DATE DUE

D0948953

Contemporary practice of chromatography

Contemporary practice of chromatography

Colin F. Poole and Sheila A. Schuette

Department of Chemistry, Wayne State University, Detroit, MI 48202, U.S.A.

ELSEVIER
Amsterdam — Oxford — New York — Tokyo   1984

ELSEVIER SCIENCE PUBLISHERS B.V.
Molenwerf 1
P.O. Box 211, 1000 AE Amsterdam, The Netherlands

Distributors for the United States and Canada:

ELSEVIER SCIENCE PUBLISHING COMPANY INC.
52, Vanderbilt Avenue
New York, NY 10017

ISBN 0-444-42410-5

© Elsevier Science Publishers B.V., 1984
All rights reserved. No part of this publication may be reproduced, stored in a retrieval system or
transmitted in any form or by any means, electronic, mechanical, photocopying, recording or otherwise, without the prior written permission of the publisher, Elsevier Science Publishers B.V./Scinece
& Technology Division, P.O. Box 330, 1000 AH Amsterdam, The Netherlands.

Special regulations for readers in the USA — This publication has been registered with the Copyright
Clearance Center Inc. (CCC), Salem, Massachusetts. Information can be obtained from the CCC about
conditions under which photocopies of parts of this publication may be made in the USA. All other
copyright questions, including photocopying outside of the USA, should be referred to the publisher.

Printed in The Netherlands

VANDERBILT UNIVERSITY
LIBRARY
NASHVILLE, TENNESSEE

CONTENTS

VIII

Preface

Chromatography is a widely used analytical tool that is often at the forefront of new discoveries in chemistry, biology, medicine, pharmacy, clinical chemistry, and the environmental sciences. As the technique is represented by a buoyant literature there has been a trend in recent years to divide its subject matter into a series of subsections or specialities. For some time we have felt the need for a book with the opposite philosophical approach, if only to provide a balanced perspective of the principal chromatographic methods used by analytical chemists.

In writing this book we had in mind that it should present a review of modern gas, liquid, and thin-layer chromatography at a level commensurate with the needs of a text book for teaching post-baccalaureate courses in the separation sciences and as a self-study guide for practicing chromatographers who completed their formal training several years ago. It was also decided that this book should take the practice of chromatography as its unifying theme, as there are already several excellent texts devoted to the physicochemical aspects of chromatography. For the above reason, we have treated sample preparation methods in a fairly comprehensive manner and included a chapter on instrumental methods of sample identification.

Even within the above restrictions, this has resulted in the production of a fairly long book. When using this material to teach graduate level courses, it is necessary to select a group of topics to be covered. The book is written in a modular fashion to accommodate this approach without loss in continuity. The material can be presented as two one-semester courses in separation science (chapters 1-5 and 9) and sample preparation (parts of chapters 4, 5, and 9, and chapters 6-8). The material has also been used as the basis for six modular one-day short courses taught to groups of indrustrial chemists.

Finally, but not least, it is a pleasure to acknowledge the friendly cooperation of our colleagues who unselfishly gave their time to proofread and comment on various portions of the text. In this regard, Dr. Roswitha Brazell, Dr. Hal Butler, Ms. Myra E. Coddens, and Dr. Frank Pacholec deserve a special thank you. We would also like to thank Ms. Coddens for patiently accommodating the many changes made to the manuscript during its preparation and for formatting the final camera-ready copy version.

Colin F. Poole
Sheila A. Schuette

Detroit, June 1984

Chapter 1

FUNDAMENTAL RELATIONSHIPS OF CHROMATOGRAPHY

1.1   <u>Introduction</u>

Chromatographic methods have had an enormous impact on the practice of analytical chemistry. Discovered about 80 years ago, reborn and augmented many times since, chromatography provides the present-day scientist with a powerful separation tool and a wealth of both theoretical and practical information to help him/her apply that tool in the most efficient manner [1].

Chromatography is essentially a physical method of separation in which the components to be separated are distributed between two phases; one of these is a stationary phase bed and the other is a mobile phase which percolates through this bed [2]. The chromatographic process occurs as a result of repeated sorption/desorption acts during the movement of the sample components along the stationary bed, and the separation is due to differences in the distribution constants of the individual sample components. In gas chromatography the mobile phase is an inert gas and the stationary phase is either an adsorbent or a liquid distributed over the surface of a porous, inert support. In liquid chromatography the mobile phase is a liquid of low viscosity that is caused to flow through a bed of sorbent. The sorbent may be an immiscible liquid coated onto a porous support, a thin-film of liquid phase bonded to the surface of a sorbent, or an inert sorbent of controlled pore size. In thin-layer chromatography a liquid mobile phase moves through a layer of sorbent by the action of capillary forces. Thin-layer chromatography is an open bed technique as pressure is not required to cause movement of the mobile phase. Gas and liquid chromatography are closed bed techniques and employ pressure gradients across the column ends to induce movement of the mobile phase. The column is a glass or metal tube of sufficient mechanical strength to withstand the column operating pressure. In packed column chromatography the sorbent, in granular form, is packed into a homogeneous bed that completely fills the column. In open tubular columns the stationary phase is distributed as a thin film or layer on the column wall, leaving an open passageway through the center of the column.

The information obtained from a chromatographic experiment is contained in the chromatogram, a record of the concentration or mass profile of the sample components as a function of the movement of the mobile phase. Information readily extracted from the chromatogram includes an indication of sample complexity based on the number of observed peaks [3], qualitative identification of sample components based on the accurate determination of peak position, quantitative assessment of the relative concentration or amount of each peak, and an indication of column performance. The fundamental information of the chromatographic process which can be extracted from the chromatogram and its associated vocabulary form the subject of this chapter [2,4-6]. A more detailed discussion of the theoretical basis and thermodynamic principles of the chromatographic process can be found in standard texts [7-22].

1.2 Retention

During their passage through the column, sample molecules spend part of the time in the mobile phase and part in the stationary phase. All molecules spend the same amount of time in the mobile phase. This time is called the column dead time or holdup time (t_m) and is equivalent to the time required for an unretained solute to reach the detector from the point of injection. The solute retention time is the time between the instant of sample introduction and when the detector senses the maximum of the retained peak. This value is greater than the column dead time by the amount of time the solute spends in the stationary phase and is called the adjusted retention time (t_R'). These values lead to the fundamental relationship, equation (1.1), describing retention in gas and liquid chromatography.

$$t_R = t_R' + t_m \qquad\qquad (1.1)$$

t_R = solute retention time
t_R' = adjusted retention time
t_m = column dead time

Retention is usually measured in units of time although volume units may also be used. At constant mobile phase flow rates the two measures of retention are easily related, as indicated in Table 1.1. Under average HPLC conditions liquids are considered to be incompressible. This is not the case for gases, and in gas chromatography elution volumes are corrected to a mean column pressure by multiplying them by the pressure gradient correction factor, j, equation (1.2).

$$j = \frac{3}{2} \left(\frac{P^2 - 1}{P^3 - 1} \right)$$

(1.2)

j = gas compressibility correction factor
P = relative pressure (P_i/P_o)
P_i = column inlet pressure
P_o = column outlet pressure

The column inlet pressure is usually measured with a pressure gauge at the head of the column. The gauge actually reads the pressure drop across the column; thus the inlet pressure used for calculating P in equation (1.2) is the value read from the gauge plus the value for P_o. It is also common practice to measure flow rates in gas chromatography with a soap-film meter. For accurate measurements it is necessary to correct the measured value of the flow rate for the vapor pressure of the soap film (assumed to be the same as that of pure water) and also for the difference in temperature between the column and flowmeter, as indicated in equation (1.3).

$$F_c = F_a \left(\frac{T_c}{T_a} \right) \left(1 - \frac{P_w}{P_a} \right)$$

(1.3)

F_c = corrected value of the carrier gas flow rate
F_a = flow rate at the column outlet
T_c = column temperature (K)
T_a = ambient temperature (K)
P_a = ambient pressure (Torr)
P_w = vapor pressure of water (Torr)

Appropriate values for P_w over a temperature range of 16-25.8°C are given in Table 1.2. The net retention volume and the specific retention volume, defined in Table 1.1, are important parameters for determining physicochemical constants from gas chromatographic data [14,15].

A more fundamentally important parameter than the absolute retention time is the ratio of the time spent by the solute in the stationary phase to the time it spends in the mobile phase. This ratio is called the solute capacity factor and is given by equation (1.4).

TABLE 1.1

RETENTION EXPRESSED IN TERMS OF VOLUME

Term	Symbol	Definition	Method of Calculation[+]
Column Dead Volume	V_m	Retention volume corresponding to the column dead time	$V_m = t_m F_c$
Retention Volume	V_R	Retention volume corresponding to the retention time	$V_R = t_R F_c$
Adjusted Retention Volume	V_R'	Retention volume corresponding to the adjusted retention time	$V_R' = t_R F_c = V_R - V_m$
Corrected Dead Volume	V_m°	V_m corrected for mobile phase compressibility; equivalent to the column volume	$V_m^\circ = j V_M$
Corrected Retention Volume	V_R°	Retention volume corrected for mobile phase compressibility	$V_R^\circ = j V_R$
Net Retention Volume	V_N	Adjusted retention volume corrected for mobile phase compressibility	$V_N = j V_R'$ $V_N = V_R^\circ - V_m^\circ$
Specific Retention Volume	V_g	The volume of mobile phase per gram of stationary phase corrected to 0°C	$V_g = V_N \dfrac{273.16}{W_L \, T_c}$

[+] F_c = column flow rate (ml/min); T_c = column temperature in K;
W_L = weight of liquid phase in the column

$$k = \frac{t_R'}{t_m} = \left(\frac{t_R - t_m}{t_m} \right) \qquad (1.4)$$

k = capacity factor
From its capacity factor, the retention time of any solute can be calculated

from equation (1.5).

$$t_R = t_m (1 + k) = \frac{L}{\bar{u}} (1 + k) \qquad (1.5)$$

L = column length
$\bar{u}$ = average mobile phase velocity

The relative retention of two adjacent peaks in the chromatogram is

described by the separation factor, α , given by equation (1.6).

$$a = \frac{t'_{R(B)}}{t'_{R(A)}} = \frac{k_B}{k_A} \tag{1.6}$$

a = separation factor

By convention, the adjusted retention time or the capacity factor of the later of the two eluting peaks is made the numerator in equation (1.6); a thus has values greater than or equal to 1.0. The separation factor is a measure of the selectivity of a chromatographic system.

TABLE 1.2

VAPOR PRESSURE OF WATER IN TORR (mm Hg)

Temperature (°C)	0.0	0.2	0.4	0.6	0.8
16	13.634	13.809	13.987	14.166	14.347
17	14.530	14.715	14.903	15.092·	15.284
18	15.477	15.673	15.871	16.071	16.272
19	16.477	16.685	16.894	17.105	17.319
20	17.535	17.735	17.974	18.197	18.422
21	18.650	18.880	19.113	19.349	19.587
22	19.827	20.070	20.316	20.565	20.815
23	21.068	21.324	21.583	21.845	22.110
24	22.377	22.648	22.922	23.198	23.476
25	23.756	24.039	24.826	24.617	24.912

The distribution constant (the ratio of the concentration of the solute in the stationary phase to its concentration in the mobile phase) is related to the capacity factor by equation (1.7).

$$K_D = \beta k \tag{1.7}$$

K_D = distribution constant

β = phase ratio

In gas chromatography the value of the distribution constant depends only on the type of stationary phase and the column temperature. It is independent of column type and instrumental parameters. The proportionality factor in equation (1.7) is called the phase ratio and is equal to the ratio of the volume of the gas (V_G) and liquid (V_L) phases in the column. The relationship between retention and the distribution constant is given by equations (1.8) and (1.9).

$$V_R = V_m + K_D V_s \tag{1.8}$$

$$t_R = \frac{L}{\bar{u}} \left(1 - \frac{V_s}{V_m} \right) K_D \tag{1.9}$$

V_m = retention volume of an unretained solute

V_s = volume of stationary phase

1.3 Flow in Porous Media

For an understanding of band broadening in chromatographic systems, the linear velocity of the mobile phase is more important than the column volumetric flow rate. The mobile phase velocity and flow rate in an open tubular column are simply related to one another by equation (1.10).

$$u_o = F_c / A_c \tag{1.10}$$

u_o = mobile phase velocity at the column outlet
F_c = column volumetric flow rate
A_c = column cross-sectional area available to the mobile phase

In a packed bed only a fraction of the column geometric cross-sectional area is available to the mobile phase, the rest is occupied by the solid (support) particles. The flow of mobile phase in a packed bed occurs predominantly through the interstitial spaces; the mobile phase trapped within the porous particles is largely stagnant [23-25]. The mobile phase velocity at the column outlet is thus described by equation (1.11).

$$u_o = \frac{F_c}{\pi r^2 \varepsilon_e} \tag{1.11}$$

r = column radius
ε_e = interparticle porosity

Typical values for the inter- and intraparticle porosity are 0.4 and 0.5, respectively.

By definition, the experimentally determined average mobile phase velocity is equal to the ratio of the column length to the retention time of an unretained solute. The value obtained will depend on the ability of the unretained solute to probe the pore volume. In liquid chromatography, a value for the intersticial velocity can be obtained by using an unretained solute that is excluded from the pore volume for the measurement. The interstitial velocity is probably more fundamentally significant than the chromatographic velocity in liquid chromatography [26].

Under chromatographic conditions, the flow profile is laminar and therefore the mobile phase velocity can be described by Darcy's law, equation (1.12).

$$u(x) = \left(\frac{-K}{\eta}\right)\left(\frac{dP}{dx}\right) \tag{1.12}$$

$u(x)$ = velocity at some point x
K = column permeability
η = mobile phase viscosity

As gases are compressible and liquids are not under average chromatographic conditions, equation (1.12) must be integrated differently for gases and liquids. For gas chromatography, the mobile phase velocity at the column outlet is given by equation (1.13).

$$u_o = \frac{KP_o (P^2-1)}{2\eta L} \tag{1.13}$$

Equation (1.13) is valid for open tubular columns under all normal conditions and for packed columns at low mobile phase velocities (less than 100-200 cm/s). However, it is the average, and not the outlet carrier gas velocity, that is of fundamental importance. The average carrier gas velocity is calculated from the outlet velocity by correcting the latter for the pressure drop across the column, equation (1.14).

$$\bar{u} = ju_o = \frac{3}{4}\left(\frac{KP_o}{\eta L}\right)\frac{(P^2 - 1)^2}{P^3 - 1} \tag{1.14}$$

In liquid chromatography, equation (1.12) can be integrated directly, neglecting the variation of viscosity with pressure and the compressibility of the mobile phase, equation (1.15).

$$u = \frac{\Delta P K_o d_p^2}{L} \tag{1.15}$$

K_o = specific permeability coefficient
d_p = particle diameter
ΔP = column pressure drop

These assumptions are valid up to ΔP values of about 600 atmospheres. The specific permeability coefficient, K_o, has a value of ca. 1000. Accurate values can be obtained from the Kozeny-Carman equation (1.16).

$$K_o = \frac{\varepsilon_e^3}{180 (1 - \varepsilon_e)^2} \tag{1.16}$$

The product $K_o d_p^2$ is the column permeability.

1.4 Band Broadening Mechanisms

As a sample traverses a column its distribution about the zone center increases in proportion to its migration distance or time in the column. The extent of zone broadening determines the chromatographic efficiency, which can

be expressed as either the number of theoretical plates (n) or the height
equivalent to a theoretical plate (H or HETP). Column efficiency is readily
determined from the peak profile, Figure 1.1, using any of equations (1.17) to
(1.21).

$$n = \left(\frac{t_R}{\sigma_t}\right)^2 \qquad (1.17)$$

$$n = \left(\frac{t_R}{w_b}\right)^2 \qquad (1.18)$$

$$n = 5.54\left(\frac{t_R}{w_h}\right)^2 \qquad (1.19)$$

$$n = 4\left(\frac{t_R}{w_i}\right)^2 \qquad (1.20)$$

$$n = 2\pi\left(\frac{t_R h}{A}\right)^2 \qquad (1.21)$$

$$H = L/n \qquad (1.22)$$

σ_t = band variance in time units
w_h = peak width at half height
w_b = peak width at the base
w_i = peak width at the inflection point
h = peak height
A = peak area

The height equivalent to a theoretical plate is given by the ratio of the
column length to the column plate count, equation (1.22). Column efficiency can
also be measured as the number of effective theoretical plates (N) by
substituting the adjusted retention time ($t_R - t_m$) for the retention time (t_R)
in equations (1.17) through (1.21). N is considered more fundamentally
important than n since it measures the band broadening that occurs only in the
stationary phase. For many of the relationships discussed in this chapter, n
and N can be used interchangeably; at sufficiently high values of the capacity
factor, N and n converge to a similar value.

The terms plate number and plate height have their origin in the plate
model of the chromatographic process [8,11,12,21,27,28]. The plate model
assumes that the column can be visualized as being divided into a number of
volume elements or imaginary sections called plates. At each plate the
partitioning of the solute between the mobile and stationary phase is rapid and
equilibrium is reached before the solute moves onto the next plate. The
distribution coefficient of the solute is the same in all plates and is
independent of the solute concentration. The mobile phase flow is assumed to
occur in a discontinuous manner between plates and diffusion of the solute in
the axial direction is negligible (or confined to the volume element of the
plate occupied by the solute). The plate model is useful for characterizing the

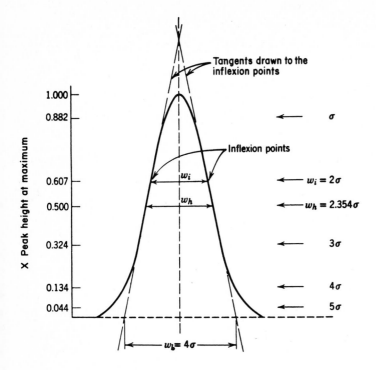

Figure 1.1 Characteristic properties of a Gaussian peak profile

efficiency of distillation columns and liquid extractors but its physical
significance in chromatography is questionable. Axial diffusion contributes
significantly to band broadening, the distribution constant is independent of
concentration only over a narrow concentration range, and, quite obviously, the
assumption that flow occurs in a discontinuous manner is false. Perhaps the
largest shortcoming of the plate model is that it fails to relate the band
broadening process to the experimental parameters (e.g., particle size,
stationary phase thickness, mobile phase velocity, etc.) that are open to
manipulation by the investigator. Nevertheless, the measured quantities n and H
are useful parameters for characterizing chromatographic efficiency and are not
limited by any of the deficiencies in the plate model. The various rate models
of the chromatographic process enable a similar expression for the theoretical
plate to be derived [8,27].

The rate theory makes the following assumptions in its explanation of band
broadening:

1. Resistance to mass transfer in both the stationary and mobile phase
    prevents the existence of an instantaneous equilibrium. Under most
    practical conditions this is the dominant cause of band broadening.

2. The flow velocity through a packed column varies widely with radial
    position in the column. Some molecules will travel more rapidly by

following open pathways (channelling); others will diffuse into restricted areas and lag behind the zone center (eddy diffusion). These differing flow velocities will cause zone dispersion about the average velocity. For open tubular columns, the contribution of unequal flow velocities to the plate height is zero.

3. Longitudinal diffusion (molecular diffusion in the axial direction) leads to band broadening that is independent of the mobile phase velocity. Its contribution to band broadening increases with the amount of time the solute spends in the column.

The individual contributions to the band broadening mechanism are considered as independent variables except under some circumstances when the eddy diffusion term is coupled to the mobile phase mass transfer term. The above approach can be applied to gas or liquid mobile phases, although it is necessary to make allowances for the different physical properties of gases and liquids, Table 1.3. For liquid chromatography, the mobile phase is assumed to be incompressible. This is generally true in most practical situations, although at large pressure drops the solute diffusion coefficients, capacity factors, and the column plate height are all influenced by the column pressure drop [29]. Instrumental contributions to band broadening are considered is sections 3.1 and 5.2 and will not be discussed here. General reviews of the band broadening process which expand on the treatment presented below are available [8,11,12,19,30-37].

TABLE 1.3

APPROXIMATE VALUES OF CHARACTERISTIC PARAMETERS IMPORTANT IN PREDICTING BAND BROADENING

Parameter	Gas	Liquid
Diffusion Coefficients (cm^2/s)	10^{-1}	10^{-5}
Density (g/cm^3)	10^{-3}	1
Viscosity (poise)	10^{-4}	10^{-2}
Reynold's Number	10	10^2

When a sample band migrates through a packed bed, the individual flow paths around the packing particles are of different lengths. These variations in the flow direction and rate lead to band broadening that should depend only on the density and homogeneity of the column packing. Its contribution to the total plate height is proportional to the particle size and can be described by equation (1.23).

$$H_E = 2 \lambda d_p \qquad\qquad (1.23)$$

H_E = contribution to the total plate height from eddy diffusion
λ = packing factor
d_p = average particle diameter

The packing factor is a dimensionless constant and usually has a value between 1 and 2.

The contribution to the plate height from molecular diffusion in the mobile phase arises from the natural tendency of the solute band to diffuse away from the zone center as it moves through the column [38]. Its value is proportional to the diffusion coefficient and the time the sample spends in the column. Its contribution to the total plate height is given by equation (1.24)

$$H_L = \frac{2 \, Y \, D_m}{\bar{u}} \qquad\qquad (1.24)$$

H_L = contribution to the plate height from longitudinal molecular
        diffusion in the mobile phase
Y = obstruction or tortuosity factor
D_m = solute diffusion coefficient in the mobile phase
$\bar{u}$ = average mobile phase velocity

The obstruction factor is a dimensionless quasi-constant that is not totally independent of the mobile phase velocity [39]. This dependence arises from the fact that the lowest flow resistance is offered by gaps or wide paths in the packing structure. Thus, at low velocities the value of the obstruction factor is averaged over tightly-packed and loosely-packed domains, while at high velocities it is weighted in favor of the loosely-packed domains where more flow occurs. Typical values for the obstruction factor are 0.6 to 0.8 in a packed bed and 1.0 for an open tubular column.

An alternative argument for the existence of the obstruction factor has been proposed by Deininger [40]. In a bed comprised of a porous packing, a sample spends part of its time within the pores of the packing where longitudinal diffusion is negligible. The obstruction factor then arises to correct the diffusion time for the fraction of the column residence time the sample spends in the stagnant mobile phase. It is proportional to some function of the specific pore volume of the packing and the interparticle volume.

In liquid chromatography where diffusion coefficients are small, the contribution of H_L to the plate height is often negligible. Diffusion coefficients are much larger in gases and hence H_L is more important, particularly at low mobile phase velocities.

Mass transfer in either the stationary or mobile phase is not instantaneous and, consequently, complete equilibrium is not established under normal separation conditions. The result is that the solute concentration profile in the stationary phase is always displaced slightly behind the equilibrium position and the mobile phase profile is similarly slightly in advance of the equilibrium position. The combined peak observed at the column outlet is broadened about its band center, which is located where it would have been for instantaneous equilibrium, provided the degree of nonequilibrium is small. The stationary phase contribution to mass transfer is given by equation (1.25)

$$H_S = \frac{2kd_f^2 \bar{u}}{3D_s(1 + k)^2} \qquad (1.25)$$

H_S = contribution to the plate height from the resistance to mass transfer in the stationary phase

d_f = stationary phase film thickness

D_s = diffusion coefficient in the stationary phase

Equation (1.25) applies exactly to thin-film open tubular columns and is a reasonable approximation for packed column gas chromatography. For liquid chromatography the agreement is poor as there is no allowance for the contribution of diffusion resistance in the stagnant mobile phase [31,32].

When a liquid flows through a packed bed an appreciable fraction of the interstitial fluid is essentially stagnant with respect to the actual stream in the center region of the interparticle channels. The fluid space in the column is depicted as consisting of three domains: the free, streaming fluid space, the stagnant interstitial fluid space, and the intraparticulate fluid space, which is also assumed to be stagnant. Diffusion is relatively slow in the stagnant mobile phase and its influence on band broadening in liquid chromatography is often significant. Thus equation (1.26) more adequately accounts for the contribution of mass transfer in the stationary phase to the total plate height in liquid chromatography than does equation (1.25).

$$H_S = \frac{\theta(k_o + k + k_o k)^2 d_p^2 u_e}{30 D_m k_o (1 + k_o)^2 (1 + k)^2} \qquad (1.26)$$

θ = tortuosity factor for the pore structure of the particles

k_o = ratio of the intraparticle void volume to the interstitial void space

u_e = interstitial mobile phase velocity

With multicomponent eluents the value of k_o will vary with the mobile phase composition since the intraparticulate space occupied by the stagnant mobile phase may change due to solvation of the stationary phase surface. In the derivation of equation (1.26), the influence of diffusion through the interparticle stagnant mobile phase has been neglected as it is generally very small compared to the value for the intraparticle stagnant mobile phase contribution. If a liquid, coated on a support, were used rather than a bonded-phase column packing, it would also be necessary to modify equation (1.26) to allow for mass transfer resistance at the liquid-liquid interface [30].

Mass transfer resistance in the mobile phase is more difficult to calculate because it requires an exact knowledge of the flow profile of the mobile phase. This is only known exactly for open tubular columns for which the contribution of mass transfer resistance in the mobile phase to the total plate height can be described by equation (1.27).

$$H_M = \frac{1 + 6k + 11k^2}{96 (1 + k)^2} \cdot \frac{d_c^2 \bar{u}}{D_m} \qquad (1.27)$$

H_M = contribution to the plate height from the resistance to mass transfer in the mobile phase
d_c = column diameter

In a packed bed the mobile phase flows through a tortuous channel system and lateral mass transfer can take place by a combination of diffusion and convection. The diffusion contribution can be approximated by equation (1.28).

$$H_{M,D} = \frac{w \, \bar{u} \, d_p^2}{D_m} \qquad (1.28)$$

$H_{M,D}$ = contribution to the plate height from diffusion-controlled resistance to mass transfer in the mobile phase
w = packing factor function to correct for radial diffusion (ca. 0.02 to 5)

To account for the influence of convection, that is, band broadening resulting from the exchange of solute between flow streams moving at different velocities, the eddy diffusion term must be coupled to the mobile phase mass transfer term, as indicated in equation (1.29).

$$H_{M,C} = \frac{1}{(1/H_E + 1/H_{M,D})} = \frac{1}{(1/2\lambda d_p + D_m/w\bar{u}d_p^2)} \qquad (1.29)$$

$H_{M,C}$ = contribution to the plate height resulting from the coupling of eddy diffusion and mobile phase mass transfer terms

In general, $H_{M,C}$ increases with increasing particle diameter and flow velocity and decreases with solute diffusivity. The packing structure, the velocity range, and the capacity factor value can significantly influence the exact form

14

of the relationship.

In gas chromatography, the coupled plate height equation flattens out the ascending portion of the van Deemter curve at high mobile phase velocities and generally provides a better fit to experimental data than the classic van Deemter equation. The coupling effect is most apparent in liquid chromatography where the contribution from mass transfer resistance in the mobile phase at high flow rates is generally the predominant term in the overall plate height expression.

Although the above listing of contributions to the column plate height is not comprehensive, it encompasses the major band-broadening factors and the overall plate height can be expressed as their sum, equation (1.30).

$$HETP = H_E + H_L + H_S + H_M \tag{1.30}$$

A plot of HETP as a function of mobile phase velocity is a hyperbolic function (Figure 1.2) most generally described by the van Deemter equation (1.31).

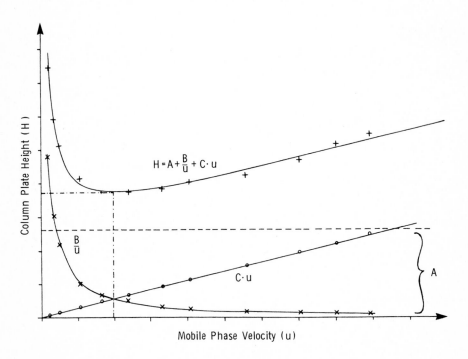

Figure 1.2 Relationship between band broadening and flow rate (van Deemter equation)

$$\text{HETP} = A + \frac{B}{\bar{u}} + (C_S + C_M)\,\bar{u} \tag{1.31}$$

The A term represents the contribution from eddy diffusion, the B term the contribution from longitudinal diffusion, and the C terms the contributions from mass transfer in the mobile and stationary phases to the total column plate height. By differentiating equation (1.31) with respect to the mobile phase velocity and setting the result equal to zero, the optimum values of mobile phase velocity ($\bar{u}_{opt}$) and plate height (HETP_{opt}) can be obtained.

$$\bar{u}_{opt} = \sqrt{B/(C_M + C_S)} \tag{1.32}$$

$$(\text{HETP})_{min} = A + 2\sqrt{B(C_M + C_S)} \tag{1.33}$$

The highest column effficieny will be obtained at $\bar{u}_{opt}$. In practice, higher values are frequently used to minimize analysis times. For gas chromatography a velocity of about $2\,\bar{u}_{opt}$, called the practical operating gas velocity, is frequently recommended [41]. Provided that the ascending portion of the van Deemter curve is fairly flat at higher velocities than $\bar{u}_{opt}$, then the loss in efficiency will be small and well worth the gain in analysis time.

The van Deemter equation (and similar functions) has made an important contribution to column technology. Evaluation of the coefficients of equation (1.31) by curve fitting techniques provides a link between chromatographic efficiency and the variables that can be manipulated to maximize column efficiency [34,37,42]. This approach can be used to establish the optimum operating conditions of the various column types used in gas and liquid chromatography.

Beginning with the most favorable case, band broadening in open tubular columns is satisfactorily described by the Golay equation, extended to situations of appreciable pressure drop by Giddings, equation (1.34) [43,44].

$$\text{HETP} = \left(\frac{2D_{m,o}}{u_o} + \frac{11k^2 + 6k + 1}{96\,(1+k)^2} \cdot \frac{d_c^2 u_o}{D_{m,o}} \right) f_1 + \left(\frac{2\,kd_f^2 u_o}{3D_s(1+k)^2} \right) f_2 \tag{1.34}$$

$D_{m,o}$ = mobile phase diffusion coefficient at the column outlet pressure

u_o = mobile phase velocity at the column outlet

d_f = stationary phase film thickness

$$f_1 = \frac{9}{8} \cdot \frac{(P^4 - 1)\,(P^2 - 1)}{(P^3 - 1)^2} \qquad\qquad f_2 = \frac{3}{2} \cdot \frac{P^2 - 1}{P^3 - 1}$$

P = column inlet/column outlet pressure

Open tubular columns in current use have thin films of stationary phase (ca. 0.1 to 0.5 micrometers) to minimize the contribution of mass transfer in the stationary phase to band broadening [i.e., the last term in equation (1.34) approaches zero]. They have column diameters, d_c, in the range 0.2-0.5 mm to minimize the HETP while still maintaining an acceptable pressure drop across the column. Gases of high diffusivity, hydrogen or helium, are used to minimize mass transfer resistance in the mobile phase and, at the same time, minimize analysis time [(HETP)$_{min}$ occurs at higher values of $\bar{u}_{opt}$ for gases of high diffusivity]. Because the stationary phase film is thin, the maximum sample size which can be injected without overloading the column is low, typically in the nanogram range. The film thickness can be increased to enhance sample detectability and columns with films up to about 6 micrometers thick are commercially available [45]. It should be noted though that as the film thickness is increased the resistance to mass transfer in the stationary phase increases much more rapidly than the mobile phase term and eventually becomes the dominant term in the plate height equation. Thus an increase in sample capacity is gained only at the expense of a decrease in column efficiency. Equation (1.34) also indicates that column efficiency can be expected to increase with a decrease in column diameter [46-49]. With columns of conventional diameters, ca. 0.2 mm, it is difficult to exceed a column plate count of about 300,000 without resorting to unacceptably long analysis times. Recent work has shown that a maximum of 2.7 million theoretical plates should be attainable from a column with a diameter of 35 micrometers without resorting to unreasonable column pressures [48]. For very fast analysis, instrumental constraints remains the major hurdle [46,47]. Commercially available instruments are not generally suitable for separations employing microbore open tubular columns without extensive modification of injectors, detectors, pneumatic systems, extracolumn dead volumes, and electronic time constants.

The Golay equation is strictly applicable to open tubular columns with smooth walls but, with certain approximations, it can be extended to include support-coated [50] and whisker-walled [51] open tubular columns. It can also be used to predict optimum separation conditions in open tubular liquid chromatography [52].

As the exact profile of the mobile phase flow through a packed bed is unknown, only an approximate description of the band broadening process can be attained. For packed column gas chromatography at low mobile phase velocities, equation (1.35) provides a reasonable description of the band broadening process [35,53,54].

$$\text{HETP} = 2\lambda d_p + \frac{2\gamma D_m}{\bar{u}} + \frac{2kd_f^2\bar{u}}{3D_s(1+k)^2} + \frac{wd_p^2\bar{u}}{D_m} \tag{1.35}$$

Small particles having a narrow size distribution and coated with a thin, homogeneous film of liquid phase are required for high efficiency columns. The particle size is limited by the need to remain within limited pressure constraints; this results in the use of column packings with diameters of 120-180 micrometers in columns less than ca. 5 meters long. For heavily loaded columns, liquid phase loading of 25-35% w/w, slow diffusion in the stationary phase film is the principal cause of band broadening. With lightly loaded columns (less than 5% w/w), resistance to mass transfer in the mobile phase is no longer negligible. At high mobile phase velocities the coupled form of the plate height equation is used to describe band broadening.

When the mobile phase is a liquid a variety of equations can be used in addition to the van Deemter equation (1.31) to describe band broadening as a function of the mobile phase velocity, equations (1.36) to (1.39) [26,32,55-58].

$$\text{HETP} = \frac{A}{1 + E/u} + \frac{B}{u} + Cu \qquad (1.36)$$

$$\text{HETP} = \frac{A}{1 + E/u^{1/2}} + \frac{B}{u} + Cu + Du^{1/2} \qquad (1.37)$$

$$\text{HETP} = Au^{1/3} + \frac{B}{u} + Cu \qquad (1.38)$$

$$\text{HETP} = \frac{A}{1 + E/u^{1/3}} + \frac{B}{u} + Cu + Du^{2/3} \qquad (1.39)$$

A, B, C, D, and E are appropriate constants for a given solute chromatographed on a given column system. Scott's [26] comparison of these equations indicated a good fit with experimental data for all equations, but only equations (1.31), (1.36), and (1.38) consistantly gave physically meaningful values for coefficients A through E. The van Deemter equation, expressed in form (1.40), was found to give the most reasonable fit for porous silica packings over the mobile phase velocity range of 0.02 to 1.0 cm/s.

$$\text{HETP} = 2\lambda d_p + \frac{2\gamma}{u_e} \cdot D_m + \frac{(a + bk_e + ck_e^2)}{24 (1 + k_e)^2 D_m} \cdot d_p^2 u_e \qquad (1.40)$$

k_e = capacity factor determined for the interstital column volume

u_e = interstitial mobile phase velocity

The values of λ and γ may vary with the quality of the packing and, for a reasonably well-packed column, can be assumed to be 0.5 and 0.8, respectively;

a, b, and c can be assigned values of 0.37, 4.69, and 4.04, respectively. At mobile phase velocities higher than those investigated by Scott, the coupled form of the plate height equation may be more appropriate. Equation (1.40) was derived with the assumption that diffusion in the mobile and stationary phases was similar and could therefore be represented by a single mass transfer term.

The highest efficiency in liquid chromatography is obtained using columns packed with particles of small diameter, operated at high pressures, with mobile phases of low viscosity. Both solute diffusivity and column permeability decrease as the mobile phase viscosity increases. For a fixed column pressure, the separation time will increase as the viscosity of the mobile phase is increased. Diffusion coeffcients are much smaller in liquids than in gases and, although this means that longitudinal diffusion can often be neglected in liquid chromatography, the importance of mass transfer resistance in the mobile phase is now of much greater significance. The adverse effect of slow solute diffusion in liquid chromatography can be partially overcome by operating at much lower mobile phase velocities than is common for gas chromatography. This increase in efficiency is, however, paid for by an extended analysis time.

Efficiency in itself can be a false god; the analyst is generally more interested in the ability of a chromatographic system to deliver a certain efficiency per unit time under experimentally realizable conditions [17,25,59-66]. Guiochon has developed limiting equations for the maximum number of theoretical plates of a column and for the time required to deliver a theoretical plate [65,66]. In liquid chromatography the velocity diminishes rapidly as the column length is progressively increased, and the efficiency tends towards a limit while the analysis time becomes infinite, equation (1.41).

$$(n)_{lim} = \frac{k_o}{\eta \, D_m} \cdot \frac{d_p^2}{2\gamma} \cdot \Delta P \tag{1.41}$$

ΔP = column pressure drop

At high mobile phase velocities both the efficiency and analysis time tend toward zero and the ratio t/n approaches a limiting value, given by equation (1.42).

$$(t/n)_{lim} = C \, \frac{1 + k}{D_m} \cdot d_p^2 \tag{1.42}$$

C = a constant for mass transfer obtained from the Knox equation (see Section 4.6.2). It has a value of ca. 0.01

If a fast analysis is desired then a column packed with fine particles and solutes with large diffusion coefficients are required ($d_p^2/D_m \longrightarrow$ minimum).

This will provide moderate plate numbers in short times, but the maximum attainable plate number will be rather small. On the other hand, if the emphasis is placed on obtaining very high efficiency, then coarse particles and low diffusion coefficients are required ($d_p^2/D_m \longrightarrow$ maximum). This permits the use of very long columns with reasonable pressure drops ($\mathbf{\Delta}P/L$) but leads to very long analysis times. The current practice in HPLC is a compromise between the two extremes, with the emphasis on maintaining acceptable analysis times (less than 1 h). Thus, short columns packed with fine particles and operated at high inlet pressures are routinely used.

If the pressure gradient ($\mathbf{\Delta}P/L$) is kept constant in liquid chromatography, then the mobile phase will also be constant; hence, the plate height and the analysis time increase proportionately with the plate number. This is not the case in gas chromatography, where the outlet velocity will increase in proportion to the inlet pressure at a constant outlet pressure. With increasing column length the plate height increases more rapidly than the change in column length. On the other hand, the average velocity remains constant and the analysis time increases in proportion to the column length. The limiting relationships for n and t/n are therefore given by equations (1.43) and (1.44).

$$(n)_{lim} = \frac{k_o}{4Y \, \eta \, D_m} \cdot \frac{P_i^2}{P_o} \cdot d_p^2 \tag{1.43}$$

$$(t/n)_{lim} = \frac{2C}{3} \cdot \frac{1 + k}{D_m} \cdot \frac{P_i}{P_o} \cdot d_p^2 \tag{1.44}$$

C = mass transfer coefficient with values from 0.01 to 0.2
P_i = column inlet pressure
P_o = column outlet pressure

The optimum conditions derived from equations (1.43) and (1.44) are similar to those indicated for liquid chromatography, except that in gas chromatography the limit (t/n) is a function of the inlet pressure at high carrier gas velocities.

The column length required to achieve a given plate number and the optimum carrier gas velocity are similar for packed and open tubular columns. Open tubular columns, however, have much higher permeabilities than packed columns and therefore the inlet pressure required to achieve the optimum flow velocity is much lower. Also, the average mobile phase velocity corresponding to a given outlet flow velocity is much larger for an open tubular column than a packed column and, consequently, the analysis time is much shorter. Thus the highest column efficiency and the shortest analysis time will always be obtained with open tubular columns, assuming reasonable instrument operating parameters. With the present column technology, open tubular capillary gas chromatography is clearly superior in terms of the performance criteria discussed above. The efficiencies of packed column gas and liquid chromatography are very similar, although gas chromatography is faster. This difference in speed is directly related to the physical and mechanical properties of liquids and gases, and is therefore probably an unsurmountable barrier.

1.5 Parameters Affecting Resolution

The separation factor, α, introduced in section 1.2 is a useful measure of relative peak separation. It is a constant for a given set of analytical conditions (stationary phase, temperature, etc.) and is independent of the column type and dimensions. The actual separation of two peaks in a chromatogram is not adequately described by the separation factor alone, however, since it does not contain any information about peak widths. The degree of separation between two peaks is defined by their resolution, R_s, the

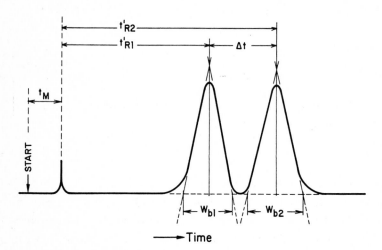

Figure 1.3 Measurement of resolution for two closely spaced peaks

ratio between the two peak maxima (Δt) and the average base width of the two peaks, Figure 1.3 and equation(1.45).

$$R_s = \frac{2 \Delta t}{w_{b1} + w_{b2}} \tag{1.45}$$

$$t = t_{R2} - t_{R1} = t_{R2}' - t_{R1}'$$

A value of $R_s = 1.0$ corresponds to a peak separation on the order of 94% and is generally considered an adequate goal for an optimized separation. Baseline resolution corresponds to an R_s value of 1.5. Resolution is related to the adjustable variables of a chromatographic experiment by equation (1.46).

$$R_s = \frac{\sqrt{n}}{4} \cdot \left(\frac{a-1}{a}\right)\left(\frac{k_2}{1+k_2}\right) \tag{1.46}$$

k_2 = capacity factor value of the later eluting peak of a peak pair

To a first approximation the three terms in equation (1.46) can be considered independent variables. Resolution is strongly dependent on a; when $a = 1$, a separation is not possible in that chromatographic system. The larger the value of a, the easier a separation becomes. Resolution increases only as the square root of column efficiency. Thus, the column length must be increased four-fold to increase resolution by a factor of two. Resolution is generally low for small values of the capacity factor, k_2. Although k_2 can be increased resolution will not be significantly improved beyond a certain range while the time required to carry out the separation may be considerably increased. The optimum resolution zone for most separations falls in the range k_2 = 3-10. Equation (1.46) can be arranged to predict the number of theoretical plates required to give a certain separation, equation (1.47).

$$n_{req} = 16 \, R_s^2 \cdot \left(\frac{a}{a-1}\right)^2 \left(\frac{k_2 + 1}{k_2}\right)^2 \tag{1.47}$$

Equations (1.46) and (1.47) can be strictly applied only to two peaks that are close together in the chromatogram.

1.6 Peak Shape Models

The column is usually assumed to function as a Gaussian operator, broadening the sample plug into a Gaussian distrubution as it passes through the column. In practice, chromatographic peaks are rarely Gaussian and significant errors might result from the calculation of chromatogaphic parameters based on this false assumption [67,68]. The Gaussian model is only appropriate when the degree of peak asymmetry is slight.

Peak asymmetry can arise from a variety of instrumental and chromatographic sources, Table 1.4 [15,69,70]. Meaningful chromatographic data can be extracted from asymmetric peaks by digital integration or curve fitting routines applied to the chromatographic peak profile. Digital acquisition of chromatographic data by computer permits the rapid and easy calculation of the statistical moments of the peak profile by direct integration [71-76]. The nth statistical moment (M_n) is defined by equation (1.48)

$$M_n = \frac{\int_0^\infty t^n c(t) \, dt}{\int_0^\infty c(t) \, dt} \qquad\qquad (1.48)$$

where $c(t)$ is the concentration of the solute observed at the detector at time t after injection. The first moment ($n = 1$) corresponds to the elution time of the center of gravity of the peak since it is obtained by weighting the elution time of each point in the peak by its concentration, Table 1.5. Higher moments are more usefully defined as central moments, obtained from the distribution of the peak around the first moment. The second central moment (M_2) is the variance (the square of the standard deviation of the peak). This is the parameter used to calculate the column plate height. The third and fourth central moments measure the peak asymmetry (skew) and the extent of vertical

TABLE 1.4

CHROMATOGRAPHIC PROCESSES LEADING TO PEAK ASYMMETRY

Incomplete Resolution	Distorted peak shapes are sometimes due to the presence of unresolved solute components
Slow Kinetic Processes	The model proposed by Giddings assumes that the stationary phase contains two types of sites. Fast solute exchange between the mobile and stationary phases occurs at normal sites. On the second type of site solute molecules are only slowly sorbed and desorbed. If the time constant for the desorption step of the slow process is greater than half the standard deviation of the peak, the peak will not only be broadened but will also carry an exponential tail. Examples of slow mass transfer processes include diffusion of the solute in microporous solids, polymers, organic gel matrices, and deep pores holding liquid droplets; interactions involving surfaces with a heterogeneous energy distribution; and, in liquid chromatography, interfacial mass transfer resistance caused by poor solvation of bonded phases.
Chemical Reactions	If the solute undergoes a chemical reaction on the column, peak tailing or anomalous peak broadening may result; more likely, the effect will be complex with multiple peaking, often involving sharp peaks superimposed on a strongly tailed peak profile.
Column Voids	Bed shrinkage is usually gradual and results in progressive peak broadening and/or distortion. A void over the entire cross-section of the column near the inlet produces more peak broadening than asymmetry. However, voids occupying only part of the cross-section along the length of the bed can produce pronounced tailing or fronting, or even split all peaks into resolved or unresolved doublets. Partial voidage effects are due to channelling, i.e., different residence times in the two different flow paths, formed by the void and packed regions. Poor radial diffusion in liquids fails to relax the radial concentration profiles fast enough to avoid asymmetry or split peaks. In gas chromatography the phenomenon is far less significant because diffusion in gases in much faster.

flattening (excess), repsectively. For a Gaussian distribution, statistical
moments higher than the second have a value of zero. A positive value for the
skew indicates a tailing peak. A positive value for the excess indicates a
sharpening of the peak profile relative to a Gaussian peak, while a negative
value indicates a relative flattening of the upper portions of the peak
profile.

Direct numerical integration of the peak profile may lead to many errors
and uncertainties arising from the limits used in the integration, baseline
drift, noise, and extracolumn contributions [77,78]. A more accurate method for
calculating moments of asymmetric peaks is curve fitting to an appropriate
model, such as the exponentially modified Gaussian model [78-84]. The peak
profile is fitted by computer or by manual methods to a Gaussian peak function,
onto which is imposed an exponential decay function. The peak moments are now
described by two components: a symmetrical component, σ_t, due to the original
Gaussian distribution and τ, a nonsymmetrical contribution due to the
exponential decay, Table 1.5. The mathematics involved in applying the
exponentially modified Gaussian function to chromatographic peaks are fairly
complex and the reader is referred to the reviews by Dorsey for further details
[83,84].

TABLE 1.5

STATISTICAL MOMENTS OF CHROMATOGRAPHIC PEAKS
τ = time constant of the exponential modifier; A = peak area (M_o);
σ_t = peak standard deviation.

Statistical Moment	Order	Equation	Chromatographic Parameter Measured	Exponentially Modified Gaussian
M_o	0	$\int_o^\infty c(t)\,dt$	area	
M_1	1	$1/A \int_o^\infty c(t)\,dt$	retention time (t_R)	$t_R + \tau$
M_2	2	$1/A \int_o^\infty c(t-t_R)^2\,dt$	variance	$\sigma_t^2 + \tau$
M_3	3	$1/(A\sigma_t^3) \int_o^\infty c(t-t_R)^3\,dt$ $S = M_3/M_2^{3/2}$	skew (S)	$2\tau^3$
M_4	4	$1/(A\sigma_t^4) \int_o^\infty c(t-t_R)^4\,dt - 3$ $E = (M_4/M_2^2) - 3$	excess (E)	$3\sigma_t^4 + 6\sigma_t^2\tau^2 + 9\tau^4$

1.7 Retention Index Systems

Methods based on relative retention are inaccurate for reporting
chromatographic data which is to be used for interlaboratory substance

identification. In gas chromatography, retention index systems are used for
this purpose. The retention behavior is expressed on a uniform scale determined
by a series of closely related standard substances. It is known that for a
homologous series of compounds, the logarithms of the adjusted retention time
values obtained on a given column at a given temperature are linearly
proportional to the carbon number. In the retention index system devised by
Kovats, the system most frequently used, the retention index value of a
substance is defined according to equation (1.49) [85-88].

$$I = 100 \left(\frac{\log t'_{R(x)} - \log t'_{R(z)}}{\log t'_{R(z+1)} - \log t'_{R(z)}} + z \right) \qquad (1.49)$$

I = retention index
x = substance of interest
z = normal alkane with x carbon atoms emerging before the substance of
     interest
z+1 = normal alkane with z+1 carbon atoms emerging after the substance of
     interest

As defined in equation (1.49), the retention index of a substance is equivalent
to 100-times the carbon number of a hypothetical n-alkane with the same adjusted
retention time. By convention, the retention index has the column temperature
and the stationary phase as subscript and superscript, respectively. Ideally,
the retention index should depend only on the interaction of the solute with the
stationary phase and the column temperature. It is independent of other
instrumental and experimental parameters.

The temperature dependence of the retention index value is a hyperbolic
function described by an Antoine-type function, equation (1.50).

$$I(T) = A + \frac{B}{T + C} \qquad (1.50)$$

I(T) = retention index at temperature T
T   = column temperature (K)

where A, B, and C are experimentally derived constants. The curve can
nevertheless have a significant linear portion, particularly for substances of
low polarity on nonpolar stationary phases.

For mixtures of wide boiling point range the determination of retention
indices under isothermal conditions would be time consuming and unnecessarily
restrictive. Under temperature program conditions an approximately linear
relationship exists between the elution temperature of n-alkanes and their
carbon number, provided that the initial column temperature is low and that only
a relatively limited range of carbon numbers is considered. An expression
equivalent to equation (1.49) can be given for linear temperature program

conditions by replacing the adjusted retention time in equation (1.49) with the time required for elution or the column temperature at this time, equation (1.51) [89,90].

$$I = 100 \left(\frac{T_{R(s)} - T_{R(z)}}{T_{R(z+1)} - T_{R(z)}} + z \right) \qquad (1.51)$$

T_R = elution temperature (K)

Sources of error in determining retention indices are due primarily to instrumental variations in temperature and carrier gas flow rates, inaccuracies in the measurement of retention times and the column dead time, and effects caused by support activity and impurities in the stationary phase [85,91]. Under normal circumstances, adsorption by the support and poorly defined stationary phases are the principal causes of error. Many stationary phases used in gas chromatography are commercial products of poorly defined structure and are often of less than adequate purity. Whenever possible, single substances or specially purified and fractionated polymeric phases for gas chromatography should be used. In theory, the retention index should be independent of the stationary phase loading. In practice this is true within the normal range of phase loadings only if support interactions can be neglected and if the solute is retained entirely by partition (i.e., solute adsorption at the gas-liquid interface is negligible). The sample size may also influence retention. Polar phases show poor solvation properties for alkanes, which consequently exhibit variable retention times at moderate to large sample sizes. The smallest practical sample size commensurate with obtaining a reasonable detector response should be used.

The reproducibility of retention indices within the same laboratory is generally on the order of ± 0.2 or ± 0.05-0.1 index units, depending on whether measurements are made manually or by computer acquisition of the experimental data, respectively. Interlaboratory variations of about 0.2 index units are considered acceptable but wider variations (1-3 index units) are often observed in the literature.

When the retention index of a substance is unknown and the substance is unavailable for measurement, its retention index can be estimated. The partial indices for the carbon skeleton and functional groups are summed, making allowances where necessary for second order interactions between functional groups [86-88,92]. The partial retention indices are measured by analysis of selected standards. For nonpolar compounds on stationary phases of low polarity, good agreement between estimated and measured retention indices is obtained. As the polarity of the solute and/or the stationary phase increases, the accuracy of estimated retention indicies declines.

Alternative retention index scales based on standards other than alkanes
are occasionally used. For example, fatty acid methyl esters are used in lipid
analysis; androstane and cholestane for the calculation of steroid numbers; and
naphthalene, phenanthrene, chrysene, and picene for polycyclic aromatic
compounds [85,87,93,94]. One reason for selecting these new standards is that
they are often already present in the samples being analyzed. Detector
compatibility can be another problem. Selective detectors such as the
electron-capture and flame photometric detectors do not respond significantly to
n-alkanes. n-Alkyl bromides [95] or n-alkyl trichloroacetates [96,97] are used
with the electron-capture detector while n-alkyl sulfides are used with the
flame photometric detector [98]. The retention index is calculated in the same
way as for the alkanes using equation (1.49) or (1.51).

Retention index systems are not commonly used in liquid chromatography.
For reversed-phase liquid chromatography, 2-keto alkanes [99], alkyl aryl
ketones [100], and a series of polycyclic aromatic hydrocarbons (benzene,
naphthalene, phenanthrene, benzo[a]anthracene, and benzo[b]chrysene) [101] have
been used as retention index standards. In gas chromatography the mobile phase
is inert and is assumed to have no influence on retention. This is not the case
in liquid chromatography where retention is generally manipulated by adjusting
the mobile phase composition. The relationship between retention, mobile phase
composition, and temperature is complex and not adequately described by current
theory [102-104]. Further work will be required to develop a general and
accurate retention index system for liquid chromatography.

1.8 References

1. L. S. Ettre and A. Zlatkis (Eds.), "75 Years of Chromatography. A
    Historical Dialogue", Elsevier, Amsterdam, 1979.
2. L. S. Ettre, "Basic Relationships of Gas Chromatography", Perkin-Elmer,
    Norwalk, 1977.
3. J. M Davis and J. C. Giddings, Anal. Chem., 55 (1983) 418.
4. L. S. Ettre, J. Chromatogr., 165 (1979) 235.
5. L. S. Ettre, J. Chromatogr., 220 (1981) 29.
6. L. S. Ettre, J. Chromatogr., 220 (1981) 65.
7. J. H. Purnell, "Gas Chromatography", Wiley, New York, 1962.
8. J. C. Giddings, "Dynamics of Chromatography", Dekker, New York, 1965.
9. L. S. Ettre and A. Zlatkis (Eds.), "The Practice of Gas Chromatography",
    Interscience, New York, 1967.
10. A. B. Littlewood, "Gas Chromatography. Principles, Techniques, and
    Applications", Academic Press, New York, 2nd Edn., 1970.
11. B. L. Karger, L. R. Snyder, and C. Horvath, "An Introduction to
    Separation Science", Wiley, New York, 1973.
12. R. P. W. Scott, "Contemporary Liquid Chromatography", Wiley, New York,
    1976.
13. C. J. O. R. Morris and P. Morris, "Separation Methods in Biochemistry",
    Wiley, New York, 2nd Edn., 1976.
14. R. J. Laub and R. L. Pecsok, "Physiochemical Applications of Gas
    Chromatography", Wiley, New York, 1978.
15. J. R. Conder and C. L. Young, "Physicochemical Measurement by Gas
    Chromatography", Wiley, New York, 1979.

16. O. Mikes (Ed.), "Laboratory Handbook of Chromatographic and Allied Methods", Wllis Horwood, Chichester, 1979.
17. L. S. Ettre, "Introduction to Open Tubular Columns", Perkin-Elmer, Norwalk, 1979.
18. L. R. Snyder and J. J. Kirkland, "Introduction to Modern Liquid Chromatography", Wiley, New York, 2nd Edn., 1979.
19. J. A. Perry, "Introduction to Analytical Gas Chromatography. History, Principles, and Practice", Dekker, New York, 1981.
20. R. E. Kaiser and E. Oelrich, "Optimization in HPLC", Huthig, Heidelberg, 1981.
21. A. S. Said, "Theory and Mathematics of Chromatography", Huthig, Heidelberg, 1981.
22. E. Heftmann (Ed.), "Chromatography Part A: Fundamentals and Techniques", Elsevier, Amsterdam, 1983.
23. C. A. Cramers, J. A. Rijks, and C. P. M. Schutjes, Chromatographia, 14 (1981) 439.
24. G. Guiochon, J. Chromatogr. Revs., 8 (1966) 1.
25. G. Guiochon, J. Chromatogr., 185 (1979) 3.
26. E. Katz, K. L. Ogan, and R. P. W. Scott, J. Chromatogr., 270 (1983) 51.
27. E. Grushka, L. R. Snyder, and J. H. Knox, J. Chromatogr. Sci., 13 (1975) 25.
28. J. S. Fritz and D. M. Scott, J. Chromatogr., 271 (1983) 193.
29. M. Martin and G. Guiochon, Anal. Chem., 55 (1983) 2302.
30. C. Horvath and W. E. Melander, in E. Heftmann (Ed.), "Chromatography Part A: Fundamentals and Techniques", Elsevier, Amsterdam, 1983, p. A27.
31. C. Horvath and H.-J. Lin, J. Chromatogr., 126 (1976) 401.
32. C. Horvath and H.-J. Lin, J. Chromatogr., 149 (1978) 43.
33. J. C. Sternberg, Adv. Chromatogr., 2 (1966) 205.
34. V. Batu, J. Chromatogr., 260 (1983) 255.
35. J. F. K. Huber, H. H. Lauer, and H. Poppe, J. Chromatogr., 112 (1975) 377.
36. J. H. Knox, J. Chromatogr. Sci., 18 (1980) 453.
37. J. H. Knox and H. P. Scott, J. Chromatogr., 282 (1983) 297.
38. V. R. Maynard and E. Grushka, Adv. Chromatogr., 12 (1975) 99.
39. R. Thumneum and S. Hawkes, J. Chromatogr. Sci., 14 (1981) 576.
40. G. Deininger, Chromatographia, 9 (1976) 251.
41. R. P. W. Scott, Adv. Chromatogr., 9 (1970) 193.
42. E. Bottari and G. Goretti, J. Chromatogr., 154 (1978) 228.
43. C. A. Cramers, F. A. Wijnheymer, and J. A. Rijks, J. High Resolut. Chromatogr. Chromatogr. Commun., 2 (1979) 329.
44. D. F. Ingraham, C. F. Shoemaker, and W. Jennings, J. High Resolut. Chromatogr. Chromatogr. Commun., 5 (1982) 227.
45. L. S. Ettre, Chromatographia, 17 (1983) 553.
46. G. Guiochon, Anal. Chem., 50 (1978) 1812.
47. G. Gasper, C. Vidal-Madjar, and G. Guiochon, Chromatographia, 15 (1982) 125.
48. C. P. M. Schutjes, E. A. Vermer, J. A. Rijks, and C. A. Cramers, J. Chromatogr., 253 (1982) 1.
49. M. F. Gonnord, G. Guiochon, and F. I. Onuska, Anal. Chem., 55 (1983) 2115.
50. I. Brown, Chromatographia, 12 (1979) 467.
51. J. D. Schieke and V. Pretorius, J. Chromatogr., 132 (1977) 217.
52. J. W. Jorgenson and E. J. Guthrie, J. Chromatogr., 255 (1983) 335.
53. G. Guiochon, Adv. Chromatogr., 8 (1969) 179.
54. R. J. Jonker, H. Poppe, and J. F. K. Huber, Anal. Chem., 54 (1982) 2447.
55. J. C. Giddings, J. Chromatogr., 5 (1961) 46.
56. J. F. K. Huber and J. A. R. J. Hulsman, Anal. Chim. Acta, 38 (1967) 305.
57. G. J. Kennedy and J. H. Knox, J. Chromatogr. Sci., 10 (1972) 549.
58. E. D. Katz and R. P. W. Scott, J. Chromatogr., 270 (1983) 29.
59. L. S. Ettre and E. W. March, J. Chromatogr., 91 (1974) 5.
60. M. Martin, C. Eon, and G. Guiochon, J. Chromatogr., 99 (1974) 357.

61. I. Halasz, R. Endele, and J. Asshauer, J. Chromatogr., 112 (1975) 37.
62. I. Halasz, H. Schmidt, and P. Vogtel, J. Chromatogr., 126 (1976) 19.
63. E. Katz and R. P. W. Scott, J. Chromatogr., 253 (1982) 159.
64. G. Guiochon, Adv. Chromatogr., 8 (1969) 179.
65. G. Guiochon, in C. Horvath (Ed.), "High-Performance Liquid Chromatography. Advances and Perspectives", Academic Press, New York, 2 (1980) 1.
66. G. Guiochon, Anal. Chem., 52 (1980) 2002.
67. J. J. Kirkland, W. W. Yau, H. J. Stoklosa, and C. J. Dilks, J. Chromatogr. Sci., 15 (1977) 303.
68. R. E. Pauls and L. B. Rogers, Sepn. Sci., 12 (1977) 395.
69. J. R. Conder, J. High Resolut. Chromatogr. Chromatogr. Commun., 5 (1982) 341.
70. J. R. Conder, J. High Resolut. Chromatogr. Chromatogr. Commun., 5 (1982) 397.
71. O. Grubner and D. Underhill, J. Chromatogr., 73 (1972) 1.
72. K. Yamaoka and T. Nakagawa, Anal. Chem., 47 (1975) 2051.
73. K.-P. Li and Y.-Y. Hwa Li, Anal. Chem., 48 (1976) 737.
74. O. Pazdernik and P. Schneider, J. Chromatogr., 207 (1981) 181.
75. A. S. Said, H. Al-Ali, and E. Hamad, J. High Resolut. Chromatogr. Chromatogr. Commun., 5 (1982) 306.
76. J. R. Conder, G. J. Rees, and S. McHale, J. Chromatogr., 258 (1983) 1.
77. S. N. Chesler and S. P. Cram, Anal. Chem., 43 (1971) 1922.
78. C. Vidal-Madjar and G. Guiochon, J. Chromatogr., 142 (1977) 61.
79. T. Peticlerc and G. Guiochon, J. Chromatogr. Sci., 14 (1976) 531.
80. R. E. Pauls and L. B. Rogers, Anal. Chem., 49 (1977) 625.
81. W. W. Yau, Anal. Chem., 49 (1977) 395.
82. W. E. Barber and P. W. Carr, Anal. Chem., 53 (1981) 1939.
83. J. P. Foley and J. G. Dorsey, Anal. Chem., 55 (1983) 730.
84. J. P. Foley and J. G. Dorsey, J. Chromatogr. Sci., 22 (1984) 40.
85. L. S. Ettre, Chromatographia, 6 (1973) 489.
86. L. S. Ettre, Chromatographia, 7 (1974) 39.
87. J. K. Haken, Adv. Chromatogr., 14 (1976) 367.
88. M. V. Budahegyi, E. R. Lombosi, T. S. Lombosi, S. Y. Meszaros, Sz. Nyiredy, G. Tarjan, I. Timar, and J. M. Takacs, J. Chromatogr., 271 (1983) 213.
89. H. van den Dool and P. Dec. Kratz, J. Chromatogr., 11 (1963) 463.
90. J. Lee and D. R. Taylor, Chromatographia, 16 (1983) 286.
91. F. Vernon and J. B. Suratman, Chromatographia, 17 (1983) 597.
92. C. M. White, A. Robbat, and R. M. Hoes, Chromatographia, 17 (1983) 605.
93. M. L. Lee, D. L. Vassilaros, C. M. White, and M. Novotny, Anal. Chem., 51 (1979) 768.
94. D. L. Vassilaros, R. C. Kong, D. W. Later, and M. L. Lee, J. Chromatogr., 252 (1982) 1.
95. F. Pacholec and C. F. Poole, Anal. Chem., 54 (1982) 1019.
96. K. Ballschmiter and M. Zell, Z. Anal. Chem., 293 (1978) 193.
97. T. Schwartz, J. Petty, and R. Kaiser, Anal. Chem., 55 (1983) 1839.
98. L. N. Zotov, G. V. Golovkin, and R. V. Golovnya, J. High Resolut. Chromatogr. Chromatogr. Commun., 4 (1981) 6.
99. J. K. Baker and C.-Y. Ma, J. Chromatogr., 169 (1979) 107.
100. R. M. Smith, Anal. Chem., 56 (1984) 256.
101. M. N. Hasan and P. C. Jurs, Anal. Chem., 55 (1983) 263.
102. H. Colin, G. Guiochon, and P. Jandera, Chromatographia, 17 (1983) 83.
103. H. Colin, A. M. Krstulovic, M.-F. Gonnord, G. Guiochon, Z. Yun, and P. Jandera, Chromatographia, 17 (1983) 9.
104. A. M. Krstulovic, H. Colin, A. Tchapla, and G. Guiochon, Chromatographia, 17 (1983) 228.

Chapter 2

THE COLUMN IN GAS CHROMATOGRAPHY

2.1 Introduction

Numerous papers have been written about gas–liquid chromatography since the
first description by James and Martin [1]. Its impact on modern analytical
chemistry has clearly been immense; this technique has been exploited to solve a
range of problems that encompass medical, biological, and environmental
sciences, as well as industrial applications. In spite of developments in
spectroscopy and liquid chromatography, gas chromatography remains the most
widely used separation tool in analytical chemistry and is likely to remain so
in the foreseeable future. No other analytical technique can provide equivalent
resolving power or sensitivity for volatile organic compounds. The limitations
of the technique are established primarily by the thermal stability of the
samples and chromatographic substrates. Generally, one is restricted to an
upper temperature of around 400°C and a molecular weight less than 1000,
although higher temperatures have been used and larger molecular weight samples
have been separated in a few instances. To be suitable for gas chromatographic
analysis the sample, or some convenient derivative of it, must be thermally
stable at the temperature required for volatilization.

In gas chromatography samples are separated by distribution between a
stationary phase and a mobile phase by adsorption, partition, or a combination
of the two. When a solid adsorbent serves as the stationary phase the technique
is called gas–solid chromatography (GSC). When a liquid phase is supported on
an inert support or coated as a thin film onto the wall of a capillary column
the technique is designated gas–liquid chromatography (GLC). The separation
medium may be a coarse powder, coated with a liquid phase through which the
carrier gas flows. This is an example of packed column gas chromatography. If
the adsorbent, liquid phase, or both, are coated onto the wall of a narrow bore
column of capillary dimensions, the technique is called wall–coated open tubular
(WCOT), support–coated open tubular (SCOT), or porous–layer open tubular (PLOT)
column gas chromatography. A distinction between these column types will be
made later.

In this chapter we will adhere closely to the historical development of gas
chromatography, describing packed columns before open tubular columns. In
recent years spectacular developments in column fabrication and instrument
design have diminished the importance of packed columns even for simple mixture
analysis. This trend will surely continue. However, many of the theoretical
and practical developments in gas chromatography have employed packed columns
and it would be inappropriate to ignore them entirely.

2.2 Gas-Liquid Chromatography

Separations occur in GLC because of selective interactions between the solute and the stationary liquid phase. All solutes spend the same length of time in the gas phase. On the molecular level the principal intermolecular forces that occur between a solute and a solvent are dispersion, induction, orientation, and donor-acceptor interactions, including hydrogen bonding [2-8]. The physical descriptions of these interactions are reviewed in Table 2.1. Although these forces provide a suitable framework for the qualitative understanding of the separation process, the concert of interacting forces between polyatomic molecules is far too complex for a quantitative description.

For nonpolar solutes (e.g., hydrocarbons) dispersive forces are the sole forces of interaction in the pure liquid state. Dispersive forces are non-selective and do not vary greatly among simple organic molecules. The elution of nonpolar species from any chromatographic column occurs, therefore, in order of increasing boiling point. Dispersive forces are not considered important for the separation of polar solutes. Here the important interactions are orientation and induction forces, which depend on the polarizabilities, ionization potentials, and dipole moments of the solute and liquid phase, and specific electron donor-acceptor interactions of a chemical nature. Orientation and induction forces are usually in the range of 1-10 kcal/mole while donor-acceptor interactions, including hydrogen bonding, are about 1-8 kcal/mole [8]. The sum of the various intermolecular forces between a particular solute and liquid phase is a measure of the polarity of that phase with respect to that solute. The magnitude of the individual interaction energies is a measure of phase selectivity. Differences in selectivity are important chromatographically, for these interactions enable two solutes of equal polarity to be separated by a selective liquid phase. Although not specifically discussed above, when fine tuning a separation it may be necessary to also take molecular size and shape into account.

2.2.1 Frequently Used Liquid Phases

The properties desired of an ideal liquid phase are contradictory and a compromise must be reached between theoretical and practical considerations [9]. It is generally desirable for the liquid phase to have a wide temperature operating range. Ideally, this range would include all temperatures from the lowest to the highest used in GLC, approximately -60°C to 400°C. No phase meets this requirement but several popular phases have stable liquid temperature ranges extending from 100 to 300°C. The liquid phase should have a low vapor pressure, be chemically stable, and have a low viscosity (for packed column use)

TABLE 2.1

PRINCIPAL INTERMOLECULAR FORCES CHARACTERIZING SOLUTE/STATIONARY PHASE INTERACTIONS

α_A and α_S = polarizabilities of solute and solvent molecules; I_A and I_S = ionization energies of solute and solvent molecules; r = distance between dipoles; μ_A and μ_S = dipole or induced dipole moments in solute and solvent molecules; K = Boltzmann constant; T = temperature (K).

Type	Description	Representation	Comments
Dispersive (or London) Forces	Arise from the electric field produced by the very rapidly varying dipoles between nuclei and electrons at in molecules with zero-point motion. Induced dipoles are formed in phase with the instantaneous dipoles producing them.	$E_D = \dfrac{-3\alpha_A\alpha_S I_A I_S}{2r^6(I_A+I_S)}$	Present in all solute/solvent systems. Only source of attraction between nonpolar molecules. Independent of temperature.
Induction (or Debye) Forces	Arise from the interaction of a permanent dipole with a polarizable molecule.	$E_i = 1/r^6(\alpha_S\mu_A^2 + \alpha_A\mu_S^2)$	Generally weak and decrease with increasing temperature. Dipole-induced dipole interactions are not the same in all directions and depend on relative molecular orientation.
Orientation (Keesom) Forces	Arise from the net attraction between molecules or portions of molecules possessing a permanent dipole moment.	$E_o = \dfrac{-2}{3}\dfrac{\mu_A^2\mu_S^2}{r^6 KT}$	Decrease with increasing temperature and approach zero at very high temperatures when all orientations are equally probable.
Donor-Acceptor Complexes	Special chemical bonding interactions which arise from the partial transfer of electrons from a filled orbital on the donor to a vacant orbital on the acceptor molecule.		Examples include coordination forces between metal ions and olefins, charge transfer forces, and hydrogen bonding interactions.

at the desired operating temperature. For high temperature operation this
generally dictates the use of high molecular weight polymeric liquids. As
polymers lack a well-defined chemical structure their composition is subject to
batch-to-batch variations and they cannot be considered ideal liquid phases.
Practical considerations dictate that the liquid phase must have reasonable
solubility in some common volatile organic solvent and adequately wet the
supports used in column fabrication. The phases themselves must show reasonable
solvent properties (i.e., dissolving power) for the solutes to be separated and
phases with varying selectivities are needed to effect difficult separations of
polar samples.

Prior to the development of suitable techniques for liquid phase
characterization numerous phases were in general use, leading to what has been
called, "stationary phase pollution" [10]. Gradually, many of those phases with
poor chromatographic properties, those that simply duplicated the properties of
others, and those available as general industrial materials whose composition
varied from the same or different sources, have passed into disuse. Today most
separations can be performed on a handful of preferred phases, often specially
synthesized for GLC purposes [10-12]. These and some additional popular or
novel phases will be described in the following paragraphs.

From the inception of GLC high molecular weight hydrocarbons such as
squalane (I) and Apiezon greases have been used as nonpolar stationary phases
[13]. Dispersion is the sole type of interaction possible with these phases.

$$CH_3 \quad CH_3 \quad CH_3 \quad CH_3 \quad CH_3 \quad CH_3$$
$$CH-(CH_2)_3-CH-(CH_2)_3-CH-(CH_2)_4-CH-(CH_2)_3-CH-(CH_2)_3-CH$$
$$CH_3 \qquad\qquad\qquad\qquad\qquad\qquad\qquad\qquad CH_3$$

2,6,10,15,19,23-hexamethyltetracosane

squalane (I)

Accordingly, nonpolar solutes are eluted in order of volatility while polar
solutes elute largely according to their hydrophobicity (size of nonpolar
portion), as their inherent polarity is not a factor. The principal limitation
of squalane is its low maximum allowable operating temperature limit of about
120°C. The Apiezon greases, prepared by the high temperature treatment and
molecular distillation of lubricating oils, provide higher maximum allowable
operating temperature limits, up to about 300°C. Commercial materials with a
characteristic yellow-brown color contain residual carbonyl and carboxylic acid
groups which may cause tailing of polar solutes. Prior to use the greases

should be purified by liquid chromatography over alumina [13]. The structure of
the Apiezon greases has not clearly been established, although they are known to
contain both olefinic and aromatic unsaturated groups. Hydrogenation removes
any residual unsaturation and produces superior high temperature phases
[14-16]. However, discrepancies exist in the values obtained for the average
molecular weight. By vapor-phase osmometry a value of 3970 was obtained for
Apiezon L and 2340 for Apiezon M [14]. Values of 1300 and 950, respectively,
were obtained by size-exclusion chromatography [15]. Kovats has prepared a
synthetic, high temperature hydrocarbon phase ($C_{87}H_{176}$, Apolane-87) (II) with a
molecular weight of 1222 and a maximum operating temperature of 280-300°C
[17-19].

$$\begin{array}{ccc}
H_{37}C_{18} & C_2H_5 & C_{18}H_{37} \\
| & | & | \\
CH-(CH_2)_4-C-(CH_2)_4-CH \\
| & | & | \\
H_{37}C_{18} & C_2H_5 & C_{18}H_{37}
\end{array}$$

24,24-diethyl-19,29-dioctadecylheptatetracontane

Apolane - 87 (II)

Hydrocarbon phases are widely used as nonpolar reference phases in schemes
designed to measure the selectivity of polar phases. Both Rohrschneider and
McReynolds selected squalane for this purpose. Apolane-87 and hydrogenated
Apiezon, Apiezon MH, have been suggested as substitutes due to their greater
thermal stability. Retention index values on Apiezon MH and Apolane-87 are
virtually identical and, in both instances, slightly larger than on squalane;
this difference is attributed to the molecular weight difference between the
high temperature phases and squalane. Schemes for measuring stationary phase
selectivity are discussed in the next section.

All hydrocarbon phases are susceptible to oxidation, which alters their
chromatographic properties by the introduction of polar, oxygenated functional
groups [20-22]. In extreme cases scission of the carbon backbone may occur,
resulting in a high level of column bleed. Oxidation of squalane (I), which is
fully hydrogenated, arises from the presence of tertiary hydrogen atoms that
react with oxygen to form thermally unstable hydroperoxides that in turn yield
hydroxylic and carbonyl derivatives. Apolane-87 has a much lower concentration
of tertiary hydrogens and is therefore more resistant to oxidation.

Sporadic accounts of the use of fluorocarbon liquid phases have appeared
almost from the inception of GLC. Early reports employed perfluoroalkanes,
Kel-F oils, and fluoroalkyl esters for the analysis of corrosive and reactive
chemicals (e.g., UF_6, ClO_2F, SF_5Cl, etc.) which could not be separated on

conventional phases [23,24]. Attempts to use these phases for more general applications were less successful due to a combination of poor support wetting characteristics, low column efficiencies, and low upper operating temperature limits [25]. Intermolecular forces in perfluorocarbon liquids are weak, giving rise to diminished retention in GLC, thus making the separation of high molecular weight or thermally labile samples possible at lower temperatures than on conventional nonfluorinated phases [26,27]. The poly(perfluoroalkyl ether) liquid phase, Fomblin YR (III), produces columns of high efficiency and may be used at temperatures of up to 255°C.

$$\left[\begin{array}{c} CF_3 \\ | \\ (O\text{-}CF\text{-}CF_2)_n\text{-}(O\text{-}CF_2)_m \end{array} \right]$$

Fomblin YR (MW = 6000-7000)

(III)

This temperature is well below that at which column bleed is significant and is established by the stability of the liquid film; above 255°C the liquid film "beads up", lowering column efficiency. The fluorinated alkyl esters, Fluorad FC-430 and FC-431, are generally useful, medium polarity liquid phases that strongly deactivate diatomaceous earth supports [28,29]. Low concentrations of underivatized amines, phenols, and carboxylic acids can be separated on these phases. The absolute retention of organic compounds on Fluorad FC-430 is approximately 2 to 6 times lower than on a nonfluorinated phase with similar polarity.

By far the most important group of liquid phases are the polysiloxanes [30-33]. Their high thermal stability, wide liquid range, and availability in different polarities, spanning the range from nonpolar to some of the most polar phases presently available, contribute to their popularity. The basic siloxane backbone can be represented by structure (IV), in which R can be either methyl, vinyl, phenyl, 3,3,3-trifluoropropyl, cyanoethyl, or cyanopropyl groups. Many polymers contain mixtures of the above functional groups, as indicated in Table 2.2. The polysiloxane polymers are prepared by the acid hydrolysis of appropriately substituted dichlorosilanes or by the catalytic polymerization of

$$\left[\begin{array}{cc} R & R \\ | & | \\ Si\text{-}O\text{-}Si\text{-}O \\ | & | \\ R & R \end{array} \right]$$

(IV)

cyclosiloxanes in the presence of a small amount of hexamethyldisiloxane as a chain stopper. Materials of high purity, particularly with regard to the presence of residual catalyst, and narrow molecular weight range are required for chromatographic purposes [34]. In addition, terminal silanol groups must be endcapped to maximize thermal stability. Polysiloxanes containing α-cyano or α- or β-fluoro groups are thermally labile and not useful chromatographically. The poly(3,3,3-trifluoropropylsiloxanes) are particularly sensitive to strong bases, undergoing depolymerization to form volatile cyclotetrasiloxanes. The polysiloxanes otherwise exhibit exceptional chemical, thermal, and oxidative stability.

The phenyl-containing polysiloxanes are more polarizable than the poly(methylsiloxanes) and exhibit some benzene-like solvent properties. The 3,3,3-trifluoropropyl-containing polymers have an unusually high affinity for carbonyl-containing solutes due to induction of some positive charge character at the carbon atom adjacent to the electronegative trifluoromethyl group. The cyanopropyl-containing polysiloxanes interact strongly with dipolar solutes.

TABLE 2.2

PROPERTIES OF SOME COMMERCIALLY AVAILABLE POLYSILOXANE PHASES

Name	Type	Structure (R in IV)	Density (g/ml)	Viscosity (cP)	Average Molecular Weight
OV-1	Dimethylsilicone gum	CH_3	0.975		$>10^6$
OV-101	Dimethylsilicone fluid	CH_3		1500	3×10^4
OV-7	Phenylmethyldimethyl-silicone	80% CH_3 20% C_6H_5	1.021	500	1×10^4
OV-17	Phenylmethylsilicone	50% CH_3 50% C_6H_5	1.092	1300	4×10^3
OV-25	Phenylmethyldiphenyl-silicone	25% CH_3 75% C_6H_5	1.150	$>100,000$	1×10^4
OV-210	Trifluoropropyl-methylsilicone	50% CH_3 50% $CH_2CH_2CF_3$	1.284	10,000	2×10^5
OV-225	Cyanopropylmethyl-phenylmethylsilicone	50% CH_3 25% C_6H_5 25% C_3H_6CN	1.096	9000	8×10^3
OV-275	Dicyanoethylsilicone	C_2H_4CN		20,000	5×10^3

A series of carborane-siloxane polymers with exceptional thermal stability have been described [32]. Dexsil 300GC (V) contains one carborane group attached in the meta position as part of a chain with dimethylsiloxane groups as the repeating unit. Dexsil 400GC and 410GC have as the repeating unit one carborane group and five dimethysiloxane units, with the central silicon atom substituted with a phenyl and a 2-cyanoethyl group, respectively. These phases are of modest polarity but can be operated at temperatures up to 450°C for Dexsil 300GC and 375°C for Dexsil 400GC and 410GC.

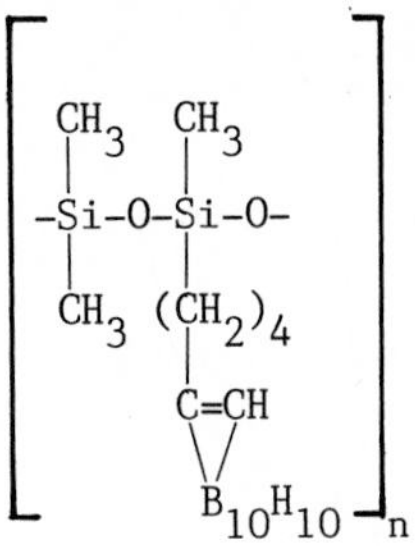

$$\left[\begin{array}{c} \underset{\underset{CH_3}{|}}{\overset{\overset{CH_3}{|}}{Si}}-O-\underset{\underset{(CH_2)_4}{|}}{\overset{\overset{CH_3}{|}}{Si}}-O- \\ \\ C=CH \\ \diagdown \diagup \\ B_{10}H_{10} \end{array} \right]_n$$

Dexsil 300GC

(V)

The meta-linked poly(phenyl ethers) (VI) are useful, chemically well-defined, moderately polar liquid phases. Their low volatility is exceptional

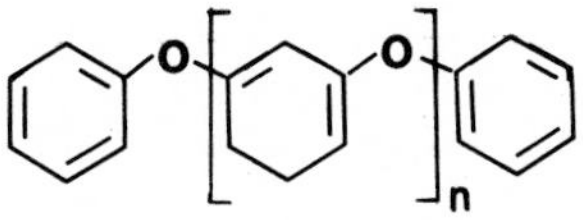

poly(phenyl ether)   n = 3 or n = 4

(VI)

for their molecular weight. The five and six ring poly(phenyl ethers) are stable to 200°C and 250°C, respectively. The flexibility imparted to the chain by the ether linkage is responsible for their low viscosity. A high molecular weight poly(phenyl ether), in this instance a true polymer containing an average of 20 rings, was used to separate solutes of high boiling point with column temperatures in the range 125–400°C [35]. Poly(phenyl ethers) with polar functional groups have been investigated for use as polar reference phases [36,37]. In general these phases have low column efficiencies and poor thermal stability [36,38]. The five and six ring poly(phenyl ethers) can be copolymerized with diphenyl ether-4,4'-disulfonyl chloride to produce a polymer of undefined molecular weight and chemical structure, containing repeating units of poly(phenyl ether) joined together by diphenyl ether sulfonyl bridging groups [39,40]. Commercially available as polyphenyl ether sulfone or, Poly-S-179, it has been used to separate a wide range of high-boiling polar solutes at temperatures in the range 200-400°C.

Phthalate esters (VII) are well-characterized, moderately polar liquid phases [41]. As might be expected, the polarity of the phases declines as the

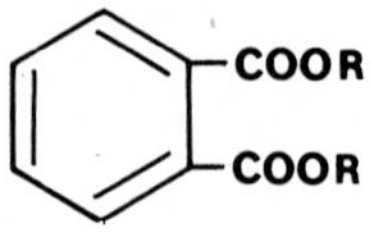

Phthalate ester

(VII)

alkyl (R) group increases in size, while their volatility decreases. Tetrachlorophthalate esters show selective charge transfer interactions with electron donor solutes. Poor thermal stability compared to other available liquid phases has reduced their importance in recent years.

A more important group of moderately polar liquid phases is the polyesters [42]. The term polyester is used to describe a wide range of resinous composites derived from the reaction of a polybasic acid with a polyhydric alcohol [42,43]. The most frequently used polyester phases for GLC are the succinates and adipates of ethylene glycol, diethylene glycol, and butanediol. In particular, poly(ethylene glycol succinate) (VIII) and poly(diethylene glycol succinate) (IX) have been frequently used.

Poly(ethylene glycol succinate)   n = 17 to 21

EGS   (VIII)

Poor column stability and column-to-column variations in chromatographic properties have recently diminished their use. Upon conditioning the molecular weights of EGS (VIII) and DEGS (IX) are increased from about 3000 to about 9000-16,000 [44]. The higher molecular weight polymer is a superpolyester, formed with the loss of volatile stationary phase components upon heating to

high temperatures. These phases show poor hydrolytic stability. The ester
linkages along the polymer backbone break upon hydrolysis, resulting in the
formation of additional hydroxyl and carboxylic acid functionalities. The
presence of residual acid and hydroxyl groups precludes the use of these phases
in the analysis of isocyanates, epoxides, anhydrides, halo acids, etc., which
may react with these groups. Amines may also cause aminolysis of the ester
linkage. At high temperatures polyesters are also susceptible to oxidative
degradation, which occurs through the formation of hydroperoxides that later
decompose to form free radicals. The free radicals decompose the polyester
chains, producing vinyl-type unsaturation, and may also cause polymerization.

$$HO-(CH_2)_2-O-(CH_2)_2\left[-O-\overset{\overset{O}{\|}}{C}-CH_2-CH_2-\overset{\overset{O}{\|}}{C}-O-(CH_2)_2-O-(CH_2)_2\right]_n OH$$

Poly(diethylene glycol succinate)   n = 12 to 18

DEGS (IX)

Poly(ethylene glycols) of general formula (X) are prepared by the
polymerization of ethylene oxide [45]. The products of the reaction are

$$HO(CH_2CH_2O)_n CH_2CH_2OH$$

Poly(ethylene glycol)

(X)

separated into a series of fractions having different nominal average molecular
weights. Carbowax 20M for example, has an average molecular weight of 14,000
[46] while that of Superox-4 is approximately four million [47]. This molecular
weight difference is reflected in their thermal stability; Superox-4 may be used
at temperatures up to 300°C while Carbowax 20M has an upper temperature limit of
only 225°C. Superox-4 is somewhat difficult to use due to its low solubility in
common solvents and its rapid decomposition by oxygen or moisture at high
temperatures [47,48]. Oxidative degradation can be a problem with the lower
molecular weight Carbowaxes as well.

The pluronics are related to the poly(ethylene glycols) but have better
defined molecular weights and a narrower polydispersity [46,49]. They are
prepared by condensing proplyene oxide with proplyene glycol. The resulting
chain is then extended on both sides by the addition of controlled amounts of
ethylene oxide until the desired molecular weight is obtained. The pluronics
most useful as liquid phases have average molecular weights in the range 2000 to
about 8000, with maximum allowable operating temperatures in the range

220–260°C.

The poly(ethylene glycols) are useful phases for the separation of a wide range of polar and hydrogen bonding solutes. The principal factors determining retention are the concentration of hydroxyl groups and, to a much lesser degree, the molecular weight distribution of the liquid phase.

Fused inorganic salt eutectic mixtures were used by Juvet to separate metal halides [50]. Samples of this type are difficult to separate using organic liquid phases because they react chemically with the phase and require high temperatures for volatilization. The solutes studied were capable of forming chloro-complexes with the chloride ions present in the melt. The general application of inorganic molten salts to the separation of organic compounds was far less successful due to the limited solubility of organic compounds in these melts [51]. The affinity of organic compounds for organic molten salts should be much greater and the availability of low temperature organic melts is a further advantage. This has been borne out in practice [52–54]; in particular, ethylpyridinium bromide and tetrabutylammonium tetrafluoroborate have been shown to be generally useful polar liquid phases. They exhibit high selectivity for compounds possessing large dipole or hydrogen bonding functional groups. Retention is believed to depend upon a combination of solute partitioning and adsorption interactions; the relative magnitude of each depends on the nature of the solute's functional groups. Organic molten salts are relatively new liquid phases and further study is needed to better characterize their properties [55].

2.2.2 Selective Liquid Phases

Certain liquid phases employing suitably selective interactions are occasionally used to perform specific separations. The separation of donor solutes by phases containing coordination complexes and of enantiomers by chiral liquid phases are described in Chapter 7.17. Electron-acceptor compounds such as dialkyltetrachlorophthalates and 2,4,7-trinitro-9-fluorenone have been used to separate unsaturated hydrocarbons and aromatic hydrocarbons (electron donors) by π-charge transfer complex formation [56–61]. As such interactions are perturbed by steric influences, these phases are often able to resolve isomeric mixtures that are difficult to resolve on other phases.

Liquid crystal phases are widely used for separating mixtures of rigid isomeric solutes such as substituted benzenes, polycyclic aromatic hydrocarbons, polychlorinated biphenyls, and steroids [62–64]. The liquid crystalline state represents a specific state of matter intermediate between a crystalline solid and an isotropic liquid. Thus in the liquid crystalline state the phase exhibits the mechanical properties characteristic of a liquid while some of the anisotropic properties of the solid are maintained due to the preservation of a higher degree of ordering than in liquids. The liquid crystals used as stationary phases are of the thermotropic type, that is, the liquid crystalline state commences at the melt and is thermally stable until, at some higher temperature (clearing temperature), a transition to an isotropic liquid takes place. Depending on the way the molecules are ordered, thermotropic liqid crystals are classified into nematic, cholesteric, and smectic types. Smectic phases possess a two-dimensional array in which rod-like molecules are arranged in parallel to give a layered structure whose thickness is approximately equal to the length of the molecules comprising the layer. In the nematic phase the parallel orientation is maintained but the layered structure does not exist and liquid molecules are free to move within the limits of the parallel configuration. The cholesteric phase is a twisted nematic structure and occurs in compounds possessing chiral centers. Most liquid crystalline phases used in GLC are of the nematic type as these usually provide the widest temperature operating range.

Well over two hundred liquid crystalline phases have been used in GLC [62-64]. They are of a variety of chemical types but all have in common a markedly elongated rigid rod-like or lath-like structure. The most common types are Schiff's bases, esters, and azoxybenzenes as shown in Table 2.3. When coated onto solid supports their column efficiency is usually less than that found with conventional polymeric phases due to their high viscosity and poor mass transfer characteristics [65]. Retention volumes may vary unpredicatably with the amount of liquid phase coated and the type of support used [65,66]. Reproducible retention data can only be expected with high loadings on low surface area supports. At low loadings most of the stationary phase is present as a surface film and solute retention is due largely to adsorption at the gas-liqiuid and gas-solid interfaces. As the amount of liquid phase increases, the support pores are filled and solute partition with the bulk liquid dominates the retention process. The efficiency of liquid crystalline phases can be improved by blending them with conventional stationary phases [67] or by using eutectic mixtures of two or more liquid crystalline phases [68]. Polysiloxanes containing liquid crystalline functional groups have been shown to provide higher efficiency and wider temperature operating ranges than conventional liquid crystalline phases [69].

In contrast to conventional isotropic liquid phases where the separation mechanism is mainly determined by solute volatility and polarity, anisotropic liquid crystals discriminate on the basis of solute shape differences. The remarkable shape selectivity exhibited by such phases is closely related to the parallel molecular alignment of the mesophase molecules that exercise preferential solubility for rigid linear molecules and steric discrimination against bulky molecules. The order of elution is generally in accord with the solute length-to-breadth ratios. Therefore, planar molecules are generally retained longer than nonplanar molecules.

TABLE 2.3

MOST COMMON STRUCTURES OF LIQUID CRYSTALLINE STATIONARY PHASES

$$Y_1 \left[\langle \bigcirc \rangle \right]_n -X- \left[\langle \bigcirc \rangle \right]_n -Y_2 \qquad n = 1 \text{ or more}$$

X	Y_1 and Y_2
$(-CH_2-CH_2-)_n$	H
$(-CH=CH-)_n$	H
$-OCH_2CH_2O-$	R
$(-OCH_2-)_n$	RO
$-CH=N-N=CH-$	CN
$-C\equiv C-$	Cl, Br, F
$-N\equiv C-$	$R-C-O-$ $\overset{\parallel}{O}$
$-N=N-$ $\overset{\parallel}{O}$	$RO-(CH_2)_n-O-$
$-CH=N-$ $\overset{\parallel}{O}$	$(CH_3)_2N-$
$-C-O-$ $\overset{\parallel}{O}$	R = alkyl or aryl

2.2.3 Liquid Phase Characterization

    The liquid phase characteristics of primary interest to the chromatographer are its temperature operating range, its viscosity within this temperature range, and the ability of the phase to selectively interact with different solutes. The minimum operating temperature is usually established by the melting point of the phase. Some phases have been used below their melting point as selective adsorbents for the separation of spatial and positional isomers which could not be resolved at temperatures above the melt [70,71]; however, this is the exception rather than the rule. As shown in Figure 2.1, the performance of columns below the liquid phase melting point is generally low and declines rapidly as lower temperatures are reached [54]. The retention characteristics of the phase may also change dramatically on melting,

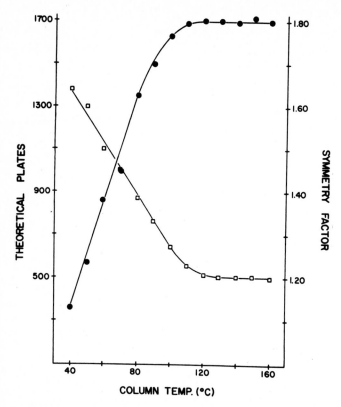

Figure 2.1   The effect of column temperature on column efficiency and peak
symmetry for the organic molten salt ethylpyridinium bromide
(m.p. 110°C). (Reproduced with permission from ref. 54.   Copyright
Friedr. Vieweg & Sohn).

particularly for solutes retained by a partition mechanism, Figure 2.2 [54].
Some polymers and liquid crystals exhibit multiple phase transitions, each one
associated with a change in retention corresponding to a change in surface area,
volume, or viscosity above the transition temperature [70].

The upper temperature limit for a liquid phase is established as the
highest temperature at which the phase can be maintained without decomposition
or significant bleed from the column. Many phase are polydisperse substances
containing low molecular weight oligomers. These oligomers may be selectively
evaporated away during column conditioning, leaving a much lower amount of phase
on the column than expected. This is less of a problem for chemically defined
phases and for phases of low polydispersity. In this case the maximum allowable
operating temperature of the phase is defined as the highest temperature at
which the phase can be maintained for 24 h without changing the retention and
performance characteristics of the components in a test chromtogram [29,36,72].
Alternatively, the maximum allowable operating temperature can be assumed to be

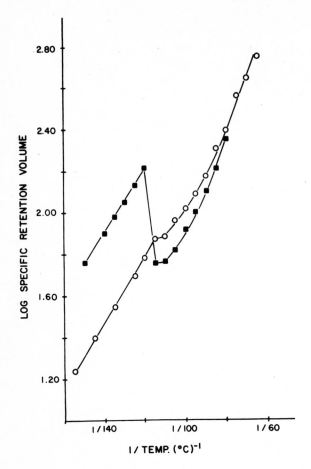

Figure 2.2   Change in retention for naphthalene (■) and n-butanol (o) at the
            phase transition point for ethylpyridinium bromide. Above the
            m.p. of the phase naphthalene is retained largely by partitioning
            while n-butanol is retained largely by an adsorption mechanism.
            (Reproduced with permission from ref. 54. Copyright Friedr. Vieweg
            & Sohn).

that temperature at which the liquid phase exhibits a vapor pressure of 0.5
Torr.

     The general polarity and selectivity of a liquid phase is determined by the
method of McReynolds [73] which is based on theoretical considerations proposed
by Rohrschneider [74,75]. The founding principle of the Rohrschneider/
McReynolds system is the additivity of intermolecular forces evaluated from the
differences in retention index values of a series of test probes measured on the
liquid phase to be characterized and on squalane, a nonpolar reference phase
[76,77]. The set of test probes must adequately characterize the principal
molecular interactions responsible for retention in gas chromatography:

dispersion, orientation, induction, and donor–acceptor complexation (including hydrogen bonding). Rohrschneider suggested that benzene, ethanol, methyl ethyl ketone, nitromethane, and pyridine be used for this purpose. McReynolds suggested ten probes, Table 2.4, including the five of Rohrschneider, except that butanol, 2–pentanone, and nitropropane were proposed for the lower molecular weight homologs. The higher molecular weight homologs were selected for practical rather than theoretical considerations. The greater retention of the higher molecular weight homologs improves the accuracy with which the retention index values can be measured. The number of probes required to completely characterize the properties of a liquid phase has been the subject of some controversy [76–80]. Conventional practice is to use the first five probes, x' through s' in Table 2.4, suggested by McReynolds. Virtually all commercially available liquid phases have been characterized this way. The system is thus well established and unlikely to be changed in the near future.

McReynolds assumed that for each type of polar intermolecular interaction, the interaction energy is proportional to values a,b,...,e characteristic of each test probe and values x',y',...,s' characteristic of the liquid phase. The retention index difference I is thus compiled of products as shown in equation (2.1).

$$\Delta I = ax' + by' + cz' + du' + es' \qquad (2.1)$$

The liquid phase constants x',y',....s' are determined by injecting each of the test probes and hydrocarbon retention index standards on the phase to be characterized. The phase specific constants are calculated by subtracting the retention index value measured on squalane from that measured on the phase to be characterized. This calculation is indicated for benzene in equation (2.2).

$$x' = \Delta I = I^{phase}(benzene) - I^{squalane}(benzene) \qquad (2.2)$$

The calculation is repeated for each probe. Liquid phase loadings of 10–20% and column temperatures of 100°C or 120°C are commonly used. The phase constants are temperature dependent, but the deviations are small and can generally be neglected within the temperature range normally used. A disadvantage of the McReynolds method is that it cannot be used to characterize those liquid phases with minimum operating temperatures above 120°C, the maximum operating temperature of squalane. Suggested solutions to this problem include the use of alternate thermally–stable reference phases [14,15,17,19] and a calculation method that does not require a reference phase [81]. The original probes proposed by McReynolds, however, are usually too volatile to provide accurate retention index values at temperatures much above 120°C. Alternative probes, such as benzene and naphthalene derivatives [14,55,82] have been used at higher temperatures, but there is as yet no well–recognized system for characterizing

TABLE 2.4

INTERACTIONS CHARACTERIZED BY McREYNOLDS' PROBES

Symbol	Test Substance	Interactions Measured	Characteristic Substance Group
X'	Benzene	Primarily dispersion with some weak proton acceptor properties.	Aromatics, olefins
Y'	Butanol	Orientation properties with both proton donor and proton acceptor capabilities.	Alcohols, nitriles, acids
Z'	2-Pentanone	Orientation properties with proton acceptor but not proton donor capabilities.	Ketones, ethers, aldehydes, esters, epoxides, dimethylamino derivatives
U'	Nitropropane	Dipole orientation properties.	Nitro and nitrile derivatives
S'	Pyridine	Weak dipole orientation with strong proton acceptor capabilities. Proton donor properties are absent.	Aromatic bases
H'	2-Methyl-2-pentanol		Branched chain compounds particularly alcohols
J'	Iodobutane		Halogenated compounds
K'	2-Octyne		
L'	1,4-Dioxane	Orientation properties with proton acceptor but not proton donor capabilities.	
M'	cis-Hydrindane		

liquid phases with high melting points.

During the early development of GLC literally thousands of liquid phases were employed by various researchers. Use of the McReynolds system enabled phases to be characterized quantitatively. This led to the discontinuation of many phases with duplicate separation properties; only those liquid phases with superior chromatographic properties and useful selectivity have endured. Some examples of popular liquid phases, along with their chromatographic properties, are given in Table 2.5.

McReynolds' constants can also be used to select a liquid phase for a particular application [76]. The sample is first considered in terms of the functional groups present. For example, consider the separation of a sample containing a series of components with a common functional group. Using Table 2.4, the McReynolds' constant which is most representative of the sample is selected. Then from Table 2.5 a phase with a large value for that constant would be chosen to provide the highest selectivity for the sample. Many samples contain components with more than one functional group and to achieve the desired separation it may be necessary to consider more than one McReynolds' constant. For example, to separate a mixture of saturated and unsaturated fatty acid esters, a phase with a large z' value (selectivity for esters) and a large x' value (selectivity for unsaturated systems) would be selected. Similar arguments can be used to select a phase for the separation of samples containing components with different functional groups; for example, a phase with a high y' value would be selected to separate alcohols from nitriles. In reaching a final decision, it is also necessary to consider the relationship between the thermal stability of the phase and the volatility of the sample.

2.2.4 Predicting Separations with Binary Liquid Phase Mixtures

Consider the situation where a sample is incompletely resolved on two different liquid phases and the components unseparated on each phase are not the same. The natural conclusion would be that a complete separation could probably be obtained if the two phases were combined in appropriate proportions. Given the separations on the two pure liquids how can one calculate the exact composition of a binary liquid phase mixture that will provide complete resolution of the mixture? A method for this purpose, based on the theory of diachoric solutions, was developed by Purnell and Laub [83]. It is commonly known as the window diagram method and has been applied with spectacular success to a number of practical problems [83-89]. Possible limitations of the theory have been discussed by others [90,91].

Diachoric solution theory predicts that retentions on mixed phases can be calculated from those on each of the pure phases according to equation (2.3).

TABLE 2.5

CHROMATOGRAPHIC PROPERTIES OF SOME TYPICAL LIQUID PHASES

Name	Type	Minimum Operating Temperature	Maximum Operating Temperature	McReynolds' Constants				
				X'	Y'	Z'	U'	S'
Squalane	Hydrocarbon	20	120	0	0	0	0	0
Apiezon MH	Hydrocarbon	40	200	18	9	5	15	22
Apiezon L	Hydrocarbon	50	250	32	22	15	32	42
Apolane-87	Hydrocarbon	30	280	21	10	3	12	25
Fomblin YR	Poly(perfluoroalkyl ether)	a	255	-15	138	88	141	51
Fluorad FC-430	Fluorinated alkyl ester	a	250	178	466	381	462	460
Fluorad FC-431	Fluorinated alkyl ester	a	200	281	423	297	509	360
OV-101	Dimethylsilicone	0	350	17	57	45	67	43
OV-7	Phenylmethyldimethylsilicone	0	350	69	113	111	171	128
OV-17	Phenylmethylsilicone	0	350	160	188	191	283	253
OV-25	Phenylmethyldiphenylsilicone	0	350	178	204	208	305	280
OV-210	Trifluoropropylmethylsilicone	0	275	146	238	358	468	310
OV-225	Cyanopropylmethylphenylmethylsilicone	0	265	228	369	338	492	386
OV-275	Dicyanoethylsilicone	25	250	629	872	763	1106	849

Dexsil 300	Methylsiliconecarborane	50	450	47	80	103	148	96
Dexsil 400	Methylphenylsiliconecarborane	50	400	72	108	118	166	123
Dexsil 410	Methylcyanoethylsiliconecarborane	50	400	72	286	174	249	171
OS-124	Polyphenyl ether (5 rings)	0	200	176	227	224	306	283
OS-138	Polyphenyl ether (6 rings)	0	250	182	233	228	313	293
Poly-S 179	Polyphenyl ether sulfone	200	400	---	---	---	---	---
	Dionyl phthalate	20	150	83	183	147	231	159
	Didecyl phthalate	10	175	136	255	213	320	235
EGS	Poly(ethylene glycol succinate)	100	200	537	787	643	903	889
EGA	Poly(thylene glycol adipate)	100	225	372	576	453	655	617
DEGS	Poly(diethylene glycol succinate)	20	200	496	746	590	837	835
DEGA	Poly(diethylene glycol adipate)	0	200	378	603	460	665	658
Carbowax 20M	Poly(ethylene glycol)	60	225	322	536	368	572	510
Superox-4	Poly(ethylene glycol)	65	300	322	536	368	572	510
Pluronic L64	Poly(ethylene glycol)	0	200	214	418	278	421	375
	Ethylpyridinium bromide	110	160	---	678	613	580	485
	Tetraheptylammonium chloride	a	130	159	774	248	440	327
	Tetrahexylammonium benzoate	a	110	154	799	384	427	330
	Tetrabutylammonium tetrafluoroborate	170	280	---	---	---	---	---
TCEP	1,2,3-Tris(2-cyanoethoxy)propane	0	125	594	857	759	1031	917

a Liquids at room temperature

$$K_R = \Theta_A K_{R(A)}{}^{\circ} + \Theta_S K_{R(S)}{}^{\circ} \qquad (2.3)$$

K_R = infinite dilution partition coefficient between the liquid mixture (A+S)

$K_{R(A)}{}^{\circ}$ = infinite dilution partition coefficient in pure liquid A

$K_{R(S)}{}^{\circ}$ = infinite dilution partition coefficient in pure liquid B

Θ_A = volume fraction composition of A

Θ_S = volume fraction composition of S

Rearranging equation (2.3) to equation (2.4), it can be seen that a plot of K_R against Θ_A will be linear, Figure 2.3A.

$$K_R = K_{R(S)}{}^{\circ} + \Theta_A (K_{R(A)}{}^{\circ} - K_{R(S)}{}^{\circ}) \qquad (2.4)$$

Furthermore, for any pair of solutes (1 and 2) whose retention behavior is described by equation (2.3), the relative retention, a, is defined by equation (2.5).

$$a = \frac{K_{R1}}{K_{R2}} = \frac{\left[\Theta_A K_{R(A)}{}^{\circ} + \Theta_S K_{R(S)}{}^{\circ}\right] \text{solute 1}}{\left[\Theta_A K_{R(A)}{}^{\circ} + \Theta_S K_{R(S)}{}^{\circ}\right] \text{solute 2}} \qquad (2.5)$$

For computational purposes a more useful form of equation (2.5) is given by equation (2.6).

$$\Theta_A = \frac{[K_{R(S)1}{}^{\circ} - a K_{R(S)2}{}^{\circ}]}{a \Delta K_{R2}{}^{\circ} - \Delta K_{R1}{}^{\circ}} \qquad (2.6)$$

$$\Delta K_R{}^{\circ} = K_{R(A)}{}^{\circ} - K_{R(S)}{}^{\circ}$$

A plot of a versus Θ_A for all solute pairs provides a window diagram, Figure 2.3B. For simplification all solute pairs with $a > 1.3$ are ignored, as their separation represents a relatively trivial problem. Also, for the purpose of visualizing the data, when plots of K_R versus Θ_A cross for any solute, the calculated value of a is inverted so that $a \geq 1$ at all times. The resultant window diagram is a series of approximate triangles rising from the Θ_A axis that constitute the windows within which complete separtion for all solutes can be achieved. The optimum operating conditions (a, Θ_A) correspond to the peak of the highest window. Further, the number of plates required for complete separation can be calculated from equation (2.7).

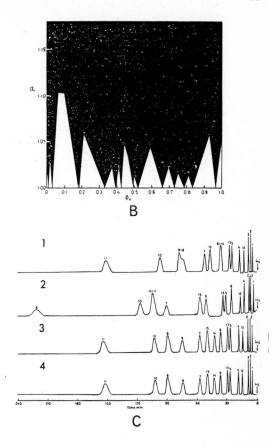

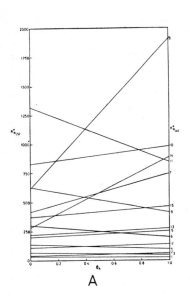

Figure 2.3   An example of the use of the window diagram method to select an
             optimum binary liquid phase mixture. A, retention data for all
             solutes on (S) squalane and (A) dinonylphthalate; B, window
             diagram; C, 1. separation on squalane, 2. separation on
             dinonylphthalate, 3. separation on the optimum binary phase
             mixture, 4. separtion obtained by mechanically combining individual
             packings to give the predicted optimum binary phase.

$$n_{req} = 36\left[\frac{a}{a-1}\right]^2\left[\frac{k+1}{k}\right]^2 \qquad (2.7)$$

n_{req} = number of theoretical plates required to give baseline resolution

a = selectivity factor
k = capacity factor

This simplifies to equation (2.8) for mixtures which are difficult to separate

and $k \geq 10$.

$$n_{req} = 36 \left[\frac{a}{a-1} \right]^2 \tag{2.8}$$

Thus, knowing the values of n attainable with the pure liquid phase columns, the required length of the mixed-phase column can be calculated with reasonable accuracy.

Consider an example based on the above discussion. Figure 2.3A shows the K_R° values measured on squalane and dinonyl phthalate for a 15-component mixture. Equation (2.3) is assumed to hold true and the K_R° values for each component on the pure phases are connected by straight lines. This information is used to construct the window diagram in Figure 2.3B. It can be seen that of the 14 windows available all but three are valueless since the highest a value available for the most difficult to separate pair is less than the value for pure dinonyl phthalate. Of the other three options, that at $\Theta_A = 0.075$ is unquestionably the best. This value of a is then used to predict the column length required for the separation using equation (2.8). Figure 2.3C presents the data in terms of the actual chromatograms. The chromatograms in Figure 2.3(C1) and (C2) illustrate the separation obtained on pure squalane and dinonyl phthalate, respectively. Neither phase provides a complete separation of the mixture. Figure 2.3(C3) illustrates the separation on the binary phase mixture predicted by the window diagram method. The separation is complete and, furthermore, the elution order is the same as that predicted by reading up the window lines in Figure 2.3A. Figure 2.3(C4) indicates that identical results can be obtained by mechanically combining the correct proportions of squalane and dinonyl phthalate packings.

The window diagram method can also be used to optimize the separation of mixtures when the number and identity of the components are unknown [86]. Two liquid phases, A and S, of different selectivity are chosen. Trial chromatograms are run on the two columns at some common temperature arrived at by visual optimization. A standard solute for which $K_{R(A)}^{\circ}$ and $K_{R(S)}^{\circ}$ are accurately known is included in the trial chromatograms to allow retention time data to be converted into K_R° values. The sample, including standard, is then run at the selected temperature on columns containing pure A, pure S, and on 2:1, 1:1, and 1:2 v/v mixtures of the two liquid phases. For any given solute eluted from A, S, and A/S phases the corresponding K_R/Θ_A data must lie on a straight line. This consideration enables the corresponding data points on each column to be distinguished and a plot similar to Figure 2.3A to be constructed. Once this has been completed the window diagram can be constructed as described previously. With this approach the presence of overlapping peaks on either column or changes in the elution order between the two columns does not present problems.

The window diagram method has been developed for isothermal packed column gas chromatography, although extension to temperature programmed conditions should be possible [92]. The approach is less frequently used with open tubular columns where mixed phases are rarely used [93,94]. Because of the poor efficiency of packed columns it is essential to optimize α to provide maximum resolution. With open tubular columns very large values of n can be obtained, making feasible the separation of mixtures with low α values. In addition, the number of calculations required for an average capillary column run is prohibitive.

2.2.5 Supports for Gas-Liquid Chromatography

An ideal support would have a large surface area capable of holding the stationary liquid phase as a thin film and would play no further role in the separation process. To eliminate solute interaction with the support its surface activity should be zero or at least very small. From the practical point of view, this is obviously impossible as the surface must have sufficient energy to both hold the liquid phase stationary and to cause it to wet the surface in the form of a thin film. The ideal support would have a large surface area to weight ratio to allow for the use of high liquid phase loadings and have a regular shape with a narrow range of cross-sectional diameters so that columns of high efficiency could be prepared. It should also be a good conductor of heat and mechanically and thermally stable to avoid changes in properties on handling or while in use. No such ideal support exists and a compromise between those properties which are desirable and those which can be obtained practically must be reached. By far the most important support materials are the diatomaceous earths, which provide a favorable balance between support area, mechanical and thermal properties, support activity and shape.

Diatomite (diatomaceous earth, Kieselguhr) is composed of the skeletons of diatoms, single-celled algae, which have accumulated in large beds in various parts of the world. The skeletal material is essentially microamorphous silica with small amounts of alumina and metallic oxide impurities, Table 2.6 [95]. The porous nature of the diatomite with its associated secondary structure gives the material a high surface area to weight ratio, approximately 20 m^2/g. As mined, the diatomite skeletons are too small and fragile to be used as chromatographic supports; further processing agglomerates and strengthens the natural material. This is achieved by calcining at temperatures in excess of 900°C. The crude material is ground to a powder and shaped into the form of a fire brick (perhaps with the addition of a small amount of clay binder) before being calcined. The calcination serves to fuse the ground particles together with some of the silica being converted to cristobalite and the mineral impurities forming complex oxides or silicates. The presence of complex iron oxides gives this material its characteristic pink color. Close inspection of the processed diatomite material reveals a compact mass of broken diatomite fragments with a portion of the secondary pore structure still intact so that the material retains a relatively large surface area. If the diatomite is first broken into chunks and then calcined in the presence of a small amount of sodium carbonate flux, then a useful modification of its properties can be achieved. This material is white, as the metal impurities are converted to colorless sodium metal silicates, and close inspection reveals a mass of large diatomite fragments held together by sodium silicate glass with most of the secondary pore structure destroyed. Thus the white material has a smaller surface area when compared to the pink and the use of sodium carbonate flux imparts a slightly basic character to the material compared to the natural acidic character of the pink material [96-98].

TABLE 2.6

THE COMPOSITION OF CALCINED DIATOMACEOUS EARTH

Component	Percent Composition
SiO_2	91.0
Al_2O_3	4.6
Fe_2O_3	1.9
CaO	1.4
MgO	0.4
Volatile	0.3
Other	0.4

The physical properties of the popular series of Chromosorb supports are summarized in Table 2.7. The pink supports are relatively hard, have a high packing density, large surface area, and high capacity. They are used in both analytical and preparative gas chromatography. The white supports are more friable, less dense, have a smaller surface area and capacity, and are used in analytical gas chromatography. Chromosorb G is a specially prepared white support which is harder, more robust, and inert than Chromosorb W. It has a low capacity but is more dense than the ordinary white supports so that a given column contains approximately 2-1/2 times the amount of Chromosorb G, and therefore liquid phase, as Chromosorb W of the same nominal percentage weight per weight coating. Chromosorb A, a pink support developed specifically for preparative gas chromatography, possess the mechanical strength and high liquid

phase capacity of pink supports with a reduced surface activity approaching that
of white supports. Chromosorb 750 is a white support which is chemically
treated after processing and is the most inert of the Chromosorb series. As
well as the Chromosorb supports manufactured by Johns-Manville, numerous other
supply companies either manufacture or resell similar materials under various
brand and house names.

TABLE 2.7

PHYSICAL PROPERTIES OF CHROMOSORB SUPPORTS

Property	P	W	G	A	750	T
Color	pink	white	oyster white	pink	white	white
pH	6.5	8.5	8.5	7.1	– – –	– – –
Free fall density (g/ml)	0.38	0.18	0.47	0.40	0.32	0.42
Packed density (g/ml)	0.47	0.24	0.58	0.48	0.36	0.49
Surface area (m^2/g)	4.0	1.0	0.5	2.7	0.5-1.0	7-8
Surface area (m^2/ml)	1.9	0.3	0.3	1.3	– – –	– – –
Maximum useful liquid phase loading (%)	30	20	5	25	7	20

Theoretical studies predict that the highest column efficiencies will be
obtained with the minimum possible support particle size. From the practical
point of view, this particle size optimization is limited by the flow resistance
of the finished column. The pressure differential across the column is
inversely proportional to the square of the particle diameter, whereas the
column efficiency (as measured by the HETP) is directly proportional to the
particle size. Gas chromatographs are designed to operate at differential
column pressures of a few atmospheres and, for analytical columns, supports with
a mesh range of 80-100 or 100-120 are most frequently used. The highest
efficiencies are obtained with particles of a narrow mesh range, as upon packing
the larger particles will segregate near the wall. Traditionally, particle size
range is quoted in terms of mesh and the designation 100-120 mesh means that the
particles passed through a sieve of 100 mesh but not a sieve of 120 mesh. The
mesh range in units of micrometers is given in Table 2.8.

In the early development of gas chromatography, the crude diatomaceous supports performed successfully as long as experimental work was limited to hydrocarbons and other such nonpolar samples. The extension of gas chromatography into the biomedical and environmental fields proved that the crude diatomaceous supports were not sufficiently inactive. Severe tailing of polar molecules, sample decomposition, structural rearrangements, and even complete sample adsorption were often observed [96,97,99]. These undesirable interactions were associated with the presence of metallic impurities and silanol groups at the surface of the diatomaceous supports. Early attempts to reduce the surface activity of the support without diminishing its surface area used an overcoat of an inert material such as silver [100], Teflon [101], or a surface active polymer such as Epikote 100 or polyvinylpyrrolidone [102]. Acid and/or base washing to remove metallic impurities and silanization of surface silanol groups are presently the most widely used methods of support deactivation. It has been suggested that coating diatomaceous earths with pyrocarbon strengthens the particles towards attrition and reduces their surface activity [103].

TABLE 2.8

PARTICLE SIZE RELATIONSHIPS FOR DIATOMACEOUS SUPPORTS

Mesh Range	Top Screen Opening(μm)	Bottom Screen Opening(μm)	Spread (μm)	Range Ratio
10-20	2000	841	1159	2.38
10-30	2000	595	1405	3.36
20-30	841	595	246	1.41
30-40	595	420	175	1.41
35-80	500	177	323	2.82
45-60	354	250	104	1.41
60-70	250	210	40	1.19
60-80	250	177	73	1.41
60-100	250	149	101	1.68
70-80	210	177	33	1.19
80-100	177	149	28	1.19
100-120	149	125	24	1.19
100-140	149	105	44	1.42
120-140	125	105	20	1.19
140-170	105	88	17	1.19
170-200	88	74	14	1.19
200-230	74	63	11	1.19
230-270	63	53	10	1.19
270-325	53	44	9	1.20
325-400	44	37	7	1.19

Surface metallic impurities, principally iron, are removed by soaking for several hours in 3 N HCl [104], by Soxhlet extraction with 6 N HCl until no further coloration is extracted [105,106], or by heating for one to three days at 850-900°C while passing a mixture of nitrogen and hydrogen chloride gases

through the support [106,107]. The above treatments remove 95-98% of the iron
present at the surface of the support and reduce the concentrations of sodium
and aluminum as well. Reduction of metal oxides, particularly iron oxide, by
high temperature treatment with hydrogen has also been claimed to improve the
properties of diatomaceous supports [108].

Surface silanol groups are converted to silyl ethers by reaction with
dimethyldichlorosilane (DMCS), hexamethyldisilazane (HMDS), trimethylchlorosilane (TMCS), or a combination of these reagents, Figure 2.4 [109-112].
However, some silanol groups remain reactive even with exhaustive silylation.
Upon silylation the hydrophilic character of the support surface becomes
hydrophobic. Polar liquid phases are no longer capable of wetting the support
surface and silanized supports should not be used to prepare packings with polar
liquids.

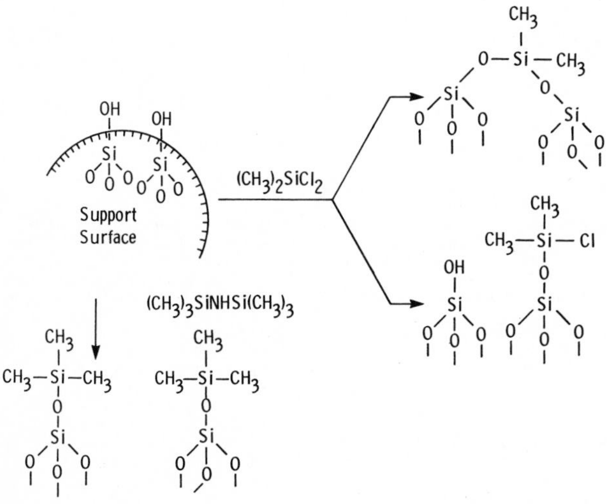

Figure 2.4 Support deactivation by silanization.

The number of hydroxyl centers is 12-24 times greater on the surface of the
pink compared with the white supports, so that even after silylation deactivated
pink supports show approximately the same adsorptive activity as untreated white
supports. The treatment that the support has undergone is normally stated on
the manufacturer's label, for example, Chromosorb W-AW-DMCS would indicate that
the white diatomite support was deactivated by acid washing and silylated with
dimethyldichlorosilane.

Although silanized supports show less adsorptive activity than unsilanized
supports, their coating characteristics are not necessarily ideal, even for
nonpolar phases [18,113]. Deposition of a nonextractable film of Carbowax 20M

[113-116], surface treatment with cyclic siloxanes [117,118], and silylation with a bulky reagent such as octadecyldimethylchlorosilane [18,113] have been used to modify the wetting characteristics of diatomaceous supports. None of these treatments yields ideal coating characteristics for nonpolar phases; however, they all provide higher column efficiencies and improved mass transfer characteristics compared with conventional silylation, indicating that some improvement in the support wetting characteristics has been realized.

The most extensive chemical treatment and silylation of diatomaceous materials cannot completely remove those active centers which cause tailing of strongly basic or acidic components. When compounds of this type are analyzed, the addition of small quantities of "tailing reducers" is necessary. Tailing reducers are coated onto the support in a manner similar to that of the liquid phase and, to be effective, they must be stronger acids or bases than the compounds to be chromatographed. For amines, the tailing reducer could be a few percent of potassium hydroxide or another strongly basic substance such as polyethyleneimine, polypropyleneimine, N,N'-bis-1-methylheptyl-p-phenylene-diamine, or sodium metanilate [96,119,120]. For acidic compounds, suitable tailing reducers are phosphoric acid, the stationary phase FFAP, or commercial corrosion inhibitors such as trimer acid [96]. It must be remembered that these active substances will also act as subtractive agents and thus an acidic tailing reducer will remove basic substances from the chromatogram and vice versa. As well as the most obvious cases, FFAP and polypropyleneimine will selectively abstract aldehydes, while polyethyleneimine abstracts ketones and phosphoric acid may esterify or dehydrate alcohols. In addition, phosphoric acid will convert amides to the nitrile of the same carbon number and desulfonate sulfur-containing compounds [96]. The phase itself must also be compatible with reagents. For example, potassium hydroxide and phosphoric acid catalyze the depolymerization of polyesters and polysiloxane phases.

More than 90% of the separations using packed column GLC have probably been performed on columns prepared with diatomaceous supports. The only other support materials of any importance are various fluorocarbon powders, glass beads, graphitized carbon black, and dendrite salt [96,98,121]. The most important of the fluorocarbon supports are: Teflon-6, Chromosorb T, Holoport, Fluoropak 80, and Kel-F. Teflon-6 is a polymerized tetrafluoroethylene prepared by the agglomeration of Teflon emulsions which are sieved to a 40-60 mesh fraction. The individual particles are composed of fragile aggregates possessing a porous structure, providing a large surface area with little surface energy. This results in a very inert surface but presents problems with polar liquid phases, as these cannot wet the surface of the particles. The Teflon supports also present problems in packing and handling since the particles develop electrostatic charges easily. For high column efficiencies the Teflon particles should be handled at temperatures below 19°C (solid transition point), at which temperature they are rigid and free flowing. Kel-F is a hard, granular chlorofluorocarbon molding powder that can be handled like diatomaceous supports but generally gives poor column efficiencies. Fluoropak-80 is a granular type of fluorocarbon resin with a sponge-like structure of low surface area and therefore low capacity. The properties of the fluorocarbon supports are summarized in Table 2.9 [122].

TABLE 2.9

CHROMATOGRAPHIC PROPERTIES OF FLUOROCARBON SUPPORTS

Property	Kel-F	Fluoropak-80	Teflon-6
Surface area (m^2/g)	2.2	1.3	10.5
Optimum liquid phase loading (%)	15-20	2-5	15-20
Maximum temperature (°C)	160	275	250

Glass beads are available as spheroids in narrow mesh ranges. They can be used to prepare very efficient columns. As the surface energy of the beads is low, liquid phase loadings are limited to less than 0.5% w/w. At higher loadings the column efficiency is reduced due to the formation of pools of liquid phase at the contact points of the beads [123]. The glass surface is not completely inert and tailing is observed with polar solutes. In addition to the silanol groups on the surface of the bead, Na^+ and Ca^{2+} ions are also present. These ions may function as Lewis acids and provide adsorption sites for lone pair donor molecules such as ketones and amines [124]. Silylation of the beads reduces the tailing caused by silanol groups but also diminishes the support capacity. To increase the capacity of the beads, chemical etching of the surface [124] or coating with a thin layer of an intermediate material of high surface area (such as boehmite or graphitized carbon black) have been used [125]. The liquid phase is then coated onto the intermediate support layer covering the bead. Glass beads (Cerabeads) were developed with a sponge-like outer surface of high-content alumina soda-lime glass that could be coated with up to 4% w/w of liquid phase. These beads exhibited severe tailing of polar solutes, which could be reduced by silylation but only at the expense of capacity. Glass beads are used mainly in theoretical studies because of their controlled shape and size and for very rapid analyses where their low capacity is not a disadvantage.

2.3 Preparation of Packed Columns

Column preparation is a very simple task if due care is taken and proper attention is paid to the various steps. The liquid phase, in a suitable solvent, is mixed with the support, the solvent is removed, and the dried packing is then added to the empty column with the aid of pressure or vacuum. The solvent should be a good solvent for the phase and of sufficient volatility to be removed conveniently when using one of the evaporation methods. Conventionally, column packings are prepared on a weight per weight basis. Thus, to prepare 10 g of a 10% w/w phase loaded column, 9 g of support and 1 g of liquid phase would be used. Typical phase loadings range from 0.5 to 30% w/w depending upon the type of separation. The maximum useful phase loadings for various supports were given in Tables 2.7 and 2.9. The support is coated by adding it slowly to a flask containing the liquid phase solution. The support and solution are slurried together and then filtered on a Buchner funnel. Alternatively, the support can be packed into a glass column that has a glass frit at its base; a slight vacuum or pressure is then used to force the solution through it [127-129]. Excess solvent is removed either by filtration or evaporation. The rotary evaporator or the pan-dry method, which employs a stream of nitrogen or an infrared lamp to aid evaporation, is most frequently used. The rotary evaporator method is very convenient but, because of the fragile nature of diatomaceous supports, it is necessary to avoid rapid rotation of the flask or violent bumping of the damp solid. After coating the damp support may be air dried, oven dried, or dried in a fluidized bed dryer. Fluidized bed drying is faster and may lead to more efficient column packings as fines developed during the coating procedure are carried away by the gas passing through the packing [128-131]. The dried packing may be dry sieved or classified by suspension in fluids prior to packing in the empty column [132]. Dry sieving is easy to perform but may also cause extensive attrition of the fragile diatomaceous supports. Mechanical sieving prior to coating and fluidized bed drying after coating are the preferred methods of minimizing the deleterious effects of fines on column performance. Unopened bottles of support may contain from 30 to 50% fines by volume, thus the need for a preliminary screening process.

For general purposes it is assumed that the ratio of liquid phase to support does not change during the coating process. If an accurate determination of the liquid phase loading is required, for example in physicochemical studies, this can be performed by either Soxhlet extraction or high temperature evaporation of the dried packing [133].

Copper, aluminum, stainless steel, nickel, or glass tubes bent into various shapes to fit the dimensions of the column oven provide the container for column packings. Neither copper nor aluminum tubing is recommended as both metals are readily oxidized; active, oxide-coated films formed on the inner walls promote decomposition or tailing of labile and polar solutes [134,135]. Stainless steel is adequate for nonpolar samples but its catalytic activity precludes the analysis of labile solutes. Nickel, after acid passivation, and glass are the most inert column materials, and are the preferred materials for the analysis of labile samples [135]. Teflon or plastic tubing is also used occassionally, often in the separation of chemically-reactive substances for which Teflon supports are also required. Low upper temperature limits and permeability to some solutes restricts their general use.

Columns of 0.5-3.0 m in length and internal diameters of 2-4 mm can be conveniently packed by the tap-and-fill method, aided by vacuum suction. One end of the column is terminated with a short glass wool plug and attached via a hose to a water pump aspirator. A small filter funnel is attached to the other end and the packing material is added in small aliquots. The column bed is consolidated as it forms by gentle tapping of the column sides with a rod or with the aid of an electric vibrator. Over vibration is detrimental to the preparation of an efficient column and should be avoided. When sufficient packing has been added the funnel is removed and a glass wool plug is inserted. The length of the plug is usually dictated by the design of the injector. Columns longer than 3.0 m are difficult to pack and require a combination of vacuum suction at the free end and pressure applied to the packing reservoir at the other end [137]. Columns containing segmented liquid phase loadings have been shown to have some advantages over uniformly loaded columns [138]. The column is packed in segments, with different percentage loadings of the same phase, generally in the order low-to-high or high-to-low. Using high-to-low loaded segments the resolution of the column is enhanced compared to a uniformly loaded column containing an amount of liquid phase corresponding to the mean loading of the segmented column. Reduced analysis time is claimed to be the advantage of the low-to-high series packings.

The packed column requires a temperature conditioning period before use. The column is installed in the oven with the detector end disconnected. The column temperature is then raised to a value about 20°C higher than its intended operating temperature while a slow flow of nitrogen is passed through it. In all cases the temperature used for conditioning should not exceed the maximum operating temperature of the phase. The conditioning period might last from several hours to several days, depending on the quality of the liquid phase. If the column bed should shrink during conditioning, additional amounts of packing should be added to restore it to the original level. The conditioning period is complete when a stable detector baseline is obtained.

2.4 Performance Evaluation of Packed Columns

After conditioning the column a few preliminary tests are performed to ensure that the performance of the column is adequate. A measure of column efficiency is made with a test sample, for example a mixture of hydrocarbons, Figure 2.5 [36]. An average value of about 2000 theoretical plate per meter is acceptable for a 10% w/w loaded column on a 80-100 mesh support. Higher values can be expected for columns with lighter loadings and with finer mesh supports. Columns with less than 1500 theoretical plates per meter are of doubtful quality. An unusually high column pressure differential also signals problems which may be responsible for the loss in efficiency. A high column back pressure is usually associated with attrition of the column packing.

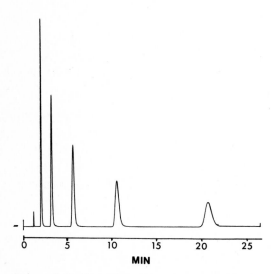

Figure 2.5 Test chromatogram to establish column efficiency. Test probes are
n-hydrocarbons separated isothermally.

After testing the efficiency of the column, some test of column activity and resolving power may be made. A polarity test mixture, such as the one illustrated in Figure 2.6, can be used to evaluate column adsorptivity. Excessive tailing or adsorption of the test probes indicates residual column activity. This activity can often be reduced by on-column silylation using one of the many commercially available column conditioning reagents. Residual sorptive activity is not unusual for nonpolar phases on diatomaceous supports and a modified polarity test mixture is more appropriate for testing them [140]. To test the resolving power a mixture that reflects the intended use of the column should be selected.

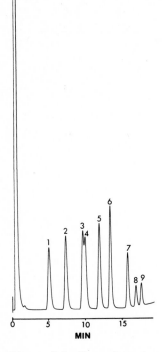

Figure 2.6 Test chromatogram to evaluate column adsorptive behavior.
1. decane; 2. undecane; 3. dodecane; 4. 5-nonanone; 5. tridecane;
6. 1-octanol; 7. naphthalene; 8. 2,6-dimethylaniline;
9. 2,6-dimethylphenol.

The original test chromatograms for the column should be retained. Routine
retesting at intervals allows a quantitative assessment of column aging to be
made.

2.5 Contribution of Interfacial Adsorption to Retention in Gas-Liquid Chromatography

In many instances retention in GLC cannot be explained entirely by a
gas-liquid partitioning mechanism. The possibility that adsorption at the
gas-liquid and gas-solid interfaces may contribute significantly to retention
must also be considered [139-142]. In this case solute retention is described
by equation (2.9).

$$V_N = V_L K_L + A_{GL} K_{GL} + A_{LS} K_{GLS} \qquad (2.9)$$

V_N = net retention volume
V_L = volume of liquid phase
K_L = gas-liquid partition coefficient
A_{GL} = coefficient for adsorption at the gas-liquid interface
A_{LS} = gas-solid interfacial area
K_{GLS} = coefficient for adsorption at the support surface

When the terms that contain K_{GL} and K_{GLS} are not negligible then retention can be expected to vary with the properties of the support, the support-wetting characteristics of the liquid phase and the percent liquid phase loading. The nature of the solute must also be considered, for example, a polar solute such as an alcohol chromatographed on a nonpolar liquid phase, is more likely to displace liquid phase from the support surface and have a large $A_{LS} K_{GLS}$ term than would a hydrocarbon. For a polar liquid phase, displacement of phase molecules by the solute is less likely but the possibility of adsorption at the gas-liquid phase interface becomes more probable. Gas-liquid phase adsorption should become increasingly significant as the dissimilarity between the liquid phase and solute increases, and is likely to be very important in the retention of hydrocarbons on polar liquid phases. Only when retention is determined solely by partitioning ($V_L K_L$) will relative retention and retention index values be independent of the liquid phase loading.

A model devised to explain retention in general terms must take into account the distribution of the liquid phase on the support. It cannot be assumed that the liquid phase always distributes itself as an even film over the entire support surface. This will only occur with those liquids that easily wet the support [143,144]. When the support is not readily wetted by the liquid phase, the liquid phase will be present in the form of individual droplets, located primarily at the outer grain surface with little penetration into the pores. Even when a liquid wets the support surface the liquid film will exist in two separate states. Liquid phase molecules near the support surface exist in a regular arrangement with an entropy of solution that is lower than that of the pure liquid. This structured layer adjacent to the support surface may be of considerable thickness. For supports coated with low volumes of stationary phase this structured layer is likely to predominate. For polar phases, in which long range forces exist, a structured multimolecular layer may exist up to several percent liquid phase loadings. At still higher phase loadings a bulk liquid phase layer will be formed over the structured layer. The bulk liquid phase layer will exhibit normal solution behavior and its properties can be predicted from those of the bulk liquid. To take account of the presence of the structured layer, equation (2.9) can be modified to equation (2.10) [145].

$$V_N = V_L K_L + V_L K_{SL} + (1-\delta)A_{LS}K_{DSL} + A_{GL}K_{GL} + A_{LS}K_{GLS} \qquad (2.10)$$

K_{SL} = partition coefficient for the structured liquid phase layer
K_{DSL} = coefficient for adsorption at the gas-structured liquid phase interface
δ = constant equal to 1 when $d < d_s$ and 0 when $d \geq d_s$
d_s = thickness of the structured liquid phase layer
d = film thickness

An exact solution of equation (2.10) is difficult both mathematically and experimentally. Equation (2.10) contains five unknowns (K_L, K_{SL}, K_{DSL}, K_{GL}, K_{GLS}). To solve the equation, the phase characteristics V_L, A_{GL}, and A_{LS} must be known for a minimum of five column packings prepared with different phase loadings. A value for δ must also be found by a combination of intuition and iteration.

Some representative data for different solutes on Apiezon M and Carbowax 4000 coated on the same support are given in Table 2.10. The interaction of polar solutes with the support contributes substantially to retention on a nonpolar Apiezon column. Even at high liquid phase loadings, adsorption at the support surface may contribute to between 4 and 25% of solute retention. For Carbowax 4000, gas-liquid phase adsorption contributes only moderately to the retention of all solutes, decreasing rapidly with increasing liquid phase loadings. At low liquid phase loadings adsorption at the gas-liquid phase interface may contribute to between 3 and 42% of the retention of the solutes investigated.

2.6 Bonded Liquid Phases

Three approaches have been used to attach a liquid phase to a support by either chemical or physical means. Halasz described the preparation of estersils, support bonded alcohols, by the condensation of a normal, branched, or substituted alcohol with porous silica beads at high temperatures in an autoclave [145,147]. These bonded phases are available commercially under the tradename Durapak [148]. Similar materials have been prepared by reacting substituted isocyanate or isothiocyanate reagents with porous silica beads, forming urethane bonds [149].

The second approach involves the reaction of a monofunctional or multifunctional chlorosilane, disilazane, or cyclosiloxane reagent with a porous silica or diatomaceous support [150-154]. Chemical attachment via formation of

TABLE 2.10

RELATIVE CONTRIBUTION (%) TO THE NET RETENTION VOLUME BY DIFFERENT SOLUTE
INTERACTIONS

Solute	Phase	Interaction[a]	Percent Liquid Phase					
			1	2.9	4.7	9.1	13	17
Octane	Apiezon M	$V_L K_L$	100	100	100	100	100	100
	Carbowax 4000	$V_L K_L$	58	81	88	93	95	97
		$A_{GL} K_{GL}$	42	19	12	7	5	3
Benzene	Apiezon M	$V_L K_L$	100	100	100	100	100	100
	Carbowax 4000	$V_S K_S$	89	96	98	56	37.5	28.2
		$V_L K_L$				43	61.7	71.3
		$A_{GL} K_{GL}$	11	4	2	1	0.8	0.5
1-Butanol	Apiezon M	$V_L K_L$	13	30	42	59	67	75
		$A_{LS} K_{GLS}$	87	70	58	41	33	25
	Carbowax 4000	$V_S K_S$	86	95	97	60	40	30
		$V_L K_L$				38	59	69
		$A_{GL} K_{GL}$	14	5	3	2	1	1
2-Butanone	Apiezon M	$V_L K_L$	52	77	84	92	94	96
		$A_{LS} K_{GLS}$	48	23	16	8	6	4
	Carbowax 4000	$V_S K_S$	82	93	96	60	40	30
		$V_L K_L$				38	58	69
		$A_{GL} K_{GL}$	18	7	4	2	2	1
Chloro-	Apiezon M	$V_L K_L$	100	100	100	100	100	100
Benzene	Carbowax 4000	$V_S K_S$	96	99	99	62.8	41.8	31.4
		$V_L K_L$				36.8	57.9	68.4
		$A_{GL} K_{GL}$	4	1	1	0.4	0.3	0.2

[a] $V_L K_L$ = contribution due to partition with the liquid phase

$A_{GL} K_{GL}$ = contribution due to adsorption at the gas-liquid interface

$V_S K_S$ = contribution due to partition with the structured liquid
phase layer

$A_{LS} K_{GLS}$ = contribution due to adsorption at the support surface

siloxane bonds and polymerization are probably both involved in bonding the
liquid phase to the support.

The third approach involves bonding the liquid phase to the support by the
high temperature treatment of a coated support in the presence of a low flow of
carrier gas. After treatment the packing is exhaustively extracted with
appropriate solvents to leave a fraction of the original liquid phase
permanently attached to the support [155-159]. The nonextractable liquid phase
loading corresponds to about 0.2% w/w. Phases which can be bonded in this way

include Carbowax 20M, polyesters, and polysiloxanes. They are commercially available under the tradename Ultra-bond. The chromatographic properties of these bonded liquid phases have been reviewed [160-162].

Due to the specific reaction between an alcohol or isocyanate with silanol groups on the support surface, only a monomeric film is possible. As long as the alkyl groups are relatively short they are probably orientated normal to the surface and closely packed. Consequently the speed of mass transfer is considerably higher than with conventional liquid phases [147,148,163]. The relatively flat van Deemter curve enables columns to be operated at high mobile phase velocities without loss in efficiency. The bonded phases are thermally stable and can be operated at temperatures 80 to 90°C higher than when the same liquid phases are conventionally coated onto supports. The estersils are easily hydrolyzed by water, alcohols, and other protonic substances. At high temperatures oxidative degradation by traces of oxygen in the carrier gas can be a problem.

The extent of solute retention on estersils depends on the size and functionality of the organic chain, the degree of surface bonding, and the size and polarity of the solute [148,164-167]. The most important mechanism of solute retention is adsorption on either the solid support or the bonded organic group. These groups dominate the retention mechanism for those bonded phases containing short-chain alkyl groups. The extent of interaction between the solute and silanol groups appears to decrease with increasing size of either the solute or the bonded organic group. For bonded phases containing polar groups, adsorption at the surface of the bonded organic group tends to dominate the retention mechanism. This accounts for the contrary observation that the octane-bonded phase (prepared by reaction with octanol) is more polar than the Carbowax-bonded phase, Table 2.11, the exact opposite of results obtained with conventionally-coated liquid phases. Long-chain bonded alkyl groups show less spatial ordering than their short chain analogs. Their structure shows more liquid-like behavior and the possibility that partitioning contributes to the retention mechanism is more likely.

Less is known about the retention mechanism of solutes on the nonextractable phases. It is probably similar to the estersil phases except that the liquid film may be present, in part, as "dense patches", and therefore a greater contribution from partitioning might be anticipated. The principal chromatographic properties of the nonextractable phases are low retention, very low column bleed, and higher column efficiencies and maximum operating temperatures than conventionally coated phases. The packings are also more inert and can be used for the separation of polar solutes. Oxidative degradation can occur at high temperatures.

TABLE 2.11

PERMANENTLY BONDED LIQUID PHASES (ESTERSILS)

Support[a]	Stationary Phase(%w/w)	Polarity	Isothermal Temperature Limit (°C)	Recommended for the Separation of
Porasil C	3-Hydroxy-propionitrile (3.60)	Medium	135	Low molecular weight hydrocarbons, aromatics, and oxygenated compounds
Porasil C	Carbowax 400 (7.86)	Nonpolar	175	Low molecular weight alcohols
Porasil C	n-Octanol	Polar	175	Low molecular weight alcohols, hydrocarbons
Porasil C	Phenyl isocyanate	Polar	60	C_1-C_3 hydrocarbons and their isomers
Porasil S	Carbowax 400 (16.75)	Nonpolar	230	
Porasil S	Carbowax 4000	Polar	230	Aromatics and chloro-aromatics
Porasil F	Carbowax 400 (1.41)	Nonpolar	230	Higher molecular weight substances - waxes, steroids, polynuclear aromatics

[a] The Porasils are spherical porous silicas that differ in surface area and pore diameter. Porasil C (100 m^2/g, 300 Å), Porasil F (10 m^2/g, 3000 Å), and Porasil S (300 m^2/g)

2.7 Porous Polymer Beads

Porous polymer beads are prepared by the process of suspension polymerization, in which a mixture of monomers and crosslinking reagents is polymerized in the presence of an inert solvent. The microstructure of the gel, formed within the droplet in the early stages, gradually grows into a sponge-like structure, with inert solvent filling the space between the microstructures [168]. Upon drying and evacuation, the porous structure remains, producing a uniform bead that is sufficiently rigid to be dry packed into a column. By adroit selection of monomer, crosslinking reagent, and inert solvent, porous polymer beads of different properties can be prepared. Some examples are given in Table 2.12. The properties of these beads vary with the chemical nature, pore structure, surface area, and particle size of the material [169,170]. As a general rule, the larger the average pore diameter the faster the analysis. Thus porous polymeric beads with pore diameters less than 0.01 micrometers are used primarily for the separation of gases while those with pore diameters greater than 0.01 micrometers are employed for higher-boiling organics. The porous polymeric beads have found many applications in the analysis of volatile inorganic and organic compounds, particularly for samples with poor peak shapes on support-coated diatomaceous column packings. Notable examples are water, formaldehyde, carboxylic acids, and inorganic gases. Some general indications of the type of samples which can be separated on porous polymer beads are given in Table 2.13 [171,172].

TABLE 2.12

PHYSICAL PROPERTIES OF POROUS POLYMER BEADS

Porous Polymer	Type[a]	Physical Property			
		Free-fall Density (g/cm^3)	Surface Area (m^2/g)[b]	Average Pore Diameter (micrometers)	Temperature Limit (°C)
Chromosorb 101	STY-DVB	0.30	50	0.3-0.4	275
Chromosorb 102	STY-DVB	0.29	300-500	0.0085	250
Chromosorb 103	Polystyrene	0.32	15-25	0.3-0.4	275
Chromosorb 104	ACN-DVB	0.32	100-200	0.06-0.08	250
Chromosorb 105	Acrylic Ester	0.34	600-700	0.04-0.06	250
Chromosorb 106	Polystyrene	0.28	700-800	0.5	250
Chromosorb 107	Acrylic Ester	0.30	400-500	0.8	250
Chromosorb 108	Acrylic Ester		100-200	0.25	250
Porapak N	Vinylpyrolidone	0.39	225-350		200
Porapak P	STY-DVB	0.28	100-200		250
Porapak Q	EVB-DVB	0.35	500-700	0.0075	250
Porapak R	Vinylpyrolidone	0.33	450-600	0.0076	250
Porapak S	Vinylpyrolidone	0.35	300-450	0.0076	250
Porapak T	EGDMA	0.44	250-300	0.009	200
Porapak PS	Silanized P				
Porapak QS	Silanized Q				250
Tenax-GC		0.37	18.6		375

[a] STY = styrene; DVB = divinylbenzene; ACN = acrylonitrile;
EVB = ethylvinylbenzene; EGDMA = ethylene glycol dimethacrylate

[b] Values for surface area vary widely in the literature

A substantial amount of literature describes the application of porous polymer beads to various problems; far less information is available concerning their retention mechanism. The consensus of opinion seems to be that at low temperatures solutes are retained principally by adsorption. At higher temperatures the surface of the polymer behaves as a highly extended liquid and partition becomes more significant [173-175]. In many instances both mechanisms may be simultaneously involved.

TABLE 2.13

SAMPLES SUITABLE FOR SEPARATION ON POROUS POLYMER BEADS

Porous Polymer	Recommended for the Separation of	Not Recommended For
Chromosorb 101 Porapak P and PS	Esters, ethers, ketones, alcohols, hydrocarbons, fatty acids, aldehydes, and glycols	Amines, anilines
Chromosorb 102 Porapak Q	Light and permanent gases, low molecular weight acids, alcohols, glycols, ketones, hydrocarbons, esters, nitriles, and nitroalkanes	Amines, anilines
Chromosorb 103	Amines, amides, alcohols, aldehydes hydrazines, and ketones	Acidic substances, glycols, nitriles, and nitroalkanes
Chromosorb 104	Nitriles, nitro compounds, sulfur gases, oxides of nitrogen, ammonia, and xylenols	Amines and glycols
Chromosorb 105 Porapak N	Aqueous mixtures of formaldehyde, acetylene from lower hydrocarbons, and most gases	Glycols, acids, and amines
Chromosorb 106 Porapak QS	Alcohols, C_2–C_5 carboxylic acids, alcohols, and sulfur gases	Glycols and amines
Chromosorb 107 Porapak T	Formaldehyde from water, acetylene from lower hydrocarbons	Glycols and amines
Chromosorb 108	Gases, polar materials such as water, alcohols, aldehydes, and glycols	
Porapak S	Normal and branched alcohols, ketones, and halocarbon compounds	Acids and amines
Porapak R	Esters and ethers, nitriles, and nitro compounds	Glycols and amines
Tenax-GC	High boiling polar compounds, diols, phenols, methyl esters of dicarboxylic acids, amines, diamines, ethanolamines, amides, aldehydes, and ketones	

With the exception of Tenax-GC, the porous polymers are limited in application to those compounds with boiling points less than 300°C; otherwise, retention times may be excessively long. Tenax-GC is unique in terms of its thermal stability. It may be used isothermally at temperatures up to 375°C and in temperature programmed separations it has been used to temperatures in excess of 400°C. Tenax-GC is a linear polymer of p-2,6-diphenyl-phenylene oxide (XI) having a molecular weight of 5×10^5 to 10^6. A granulated powder of low surface area, it differs physically from the Porapak and Chromosorb polymers [176,177]. Its separation properties are not very different from Porapak Q and similar porous polymers; its principal advantage is its greater thermal stability.

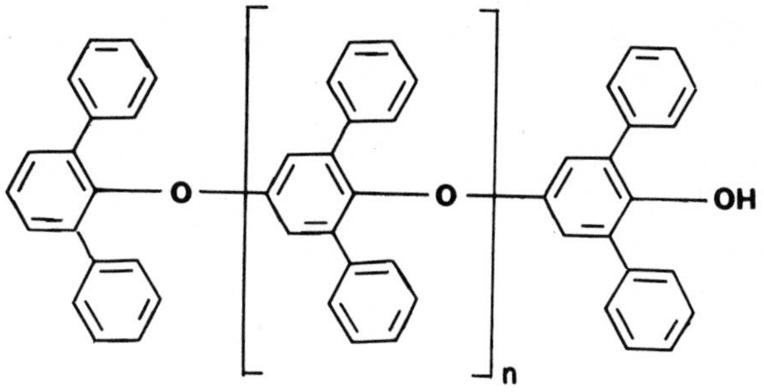

Tenax-GC

(XI)

Several methods have been described to characterize the separation properties of the porous polymer beads. For low temperature operation the relative retention of ethylene, acetylene, and carbon dioxide on a porous polymer compared to their values on Porapak Q as a reference sorbent has been suggested as a measure of general polarity [178-180]. Using this measure the order of polarity of the porous polymer beads was established as Porapak Q < Chromosorb 106 and 102 < Porapak P < Chromosorb 101 < Porapak S , Chromosorb 103 and 105 < Porapak K and N < Chromosorb 107 and 108 < Porapak T < Chromosorb 104. For characterization at higher temperatures several versions of the Rohrschneider/McReynolds system have been used. Squalane [181], Carbopak B [182], and Chromosorb 106 [183] were employed as the standard, nonpolar reference phases. However, the applicability of these methods has not been thoroughly tested. Table 2.14 provides values for the McReynolds constants at 200°C using Chromosorb 106 as the reference phase. The figures in parentheses were obtained after aging the columns for 22 days. When some of the porous polymer bead packings are operated at high temperatures, continuous changes in retention characteristics occur [183,184].

Some of the porous polymer beads have been shown to react with certain samples. Porapak QS reacts with amines, Porapak Q and Chromosorb 102 are nitrated by nitrogen dioxide [185], and Porapak S and Chromosorb 103 react with nitroalkanes to give multiple or very broad peaks [183]. The porous polymer beads are also used as trapping agents for organic volatiles in air. This very important application is discussed in Chapter 7.9.

TABLE 2.14

McREYNOLDS' SELECTIVITY CONSTANTS FOR POROUS POLYMER BEADS[a]

		Benzene	Butanol	2-Pentanone	Nitropropane	Pyridine
Porapak	N	50	108	83	126	98 (176)
	P	130	99	85	150	196
	Q	7	10	4	3	15
	R	32	77	46	74	69
	S	18	53	33	b	41
	T	114	183	153	293	187
		(162)	(245)	(189)	(318)	(273)
Chromosorb	101	107	76	66	122	144
		(100)			(114)	(138)
	102	32	25	21	40	50
	103	147	160	122	b	212
		(140)	(167)			(203)
	104	246	330	314	467	405
					(476)	(415)
	105	31	62	45	72	74
			(71)	(55)		(98)
	106	0	0	0	0	0
	107	54	139	106	181	140
		(92)	(168)	(140)	(214)	(172)
	108	132	215	161	265	231
		(155)	(235)	(182)	(292)	(262)

[a] Values in parentheses refer to the value measured after aging the column for 22 days at 200°C.

[b] Reacts with the phase, giving multiple or very broad peaks.

2.8 Gas-Solid Chromatography

Although it preceeded gas-liquid chromatography, gas-solid chromatography (GSC) has never attained the same status as a separation technique. There are a number of reasons for this. First, adsorption isotherms in GSC are often non-linear, even with small sample amounts. This produces several detrimental effects: solute retention volumes vary with sample size, peaks are asymmetric, and recovery of some samples from the column is incomplete. Second, the comparatively large surface area and energy result in excessively long retention times, particularly for large, polar molecules. This limits the application of GSC to the separation of relatively low molecular weight solutes at normal temperatures, and slightly larger molecules at higher temperatures. The temperature required to obtain reasonable retention volumes may exceed the thermal stability of some solutes; in addition, some adsorbents show considerable catalytic activity at high temperatures. The sample molecular weight range for which GSC is suitable is thus more restricted than for GLC.

Third, adsorbents are generally more difficult to standardize and to prepare reproducibly than liquid phases. Finally, the number of useful commercially available adsorbents is relatively small compared with the number of liquid phases available for GLC.

GSC does enjoy some advantages over GLC and these have been sufficient to maintain some interest in the technique. Adsorbents are generally stable over a wide temperature range and are often insensitive to attack by oxygen. Column bleed is virtually nonexistent, so high sensitivity detectors such as the helium ionization detector can be used. The selectivity of GSC is usually much greater then GLC for the separation of geometric and isotopic isomers. GSC is also suitable for the separation of inorganic gases and low molecular weight hydrocarbons, for which GLC generally shows little selectivity.

The principal adsorbents used in GSC are silica, alumina, graphitized carbon balcks, and molecular sieves [9,186]. Silica gel adsorbents are available for GSC in the form of rigid, incompressible, spherical beads that can be dry packed into empty columns. These macroporous adsorbents, Table 2.15, are prepared from ordinary silica gel by either simple high temperature treatment (700 to 950°C) or by hydrothermal treatment with steam in an autoclave. Hydrothermal treatment causes pronounced pore widening, leading to a reduction in the specific surface area. Thus silica adsorbents with a wide range of specific surface areas, pore diameters, and pore volumes can be readily prepared [187-189]. Silica and alumina adsorbents are recommended for the separation of low molecular weight saturated and unsaturated hydrocarbons, halogenated hydrocarbons, and derivatives of benzene. Polar solutes interact too strongly with the highly adsorptive surface to be separated in reasonable analysis times or with reasonable peak shapes.

Retention on silica or alumina adsorbents is a function of the specific surface area, the degree of surface contamination (particularly that of water), the prior thermal conditioning of the adsorbent, and the ability of the solute to undergo specific interactions, such as hydrogen bonding, with surface silanol groups. Alumina adsorbents with specific surface areas comparable to silica adsorbents show similar retention properties but different selectivities. This is due to the presence of Lewis acid sites associated with surface aluminum ions. In both cases a dramatic reduction in the retention of nonpolar solutes can be obtained by the uptake of water vapor by the adsorbents. Mixing the adsorbents with an inert solid diluent such as glass beads or a diatomaceous support reduces retention and simultaneously improves column efficiency [190,191]. These columns were called "dusted columns", as the quantity of adsorbent was generally only a few percent of the total mixture. Rapid isothermal analyses of samples with wide boiling ranges and column efficiencies similar to GLC columns were possible with these columns. Short columns packed with microparticulate adsorbents (7-10 or 25-35 micrometer average particle diameters) have been used for high speed GSC [192,193]. As mass transfer for these columns is very fast, high carrier gas flow rates can be used without substantial reductions in column efficiency. Column efficiencies on the order of 2400 theoretical plates per second can be obtained. However, operation of these columns requires very high head pressures and fast response detectors.

TABLE 2.15

ADSORBENTS FOR GAS-SOLID CHROMATOGRAPHY

Type	Name		Specific Surface Area (m^2/g)	Pore Diameter (nm)
Silica	Spherosil	XOA 400	300-500	8
		XOA 200	140-230	15
		XOB 075	75-125	30
		XOB 030	37-62	60
		XOB 015	18-31	125
		XOC 005	5-15	300
	Porasil	B	125-250	10-20
		C	50-100	20-40
Graphitized Carbon Black	Carbopack	C	12	
		B	100	
	Carbosieve		1000	1.3
	Spherocarb		1200	1.5
Sodium Aluminum Silicate	Molecular Sieve	4A		0.4
		13X		1.0
Calcium Aluminum Silicate	Molecular Sieve 5A			0.5

To reduce retention volumes, modify column selectivity, and improve the efficiency of silica and alumina adsorbents, they can be coated with a small quantity of a nonvolatile liquid [189,194,195] or an inorganic salt [196,197]. It is assumed that the liquid film is selectively adsorbed by the most energetic adsorption sites, which are then no longer available to the solute. The heterogeneous distribution of active sites is consequently reduced, providing a more homogeneous surface for interaction with the solute. As a result, both retention and peak asymmetry are dramatically reduced. As the amount of liquid phase is increased gas-liquid adsorption and, at still higher levels, gas-liquid partitioning can be expected to contribute to the retention mechanism. For the low levels of liquid phase commonly used in GSC it is assumed that gas-solid interactions dominate the retention mechanism.

Commonly used inorganic salt modifiers generally fall into two categories.

Alkali metal salts are used to reduce the surface heterogeneity of the adsorbent
in a manner somewhat similar to polar liquid phases. Complex-forming salts such
as silver nitrate and cuprous chloride deactivate the surface and can also
adjust column selectivity by their ability to selectively retain some solutes.
With low salt loadings, active sites on the original adsorbent surface are
covered by the salt in the form of ions and/or ion pairs. With increasing salt
loadings a new surface layer of salt gradually covers the original adsorbent
surface. After a complete monolayer is formed additional salt is deposited as a
thin, crystalline film whose thickness depends on the amount of salt added. At
this point the specific surface area will decline due to a reduction in the pore
volume now occupied by salt crystals. Therefore, at high salt loadings, solute
retention is determined chiefly by adsorption onto the surface of the salt
film. At low salt loadings retention is determined primarily by interactions
with the adsorbent surface. At intermediate salt loadings both retention
mechanisms may be important, depending on the nature of the solute and its
ability to displace the salt modifier from active sites on the adsorbent
surface. Thus optimum salt loadings for a particular sample are generally found
by trial and error and may vary from about 0.5% w/w up to about 30% w/w.

   In addition to coating salts onto silica, alkali-metal modified adsorbents
have been prepared by the coprecipitation of silica gel with the desired alkali
salt [198]. These synthetically modified adsorbents have properties similar to
those of the coated adsorbents.

   Graphitized carbon blacks are prepard by heating ordinary carbon blacks to
about 3000°C in an inert gas atmosphere [186,199]. This drives off volatile,
tarry residues and induces the growth of graphite crystallites, particles in
which graphite crystals are arranged in the form of polyhedrons. At the same
time various functional groups originally present on the carbon black surface
are destroyed. The surface of the graphitized carbon blacks is almost
completely free of unsaturated bonds, lone electron pairs, free radicals, and
ions. Because of the high surface homogeneity of the adsorbent, solute-solute
interactions on the adsorbent surface can be important. At moderate sample
sizes this can result in a decreased linearity of the adsorption isotherm. The
flat surfaces of graphitized carbon blacks are particularly well suited for the
separation of molecules which differ only in geometric structure, i.e.,
structural isomers and geometric stereoisomers.

   The retention behavior of solutes on graphitized carbon blacks can be
explained on the basis of the availability of two types of adsorption sites.
The vast majority of the surface sites are nonpolar and correspond to the
graphite-like array of carbon atoms. These sites show no tendency to interact
preferentially with molecules carrying functional groups and dispersive
interactions dominate retention behavior. Polar adsorption sites are few in

number but they can establish specific, strong interactions with polar solutes. Preliminary treatment of the adsorbent may be used to reduce their number. For example, heating to 1000°C in a stream of hydrogen has been used to minimize those active sites associated with the presence of surface oxygen complexes [9,199] while washing with perchloric or phosphoric acid eliminates basic carbonium-oxygen complexes and sulfur present as sulfide [200,201]. However, even after such treatment further modification of the surface by the addition of small quantities of liquids that are capable of establishing strong interactions with residual active sites is generally required [199,201,202]. At low surface concentrations the modifying agents act primarily as tailing inhibitors and retention volume reducers. Liquids which are capable of entering into specific interactions with the sample may be used as selectivity modifiers. An example of this is shown in Figure 2.7, in which a low concentration of picric acid is added to the adsorbent to enhance the separation of a mixture of saturated and unsaturated hydrocarbons [203]. At the surface concentration of liquid modifier required to build up a densely-packed monolayer, retention will occur primarily by adsorption on this layer and the selectivity of the column will be different to that of a partially complete monolayer. This film, however, will not show properties characteristic of the bulk liquid as its properties are influenced by those of the adsorbent with which it is in contact. At high liquid phase loadings the adsorbent can be considered to behave as a support for the bulk liquid [204,205]. Thus quite different column selectivities can be expected over the range of the liquid phase modifier added to the adsorbent. In most cases quite low levels are used to exploit the full advantages of gas-solid chromatographic interactions.

Very efficient packed columns have been prepared from graphitized carbon blacks having average particle diameters in the range 25-33 micrometers [206,207]. Values in excess of 10,000 plates per meter were achieved. As might be anticipated the principal disadvantage of these columns is the large pressure differential required for their operation.

Molecular sieves (zeolites) of general formula $M_{2/m}O.Al_2O_3.nSiO_2.xH_2O$ (M = alkali or alkaline earth cation) occur naturally or can be synthesized. The structural units of the aluminosilicate crystalline lattice consist of tetrahedral anions $[SiO_4]^{4-}$ and $[AlO_4]^{5-}$ bound through oxygen atoms. The excess negative charge arising from the replacement of silicon by aluminum is compensated for by the alkali or alkaline earth metal ion. Molecular sieves, therefore, are porous bodies which have cavities or holes, the diameter of which can be varied by changing the cation. The pore structure of the zeolites is microporous as the cross-sectional diameter of the channels is on the same order as the dimensions of small molecules. As only those molecules which can penetrate the micropores are retained by molecular sieve columns, they are used

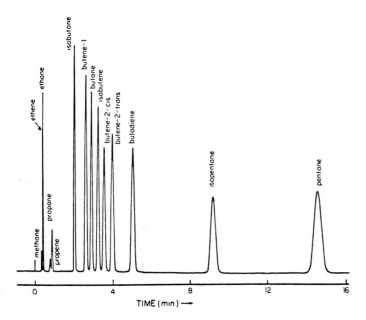

Figure 2.7   Isothermal separation of C_1 –C_5 hydrocarbons on a 2.2 m x 2 mm I.D.
column Carbopak + 0.19% picric acid at 50°C. (Reproduced with
permission from ref. 203. Copyright American Chemical Society).

primarily for the separation of permanent gases (e.g., H_2, O_2, N_2, CH_4, and CO),
and for the separation of low molecular weight linear hydrocarbons from branched
chain and cyclic hydrocarbons [208,209]. Those solutes which can penetrate the
cavities are further retarded by surface adsorption and interactions with the
charge field of the zeolite framework and the counter–charged cation. It has
also been shown that the degree of hydration of the cavities can influence
solute retention through its modification of polar interactions.

Unlike the molecular sieves, carbon sieves are not highly crystalline but
are composed of very small crystallites in which the carbon atoms are trigonally
bound [210]. The crystallites, in turn, are crosslinked to yield a disordered
cavity–aperture structure. The apertures, produced by the more or less close
approach of basal planes from adjacent crystallites, are slit shaped and of
variable size, depending upon the organic precursor and the method of thermal
treatment. Kaiser prepared microporous carbon sieves having surface areas of
1000 m^2/g and average pore radii of 1.24 nm by the pyrolysis of polyvinylidene
chloride [211]. Carbon molecular sieves show very pronounced retention of
organic compounds and are used primarily for the separation of inorganic gases
and, in particular, the separation of small polar molecules such as water,
formaldehyde, and hydrogen sulfide [211,212].

Other adsorbents which have been used in GSC include bentonite, clathrate

compounds, chromium(III) oxide, iron(III) oxide, barium sulfate, alkali metal
tetraphenylborates, and molybdenum sulfide [9,213-216]. Although they may show
specific advantages for certain separations, they have not been adopted for
general use and will not be discussed further.

2.9 Micropacked Columns

Micropacked columns, sometimes referred to as packed capillary columns, are
characterized by small internal column diameters, usually less than 1.0 mm, and
packing densities comparable to classical packed columns [217,218]. Two types
of micropacked columns can be distinguished according to their particle diameter
to column diameter ratio (d_p/d_c): columns packed with particles of a fine
grain, for example 30-35 micrometers, with a d_p/d_c ratio less than 0.1 and
columns packed with coarser particles such that the d_p/d_c ratio is usually
between 0.1 and 0.3. Both column types are capable of generating high
efficiencies, 30,000 to 60,000 theoretical plates. They differ primarily in
terms of their permeability. Micropacked columns filled with coarse packings in
column lengths up to 20 m can be used at reasonable inlet pressures. Columns
packed with fine grain particles and only a few meters in length may require
inlet pressures from 25 to 50 atmospheres for operation at optimal carrier gas
flow velocities [219]. These large pressure drops lead to a number of
instrumental problems, requiring extensive modification to instruments designed
for use with classical packed columns.

Micropacked columns are prepared by simply modifying those packing
techniques used to prepare classical packed columns. In general, higher
pressures are required and constant vibration or ultrasonic treatment is needed
to achieve the necessary packing density [217,218]. Virtually all packings used
in classical packed columns can be used to prepare micropacked columns.

Micropacked columns combine high efficiency with a relatively high phase
ratio. Compared to open tubular capillary columns they are particularly
suitable for the separation of weakly retained solutes such as gases and low
molecular weight hydrocarbons without resorting to subambient temperature
control. They are also useful for the determination of trace components in
complex mixtures where the limited sample capacity of open tubular columns makes
direct analysis difficult and classical packed columns lack the resolving power
required for the separation. Practical problems, particularly with injection at
high back pressures, have limited the use of micropacked columns to a few
chromatographic laboratories.

2.10 Open Tubular Columns

In his seminal paper published in 1957, Marcel Golay showed theoretically that an open tube of small internal diameter coated with a thin liquid film would be capable of realizing efficiencies a hundred-fold or so greater than a packed column [220]. Thus the concept of open tubular column gas chromatography was born and the next quarter century was an era of struggles and termoil devoted to developing the column technology necessary to make the technique widely accepted [221,222]. In the past column preparation was often described as an art rather than a science, with many practitioners but only a few grand masters. Today it is possible to purchase excellent open tubular columns whose efficiency, thermal stability, and inertness surpass the capabilities of packed columns.

The theoretical efficiency of packed columns is limited by the intrinsic dispersion of solute molecules associated with the multiplicity of flow paths through the packing, and by differences in retention resulting from the uneven distribution of liquid phase within the particles and at the contact points between individual particles. By contrast, the efficiency of an open tubular column is governed solely by the rate of mass transfer in the liquid phase; flow heterogeneities due to the presence of packing are, of course, eliminated. The general improvement in efficiency of an open tubular column over a packed column can best be demonstrated by example. Consider the separation of the solvent extract of a river water sample in Figure 2.8 [223]. Clearly the complexity of the sample is better represented in the open tubular column chromatogram than in the packed column case. Both theory and practice indicate that open tubular column gas chromatography is capable of better separation efficiencies and greatly improved sample detectability for a given analysis time than any packed column [224]. The principal disadvantages of the general use of open tubular columns are that they are more demanding of instrument performance, less forgiving of poor operator technique, and possess a lower sample capacity than packed columns.

Before discussing column preparation a few comments on nomenclature are in order. Open tubular columns are also widely known as capillary columns. The characteristic feature of these columns is their openness, which provides an unrestricted gas path through the column. Thus open tubular column rather than capillary column is a more apt description. However, both descriptions appear frequently in the literature and can be considered interchangeable. The type of columns discussed so far are also known as wall-coated open tubular columns (WCOT). Here the liquid phase is deposited directly onto the column wall without the inclusion of any additive that might be considered solid support. Alternatives to the WCOT columns are the porous-layer open tubular columns

80

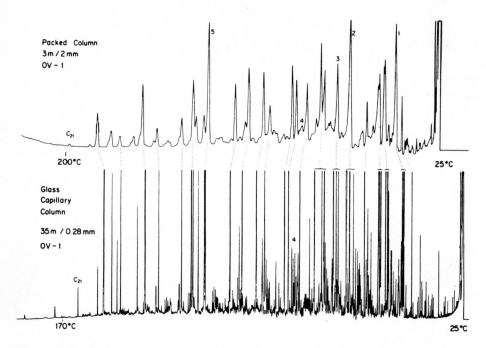

Figure 2.8 Comparison of a typical separation on a packed and a capillary
column. The sample (solvent extract of river water) is the same in
both cases. (Reproduced with permission from ref. 223. Copyright
Ann Arbor Science Publishers).

(PLOT) and support-coated open tubular columns (SCOT). PLOT columns are
prepared by extending the inner wall of the column by such substances as fused
silica or elongated crystal deposits. In SCOT columns the liquid phase is
coated on a surface that has been covered with some type of solid support
material. Every SCOT column is a PLOT column but not every PLOT column is a
SCOT column.

Recent developments in column technology have made much of the previous
work on column preparation obsolete. We will present only a brief account of
these older methods in the following sections. More detailed accounts are
available in standard reference works [225-230].

2.10.1 Drawing Columns With Capillary Dimensions

Numerous materials have been used to fabricate open tubular columns. Metal
columns of capillary dimensions made from stainless steel, nickel, aluminum,
copper, etc. are commercially available. In general, metal columns have poor
coating characteristics and exhibit catalytic and adsorptive activity,
precluding their use with many polar samples [231,232]. Plastic and nylon

capillary tubing is difficult to coat and has limited thermal stability. Teflon columns modified by brief reaction with a solution of sodium naphthalate can be coated with polar liquid phases [233,234]. Chemical pretreatment is necessary to create an adsorptive layer, largely nodules of carbon, to support the liquid phase. These columns have a high surface activity and a high specific area (up to 4000 m^2/g). None of these materials has found wide application.

At present, virtually all open tubular columns are prepared from either soda-lime, borosilicate, or fused silica glasses. Desty devised a simple glass drawing machine which made available soda-lime and borosilicate capillary columns with any desired internal diameter to early workers in the field [235]. These columns remained the standard for capillary column work until the introduction of flexible fused silica columns by Dandeneau and Zerenner in 1979 [236]. Dandeneau and Zerenner also investigated the properties of uranium and lead glasses [236]. Uranium glasses were found to be very adsorptive towards amines, phenols, and alcohols, whereas lead glasses showed little adsorption of phenols or fatty acids but could not be used for the separation of sulfur-containing compounds. However, neither of these glasses has been widely used in gas chromatography. The chemistry of glasses and their influence on chromatographic properties have been lucidly reviewed by Jennings [229] and Pretorius [237].

The major component of the glasses used to prepare open tubular columns is silica. Fused silica, prepared by the flame hydrolysis of silicon tetrachloride, is essentially pure silica containing less than 1.0 ppm of metal impurities [236]. It has a relatively high melting point and both special equipment and precautions are required for fabricating columns from this material. The surface of fused silica glass is relatively inert, containing primarily silanol and siloxane groups. The presence of silanol groups is responsible for the residual acidic character of the glass. High temperature treatment in an inert atmosphere can reduce the concentration of silanol groups, and hence activity, to very low levels [238,239].

Glasses with lower working temperatures than silica are prepared by adding metal oxides to the silica during manufacture. Thousands of glasses with a variety of compositions are known, but for chromatographic purposes only soda-lime (soft) and borosilicate (hard, Pyrex, Duran) glasses are important. Typical values of the bulk compositions of soda-lime and borosilicate glasses are given in Table 2.16. Soda-lime glasses are slightly alkaline due to the high content of Na_2O while the borosilicate glasses are somewhat acidic as a result of the presence of B_2O_3. Their surface composition is often very different than the bulk composition of the glass, Table 2.16 [240]. The adsorptive and catalytic activity of the column materials can be attributed to the silica surface (silanol and siloxane groups) and to the presence of metal

impurities at the surface which can function as Lewis acid sites. The metal
impurities act as adsorption sites for lone-pair donor molecules such as ketones
and amines. The types of surface adsorption interactions experienced with glass
open tubular columns are summarized in Table 2.17 [226].

TABLE 2.16

COMPOSITION OF SODA-LIME AND BOROSILICATE GLASSES

Component	Soda-lime	Borosilicate
SiO_2	67.7	81.0
Na_2O	15.6	4.0
CaO	5.7	0.5
Al_2O_3	2.8	2.0
B_2O_3	---	13.0
MgO	3.9	---
BaO	0.8	---
K_2O	0.6	---

Composition by Element (atomic %)

Element	Bulk	Surface	Bulk	Surface
Si	23.4	11.4	25.5	24.0
O	59.4	57.8	64.0	69.0
Na	10.4	16.0	2.5	– – –
Ca	2.1	12.0	0.2	– – –
B	1.0	1.6	7.0	7.0
Al	1.2	– – –	0.8	– – –
K	0.3	1.2	– – –	– – –

Open tubular glass capillary columns with internal diameters of 0.05 to 1.0
mm and outer diameters of 0.3 to 1.5 mm can be prepared from glass stock,
typically 4.0 to 10.0 mm outer diameter and 2.0 to 6.0 mm internal diameter,
with a glass drawing machine, Figure 2.9 [226,228,241]. The glass stock tube is
fed into a softening furnace that is electrically heated to a temperature of 650
to 850°C and drawn out at the lower end by passage through a pair of
motor-driven rollers. Capillary tubes of different dimensions are drawn by
maintaining a fixed differential between the feed rate of the stock tube to the
furnace and the pull rate of the softened tube, at the exit of the furnace.

TABLE 2.17

SOLUTE-SURFACE INTERACTIONS IN GLASS OPEN TUBULAR COLUMNS

1. The metallic oxides used in the manufacture of glass give rise to cationic charge locations that function as Lewis acid sites. These sites adsorb molecules having regions of localized high electron density such as alcohols, ketones, amines, and π-bond-containing molecules.

2. The surface hydroxyl groups act as proton donors in hydrogen bond interactions and can function as very strong adsorptive sites for molecules having localized high electron density.

3. The surface siloxane bridges act as proton acceptors in hydrogen bond interactions and can function as adsorptive sites for molecules like alcohols. In addition, these sites give rise to significant van der Waals interactions.

4. Weak dispersive interactions can arise from the silicon-oxygen network and other functionalities present on the glass surface.

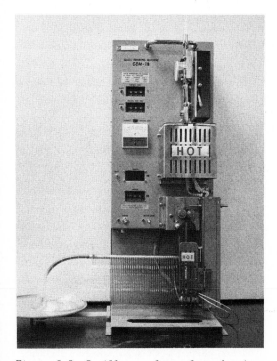

Figure 2.9 Capillary column glass drawing machine.

The capillary tubing emerging from the softening oven is coiled by passage through a heated bent pipe maintained at 550 to 625°C. The actual temperatures used for the softening oven and bent pipe depend on the wall thickness and composition of the glass. Higher temperatures are required for borosilicate than soda-lime glasses.

Non-uniformity of the column's inner diameter can affect the performance of high efficiency open tubular columns [242-245]. Variations can arise from irregularities in the dimensions of the stock tube, fluctuations in the temperature of the softening oven and bent pipe, and shifting of the glass tube between the feed rollers. Commercial glass drawing machines will usually provide a tolerance of 3-5% RSD for the internal capillary tube diameter; modification of the power supply and temperature control circuit might improve this to 1.5% RSD [246].

Fused silica and quartz capillary tubes cannot be drawn on standard glass drawing machines. Quartz capillary columns can be drawn by replacing the softening furnace with a hydrogen-oxygen burner or graphite furnace and making the bent pipe from a platinum alloy that is air stable when heated to 1250-1350°C [247]. However, this is not the process used to fabricate fused silica open tubular columns widely used in gas chromatography today. These thin-walled polymer-coated columns are drawn by techniques similar to those used for manufacturing optical fibers [247,248]. This type of equipment is not normally found in analytical laboratories. The tubing, as drawn, is inherently straight but sufficiently flexible to be coiled on a spool for collection purposes. A protective polymeric sheath is required to protect the newly drawn capillary from atmospheric corrosion. Moist air causes the hydration of strained siloxyl bonds in the freshly drawn tube, promoting the growth of fissures or cracks which make the capillary weak and friable [249]. For this reason the capillary emerging from the softening oven is immediately passed through a reservoir containing a solution of a high temperature polymer, usually a polyimide, and then through an electrically-heated drying oven prior to collection on a spool. Typical drawing rates for flexible thin-walled fused silica capillary columns are around 1 m/s, substantially higher than the rate used to draw glass capillaries.

The desirable flexibility and inherent straightness of fused silica columns are not specific properties of silica but, rather, are due to the thinness of the wall and the protective polymeric coating. Flexible soda-lime glass capillary columns can also be prepared by modifying a standard glass drawing machine to draw thin-walled columns which are coated with a polymeric film in a manner similar to the fused silica columns [248,250]. However, the chromatographic inertness of soda-lime glass is not comparable to fused silica and it remains to be seen whether these new columns will become popular [251].

Nearly all open tubular columns in current use have circular cross-sections with internal diameters in the range 0.05 to 1.0 mm. Theory predicts that the

efficiency of such columns could be increased by inducing radial mixing along
the flow path to minimize the gas phase mass transfer contribution to the column
HETP value. Geometrically deformed helical, oval, square, and crinkled open
tubular columns have been prepared [252-255]. These columns are difficult to
coat efficiently and practical demonstrations of their superiority over circular
columns have not been forthcoming.

2.10.2 Film Formation on Glass Surfaces

To achieve a high separation efficiency in any type of open tubular column
it is essential that the stationary phase be deposited as a smooth, thin, and
homogeneous film. This film must also maintain its integrity without forming
droplets when the column temperature is varied.

The ability of a liquid to wet a solid surface is determined by the contact
angle, that is, the angle between the tangent to the liquid drop and the solid
surface. When the contact angle is zero the liquid wets the surface
completely. As the contact angle increases the tendency for a liquid to coat a
surface decreases. Whether a liquid will wet a surface depends on the
difference between the cohesion forces inside the liquid and the energy of the
solid surface; spreading generally occurs when the specific free energy of the
liquid is less than that of the solid. The critical surface tension of a
smooth, clean glass is generally less than 30 dyne/cm while the surface tensions
of typical stationary phases are in the range of 30-50 dyne/cm [226,256-259].
Consequently, most stationary phases exhibit large contact angles on glass and
do not form uniform films.

Solutions to the above problem are required if efficient open tubular
columns are to be prepared. The energy of the smooth glass surface can be
increased by roughening or chemical modification, or the surface tension of the
stationary phase can be lowered by the addition of a surfactant. Roughening
and/or chemical modification are the most widely used techniques for column
preparation; the addition of a surfactant, although effective, modifies the
separation properties of the stationary phase and may also limit the thermal
stability of columns prepared with high temperature stable phases.

Roughening, achieved by chemical treatment of the glass surface, enhances
the wettability of the glass surface by increasing the surface area over which
interfacial forces can act and dissipate the cohesive energy of the drop. The
surface tension of the stationary phase can also be lowered by creating a thin
film on the glass surface. This film can be either a bonded organic layer,
produced by silylation of the glass surface, or a non-extractable polymeric film
formed by the thermal degradation of a polymeric substrate. In subsequent
sections the chemistry of the above processes will be discussed in more detail.

Although the stationary phases used to prepare packed columns might be used
to prepare open tubular columns, all of these phases are not necessarily optimal
for this purpose. Open tubular columns are generally coated with films that are
sufficiently thin that diffusion in the liquid phase is not limited by the
viscosity of the stationary phase. Efficient open tubular columns can therefore
be coated with specially prepared gum or crosslinked phases of very high
viscosity [33,260,261]. At high temperatures viscous phases resist droplet
formation and can remain indefinitely stable as smooth films. We will return to
this point later.

2.10.3 Surface Modification Reactions

Most stationary phases do not form thin, uniform, stable films when coated
onto smooth glass surfaces. To improve the wetting characteristics of common
stationary phases, surface roughening techniques are used [226,229,241,251,
262]. Numerous roughening methods have been investigated but only a few are
used frequently. The most important reactions are etching by gaseous hydrogen
chloride, leaching with aqueous hydrochloric acid, formation of whiskers, and
solution deposition of a film or layer of solid particles.

The static or dynamic etching of soda-lime glass capillaries results in the
formation of a regularly spaced array of sodium chloride crystals on the inner
surface of the column. Borosilicate glasses contain a low concentration of
alkali ions and are little changed by hydrogen chloride treatment, unlike
soda-lime glasses which become opaque in appearance. Hydrogen chloride etching
is considered essentially a technique for roughening the surface of soda-lime
capillaries, although similar treatment of borosilicate glass often improves its
wetting and deactivation characteristics as well.

To etch soda-lime capillary columns the dry column is filled with hydrogen
chloride gas and then sealed at both ends with a microtorch. The column is then
heated at 300 to 400°C for between 3 to 12 h depending on the method followed
[256,263,264]. Dynamic etching with a continuous flow of hydrogen chloride
yields a surface covered with a large number of relatively small particles of
dissimilar size and shape. In the presence of excess hydrogen chloride (the
static method) recrystallization is relatively slow compared to particle growth,
accounting for the differences in the surfaces generated by the two methods.
The static method provides the most uniform surface and is preferred for column
preparation. A detailed study of the mechanism of sodium chloride crystal
formation has been given by Franken et al. [263] and is summarized in Table
2.18. Badings et al. noted that capillary columns sealed and stored after
hydrogen chloride treatment developed surface irregularities compared to freshly
prepared columns [264]. Residual water in the capillaries caused migration and

recrystallization of the sodium chloride particles. Columns should be either coated immediately after etching or evacuated prior to sealing for storage. Hydrogen chloride etching has certain limitations, the most important of which are [226]:

- The reproducibility of the microroughened surface depends on the surface composition of the glass.
- The solubility of sodium chloride in various solvents can present a problem during the rinsing and coating of columns.
- Thin film columns show adsorptive behavior resulting from the weak Lewis acid character of the sodium ions.
- The high alkali concentration on capillary surfaces increases the catalytic decomposition of some stationary phases.

Sequential treatment of soft glass capillaries with a continuous flow of hydrogen chloride at 450°C, followed by treatment with hydrogen fluoride-nitrogen (1:1) for 1 to 2 h at the same temperature, results in a less adsorptive surface. After coating these columns are more thermally stable and show less tailing and stationary phase bleed than statically-etched soda-lime columns [265]. The experimental conditions must be carefully selected to avoid whisker formation, which takes place under static etching conditions [266,267].

TABLE 2.18

MECHANISM FOR THE FORMATION OF SODIUM CHLORIDE CRYSTALS DURING STATIC ETCHING OF SODA-LIME CAPILLARIES WITH GASEOUS HYDROGEN CHLORIDE

---

1. During the drawing of capillary columns, vapor-phase Na_2O condenses on the column surface during cooling, forming an alkali-rich surface layer which serves as sites for nucleation.

2. The sodium ions become more mobile at the high temperatures needed for reaction. Because of their small diameters, these ions move rather freely through the lattice toward the surface. At the same time H^+ ions from the HCl gas diffuse into the glass surface.

3. At the surface, sodium ions exchange with hydrogen ions according to the following equation:

$$\begin{array}{ccc} | & & | \\ -Si-ONa & + \; H^+ \longrightarrow \; -Si-OH & + \; Na^+ \\ | & & | \end{array}$$

4. The Na^+ ions move across the glass surface, associate with Cl^- ions, and randomly locate on a nucleation site. Consequently, the initial NaCl particles formed tend to have circular, flattened-convex shapes.

5. As the HCl gas is consumed, particle growth slows down and recrystallization becomes competitive. Ultimately, rectangular crystals are formed.

6. If the heat treatment is continued for long periods, large NaCl crystals grow at the expense of smaller ones.

Hydrogen fluoride gas is highly toxic and this has probably limited the general appeal of the method.

Acid leaching with solutions of hydrochloric acid produces a different effect than acid etching in the gas phase. Controlled acid leaching removes metallic cations from the glass surface to form a silica-rich surface. The new surface layer greatly minimizes the effects of glass variety on subsequent treatments and lends a higher degree of reproducibility to column preparation. Lewis acid sites are removed, leaving highly adsorptive silanol groups; the latter are easily deactivated, for example, by silylation. Thus acid-leached borosilicate and soda-lime glasses are very inert after deactivation and can be coated with thin films of nonpolar phases. As fused silica contains only traces of metal impurities, acid leaching prior to coating with nonpolar phases is not normally required.

Borosilicate and soda-lime glass capillary columns can be leached with aqueous hydrochloric acid using the static [268,269] or dynamic [270-272] method. In the static method the column is filled with 20% v/v hydrochloric acid, a small portion of the column is emptied and then evacuated, the ends are sealed and the column is heated overnight at 140 to 170°C. The column is partially emptied and evacuated to provide an expansion zone for the heated liquid. For borosilicate columns it is recommended that the evacuated portion of the column should correspond to 7% of the column length and the column should be heated at 170°C for 12 to 16 h. For soda-lime columns the evacuated portion of the column should correspond to 4% of the column length and it should be heated at 140°C for 12 to 16 h. After leaching the columns are cooled and then rinsed with 1% v/v hydrochloric acid to remove the dissolved metal ions. One to two column volumes at a flow rate not greater than 2 cm/s are usually sufficient for the rinse step. To avoid readsorption of dissolved metal ions by the surface hydrogel, the entire rinsing process should occur under strong acid conditions (distilled water should not be used). Also, short (20 meter) columns are much easier to rinse out without readsorption of metal ions than longer columns. After rinsing the swollen surface hydrogel is dehydrated by applying a vacuum to both ends of the column simultaneously, or by applying the vacuum to either end of the column for 10 to 20 minutes alternately, while the column is heated. Recommended conditions for 15 to 20 m column lengths are 2 h at 300°C, and 3 h at 300°C for 30 to 50 m column lengths.

For dynamic leaching the column is mounted in an oven with both ends extending to the outside. One end is connected to a reservoir and a solution of 20% v/v hydrochloric acid is forced through the column at a temperature of 100°C. Generally, long lengths (e.g., 90 m) of capillary tubing are leached with 50 ml of acid for approximately 48 h. The leaching step may need to be repeated after an intermediate drying stage to produce the best results. After leaching the column is rinsed with distilled water and then dried by purging with dry nitrogen at 150°C for about 12 h. Venema et al. [272] have recommended slightly different conditions for the dynamic leaching of soda-lime capillary columns.

A variety of reagents and methods have been used for extending the surface area of borosilicate or soda-lime glasses by the preparation of whisker surfaces. A high density light surface etch can be obtained by flushing the capillary column with either an aqueous solution of potassium bifluoride [273] or a methanolic solution of ammonium bifluoride [274]. Treatment of borosilicate glass with ammonium bifluoride results in a thin, uniform etch with a depth of 0.05 to 0.4 micrometers and a distance of 2 to 6 micrometers between whiskers. This is a relatively light surface modification compared to whiskers grown at high temperatures. However, the above methods are simple to carry out and, after acid leaching and deactivation, are suitable for preparing columns coated with nonpolar and moderately polar stationary phases.

Whisker columns of high density are prepared by the action of hydrogen fluoride gas on the glass surface at high temperatures. The gas itself is rarely used because of safety considerations (toxicity) and the often observed non-uniformity of whisker growth [266,267]. The preferred method is to generate hydrogen fluoride in situ by the thermal decomposition of 2-chloro-1,1,2-tri-fluoroethyl methyl ether [267,275-279] or ammonium bifluoride [280]. The mechanism of whisker formation is not known with certainty but it is assumed that the hydrogen fluoride generated reacts with the glass to form silicon tetrafluoride, which is then converted into silicon dioxide and deposited in the form of whiskers. The whisker surface has a surface area a thousand times or so higher than the smooth glass surface and consists of a densely packed layer of filamentary projections whose size and length depend on the experimental conditions. Under optimum conditions whiskers of 4 to 5 micrometers in length are obtained. With the 2-chloro-1,1,2-trifluoroethyl methyl ether, the important experimental variables are reagent concentration, temperature, and reaction time. At temperatures below 250°C whisker formation does not occur. The optimum temperature for whisker formation is in the region of 400°C. Whisker length and surface density increase with the growth period and a time of 24 h is considered optimum. Most variations in methodology differ in the amount of reagent used for whisker growth. Pretorius recommended a concentration of 2.5 to 10% of the column volume [275]; Sandra [277] suggested that it was preferable to calculate the amount of ether as a function of the column surface area and recommended a concentration of approximately 0.3 microliters/cm^2; while Gates [279] found mastering the technique of Pretorius (injection of sample through a septum into a column evacuated to 10^{-4} mmHg) difficult and recommended that the reagent be dynamically coated onto the column wall and the column then sealed. After etching by any of the above methods carbon deposits formed by decomposition of the ether must be removed by flushing with oxygen. Typical conditions involve passing a continuous flow of oxygen through the column at 450°C for 8 to 12 h. The ammonium bifluoride method has the advantage that only volatile products (HF and NH_3) are produced by thermal decomposition. Column preparation time is also considerably reduced. For optimum whisker growth the reagent is dynamically coated and the sealed column is then heated at 450°C for 3 h. Soda-lime glasses are unsuitable for preparing whisker columns due to their low softening temperature.

The large surface area of whisker columns enables any stationary phase to be coated efficiently without droplet formation and their sample capacity is much higher than WCOT columns. Another advantage is that the roughened surface improves the thermal stability of the stationary phase film, allowing high temperature column operation. On the debit side, whisker surfaces are extremely active and very difficult to deactivate, and, because of the high degree of roughening, they are less efficient than WCOT columns.

Glass surfaces can also be roughened by deposition of a thin homogeneous film of particle material from solution. Since hydrogen chloride etching of soda-lime glass results in a crystalline layer of sodium chloride then a logical extension would be to form a similar layer by deposition of sodium chloride from solution. Soda-lime glass capillaries coated with sodium chloride dendrites were prepared by dynamic coating with an aqueous sodium chloride solution that also contained a surfactant [281]. Although these columns provided adequate roughening for coating with moderately polar phases, the degree of roughness was inadequate to stabilize the more difficult to coat polar phases. In an alternative approach a dense layer of sodium chloride crystals was formed by dynamically coating the columns with a stable sodium chloride suspension that was prepared by the addition of a saturated solution of sodium chloride in methanol to 1,1,1-trichloroethane [282]. During the passage of the suspension through the column, particles of sodium chloride deposit spontaneously on the column wall, building up a layer of sodium chloride which gradually reaches a maximum density. After deactivation, coating with polar stationary phases produces efficient and thermally stable columns. Surface roughening with a thin film of amorphous silica can be achieved by the dynamic coating of capillary columns with an aqueous solution of water glass followed by reaction with hydrogen chloride [283] or by allowing silicon tetrachloride vapor to react with a humid glass surface [284]. After deactivation, coating with nonpolar and medium polar phases produces efficient columns. Neither the sodium chloride nor silica deposition methods have been widely adopted.

Of the deposition methods, the barium carbonate procedure developed by Grob has been the most widely evaluated [285-288]. The general procedure consists of dynamically coating the glass surface with barium hydroxide solution using carbon dioxide gas to push the plug of hydroxide solution through the column and generate the barium carbonate layer. From a saturated solution of barium hydroxide, a rather dense crystal layer is obtained that is suitable for coating with polar phases. From gradually diluted coating solutions the crystals become more distant while the glass surface between the crystals remains covered by a very thin and smooth layer of barium carbonate. When the solution is diluted 50- to 100-fold the crystals disappear, leaving the smooth surface cover only. This thin film is suitable for coating with nonpolar phases where the barium carbonate layer functions more as a deactivating agent for surface active sites than as a roughening agent. The preparation of the barium carbonate layer is influenced by a large number of experimental variables including the glass surface structure, the crystallization temperature, the addition of stationary phase modifiers (surfactants), and the method used for deactivation. Careful attention to the experimental details enables columns to be prepared with a fairly high success rate, although column efficiency is usually lower than with other methods. General problems are the difficulty of completely deactivating the barium carbonate layer and the poor thermal stability of some phases when coated onto this layer.

For the purpose of perspective it should be noted that the roughening methods discussed above were important developments in capillary column technology during the mid-1970's to the 1980's when borosilicate and soda-lime glasses were used predominantly. The shift in emphasis in column technology from these glasses to the use of fused silica capillaries coated with immobilized phases has diminished their general importance, even for polar phases, which at first were thought to be impossible to coat on the smooth, inactive fused silica surface. Improvements in stationary phase technology have rendered this concern virtually obsolete; many now consider the soda-lime/borosilicate glass capillary column extinct in modern column technology.

2.10.4 Surface Deactivation Methods

Roughening the surfaces of soda-lime and borosilicate glasses to improve their wetting characteristics results in a concomitant increase in the adsorptive activity of the columns. Without deactivation, columns coated with nonpolar and moderately polar stationary phases exhibit undesirable chromatographic properties such as peak tailing and incomplete elution of polar solutes [289]. Although polar stationary phases may act as their own deactivating agents, column pretreatment is still advisable to ensure complete

deactivation and to improve the thermal stability of the stationary phase film
[290]. Thus, in general, the processes of surface roughening and deactivation
are complementary steps in the column preparation sequence, especially since it
is often difficult to completely deactivate glass surfaces without first
including an activation procedure. In this respect, acid leaching may be
considered as both an activation and deactivation process as it denudes the
glass surface of Lewis acid sites while increasing the concentration of silanol
groups. Deactivating silanol groups is more straightforward than Lewis acid
sites, thus acid leaching is now widely used as a preliminary step in the
preparation of inert, nonpolar open tubular columns.

The effectiveness of a particular deactivating reagent depends on the
properties of the glass, the stationary phase, and the sample to be analyzed, as
well as on the method of surface roughening. No universal method of
deactivation exists but some techniques have emerged as more useful than
others. The most widely used methods include coating with a surface active
reagent, formation of a nonextractable polymeric film, or silylation prior to
coating with the stationary phase.

Precoating the column with a thin film of a surface active reagent such as
benzyltriphenylphosphonium chloride (BTPPC), di-iso-butylphenoxyethoxy-
ethyldimethylbenzylammonium chloride (BAC), or trioctadecylmethylammonium
bromide is notable from an historic point of view [277,291-293]. However,
deactivation using the above reagents was rarely complete, was often difficult
to reproduce exactly from column to column, and the finished columns usually
exhibited poor thermal stability (generally ineffective at temperatures above
220-250°C). Also, the reagents themselves contributed to the retention of polar
solutes, particularly if columns were subsequently coated with nonpolar
stationary phases. A more widely used method is to precoat the column with a
nonextractable film of a polymeric substance. The thermal degradation products
of Carbowax 20M [294-298], the polymethylsiloxanes [298,299], and
N-cyclohexyl-3-azetidinol [277,278] are the most widely used. The Carbowax
treatment can be performed in either of two ways. The column can be dynamically
coated with a solution of Carbowax 20M in a volatile solvent, excess solvent is
then evaporated with a stream of nitrogen, the column ends are sealed, and the
column is heated at about 260-280°C for 12-16 h. After the column is cooled,
excess Carbowax and other by-products are removed by rinsing with suitable
solvents. The column may also be deactivated by vapor phase treatment [297].
For this purpose a short precolumn, containing Carbowax 20M coated onto a
support, is inserted into a separate heater or the injection heater of a gas
chromatograph and connected directly to the column to be deactivated, which is
housed in a separate oven. The precolumn in maintained at a temperature of
about 260°C, 5-10°C higher than the capillary column. Deactivation is completed

by allowing the degradation products to bleed through the capillary column for about 16 h with a slow flow of carrier gas. The mechanism of Carbowax deactivation remains unknown, although it is believed that the thermal degradation products bond chemically to the surface silanol groups at the high temperatures employed. Carbowax pretreatment is reasonably successful in masking the influence of silanol groups but less so in masking Lewis acid sites associated with metal ions; the latter are better removed by acid leaching. Carbowax deactivation methods have been used to prepare both nonpolar and polar columns and reduced activity toward many sensitive compounds has been demonstrated. The deactivated layer is thermally stable under continuous use at temperatures up to about 220°C and for intermittent use up to about 250°C. Columns coated with nonpolar stationary phases may show mixed liquid phase retention behavior. Carbowax pretreatment is inadequate for deactivating whisker columns.

The effectiveness of deactivation techniques in reducing the activity of glass columns can be judged from the analysis of polarity test mixtures [277]. On bare Pyrex glass, Figure 2.10A, dibutylketone exhibits tailing while 2-propylcyclohexanol and 2,6-dimethylaniline are completely adsorbed. Etching the column with hydrogen chloride gas reduces the activity slightly, Figure 2.10B. Further deactivation of the etched Pyrex column with benzyltriphenylphosphonium chloride diminishes the column activity, although 2-propylcyclohexanol and 2,6-dimethylaniline still display some tailing, Figure 2.10C. Carbowax 20M is more successful than BTPPC, Figure 2.10D. The higher adsorptive activity of Pyrex glass compared to soda-lime glass can be seen by comparing the chromatograms for hydrogen chloride etched and Carbowax 20M deactivated columns, Figures 2.10D and 2.10E. As the adsorptive activity is an important consideration in judging the quality of an open tubular column we will return to this subject in a subsequent section.

The thermal degradation products of polymethylsiloxane stationary phases have been used by Schomburg to deactivate glass and fused silica columns prior to coating the columns with those same phases [298,299]. An acid-leached column is dynamically coated with a solution of the polymethylsiloxane phase, the solvent is evaporated, and the column is then filled with an inert gas, sealed, and heated to between 300 and 450°C (depending on the glass used) for several hours. Afterwards, excess material is removed by solvent extraction or by high temperature conditioning with a continuous flow of inert gas. It is postulated that under the experimental conditions described above the polymethylsiloxane phase decomposes, releasing degradation products which bond chemically to the glass surface. The mechanism may be very similar to that of silylation with cyclic siloxanes described later. Columns prepared in the above manner exhibit good deactivation of silanol groups with thermal stabilities that endure above

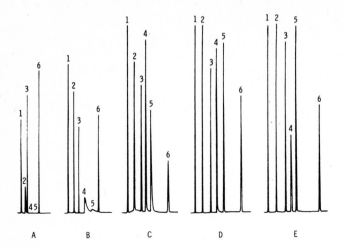

Figure 2.10 The influence of column activity on the separation of a polarity
test mixture containing 1. n-decane; 2. dibutylketone;
3. undecane; 4. 2-propylcyclohexanol; 5. 2,6-dimethylaniline; and
6. n-dodecane on A, bare Pyrex glass; B, Pyrex glass etched with
hydrogen chloride gas; C, column B after deactivation with BTPPC;
D, column B after deactivation with Carbowax 20M; and column E
soda-lime capillary treated as in D. (Reproduced with permission
from ref. 277. Copyright Friedr. Vieweg & Sohn).

300°C.

To reduce the activity of both Lewis acid sites and silanol groups on
borosilicate glass, and in particular for whisker columns, Verzele has suggested
the use of cyclic alkyl-azetidinol reagents [277,278]. These reagents are heat
polymerized to form a nonextractable polymeric film as shown below.

$$R-N\bigcirc-OH \xrightarrow{\Delta H} \left[-\underset{R}{\overset{|}{N}}-\underset{OH}{\overset{|}{}} \right]_n$$

R = cyclohexyl (CHAZ)

The column is coated dynamically with a 5% w/v solution of the reagent, sealed
after evaporation of the solvent, and then heated overnight at 125°C. Afterwards
the extractable portion of the polymer is removed by washing the column
successively with dry methanol and dichloromethane. The deactivated column is
thermally stable up to about 240°C with occasional use to about 280°C. Verzele
has also shown that coated columns may be deactivated in situ by several
injections (0.1-0.2 microliters) of a 1:1 solution of triethanolamine, or
preferably tri-isopropylamine, in dichloromethane. This method is restricted to
those stationary phases which are able to elute the reagent below their maximum
operating temperature.

In recent years, silylation by pyrolysis of cyclic siloxanes [300-304] or by high temperature reaction with a disilazane [269,270,305-309] has emerged as the preferred technique for deactivating acid leached soda-lime and borosilicate glasses or fused silica capillary columns. Using reagents that contain polar functional groups simultaneously improves the coating characteristics of polar liquid phases. This is significant since silylation by conventional alkylsilylating reagents results in a surface which can only be coated effectively with nonpolar liquid phases.

Deactivation with octamethylcyclotetrasiloxane (D_4) is performed by dynamically coating the column with the reagent, sealing the ends, and then heating the column at 400-420°C for about 2 h. The column ends are then broken and the column is purged with nitrogen for about 30 min at 350°C to remove any residual reagent. Slight modification of the experimental conditions enables deactivation to be carried out with hexaphenylcyclotrisiloxane [300], octaphenylcyclotetrasiloxane [301], cyanopropyl(methyl)cyclotetrasiloxane [303], and cyclic (3,3,3-trifluoropropyl)methylsiloxane, creating inert surfaces which can be coated with liquid phases containing the same substitutent group as the deactivating reagent.

High-temperature silylation of leached glass or fused silica columns using hexamethyldisilazane, either alone or mixed with trimethylchlorosilane, can be performed statically [305,306] or dynamically [270]. In the static method a plug of hexamethyldisilazane is quickly pushed through the column at 4-5 cm/s. After the reagent has been expelled the column ends are sealed and the entire column is slowly heated to 300-400°C and maintained there for 4-20 h, depending on the method followed. In the dynamic method the column is placed in an oven at 400°C with both ends extending out of the oven. One is connected to a bubbler containing the reagent, the vapors from which are circulated through the column by a stream of dry nitrogen gas for about 48 h. After silylation, the oven is cooled to 200°C, the bubbler is removed, and dry nitrogen is passed directly through the column for about 30 min to flush out any unreacted reagent.

An unexpected problem has been observed in the static silylation of glass capillary columns with hexamethyldisilazane [307]. Silylation would be expected to proceed to completion, provided an excess of reagent were present. However, completion is not reached. The probable reason is that the ammonia generated during the silylation reaction continues to open up the silica network, thus constantly liberating more reactive silanol groups. With time a sponge-like surface layer forms and becomes thoroughly silylated. This layer is an effective adsorbent for nonpolar substances but repels polar materials. Consequently, excessive silylation results in columns which produce excellent peak shapes for polar solutes while peaks for alkanes are broadened and possibly retained longer than expected. Thus the optimized conditions for silylation discussed above should be carefully followed in order to avoid excessive silylation.

After high-temperature silylation with hexamethyldisilazane the surface energy of the glass column surface is too low to allow coating with other than nonpolar stationary phases. Silylation with diphenyltetramethyldisilazane and tetraphenyldimethyldisilazane yields inactive surfaces with better wettability for medium-polar stationary phases [269,270,305,308]. The large size of the phenyl-substituted alkylsilyl groups may prevent complete reaction of all silanol groups but, by the same token, it may provide a better shielding layer (umbrella effect) to cover unreactive silanol groups. The preliminary data available suggests that phenyl-substituted silylating reagents produce a more stable deactivated surface than treatment with hexamethyldisilazane.

2.10.5 Procedures for Coating Open Tubular Columns

The object of all coating procedures is the distribution of the stationary phase as a thin, even film that completely covers the glass surface. Three methods are currently in use for this purpose: dynamic coating, the mercury plug dynamic method, and static coating.

For dynamic coating, a suitable coating reservoir is charged with a solution of the stationary phase, 5-15% w/v, which is forced by gas pressure into the capillary column to be coated [226,228,259,277]. When about 25% of the column volume is filled the column is withdrawn from the coating solution and gas pressure is used to force the liquid plug through the column at a linear velocity of 1-2 cm/s. To maintain a constant coating velocity a buffer capillary column, which is about 25% of the length of the column being coated, is attached to the outlet end of the capillary column. When all the coating solution has left the capillary column it is disconnected from the buffer column and the gas pressure is increased to evaporate the solvent. A suitable experimental arrangement for dynamic coating is shown in Figure 2.11A [277]. For coating long lengths of narrow bore capillary columns a high pressure glass-lined stainless steel reservoir is required [310,311]; otherwise, coating reservoirs are usually made from any convenient screw-capped PTFE sealed vial, bottle, or test tube [312].

The thickness and uniformity of the stationary phase film produced by dynamic coating is dependent on the concentration and viscosity of the coating solution, the coating velocity, and the rate of solvent evaporation. A weakness of the dynamic coating procedure is the difficulty in accurately predicting the stationary phase film thickness prior to preparing the column. Methods used to determine the stationary phase thickness of coated columns are summarized in Table 2.19. When trying to prepare thick-film columns by depositing a fairly

TABLE 2.19

CALCULATION OF STATIONARY PHASE FILM THICKNESS FOR OPEN TUBULAR COLUMNS

Dynamically-Coated Columns

A. Relationship between film thickness and coating conditions [243,313-315]:

$$d_f = \frac{rc}{200} \left[\frac{u\eta}{\tau} \right]^{1/2} \qquad \text{(i)}$$

$$d_f = \frac{1.34rc}{100} \left[\frac{u\eta}{\tau} \right]^{2/3} \qquad \text{(ii)}$$

d_f = film thickness; r = column radius; c = concentration of coating solution; u = velocity of coating plug; η = viscosity of coating solution; and τ = surface tension of coating solution.

Equation (i) applies when $u\eta/\tau > 0.001$. Equation (ii) is applicable when $u\eta/\tau < 0.001$ (thin films or low solution viscosities)

B. Determining the film thickness of coated columns:

1. From the specific retention volume of a solute measured on a packed column with the same stationary phase [243,315-317]

$$d_f = \frac{273\ rk}{2V_g T_c \rho_T} \qquad \text{(iii)}$$

k = capacity factor; V_g = specific retention volume; T_c = column temperature and ρ_T = phase density at the column temperature.

2. From measurement of capacity ratios on a statically-coated capillary column of known phase ratio and a dynamically-coated column [243,313,318]

$$d_f = \frac{rk_2}{2\beta_1 k_1} \qquad \text{(iv)}$$

k_1 = capacity factor for statically-coated column; k_2 = capacity factor for dynamically-coated column; and β_1 = phase volume ratio for the statically-coated column.

C. Estimation of film thickness from the conditions used for column preparation:

1. By measuring the difference in volume of the coating solution before and after coating [319]

2. From the weight of stationary phase that can be rinsed out of the column [243,320]

3. By measuring the decrease in size of the plug of coating solution over a given column length [319]

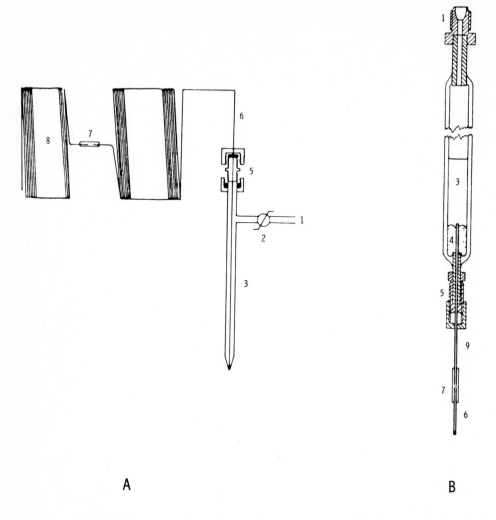

A   B

Figure 2.11 Coating resrvoirs for dynamic coating (A) and mercury plug dynamic
            coating (B). 1. connection for nitrogen pressure; 2. pressure
            regulator; 3. solution of stationary phase; 4. mercury;
            5. connection to capillary column; 6. capillary column; 7. Teflon
            shrink tubing; 8. buffer capillary column; and 9. movable
            stainless steel capillary.

large amount of solution on the column wall, drainage of the solution to the
lowest parts of the coil can be a problem. This results in inefficient columns
having localized thick patches of stationary phase. Drainage can be minimized
by orienting the coils of the column horizontally when coating and drying.
Phase rearrangement can also occur in the drying step which, by necessity,
requires several hours for completion. Fast column flow rates are preferred for
drying as this minimizes the column preparation time and results in columns of

higher efficiency. This also leads to the formation of thicker films
[317,321]. Slow drying speeds do not disturb droplets which are pushed out of
the column, while high drying speeds destroy droplets and prevents stationary
phase transportation. In recording experimental details in the literature
insufficient attention is paid to the flow rate at which columns are dried.
Most authors use flow rates between 4-10 ml/min, although higher flow rates may
be beneficial. Discarding the first 5% of the column length after drying may
produce higher overall efficiency in the final column. According to Parker and
Marshall [259], the principal cause of low column efficiency provided, that the
stationary phase wets the column wall, is the formation of "lenses" (small plugs
of liquid left behind in the main plug during coating). Small temperature
fluctuations may cause condensation of solvent vapors, forming a lens which is
able to bridge the bore of the capillary column. Since the lens is pure solvent
it will dissolve the film already laid down. Lens formation can be eliminated
by raising the temperature of the column during coating, although this is rarely
done in practice.

To improve the success rate of the dynamic coating method, and to minimize
the problems due to drainage of the coating solution, Schomburg suggested a
modification that is known as the mercury plug dynamic method [277,321-326]. In
this procedure a mercury plug interposed between the solvent plug and the
driving gas wipes most of the coating solution off the surface as it moves
through the column due to its high surface tension. More concentrated coating
solutions (10-50% v/v) are used in this procedure, producing films which are
more resistant to drainage. A suitable reservoir for coating by the mercury
plug method is shown in Figure 2.11B [322]. The empty capillary column, with
its coils oriented horizontally, is attached via a flexible length of shrinkable
Teflon tubing to the stainless steel tube. The metal tube can be raised or
lowered to feed from either the coating solution, the mercury, or the gas
atmosphere. Coating solution is forced into the column by nitrogen pressure
until about 15% of the column volume is filled. The feed tube is lowered and a
1-5 cm mercury plug is introduced into the column, carefully avoiding the
formation of a gas zone between the mercury plug and the coating solution. The
pressure in the coating reservoir is released, the feed tube is raised to the
nitrogen atmosphere above the coating solution, and the reservoir is
repressurized to force both the coating solution and the mercury plug through
the column. Depressurization is required to prevent the formation of a second
liquid plug behind the mercury plug as the feed tube is raised to the gas zone.
After the coating solution and the mercury plug are discharged from the column,
the column is disconnected from the coating reservoir and residual solvent is
evaporated with a stream of nitrogen. Optimization of all parameters which
affect the film thickness and column efficiency has not been studied in detail.

For columns of high efficiency it has been recommended that the viscosity of the coating solution should fall in the range 0.4-1.0 cP and that the coating velocity should be selected so that the square root of the product of velocity and viscosity is equal to about 0.5, or in the range of 0.5-0.7 $g^{1/2}/s$ [325].

The static coating method has the advantage of producing efficient columns of predictable film thickness with gum or solid phases [277,327-331]. Phases of low viscosity, however, cannot be successfully coated by the static method due to their ability to flow and accummulate in the lowest portions of the capillary column during the relatively long time required for solvent evaporation. For static coating the column is filled entirely with a dilute solution, 0.02-4.0% v/v, of the stationary phase in a volatile solvent such as pentane, dichloromethane, or diethyl ether. One end of the column is sealed and the other is connected to a vacuum pump (ca. 0.2 Torr). The solvent in the thermostatted column is then evaporated under quiescent conditions, leaving behind a thin film of liquid phase. The thickness of the stationary phase film can be calculated from equation (2.11)

$$d_f = \frac{rc}{200} \qquad (2.11)$$

r   = capillary column radius
c   = coating solution concentration v/v
d_f = stationary phase film thickness

The principal problem with the static coating method is the breakthrough of gas bubbles when the column is placed under vacuum. These bubbles generally originate at the solvent/seal interface or are trapped in microcavities in the glass surface. Breakthrough can eliminate all or part of the coating solvent from the capillary column. The creation of a solvent/seal air-free interface is very important. Numerous methods have been described for this purpose and can be grouped into two categories: mechanical methods include clamping, crimping, or placing a stopper in a short length of Teflon or silicone tubing attached to the column end [329,332-334]; and chemical sealants such as waxes [335,336], Apiezon grease [337], epoxy resins [277,338-340], or water glass [327,331], etc., which are sucked into the end of the column and allowed to harden before evacuation is begun. Some sealants may dissolve in the coating solution, which must be shielded from it by a short aqueous plug [340]. It may be necessary to degas some sealants prior to use by placing them under vacuum.

Air bubbles remain trapped in microcavities or scratches, etc. in the glass surface. If evacuation is commenced immediately after filling, the enclosed gas will produce a vapor bubble which will cause breakthrough [312,341]. If the filled column is allowed to stand several hours before evacuation, the trapped gas will dissolve in the coating solution and evaporation then proceeds smoothly.

Only volatile solvents such as pentane and dichloromethane provide reasonable column preparation times at ambient temperatures. For example, a 20 m x 0.3 mm I.D. capillary column filled with dichloromethane requires about 15 h for complete evacuation. Pentane requires only about half this time and is the preferred solvent for those phases having sufficient solubility [337,341,342]. Mixtures of pentane-dichloromethane can also be used to decrease the column preparation time of medium polar stationary phases. Static coating of narrow bore capillary columns at temperatures above the solvent boiling point has been used successfully to minimize column preparation time [337]. To ensure even coating, columns should generally be thermostatted during the coating process by immersion in a large volume water or oil bath.

2.10.6 Immobilized Stationary Phases

It is difficult to prepare thermally stable capillary columns with phases of low to medium viscosity. As the temperature increases the viscosity declines and the mobility of the liquid phase on the glass surface increases, eventually resulting in a breakup of the stationary phase film and the formation of droplets. The use of gum phases and roughened surfaces to counteract this tendency has already been discussed. An alternative approach is the use of immobilized phases which are created by bonding some functional group of the liquid phase to the column wall, or by crosslinking the stationary phase in situ to form a nonextractable rubber, or a combination of both techniques [342]. Immobilized phases are also known as "bonded" or "nonextractable" phases. Immobilized phases is the preferred term as the mechanism of stationary phase fixation remains unclear and, as currently practiced, quite probably involves a combination of surface bonding and crosslinking reactions [344]. The term "nonextractable phase" is a reasonable description of at least one of the characteristic properties of immobilized phases - their insolubility in and resistance to extraction by common laboratory solvents. Other desirable properties of immobilized stationary phase films are the formation of physically stable films on surfaces not completely wet by the stationary phase before immobilization; the creation of films having exceedingly low levels of bleed at elevated temperatures; the ability to prepare columns with thicker films than is possible using conventional phases and coating techniques; the formation of nonextractable coatings resistant to phase stripping by large volume splitless or on-column injection, and which permit solvent rinsing to free the column from non-volatile sample by-products or from active breakdown products of the liquid phase; and the preparation of capillary columns suitable for use in supercritical fluid chromatography [33,342-346].

Stationary phase immobilization has been achieved by two different mechanisms. In early work columns were coated with an **α,ω**-hydroxypolysiloxane prepolymer and then polymerized in situ by thermosetting in the presence of ammonia gas [347-350]. In some studies the prepolymer was first conditioned with silicon tetrachloride to introduce some crosslinking into the finished polymer [345,351-354]. The prepolymers were prepared by the hydrolysis of suitable mixtures of chlorosilanes and, after removal of the low molecular weight fraction of the mixture, coated dynamically or statically by heating in the usual way. Polymerization is completed by heating the sealed column, filled with ammonia gas, to about 320°C and maintaining it at that temperature for about 20 h. Si-O-Si bonds are formed by condensation of the prepolymer chains and possibly also by reaction of the prepolymer with silanol groups on the column wall. To encourage bond formation with the column wall, the leached glass surface can be treated with silicon tetrachloride prior to coating. Likewise, the coated prepolymer can be treated with nitrogen gas that is saturated with silicon tetrachloride to introduce a low concentration of crosslinks into the phase after ammonia-catalyzed thermosetting as described above. The prepolymers contain an average of 44 monomer units and thus the number of crosslinks is fairly low. If too many crosslinks are introduced a hard, brittle polymer with poor chromatographic performance will result.

The silicone phases immobilized as described above showed excellent thermal stability but often only attained about 70% of the expected column efficiency. These columns also often exhibited undesirably high column activity, caused by residual silanol or alkoxy groups that were left in the phase after crosslinking and could not be chemically deactivated or thermally removed due to steric problems.

An alternative approach to stationary phase immobilization, and the one most frequently used, is the free radical crosslinking of the polymer chains using peroxides [303,344,346,355-362], azo-compounds [302,346,363,364], or gamma radiation [365-368] as free radical generators. In this case crosslinking occurs through the formation of Si-C-C-Si bonds as shown below:

$$
2\begin{bmatrix} \quad CH_3 \\ -Si-O- \\ \quad CH_3 \end{bmatrix}_n \ \xrightarrow{\ 2R^\cdot\ } \
\begin{array}{l} CH_3 \\ -Si-O- \\ \cdot CH_2 \\ \cdot CH_2 \\ -Si-O- \\ CH_3 \end{array}
\ \longrightarrow \
\begin{array}{l} CH_3 \\ -Si-O- \\ CH_2 \\ CH_2 \\ -Si-O- \\ CH_3 \end{array} \ + \ 2RH
$$

Very little crosslinking (0.1-1.0%) is required to insolubilize polysiloxanes with long polymer chains. An advantage of the method is the relatively simple column preparation procedure. The column, which may be deactivated by silylation, is statically coated with a freshly prepared solution of the stationary phase that contains the free radical generator at a concentration of between 0.2 and 5% of the stationary phase weight. It is very important that the stationary phase is deposited as an even film upon coating since the film will be fixed in position upon crosslinking and no improvement in film homogeneity can be obtained after fixation. The coated column is sealed and slowly raised to the curing temperature for static curing. For dynamic curing a slow flow of carrier gas is used. The curing temperature is a function of the thermal stability of the free radical generator and is selected to give a reasonable half-life for the crosslinking reaction. Temperatures in the range 80 to 250°C are common with the reagents presently used. Once the column reaches the curing temperature, remaining at that temperature a short while is usually all that is required to complete the curing reaction. After curing, the phase which was not immobilized is rinsed from the column by flushing with appropriate organic solvents.

Peroxides have been the most widely used free radical generators. The principal problem with their use is the formation of by-products which cannot be rinsed from the column and give rise to stationary phase degradation and column activity. For example, benzoic acid generated as a by-product of benzoyl peroxide catalyzes silicone degradation. A further problem with the use of peroxides is that they may cause oxidation of susceptible functional groups, such as tolyl or cyanopropyl groups, during the crosslinking process, again leading to undesirable column activity [362]. Azo-compounds such as azo-t-alkanes [363,364] or azoisobutylnitrile [346] have been investigated recently. They produce neutral, volatile by-products which do not oxidize susceptible functional groups. Of course, none of the above problems exist when crosslinking is induced by physical means. This can be achieved by exposing the coated column to gamma radiation from a cobalt-60 source [365-368]. The method has been shown to be very reliable but, as few analytical laboratories have easy access to a coablt-60 source, this method is not widely applied. Irradiation of fused silica capillary columns causes some loss in flexibility due to deformation of the protective polyimide coating but this is not sufficient to preclude their use in gas chromatography [368].

Polymethylsiloxane gum phases are relatively easy to crosslink by treatment with peroxides and azo-compounds. High molecular weight polymers with long polymer chains increase the probability of crosslinking. Many of the conventional phases preferred for packed column gas chromatography are comprised of short chains which cannot be easily crosslinked. For this reason, these phases are not suitable for preparing immobilized phases. The introduction of phenyl and cyanopropyl groups into the silicone backbone also considerably reduces the effectiveness of peroxide crosslinking. The synthesis of new gum phases containing vinyl or tolyl groups, which are easy to crosslink, greatly facilitates the immobilization of silicone phases containing phenyl, trifluoropropyl, or cyanopropyl groups [303,361,364]. The di-tolylsiloxane group was shown to be the most useful for crosslinking phases containing a high percentage of cyanopropyl groups. When two p-tolyl groups are attached to the same silicon atom the polysiloxane chain bends away from them, leaving the tolyl groups more exposed and available for crosslinking with another chain. The vinyl group, because of its small size, is not as effective in this respect, although it is very useful for producing crosslinking in methylsiloxanes containing moderate amounts of phenyl groups [360,361]. Dicumyl peroxide shows a higher reactivity with vinyl groups than methyl groups and, therefore, a smaller amount of peroxide can be used to crosslink vinyl-containing siloxanes than methylsiloxanes. Siloxane gum phases that contain substitutents to enhance their crosslinking ability have recently become commercially available.

Another general advantage of immobilization is that it permits the preparation of open tubular columns having very thick films [369]. Normal film thicknesses in open tubular gas chromatography are 0.1-0.3 micrometers. It is very difficult to prepare conventionally-coated columns with stable films thicker than about 0.5 micrometers. However, thick-film columns of 1.0-8.0 micrometers can easily be prepared by immobilization [369-373]. Thick-film columns permit the analysis of volatile substances with reasonable retention times without resorting to subambient temperatures. They also permit the injection of much larger sample sizes without loss of resolution. This factor is important in identifying trace components using hyphenated techniques such as GC-FTIR and GC-MS, where the small sample sizes tolerated by conventional columns may preclude positive identification. The disadvantage of thick film columns is the large increase in retention time or elution temperature observed for substances of moderate to high molecular weight. The influence of film thickness on retention can be seen in Figure 2.12 [374]. Thick-film columns are thus best reserved for the separation of volatile compounds or as short columns for substance identification by spectroscopic techniques, where sample capacity is as important as resolution.

2.10.7 Porous-Layer Open Tubular (PLOT) and Support-Coated Open Tubular (SCOT) Columns

Porous-layer open tubular (PLOT) and support-coated open tubular (SCOT) columns are prepared by extending the inner surface area of the capillary tube.

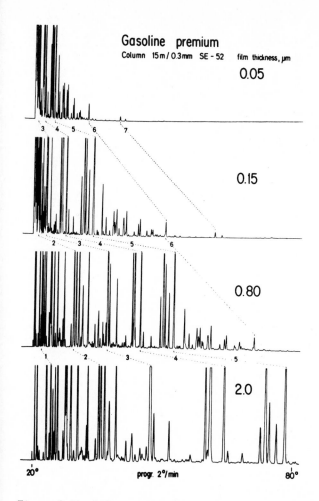

Figure 2.12   Influence of film thickness on retention for columns of identical
geometry and operatig conditions. 1. n-pentane; 2. benzene;
3. toluene; 4. o-xylene; 5. 1,2,4-trimethylbenzene;
6. naphthalene; and 7. 2-methylnaphthalene. (Reproduced with
permission from ref. 374. Copyright Dr. Alfred Huethig
Publishers).

A layer of particles can be deposited on the surface or the column wall can be
chemically treated to create a porous adsorbent layer. Obviously some of the
wall-modified open-tubular columns discussed in the previous section, for
example those prepared by barium carbonate deposition, sodium chloride dendrite
formation, and whisker formation, could also be considered examples of SCOT and
PLOT columns. In fact, distinction between certain kinds of WCOT, PLOT, and
SCOT columns is quite blurred and we will attempt no such decisive division
here. SCOT columns contain a liquid phase supported on a surface that is
covered with some type of solid support material. Thus every SCOT column could

certainly be called a PLOT column, but not vice versa, as certain PLOT columns, for example those used in adsorption chromatography, do not necessarily contain any liquid phase. However, adsorption columns are little used in open tubular column chromatography while, until quite recently, SCOT columns enjoyed modest interest. Thus the use of the term SCOT column has been more frequent although it could be argued that the term is redundant.

The advantages of PLOT and SCOT columns are that they provide higher sample capacity than WCOT columns with efficiencies and resolution intermediate between those of packed and WCOT columns [375-377]. Having a lower phase ratio than WCOT columns, they are particularly useful for the analysis of volatile organic compounds at temperatures above ambient. These are advantages and areas of application which PLOT and SCOT columns must share with micropacked columns (section 2.9) and thick-filmed immobilized WCOT columns (section 2.10.1). As micropacked columns require special instrumentation designed to operate at high pressures, PLOT and SCOT columns have generally been preferred over them. Thick-film WCOT columns were only recently introduced and have not been fully evaluated. However, they have many potential advantages, not least of which are inertness (they do not contain an active support) and robustness (involatile and active sample components can be rinsed out and column bleed is very low). These columns could eventually replace SCOT columns entirely.

PLOT and SCOT columns are usually prepared by an adaptation of one of three general methods [376]. The porous layer can be formed by chemical treatment of the inside tube wall, the inside tube wall can be coated with a layer of porous particles, or the porous particles can be partially embedded in the tube's inside wall as the capillary tube is drawn from the tube of larger diameter. The latter method generally yields columns of the lowest efficiency. A large diameter tube is loosely packed with sorbent [375,376] or coated with a thin stable film of sorbent [378,379] and then drawn out to a capillary tube in the conventional manner. During the drawing and coiling process the solid particles are partially embedded in the inside wall of the capillary tube. Various adsorbents including diatomaceous earths, graphitized carbon black, and powdered quartz have been used as support materials; the basic requirements of the support material are that it should be available in a narrow size range and have a melting point higher than the softening point of the glass. SCOT columns can then be prepared by coating the wall-modified capillary columns in the usual way. A wire or rod, suspended centrally in the packed column blank, enables drawn columns of very high permeability to be produced. The water soluble glue, poly(diethylaminoethyl methylacrylate) can be used to stabilize a thin film of adsorbent on the surface of the column blank. During the column drawing process the glue is decomposed to carbon dioxide, water, and ammonia and is thus eliminated from the final column.

Open tubular columns for gas-solid chromatography are generally prepared by chemical etching of the inside column wall (glass for silica and aluminum for alumina) or by dynamic or static coating of the capillary column with a suspension of micrometer or sub-micrometer-sized adsorbent particles [376,380-388]. Glass that is etched by heating with aqueous ammonia, hydrochloric acid, or sodium hydroxide for various lengths of time produces a porous silica layer of average thickness 5-100 micrometers, depending on the reaction conditions. Dehydration of this layer by conditioning at a high temperature produces a surface suitable for gas-solid chromatography or as a support for metal salts or organic modifiers [376,380]. Whisker-walled columns have also been used to prepare salt-modified adsorption columns [381]. Good quality alumina or silica columns can be prepared by dynamic coating of the capillary column with an aqueous or organic suspension of colloidal adsorbent [382,282,288] or with a stabilized suspension of micrometer-sized particles [384-388]. Stable suspensions are prepared using a balanced density slurry, by the addition of a surfactant to the coating solution, or by adding colloidal particles of the same material to the suspension. Surface-modified porous silicas containing a bonded film of liquid phase can be readily dispersed in most common organic solvents and present no particular problems in column preparation [384,385]. The activity and selectivity of the adsorbent columns can be controlled by thermal conditioning or by the addition of inorganic salts or polymeric liquids as modifiers in a manner analogous to that described for packed columns (section 2.8).

Adsorbent columns generally show high retention of organic compounds, particularly those containing polar functional groups. They are often used for the separation of volatile hydrocarbons where their high selectivity for the separation of isomeric and isotopic compounds is very useful and is often difficult to obtain using conventional partition columns [376,385,386]. Other applications include the separation of inorganic gases.

The popularity of SCOT columns in the early 1970's had much to do with the simple and reliable two-step column preparation procedure described by Horning [389,390]. Stabilization of the liquid phase film was achieved by adding Silanox 101 (trimethylsilylated silica with a primary particle size of 7 nanometers) to the coating solution. As these particles do not adhere to the glass surface or self-aggregate, they must be coated with a binder. The stationary phase itself and, in some instances, a surfactant such as benzyltriphenylphosphonium chloride are used as binders. The optimum concentration of stationary phase and Silanox 101 often produced solutions of too high a viscosity for effective movement through the column. Thus a two-step coating procedure was developed. In the first step, a chloroform suspension containing SE-30, Silanox 101, and surfactant was sonicated and passed through a silanized capillary column behind a short plug of solvent that was used to prewet the glass. The solvent was then evaporated by increasing the nitrogen flow. To increase the loading of stationary phase, the dried column was coated a second time with a more concentrated solution of SE-30 in isooctane. The solvent was then evaporated by nitrogen flow and the column conditioned in the usual way before use. Isooctane was selected as the second coating solvent because it did not disturb the initial layer due to its low viscosity. To prepare columns coated with polar phases, the surfactant was omitted in the first coating step and static coating with an acetone solution of the stationary phase was used for the second step [392]. Subsequently, several investigators have studied the various critical steps in the coating procedure. Sedimentation to remove agglomerates from the suspension, a change to chloroform-acetone (9:1) as the coating solvent, and the recommendation of specific coating and drying speeds to produce films of different thicknesses increased the reliability of column preparation [391-393]. The chloroform-acetone solvent combination had adequate density to stabilize the Silanox suspension and sufficient polarity to prevent formation of thixotropic gels during the coating process.

Although columns with polar phases can be prepared using Silanox 101 [390,393] most workers do not recommend its use. Silanox 101 is hydrophobic and, therefore, not generally wet by polar phases. Unsilanized fumed silica (e.g., Cab-O-Sil, Silica T40) [394,395], diatomaceous earth (Chromsorb R6470-1) [396,397], graphite [398], and kaolin [398,399] have been used to prepare SCOT columns with most common polar stationary phases. Other details of column preparation are similar to those discussed for Silanox 101. A one-step dynamic coating procedure using acetone [395] or chloroform-isooctane (1:1) [397] as the coating solvent was recently described. About 20-25% of the column length was filled with the coating solution, which consisted of the stationary phase (5% w/v) and Chromosorb R6470-1 (5% w/v) in the chloroform-isooctane (1:1) solvent. This liquid plug was forced through the column, which was suspended in a water bath at 40°C, at a velocity of 4 cm/s. As soon as the plug left the capillary column, the dummy column was disconnected and the flow of carrier gas was increased to about 15 ml/min to evaporate all the solvent.

The mechanism of film stabilization in columns prepared by the above methods has been discussed by Cramers [394]. He comments that as the solid material is dispersed in the liquid phase (rather than the liquid phase being dispersed in a porous layer as is the case for PLOT columns) it cannot adhere to the wall through surface tension forces, nor can the stability of the film-surface relationship be explained in terms of surface roughness. Cramers suggests that the film stability can be explained by the rheological behavior of structured dispersions.

2.11 Evaluation of the Quality of Open Tubular Columns

The quality of a column is a rather loosely defined concept but embodies such criteria as efficiency, inertness, and the thermal stability of the stationary phase film. All these properties depend upon the method and materials chosen for column preparation as well as the care taken to exactly reproduce the critical steps in that procedure. Whether using homemade or commercial columns, the assessment of column quality is of importance to all chromatographers. In many areas of column testing there is little standardization and therefore it is not possible to treat the subject didactically. Instead we will collect together some of the more frequently used methods and comment on their general applicability.

2.11.1 Measurement of Column Dead Time

The measurement of the column dead time or gas holdup time is an important parameter in gas chromatography. It influences the accuracy with which the retention due to pure chromatographic processes (the adjusted retention time) can be determined and, consequently, the accuracy with which all chromatographic parameters that depend on the absolute measure of retention can be calculated. Experimentally, the dead time of a chromatographic system is taken to be the retention time of an unretained substance such as methane, hydrogen, air, an inert gas, etc. [400-403]. Methane is frequently used with the popular FID detector, although it is apparent from many studies that methane is retarded, albeit slightly, by many common stationary phases, leading to a slightly high value for the dead time. Regression analysis of the retention data obtained from the separation of a homologous series of alkanes or other standards has been used as a more precise method [404-406]. Even here slight discrepancies exist between values obtained by different methods of calculation. The procedure used to measure retention, for instance an on-line data system or a stopwatch and graphical treatment, can influence the accuracy of the calculated dead time value through the choice of mathematical technique. For further details and a comparison of available methods, references [407,408] are recommended. For the purpose of evaluating column quality, the injection of methane provides sufficient accuracy and is easy to perform.

2.11.2 Activity Tests for Uncoated Columns

Activity tests for uncoated columns are important for studying the effectiveness of surface preparation and, in particular, column deactivation procedures [265,289,409,410]. Even when using a well-tried method something may go wrong or the deactivation may be incomplete, and an intermediate test prior

to coating is very useful for recognizing problems of this nature. The test
method is based on the principle that if a symmetrical solute band passes
through an inert and uncoated capillary column it should emerge from that column
unchanged except for symmetrical band broadening, and its retention time in that
column should be equivalent to the column dead time. To perform the test a
short capillary precolumn of high quality, coated with a moderately polar phase
such as Carbowax 20M, is connected to the injector at one end and a short length
of the test capillary column at the other by a piece of shrinkable Teflon
tubing. The other end of the column to be tested is connected to the detector.
The test substances, after separation by the precolumn, enter the test capillary
as perfectly shaped elution bands. The procedure generates nearly Gaussian
peaks, from which symmetry distortions may be mathematically evaluated using
symmetry factors, peak area/peak height ratios, etc. A practical example is
shown in Figure 2.13, where the activity of four acid-leached soda-lime
capillary columns, deactivated with hexamethyldisilazane under different
conditions, is compared [307]. Almost any of the popular polarity test mixtures
can be used to estimate the nature and extent of the activity of the test
capillary column.

2.11.3 Efficiency as a Column Quality Test

The efficiency of an open tubular column can be measured in several ways;
the most widely used methods are the number of theoretical plates (n), the
number of effective theoretical plates (N), the height equivalent to a
theoretical plate (HETP) or effective plate (HEETP), the coating efficiency
(CE), and the separation number (SN). No single method is ideal,
standardization is lacking, and the terms are often used loosely in the
literature without an adequate definition of the experimental conditions. All
the above methods measure the total efficiency of both the column and
instrument; if the values obtained are to be meaningful, it must be assumed that
the instrumental contribution to band broadening is negligible.

The number of theoretical or effective plates is easily measured from a
test chromatogram. For comparative purposes they are usually normalized per
meter of column length. However, the values measured will vary with retention,
column dimensions, carrier gas velocity, and stationary phase film thickness
[228,411,412]. If the column is very long, as open tubular capillary columns
often are, then the dead time may become a sizeable fraction of the total
retention time and the resultant number of theoretical plates speciously high.
For this reason some authors consider the number of effective theoretical plates
to be a better measure of column efficiency than the total number of theoretical
plates, as the former is less influenced by the column dead time. The two

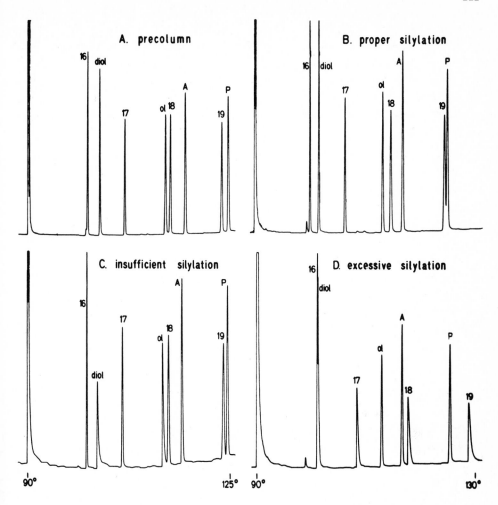

Figure 2.13 Activity test of uncoated silylated capillary columns according to the method of Schomburg. Test mixture: 16,17,18, and 19 are n-alkanes; diol, 2,3-butanediol; A, 2,6-dimethylaniline; P, 2,6-dimethylphenol; and ol, 1-decanol. This test was performed under temperature programmed conditions. (Reproduced with permission from ref. 307. Copyright Dr. Alfred Huethig Publishers).

measures of column efficiency are related to one another by equation (2.12).

$$N = n \left[\frac{k}{1 + k} \right]^2 \qquad (2.12)$$

N = number of effective plates
n = number of theoretical plates
k = capacity factor of the test solute

For a weakly retained solute, for example one with k = 1, N will be only 25% of
the value of n; however, for well retained solutes, k > 10, N and n will be
approximately equivalent. If a well-retained solute is used to measure column
efficiency then either n or N would be an adequate measure. This can be seen
from Figure 2.14 where both terms converge to an approximately constant value at
a sufficiently large value of the capacity factor [411]. In fact, as n levels
off faster than N with increasing values of k, this might be considered as the
more reliable measure of column efficiency.

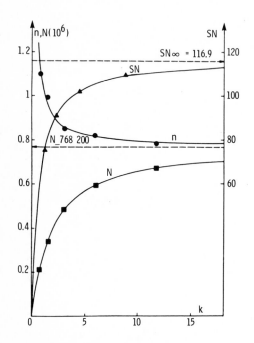

Figure 2.14   Plot of theoretical plate number (n), effective plate number (N),
              and separation number (SN) versus the capacity factor for an open
              tubular capillary column operated under isothermal conditions.
              (Reproduced with permission from ref. 411. Copyright Friedr.
              Vieweg & Sohn).

     The plate height, and thus the total number of theoretical or effective
plates, depends on the average linear carrier gas velocity (van Deemter
relationship) and, for a particular carrier gas, the efficiency will maximize at
a particular flow rate. Only at the optimum carrier gas flow rate are n, N, and
HETP independent of the column length. For thin-film columns, the efficiency
will also depend on the column diameter. From the Golay equation the maximum
number of theoretical plates for a WCOT column is given by equation (2.13).
Note that this derivation assumes that the liquid phase mass transfer term is

negligible and that the column pressure drop can be ignored, assumptions which are unlikely to hold for all situations.

$$(HETP)_{min} = r\left[\frac{1 + 6k + 11k^2}{3 (1 + k)^2}\right]^{1/2}$$ (2.13)

r = column radius

Typical values for columns with different diameters and solutes with capacity factors of 4 and 10 are given in Table 2.20. Values for n, N, and HETP are reasonably independent of temperature but may vary with the substance used for their determination, particularly if the test substance and stationary phase are not totally compatible.

Brown suggested that the mean specific plate number, M, is a more reliable parameter for measuring column efficiency as it is less influenced by partition ratios [413]. The mean specific plate parameter is obtained by normalizing the arithmetic mean of n and N for the column length L and the diameter of the free gas passage, equation (2.14)

$$M = N \frac{2(r - d_f)}{L} = n (r - d_f)\left[1 + \left(\frac{k}{1 + k}\right)^2\right]/L$$ (2.14)

M = mean specific plate number
L = column length
d_f = film thickness

This value has been little used in practice. Desty suggested that the rate of

TABLE 2.20

THE INFLUENCE OF COLUMN DIAMETER ON EFFICIENCY FOR SUBSTANCES OF DIFFERENT RETENTION

Column Diameter (mm)	Capacity Factor				
	k = 4			k = 10	
	$(HETP)_{min}$	n_{max}/m	N_{max}/m	$(HETP)_{min}$	n_{max}/m
0.1	0.08 (mm)	12,500	7700	0.09 (mm)	11,100
0.2	0.16	6250	3850	0.18	5550
0.25	0.20	5000	3125	0.22	4470
0.35	0.29	3450	- - -	0.31	3190
0.5	0.41	2440	1560	0.45	2240
0.75	0.61	1640	1040	0.67	1490
1.0	0.82	1220	780	0.89	1120

production of effective plates per retention time (N/t_R) would give a better measure of column performance than the number of effective plates itself [412]. However, this term is also dependent on the partition ratio and, except at high linear carrier gas velocities, the column flow rate as well. It finds occasional use in the literature.

The coating efficiency (CE), also known as the utilization of theoretical efficiency (UTE), is defined as the ratio of the theoretical to experimental plate height, expressed as a percentage, under optimum conditions, equation (2.15).

$$CE = \frac{(HETP)_{Theor}}{(HETP)_{Exp}} \times 100\% \qquad (2.15)$$

A value for $(HETP)_{Theor}$ can be calculated from equation (2.13) which, as it is derived for WCOT columns, is not strictly applicable to PLOT and whisker-wall columns [414]. The coating efficiency corrects for the effects of retention and the diffusion coefficient of the test compound on the measured efficiency and should be constant for a given column. However, the accuracy of this measure can be no greater than the error introduced in calculating $(HETP)_{Theor}$ [415], and this is reflected in the fact that CE values greater than 100% can be found in the literature. A column with a continuous, even, thin film should have CE values close to 100%. Normal values fall into the range 80-100% for columns of superior quality.

The separation number (SN), also know as the Trennzahl (TZ), is defined as the number of component peaks which can be placed between the peaks of two consecutive homologous standards with z and z+1 carbon atoms and separated with a resolution of $R_s = 1.177$, equation (2.16).

$$SN = \left[\frac{t_{R(z+1)} - t_{R(z)}}{w_{h(z)} + w_{h(z+1)}} \right] - 1 \qquad (2.16)$$

t_R = retention time

w_h = peak width at half height

Normal alkanes or fatty acid methyl esters are most frequently used as the standard homologous compounds. The column separation number is dependent on the nature of the stationary phase, the column length, column temperature, and carrier gas flow rate [411,412,416-418]. The relationship between the separation number and efficiency can be expressed by equation (2.17) (temperture also enters into this relationship and is emboddied in the value of the separation factor).

$$SN = 0.425 \left[\frac{a-1}{a+1}\right]\sqrt{N} - 1 \qquad\qquad (2.17)$$

a = separation factor
N = number of effective plates

The separation factor tends to increase as the column temperature decreases, causing an increase in the separation number. At constant temperature, equation (2.17) also predicts that the separation number will approach a maximum and remain nearly constant as the capacity factor increases. This is because the separation factor between neighbors of a homologous series and the number of effective plates becomes nearly constant. Referring to Figure 2.14, at a sufficiently high capacity factor value either n, N, or SN provides a reasonable value for comparing the efficiencies of columns of constant length. The disadvantage of the separation number is that it varies with temperature, whereas n and N are reasonably temperature independent.

The separation number is the only column efficiency parameter that can be determined under temperature programmed conditions [418,419]. The critical parameters that must be standardized to obtain reproducible SN values for columns of different length are the carrier gas flow rate and the temperature program (details are given in the next section). A maximum SN value is reached when the first compound of a homologous pair elutes with a capacity factor value of at least ten under optimized carrier gas flow and temperature program rates.

2.11.4 Standardized Quality Test for Coated Open Tubular Columns

The comprehensive column test procedure devised by Grob is now universally used by both column producers and column users [420,421]. It supplants the use of the various polarity test mixtures used previously [251,265,277,307, 422-425]. The principal advantage of the Grob test is that it provides quantitative information about four important aspects of column quality: separation efficiency, adsorptive activity, acidity/basicity, and the stationary phase film thickness. The standard test conditions are optimized for the measurement of efficiency and, as it is not feasible to standardize all parameters simultaneously in a multi-purpose test, these conditions are not necessarily optimum for measuring column activity. The experimental conditions are optimized for columns with a medium range of film thicknesses (0.08-0.4 micrometers) and column internal diameters of 0.25-0.35 mm. Although the separation number values obtained for wide bore or thick film columns may be slightly below their maximum values, for convenience the same experimental conditions are generally used for all column types.

The composition of the Grob test mixture is given in Table 2.21. To prepare a working solution, 1.0 ml of each standard solution is added to a 10.0

ml vial and 1.0 ml of this secondary standard is diluted to 20.0 ml with hexane. The resultant mixture can be stored in a refrigerator and is normally stable for many months. The concentrated standards should be stored in a freezer and are stable indefinitely.

A stepwise guide to the test method is given in Table 2.22. The test is performed under optimized conditions of carrier gas flow rate and rate of temperature programming, which are adjusted for column length and carrier gas viscosity, Table 2.23. To obtain a correct dead time value for thick-film columns (d_f ca. > 0.7 micrometers) it should be measured at 100°C (methane is considerably retained at 25°C). As the viscosity of the carrier gas is greater at this temperature, the dead time should be corrected by adding 10% for hydrogen and 15% for helium. Some instruments do not allow the temperature program rate to be changed continuously to meet all conditions given in Table 2.23. In this case it is necessary to correct the speed of the run to an available temperature program rate by selecting a dead time corresponding to this program rate.

TABLE 2.21

COMPONENTS OF THE CONCENTRATED GROB TEST MIXTURE

Test Substance	Abbreviation	Amount Dissolved in 20.0 ml solvent[a] (mg)
Methyl decanoate	E_{10}	242
Methyl undecanoate	E_{11}	236
Methyl dodecanoate	E_{12}	230
n-Decane	10	172
n-Undecane	11	174
n-Dodecane[b]	12	176
1-Octanol	ol	222
Nonanal	al	250
2,3-Butanediol	D	380
2,6-Dimethylaniline	A	205
2,6-Dimethylphenol	P	194
Dicyclohexylamine	am	204
2-Ethylhexanoic Acid	S	242

[a] Hexane, except for 2,3-butanediol, which is dissolved in chloroform;
[b] Used in place of n-undecane to reduce the possibility of peak coincidences on nonpolar stationary phases.

The column efficiency is determined by the separation number obtained for the methyl esters of decanoic, undecanoic, and dodecanoic acid (E_{10}, E_{11}, E_{12}). As the relative difference in the molecular sizes of the homologous pairs decreases with increasing molecular size, the first pair of methyl esters (E_{10}/E_{11}) provides a separation number value that is about 8% higher than that of the second pair (E_{11}/E_{12}). The average of the two values is normally used as a measure of the column separation efficiency.

TABLE 2.22

STEPWISE PROCEDURE FOR PERFORMING THE GROB TEST

- Cool the column oven to below 40°C.
- Adjust the flow rate by measuring the dead time of methane. For most instruments it is advisable to set the split flow prior to the flow measurement because changes in the split flow usually change the pressure at the column inlet. Adjust the time to the standard time (± 5%) Table 2.23.
- Adjust the temperature program rate to the appropriate value given in Table 2.23.
- Inject the test mixture under conditions that allow ca. 2 ng of a single test substance to enter the column (e.g., 1 microliter with a split ratio of 1:20 to 1:50, depending on injector design).
- Immediately after injection, heat the oven to 40°C (for very thin films, to 30°C) and start the temperature program.
- Within the temperature range in which the third ester is eluted (on most columns, 110-140°C), make two marks on the recorder chart noting the actual oven temperature.
- At the end of the run, inter- or extrapolate the elution temperature of the third ester.
- Draw the "100% line" over the two alkanes and the three esters.
- Express the height of the remaining peaks as a percentage of the distance between the baseline and the 100% line.
- Determine SN as an average of SN E_{10}/E_{11} and SN E_{11}/E_{12}.
- Determine the film thickness using the nonogram (see text).

TABLE 2.23

STANDARD EXPERIMENTAL CONDITIONS FOR PERFORMING THE GROB TEST

Column	Hydrogen		Helium	
Length (m)	CH_4 elution (sec)	Temperature Program (°C/min)	CH_4 elution (sec)	Temperature Program (°C/min)
10	20	5.0	35	2.5
15	30	3.3	53	1.65
20	40	2.5	70	1.25
30	60	1.67	105	0.84
40	80	1.25	140	0.63
50	100	1.0	175	0.5

An easy way to quantify the adsorptive and acid/base characteristics of the column is to measure the peak height as a percentage of that expected for complete and undisturbed elution. The two alkanes and three fatty acid methyl esters (non-adsorbing peaks) are connected at their apices to provide the 100% line as shown in Figure 2.15 [287]. The column activity is then quantified by expressing the height of the remaining peaks as a percentage of the distance between the baseline and the 100% line. It accounts for all types of peak distortion that occur in practice: peak broadening, peak tailing, irreversible adsorption, and degradation. The alcohols octanol and 2,3-butanediol are used to measure adsorption by a hydrogen bonding mechanism. Acid/base interactions are assessed from the adsorptive behavior of 2,6-dimethylaniline, 2,6-dimethylphenol, dicyclohexylamine, and 2-ethylhexanoic acid. Probes with sterically-hindered functional groups are used for this purpose in order to avoid adsorption by hydrogen bonding, which can complicate the interpretation of the interaction. 2-Ethylhexanoic acid and dicyclohexylamine provide much more stringent tests of acid/base behavior than the phenol and aniline. On nonpolar phases, even less than 1.0 ng of 2-ethylhexanoic acid may cause column overloading, resulting in a distorted (leading) peak. In this case, peak area as opposed to peak height must be used to quantify column interactions. Nonanal is used to investigate the capacity of the column to adsorb saturated aldehydes. This is the most specific of the polar test probes and shows

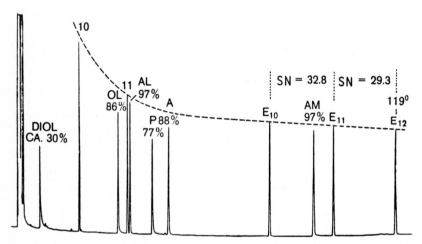

Figure 2.15   Test chromatogram of an open tubular column according to the
method of Grob. A line is drawn over the peaks of the
non-adsorbed solutes. The peak height of the remaining peaks is
determined as a percentage of the ideal peak height. In the
absence of adsorption all peaks should reach the dotted line.

the least variation on most columns of at least moderate quality. On very inert, nonpolar columns the alkane standards may be partially adsorbed. In this case the aldehyde and/or alcohol peaks will be larger than those of the alkanes; quantitation of adsorption is, of course, not possible in such cases.

The film thickness is calculated from the elution temperature of methyl dodecanoate (E_{12}). However, quantitation of film thickness requires a calibration of elution temperature against stationary phase thickness for all stationary phases of interest. A nonogram can then be constructed and the stationary phase film thickness can be obtained from the elution temperature of E_{12} [420]. The elution temperature of E_{12} is reproducible to within ± 1°C if the standard conditions are kept within reasonable limits; in terms of film thickness this corresponds to a variation of 5% or less.

Three problems with the Grob test are worthy of mention. It cannot be used to test columns coated with liquid phases of high melting point. The elution order of the test mixture is not the same on all stationary phases and the occurrence of peak co-elution cannot be entirely eliminated. The elution order of the test mixture on many common phases has been given by Grob [420,421]. For phases not previously tested, or in cases where the elution order is in doubt, individual standards must be injected for peak identification. In the case of peak co-elution, the test mixture should be divided into groups of separated components and each group injected independently. Finally, the test is biased towards the measurement of adsorptive as opposed to catalytic activity [426,427]. Catalytic activity causes time-dependent, concentration-independent losses of sample components. With increasing column temperature, adsorption decreases whilst catalytic/thermal decomposition increases. Although at high column temperatures adsorption may be of little importance compared to catalytic activity, the latter may not be observed in the Grob test. As catalytic activity is a fairly specific influence on particular solutes it may be necessary to customize an activity test for it based on the intended use of the column.

2.11.5 Column Thermal Stability

In quite general terms the amount of volatile decomposition products formed in the column per unit time depends on the stationary phase, temperature, surface properties of the support, surface area covered by the stationary phase, and film thickness [226,290,299]. The mass flow of volatile products at the column outlet depends on the column length, the carrier gas flow rate, and the rate of formation of the volatile products.

Standardization of the measurement of column bleed from different columns presents several problems. The influence of detector sensitivity on the results

120

obtained can be eliminated by injecting a standard compound as part of the test,
Figure 2.16. The amount of stationary phase bleed is represented by a rectangle
whose area is defined by the change in recorder response between two set
temperatures, one of which is selected to represent conditions of low column
bleed, and the section of the abscissa that corresponds to unit time (minutes).
By comparison of this area with that of a test compound, usually an n-alkane,
the loss of liquid phase can be expressed as micrograms of n-alkane per minute.
It is assumed that the response factors of the column bleed products and the
test compound are identical. Although this is required as a working hypothesis,
it need not necessarily be the case.

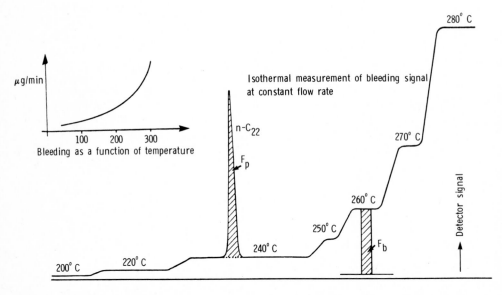

Figure 2.16 Standardized bleed test for gas chromatographic columns (see text
for details). (Reproduced with permission from ref. 290.
Copyright Elsevier Scientific Publishing Co.)

The above method can be used to study the influence of the column
preparation procedure on the stability of the liquid phase film. Surface
effects are particularly noticeable in capillary columns since the ratio of the
surface area to the amount of stationary phase is high. An example is shown in
Figure 2.17 for a poly(methyl)siloxane phase [299]. In general, glass capillary
columns made from alkali glass exhibit much greater column bleed than
borosilicate glasses unless the concentration of alkali ions is reduced by acid
leaching. Fused silica columns show very low levels of thermally-induced
catalytic phase decomposition. The method is also suitable for observing the
influence of different methods of deactivation on stationary phase

decomposition.

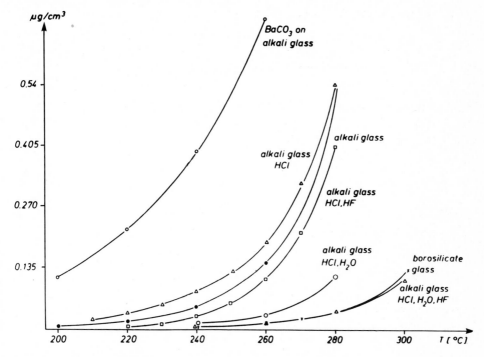

Figure 2.17 Column bleed of OV-101 coated onto different glass surfaces.
(Reproduced with permission from ref. 299. Copyright Friedr.
Vieweg & Sohn).

2.12 The Mobile Phase in Gas Chromatography

The mobile phase is gas chromatography is generally considered to be inert
in the sense that it does not react chemically with the sample or stationary
phase, nor does it influence the sorption-desorption or partitioning processes
that occur within the column. Thus the choice of carrier gas does not influence
selectivity. It can influence resolution through its effect on column
efficiency which arises from differences in solute diffusion rates. It can also
affect analysis time because the optimum carrier gas velocity decreases as the
solute diffusivity decreases, and also plays a role in pressure-limiting
situations due to differences in gas viscosities. Other considerations for
choosing a particular carrier gas might be cost, purity, safety, and detector
compatibility. Taking all these considerations into account, hydrogen, helium,
nitrogen, and argon are the principal carrier gases used in gas chromatography.

For packed column gas chromatography, little account is taken of the choice
of carrier gas when using columns of normal length and supports of average
particle size. In situations where the pressure drop is the limiting feature,

hydrogen would be preferred because its viscosity is only about half that of
helium and nitrogen. In terms of efficiency nitrogen is slightly superior at
low temperatures and flow rates while hydrogen is perferred at higher
temperatures and at above optimum flow rates [428,429].

For open tubular columns a more clear-cut preference for hydrogen can be
made [228,374,430,431]. Comparing the van Deemter curves in Figure 2.18,
nitrogen provides the lowest plate height but it occurs at an optimum gas
velocity which is rather low and hence leads to long analysis times. In the
optimum plate height region the ascending portions of the curve are much
shallower for hydrogen and helium. Thus for operation at carrier gas flow rates
above the optimum in the van Deemter curves, the situation which applies in
practice, hydrogen, and to a lesser extent helium, show only a modest sacrifice
in column efficiency compared to nitrogen. Expressed as the rate of production
of effective theoretical plates, Figure 2.18, it is clear that hydrogen provides
the highest column efficiency per unit time. In addition, because equivalent
column efficiencies are obtained at higher average linear gas velocities,

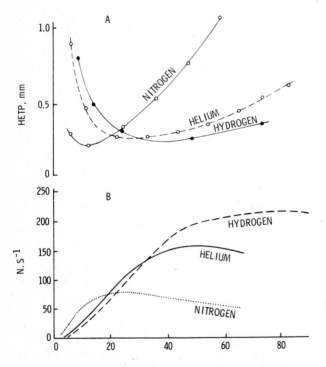

Figure 2.18 A, van Deemter curves for three different carrier gases. The test
solute has a partition ratio of 7.95. In B, the same data is
replotted as the rate of production of effective plates as a
function of the average linear carrier gas velocity.

compounds exhibit lower elution temperatures during temperature programmed runs and, as the peaks are taller and narrower, sample detectability is improved. Figure 2.19 provides an example of the separation of a complex mixture optimized to produce similar elution temperatues with nitrogen and hydrogen as the carrier gases. When nitrogen is used the resolution is slightly poorer and the sensitivity is reduced, as indicated by the larger sample size required to provide approximately the same signal. However, the most dramatic difference is in the analysis time, which is 2.5 times longer in the case of nitrogen. The only disadvantage to the use of hydrogen as a carrier gas is the real or perceived explosion hazzard from leaks within the column oven. Experience has shown that the conditions required for a catastrophic explosion may never be achieved in practice. However, commercially available gas sensors will automatically switch off the column oven and carrier gas flow at air-hydrogen

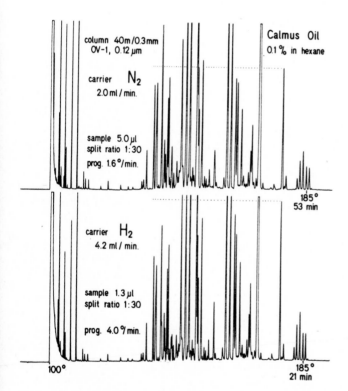

Figure 2.19   Comparison of nitrogen and hydrogen as carrier gases for open tubular columns. For nitrogen the conditions (temperature and flow rate) were selected to yield the optimum resolution. For hydrogen the conditions were set to yield elution temperatures similar to those of nitrogen, and for high, but not maximum, resolution. (Reproduced with permission from ref 374. Copyright Dr. Alfred Huethig Publishers).

mixtures well below the explosion threshold limit [432].

Several studies have been made on the influence of vapors or gases added to the carrier gas to modify the column separation characteristics [433]. Steam, ammonia, formic acid, etc. have been used to adjust the retention, improve the peak shape, and diminish the adsorption of polar solutes in gas–liquid and gas–solid chromatography [434]. Organic vapors with appreciable stationary phase solubility have been used to create dynamic binary liquid phase mixtures whose selectivity can be adjusted by changing the partial pressure of the organic vapor in the carrier gas [435,436]. In general, problems with detector incompatibility, the limited available temperature range for the desired effect to apply, and other inconveniences related to safety and equipment requirements have limited studies in this area.

2.13 Supercritical Fluid Chromatography

Technical difficulties in establishing microcolumn liquid chromatography as a routine analytical technique and recent developments in gas chromatography, have produced a rebirth of interest in supercritical fluid chromatography (SFC). SFC should not be viewed as a replacement for either gas or liquid chromatography but rather as a technique which is complementary to both. SFC can be used to separate thermally labile and high molecular weight samples unsuited to gas chromatographic analysis, and, when compared to HPLC, it offers the possibility of higher separation efficiencies, shorter analysis times, and a wider range of detection possibilities. There are of course limitations as well as advantages. SFC cannot provide comparable separating power for those compounds which are stable to gas chromatographic conditions and it lacks the flexibility of the many different retention mechanisms available in HPLC. Additionally, as SFC separations are performed at temperatures above the critical temperature of the mobile phase, the compounds analyzed must have sufficient thermal stability to meet this requirement. A further limitation to the current exploitation of SFC is a lack of commercially available instrumentation.

A supercritical fluid is a gas which has been heated to a temperature above its critical temperture while simultaneously compressed to a pressure exceeding its critical pressure. Under these conditions the gas is converted to a single phase, dense fluid whose solvent properties differ from both gas and liquid phases. Gases and supercritical fluids have similar viscosities and are about 100–fold lower than those of liquids. Diffusion coefficients are higher in supercritical fluids than in liquids but generally lower than in gases. The solubilizing power of supercritical fluids is appreciably higher than that of gases at similar temperatures and, in fact, approaches that of normal liquids.

A number of recent reviews have dealt with the physical properties and chromatographic uses of supercritical fluids [437-443].

Most of the early research in SFC was based on packed column methods. However, the resurgence of interest in the technique can be traced to the development of open tubular columns with immobilized phases for gas chromatography. These columns eliminated the problem of phase stripping by the supercritical fluid which limited the usefulness of conventional wall-coated open tubular columns [443-448]. Open tubular columns have certain other features which make them desirable for SFC. The pressure drop across a capillary column is lower than that for a packed column, allowing the use of longer columns that can attain higher efficiencies [441,446]. Additionally, the contribution to the column plate height from alternate fluid flow paths found in packed columns is absent in open tubular columns. The low pressure drop in capillary SFC is moreover beneficial because the density of the mobile phase is more uniform throughout the length of the column. As the mobile phase density has an important influence on retention in SFC, this parameter assumes greater importance than in gas or liquid chromatography. The principal problems in capillary SFC relate to instrumental constraints resulting from the high pressures and temperatures (in some cases) required to maintain the mobile phase at or above its critical point, and the low tolerance of extracolumn volume in sample introduction and detection systems. These difficulties have been sufficient to maintain an interest in packed column SFC, employing columns and packings similar to those used in HPLC [438,439,449-451]. In packed column SFC, instrument requirements can be simplified by operating in the near supercritical region with inorganic gases compressed to liquid densities but maintained below their critical temperatures [451].

Some suitable mobile phases for SFC are given in Table 2.24 [438,452]. With equipment not specifically designed for SFC, mobile phases with low critical constants, such as carbon dioxide, are preferred [447]. When equipment constraints are not limiting, then mobile phases with higher critical constants offer certain advantages, particularly for the separation of high molecular weight samples. This is because retention in SFC is controlled by a combination of extraction and evaporation mechanisms. To elute high molecular weight samples at low temperatures, a compensating increase in the mobile phase density (pressure) is required, which may not be convenient in practice. Of course, the overriding consideration with respect to temperature is the thermal stability of the sample itself. Some of the organic mobile phases listed in Table 2.24 may constitute a safety hazard if a leak develops at high temperatures and pressures [447]; still others are corrosive or toxic agents that present their own particular problems. Degradation of some solvents occurs under useful operating conditions [452]. For the above reasons the most widely used mobile phases have

been carbon dioxide and pentane. As neither of these fluids is very selective
there is still a strong interest in evaluating alternative phases to improve
resolution by manipulation of the mobile phase selectivity. The availability of
immobilized stationary phases with different selectivities may partially solve
this problem.

TABLE 2.24

PROPERTIES OF POSSIBLE MOBILE PHASES FOR SUPERCRITICAL FLUID CHROMATOGRAPHY

Compound	Atm	Critical Point Data		
	b.p. (°C)	T_c (°C)	P_c (atm)	d_c (g/ml)
Nitrous oxide	-89	36.5	71.4	0.457
Carbon Dioxide	-78.5[a]	31.3	72.9	0.448
Sulfur dioxide	-10	157.5	77.6	0.524
Sulfur hexafluoride	-63.8[a]	45.6	37.1	0.752
Ammonia	-33.4	132.3	111.3	0.24
Water	100	374.4	226.8	0.344
Methanol	64.7	240.5	78.9	0.272
Ethanol	78.4	243.4	63.0	0.276
Isopropanol	82.5	235.3	47.0	0.273
Ethane	-88	32.4	48.3	0.203
n-Propane	-44.5	96.8	42.0	0.220
n-Butane	- 0.5	152.0	37.5	0.228
n-Pentane	36.3	196.6	33.3	0.232
n-Hexane	69.0	234.2	29.6	0.234
n-Heptane	98.4	267.0	27.0	0.235
2,3-Dimethylbutane	58.0	226.8	31.0	0.241
Benzene	80.1	288.9	48.3	0.302
Diethyl ether	34.6	193.6	36.3	0.267
Methyl ethyl ether	7.6	164.7	43.4	0.272
Dichlorodifluoromethane	-29.8	111.7	39.4	0.558
Dichlorofluoromethane	8.9	178.5	51.0	0.522
Trichlorofluoromethane	23.7	196.6	41.7	0.544
Dichlorotetrafluoroethane	3.5	146.1	35.5	0.582

[a] Sublimation point

Two properties, the volatility of the solute and its solubility in the
mobile phase, are responsible for determining the retention of solutes in SFC.
For in-volatile compounds the mobile phase density plays the most important role
in determining capacity factor values. For open tubular columns the pressure
drop across the column is approximately linear and, therefore, the density of
the mobile phase in the column is given by equation (2.18).

$$\rho = D - wl \qquad\qquad (2.18)$$

ρ = density of the mobile phase
D = density of the mobile phase at the head of the column
w = rate at which the density changes
l = length along the column

The effect of mobile phase density on selectivity and retention can be calculated from equations (2.19) and (2.20).

$$\ln a = B_o - m\rho \qquad (2.19)$$

$$\ln k = a - b\rho \qquad (2.20)$$

a = selectivity coefficient
k = capacity factor
a = log k of a compound under gas chromatographic conditions at the same temperature
b = rate of change of log k with respect to the mobile phase density

B_o, m, a, and b are constants whose values depend on the solute, the nature of the stationary phase, and the temperature of the mobile phase. Equations (2.17)-(2.19) can be combined to give expressions for k and a as a function of the distance a solute has travelled along the column. The integration of each expression over the column length, divided by the column length, provides average values for k and a. This enables observed values for k and a to be calculated according to equations (2.21) and (2.22).

$$k_{OBS} = e^{(a-bD)} (e^{bwL} - 1)/bwL \qquad (2.21)$$

$$a_{OBS} = e^{(B-mD)} (e^{mwL} - 1/mwL \qquad (2.22)$$

L = column length

In both equations wL is simply the density change, $\Delta\rho$, across the column. The following general observations can be made from equations (2.21) and (2.22):

1. Observed selectivity and retention variations caused by linear changes in density depend only on the column total density drop.

2. If the density at the head of the column is held constant while $\Delta\rho$ is allowed to increase, then both selectivity and retention will increase.

3. Increasing the density drop affects both retention and selectivity; if k_{OBS} remains constant, selectivity will always increase, and vice versa.

Resolution depends on the column efficiency as well as on k and a. The pressure drop across an open tubular column has only a limited effect on column efficiency since it is small. For a fixed column length, the number of theoretical plates in capillary SFC is given by equation (2.23) [444,445].

$$n = \frac{24 \, Lu \, (1 + k)^2 \, D_m}{48 \, D_m^{\,2} \, (1 + k)^2 + (1 + 6k + 11k^2) \, r^2 u^2} \qquad (2.23)$$

The maximum number of theoretical plates that can be attained at a given mobile phase velocity is given by equation (2.24)

$$n_{max} = \frac{3 \, \Delta P_{max} \, r^2 D_m \, (1 + k)^2}{\eta [48 \, D_m^2 \, (1 + k)^2 + (1 + 6k + 11k^2) \, r^2 u^2]} \qquad (2.24)$$

and the longest column length that can be used is given by equation (2.25).

$$L_{max} = \frac{\Delta P_{max} \, r^2}{8 \, \eta \, u} \qquad (2.25)$$

η = mobile phase viscosity
u = mobile phase linear velocity at the head of the column
r = column radius
ΔP = column pressure drop
L = column length
D_m = mobile phase solute diffusion coefficient

To maximize resolution the column efficiency should be maximized while simultaneously minimizing the selectivity loss caused by the column pressure drop. This favors the use of fairly short capillary columns (ca. 20 m in length) of small internal diameter (less than 50 micrometers). For technical reasons, internal column diameters are currently limited to the range of 50-100 micrometers.

Two gradient techniques are applicable to SFC: pressure gradients, where the mobile phase density is altered to shorten the analysis time and increase resolution, and eluent gradients, whereby the eluent composition is altered. Pressure programming has been more widely used than eluent programming and will be described below [438,450]. Temperature programming is not generally useful in SFC; most separations are performed isothermally at a temperature a few degrees higher than the critical temperature.

By gradually increasing the pressure during a chromatographic run it is possible to elute compounds over a wide molecular weight range. Here, it is not the pressure programming rate which is important, but rather the rate at which the mobile phase density increases relative to its flow rate. It is generally recognized that slow programming rates give the best results, that increasing the column temperature may lead to better results since the pressure-density isotherms become more linear, and that the pressure drop across the column should be as small as possible. Since density is not directly proportional to pressure near the critical point, linear pressure programming will produce inferior results for the separation of mixtures having a wide molecular weight range [443,453,454]. The elution time of compounds in a homologous series is described by equation (2.26)

$$\ln k = A + B_o n - mn\rho \tag{2.26}$$

where A, B_o, and m are constants and n is the number of monomeric units in the eluted oligomers. For linear pressure programming, the elution time for each oligomer is given by equation (2.27).

$$t_n = t_\infty - c/n \tag{2.27}$$

t_n = elution time for an oligomer containing n monomer units

t_∞ = elution time as n $\longrightarrow \infty$

c = constant whose value depends on the density programming rate

Equation (2.27) predicts that with linear density programming, successive oligomers will be eluted close together and all oligomers will be eluted by the time t_∞ is reached, in agreement with experimental data, Figure 2.20A. It would be preferable for members of a homologous series to elute at regular time intervals. In this case n becomes equal to some constant multiplied by the sum of the time, t, and a reference time, t'. The reference time is required since the oligomers cannot elute before the dead time. Under these conditions the elution time of a particular oligomer can be described by equation (2.28).

$$\rho = \rho_a - \frac{c}{t + t'} \tag{2.28}$$

t = elution time

t' = reference time

ρ_a = limiting density (ρ at which $\ln a = 0$)

Both c and t' determine the spacing and retention of components as they elute. When selecting a suitable density (pressure) program to separate a homologous series, it is important to take the limiting density value into account, since further separation cannot be expected at higher densities. A value for the limiting density is easily established by a few preliminary experiments, as described elsewhere [453]. The applicability of equation (2.28) is demonstrated by the separation in Figure 2.20B.

The instrumental requirements of SFC resemble those of HPLC, with some characteristic differences. Compared to liquid chromatography, the entire column must be operated under high pressure rather than just the inlet section, pressure rather than flow control of the mobile phase is required, and detectors must be designed to operate either at high pressures or provide a means to handle eluted compounds after the mobile phase pressure is reduced to atmospheric values [438,439,443]. The modifications of a standard high pressure liquid chromatograph for packed column SFC [449-451] and a gas chromatograph for capillary SFC [446,447,453] have been described.

For mobile phases that are gases at ambient pressure the simplest delivery

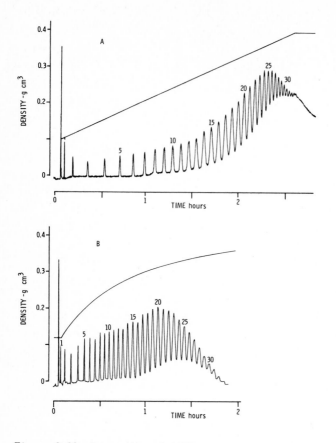

Figure 2.20   Separation of 2000 MW polystyrene oligomer mixture by capillary
SFC.   A, linear density program; B, asumptotic density program.
(Reproduced with permission from ref. 453.   Copyright Preston
Publications, Inc.)

system is a pressurized gas cylinder connected to a high pressure reducing
valve.   The cylinder can be heated electrically to provide higher pressures than
those normally available.   Alternatively, a high pressure pump can be used for
either gases or liquids at room temperature.   These are usually HPLC pumps,
modified for pressure rather than flow control.   Although any type of pump
capable of delivering a continuous pulse-free flow is adequate, syringe-type
pumps have been generally used.   These pumps can be readily modified for
pressure programming [455].

The fluid is converted to the supercritical state by passage through a
preheating or conditioning coil, often located in the column oven, that is
maintained at the column temperature.   The supercritical fluid then flows
through an injection device which should be heated in a separate oven or located
in the column oven.   Injection is made by a high pressure syringe through a

septum injector or, for capillary columns, by on-column injection (similar to gas chromatography) or by a valve injector connected to a flow splitter. The latter can be fabricated from a one-sixteenth inch Swagelok T-piece and a short length of capillary tubing that acts as the balancing restrictor [446]. Injection dead volume and column sample overload are severe problems in the operation of the small internal diameter open tubular columns currently in favor for SFC.

Detection remains a problem in open tubular SFC because of the need for low dead volume detectors with high sensitivity. The most commonly used detectors are those already in use for gas and liquid chromatography [438,439,441,443, 456,457]. The UV and fluorescence detectors can be used after modification for high pressure and high temperature operation. This usually involves replacing the cell used for HPLC with one designed for SFC, or carrying out detection on-line by using a section of the column itself as the cell. Alternatively, detection can be performed after decompression. However, if the effluent is gaseous at ambient conditions, high molecular weight solutes tend to form droplets during the decompression stage; detection then becomes erratic and unpredictable. Flame-based gas chromatographic detectors are limited to mobile phases which do not generate a detector response. In addition, mobile phases which corrode the detector parts must be avoided.

To maintain a sufficient pressure and to control the flow rate through the column, a pressure regulator is required after or prior to the detector. For open tubular SFC this is usually a length of narrow bore capillary tubing having an internal diameter less than that of the column, or a short length of capillary tubing connected to a backpressure regulator or a flow metering valve.

Coupling a supercritical fluid chromatograph to a mass spectrometer [458-461] and a Fourier transform infrared spectrometer [462,463] has been achieved with fewer problems than encountered in HPLC. Open tubular columns can be coupled to a mass spectrometer using a the direct fluid injection interface. This is essentially the same as direct injection interface for HPLC-MS (Chapter 8.4) except that a constant temperature is maintained throughout the length of the probe, a desolvation region is unnecessary, and the pressure restrictor at the end of the probe is chosen more to meet chromatographic requirements (flow through the column) and less to protect the mass spectrometer vacuum system. Typical liquid flow rates in capillary SFC are on the order of 50-80 microliters per minute, well within the capacity of the pumping system on most modern mass spectrometers. Most mass spectra have been recorded in the chemical ionization mode where the mobile phase also acts as the reagent gas. On-line SFC-FTIR provides higher sensitivity and greater spectral information than HPLC-FTIR. The common mobile phases used in SFC have wider spectral windows and greater IR transparencies than those used in HPLC. However, pressure programming produces an appreciable change in the background spectrum as neither the intensity nor the width of absorption bands for the mobile phase are independent of pressure. Spectral subtraction must therefore be performed from previously recorded reference spectra measured at different pressures.

2.14 References

1. A. T. James and A. J. P. Martin, Biochem. J., 50 (1952) 679.
2. I. Brown, J. Chromatogr., 10 (1963) 284.
3. S. H. Langer and R. J. Sheehan, in J. H. Purnell (Ed.), "Progress in Gas Chromatography", Wiley, New York, 1968, p. 289.
4. R. A. Keller, J. Chromatogr. Sci., 11 (1973) 49.
5. B. L. Karger, L. R. Snyder, and C. Eon, J. Chromatogr., 125 (1976) 71.
6. F. Vernon, in C. E. H. Knapman (Ed.), "Developments in Chromatography", Applied Science Publishers, London, Vol. 1, 1978, p. 1.
7. R. V. Golovnya and T. A. Misharina, J. High Resolut. Chromatogr. Chromatogr. Commun., 3 (1980) 4.
8. R. V. Golovnya and T. A. Misharina, J. High Resolut. Chromatogr. Chromatogr. Commun., 3 (1980) 51.
9. G. E. Bailescu and V. A. Ilie, "Stationary Phases in Gas Chromatography", Pergamon Press, Oxford, 1975.
10. J. R. Mann and S. T. Preston, J. Chromatogr. Sci., 11 (1973) 216.
11. R. A. Keller, J. Chromatogr. Sci., 11 (1973) 188.
12. L. V. Semenchenko and M. S. Vigdergauz, J. Chromatogr., 245 (1982) 177.
13. G. Castello, J. Chromatogr., 66 (1972) 213.
14. F. Vernon and C. O. E. Ogundipe, J. Chromatogr., 132 (1977) 181.
15. J. K. Haken and F. Vernon, J. Chromatogr., 186 (1979) 89.
16. F. I. Onusca, J. Chromatogr., 186 (1979) 259.
17. F. Reido, D. Fritz, G. Tarjan, and E. Sz. Kovats, J. Chromatogr., 126 (1976) 63.
18. L. Boksanyi and E. Sz. Kovats, J. Chromatogr., 126 (1976) 87.
19. J. K. Haken and D. K. M. Ho, J. Chromatogr., 142 (1977) 203.
20. M. B. Evans, J. Chromatogr., 160 (1978) 277.
21. M. B. Evans and M. J. Osborn, Chromatographia, 13 (1980) 177.
22. M. B. Evans, M. I. Kawar, and R. Newton, Chromatographia, 14 (1981) 398.
23. R. S. Juvet and R. L. Fisher, Anal. Chem., 38 (1966) 1860.
24. W. S. Pappas and J. G. Million, Anal. Chem., 40 (1968) 2176.
25. U. Muller, P. Deitrich, and D. Prescher, J. Chromatogr., 259 (1983) 243.
26. S. C. Dhanesar and C. F. Poole, J. Chromatogr., 267 (1983) 388.
27. S. C. Dhanesar and C. F. Poole, Anal. Chem., 55 (1983) 1462.
28. W. W. Blaser and W. R. Kracht, J. Chromatogr. Sci., 16 (1978) 111.
29. S. C. Dhanesar and C. F. Poole, Anal. Chem., 55 (1983) 2148.
30. C. R. Trash, J. Chromatogr. Sci., 11 (1973) 196.
31. J. K. Haken, J. Chromatogr., 300 (1984) 1.
32. J. K. Haken, J. Chromatogr., 141 (1977) 247.
33. L. Blomberg, J. High Resolut. Chromatogr. Chromatogr. Commun., 5 (1982) 520.
34. H. Heckers, K. Dittmar, F. W. Melcher, and H. O. Kalinowski, J. Chromatogr., 135 (1977) 93.
35. J. H. Beeson and R. E. Pecsar, Anal. Chem., 41 (1969) 1678.
36. S. C. Dhanesar and C. F. Poole, J. Chromatogr., 267 (1983) 293.
37. C. F. Poole, H. T. Butler, S. A. Agnello, W.-F. Sye, and A. Zlatkis, J. Chromatogr., 217 (1981) 39.
38. S. C. Dhanesar and C. F. Poole, J. Chromatogr., 253 (1982) 255.
39. R. G. Matthews, R. D. Schwartz, C. D. Pfaffenberger, S.-N. Lin, and E. C. Horning, J. Chromatogr., 99 (1974) 51.
40. R. D. Schwartz, R. G. Matthews, S. Ramachandran, R. S. Henly, and J. E. Doyle, J. Chromatogr., 112 (1975) 111.

41. P. Alessi, I. Kikic, and A. Papo, J. Chromatogr., 131 (1977) 31.
42. D. G. Anderson and R. E. Ansel, J. Chromatogr. Sci., 11 (1973) 192.
43. F. Vernon, J. Chromatogr., 148 (1978) 397.
44. R. F. Kruppa and R. S. Henly, J. Chromatogr. Sci., 12 (1974) 127.
45. M. B. Evans and J. F. Smith, J. Chromaotgr., 36 (1968) 489.
46. Gy. Vigh, A. Bartha, and J. Hlavay, J. High Resolut. Chromatogr. Chromatogr. Commun., 4 (1981) 3.
47. M. Verzele and P. Sandra, J. Chromatogr., 158 (1978) 111.
48. P. Sandra, M. Verzele, M. Verstappe, and J. Verzele, J. High Resolut. Chromatogr. Chromatogr. Commun., 2 (1979) 288.
49. K. Grob and K. Grob, J. Chromatogr., 140 (1977) 257.
50. R. S. Juvet and F. M. Wachi, Anal. Chem., 32 (1960) 290.
51. W. W. Hanneman, C. F. Spencer, and J. F. Johnson, Anal. Chem., 32 (1960) 1386.
52. J. E. Gordon, J. E. Selwyn, and R. J. Thorpe, J. Org. Chem., 31 (1966) 1925.
53. F. Pacholec, H. T. Butler, and C. F. Poole, Anal. Chem., 54 (1982) 1938.
54. F. Pacholec and C. F. Poole, Chromatographia, 17 (1983) 370.
55. C. F. Poole, H. T. Butler, M. E. Coddens, S. C. Dhanesar, and F. Pacholec, J. Chromatogr., 289 (1984) 299.
56. D. L. Meen, F. Morris, and J. H. Purnell, J. Chromatogr. Sci., 9 (1971) 281.
57. M. Ryba, Chromatographia, 5 (1972) 23.
58. S. H. Langer, Anal. Chem., 44 (1972) 1915.
59. J. H. Purnell and O. P. Srivastava, Anal. Chem., 45 (1973) 1111.
60. R. J. Laub and P. L. Pecsok, J. Chromatogr., 113 (1975) 47.
61. C. L. de Ligny, Adv. Chromatogr., 14 (1976) 265.
62. H. Kelker, Adv. Liq. Cryst., 3 (1978) 237.
63. G. M. Janini, Adv. Chromatogr., 17 (1979) 231.
64. Z. Witkiewicz, J. Chromatogr., 251 (1982) 311.
65. J. Szulc and Z. Witkiewicz, J. Chromatogr., 262 (1983) 141.
66. G. M. Janini and M. T. Ubeid, J. Chromatogr., 248 (1982) 217.
67. R. J. Laub, W. L. Roberts, and C. A. Smith, J. High Resolut. Chromatogr. Chromatogr. Commun., 3 (1980) 355.
68. J. Szulc, Z. Witkiewicz, and A. Ziolek, J. Chromatogr., 262 (1983) 161.
69. R. C. Kong, M. L. Lee, Y. Tominaga, R. Pratap, M. Iwao, and R. C. Castle, Anal. Chem., 54 (1982) 1802.
70. P. R. McCrea, in J. H. Purnell (Ed.), "New Developments in Gas Chromatography", Wiley, New York, 1973, p. 87.
71. L. Sojak and J. Krupcik, J. Chromatogr., 190 (1980) 283.
72. A.-M. Olsson, L. Mathiasson, J. A. Jonsson, and L. Haraldson, J. Chromatogr., 128 (1976) 35.
73. W. O. McReynolds, J. Chromatogr. Sci., 8 (1970) 685.
74. L. Rohrschneider, J. Chromatogr. Sci., 11 (1973) 160.
75. W. R. Supina and L. P. Rose, J. Chromatogr. Sci., 8 (1970) 214.
76. L. S. Ettre, Chromatographia, 7 (1974) 261.
77. J. K. Haken, Adv. Chromatogr., 14 (1976) 366.
78. A. Hartkopf, J. Chromatogr. Sci., 12 (1974) 113.
79. A. Hartkopf, S. Grunfeld, and R. DeLumyea, J. Chromatogr. Sci., 12 (1974) 119.
80. S. R. Lowry, S. Tsuge, J. J. Leary, and T. L. Isenhour, J. Chromatogr. Sci., 12 (1974) 124.
81. J. K. Haken and D. Srisukh, J. Chromatogr., 199 (1980) 199.
82. R. D. Schwartz and R. G. Matthews, J. Chromatogr., 126 (1976) 113.
83. R. J. Laub, in T. Kuwana (Ed.), "Physical Methods of Modern Chemical Analysis", Academic Press, New York, Vol. 3, 1983, p. 249.
84. C.-F. Chien, R. J. Laub, and M. M. Kopecni, Anal. Chem., 52 (1980) 1402.
85. R. J. Laub and J. H. Purnell, Anal. Chem., 48 (1976) 799.
86. R. J. Laub and J. H. Purnell, Anal. Chem., 48 (1976) 1720.
87. R. J. Laub, J. H. Purnell, and P. S. Williams, J. Chromatogr., 134 (1977) 249.

88. R. J. Laub and J. H. Purnell, J. Chromatogr., 112 (1975) 71.
89. R. J. Laub, J. H. Purnell, D. M. Summers, and P. S. Williams, J. Chromatogr., 155 (1978) 1.
90. J. F. K. Huber, E. Kenndler, and H. Markens, J. Chromatogr., 167 (1978) 291.
91. P. Tiley, J. Chromatogr., 179 (1979) 247.
92. P. Hurber and T. van Ree, J. Chromatogr., 169 (1979) 93.
93. P. Sandra and M. van Roelenbosch, Chromatographia, 14 (1981) 345.
94. T. Czajkowska, J. Chromatogr., 243 (1982) 35.
95. W. Q. Hull, H. Kenneys, and B. W. Garson, Ind. End. Chem., 45 (1959) 256.
96. D. M. Ottenstein, J. Chromatogr. Sci., 11 (1973) 136.
97. D. M. Ottenstein, Adv. Chromatogr., 3 (1966) 137.
98. J. F. Palframan and E. A. Walker, Analyst, 92 (1967) 71.
99. W. Kusy, Anal. Chem., 37 (1965) 1749.
100. E. C. Ormerod and R. P. W. Scott, J. Chromatogr., 2 (1955) 65.
101. T. Onaka and T. Okamoto, Chem. Pharm. Bull. (JPN), 10 (1962) 757.
102. W. J. A. Vanden Heuvel, W. L. Gardiner, and E. C. Horning, Anal. Chem., 35 (1963) 1745.
103. V. Pretorius and D. H. Desty, J. High Resolut. Chromatogr. Chromatogr. Commun., 4 (1981) 181.
104. M. A. Kaiser and C. D. Batich, J. Chromatogr., 175 (1979) 174.
105. M. M. Daniewski and W. A. Aue, J. Chromatogr., 150 (1978) 506.
106. W. A. Aue, M. M. Daniewski, E. E. Pickett, and P. R. McCullough, J. Chromatogr., 111 (1975) 37.
107. J. A. Jonsson, L. Mathiasson, and Z. Suprynowicz, J. Chromatogr., 207 (1981) 69.
108. P. P. Wickramanayake and W. A. Aue, J. Chromatogr., 210 (1981) 133.
109. J. Bohemen, S. H. Langer, R. H. Perrett, and J. H. Purnell, J. Chem. Soc., (1960) 244.
110. D. M. Ottenstein, J. Gas Chromatogr., 6 (1968) 129.
111. Z. Suprynowicz, A. Waksmundzi, and W. Rudzinski, J. Chromatogr., 72 (1972) 5.
112. Y. Takayamo, J. Chromatogr., 178 (1979) 63.
113. C. Gaget, D. Morel, and J. Serpinet, J. Chromatogr., 244 (1982) 209.
114. Z. Suprynowicz and J. A. Jonsson, Chromatographia, 14 (1981) 455.
115. A. N. Korol, G. M. Belokleytseva, and G. V. Filonenka, J. Chromatogr., 194 (1980) 145.
116. S. Kapila, W. A. Aue, and J. M. Augl, J. Chromatogr., 87 (1973) 35.
117. G. V. Filonenka, V. A. Tertykh, V. V. Pavlov, G. Ya Guba, and A. N. Korol, J. Chromatogr., 209 (1981) 385.
118. W. A. Aue and D. R. Younker, J. Chromatogr., 88 (1974) 7.
119. W. Biernacki, J. Chromatogr., 50 (1970) 135.
120. M. B. Evans, R. Newton, and J. D. Carni, J. Chromatogr., 166 (1978) 101.
121. R. D. Schwartz, R. G. Mathews, and N. A. Pedro, J. Chromatogr., 142 (1977) 103.
122. J. J. Kirkland, Anal. Chem., 35 (1963) 2003.
123. R. W. Ohline and R. Jojola, Anal. Chem., 36 (1964) 1681.
124. C. Hista, J. P. Messely, and R. F. Reschke, Anal. Chem., 32 (1960) 1930.
125. A. M. Filbert, in I. I. Domsky and J. A. Perry (Eds.), "Advances in Gas Chromatography", Dekker, New York, 1971, p. 49.
126. C. Hishta, J. Bomstein, and W. D. Cooke, Adv. Chromatogr., 9 (1970) 215.
127. W. Averill, J. Gas Chromatogr., 1 (1963) 34.
128. J. F. Parcher and P. Urone, J. Gas Chromatogr., 2 (1964) 184.
129. A. A. Spark, J. High Resolut. Chromatogr. Chromatogr. Commun., 2 (1979) 577.
130. R. F. Kruppa, R. S. Henly, and D. L. Smead, Anal. Chem., 39 (1967) 851.
131. S. Melendez-R and W. C. Parker, J. High Resolut. Chromatogr. Chromatogr. Commun., 2 (1979) 580.
132. K. Tesarik and M. Necasova, J. Chromatogr., 75 (1973) 1.

133. R. N. Nikolov, N. D. Petsev, and A. D. Stefanova, Chromatographia, 9 (1976) 81.
134. D. H. Desty, A. Goldup, and B. H. F. Whyman, J. Inst. Petrol., 45 (1959) 287.
135. D. J. Petitjean and C. J. Hiftoult, J. Gas Chromatogr., 3 (1963) 18.
136. D. C. Fenimore, J. H. Whitford, C. M. Davis, and A. Zlatkis, J. Chromatogr., 140 (1977) 9.
137. R. J. Laub and J. H. Purnell, J. High Resolut. Chromatogr. Chromatogr. Commun., 3 (1980) 195.
138. E. F. Barry, K. P. Li, and C. Merritt, J. Chromatogr. Sci., 20 (1982) 357.
139. F. Vernon and M. Rayanakorn, Chromatographia, 13 (1980) 611.
140. D. E. Martire, in J. H. Purnell (Ed.), "Progress in Gas Chromatography", Wiley, New York, 1968, p. 93.
141. J. Klein and H. Widdecke, J. Chromatogr., 129 (1976) 375.
142. E. Grushka and T. A. Goodwin, Chromatographia, 10 (1977) 549.
143. A. N. Korol, J. Chromatogr., 67 (1972) 213.
144. V. G. Berezkin, E. Ju. Sorokina, G. F. Shaligin, I. A. Litvinov, and M. N. Budantseva, J. Chromatogr., 193 (1980) 132.
145. R. N. Nikolov, J. Chromatogr., 241 (1982) 237.
146. I. Halasz and I. Sebastian, Angew. Chem. Int. Edn., 8 (1969) 453.
147. I. Halasz and I. Sebastian, J. Chromatogr. Sci., 12 (1974) 161.
148. J. N. Little, W. A. Dark, P. W. Farlinger, and K. J. Bombaugh, J. Chromatogr. Sci., 8 (1970) 647.
149. G. E. Pollock, D. R. Kojiro, and F. H. Woeller, J. Chromatogr. Sci., 20 (1982) 176.
150. E. W. Abel, F. H. Pollard, P. C. Uden, and G. Nickless, J. Chromatogr., 22 (1966) 23.
151. W. A. Aue and C. R. Hastings, J. Chromatogr., 42 (1969) 319.
152. M. M. Daniewski and W. A. Aue, J. Chromatogr., 147 (1978) 119.
153. P. P. Wickramanayake and W. A. Aue, J. Chromatogr., 195 (1980) 25.
154. W. A. Aue and P. P. Wickramanayake, J. Chromatogr., 200 (1980) 3.
155. W. A. Aue, C. R. Hastings, and S. Kapila, J. Chromatogr., 77 (1973) 299.
156. W. A. Aue, C. R. Hastings, and S. Kapila, Anal. Chem., 45 (1973) 725.
157. W. A. Aue, C. R. Hastings, J. M. Augl, M. K. Norr, and J. V. Larsen, J. Chromatogr., 56 (1971) 295.
158. R. F. Moseman, J. Chromatogr., 166 (1978) 397.
159. T. R. Edgerton and R. F. Moseman, J. Chromatogr. Sci., 18 (1980) 25.
160. E. Grushka (Ed.), "Bonded Phases in Chromatography", Ann Arbor Science, Ann Arbor, 1974.
161. V. Rehak and E. Smolkova, Chromatographia, 9 (1976) 219.
162. D. C. Locke, J. Chromatogr. Sci., 11 (1973) 120.
163. S. Mori, J. Chromatogr., 135 (1977) 261.
164. B. L. Karger and E. Sibley, Anal. Chem., 45 (1973) 740.
165. J. J. Pesek and J. E. Daniels, J. Chromatogr. Sci., 14 (1976) 288.
166. J. J. Pesek and J. A. Graham, Anal. Chem., 49 (1977) 133.
167. Z. Kessaissia, E. Papier, and J.-B. Donnet, J. Chromatogr., 196 (1980) 481.
168. O. L. Hollis, J. Chromatogr. Sci., 11 (1973) 335.
169. M. Kraus and H. Kopecka, J. Chromatogr., 124 (1976) 360.
170. H. L. Gearhart and M. F. Burke, J. Chromatogr. Sci., 15 (1977) 1.
171. O. L. Hollis, Anal. Chem., 38 (1966) 309.
172. S. B. Dave, J. Chromatogr. Sci., 7 (1969) 389.
173. F. M. Zado and J. Fabecic, J. Chromatogr., 51 (1970) 37.
174. E. Bayer and G. Nikolson, J. Chromatogr. Sci., 8 (1970) 467.
175. W. Hertl and M. G. Neumann, J. Chromatogr., 60 (1971) 319.
176. R. van Wijk, J. Chromatogr. Sci., 8 (1970) 418.
177. J. M. H. Daeman, W. Kankelman, and M. E. Hendricks, J. Chromatogr. Sci., 13 (1975) 79.
178. G. Castello and G. D'Amato, J. Chromatogr., 243 (1982) 25.

136

179. G. Castello and G. D'Amato, J. Chromatogr., 212 (1981) 261.
180. G. Castello and G. D'Amato, J. Chromatogr., 196 (1980) 245.
181. R. Komers, H. Kopecka, and M. Kraus, J. Chromatogr., 148 (1978) 45.
182. J. Lukas, J. Chromatogr., 190(1980) 13.
183. G. Castello and G. D'Amato, J. Chromatogr., 254 (1983) 69.
184. G. Castello and G. D'Amato, J. Chromatogr., 248 (1982) 391.
185. J. M. Trowell, J. Chromatogr. Sci., 9 (1971) 253.
186. A. V. Kiselev, Adv. Chromatogr., 4 (1967) 113.
187. E. Smolkova, L. Feltl, and J. Zima, Chromatographia, 12 (1979) 463.
188. C. L. Guillemin, J. Chromatogr., 158 (1978) 21.
189. C. L. Guillemin, M. Deleuil, S. Cirendini, and J. Vermont, Anal. Chem., 43 (1971) 2015.
190. W. K. Al.-Thamir, J. H. Purnell, and R. J. Laub, J. Chromatogr., 176 (1979) 232.
191. W. K. Al.-Thamir, J. H. Purnell, and R. J. Laub, J. Chromatogr., 188 (1980) 79.
192. L. Peichang, Z. Liangmo, W. Chinghai, W. Guanghua, X. Aizu, and X. Fangbao, J. Chromatogr., 186 (1979) 25.
193. R. Dandeneau and S. J. Hawkes, Chromatographia, 13 (1980) 686.
194. K. Naito, R. Kurita, S. Moriguchi, and S. Takei, J. Chromatogr., 246 (1982) 199.
195. W. K. Al.-Thamir, R. J. Laub, and J. H. Purnell, J. Chromatogr., 142 (1977) 3.
196. C. S. G. Phillips and C. D. Scott, in J. H. Purnell (Ed.), "Progress in Gas Chromatography", Wiley, New York, 1968, p. 121.
197. K. Naito, M. Endo, S. Moriguchi, and S. Takei, J. Chromatogr., 253 (1982) 205.
198. M. M. Kopecni, S. K. Milonjic, and R. J. Laub, Anal. Chem., 52 (1980) 1032.
199. A. Di Corcia and A. Liberti, Adv. Chromatogr., 14 (1976) 305.
200. A. Di Corcia, R. Samperi, E. Sabestaini, and C. Severini, Anal. Chem., 52 (1980) 1345.
201. A. Di Corcia, R. Samperi, E Sabestiani, and C. Sererini, Chromatographia, 14 (1981) 86.
202. A. Di Corcia, A. Liberti, and R. Samperi, Anal. Chem., 45 (1973) 1228.
203. A. Di Corcia and R. Samperi, Anal. Chem., 47 (1975) 1853.
204. A. Di Corcia, A. Liberti, and R. Samperi, J. Chromatogr., 122 (1976) 459.
205. F. Bruner, G. Bertoni, and P. Ciccioli, J. Chromatogr., 120 (1976) 307.
206. A. Di Corcia and M. Giabbai, Anal. Chem., 50 (1978) 1000.
207. A. Di Corcia, A. Liberti, and R. Samperi, J. Chromtogr., 167 (1978) 243.
208. N. Brenner, E. Cieplinski, L. S. Ettre, and V. J. Coates., J. Chromatogr., 3 (1960) 230.
209. R. M. Peterson and J. Rogers, Chromatographia, 5 (1972) 13.
210. S. P. Nandi and P. L. Walker, Sepn. Sci., 11 (1976) 441.
211. R. E. Kaiser, Chromatographia, 3 (1970) 38.
212. A. Zlatkis, H. R. Kaufman, and D. E. Durbin, J. Chromatogr. Sci., 8 (1970) 416.
213. W. E. Sharples, in C. E. H. Knapman (Ed.), "Devlopments in Chromatography", Applied Science Publishers, London, Vol. 1, 1978, p. 96.
214. T. B. Gavrilova, A. V. Kiselev, and T. M. Roshchima, J. Chromatogr., 192 (1980) 323.
215. F. Vernon and A. N. Khakoo, J. Chromatogr., 157 (1978) 412.
216. E. Smolkova, L. Feltl, and J. Vsetecka, J. Chromatogr., 148 (1978) 3.
217. V. G. Berezkin, L. A. Shkolina, V. N. Lipavsky, A. A. Serdan, and V. A. Barnov, J. Chromatogr., 141 (1977) 197.
218. C. A. Cramers and J. A. Rijks, Adv. Chromatogr., 17 (1979) 101.
219. J. F. K. Huber, H. H. Lauer, and H. Poppe, J. Chromatogr., 112 (1975) 377.
220. M. J. E. Golay, Anal. Chem., 29 (1957) 928.

221. L. S. Ettre in W. G. Jennings (Ed.), "Applications of Glass Capillary Gas Chromatography", Dekker, New York, 1981, p. 1.
222. L. S. Ettre, Chromatographia, 16 (1982) 18.
223. K. Grob and G. Grob, in L. H. Keith (Ed.), "Identification and Analysis of Organic Pollutants in Water", Ann Arbor Science, Ann Arbor, 1976, p. 75.
224. L. S. Ettre and E. W. March, J. Chromatogr., 91 (1974) 5.
225. M. Novotny and A. Zlatkis, J. Chromatogr., 14 (1971) 1.
226. M. L. Lee and B. W. Wright, J. Chromatogr., 184 (1980) 235.
227. L. S. Ettre, "Introduction to Open Tubular Columns", Perkin-Elmer, Norwalk, 1979.
228. W. G. Jennings, "Gas Chromatography with Glass Capillary Columns", Academic Press, New York, 1980.
229. W. G. Jennings, "Comparison of Fused Silica and Other Glass Columns in Gas Chromatography", Huthig, Heidelberg, 1981.
230. M. L. Lee, F. J. Yang, and K. D. Bartle, "Open Tubular Column Gas Chromatography. Theory and Practice", Wiley, New York, 1984.
231. W. Bertsch, F. Shunbo, R. C. Chang, and A. Zlatkis, Chromatographia, 7 (1974) 128.
232. A. Zlatkis, C. F. Poole, R. S. Brazell, and S. Singhawangcha, J. High Resolut. Chromatogr. Chromatogr. Commun., 2 (1979) 423.
233. J. E. DiNunzio, D. D. Bombick, and G. Doebler, Chromatographia, 15 (1982) 641.
234. D. D. Bombick and J. DiNunzio, Chromatographia, 14 (1981) 19.
235. D. H. Desty, J. N. Haresnape, and B. H. F. Whyman, Anal. Chem., 32 (1960) 302.
236. R. D. Dandeneau and E. H. Zerenner, J. High Resolut. Chromatogr. Chromatogr. Commun., 2 (1979) 351.
237. V. Pretorius and J. C. Davidtz, J. High Resolut. Chromatogr. Chromatogr. Commun., 2 (1979) 703.
238. S. R. Lipsky and W. J. McMurray, J. Chromatogr., 217 (1981) 3.
239. H. Saito, J. Chromatogr., 243 (1982) 189.
240. M. L. Lee, D. L. Vassilaros, L. V. Phillips, D. M. Hercules, H. E. Azumaya, J. W. Jorgenson, M. P. Maskorinec, and M. Novotny, Anal. Letts., 12 (1979) 191.
241. W. G. Jennings, Adv. Chromatogr., 20 (1982) 197.
242. M. J. Hartigan and L. S. Ettre, J. Chromatogr., 119 (1976) 187.
243. L. Blomberg, J. Chromatogr., 138 (1977) 7.
244. J. L. Marshall and D. A. Parker, J. Chromatogr., 122 (1976) 425.
245. W. G. Jennings, R. H. Wohleb, and R. G. Jenkins, Chromatographia, 14 (1981) 484.
246. J. G. Schenning, L. G. van der Ven, and A. Venema, J. High Resolut. Chromatogr. Chromatogr. Commun., 1 (1978) 101.
247. S. R. Lipsky, W. J. McMurray, M. Hernandez, J. E. Purcell, and K. A. Billeb, J. Chromatogr. Sci., 18 (1980) 1.
248. V. Pretorius and D. H. Desty, Chromatographia, 15 (1982) 569.
249. T. A. Michalske and S. W. Freeman, Nature 295 (1982) 511.
250. K. L Ogan, C. Reese, and R. P. W. Scott, J. Chromatogr. Sci., 20 (1982) 425.
251. S. R. Lipsky, J. High Resolut. Chromatogr. Chromatogr. Commun., 6 (1983) 359.
252. D. H. Desty and A. A. Douglas, J. Chromatogr., 142 (1977) 39.
253. D. H. Desty and A. A. Douglas, J. Chromatogr., 158 (1978) 73.
254. H. D. Papendick and J. Budisch, J. Chromatogr., 122 (1976) 443.
255. G. Du Plessis, P. Torline, and N. Kozma, Chromatographia, 10 (1977) 624.
256. G. Alexander and G. A. F. M. Rutten, J. Chromatogr., 99 (1974) 81.
257. K. D. Bartle, B. W. Wright, and M. L. Lee, Chromatographia, 14 (1981) 387.
258. K. Grob, J. High Resolut. Chromatogr. Chromatogr. Commun., 2 (1979) 599.
259. D. A. Parker and J. L. Marshall, Chromatographia, 11 (1978) 426.

138

260. K. Grob, Chromatographia, 10 (1977) 625.
261. M. Verzele, J. High Resolut. Chromatogr. Chromatogr. Commun., 1 (1978) 288.
262. M. Verzele, J. High Resolut. Chromatogr. Chromatogr. Commun., 2 (1979) 647.
263. J. J. Franken, G. A. F. M. Rutten, and J. A. Rijks, J. Chromatogr., 126 (1976) 117.
264. H. T. Badings, J. J. G. van der Pol, and D. G. Schmidt, Chromatographia, 10 (1977) 404.
265. G. Schomburg, H. Husmann, and F. Weeke, Chromatographia, 10 (1977) 580.
266. F. I. Onuska and M. E. Comba, J. Chromatogr., 126 (1976) 133.
267. F. I. Onuska and M. E. Comba, Chromatographia, 10 (1977) 498.
268. K. Grob, G. Grob, and K. Grob, J. High Resolut. Chromatogr. Chromatogr. Commun., 2 (1979) 677.
269. K. Grob, G. Grob, W. Blum, and W. Walther, J. Chromatogr., 244 (1982) 197.
270. B. W. Wright, M. L. Lee, S. W. Graham, L. V. Phillips, and D. M. Hercules, J. Chromatogr., 199(1980) 355.
271. B. W. Wright, P. A. Peaden, M. L. Lee, and G. M. Booth, Chromatographia, 15 (1982) 584.
272. A. Venema, J. T. Sukkel, and N. Kampstra, J. High Resolut. Chromatogr. Chromatogr. Commun., 6 (1983) 236.
273. R. A. Heckman, C. R. Green, and F. W. Best, Anal. Chem., 50 (1978) 2157.
274. T. L. Petters, T. J. Nestrick, and L. L. Lamparski, Anal. Chem., 54 (1982) 2397.
275. J. D. Schieke, N. R. Comins, and V. Pretorius, J. Chromatogr., 112 (1975) 97.
276. J. D. Schieke, N. R. Comins, and V. Pretorius, Chromatographia, 8 (1975) 354.
277. P. Sandra and M. Verzele, Chromatographia, 10 (1977) 419.
278. P. Sandra, M. Verstappe, and M. Verzele, J. High Resolut. Chromatogr. Chromatogr. Commun., 1 (1978) 28.
279. J. F. G. Clarke, J. High Resolut. Chromatogr. Chromatogr. Commun., 2 (1979) 357.
280. F. I. Onuska, M. E. Comba, T. Bistricki, and R. J. Wilkinson, J. Chromatogr., 142 (1977) 117.
281. P. Sandra, M. Verstappe, and M. Verzele, Chromatographia, 11 (1978) 223.
282. R. C. M. de Nijs, G. A. F. M. Rutten, J. J. Franken, R. P. M. Dooper, and J. A. Rijks, J. High Resolut. Chromatogr. Chromatogr. Commun., 2 (1979) 447.
283. H. T. Badings, J. J. G. van der Pol, and J. G. Wassink, J. High Resolut. Chromatogr. Chromatogr. Commun., 2 (1979) 297.
284. H. T. Badings, J. J. G. van der Pol, and J. G. Wassink, J. Chromatogr., 203 (1981) 227.
285. K. Grob and G. Grob, J. Chromatogr., 125 (1976) 471.
286. K. Grob, G. Grob, and K. Grob, Chromatographia, 10 (1977) 181.
287. K. Grob, G. Grob, and K. Grob, J. High Resolut. Chromatogr. Chromatogr. Commun., 1 (1978) 149.
288. R. F. Arrendale, L. B. Smith, and L. B. Rogers, J. High Resolut. Chromatogr. Chromatogr. Commun., 3 (1980) 115.
289. G. Schomburg, H. Husmann, and H. Behlau, J. Chromatogr., 203 (1981) 179.
290. G. Schomburg, R. Dielmann, H. Borwitzky, and H. Husman, J. Chromatogr., 167 (1978) 337.
291. G. A. F. M. Rutten and J. A. Luyten, J. Chromatogr., 74 (1972) 177.
292. H. T. Badings, J. J. G. van der Pol, and J. G. Wassink, Chromatographia, 8 (1975) 440.
293. R. E. Kaiser and R. Rieder, Chromatographia, 8 (1975) 491.
294. D. A. Cronin, J. Chromatogr., 97 (1974) 263.
295. L. Blomberg and T. Wannman, J. Chromatogr., 148 (1978) 379.
296. K. Grob and G. Grob, J. Chromatogr., 125 (1976) 471.

297. R. C. M. De Nijs, J. J. Franken, R. P. M. Dooper, J. A. Rijks, H. J. J. M. De Ruwe, and F. L. Schulting, J. Chromatogr., 167 (1978) 231.

298. G. Schomburg, H. Husmann, and H. Borwitzky, Chromatographia, 13 (1980) 321.

299. G. Schomburg, H. Husmann, and H. Borwitzky, Chromatographia, 12 (1979) 651.

300. J. Buijten, L. Blomberg, K. Markides, and T. Wannman, J. Chromatogr., 237 (1982) 465.

301. R. Burrows, M. Cooke, and D. G. Gillespie, J. Chromatogr., 260 (1983) 168.

302. B. W. Wright, P. A. Peaden, M. L. Lee, and T. J. Stark, J. Chromatogr., 248 (1982) 17.

303. K. Markides, L. Blomberg, J. Buijten, and T. Wannman, J. Chromatogr., 267 (1983) 29.

304. L. Blomberg, K. Markides, and T. Wannman, J. High Resolut. Chromatogr. Chromatogr. Commun., 3 (1980) 527.

305. T. Welsch, R. Muller, W. Endewald, and G. Werner, J. Chromatogr., 241 (1982) 41.

306. M. Godefroot, M. van Roelenbosch, M. Verstappe, P. Sandra, and M. Verzele, J. High Resolut. Chromatogr. Chromatogr. Commun., 3 (1980) 337.

307. K. Grob, G. Grob, and K. Grob, J. High Resolut. Chromatogr. Chromatogr. Commun., 2 (1979) 31.

308. K. Grob and G. Grob, J. High Resolut. Chromatogr. Chromatogr. Commun., 3 (1980) 197.

309. K. Grob, J. High Resolut. Chromatogr. Chromtogr. Commun., 3 (1980) 493.

310. R. J. Laub, W. L. Roberts, and C. A. Smith, J. High Resolut. Chromatogr. Chromatogr. Commun., 6 (1983) 44.

311. C. P. Schutjes, E. A. Vermeer, and C. A. Cramers, J. Chromatogr., 279 (1983) 49.

312. K. Grob, J. High Resolut. Chromatogr. Chromatogr. Commun., 3 (1980) 525.

313. M. Novotny, L. Blomberg, and K. D. Bartle, J. Chromatogr. Sci., 8 (1970) 390.

314. G. Guiochon, J. Chromatogr. Sci., 9 (1971) 512.

315. K. D. Bartle, Anal. Chem., 45 (1973) 1831.

316. K. D. Bartle and M. Novotny, J. Chromatogr., 94 (1974) 35.

317. L. Blomberg, Chromatographia, 8 (1975) 324.

318. M. J. Hartigan and L. S. Ettre, J. Chromatogr., 119 (1976) 187.

319. J. Roeraade, Chromatographia, 8 (1975) 511.

320. K. Grob and G. Grob, Chromatographia, 4 (1971) 422.

321. G. Alexander and G. A. F. M. Rutten, Chromatographia, 6 (1975) 231.

322. G. Schomburg, H. Husmann, and F. Weeke, J. Chromatogr., 99 (1974) 63.

323. G. Schomburg and H. Husmann, Chromatographia, 8 (1975) 517.

324. G. Alexander and S. R. Lipsky, Chromatographia, 10 (1977) 487.

325. T. Czajkowska, Chromatographia, 15 (1982) 305.

326. G. Redant, P. Sandra, and M. Verzele, Chromatographia, 15 (1982) 13.

327. J. Bouche and M. Verzele, J. Gas Chromatogr., 6 (1968) 501.

328. T. Boogaerts, M. Verstappe, and M. Verzele, J. Chromatogr. Sci., 10 (1972) 217.

329. K. Grob and G. Grob, J. High Resolut. Chromatogr. Chromatogr. Commun., 5 (1982) 119.

330. M. Giabbai, M. Shoults, and W. Bertsch, J. High Resolut. Chromatogr. Chromatogr. Commun., 1 (1978) 277.

331. G. A. F. M. Rutten and J. A. Rijks, J. High Resolut. Chromatogr. Chromatogr. Commun., 1 (1978) 279.

332. P. Sandra and M. Verzele, Chromatographia, 11 (1978) 102.

333. J. C. Thompson and N. G. Schnautz, J. High Resolut. Chromatogr. Chromatogr. Commun., 3 (1980) 91.

334. G. Anders, D. Roderwald, and Th. Welsch, J. High Resolut. Chromatogr. Chromatogr. Commun., 3 (1980) 298.

335. C. H. Lochmuller and J. D. Fisk, J. High Resolut. Chromatogr. Chromatogr. Commun., 5 (1982) 232.

336. K. R. Kim, L. Ghaoui, and A. Zlatkis, J. High Resolut. Chromatogr. Chromatogr. Commun., 5 (1982) 571.

337. R. C. Kong and M. L. Lee, J. High Resolut. Chromatogr. Chromatogr. Commun., 6 (1983) 319.

338. F. Janssen and T. Kaldin, J. High Resolut. Chromatogr. Chromatogr. Commun., 5 (1982) 107.

339. C. Spagone and R. Fanelli, J. High Resolut. Chromatogr. Ghromatogr. Chromatogr. Commun., 5 (1982) 572.

340. M. K. Cueman and R. P. Hurley, J. High Resolut. Chromatogr. Chromatogr. Commun., 1 (1978) 92.

341. K. Grob, J. High Resolut. Chromatogr. Chromatogr. Commun., 1 (1978) 93.

342. R. C. Kong and M. L. Lee, Chromatographia, 17 (1983) 451.

343. L. Blomberg, J. Buijten, K. Markides, and T. Wannman, J. Chromatogr., 279 (1983) 9.

344. K. Grob, G. Grob, and K. Grob, J. Chromatogr., 211 (1981) 243.

345. S. R. Lipsky and W. J. McMurray, J. Chromatogr., 239 (1982) 61.

346. S. R. Springston, K. Melda, and M. Novotny, J. Chromatogr., 267 (1983) 395.

347. C. Madani, E. M. Chambaz, M. Rigaud, J. Durand, and P. Chebroux, J. Chromatogr., 126 (1976) 161.

348. C. Madani, E. M. Chambaz, M. Ridaud, P. Chebroux, J. C. Breton, and F. Berthou, Chromatographia, 10 (1977) 466.

349. C. Madani and E. M. Chambaz, Chromatographia, 11 (1978) 725.

350. C. Madani and E. M. Chambaz, J. Amer. Oil Chem. Soc., 58 (1981) 63.

351. L. Blomberg and T. Wannman, J. Chromatogr., 168 (1979) 81.

352. L. Blomberg and T. Wannman, J. Chromatogr., 186 (1979) 159.

353. L. Blomberg, K. Markides, and T. Wannman, J. Chromatogr., 203 (1981) 217.

354. L. Blomberg, J. Buijten, K. Markides, and T. Wannman, J. Chromatogr., 208 (1981) 231.

355. K. Grob and G. Grob, J. High Resolut. Chromatogr. Chromatogr. Commun., 5 (1982) 13.

356. K. Grob and G. Grob, J. High Resolut. Chromatogr. Chromatogr. Commun., 6 (1983) 153.

357. P. Sandra, G. Redant, E. Schacht, and M. Verzele, J. High Resolut. Chromatogr. Chromatogr. Commun., 4 (1981) 411.

358. P. Sandra, M. Van Roelenbosch, I. Temmerman, and M. Verzele, Chromatographia, 16 (1983) 63.

359. L. Blomberg, J. Buijten, K. Markides, and T. Wannman, J. High Resolut. Chromatogr. Chromatogr. Commun., 4 (1981) 578.

360. L. Blomberg, J. Buijten, K. Markides, and T. Wannman, J. Chromatogr., 239 (1982) 51.

361. J. Buijten, L. Blomberg, K. Markides, and T. Wannman, Chromatographia, 16 (1983) 183.

362. B. E. Richter, J. C. Kuel, J. I. Shelton, L. W. Castle, J. S. Bradshaw, and M. L. Lee, J. Chromatogr., 279 (1983) 21.

363. B. E. Richter, J. C. Kuel, N. J. Park, S. J. Crowley, J. S. Bradshaw, and M. L. Lee, J. High Resolut. Chromatogr. Chromtogr. Commun., 6 (1983) 371.

364. B. E. Richter, J. C. Kuel, L. W. Castle, B. A. Jones, J. S. Bradshaw, and M. L. Lee, Chromatographia, 17 (1983) 570.

365. G. Schomburg, H. Husmann, S. Ruthe, and M. Herraiz, Chromatographia, 15 (1982) 599.

366. W. Bertsch, V. Pretorius, M. Pearce, J. C. Thompson, and N. G. Schnautz, J. High Resolut. Chromatogr. Chromatogr. Commun., 5 (1982) 432.

367. J. A. Huball, P. Di Mauro, E. F. Barry, and G. E. Chabot, J. High Resolut. Chromatogr. Chromatogr. Commun., 6 (1983) 241.

368. E. F. Barry, G. E. Chabot, P. Ferioli, J. A. Huball, and E. M. Rand, J. High Resolut. Chromatogr. Chromatogr. Commun., 6 (1983) 300.
369. K. Grob and K. Grob, J. High Resolut. Chromatogr. Chromatogr. Commun., 6 (1983) 133.
370. T. Stark and P. Larson, J. Chromatogr. Sci., 20 (1982) 341.
371. L. S. Ettre, Chromatographia, 17 (1983) 553.
372. L. S. Ettre, G. L. McClure, and J. D. Walters, Chromatographia, 17 (1983) 560.
373. P. Sandra, I. Temmerman, and M. Verstappe, J. High Resolut. Chromatogr. Chromatogr. Commun., 6 (1983) 501.
374. K. Grob and G. Grob, J. High Resolut. Chromatogr. Chromatogr. Commun., 2 (1979) 109.
375. I. Halasz and E. Heine, Adv. Chromatogr., 4 (1967) 207.
376. L. S. Ettre and J. E. Purcell, Adv. Chromatogr., 10 (1974) 1.
377. J. G. Nikelly, Sepn. Purifn. Meths., 3 (1974) 423.
378. P. Torline, G. du Plessis, N. Schnautz, and J. C. Thompson, J. High Resolut. Chromatogr. Chromatogr. Commun., 2 (1979) 613.
379. L. Zu-Fong, T. Yuen-Yu, C. Rhan-Mei, O. Qing-Yu, Y. Pei-Yi, D. Kun-Nian, and Y. Wei-Lu, J. High Resolut. Chromatogr. Chromatogr. Commun., 2 (1979) 429.
380. A. G. Ober, M. Cooke, and G. Nickless, J. Chromtogr., 196 (1980) 237.
381. T. Wishousky, R. L. Grob, and A. G. Zacchei, J. Chromatogr., 249 (1982) 1.
382. E. Schulte, Chromatographia, 9 (1976) 315.
383. R. D. Schwartz, D. J. Brasseaux, and R. G. Mathews, Anal. Chem., 38 (1966) 303.
384. R. G. Mathews, J. Torres, and R. D. Schwartz, J. Chromatogr., 186 (1979) 183.
385. R. G. Mathews, J. Torres, and R. D. Schwartz, J. Chromatogr., 199 (1980) 97.
386. W. Schneider, J. C. Frohne, and H. Bruderreck, J. Chromatogr., 155 (1978) 311.
387. R. G. Mathews, J. Torres, and R. D. Schwartz, J. Chromatogr. Sci., 20 (1982) 160.
388. R. C. M. De Nijs and J. De Zeeuw, J. Chromatogr., 279 (1983) 49.
389. A. L. German and E. C. Horning, J. Chromatogr. Sci., 11 (1973) 76.
390. P. Van Hout, J. Szafranek, C. D. Pfaffenberger, and E. C. Horning, J. Chromatogr., 99 (1974) 103.
391. S. L. Mackenzie and L. R. Hogge, J. Chromatogr., 147 (1978) 388.
392. R. S. Deelder, J. J. M. Ramaekers, and J. H. M. van den Berg, J. Chromatogr., 119 (1976) 99.
393. R. G. McKeag and F. W. Hougen., J. Chromatogr., 136 (1977) 308.
394. C. A. Cramers, E. A. Vermer, and J. J. Franken, Chromatographia, 10 (1977) 412.
395. S. Kozuharov, J. Chromatogr., 198 (1980) 153.
396. J. J. Thieke, J. H. M. van den Berg, R. S. Deelder, and J. J. M. Ramaekers, J. Chromatogr., 160 (1978) 264.
397. J. Chauhan and A. Darbre, J. High Resolut. Chromatogr. Chromatogr. Commun., 4 (1981) 11.
398. A. Liberti, G. Goretti, and M. V. Russo, J. Chromatogr., 279 (1983) 1.
399. G. Goretti, F. Geraci, and M. V. Russo, Chromatographia, 14 (1981) 285.
400. W. E. Sharples and F. Vernon. J. Chromatogr., 161 (1978) 83.
401. M. S. Wainwright, J. K. Haken, and D. Srisukh, J. Chromatogr., 179 (1979) 160.
402. J. F. Parcher and D. M. Johnson, J. Chromatogr. Sci., 18 (1980) 267.
403. M. S. Wainwright and J. K. Haken, J. Chromaogr., 256 (1983) 193.
404. R. E. Kaiser, J. High Resolut. Chromatogr. Chromatogr. Commun., 1 (1978) 115.
405. J. R. Ashes, J. C. Mills, and J. K. Haken, J. Chromatogr., 166 (1978) 391.

406. L. S. Ettre, Chromatographia, 13 (1980) 73.
407. M. S. Wainwright and J. K. Haken, J. Chromatogr., 184 (1980) 1.
408. M. V. Budahegyi, E. R. Lombosi, T. S. Lombosi, I. Timar, and
     J. M. Takacs, J. Chromatogr., 271 (1983) 213.
409. G. Schomburg, J. High Resolut. Chromatogr. Chromatogr. Commun., 2 (1979)
     461.
410. K. Grob and G. Grob, J. High Resolut. Chromatogr. Chromatogr. Commun.,
     2 (1979) 302.
411. J. Krupcik, J. Garaj, G. Guiochon, and J. M. Schmitter, Chromatographia,
     14 (1981) 501.
412. L. S. Ettre, Chromatographia, 8 (1975) 291.
413. I. Brown, Chromatographia, 12 (1979) 265.
414. J. D. Schieke and V. Pretorius, J. Chromatogr., 132 (1977) 217.
415. C. A. Cramers, F. A. Wijheijmer, and J. A. Rijks, Chromatographia, 12
     (1979) 643.
416. W. Jennings and K. Yabumoto, J. High Resolut. Chromatogr. Chromatogr.
     Commun., 3 (1980) 177.
417. T. A. Rooney and M. J. Hartigan, J. High Resolut. Chromatogr.
     Chromatogr. Commun., 3 (1980) 416.
418. K. Grob and G. Grob, J. Chromatogr., 207 (1981) 291.
419. L. A. Jones, S. L. Kirby, C. L. Garganta, T. M. Gerig, and J. D. Mulik,
     Anal. Chem., 55 (1983) 1354.
420. K. Grob, G. Grob, and K. Grob, J. Chromatogr., 156 (1978) 1.
421. K. Grob, G. Grob, and K. Grob, J. Chromatogr., 219 (1981) 13.
422. K. Grob and G. Grob, Chromatographia, 4 (1971) 422.
423. Th. Welsch, W. Engewald, and Ch. Klaucke, Chromatographia, 10 (1977) 22.
424. S. P. Cram, F. J. Yang, and A. C. Brown, Chromatographia, 10 (1977) 397.
425. M. J. Hartigan, K. Billeb, and L. S. Ettre, Chromatographia, 10 (1977)
     571.
426. K. Grob, J. High Resolut. Chromatogr. Chromatogr. Commun., 3 (1980) 585.
427. M. Ahnoff and L. Johansson, J. Chromatogr., 279 (1983) 75.
428. L. Rohrschneider and E. Pelster, J. Chromatogr., 186 (1979) 249.
429. L. Kimperhaus, F. Richter, and L. Rohrschneider, Chromatographia, 15
     (1982) 577.
430. L. S. Ettre, Chromatographia, 12 (1979) 509.
431. D. F. Ingraham, C. F. Schoemaker, and W. Jennings, J. High Resolut.
     Chromatogr. Chromatogr. Commun., 5 (1982) 227.
432. B. Olufren, J. High Resolut. Chromatogr. Chromatogr. Commun., 2 (1979)
     579.
433. J. F. Parcher, J. Chromatogr. Sci., 21 (1983) 346.
434. A. Nonaka, Adv. Chromatogr., 12 (1975) 223.
435. K. W. M. Siu and W. A. Aue, J. Chromatogr., 189 (1980) 255.
436. J. F. Parcher and T. N. Westlake, J. Chromatogr. Sci., 14 (1976) 343.
437. T. H. Gouw and R. E. Jentoft, J. Chromatogr., 68 (1972) 303.
438. T. H. Gouw and R. E. Jentoft, Adv. Chromatogr., 13 (1975) 1.
439. E. Klesper, Angew. Chem. Int. Ed., 17 (1978) 738.
440. U. van Wasen, I. Swaid, and G. M. Schneider, Agnew. Chem. Int. Ed.,
     19 (1980) 575.
441. M. Novotny, S. R. Springston, P. A. Peaden, J. C. Fjeldsted, and
     M. L. Lee, Anal. Chem., 53 (1981) 407A.
442. L. G. Randall, Sep. Sci. Technol., 17 (1982) 1.
443. P. A. Peaden and M. L. Lee, J. Liq. Chromatogr., 5 (1982) 179.
444. S. R. Springston and M. Novotny, Chromatographia, 14 (1981) 679.
445. M. Novotny and S. R. Springston, J. Chromatogr., 279 (1983) 417.
446. P. A. Peaden, J. C. Fjeldsted, M. L. Lee, S. R. Springston, and
     M. Novonty, Anal. Chem., 54 (1982) 1090.
447. K. Grob, J. High Resolut. Chromatogr. Chromatogr. Commun., 6 (1983) 178.
448. P. A. Peaden and M. L. Lee, J. Chromatogr., 259 (1983) 1.
449. D. R. Gere, R. D. Board, and D. McManigill, Anal. Chem., 54 (1982) 736.
450. F P. Schmitz, H. Hilgers, and E. Klesper, J. Chromatogr., 267 (1983)
     267.

451. H. H. Lauer, D. McManigill, and R. D. Board, Anal. Chem., 55 (1983) 1370.
452. W. Asche, Chromatographia, 11 (1978) 411.
453. J. C. Fjeldsted, W. P. Jackson, P. A. Peaden, and M. L. Lee, J. Chromatogr. Sci., 21 (1983) 222.
454. J. A. Graham and L. B. Rogers, J. Chromatogr. Sci., 18 (1980) 75.
455. F. J. van Lenten and L. D. Rothman, Anal. Chem., 48 (1976) 1430.
456. J. C. Fjeldsted, B. E. Richter, W. P. Jackson, and M. L. Lee, J. Chromatogr., 279 (1983) 423.
457. J. C. Fjeldsted, R. C. Kong, and M. L. Lee, J. Chromatogr., 279 (1983) 449.
458. L. G. Randall and A. L. Wahrhaftig, Rev. Sci. Instrum., 52 (1981) 1283.
459. R. D. Smith, J. C. Fjeldsted, and M. L. Lee, J. Chromatogr., 247 (1982) 231.
460. R. D. Smith, W. D. Felix, J. C. Fjeldsted, and M. L. Lee, Anal. Chem., 54 (1982) 1883.
461. R. D. Smith and H. R. Udseth, Anal. Chem., 55 (1983) 2266.
462. K. H. Shafer and P. R. Griffiths, Anal. Chem., 55 (1983) 1939.
463. M. Novotny, S. Olesik, and S. French, J. Chromatogr., (1984) in press.

Chapter 3

INSTRUMENTAL REQUIREMENTS FOR GAS CHROMATOGRAPHY

3.1 Introduction

In gas chromatography, separations result from the interaction of a
vaporized sample in a gaseous mobile phase with a liquid or solid stationary
phase. The essential elements of a gas chromatograph are a regulated supply of
carrier gas, a device for vaporizing the sample (injector), a thermostatted oven
in which the column is housed, an on-line detector, and a recording device
[1-5]. The principal requirement of the gas chromatograph is to provide those
conditions necessary for a separation to occur without deteriorating the
performance of the column in any way. Although the basic components remain the
same, open tubular capillary columns place greater demands on instrument
peformance than packed columns. This is due to the lower sample capacity, lower
carrier gas flow rate, and faster detector response time required by the former
[6]. Older instruments designed for packed columns may be unsuitable for use
with open tubular columns; most modern instruments are designed for use with
both open tubular and packed columns, and changing from one column type to the
other requires only a few simple modifications.

A supply of gases in the form of pressurized cylinders is required for the
carrier gas and, depending on the instrument configuration, perhaps also for the
detector, for operating pneumatic controls such as switching valves, and for
providing automatic cool-down by opening the oven door on command from a
microprocessor. Each cylinder is fitted with a two-stage pressure regulator for

coarse pressure and flow control. In most instruments provision is also made
for secondary fine tuning of pressure and gas flow as well as for gas
purification. The carrier gas flow is directed through a molecular sieve trap
to remove moisture and low molecular weight hydrocarbons and possibly through an
additional trap to remove oxygen. The gas then enters a pneumatic section,
thermostatically controlled to prevent drift, in which pressure regulators, flow
controllers, and perhaps additional gas purifying traps and particle filters are
housed. The pressure regulators are usually of the diaphragm type with a
polymeric elastic membrane. This membrane can be a source of carrier gas
contamination due to its adsorptive nature and permeability to atmospheric air
[7]. The temporary accumulation of organic impurities or the ingress of
atmospheric air can be eliminated by using gauges with metallic diaphragms or by
covering the polymeric diaphragm with a thin sheet of metal foil. To maintain a
constant flow of both carrier and detector gases at a fixed pressure, a flow
impedence in the form of a calibrated constriction or choke is required [8]. In
more sophisticated systems electronic sensors, operated in a feedback loop,
provide a constant carrier gas flow independent of the column back pressure.
Another common combination is the use of a precision pressure regulator and a
needle valve to control the flow. With this arrangement the carrier gas flow
rate is not independent of the column back pressure and can be expected to vary
as the oven temperature changes. Actual flow rates are usually measured with a
soap-film meter at, or some point after, the column exit [9].

Packed column injection of the sample solution is made with a microliter
syringe through a silicone rubber septum into a heated metal block that is
continuously swept by carrier gas. When injection is made in the on-column
mode, the column is pushed right up to the septum area and the column end within
the injector is packed with glass wool. Ideally, the tip of the syringe needle
should penetrate the glass wool filling and just reach the surface of the column
packing. For flash vaporization the sample is injected into a low dead volume,
glass-lined chamber, mixed with carrier gas, and flushed directly onto the
column. Whichever technique is used, the injector must meet certain
specifications. Firstly, it must have sufficient thermal mass to rapidly
vaporize the sample. The incoming carrier gas is usually preheated by directing
its flow through a section of the injector block. This avoids possible
condensation of the vaporized sample upon mixing with the cool carrier gas. The
injector block should have variable temperature control, and a temperature range
of approximately 25-400°C. Unless dictated otherwise by the thermal instability
of the sample, an injection temperature 50°C greater than the maximum column
oven temperature used in the analysis is generally adopted. Because high
injection temperatures are frequently used, septum bleed may give rise to an
unstable baseline or the appearance of ghost peaks in the chromatograms

[10-13]. Various solutions to this problem are available; low-bleed septa with good resealability, a finned septum holder that allows cooling of the septum, or a septum purge device can be used. Many instruments incorporate a septum purge particularly for injectors used with open tubular columns. Several arrangements can be used for septum purging, but in all cases a portion of the carrier gas or an auxiliary gas is forced to flow across the face of the septum and through an adjacent orifice.

Volatile samples are injected by gas-tight syringe in the normal way or with a rotary valve [14-16]. With multi-functional valves continuous sampling is possible; the sample loops store individual samples which can be analyzed sequentially at a later time [16]. For samples of moderate volatility, the valve and transfer lines to the column may need to be thermostatted at temperatures above ambient. The sample is injected by rotating the valve body such that the carrier gas is rerouted through the sample loop and onto the column. Adsorption of the sample onto the surfaces of the valve and sample loop can lead to systematic errors in quantitative analysis [14].

Solid samples are usually dissolved in a suitable solvent and injected as described above. Alternatively, the samples can be encapsulated in glass capillaries which are then pushed or dropped into the heated injection block and crushed by a mechanical device [17]. This form of injection is particularly useful for the analysis of trace volatiles which would be hidden in the solvent front with conventional injection techniques.

The column oven is generally a forced circulation air thermostat of sufficient size to allow comfortable installation of the longest columns normally used. Temperature control, stability, and program capability are important design considerations. Commercial instruments are usually capable of maintaining a constant temperature of $\pm 0.1°C$ in time and $\pm 1.0°C$ in space. Temperature gradients within the oven volume can have a deleterious effect on column performance, particularly for open tubular columns. To minimize such gradients, some manufacturers use an oven-within-an-oven design. The utilization of microprocessors and proportional heating networks to control the initial temperature lag, the linearity of the temperature program, and the final temperature overshoot enables high precision retention data to be produced.

The detectors should be thermostatted separate from the column oven and maintained at a higher temperature to reduce contamination by column bleed or from sample condensation. Various general purpose and selective detectors are used in gas chromatography and will be described later in this chapter.

Earlier we stated that the most desirable characteristic of a gas chromatograph is that it should not adversely affect the performance of the column. Band broadening is a consequence of the chromatographic process itself but may also be influenced by the apparatus, particularly the design of the

injector/column inlet, connecting tubing and fittings, detector, electronic
amplification, and recording devices [1,18-22]. As well as extracolumn band
broadening, poorly designed instruments may also result in peak asymmetry. The
most serious problems arise from dead spaces which can behave as either mixing
or diffusion chambers, depending on whether they are swept out by the carrier
gas or not. These dead spaces create regions of turbulent, recirculating flow
outside the main flow stream. The sample is retained within these spaces and
then slowly released back into the main stream by diffusion and weak turbulent
mixing. The effect is a modification of the chromatographic peak profile
producing tailing (exponential peak broadening). Sudden diameter changes in the
carrier gas flow path are the most frequent sources of dead spaces.

In many cases it is possible to convert an older packed column gas
chromatograph to accept open tubular capillary columns [23-31]. Normally this
will involve changes to the pneumatics to provide a stable pressure-regulated
supply of carrier gas, the addition of splitter gas (if a splitter is installed)
and detector make-up gas, a new injector for split, splitless or cold on-column
injection, and additional minor changes in plumbing, fittings, etc., to
accommodate the added components. Before attempting such a conversion the
response of the detector electronics should be measured to ascertain if its time
constant is adequate [23]. The temperature stability of the oven in both time
and space should also be determined to ensure the absence of temperature
gradients and cycling or overshooting of preset or programmed temperatures,
which would result in poor column performance even if the conversion were
carried out successfully. Several companies offer kits for converting packed
column instruments for capillary column use.

3.2 Syringe Injection Techniques

In gas chromatography, samples of measured volume are usually introduced by
syringe. Various syringe handling techniques are summarized in Table 3.1
[32-36]. Packed columns accomodate relatively large sample volumes (e.g.,
approximately 1-5 microliters) and are relatively forgiving if poor injection
techniques are used. Open tubular capillary columns with their low stationary
phase volumes and carrier gas flow rates are more demanding on technique than
packed columns. Typical injection volumes vary from about 0.1 to 2.0
microliters, depending on the type of column and injection device used.

A common problem encountered with syringe injection using vaporizing
injectors is sample discrimination. An example is shown in Figure 3.1 [38].
The sample leaves the syringe and enters the vaporizer as a stream of droplets,
formed by the movement of the plunger and by evaporation of the remaining sample
from the syringe needle. It is at this evaporation stage that discrimination is

TABLE 3.1

SUMMARY OF METHODS USED FOR SAMPLE INTRODUCTION BY SYRINGE

Method	Principle
Filled needle	Sample is taken up into the syringe needle without entering the barrel. Injection is made by placing the syringe needle into the injection zone. No mechanical movement of the plunger is involved and the sample leaves the needle by evaporation.
Cold needle	Sample is drawn into the syringe barrel so that an empty syringe needle is inserted into the injection zone. Immediately the sample is injected by depressing the plunger. Sample remaining in the syringe needle leaves by evaporation.
Hot needle	Injection follows the general procedure discribed for the cold needle method except that prior to depressing the plunger the needle is allowed to heat up in the injection heater for 3-5 seconds.
Solvent flush	A solvent plug is drawn up by the syringe ahead of the sample. The solvent and sample may or may not be separated by an air barrier. The injection is usually made as indicated in the cold needle method. The solvent is used to push the sample out of the syringe.
Air flush	As for solvent flush, except that an air plug is used rather than a solvent plug.

most likely to occur; the solvent and more volatile sample components distill from the syringe needle at a greater rate than the less volatile components. As a consequence, the sample reaching the column is not identical in composition with the original sample solution; it contains more of the most volatile and less of the least volatile components of the sample. The "hot needle" and "solvent flush" techniques are about equally effective in minimizing discrimination and are generally preferrred over the other syringe handling methods denoted in Table 3.1. The influence of losses inside the syringe needle decreases as the sample size injected increases, and thus may be hardly noticed with packed columns, but may be critical with open tubular columns where the sample volume can be on the same order as that of the syringe needle. Within the injector, axial thermal gradients result in certain parts of the syringe (usually near the barrel) being cooler than other sections, thus enhancing sample discrimination. Adsorption or catalytic conversion of labile substances on the syringe surface can be a problem with some compounds [37]. Adsorption losses can be minimized by adding a carrier compound of similar polarity and in large excess to the flush solvent when using packed columns. For open tubular columns deactivated fused silica needles and cold on-column injection techniques are used to minimize thermal and catalytic degradation.

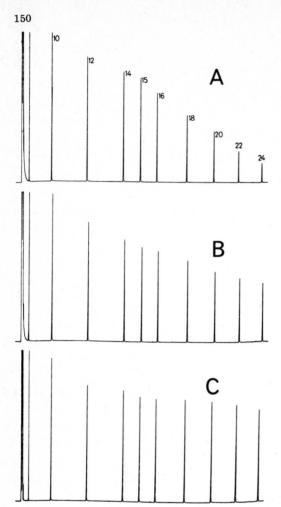

Figure 3.1   Discrimination of n-alkanes depending on injection technique.
A, filled needle; B, hot needle; C, cold on-column. A and B were
obtained using a 1:40 splitter. Sample: n-alkanes $C_{10}-C_{24}$,
1:10,000 in hexane. (Reproduced with permission from ref. 38.
Copyright Dr. Alfred Huethig Publishers).

3.3   Injection Devices for Open Tubular Capillary Columns

The limited sample capacity and the low carrier gas flow rates
characteristic of open tubular capillary gas chromatography give rise to certain
difficulties in sample introduction [39-42]. The introduction of large volume
or concentrated samples can cause column overloading, leading to a decrease in
column efficiency and/or the production of distorted peaks. The most difficult
problems arise when mixtures spanning a wide range of concentration or
volatilies must be analyzed. No single injection method will solve all of the
problems likely to be encountered and a number of different techniques are in
routine use, the most important of which are split, splitless, cold on-column,

cryogenic focussing, and falling needle injection. The sample introduction system should ideally meet all the criteria given in Table 3.2 [43], but in practice this is not usually the case and a compromise must be accepted.

TABLE 3.2

CRITERIA DESIRED IN AN IDEAL SAMPLE INTRODUCTION SYSTEM FOR OPEN TUBULAR CAPILLARY COLUMN GAS CHROMATOGRAPHY

There should be no discrimination of sample components in terms of their boiling point, polarity, or molecular weight, etc.

Thermal degradation, adsorption, rearrangements and other solute reactions should be negligible at all sample sizes.

The sampling system should introduce a negligible loss in column efficiency.

Sample recovery should be quantitative and representative for both trace and major sample components.

Changes in the column operating conditions should not influence the sampling process.

The solvent peak should not interfere with the quantitation of solute peaks.

Retention times should be reproducible to < 0.1% RSD.

Relative peak areas should be reproducible to < 1% RSD.

For many applications split injection is the most convenient sampling method, as it allows injection of mixtures virtually independent of solvent, at any column temperature with little risk of disturbing solvent effects or band broadening due to slow sample transfer from the injector to the column [39,44-48]. The apparent simplicity of the split sampling method conflicts with the many problems that arise in obtaining quantitative data for all but the simplest of mixtures. The split injector, Figure 3.2, is really a vaporizing injector in which the evaporated sample is mixed with carrier gas and divided between two streams, one exiting to the column and the other to the atmosphere. The flow through the auxiliary split line is regulated by a flow control valve. Appropriate column loads are usually achieved by injecting sample volumes of 0.1 to 2.0 microliters with split ratios between 1:10 to 1:1000; 1:20 to 1:200 being in the normal range.

The split ratio, that of column flow to split line flow, is established prior to injection. Unfortunately this preset split ratio is not representative of the sample split ratio; the latter depends in a complex way on many parameters, including the range of sample volatilities, volume of sample injected, sample solvent, syringe handling technique, injector temperature, and injector volume. For mixtures containing sample components of unequal volatility split injection discriminates against the high boiling components of

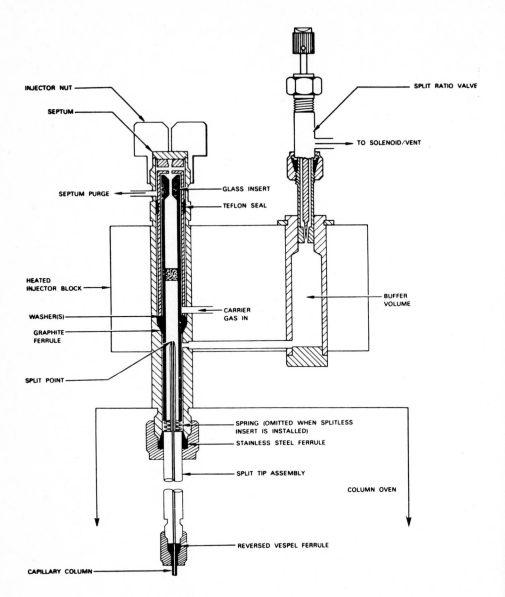

Figure 3.2 Cross-sectional view of an injection splitter assembly with a
buffer volume. (Reproduced with permission from Varian Associates).

the mixture, misrepresenting the actual sample composition. Several
interrelated processes contribute to the above observation. When the sample is
vaporized it generates an instantaneous pressure pulse, altering the flow of
carrier gas between the column and split line in a manner which is hardly
reproducible. The preset split ratio is thus altered and the portion of the
sample entering the column is different from that anticipated. However, as long

as all parts of the sample vapor plume were homogeneously mixed with carrier gas, discrimination would not occur. The length of time for sample evaporation within the injector is too short for the transfer of sufficient heat to complete the process. In most cases the sample arrives at the split point only partially evaporated as a mixture of vapor and droplets of various sizes. The sample components are likely to be unevenly distributed between the vapor and liquid phases, the latter being split to different extents, resulting in discrimination. To improve the evaporation efficiency packed precolumns containing glass wool [44,47], diatomaceous column packings [49], or baffled injector liners [50] have been used. As no dramatic improvement in sample discrimination has been demonstrated, and as a large increase in potentially catalytic and degradative surfaces is introduced, these methods are infrequently employed.

Grob has proposed an explosion-like evaporation model to further exemplify the processes occurring in a split injector [45,48]. According to this model, the amount of sample reaching the column is dependent upon the magnitude of the pressure wave, the time it takes to reach a maximum, and the column temperature. The pressure wave, caused by sample evaporation, fills the capillary column with a portion of the sample vapor followed by a period when the pressure falls back to normal. During this second period little sample enters the column and most of the sample vapor is vented through the split exit. The actual function of the preset splitter ratio is not the one expected; in actuality, it controls the sample split ratio by its influence on the magnitude of the pressure wave and the time to re-establish normal flow conditions. Column temperature influences the split ratio through a sample recondensation mechanism. Sample recondensation greatly reduces the volume of sample vapor in the cooled column inlet, creating a zone of reduced pressure that sucks in further amounts of sample vapor. This causes a decrease in the splitting ratio (i.e., increase in observed peak areas) as the column temperature is reduced and is particularly important at column temperatures near the boiling point of the solvent. At column temperatures 50–80°C below the boiling point of the solvent, recondensation is virtually complete and further decreases in the column temperature have little effect.

Although obtaining quantitative data using split injection is problematical, it is not impossible. Reproducing all aspects of the injection process and using appropriate internal standards are very important in this respect [51]. Some guidelines and further comments are given in Table 3.3.

The splitless sampling technique, which takes advantage of the solvent effect, is particularly useful for the analysis of very dilute solutions, for analyzing sample components eluting close to the tail of the solvent peak, and for the analysis of thermally-labile compounds [42,51–58]. The majority of

TABLE 3.3

FACTORS AFFECTING THE REPRODUCIBILITY OF SPLIT INJECTION

Parameter	Comments	Recommendations
Sample volume	The magnitude of the pressure wave depends on the sample size and increasing volume increases the flow into the column during the initial injection period. It may also prolong the duration of the pressure wave. Therefore, if the same amount of substance is injected in different solvent volumes, the sample peak areas may not necessarily be the same.	Reproduce the sample volume precisely for all injections.
Syringe handling	By slow movement of the plunger the pressure wave can be almost eliminated but discrimination will be very high.	Hot needle or solvent flush method. Rapid injection.
Distance between syringe needle tip and capillary inlet	Maximum amount of sample enters the column when the sample is released near the inlet of the column. This will depend on the design of the injector and the length of the syringe needle.	Reproduce penetration length of syringe needle into injector precisely.
Solvent	The molecular weight and density of the solvent influence the volume of the evaporated sample and hence the pressure wave. Solvent volatility may influence the distribution of sample between the vapor and droplet phases. The same sample amount dissolved in different solvents may produce different peak areas.	All samples should be dissolved in the same solvent for quantitative analysis.
Column temperature	Important because of the recondensation effect. Particularly important when the column temperature is near the boiling point of the principal sample component or solvent.	Reproduce starting column temperature accurately.
Standards	Internal as opposed to external standardization is recommended for quantitative analysis. If standard additions are used for calibration all the parameters listed in this table must be held constant.	

split injectors can be used in the splitless mode with minor modifications; changing the injector insert, turning off the splitter flow, and arranging for a purge flow from the base of the injector towards the top (the split line is sometimes used for this) are all that is usually required. Many commercial units, like the one shown in Figure 3.3, are designed to operate in the split/splitless mode [61]. For splitless sampling, the sample is injected into a vaporization chamber through which carrier gas is flowing. The flow rate of the carrier gas is relatively low since this is also the total flow for the column. To take advantage of the solvent effect the sample solvent must have a high boiling point with respect to the column temperature and a low boiling point with respect to the sample to be analyzed. This is critically important as the vaporized solvent is required to condense in the column inlet, forming an alternate "liquid phase". After a preselected time (rule of thumb, 1.5 times the time required for the carrier gas to sweep out the injector volume), a purge function is activated by redirecting the gas flow to the bottom of the inlet where the flow stream is divided, one portion continuing forward through the column to serve as carrier gas while the other sweeps residual volatiles from the chamber to atmosphere via a restrictor.

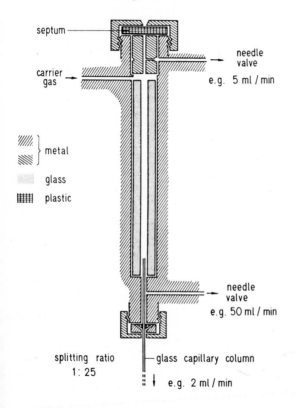

Figure 3.3   Schematic diagram of a Grob split/splitless injector. (Reproduced with permission from ref. 61. Copyright Dr. Alfred Huethig Publishers).

The splitless sample injection technique takes advantage of the solvent effect [55–61]. The condensed vaporized solvent at the column inlet behaves as a temporary thick-film stationary phase. Sample components are trapped in this layer and are later released as a sharp solute band with an extra retention time corresponding to the evaporation time of the solvent at the column inlet. The reconcentration effect of the sample band depends on the speed of the solvent evaporation from the temporary stationary phase film relative to the migration speed of the sample through this film. If the rear of the solvent band withdraws more slowly than the sample can migrate through it, then the rear of the sample band is enabled to catch up with the front. The initially wide injection band is therefore refocussed prior to separation. This refocussing is enhanced by the fact that the rear of the solvent band accelerates as it proceeds through the column. The rate of evaporation is highest at the rear of the solvent band since fresh carrier gas continuously displaces the solvent-laden gas in front of it. The solvent effect is further intensified with a ."retention gap" (a short length of uncoated column at the inlet which provides negligible retention compared to the coated section of the column). The "retention gap" accelerates the migration of the sample components and allows them to be reconcentrated at the beginning of the regular stationary phase film. The length of the retention gap zone must include the complete length of the temporary stationary phase film (approximately 60 cm for sample volumes up to 1.5 microliters).

The principal advantages of the splitless injection technique are that rapid vaporization of the sample is not required (relatively low injection temperatures can be used to minimize sample degradation), the analysis of very dilute samples is possible without preconcentration, and the injection device is easily dismantled for removing involatile sample components. Sample volumes of 0.5-1.0 microliter are usually injected, favoring its use in trace analysis. Success depends on the selection of experimental variables of which the following are the most important: sample size, sampling time, injection purge time, initial column temperature, injection temperature, and the carrier gas flow rate. Some of these parameters can be optimized only by trial and error. Some guildelines are presented in Table 3.4 but these observations should be tempered with experience gained with the injection device selected for use.

Cold on-column injection differs from the vaporization techniques discussed above in that the sample is deposited by syringe onto the column without prior heating or mixing with the carrier gas. Its advantages are the elimination of sample discrimination, excellent quantitation of individual sample components, and the elimination of sample decomposition due to thermal or catalytic effects [31,32,40,53,62-68]. A special syringe with a narrow diameter needle (e.g., 32 gauge) is required as the needle must fit inside the capillary column itself. The stainless steel or fused silica needles are very fragile and the injection device must direct the needle through a valve or septum and into the column. An example of a cold on-column injector is shown in Figure 3.4 [31,67]. A wide-bore syringe needle is used to penetrate the septum and guide the sample syringe needle into a constricted tube, ensuring alignment of the sample syringe needle with the open tubular column. The sample syringe needle enters the column to a depth of a few centimeters and the actual point of injection resides within the column oven. Some injector designs provide for secondary cooling at the injection point to refocus the sample and, in particular, to diminish the effects of sample backflushing towards the injector inlet [32,63].

TABLE 3.4

GUIDELINES FOR OPTIMIZING THE SPLITLESS INJECTION TECHNIQUE

Parameter	Comments
Solvent	Use solvents with a boiling point at least 25°C less than the most volatile sample component of interest.
Column temperature	The initial column temperature should be at least 20 to 30°C below the boiling point of the solvent.
Sample feed rate	Samples are injected slowly at about 1 microliter per second. Typical sample volumes 0.5-10 microliters. Time for injection usually less than 20 seconds.
Injector temperature	Generally lower than for split injection. Depends on the volatility of the sample and the rate of injection. A value of 200-250°C for solvents with a boiling point less than 100°C should be adequate.
Purge time	A purge flow is started to backflush excess vapors from the injector at some fixed time after injection. The delay time is determined experimentally but might be greater than 40 seconds after the start of the injection. If the purge flow is started too soon sample will be lost; too late, the solvent peak will tail into the sample. When the conditions are correct there will be no significant sample loss and the solvent peak will be rectangular.
Carrier gas	Sample transfer to the column is most effective at high carrier gas flow rates. Samples are swept from the inlet to the column more rapidly and completely and, because they are removed from the inlet more promptly, they suffer less thermal damage under these conditions.
Notes	To reproduce retention times the amount of sample injected must be identical from run to run. Solutes eluting before the solvent peak are not refocussed by the solvent effect and will be broad. Solutes that elute ahead of the solvent will not be separated efficiently or quantitatively unless there is a large difference in capacity factor values between the solutes of interest and the solvent peak.

An attractive feature of cold on-column injection is that the technique is easy to implement and can be optimized by controlling a few experimental variables. The availability of open tubular columns with immobilized phases has increased the popularity of this technique. These columns have eliminated the problem of phase stripping of the column at the inlet and also enable involatile

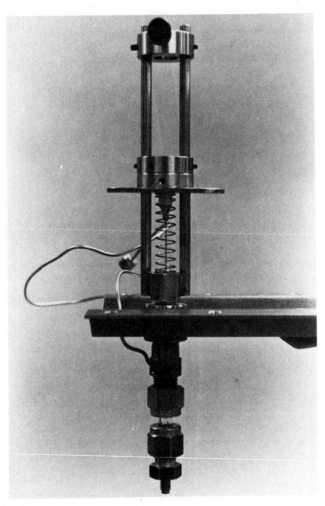

Figure 3.4 On-column injector dismounted from the gas chromatograph.

sample components to be washed out of the column by solvent backflushing.
Various studies have shown that under normal column operating conditions the
column oven temperature must be equal to or below the boiling point of the
sample solvent, the injection should be made in a continuous and rapid manner,
and that there exists a maximum in the sample volume that can be injected
(usually less than 2 microliters). If the sample volume injected is too large
or the temperature at the point of injection is incorrect, then distorted or
split-top peaks may result. Some general guidelines for cold on-column

injection are summarized in Table 3.5 [56].

TABLE 3.5

RECOMMENDATIONS FOR COLD ON-COLUMN INJECTION

The needle passage area must be cooled below the boiling point of the solvent.

If secondary cooling is not used the column oven temperature must be equal to or below the boiling point of the solvent. If secondary cooling is used, the temperature up to the point of injection should be less than the solvent boiling point and the column oven temperature slightly above the solvent boiling point (5-10°C).

Normal size injections (1-2 microliters) should be made rapidly.

After injection the column oven should be abruptly raised to operating temperature if very different from the boiling point of the solvent. Temperature programming is started from either of these two initial temperature settings depending on sample volatility.

Zlatkis has described a novel approach to the injection of large sample volumes (approximately 100 microliters) by cold on-column injection [68]. The sample is injected in the normal way, at a relatively low temperature, with the column disconnected from the detector. The column is used to strip the sample from the solvent, which is allowed to elute from the detector end of the column. The sample is then refocussed by raising the temperature of the column and condensing the sample at liquid nitrogen temperatures in a short precolumn or in a spiral made from the column itself. The sample is analyzed by installing the column in the gas chromatographic oven in the reverse position.

Besides split, splitless, and cold on-column injection several other methods of sample introduction find occasional use. Capsule-insertion methods have been used for solid sample injection [69,70]. The sample, sealed in a glass capillary or gold capsule, is dropped into the heated zone of the injector and pierced or crushed with a mechanical device. The moving-needle injector is satisfactory for the solventless injection of samples of low volatility [71-73]. The sample solution is applied to the tip of a glass plunger (needle) and the solvent is evaporated in a stream of gas. The plunger is then lowered, usually by an electromagnet, into a flash vaporization chamber and the sample is swept onto the column. Ghost peaks generated from contaminants adsorbed onto the surface of the plunger or from the gas used for solvent evaporation (charcoal prefilter recommended) are problems sometimes encountered with this injector [72]. Vogt has described a temperature programmable split/splitless injector for the analysis of samples of low volatility in large sample volumes [74]. With the split fully open, up to 250 microliters of sample solution is slowly injected into a vaporizing injector packed with glass wool and heated to a temperature above the solvent boiling point. The solvent is evaporated and vented through the split line whereupon the split valve is closed and the injector is rapidly heated to a temperature sufficient to vaporize the sample.

3.4 Detection Devices for Gas Chromatography

Numerous methods have been described for detecting organic vapors in the effluent from a gas chromatograph [75-81]. Several have passed into common use and will form the backbone of this discussion. The principal methods of detection can be grouped under four headings: ionization, bulk physical property, optical, and electrochemical detectors, according to the physical basis employed as the detection mechanism. Further division is possible based on the nature of the detector response. Detectors are broadly classified as universal, selective, or specific. These descriptions are applied loosely as no single detector exactly meets the dictionary obligations of these appellatives. The flame ionization and thermal conductivity detectors respond to the presence of nearly all organic compounds in the gas chromatographic effluent and are considered to be general or (near) universal detectors. Other detectors respond only to the presence of a particular heteroatom (e.g., the flame photometric, thermionic ionization, or emission detectors), and are considered to be specific detectors, although element-selective detectors would be more descriptive. These detectors are able to discriminate between some property of the organic compound of interest, a heteroatom, and an organic compound lacking that property. The detection process is not specific, however, as at sufficiently high concentrations a response may be obtained from a compound lacking that property being monitored. The detector response is thus selective and may be described quantitatively by a selectivity ratio, the ratio of the detector sensitivities to two different compounds, compound classes, heteroelements, etc. The response of detectors such as the electron-capture and photoionization detectors is also selective; these detectors are not element-selective but rather structure-selective. Their response range to organic compounds is broad, covering several orders of magnitude. The term "reaction detector" is occasionally applied to the electrolytic detectors, thermal energy analyzer, etc., to signify that prior to detection a chemical reaction or transformation of the organic compound into a species which can be detected is involved.

The signal from gas chromatographic detectors can be further characterized as mass or concentration dependent [82]. For concentration-dependent detectors the most notable feature is that the detector response is dependent upon the flow rate and, therefore, the sensitivity of the detector is usually defined as the product of the peak area and flow rate divided by the weight of the sample. For mass-dependent detectors sensitivity is defined as the product of the peak area divided by the sample weight in grams or moles and is independent of flow rate.

Detectors are usually compared in terms of their operational
characteristics defined by the minimum detectable quantity of standards, the
selectivity response ratio between standards of different composition or
structure, and the range of the linear portion of the detector-response
calibration curve. These terms are widely used to measure the performance of
different chromatographic detectors and are formally defined in Chapter 5 (see
particularly Table 5.6). No universal standards have been adopted for detector
characterization, which presents a problem when literature sources are used to
compare data from different detectors or for the same type of detector from
different manufacturers. Chlorpyrifos has been suggested as a suitable general
performance standard for element-selective detectors [83,84].

3.4.1 Ionization Detectors

An important property of the common carrier gases used in gas
chromatography is that they behave as perfect insulators at normal temperatures
and pressures. In the absence of conduction by the gas molecules themselves,
the increased conductivity due to the presence of very few charged molecules can
be observed, providing the high sensitivity characteristic of ionization based
gas chromatographic detectors. Examples of ionization detectors in current use
include the flame ionization detector (FID), thermionic ionization detector
(TID), photoionization detector (PID), and the electron-capture detector (ECD).
Each detector employs a different method to generate an ion current, but in all
cases the signal corresponds to the fluctuations of this ion current in the
presence of organic vapors which constitutes the quantitative basis of the
detector operation.

The flame ionization detector is as near to a universal detector as the
chromatographer has at his/her disposal. Only the fixed gases (e.g., He, Xe,
H_2, N_2), certain nitrogen oxides (e.g., N_2O, NO, etc.), compounds containing a
single carbon atom bonded to oxygen or sulfur (e.g., CO_2, CS_2, COS, etc.),
inorganic gases (e.g., NH_3, SO_2, etc.), water, and formic acid do not provide a
significant detector response. Wide applicability coupled with high
sensitivity, stability, low dead volume (approximately 1 microliter), fast
response (1 msec.), an exceptional linear response range, and simplicity of
construction and operation have made the FID the most popular detector in
current use. McWilliam has provided an interesting historic account of its
development on the occasion of the 25th anniversary of its invention (1983)
[85].

A cross-sectional view of the flame ionization detector is shown in Figure
3.5. Ions are generated by the combustion of organic compounds in a

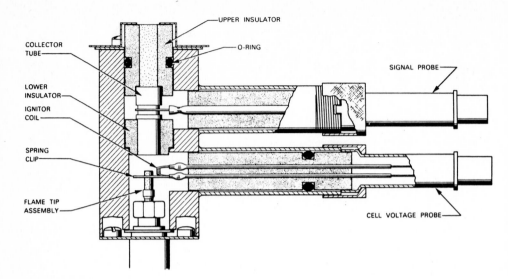

Figure 3.5 Cross-sectional view of a flame ionization detector. (Reproduced with permission from Varian Associates).

hydrogen-air diffusion flame. The carrier gas from the column is premixed with hydrogen and burned at a narrow orifice jet in a chamber through which excess air is flowing. A cylindrical collector electrode is located a few millimeters above the flame and the ion current is measured by establishing a potential between the jet tip and the collector electrode. To minimize ion recombination the potential is selected to be in the saturation region, that is, the region for which increasing the potential does not increase the ion current. Under normal conditions background currents of 2×10^{-14} to 10^{-13} Amperes, increasing to 10^{-12} to 10^{-9} Amperes in the presence of an organic vapor, are common. The small signal currents are amplified by a precision amplifier and passed to a recorder. The performance of the detector is influenced by experimental variables of which the most important are the ratio of air-to-hydrogen-to-carrier gas flow rates, the type (thermal conductance) of carrier gas, and individual detector geometry [86–88]. Methane has been suggested as a standard substance for optimizing detector operating conditions, particularly the ratio of air-to-hydrogen flow rates [89]. The optimum response plateau is usually fairly broad, permitting operation over a rather wide range of gas flow rates without incurring a large penalty in diminished response.

The mechanism of ion production in flames is complex and only poorly understood [90–93]. The thermal energy of the flame is too low to explain the production of ions and it is generally believed that organic compounds are ionized by a chemi-ionization mechanism, in which energy released in strongly exothermic chemical reactions is retained by organic molecules and leads to ionization before thermal randomization of the energy occurs. Two steps are thought to be important in this ionization process: radical formation, requiring the absence of oxygen, and chemical ionization of radicals formed by excited atomic or molecular oxygen states. At the end of a chain of reactions, in which methane or the methyl radical is thought to be a key intermediate, the dominant ion-producing reaction is shown below:

$$CH^{\cdot} + O^* \longrightarrow CHO^+ + e^- \tag{3.1}$$

As a consequence of the FID mechanism, each carbon atom capable of hydrogenation yields the same signal and the overall FID response to the analyzed substance is proportional to the sum of these "effective" carbon atoms. The FID response is the highest for hydrocarbons, being proportional to the number of carbon atoms, while substances containing nitrogen, sulfur, or halogens yield smaller responses depending upon the heteroatom-carbon character and the electron affinity of the combustion products. The reduced response can be explained by a complex series of recombination reactions and electron capture processes resulting in a lower ionization current. For a detector operating under optimum conditions the minimum detectable quantity of methane is in the range of 10^{-11} to 10^{-12} g/s. The dynamic range of the detector is 10^6 to 10^7.

Two versions of the FID which enable it to be operated as an element-selective detector have been described. The hydrogen atmosphere flame ionization detector (HAFID) can be made selective towards organometallic compounds containing, for example, aluminum, iron, tin, chromium, and lead [94,95]. The detector employs a hydrogen-oxygen flame burning in a hydrogen atmosphere doped with a reagent such as silane to improve the consistency of its response. When operated without silane doping the HAFID functions as a selective silicon detector [96]. The detector is taller in height than a conventional FID and is operated with a negative potential at the collector electrode. It has been shown that a conventional FID, operated in a hydrogen-rich mode and using oxygen to support combustion, can be made to respond as a concentration- or mass-dependent detector, as a carbon-selective detector, or as a heteroatom-selective detector simply by adjusting the hydrogen-oxygen flow rate ratios [97]. In these latter two cases the response mechanism of the detector remains unknown and the selectivity of the response has not been adequately explained.

The present generation of nitrogen and phosphorus selective thermionic ionization detectors grew out of earlier studies on the properties of alkali-metal-doped flame ionization detectors [98,99]. Adding an alkali metal salt to a flame enhanced the response of the detector to certain elements including phosphorus, nitrogen, sulfur, boron, and the halogens, as well as some

metals (e.g. Sb, As, Sn, Pb) [99,100]. The selectivity of the detector response
was dependent upon experimental variables such as flame shape, size,
temperature, and the composition of the salt tip. For this reason, the maximum
response for the selected heteroatom was achieved over a narrow range of carrier
gas, hydrogen, and air flow rates as well as salt tip composition, position, and
detector oven temperature. Accurate metering of flame gases was essential to
maintain selectivity in the region of high flame background currents in which
the alkali flame ionization detector (AFID) was operated. In this form the
detector was very noisy, the response depended critically upon the experimental
parameters, and long term stability was poor [101-103]. All these features
contributed to making the AFID difficult to use for quantitative trace
analysis.

A partial solution to the above problems was found by replacing the alkali
salt pellet with a separately heated inexhaustible alkali glass or ceramic
reservoir, reverting to a nonflame plasma for combustion, and repositioning the
collector electrode [104,105]. This detector is known as the thermionic
ionization detector (TID) or the nitrogen-phosphorus detector (NPD). Unlike the
original AFID, which responded to a wide range of heteroatoms, it is selective
for nitrogen and phosphorus only. Many commerical derivatives of the original
design by Kolb (Figure 3.6) have been described [106-110]. They differ in the
selection of the alkali source, cesium or rubidium glass/ceramic beads/
cylinders; constant current or constant temperature heating of the alkali
source; collection of negative or positive ions; availability of cold bead
operation during solvent venting, etc. We will neglect these minor operational
variables, some of which make operation more convenient, and concentrate instead
on the general principles of the TID.

The detector described by Kolb contains an electrically-heated rubidium
silicate bead situated a few millimeters above the detector jet tip and below
the collector electrode. The source is maintained at a negative potential to
prevent the loss of rubidium ions and to dampen the response of the detector in
the flame ionization mode. The temperature of the source is controlled by an
independent variable power supply to heat the bead to a dull red or orange glow
(600-800°C). A plasma is sustained in the region of the bead by hydrogen and
air support gases. The flow of hydrogen (1-5 ml/min) is too low to support a
flame in the nitrogen-phosphorus mode. A normal hydrogen flow rate
(approximately 30 ml/min) is used for the phosphorus-selective mode. To
suppress the FID signal in the flame mode the detector jet is grounded and the
negative potential of the bead deflects electrons to ground, and away from the
collector electrode. The selective response to phosphorus-containing fragments
occurs in the region of the bead and electrons generated by this reaction are
able to reach the collector electrode. In the flame mode the nitrogen response

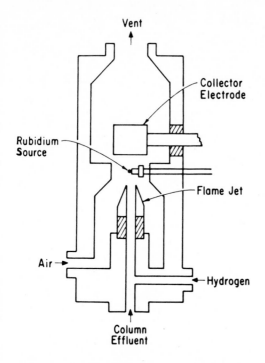

Figure 3.6   Thermionic ionization detector (Perkin-Elmer). (Reproduced with
permission from ref. 104.   Copyright Preston Publications, Inc.)

is diminished by at least an order of magnitude compared to the nonflame mode
and the detector responds mainly to phosphorus-containing compounds. Both the
selectivity and sensitivity of the detector are dependent upon experimental
variables, principally the bead heating current, jet potential, choice of
carrier gas, air and hydrogen flow rates, and the bead position [111,112].   In
particular, the detector response is influenced by surface contamination of the
bead.   Operation at a fixed bead heating current will normally give poor
long-term stability so that operation at a fixed background current is
preferred.   Optimization of the detector operating variables using the simplex
algorithm has been described [113,114].

Compared to the AFID the TID offers an order of magnitude improvement in
sensitivity and selectivity. The minimum detectable quantity of nitrogen is
about 10^{-13} g N/s and about 5×10^{-14} g P/s for phosphorus.   The response to
nitrogen-containing compounds when the nitrogen atom is not bound to a CH group
may be a good deal less than the above figure. The linear response range is
about 10^4 to 10^5 and selectivity ratios of about 4×10^4 g C/g N, 7×10^4 g C/g
P and 0.5 g N/g P may be obtained. The TID is widely used in environmental and
biomedical research for determining pesticides, drugs, and profiling, where its

high sensitivity and selectivity are useful in minimizing sample preparation. Frequent checks on calibration should be made because of the adverse effect bead contamination has on response. Injection of excess silylating reagents should be minimized for the same reason.

Several models have been proposed to account for the selectivity of the TID response to nitrogen and phosphorus. They differ principally in whether the interaction between the alkali metal atoms and organic fragments occurs as a homogeneous reaction in the gas phase or is purely a surface phenomena. According to the homogeneous mechanism theory, rubidium ions leave the bead as neutral atoms after accepting an electron from the flame [104,105,115]. While still in the vicinity of the bead, the rubidium atoms are excited and ionized by collision with plasma particles, for example:

$$Rb + 2H \longrightarrow H_2 + Rb^+ + e^- \tag{3.2}$$

The generated electrons move to the collector electrode while the positive rubidium ions return immediately to the negatively charged bead. This mechanism occurs in a cyclic fashion, producing a steady state equilibrium. If this equilibrium is disturbed, for example by a process which results in ionization of rubidium atoms, more rubidium atoms will leave the source to re-establish the equilibrium, resulting in a concomitant increase in the ion current. The selectivity of the detector owes its origin to the increase in this ion current (according to the homogeneous mechanism theory) as only those radicals with electron affinities equal to or greater than the ionization potential of rubidium atoms will contribute to the ion current. Among the many fragments generated by the pyrolysis of organic compounds in the detector plasma, this criteria is met only by the CN and PO_2 radicals.

It has proven difficult to provide any conclusive evidence for the rubidium cycle mechanism. Current thought tends to support a surface, rather than a gas phase mechanism. The surface catalytic ionization model assumes that the nitrogen compound is adsorbed onto the bead surface, pyrolyzed in situ to among other fragments the cyano radical, accepts an electron from the bead, and then departs as a cyano anion [115]. The charge-carrying species is the cyano anion in this case; the supply of electrons that supports the ionization process originates from the platinum wire used to heat the bead. The alkali glass is not a source since the rubidium atoms do not leave the glass; their role is as surface catalysts for the electron transfer reaction. An alternative surface ionization model assumes that the principal role of the alkali metal in the source is to lower the work function of the surface [110,116]. To account for the influence of experimental variables it assumes an expanded role for the hydrogen plasma compared to the surface catalytic model. The high temperature of the source is sufficient to initiate decomposition of the hydrogen and oxygen gases supplied to the detector. Consequently, in the immediate vicinity of the source a hot, chemically-reactive gas boundary layer, containing radical species, is assumed to exist. A combination of the heat from the source and the reaction of sample molecules within this boundary layer gives rise to sample decomposition. Electronegative decomposition products from phosphorus- or nitrogen-containing compounds are then ionized by extracting an electron from the surface of the thermionic source. These negative ions produce the increase in ion current measured at the collector electrode.

Photons produced by a discharge in argon, helium, or hydrogen have been used to ionize organic compounds in the carrier gas emerging from a gas chromatograph. Detectors working on this principle were called photoionization detectors, discharge detectors, or microwave emission detectors, according to whether the ionization current, the discharge current, or the intensity of the emitted light was measured. The early development of these detectors has been reviewed by Sevcik [76]. In these earlier designs the discharge and ionization compartments were not physically separated. This configuration had several drawbacks, for example, the detector and consequently the column exit had to be maintained at a reduced pressure, the partial pressure of the discharge gas varied and hence the energy distribution among the emitted photons was not constant, and metastable or excited molecules with long life times formed in the discharge space caused direct ionization of the eluted substance. Photoionization detectors of recent design eliminate the above drawbacks and make possible the use of pure photoionization processes [117,118].

Commercially available photoionization detectors use a discharge lamp containing a mixture of gases at low pressures which are excited by a high voltage placed across two electrodes positioned in the evacuated chamber. Monochromatic radiation of a specific energy is obtained by a combination of the careful choice of the filling gas mixture and the use of filters to eliminate stray radiation. Ultraviolet light sources of different energies (9.5, 10.0, 10.2, 10.9, and 11.7 eV) are available, while the 10.2 eV source is the most widely used. The discharge compartment is mechanically separated from the ionization chamber by an optically transparent window made of magnesium fluoride. The effluent from the gas chromatograph passes through the ionization chamber and between two electrodes positioned at opposite ends of the chamber. Detectors with ionization chamber volumes of 150-250 microliters are available for use with open tubular capillary columns [119-122]. An electric field is applied between the electrodes to collect the ions formed (or electrons, if preferred). The ionization chamber is capable of being heated above the column operating temperature to avoid contamination of the system by condensation of high molecular weight material.

The processes occurring within the detector can be represented by a series of equations (3.3-3.8) [123-125].

$$AB + hv \longrightarrow AB^* \tag{3.3}$$

$$AB^* \longrightarrow AB^+ + e^- \tag{3.4}$$

$$AB^+ + e^- + C \longrightarrow AB + C \tag{3.5}$$

$$AB^* + C \longrightarrow AB + C \tag{3.6}$$

$$EC + e^- \longrightarrow EC^- \tag{3.7}$$

$$EC^- + AB^+ \longrightarrow AB + EC \tag{3.8}$$

Equation (3.3) leads to reaction (3.4) representing capture of a photon by a molecule AB which leads to ionization and provides the detector signal. Equations (3.5) to (3.8) represent competing reactions involving recombination or collisional de-excitation by a carrier gas molecule C or neutralization by reaction with an electron-capturing impurity EC. These reactions should be minimized. The use of electron-capturing solvents or impurities (e.g., O_2) in the carrier gas can lead to negative peaks in the chromatogram [124]. The choice of carrier gas also influences the response of the detector through the collisional processes represented by equations (3.5) and (3.6), by its ability to influence the mobility of ions within the detector, and, in unfavorable cases, by absorption of some of the initial photon flux [123].

When competing reactions are minimized, the response of the detector can be described by equation (3.9) [123].

$$i = I F \eta \sigma N L [AB] \tag{3.9}$$

i = detector ion current
I = initial photon flux
F = Faraday constant
η = photoionization efficiency
σ = absorption cross-section
N = Avogadro's number
L = path length
[AB] = concentration of an ionizable substance

Thus for a particular detector and source, the PID signal is proportional to the ionization yield, absorption cross-section, and molar concentration. The PID is a concentration-dependent detector so its response will vary with the flow rate of the gas (carrier and make-up gas) through the detector. The product   is the photoionization cross-section, which expresses both the probability that a molecule will absorb a photon and the probability that the excited state will ionize. The calculation of photoionization cross-sections is complex but it might be intuitively expected that a direct dependence of the photoionization cross-section on the photon energy and the ionization potential of the molecule exists. In practice, a fraction of the molecules with ionization potentials up to approximately 0.4 eV above the energy of the ionizing photons will be ionized, as some of these molecules will exist in vibrationally excited states.

The PID is nondestructive, does not require an auxiliary supply of gases besides the carrier gas, is relatively inexpensive, is of rugged construction, and is easy to operate. With the 10.2 eV photon source most molecules are ionized; the exceptions are the permanent gases, C_1-C_4 hydrocarbons, methanol, acetonitrile and chloromethanes. The sensitivity for benzene is 0.3 Coulombs/g and the linear range approximately 10^7. The PID is 5 to 10 times more sensitive than the FID for alkanes and about 35 times so for aromatic compounds [126]. Freedman has shown that the ionization potential of the molecule is the most important single factor determining the PID response with the relative number of π-electrons having little significance [125]. This argument is well supported by the available experimental data although other authors claim a much greater role of π-electrons in the ionization process [126,127]. The most comprehensive collection of response data, relative to benzene, for more than a hundred compounds has been compiled by Langhorst [128]. This study led to several empirical conclusions which are summarized in Table 3.6. A general increase in sensitivity as the carbon number increases was noted and at high carbon numbers the response was attributed mainly to the carbon chain with little influence from the presence of functional groups. These observations and those in Table 3.6 generally follow the trend in ionization potentials as discussed by Freedman [125].

TABLE 3.6

RELATIONSHIP BETWEEN PHOTOIONIZATION DETECTOR RESPONSE (10.2 eV) AND MOLECULAR STRUCTURE

Sensitivity increases as the carbon number increases

Sensitivity for alkanes < alkenes < aromatics

Sensitivity for alkanes < alcohols < esters < aldehydes < ketones

Sensitivity for cyclic compounds > noncyclic compounds

Sensitivity for branched compounds > nonbranched compounds

Sensitivity for fluorine-substituted < chlorine-substituted < bromine-substituted < iodine-substituted compounds

For substituted benzenes, ring activators (electron-releasing groups) increase sensitivity and ring deactivators (electron-withdrawing groups) decrease sensitivity (exception: halogenated benzenes)

The structure-selective electron-capture detector (ECD) is the second most widely used ionization detector [129-135]. It owes much of its popularity to its unsurpassed sensitivity to a wide range of toxic and biologically active

compounds. Consequently it is widely used in trace analysis for the
determination of pesticides, herbicides, industrial chemicals in the
environment, drugs and other biologically active compounds in biological fluids,
and for determining the fate of volatile organic compounds in the upper
atmosphere. As many of these applications have an impact on commerce,
environmental quality, and health, a great deal has been written on the
properties of the ECD. However, it is not felicitous to say that the ECD is one
of the easiest to operate but least understood of the gas chromatographic
detectors in common use. Its use has also been subject to the most operator
abuse, providing both some of the most reliable and least reliable quantitative
scientific data available.

In the ECD a source of beta electrons is used to bombard the carrier gas
passing through an ionization chamber. A plasma of positive ions, radicals, and
thermal electrons is formed by a series of elastic and inelastic collisions.
The processes occurring within the detector cell can be represented
schematically as shown in Figure 3.7. Each beta electron may generate between
one hundred and one thousand thermal electrons with mean energies of 0.02 to
0.05 eV. The application of a potential difference to the electron capture
cell, either continuously or pulsed, allows the thermal electrons to be
collected. This background current, in the presence of pure carrier gas,
constitutes the detector standing current and the baseline value for all
measurements. When an electron-capturing compound enters the cell it captures a
thermal electron to produce either a negative molecular ion or a fragment ion if
dissociation accompanies capture. These ions are of larger mass than the
original electron, have a lower drift velocity, and a substantially higher rate
of recombination with positive ions. In a well-designed detector the operating
conditions are optimized such that the thermal electrons are collected but the
negative ions are not. The diminution in detector background current due to the
loss of thermal electrons constitutes the quantitative basis by which detector
response is related to solute concentration.

Commercially available electron-capture detectors use a radioisotopic
source of ionizing radiation, supported by a metallic foil; it is convenient to
use the source as one of the chamber electrodes. The ideal source would produce
a small number of ion pairs per disintegration in order for the fluctuations in
ion current and, hence, noise level to be minimal. At the same time, the total
ion pair formation should be large so that the resulting electron current during
the passage of an electron-capturing substance can be measured conveniently
without introducing other sources of noise. The best compromise among these
demands appears to be low energy beta-emitting radioisotopes (minimum number of
ion pairs per particle) at relatively high specific activities (maximum total
ion pair formation) [136]. Commerical instruments use either ^{63}Ni or ^{3}H

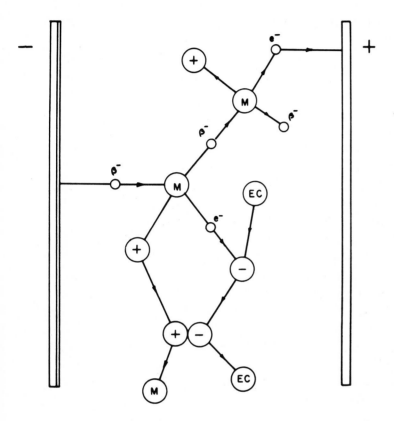

Figure 3.7 Processes occurring in an electron-capture detector. M = carrier
gas molecule and EC = electron-capturing analyte.

radioisotopic sources. Of the two types, tritium would be preferred due to its
lower energy beta emanation (0.018 MeV) compared to ^{63}Ni (0.067 MeV) and the
fact that foils with higher specific activity are less expensive to
manufacture. The principal advantage of ^{63}Ni sources is their high temperature
operation stability (to 400°C) compared to 225°C for Ti^{3}H$_2$ and 325°C for Sc^{3}H$_3$.
High temperature operation minimizes contaminiation from the chromatographic
system and enhances the response of the detector to those compounds which
capture electrons by a dissociative mechanism. Thus practical considerations
and operational convenience dictate the use of ^{63}Ni for most purposes. Other
sources of ionizing electrons include ^{55}Fe (Auger electron emitter) [138,139]
and thermionic emission from a filament housed in a separate chamber [140].

None of these alternative sources are commerically available.

Ionization chambers with parallel plate electrodes (Figure 3.8) [141–143], coaxial cylinder electrodes (Figure 3.9) [144–146], or asymmetric (e.g., coaxial displaced cylinder, Figure 3.10) [147–149] geometry have been used in commerical instruments. The parallel plate detector design was versatile and convenient for earlier experimentalists as it permitted the relative position of the source and collector electrode to be varied. It is now rarely used; most commerical detectors employ either the coaxial cylinder or asymmetric configuration. The lower specific activity of the ^{63}Ni source, compared to ^{3}H, requires the use of a larger source area to obtain the same ionization efficiency. This is more easily attained in the coaxial design. Here the minimum spacing between the source, which surrounds the centrally located collector electrode, and the collector electrode is established by the penetration depth of the beta particles [150]. This distance should be great enough to ensure that all the beta particles are deactivitated by collisions and converted to thermal energies without colliding with the anode.

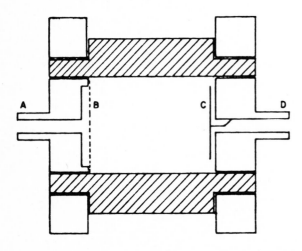

Figure 3.8 Parallel plate electron–capture detector. A, carrier gas inlet and anode; B, diffuser; C, source of ionizing radiation; D, carrier gas outlet and cathode.

By locating the anode entirely upstream from the ionized gas volume, collection of long range beta particles is minimized in the displaced coaxial cylinder design, and the direction of gas flow minimizes diffusion and convection of electrons to the collector electrode. However, the free electrons are sufficiently mobile that modest pulse voltages (e.g., 50 V) are adequate to cause the electrons to move against the gas flow and be collected during this time.

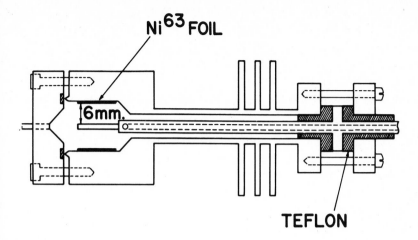

Figure 3.9   Coaxial high temperature ^{63}Ni electron–capture detector. Note
             the use of cooling fins to avoid overheating the insulation
             (Teflon) between anode and cathode (heated detector body).
             (Reproduced with permission from ref. 145. Copyright American
             Chemical Society).

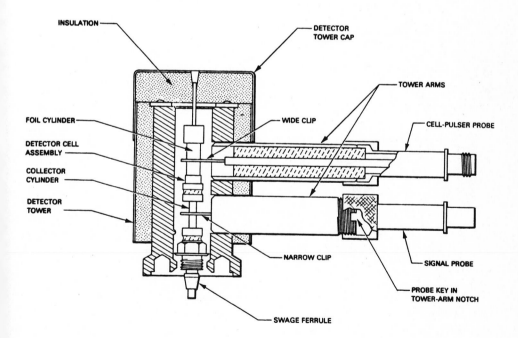

Figure 3.10 Displaced coaxial cylinder (asymmetric) electron–capture detector.
             (Reproduced with permission from Varian Associates).

Miniaturization of the ionization chamber is important for use with open tubular columns where peaks may elute in 10–100 microliter gas volumes [135]. The effective lower limit of current designs is approximately 100–400 microliters, still too large to completely eliminate extracolumn band broadening. The effective detector dead volume can be reduced by adding make-up gas at the end of the column which preserves column efficiency at the expense of some loss in detector sensitivity due to sample dilution. An equation has been proposed by De Jong which enables the flow rate of make-up gas required to maintain a certain column efficiency to be calculated, equation (3.10) [151]

$$\Phi = \frac{V}{\sigma} \left[\frac{\eta_a}{1-\eta_a} \right] \qquad (3.10)$$

Φ = necessary purge flow
V = detector volume
σ = peak width at base
η_a = actual efficiency

The actual efficiency, η_a, has a value of 1.0 for a zero dead-volume detector but a value of 0.9 is considered a reasonable, practical compromise. As some electron-capture detectors designed for packed column use have cell volumes of 2.0 to 4.0 ml, they may not be suitable for use with open tubular columns.

The thermal electron concentration in the detector cell can be measured continuously (dc voltage) or discontinuously (pulsed voltage). The dc voltage mode has several disadvantages arising from space charge effects, contact potential effects, and the promotion of non-electron-capturing ionization processes, which can result in anomalous detector operation [134,142]. Pulse sampling techniques are used in all commercial instruments to minimize the above problems. The potential is applied to the cell as a square-wave pulse of sufficient height and width to collect all the thermal electrons but of sufficient time for the concentration of thermal electrons to be replenished by the ionizing beta radiation and for their energy to reach thermal equilibrium. Additionally, in the "no field" period the electrons do not drift out of the plasma; negative ion formation occurs in the region where positive ions are also present and where recombination can most efficiently take place. The signal from an ECD, operated with a long pulse period, can be described by equation (3.11) [152].

$$\frac{(I_b - I_e)}{I_e} = K[AB] \qquad (3.11)$$

I_b = detector standing current
I_e = detector current measured at the peak height maxima
K = electron-capture coefficient
$[AB]$ = sample concentration

By analog conversion the detector output can be linearized over about four orders of magnitude [153,154]. Without analog conversion and with pulse periods less than 1 millisecond, typical operating conditions under normal circumstances, the linear response range is approximately 100 to 1000. In the dc mode shorter linear operating ranges, approximately 10 to 100, are common.

The majority of the commercially available ECD's are designed for use with a modified version of the pulsed-sampling technique termed the variable frequency constant-current mode. Rather than measuring the cell current at a constant pulse frequency, the cell current is fixed with respect to a reference value and the frequency of the pulse is modulated so that the difference between the cell current (I_{cell}) and the reference current (I_{ref}) is zero throughout the chromatographic separation. Since pulse frequency is the variable quantity in this mode of operation, the detector signal is a voltage proportional to that frequency. A block diagram of a variable frequency constant-current ECD circuit is shown in Figure 3.11. The circuit forms, in fact, a closed-loop electronic feedback network in which the cell current (I_{cell}) is combined with an external reference current (I_{ref}) such that the difference ($I_{ref} - I_{cell}$) is the input to an electrometer. The electrometer output feeds into the pulse-generating network such that the pulse frequency is determined by the magnitude of the electrometer output and the frequency of voltage pulses in turn determines the

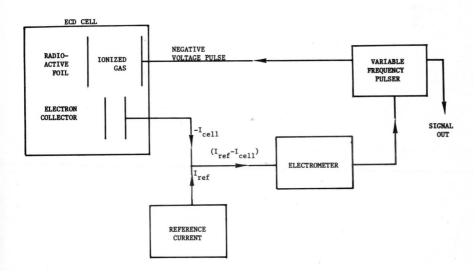

Figure 3.11   Block diagram of the electronic components of a variable frequency constant-current electron-capture detector. (Reproduced with permission from Varian Associates).

magnitude of I_{cell} (with the proviso that $I_{cell} = I_{ref}$). The two principal advantages of this method are that the linear response range is approximately 10^4 to 10^5, much greater than for the constant frequency pulse sampling mode, and secondly, the detector operation is less disturbed by traces of interferents entering the detector cell. A disadvantage of this mode of operation is that for analytes with ultrafast electron attachment rate constants (e.g., between 2.8×10^{-7} and 4.6×10^{-7} ml/molecule.s) the detector response is inherently non-linear [156-158]. Examples of compounds with ultrafast rate constants are CCl_4, SF_6, $CFCl_3$, and CH_3I. The ECD is most responsive to these types of compounds and would normally be selected for their determination. For compounds with rate constants less than these, the detector response is normally linear over the full operating range. This group includes most of the moderately strong and weak electrophores that make up the bulk of compounds determined with the ECD.

The choice of carrier gas for use with the ECD is limited to hydrogen, the noble gases, and nitrogen. Hydrogen may strip tritium from the detector at high temperatures [147]. Pure argon and helium are unsuitable as they readily form metastable ions which can transfer their excitation energy by collision with solute vapors, resulting in undesirable ionization effects (Penning reaction). The addition of 5 to 10 percent of methane to argon removes the metastable ions as quickly as they are formed (by deactivating collisions) and also serves to reduce and maintain the secondary electron energy at a constant thermal level. Argon-containing methane and oxygen-free nitrogen are the most common carrier gases used with packed columns. For open tubular capillary columns hydrogen or helium are usually used as carrier gases to maximize column efficiency and minimize analysis time while argon-methane or nitrogen are used as make-up gases. Oxygen and water vapor traps should be used to purify all gases.

The response of the ECD to organic compounds covers approximately seven orders of magnitude. The highest response is usually found among compounds having electronegative substitutents and the least among hydrocarbons and perfluoroalkanes. Some empirical observations on the molecular features contributing to a response from the ECD are summarized in Table 3.7. Some typical relative response values for organic compounds are given in Table 3.8 [130,134,135,159]. The detector responds most strongly to compounds containing halogens or nitro groups, to organometallic compounds, and to conjugated electrophores. This latter group is structurally the least well defined and is comprised of compounds containing two or more weakly electron-capturing groups connected by some specific bridge that promotes a synergistic interaction between the two groups. Examples of conjugated systems with a high detector response include conjugated carbonyl compounds (benzophenones, quinones, phthalate esters, coumarins), some polycyclic aromatic hydrocarbons, some

TABLE 3.7

MOLECULAR FEATURES GOVERNING THE RESPONSE OF THE ELECTRON-CAPTURE DETECTOR TO ORGANIC COMPOUNDS

A low response is shown by alcohols, amines, phenols, aliphatic saturated aldehydes, thioethers, fatty acid esters, hydrocarbons, aromatics and vinyl type fluorinated hydrocarbons including those containing one chlorine atom.

A high response is shown by halocarbon compounds, nitroaromatics, and conjugated compounds containing two groups which individually are not strongly electron attracting but become so when connected by specific bridges.

Compounds with a halogen atom attached to a vinyl carbon have lower responses than the corresponding saturated compounds.

Attachment of the halogen atom to an allyl carbon atom results in greater sensitivity than that obtained for the corresponding saturated compounds.

Responses towards the halogens decrease in the order I > Br > Cl > F and increase synergistically with multiple substitution on the same carbon atom.

TABLE 3.8

RELATIVE RESPONSE OF THE ELECTRON-CAPTURE DETECTOR TO ORGANIC COMPOUNDS

General Organic Compounds		Halocarbons	
Compound	Relative response	Compound	Relative response
Benzene	0.06	$CF_3CF_2CF_3$	1.0
Acetone	0.50	CF_3Cl	3.3
Di-n-butyl ether	0.60	$CF_2=CFCl$	100
Methylbutyrate	0.90	CF_3CF_2Cl	170
1-Butanol	1.00	$CF_2=CCl_2$	670
1-Chlorobutane	1.00	CF_2Cl_2	3×10^4
1,4-Dichlorobutane	15.00	$CHCl_3$	3.3×10^4
Chlorobenzene	75.00	$CHCl=CCl_2$	6.7×10^4
1,1-Dichlorobutane	111.00	CF_3Br	8.7×10^4
1-Bromobutane	280.00	$CF_2ClCFCl_2$	1.6×10^5
Bromobenzene	450.00	$CF_3CHClBr$	4.0×10^5
Chloroform	6×10^4	$CF_3CF_2CF_2I$	6.0×10^5
1-Iodobutane	9×10^4	CF_2BrCF_2Br	7.7×10^5
Carbon tetrachloride	4×10^5	$CFCl_3$	1.2×10^6

sulfonamides, and certain steroids (ecdysones, melengestrol acetate, 4-androstene-3,11,17-trione). This diverse group of electron-capturing compounds has been reviewed by Vessman [160].

The response of the ECD to halocarbon compounds decreases in the order I > Br > Cl >> F and increases synergistically with multiple substitution on the same carbon atom. The identity and number of halogen substitutents is more important than subtle variations in the geometric framework of the alkyl portion of the molecule, although even these small changes will have a measurable influence on the detector's response in many cases [161,162]. The response of the ECD to haloaromatic and nitroaromatic compounds shows similar trends to the alkyl compounds. The position of electronegative functional groups (ortho, meta, para) has a measurable influence on the detector's response but is less dramatic than the response variation due to the number and type of individual substitutents [163,164]. As the number of substitutents is increased on polychlorinated and polybrominated compounds the detector response approaches its coulometric limit and the introduction of further substituents has little effect. This law of diminishing returns is observed for all organic compounds; the introduction of the first few electronegative substituents has a large impact on detector sensitivity, but further substitution has less influence.

Much of what is known about the structure response of the detector is based on empirical observations. Clearly, the ability to correlate the response of the detector to fundamental molecular parameters would be useful. Chen and Wentworth have shown that the information required for this purpose is the electron affinity of the molecule, the rate constant for the electron attachment reaction and its activation energy, and the rate constant for the ionic recombination reaction [134,165]. In general, the direct calculation of detector response factors have rarely been carried out, as the electron affinities and rate constants for most compounds of interest are unknown.

The maximum response of the ECD to different organic compounds is markedly temperature dependent. For a single compound changes in response of 100- to 1000-fold may occur for a 100°C change in detector temperature [166,167]. This strong temperature dependence can be derived directly from the kinetic model of the ECD process [134,168]. In simple terms, an electron-capturing solute (AB) may attach an electron, forming a stable molecular ion, or the molecular ion may be formed in a sufficiently excited state to dissociate, as shown in equations (3.12) and (3.13).

$$AB + e^- \longrightarrow AB^- \qquad \text{nondissociative mechanism} \qquad (3.12)$$

$$AB + e^- \longrightarrow A^{\cdot} + B^- \qquad \text{dissociative mechanism} \qquad (3.13)$$

The favored mechanism depends on the juxtaposition of the potential energy curves for the neutral molecule and the negative ion. In the case of the nondissociative mechanism the negative molecular ion is more stable than the neutral molecule, whereas in the dissociative case the potential energy curve for the negative ion crosses that of the neutral molecule at a level corresponding to a vibrationally excited state. An increase in detector temperature would favor the populating of vibrationally excited states and thus the mechanism represented by equation (3.13). Conversely, the nondissociative mechanism is favored by low detector temperatures. The nondissociative mechanism is common among conjugated electrophores and the dissociative mechanism among halogen-containing (except fluorine) compounds. The two mechanisms can be distinguished from a plot of $\ln KT^{3/2}$ vs. $1/T$ where T is the detector temperature [134]. The calculation of K, the electron capture coefficient, is fairly involved and for routine use the peak area response for a fixed concentration of solute can be used instead. In most cases the plots will be linear with a positive slope for the nondissociative case and a negative slope for the dissociative case. In terms of maximizing the detector's response, the optimum temperature will be either the maximum operating temperature of the detector for dissociative compounds, or the lowest advisable operating temperature for nondissociative compounds. Some situations may be more complex than represented above but an optimum detector temperature in the range indicated is found easily by trial and error.

Mathematical models of the electron-capture process are based on the stirred reactor model of Lovelock [156,169] and further modifications by others [134,150,170-172]. The ionization chamber is considered to be a stirred reactor into which electrons are continuously introduced at a constant rate and electron-capturing solutes are added at a variable rate in a constant flow of carrier gas. The major consumption of cell electrons is via electron capture, recombination with positive ions, wall loss, and ventilation. The model can be expanded to allow approximately for the presence of electron-capturing contaminants and can reasonably well explain the influence of different pulse sampling conditions on the detector's response. However, an exact solution is not possible, as the importance of contaminants and the rate loss of positive ions by space charge diffusion and the corresponding magnitude of the positive ion contribution to the ECD current are unclear.

When operated in the noncoulometric mode, the response of the ECD is concentration dependent. The signal for a particular substance, when contributions from detector and carrier gas impurities can be neglected, will depend on the type and flow rate of gas through the detector, the detector operating temperature, and the magnitude of the substance specific electron-capture coefficient. The detector response is a function of the detector, the chromatographic system, and the sample itself, and response data is usually transposed into concentration terms by calibration. As the response factors cover a wide range for electron-capturing compounds, each substance to be quantified must have an individual calibration curve. Also, calibration must be repeated at frequent intervals to account for the changing conditions under which the detector is operating. A recently described calibration method uses a homologous series of alkyl bromides as calibration markers to provide both a retention index and calibration scale [173]. The molar response of the ECD to a homologous series of alkyl bromides is a constant. By adding several alkyl bromides to each sample such that the concentration of the calibration markers are related to each other in a simple, stepwise function, the area response of the bromides as a function of concentration provides an internal calibration curve. This calibration curve can be used to quantify all peaks in the chromatogram provided that the relationship between the calibration markers and each component to be quantified has been determined in an independent experiment. The alkyltrichloroacetates have also been used to provide a retention index scale for substance identification with the ECD [174].

Calibration at the extremes of the sensitivity range of the electron-capture detector is a difficult problem. If the detector could be its own absolute calibrant then the need to prepare accurate standards would be eliminated. This would require that the detector be operated in the coulometric mode, for which the time integral of the number of electrons captured during the passage of a chromatographic peak is equal, via Faraday's law, to the number of molecules ionized [175,176]. The conditions necessary for coulometry will be met only by those compounds with a high ionization efficiency and a fast rate constant for the attachment of thermal electrons. The conditions necessary for coulometric operation are difficult to achieve, can only be satisfied for a limited number of compounds, and are impractical for the analysis of multicomponent mixtures [135].

Although the ECD is normally operated with a purified carrier gas containing the minimum concentration of electron capturing impurities, it has been shown that the purposeful addition of certain electron-capturing reagents to the carrier or make-up gas provides a new and complementary method of operation. The technique is known as selective electron capture sensitization in which the detector functions as an ion-molecule reactor, using either oxygen or nitrous oxide as the reagent gas [177-179]. In the case of oxygen, it is assumed that electron attachment occurs entirely on the oxygen atom and the O_2^- ion then interacts with the analyte (AB) to generate products as shown in equations (3.14) and (3.15).

$$O_2 + e^- \longrightarrow O_2^-$$ (3.14)
$$O_2^- + AB \longrightarrow products$$ (3.15)

The instantaneous concentration of negative ions generated this way must be coupled to the electron density. The chemical sum of equations (3.14) and (3.15) is equivalent to direct electron capture. However, an ion–molecule reaction, rather than direct electron capture, determines the magnitude of this response. Both the energetics and the reaction rates are likely to be very different and, therefore, the detector would be expected to respond accordingly. This has been borne out in practice and an enhancement in sensitivity covering three or four orders of magnitude was obtained for certain weakly electron–capturing compounds such as halocarbons [134,178,180] and polycyclic aromatic hydrocarbons and their derivatives [181,182] when oxygen was added to the detector. Compounds with a high electron–capture response show only low or fractional enhancement values.

When nitrous oxide is used as a reagent gas the reactive ions formed by electron attachment are O^- and NO^-, equations (3.16)–(3.18).

$$N_2O + e^- \longrightarrow O^- + N_2 \qquad (3.16)$$
$$O^- + N_2O \longrightarrow NO^- + NO \qquad (3.17)$$
$$NO^- + N_2 \longrightarrow NO + N_2 + e^- \qquad (3.18)$$

This sequence of reactions is disrupted by any species which can remove O^- ions and a response will be observed for any analyte reacting with NO^- or O^- ions to form a stable negative ion. For a given nitrous oxide concentration and detector temperature, selectivity is determined by the reactivity of a given compound with O^- and NO^- and sensitivity by the rate constant of the reaction. The addition of a few ppm of nitrous oxide to the ECD has been shown to enhance its response to a wide range of weakly electron–capturing compounds [134,179,183].

3.4.2 Bulk Physical Property Detectors

The most important of the bulk physical property detectors, and one of the earliest routine detectors developed for gas chromatography, is the thermal conductivity detector (TCD). Also known as the hot–wire or katharometer detector, it is a universal, non–destructive, concentration–sensitive detector that responds to the difference in thermal conductivity between pure carrier gas and carrier gas containing organic vapors. The magnitude of the response will depend on the difference in thermal conductivity between the solute and carrier gas,and on the experimental variables. The TCD is generally used to detect permanent gases, light hydrocarbons, and compounds which respond only poorly to the FID. For many general applications it has been replaced by the FID, which is more sensitive (100- to 1000-fold), has a greater linear response range, and provides a more reliable signal for quantitation. Detection limits for the TCD usually fall into the range of 10^{-6} to 10^{-8} g with a response that is linear over about four orders of magnitude. For some applications the response of the TCD may be inadequate, and when the FID responds poorly to the same samples, for example, the permanent gases, then the helium ionization detector might be preferred [184-186].

In a typical TCD, the carrier gas flows through a heated thermostatted cavity that contains the sensing element, either a heated metal wire or thermistor. With pure carrier gas flowing through the cavity the heat loss from the sensor is a function of the temperature difference between the sensor and cavity and the thermal conductivity of the carrier gas. When an organic solute enters the cavity, there is a change in the thermal conductivity of the carrier gas and a resultant change in the temperature of the sensor. The sensor may be operated in a constant current, constant voltage, or constant temperature compensation circuit as part of a Wheatstone bridge network. A temperature change in the sensor results in an out-of-balance signal which is usually passed directly to a recorder without further amplification.

The TCD has appeared in several different designs, some of which have certain advantages for particular applications [75,187,188]. They usually represent some variation of the three basic geometries: the flow-through, semi-diffusion, and diffusion cells, which are shown in Figure 3.12 [189]. The diffusion cell has a slow response and is relatively insensitive; it is used mainly for preparative chromatography. The semi-diffusion cell is used for packed column analytical gas chromatography. It is less sensitive to flow variations than the flow-through cell but, with a minimum detector volume of 100 microliters (commercially available detectors have volumes from 0.01-0.1 ml), it is not suitable for use with open tubular columns unless operated at reduced pressure to overcome dead volume effects [190]. Flow-through cells with volumes of 10 to 100 microliters are easily fabricated and can be used with open tubular capillary columns [191,192]. A novel design by Hewlett-Packard employs a single-cell microvolume TCD and a microprocessor-controlled flow divider to alternately switch the reference flow and sample flow through the detector at a frequency of 10 Hz. Greater sensitivity and stability along with adequate performance is claimed for this detector when used with open tubular columns [6].

The sensing element in the TCD is usually a heated filament or thermistor. Most filaments are made from tungsten, platinum, nickel, or alloys of these with other metals such as iridium or rhenium. The desirable properties of a filament include a high temperature coefficient of resistance, relatively high electrical resistance, mechanical strength to permit forming into various shapes, a wide temperature operating range, and chemical inertness. Filaments can be coated

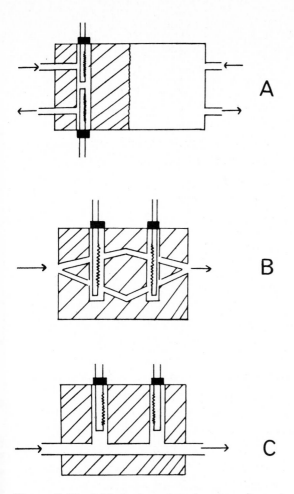

Figure 3.12 Cell designs for the thermal conductivity detector. A, flow-
through cell; B, semi-diffusion cell; C, diffusion cell.
(Reproduced with permission from ref. 189. Copyright Preston
Publications, Inc.).

with gold or PTFE to improve chemical inertness, particularly against oxidation
by air. The filament is heated by a highly-regulated power supply. The power
setting, typically in the milliampere range (200–500 mA) at about 40 V, will
affect both the sensitivity and the stability of the TCD response to different
sample amounts [193]. Thermistors can be manufactured in the very small sizes
needed to fabricate cells of low dead volume and have a much higher resistance
than filaments. They can provide greater sensitivity than filaments when
operated at low temperatures but, as they have a negative temperature
coefficient, their resistivity and hence sensitivity decline with temperature;
filament detectors are usually more sensitive above 60°C. Thermistors should
also not be used with hydrogen carrier gas to avoid reductive degradation of the

metal oxides used in their manufacture.

Temperature gradients within the detector cavity can result in poor detector performance. At the detection limit temperature changes as small as 10^{-5}°C are encountered, presenting considerable problems in cell design. It is not possible to thermostat a cell to provide the necessary absolute thermal stability to measure these small temperature changes. The sensing element must be centrally located within a detector body having a large thermal mass of controlled temperature ($\pm$ 0.01°C). A difference signal must be used by incorporating a matched reference cell in the same environment as the sample cell. Some designs use two sampling and two reference cells to improve sensitivity and stability.

In its simplest form the detector signal is measured by a Wheatstone bridge circuit containing two dummy resistances, the reference cell, and the sample cell. A change in the resistance or current drawn by the sensing element in the sample cell compared to the reference cell causes an out-of-balance potential which is fed to the recorder. The four cell configuration, now fairly common in most commercial instruments, provides better stability and does not require any external fixed resistances to complete the bridge circuit [187-189].

Carrier gases of low molecular weight and high thermal conductivity are required to maximize the response of the detector and to maintain a large linear response range. Consequently, helium and hydrogen are predominantly used. Heavier carrier gases such as nitrogen, as well as influencing sensitivity and linearity, may give rise to negative sample peaks or peaks with split tops [75,187,193].

Theoretical models have been developed to explain the operation of the TCD in the constant voltage, constant current, and constant mean temperature modes [75,187,194,195]. The fundamental relationship describing the response of the detector can be expressed by equation (3.19).

$$i^2 R_o (1 + a T) = Q_{con} + Q_{cr} + Q_{mt} \qquad (3.19)$$

i = detector current (sensor)
R_o = resistance of the sensor at 0°C
a = temperature coefficient of resistance
T = temperature of the sensor
Q_{con} = heat loss due to conduction
Q_{cr} = heat loss due to convection
Q_{mt} = heat loss due to mass transport

The mathematical solution to this model is complex and will not be dealt with here. The solutions, supported by experimental observations, indicate that there is no significant difference between constant voltage, current, or mean temperature operation on the basis of signal-to-noise ratios. The constant temperature mode will usually have a poorer signal-to-noise ratio due to a larger noise component. The constant voltage mode provides significantly better linearity than the constant current, constant temperature, or mean temperature modes.

Several compilations of relative response data for the TCD are available [75,187,189,196,197]. These values are usually expressed on a weight or molar response basis realtive to benzene. They depend on the nature of the carrier gas used for their determination but are generally sufficiently accurate for approximating sample concentrations. For precise quantitative analysis it is necessary to calibrate the detector for each substance determined. Alternatively, an internal standard can be used to determine accurate response ratios which may be subsequently used for quantitative analysis.

3.4.3 Optical Detectors

The use of flames as atom reservoirs for the spectroscopic determination of elements is a well established technique and is particularly valuable for metal analysis. Most non-metallic compounds, which account for the majority of samples analyzed by gas chromatography, have their principal emission or absorption lines in the ultraviolet region where flame background contributions are troublesome. In addition, the diffusion flames used in gas chromatography lack sufficient stability and thermal energy to be useful atom reservoirs. For direct optical emission detection, microwave induced and inductively coupled plasmas provide more appropriate atom sources for organic compounds (see Chapter 8.5). However, the determination of phosphorus and sulfur using a flame photometric detector (FPD) is widely used in gas chromatography. For these elements a hydrogen diffusion flame provides optimum excitation conditions. Since many industrial and pest control compounds contain sulfur or phosphorus, the FPD is widely used in environmental analysis. The thermal energy analyzer (TEA) is based on the photometric detection principle.

The flame photometric detector (FPD) uses a hydrogen diffusion flame to first decompose and then excite to a higher electronic state the fragments generated by the combustion of sulfur- and phosphorus-containing compounds in the effluent from a gas chromatograph. These excited molecules subsequently return to the ground state, emitting characteristic band spectra. This emission is monitored by a photomultiplier tube through either a 392 nm bandpass filter for sulfur or a 526 nm bandpass filter for phosphorus. It has been shown that by suitable changes in flame conditions the FPD will respond to elements and groups other than phosphorus and sulfur; these include nitrogen and the halogens

[198], boron [199], chromium [200], tin [201-203], selenium [203,205], tellurium [205], and germanium [203,206]. The sensitivity for the above elements is generally quite good (2 x 10^{-12} g Se/s, 1 x 10^{-14} g Ge/s, and 5 x 10^{-16} g Sn/s). Selenium exhibits properties similar to sulfur whereas tin and germanium are primarily surface-activated emitters, requiring the construction of special burners to maximize their signal. However, these applications do not represent the normal use of the detector, which is generally considered to be selective for phosphorus and sulfur [75,81,132,133,207].

In the most common detector design the carrier gas and air or oxygen are mixed, conveyed to the center orifice of a flame tip, and combusted in an atmosphere of hydrogen. With this burner and flow configuration interferring emissions from hydrocarbons occur mainly in the oxygen-rich flame regions close to the burner orifice, whereas sulfur and phosphorus emissions occur in the diffuse hydrogen-rich upper portions of the flame. To enhance the selectivity of the detector an opaque shield surrounds the base of the flame, preventing hydrocarbon emissions from reaching the photomultiplier viewing region. The extent and intensity of the various emitting regions of the flame are dependent upon the burner design and the gas flow rates. For any given burner design, the response of the FPD is critically dependent upon the ratio of hydrogen to air or oxygen flow rates, the type and flow rate of carrier gas, and the temperature of the detector block [208-211]. Different optimum conditions are usually required for sulfur and phosphorus detection, for detectors with different burner designs, and perhaps also for different compound classes determined with the same detector. The sulfur response may decrease substantially at high carrier gas flow rates, more so with nitrogen than helium, but less response variation has been noted for phosphorus. As is obvious from the above comments, careful optimization of detector flow rates is required and these optimum values should be determined on a detector-to-detector basis and, in some cases, on a compound-to-compound basis also.

Several variations of or modifications to the basic FPD design have been described since it first came into general use. Detector noise, generated by thermal radiation from the flame, may be reduced by water cooling [212] or by locating the photomultiplier tube away from the flame and transmitting the flame luminescence to it via an optical light pipe [213,214]. Dual channel FPD detectors are available for monitoring sulfur and phosphorus emissions simultaneously [215]. The injection of solvent volumes greater than a few microliters can momentarily starve the flame of oxygen and extinguish it. Solvent flameout can be solved by arranging for automatic flame reignition, using a dual-flame burner, or a oxygen-hyperventilated flame [216,217]. To reduce the hydrocarbon quenching effect in FPD's (discussed later) a new construction based on a combination of a pyrolysis furnace/single-flame burner has been described [218].

Problems with solvent flameout, hydrocarbon quenching, and structure–response variations for different sulfur– and phosphorus–containing compounds can be partially solved by using a dual–flame burner [149,219–223]. A schematic diagram of a dual–flame photometric detector is shown in Figure 3.13. The lower flame is used to decompose the incoming sample, the combustion products from which are swept into a second longitudinally–separated flame where the desired optical emission is generated. The small detector volume, 170 microliters, and the high carrier and combustion gas flow rates in the passageways between the burners provide a very small effective detector dead volume, suitable for use with open tubular capillary columns [149]. Although sensitivity for sulfur is generally higher for a single flame burner, the sample response is more likely to be dependent upon molecular structure and affected by the presence of co–eluting hydrocarbons.

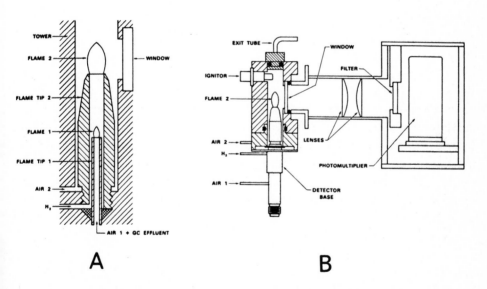

Figure 3.13   Dual–flame photometric detector (Varian). A, Schematic diagram of the dual–flame burner; B, Schematic diagram showing the relationship between the burner and the photometric viewing components. (Reproduced with permission from ref. 219. Copyright American Chemical Society).

The reponse mechanism of the FPD is known superficially, even if some of the finer details are based on surmisary evidence [133,219]. It is generally believed that incoming sulfur–containing compounds are irreversibly decomposed to yield sulfur atoms either directly or via a H_2S intermediate, which then reacts with hydrogen atoms as shown in equations (3.20) and (3.21).

$$H_2S + H \longrightarrow HS + H_2 \tag{3.20}$$

$$HS + H \longrightarrow S + H_2 \tag{3.21}$$

Excited S_2^* molecules are then formed by some mechanism in the cool, outer regions of the flame. The most probable reactions are represented by equations (3.22) to (3.25).

$$S + S \longrightarrow S_2^* \tag{3.22}$$

$$H + H + S_2 \longrightarrow S_2^* + H_2 \tag{3.23}$$

$$H + OH + S_2 \longrightarrow S_2^* + H_2O \tag{3.24}$$

$$S + S + M \longrightarrow S_2^* + M \qquad (M = gas\ molecule) \tag{3.25}$$

The emitting species is the excited S_2^* molecule and a square root dependence of the detector response on the amount of sample entering the flame is therefore expected. In the case of phosphorus, incoming phosphorus- containing compounds are first decomposed to PO molecules which are then converted into HPO^* by either of the reactions shown in equations (3.26) and (3.27).

$$PO + H + M \longrightarrow HPO^* + M \tag{3.26}$$

$$PO + OH + H_2 \longrightarrow HPO^* + H_2O \tag{3.27}$$

A linear dependence between detector response and amount of sample entering the detector is expected for phosphorus and is generally found. Deviations from the predicted detector response are more common with sulfur-containing than phosphorus-containing compounds [209,223,224]. The detector response in the sulfur mode can be described by equation (3.28). Based on the previous discussion, a value of 2 is expected for the exponential factor.

$$I_{S_2} = A\ [S]^n \tag{3.28}$$

I_{S_2} = signal intensity

A = experimental constant
[S] = mass flow rate of sulfur atoms
n = exponential factor

The value of n for a particular compound can be determined from the slope of a plot of log(response) vs. log(injected mass of sulfur) or by utilizing equation (3.29) [225].

$$n = \left[\frac{\text{peak width at base determined by FID}}{\text{peak width at base determined by FPD}} \right]^2 \tag{3.29}$$

In general, reported values for n are between 1.5 and 2.0. Non-optimized flame conditions, incomplete combustion, hydrocarbon quenching, competing flame reactions that lead to de-excitation, and sample structural effects all contribute to this deviation. As deviations from the expected value are known to occur, the use of amplifiers with a linearization function should only be used when the value of n is known to be two [226,227]. It should also be noted that when measuring column performance parameters with the sulfur FPD, the equations derived for a linear detector should be modified by incorporating the exponential factor into the denominator [225]. When n = 2, the response at one-quarter peak height yields a measure of peak width analogous to that obtained from a linear detector peak measured at half height. This discrepancy arises because the rate of elution into a detector with a square law response reaches half its maximum at one-quarter peak height. Since the aforementioned equations describing column performance contain n in the denominator, the chromatographic peaks for a non-linear detector (n > 1) will appear narrower than those obtained with a linear detector.

Minimum detection levels and selectivity values for the FPD depend on the operating conditions of the detector, burner geometry, and photomultiplier sensitivity. Typical detection limits are 5×10^{-13} g P/s and 5 to 50×10^{-12} g S/s. The linear range for phosphorus usually exceeds 1000 while the selectivity is more than 5×10^{5} g C/g P. Sulfur selectivity depends upon the amount of sulfur present; it varies from 10^{3} g C/g S at low sulfur amounts to greater than 10^{6} g C/g S at high sulfur amounts. It is not always possible to take full advantage of the detector selectivity because of quenching, particularly of the sulfur response. The co-elution of hydrocarbon compounds can quench the response of the detector due to a temporary change in flame conditions or due to de-exciting collisions or reactions in the flame-emitting zone [221,228]. The mechanism in not clearly understood but the effect, up to 80% in magnitude, has been widely observed. Therefore, when analyzing complex mixtures or quantifying compounds eluting on the tail of the solvent peak, caution is required. Quenching is less of a problem with the dual-flame detector. A retention index scale with symmetrical n-dialkyl sulfides as the fixed points has also been suggested for use with the sulfur FPD [229].

The thermal energy analyzer (TEA) is a selective detector for the determination of nitrosamines [230-232]. Low temperture catalytic pyrolysis (275-300°C) cleaves off the nitrosyl radical which is swept through a cold trap, used to condense interferring ·organic volatiles, and into an evacuated chamber (via a capillary restrictor) through which ozone is continuously bled. The nitrosyl radical reacts with ozone to form electronically-excited nitrogen dioxide, which decays to its ground state with emission in the 600 nm spectral region. The high specificity of the detector is due to the relatively low pyrolysis temperature and the selectivity of the chemiluminescent reaction for nitric oxide. Detection limits in the region of 0.5 ng for the lower alkylnitrosamines, a linear range of more than 1000, and selectivities greater than 1000 over most amines and nitro compounds have been reported. The method is not completely free of interferences, however, as has been demonstrated by several studies [232-234]. Both positive and negative response deviations have been observed from a number of unrelated and unexpected sources. The detector can be made sensitive to amines and nitro-containing compounds by employing a metal oxide catalyst for sample pyrolysis [235].

3.4.4 Electrochemical Detectors

There are two general problems in the application of electrochemical detection to gas chromatography. First of all, few electrochemical detectors are gas phase sensing devices and the sample must therefore be transferred into solution for detection. Secondly, the majority of organic compounds separated by gas chromatography are neither electrochemically active nor highly conducting. The Hall electrolytic conductivity detector and the microcoulometric detector solve both of these problems by decomposing the gas phase sample into low molecular weight electrochemically-active fragments that are readily soluble in a support solvent.

Sample decomposition is carried out either by pyrolysis or by catalytic oxidation or reduction in a low-volume flow-through tube furnace. For oxidation or reduction oxygen, air, or hydrogen is mixed with the carrier gas leaving the column and passed over a catalyst, usually a nickel or quartz tube with a nickel wire inside, maintained at a temperature of 500 to 1000°C. Organic compounds entering the furnace are decomposed into small molecular weight fragments, Table 3.9. To avoid a build-up of carbon deposits, the solvent is usually vented away from the furnace. A chemical scrubber is sometimes added to the flow system at the furnace exit to improve the selectivity of the detection process. Examples of chemical scrubbers include strontium hydroxide or potassium carbonate to selectively remove HX or SO_x, silver wire to remove HX or H_2S, alumina to remove PH_3, and aluminum silicate to remove SO_x. With careful selection of pyrolysis conditions, mode of operation, and chemical scrubber, electrochemical detection permits highly specific and sensitive element detection.

The most widely used element-selective electrochemical detector is the Hall electrolytic conductivity detector (HECD) [109,133,236]. This is an improved version of an earlier design by Coulson [237,238]. In both detectors the reaction products are swept from the furnace into a gas-liquid contactor where they are mixed with an appropriate solvent. The liquid phase is separated from

TABLE 3.9

PRODUCTS GENERATED BY THE CATALYTIC PYROLYSIS OF ORGANIC COMPOUNDS

Elemental Composition of Sample	Products Generated	
	Oxidative Mode	Reductive Mode
Carbon	CO_2	CH_4
Hydrogen	H_2O	H_2O
Halogen (X)	HX	HX
Nitrogen	NO_2 (low yield)	NH_3
Sulfur	SO_2/SO_3	H_2S
Phosphorus	P_4O_{10}	PH_3

insoluble gases in a gas-liquid separator and then passed through a conductivity cell. The Coulson detector employed a syphon arrangement for the gas-liquid separator, which had a large dead volume and would occasionally loose prime during an analysis [237-241]. The HECD employs a small-volume, concentric cylinder cell for mixing, separating, and monitoring the concentration of conducting species, Figure 3.14. The solvent and gaseous reaction products are combined in a small PTFE tee. The heterogeneous gas-liquid mixture thus formed separates into two smooth flowing phases upon contact with the inner surface of the stainless steel outer electrode. The driving forces of the cell are the downward force of the moving liquid phase, the attraction between the liquid phase and the detector surfaces, and the positive pressure of the liquid phase in the detector. Hall also improved the method of measuring sample conductance by employing an AC conductivity bridge in earlier designs [109,242] and, more recently, a bipolar, pulsed, differential conductance circuit and cell design [133,243,244]. The bipolar, pulsed, differential detector employs a series conductivity cell (Figure 3.14) in which the conductivity of the solvent is monitored in the first portion of the cell while the conductivity of the solvent plus that of the reaction products is measured in the second part of the cell. The output of the two conductivity cells are differentially summed. Changes in the concentration of conductivity solvent and temperature fluctuations, which represent the principal source of daily response variations, are thus minimized [245]. Water was employed as the conductivity solvent by most early researchers but its use is now far less common. Organic solvents such as methanol and isopropanol, either alone or in admixture with water, are in common use today. These solvents provide higher sensitivity, greater selectivity for the species to be measured, and a greater linear response range than pure water. Most detectors recirculate the conductivity solvent through a closed system; a bed of

ion-exchange resins purifies and conditions the replenished solvent.
Conditioning may involve pH modification to improve the sensitivity and
selectivity for a particular element. It has also been suggested that solutions
of dilute electroytes can improve the selectivity of the detector without use of
chemical scrubbers [239,242]. Here the detector monitors the decrease in
conductance when a highly conducting species such as H_3^+ is replaced by a
species of much lower conductance, for example, NH_4^+, in the determination of
nitrogen as ammonia.

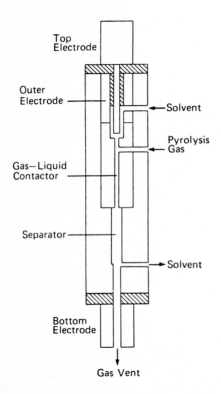

Figure 3.14   Schematic diagram of a Hall electrolytic conductivity detector
bipolar pulsed differential cell.

The practical problems most frequently encountered in operating the HECD
are poor linearity and poor peak shape. Poor linearity is usually caused by
neutralization of the conducting species when the pH of the conductivity solvent
is incorrect and/or the scrubber is exhausted. Neutralization problems are
readily recongnized by a sharp dip in the baseline just prior to the peak, and a
negative dip after the peak that gradually increases to the baseline. Peak
tailing is often due to a contaminated scrubber, contaminated transfer line from
the furnace to the cell, deactivated catalyst, or the presence of interfering,

conducting reaction products that are not removed by the scrubber. The sensitivity of the detector is also influenced by the absolute flow rate and the flow rate stability of the conductivity solvent. As it is a concentration-sensitive detector, a greater response is anticipated at lower solvent flow rates; this increase will, however, be accompanied by an increase in noise, resulting in fluctuations in the output of the solvent delivery system. Practical operating conditions are usually a compromise, in which the constancy and pulse-free nature of the solvent delivery system are extremely important.

The HECD is capable of high sensitivity and selectivity, although optimizing detector conditions and maintaining constant sensitivity at low sample levels can be troublesome. Detection limits on the order of 10^{-12} g N/s, 5×10^{-13} g Cl/s, and 10^{-12} g S/s can be obtained with a linear response range of 10^3 to 10^5. Detector selectivity values are variable, depending on the heteroelement and the detector operating conditions, but values in the range of 10^4 to 10^9 for nitrogen-, chlorine-, or sulfur- containing compounds in the presence of hydrocarbons have been obtained. Compared to the FPD in the sulfur mode, the HECD has about the same or greater sensitivity and selectivity [243,244]. In comparison with the ECD, the HECD has a greater selectivity for chlorine-containing compounds and a more predictable reponse on a per gram of chlorine basis.

The microcoulometer detector can be used to detect the same active species (e.g., SO_2, H_2S, HCl, and NH_3) as the HECD. It employs a four electrode electrochemical cell: two for generating the active species, one for sensing, and one as a reference electrode. A low concentration of a chemically active ion (e.g., Ag^+, H^+, or I_3^-) is generated in the cell and, when a reactive species enters the cell, this concentration falls. This concentration decrease is sensed by the working electrode and the resultant signal is fed via an amplifier to the generating electrodes, which then pass a current through the electrolyte until the original equilibrium ion concentration is restored. The energy consumed in regenerating the titrant ion is monitored with a precision resistor. This energy consumption is displayed as a voltage-time interval, integration of which reveals the total number of coulombs consumed in the reaction. The microcoulometer functions as an absolute detector since the total number of coulombs consumed is related to the microequivalents of sample entering the cell by Faraday's law. The microcoulometer has a sensitivity of approximately 5×10^{-11} g/s of heteroelement, a linear range of about 100, and a selectivity ratio over hydrocarbons of about 10^6. However, the detector has a rather large dead volume and a very slow response time (ca. 20 s). Therefore, it is primarily used with packed columns for the analysis of relatively simple mixtures [246-248].

3.5 Column Effluent Splitters for Parallel Detection

The column effluent may be split between two or more detectors operated in
parallel to enhance the information content of the chromatogram, or to generate
substance–characteristic detector response ratios, which aid compound
identification. Ideally, an effluent splitter should provide a fixed split
ratio that is independent of flow rate, temperature, and sample volatility. It
should also minimize extracolumn band broadening, be chemically inert, and, for
convenience, provide some mechanism for adjusting the split ratio over a
reasonable range, e.g. 1:1 to 1:100, and be easy to clean [249,250]. For packed
columns, T–splitters with fixed or value–adjustable split ratios are generally
used [251–253]. Dead volume effects are not usually a problem at packed column
flow rates, although changes in the split ratio with temperature and the
constancy of the split ratio with samples of different volatility may be. At
the low flow rates used with open tubular capillary columns, dead volume effects
can become significant unless the total volume of the splitter is reduced to the
absolute minimum and all passageways are cleanly swept out with carrier and/or
make–up gas. Effluent splitters for open tubular columns have been fabricated
from glass–lined stainless steel [249,254–256], platinum/iridium microtubing
[250,257–260], glass capillary tubing [261–265], and fused silica capillary
tubing [264]. These may be in the form of simple T- or Y–splitters for dual
detector operation or more complex roughs and stars [254,258,260,263] for
multiple detection and column switching, etc. Many of these devices can be
constructed in the laboratory without special tools, although glassblowing and
welding of capillary tubing requires practice and patience. Glass and silica
are chemically inert; platinum/iridium tubing is relatively inert, although some
catalytic activity remains even after tempering at red heat in a stream of
oxygen [250,257,265]. As effluent splitters are constructed with fixed–diameter
flow passages, the split ratio is varied by inserting wires or glass fibers of
different diameters into one of the flow paths. To maintain a constant split
ratio over a wide temperature range, as might be required during temperature
programming, it is necessary to separately thermostat the splitter from the
column oven, for example, by locating it in the oven detector block. One
commercially available effluent splitter avoids this problem by employing a
unique mixing and flow–stabilization chamber design [249]. The chromatogram in
Figure 3.15, obtained with this splitter, illustrates the increased information
content obtained by simultaneously recording the separation with two detectors
[135]. Figure 3.16 demonstrates the advantage of using a selective detector in
conjunction with a universal detector. The FID recording illustrates the
complexity of the sample while the TID records only those components that
contain nitrogen [114]. Two or more element–selective detectors can be used in

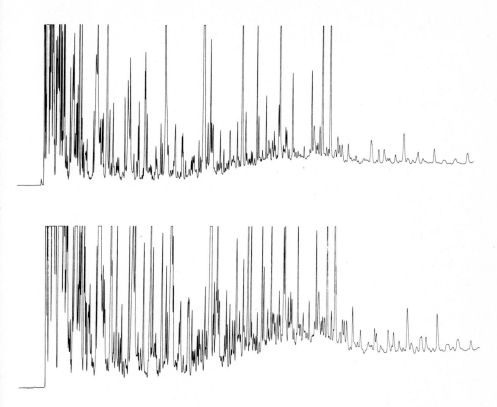

Figure 3.15   Separation of the organic volatile fraction from normal human
serum. Top, ECD and bottom, FID obtained with a 1:1 split.
(Reproduced with permission from ref 135. Copyright Dr. Alfred
Huethig Publishers).

parallel to provide heteroelement ratios sufficiently characteristic to confirm
the identity of selected peaks in a complex mixture [255,266]. Such an approach
was used to confirm the identity of polycyclic aromatic hydrocarbons by their
ECD/FID response ratios [267].

The series coupling of detectors does not require an effluent splitter.
This approach may be feasible when a non-destructive detector such as the ECD or
PID is used ahead of an FID, TID, HECD, etc. This technique has been
demonstrated in a few instances for ECD/FID [268,269] and PID/ECD detection
[139].

3.6 Multidimensional Gas Chromatography

Multidimensional gas chromatography involves the separation of a sample by
employing columns of different capacity or selectivity in series [270-272].
Some of the typical problems that arise in the analysis of complex mixtures

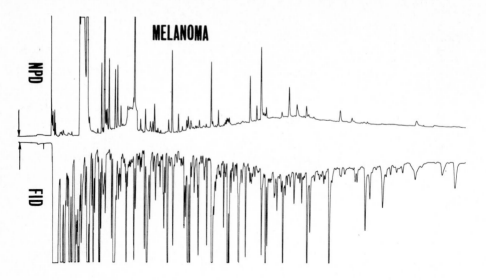

Figure 3.16 Separation of the organic volatile fraction of a sreum sample from
a melanoma patient. Top, nitrogen-phosphorus detector (TID) and
bottom, FID. (Reproduced with permission from ref. 114.
Copyright Dr. Alfred Huethig Publishers).

containing components with wide ranges of concentrations, volatilities, or
polarities can be overcome more effectively by multidimensional chromatography
than any other optimized gas chromatographic technique. Principal applications
include the separation of complex mixtures and trace enrichment of dilute
samples employing such techniques as heartcutting, backflushing, peak switching,
and intermediate trapping [273,274]. Some typical examples are given in Table
3.10. Instrumental demands vary from the simple modification of a standard gas
chromatograph to some of the most complex microprocessor-controlled gas
chromatographic equipment presently available; the application determines the
degree of sophistication required.

     Before describing some typical instrument networks in detail, some critical
instrument components deserve special attention and will be treated first. The
quantitative transfer of the effluent from one column to the next must occur
under conditions that do not degrade the resolving power of the second column.
Three transfer mechanisms are noteworthy: Deans' switches, microvolume
multifunctional valves, and cryogenic trapping with reinjection. Pneumatic
switching is based on the balance of flows at different positions in the
chromatographic system using in-line restrictors and precision pressure
regulators [277,278]. The direction of flow between columns can be changed by
opening and closing the valves located external to the column oven. Thus, no
moving parts are located inside the oven and dead volume effects are not

TABLE 3.10

APPLICATIONS FOR WHICH MULTIDIMENSIONAL GAS CHROMATOGRAPHY IS SUITABLE

Application	Comments
Solvent removal	Large amounts of solvents and excess derivatizing reagents, etc. can be excluded from the main separation column. Prevents column deterioration and improves the performance of sensitive selective detectors.
Enrichment of trace components	This can be achieved by multiple injections into a precolumn with selective trapping and storage of the fraction of interest. Often performed with packed precolumns because of their high sample capacity and because maximum resolution is not required in the pre-separation step.
Separation of trace components buried under major peaks	It is possible to vent from the system the major portion of large peaks and transfer the trace components of interest into the trap.
Heartcutting of a limited section at the beginning of the chromatogram	In a sample having a very wide boiling point range, only the components of interest are transferred into the main separation column. With a dual oven instrument high boiling components can be backflushed at an elevated temperature, thus maintaining a high sample throughput.
Switching of single peaks or selected cuts throughout a chromatographic run	Single peaks or whole areas of a chromatogram can be switched to a second column to provide better separation. With intermediate trapping and columns of different selectivity two sets of retention index data can be obtained.
Adjustment of optimal phase polarity without changing columns	Retention times of substances are individually determined on both phases and the optimal conditions (temperature or relative flow rates) are calculated by computer [275,276].

important since there are no unswept volumes. Switching between columns is controlled by pressure so that widely varying flows, as might exist in interfacing a packed column with an open tubular capillary column, are easily handled. The only critical components are the pressure controllers, which must remain reproducible under all conditions. Problems may arise from the back diffusion of sample into the switching lines, causing memory effects. In principle, Deans' switching is deceptively simple and is thus widely used, but implementation of user-built systems often requires more time than originally anticipated.

The simplest way of switching effluent from one column into another is by a mechanical valve [119,279,280]. However, technical constraints of the switching process, valve dimensions, and available materials limit this approach. Switching valves must be chemically inert, free from outgassing products, gas tight at all temperatures, have small internal dimensions, and operate without lubricants. Modern miniaturized multifunctional valves meet most of these requirements to varying extents and may be preferred to Deans' switches for less demanding chromatographic problems. Multifunctional valves are often used with cryogenic focussing when the volume of column effluent to be transferred requires a fairly lengthy transfer time.

An intermediate cryogenic trap is essential whenever a band refocussing mechanism is required as part of the switching process [278,279,281-285]. Examples include sample transfer from packed to capillary columns, preconcentration by multiple injections, and accurate determinations of retention index values on the second column. The design of the intermediate trap is of paramount importance when open tubular capillary columns are used. It is necessary to provide a temperature gradient within the trap, otherwise breakthrough is observed, due to microfog formation. A trap should not only effectively retain the substances that are directed into it but should also be of low mass so that it can be rapidly heated to introduce the trapped fraction into the second column instantaneously. The trap is usually cooled by circulating liquid nitrogen through a shroud surrounding the trap. The trap is often a short length of platinum/iridium tubing or a spiral of capillary tubing, and is heated either by forced convection with preheated nitrogen or, more generally, by resistive heating using a high amperage circuit [285-287]. For hot capillary column injection the cryogenic trap must be heated in about 20 milliseconds to retain full column efficiency. Such rapid heating is best obtained by resistive heating.

The simplest type of two-dimensional gas chromatograph for heartcutting or trace enrichment using a packed precolumn and a capillary column is shown in Figure 3.17 [279,284,286]. Almost any modern, single-oven gas chromatograph could be converted into an instrument of this kind by the addition of a few auxiliary components. It is thus a convenient design for laboratories using multidimensional chromatographic techniques on only an occasional basis and for laboratories lacking a specially-designed instrument for this purpose. Preliminary separation takes place on the packed column. The effluent from this column is directed either to a vent or to the capillary inlet by the Deans' switch. The effluent reaching the capillary inlet is split three ways. One portion passes to a detector used to monitor the preseparation, a second portion enters the capillary column and is reconcentrated in the cold trap located outside the oven, and the remaining portion passes out the split vent. Provision is made for backflushing heavy sample components from the precolumn by placing a tee leading to a second solenoid valve between the packed column and the injection port. A liquid nitrogen Dewar is used to manually cool the trap and resistive heating provides rapid reinjection onto the capillary column. An application using this instrument is shown in Figure 3.18 for the separation of the organic volatiles in a wine headspace sample by heartcutting [286].

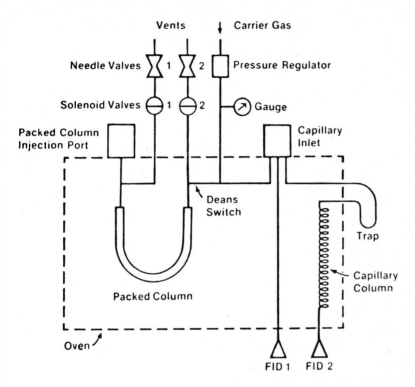

Figure 3.17   A simple two–dimensional gas chromatograph for packed column fractionation or enrichment using a Deans' switch and intermediate trap for transfer to a capillary column. (Reproduced with permission from ref. 286. Copyright Dr. Alfred Huethig Publishers).

A more sophisticated system, employing two capillary columns housed in separate ovens and connected by an intermediate trap, is shown in Figure 3.19 [273]. Complete automation by computer control and simultaneous detection with parallel detectors is possible.

For some purposes "live switching" can be used to switch column effluent between capillary columns without intermediate trapping, Figure 3.20 [285,287–295]. The critical component of the system is the double T-piece with tube connections for the adjustment of pressures and a platinum/iridium capillary for the connection of the two capillary columns. The double T-piece works with slight pressure differences between its two ends (A,B). The flows through these make–up gas lines are adjusted with two needle valves, NV2 and NV3. The pressure difference, ΔP, is indicated by the manometer. The T-piece is constructed in such a way that any gas flowing from one end to the other must

200

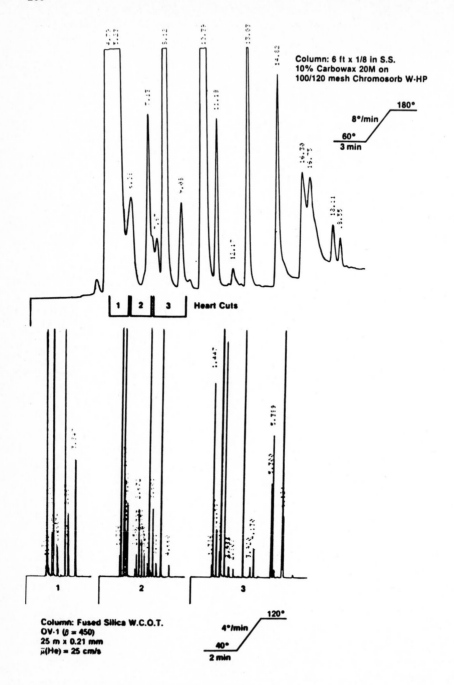

Figure 3.18   Illustration of the use of heartcutting. The top chromatogram is
a packed column separation of wine volatiles. The three labelled
fractions are switched to a capillary column for separation,
bottom chromatograms. (Reproduced with permission from ref. 286.
Copyright Dr. Alfred Huethig Publishers).

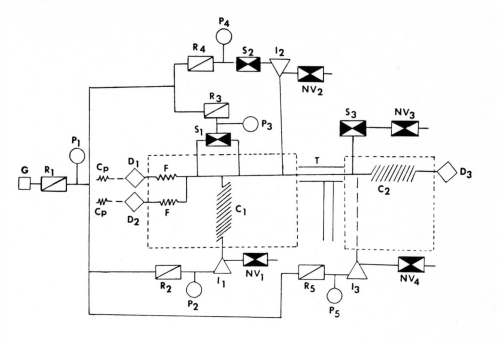

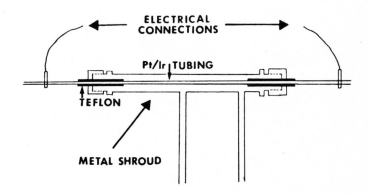

Figure 3.19   Sophisticated dual oven gas chromatograph for multidimensional gas chromatography with an expanded view of the intermediate trap. G = gas supply, I = injectors, D = detectors, T = trap, R = regulators, P = pressure gauges, S = solenoid valves, NV = needle valves, and F = restrictors. (Reproduced with permission from ref. 273. Copyright Dr. Alfred Huethig Publishers).

pass through the tiny platinum/iridium capillary, which is inserted loosely into both the first and second column. Effluent gas flow from the first column can be directed to a detector via restrictor R1 or through the second column and then to a detector. Moreover, the first column can be backflushed by activating solenoid valves SV1 and SV2.

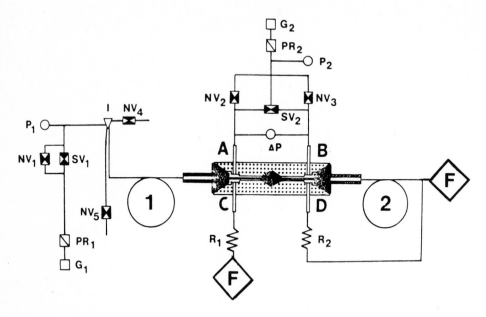

Figure 3.20   Schematic diagram of a dual oven two-dimensional gas chromatograph employing live switching between two capillary columns (Siemens). G = gas supply, I = split injector, PR = pressure regulator, P = pressure difference gauge, SV = solenoid valve, NV = needle valve, R = restrictors, and F = detectors. (Reproduced with permission from ref. 293. Copyright Dr. Alfred Huethig Publishers).

Huber has shown that the accuracy and precision of the analytical data obtained by "live switching" depends primarily on three factors: the technical characteristics of the switching device, the size of the fraction switched, and the position of the fraction with respect to overlapping matrix components [292]. Quantitative sample transfer occurs when the switch time is equal to about six times the peak standard deviation, measured symmetrically around the peak centrum. For analytical peaks that are overlapped by matrix components, a compromise must be made between the size and position of the fraction switched. If overlap is to be avoided on the second column then quantitative sample transfer may not be possible.

3.7   Temperature and Flow Programming

In isothermal gas chromatography the relationship between retention time and boiling point is logarithmic for compounds of similar chemical type, resulting in a rapid increase in retention time and peak width even for samples with narrow boiling ranges. This is commonly referred to as "the general elution problem" and is characterized by poor separation of early eluting peaks and poor detectability of late eluting peaks due to band broadening. The problem can be solved by either temperature or flow programming. Isothermal, temperature program, and flow program modes are compared in Figure 3.21 for the separation of a mixture of hydrocarbons covering a wide boiling point range [296]. Only with the programmed modes is a complete separation obtained in a reasonable time. The programmed modes tend to oppose the logarthmic dependence of retention on boiling point and, under certain optimized conditions, the relationship becomes approximately linear, with only a slow increase in peak width with increasing temperature and flow rate.

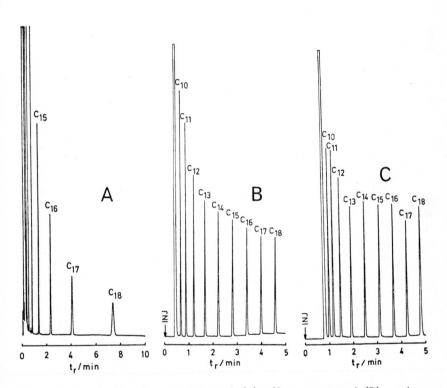

Figure 3.21   Comparison of isothermal (A), flow programmed (B), and temperature programmed (C) separation of C_{10} to C_{18} alkanes. (Reproduced with permission from ref. 296. Copyright Dr. Alfred Huethig Publishers).

The temperature program mode is the most widely used separation technique in gas chromatography [297]. A temperature program consists of a series of changes in the oven temperature and includes isothermal and controlled temperature rise segments that are selected by a mechanical or microprocessor controller. In essence, most programs are simple, consisting of an initial isothermal period, a linear temperature rise segment, a final isothermal period maintained at the temperature reached at the end of the rise segment, and a cool down period to return the oven to the starting temperature. The initial and final isothermal peroids are optional, the temperature rise segment can be selected over a broad range (0.1 to ca. 30°C/min), non-linear changes in temperature may be used (extremely rare), and for complex mixtures, several linear programs may be used in sequence to optimize the separation. The program segments are generally selected by trial and error, for although an adequate theory exists to describe the temperature program process, its implementation usually requires the input of sample data which is often unknown [297-299]. The initial oven temperature is selected with due consideration to the resolution of the earliest eluting peaks of the chromatogram. If the temperature chosen is too high the resolution of the initial peaks may be inadequate and, if it is too low, resolution may be acceptable but the analysis time will be extended needlessly. The rate of temperature rise is a compromise between the desire for a slow change to maximize resolution and a rapid change to minimize analysis time. The complexity of the sample controls this parameter. Ideally, unless the last few eluting peaks are particularly difficult to separate and require an isothermal period, elution of the last peak should coincide with the termination of the temperature rise. Peaks eluting after completion of the temperature rise will be wider than those eluted during the program.

The only practical limitations placed on the range of the temperature rise segment are the thermal stability of the stationary phase and flow changes. With high temperature stationary phases or immobilized phases the accessible upper temperature ranges have been dramatically improved. For constant-pressure-controlled instruments the flow rate will decrease with temperature, affecting the response of concentration-dependent detectors. For this reason flow-controlled instruments are preferred for temperature programmed operation. Employing identical dual columns and simultaneous detection in the difference mode, the rapid baseline rise associated with flow and stationary phase bleed problems can be highly attenuated.

Flow programming is not widely used in gas chromatography [296,300,301]. Many of its advantages tend to parallel those of temperature programmed gas chromatography, which is experimentally easier to perform. To obtain results similar to those of temperature programming an exponential increase in flow is required, limiting the dynamic range of the technique, particularly for packed columns. For short, open tubular capillary columns a dynamic operating range of approximately 0.1 to 10 ml/min provides a reasonable working range. The principal advantages of flow programming are that it shortens the analysis time for mixtures of wide volatility, while permitting the operation of the column at lower temperatures; as the analysis is performed at lower temperatures column bleed is less of a problem, and since the pressure of the gas in the column can be changed instantaneously, returning the instrument to the start conditions is very rapid. Disadvantages include a decrease in column efficiency for late eluting peaks, difficulty in calibration of flow sensitive detectors, and inconvenience of instrument operation at high pressure. Compared to temperature programming it might be preferred for the analysis of thermally labile samples.

3.8 References

1. J. R. Conder and C. L. Young, "Physicochemical Measurement by Gas Chromatography", Wiley, New York, 1979.
2. A. Zlatkis and L. S. Ettre (Eds.), "The Practice of Gas Chromatography", Wiley, New York, 1967.
3. H. M. McNair, J. Chromatogr. Sci., 16 (1978) 578.
4. F. L. Bayer, J. Chromatogr. Sci., 20 (1982) 393.
5. S. M. Sonchik, J. Chromatogr. Sci., 20 (1982) 402.
6. G. Schomburg, J. Chromatogr. Sci., 21 (1983) 97.
7. K. Grob, J. High Resolut. Chromatogr. Chromatogr. Commun., 1 (1978) 173.
8. V. Pretorius, J. High Resolut. Chromatogr. Chromatogr. Commun., 2 (1979) 186.
9. J. W. Bozzelli and S. C. Chuang, J. Chromatogr. Sci., 21 (1983) 226.
10. V. M. Oslavicky, J. Chromatogr. Sci., 16 (1978) 197.
11. V. Pretorius, J. High Resolut. Chromatogr. Chromatogr. Commun., 1 (1978) 227.
12. D. M. Ottenstein and P. H. Silvis, J. Chromatogr. Sci., 17 (1979) 389.
13. J. E. Purcell, H. D. Downs, and L. S. Ettre, Chromatographia, 8 (1975) 605.
14. J. A. Jonsson, J. Vejrosta, and J. Novak, J. Chromatogr. 236 (1982) 307.
15. S. A. Smirnova, A. G. Vitenberg, and B. V. Stolyarov, J. Chromatogr., 170 (1979) 419.
16. D. P. C. Fung and M. W. Channing, J. Chromatogr. Sci., 20 (1982) 188.
17. M. Ravey, J. Chromatogr. Sci., 19 (1981) 325.
18. J. C. Sternberg, Adv. Chromatogr. 2 (1966) 205.
19. S. P. Cram and T. H. Glenn, J. Chromatogr., 112 (1975) 329.
20. V. Maynard and E. Grushka, Anal. Chem., 44 (1972) 1427.
21. S. P. Cram, S. N., Chesler, and A. C. Brown, J. Chromatogr., 126 (1976) 279.
22. J. R. Conder, J. High Resolut. Chromatogr. Chromatogr. Commun., 5 (1982) 397.
23. W. G. Jennings, "Gas Chromatography with Glass Capillary Columns", Academic Press, New York, 2nd Ed., 1980.
24. R. F. Severson, R. F. Arrendale, and O. T. Chortyk, J. High Resolut. Chromatogr. Chromatogr. Commun., 3 (1980) 11.
25. H. T. Badings, C. De Jong, and J. G. Wassink, J. High Resolut. Chromatogr. Chromatogr. Commun., 4 (1981) 644.
26. H. J. Spencer, J. Chromatogr., 260 (1983) 164.
27. P. Baeckstrom and M. Lindstrom, J. High Resolut. Chromatogr. Chromatogr. Commun., 2 (1979) 576.
28. J. W. Hines, M. D. Erickson, D. L. Newton, M. A. Mosely, L. M. Retziaff, and E. D. Pellizzari, J. High Resolut. Chromatogr. Chromatogr. Commun., 5 (1982) 52.
29. F. Rinderknecht and B. Wenger, J. High Resolut. Chromatogr. Chromatogr. Commun., 4 (1981) 295.
30. T. L. Peters, T. J. Nestrick, and L. L. Lamparski, Anal. Chem., 54 (1982) 1893.
31. F.-S. Wang, H. Shanfield, and A. Zlatkis, Anal. Chem., 54 (1982) 1886.

206

32. M. Galli, S. Trestianu, and K. Grob, J. High Resolut. Chromatogr. Chromatogr. Commun., 2 (1979) 366.
33. K. Grob and H. P. Neukom, J. High Resolut. Chromatogr. Chromatogr. Commun., 2 (1979) 15.
34. K. Grob and H. P. Neukom, J. Chromatogr., 198 (1980) 64.
35. K. Grob and S. Rennhard, J. High Resolut. Chromatogr. Chromatogr. Commun., 3 (1980) 627.
36. J. Chauhan and A. Darbre, J. High Resolut. Chromatogr. Chromatogr. Commun., 4 (1981) 260.
37. H. Brotell, J. Chromatogr., 196 (1980) 489.
38. K. Grob and G. Grob, J. High Resolut. Chromatogr. Chromatogr. Commun., 2 (1979) 109.
39. G. Schomburg, H. Husmann, and F. Schulz, J. High Resolut. Chromatogr. Chromatogr. Commun., 5 (1982) 565.
40. K. Grob, J. High Resolut. Chromatogr. Chromatogr. Commun., 2 (1978) 263.
41. V. Pretorius and W. Bertsch, J. High Resolut. Chromatogr. Chromatogr. Commun., 6 (1983) 64.
42. V. Pretorius, K. Lawson, and W. Bertsch, J. High Resolut. Chromatogr. Chromatogr. Commun., 6 (1983) 185.
43. F. Y. Yang, A. C. Brown, and S. P. Cram, J. Chromatogr., 158 (1978) 91.
44. G. Schomburg, H. Behlau, R. Dielmann, F. Weeke, and H. Husmann, J. Chromatogr., 142 (1977) 87.
45. K. Grob and H. P. Neuknom, J. High Resolut. Chromatogr. Chromatogr. Commun., 2 (1979) 563.
46. J. Eyem, J. Chromatogr., 217 (1981) 99.
47. K. Grob, H. P. Neukom, and P. Hilling, J. High Resolut. Chromatogr. Chromatogr. Commun., 4 (1981) 203.
48. K. Grob and H. P. Neukom, J. Chromatogr., 236 (1982) 297.
49. A. L. German and E. C. Horning, Anal. Letts., 5 (1972) 619.
50. W. G. Jennings, J. Chromatogr. Sci., 13 (1975) 185.
51. D. L. Springer, D. W. Phelps, and R. E. Schirmer, J. High Resolut. Chromatogr. Chromatogr. Commun., 4 (1981) 638.
52. F. Poy, S. Visani, and F. Terrosi, J. Chromatogr., 217 (1981) 81.
53. K. Grob, J. Chromatogr., 213 (1981) 3.
54. K. Grob, J. Chromatogr., 237 (1982) 15.
55. K. Grob and B. Schilling, J. Chromatogr., 260 (1983) 265.
56. R. Jenkins and W. G. Jennings, J. High Resolut. Chromatogr. Chromatogr. Commun., 6 (1983) 228.
57. V. Pretorius, C. S. G. Phillips, and W. Bertsch, J. High Resolut. Chromatogr. Chromatogr. Commun., 6 (1983) 232.
58. W. G. Jennings, "Sample Preparation for Gas Chromatographic Analysis", Huthig, Heidelberg, 1983, p. 5.
59. W. G. Jennings, R. R. Freeman, and T. A. Rooney, J. High Resolut. Chromatogr. Chromatogr. Commun., 1 (1978) 275.
60. R. J. Miller and W. G. Jennings, J. High Resolut. Chromatogr. Chromatogr. Commun., 2 (1979) 72.
61. K. Grob and K. Grob, J. High Resolut. Chromatogr. Chromatogr. Commun., 1 (1978) 57.
62. K. Grob and P. Neukom, J. Chromatogr., 189 (1980) 109.
63. M. Galli and S. Trestianu, J. Chromatogr., 203 (1981) 193.
64. G. Schomburg, H. Husmann, and R. Rittmann, J. Chromatogr., 204 (1981) 85.
65. C. Watanabe, H. Tomita, K. Sato, Y. Masada, and K. Hashimoto, J. High Resolut. Chromatogr. Chromatogr. Commun., 5 (1982) 630.
66. F.-S. Wang, H. Shanfield, and A. Zlatkis, J. High Resolut. Chromatogr. Chromatogr. Commun., 5 (1982) 562.
67. F. Pacholec and C. F. Poole, Chromatographia, 18 (1984) in press.
68. A. Zlatkis, F.-S. Wang, and H. Shanfield, Anal. Chem., 54 (1982) 2406.
69. H. U. Buser and H. M. Widmer, J. High Resolut. Chromatogr. Chromatogr. Commun., 2 (1979) 177.

70. E. D. Morgan, R. P. Evershed, and R. C. Tyler, J. Chromatogr., 186 (1979) 555.
71. P. M. J. vanden Berg and Th. P. H. Cox, Chromatographia, 5 (1972) 301.
72. E. Evrard, C. Razzouk, M. Roberfroid, M. Mercier, and M. Bal, J. Chromatogr., 161 (1978) 97.
73. M. Verzele, G. Redant, S. Qureshi, and P. Sandra, J. Chromatogr., 199 (1980) 105.
74. W. Vogt, K. Jacob, A.-B. Ohnesorge, and H. W. Obwexer, J. Chromatogr., 186 (1979) 194.
75. D. J. David, "Gas Chromatographic Detectors", Wiley, New York, 1974.
76. J. Sevcik, "Detectors in Gas Chromatography", Elsevier, Amsterdam, 1976.
77. E. R. Adlard, CRC Crit. Revs. Anal. Chem., 5 (1975) 1.
78. E. R. Adlard, CRC Crit. Revs. Anal. Chem., 5 (1975) 13.
79. D. W. Grant, in "Developments in Chromatography", C. E. H. Knapman (Ed.), Applied Science Publishers, London, Vol. 1, 1978, p. 117.
80. L. S. Ettre, J. Chromatogr. Sci., 16 (1978) 396.
81. P. T. Holland and R. C. Greenhalgh, in "Analysis of Pesticide Residues", H. A. Moye (Ed.), Wiley, New York, 1981, p. 51.
82. J. Sevcik and J. E. Lips, Chromatographia, 12 (1979) 693.
83. R. Greenhalgh and W. P. Cochrane, J. Chromatogr., 188 (1980) 305.
84. W. P. Cochrane, R. B. Maybury, and R. Greenhalgh, J. Environ. Sci. Hlth., B14 (1979) 197.
85. I. I. G. McWilliam, Chromatographia, 17 (1983) 241.
86. A. T. Blades, J. Chromatogr. Sci., 11 (1973) 251.
87. B. A. Schaefer, J. Chromatogr. Sci., 16 (1978) 211.
88. K. Grob, J. High Resolut. Chromatogr. Chromatogr. Commun., 3 (1980) 286.
89. M. M. Thomason, W. Bertsch, P. Apps, and V. Pretorius, J. High Resolut. Chromatogr. Chromatogr. Commun., 5 (1982) 690.
90. P. Bocek and J. Janek, J. Chromatogr., 15 (1971) 111.
91. J. Sevcik and M. Klima, Chromatographia, 9 (1976) 69.
92. J. Sevcik, R. E. Kaiser, and R. Reider, J. Chromatogr., 126 (1976) 263.
93. A. J. C. Nicholson, J. Chem. Soc. Fard. Trans 1, 78 (1982) 2183.
94. J. E. Roberts and H. H. Hill, J. Chromatogr., 176 (1979) 1.
95. J. H. Wagner, C. H. Little, M. D. Dupuis, and H. H. Hill, Anal. Chem., 52 (1980) 1614.
96. M. A. Osman, H. H. Hill, M. W. Holdren, and H. H. Westberg, Anal. Chem., 51 (1979) 1286.
97. A. E. Karagozler, C. F. Simpson, T. A. Gough, and M. A. Pringuer, J. Chromatogr., 158 (1978) 139.
98. A. Karman, Adv. Chromatogr., 2 (1966) 293.
99. V. V. Brazhnikov, M. V. Gur'ev, and K. I. Sakodynsky, J. Chromatogr., 12 (1970) 1.
100. R. Greenhalgh and P. J. Wood, J. Chromatogr., 82 (1973) 410.
101. J. Sevcik, Chromatographia, 6 (1973) 139.
102. R. F. Coward and P. Smith, J. Chromatogr., 61 (1971) 329.
103. J. F. Palframan, J. McNab, and N. T. Crosby, J. Chromatogr., 76 (1973) 307.
104. B. Kolb and J. Bishoff, J. Chromatogr. Sci., 12 (1974) 625.
105. B. Kolb, M. Auer, and P. Pospisil, J. Chromatogr. Sci., 15 (1977) 53.
106. C. A. Burgett, D. H. Smith, and H. B. Bente, J. Chromatogr., 134 (1977) 65.
107. P. L. Patterson, R. Howe, V. Hornung, and C. H. Hartman, 167 (1978) 381.
108. P. L. Patterson and R. L. Howe, J. Chromatogr. Sci., 16 (1978) 275.
109. R. C. Hall, CRC Crit. Revs. Anal. Chem., 8 (1978) 323.
110. P. L. Patterson, R. A. Gatten, and C. Ontiveros, J. Chromatogr. Sci., 20 (1982) 97.
111. M. J. Hartigan, J. E. Purcell, M. Novotny, M. L. McConnell, and M. L. Lee, J. Chromatogr., 99 (1974) 339.
112. J. A. Lubkowitz, J. L. Glajch, B. P. Semonian, and L. B. Rogers, J. Chromatogr., 133 (1977) 37.

208

113. I. B. Rubin and C. K. Bayne, Anal. Chem., 51 (1979) 541.
114. F. Hsu, J. Anderson, and A. Zlatkis, J. High Resolut. Chromatogr. Chromatogr. Commun., 3 (1980) 648.
115. K. Olah, A. Szoke, and Zs. Vajta, J. Chromatogr. Sci., 17 (1979) 497.
116. P. L. Patteron, J. Chromatogr., 167 (1978) 275.
117. J. Sevcik and S. Krysl, Chromatographia, 6 (1973) 375.
118. J. N. Driscol, J. Chromatogr., 134 (1977) 49.
119. W. G. Jennings, S. G. Wyllie, and S. Alves, Chromatographia, 10 (1977) 426.
120. L. F. Jaramillo and J. N. Driscoll, J. High Resolut. Chromatogr. Chromatogr. Commun., 2 (1979) 536.
121. S. Kapila and C. R. Vogt, J. High Resolut. Chromatogr. Chromatogr. Commun., 4 (1981) 233.
122. J. N. Driscoll, J. Chromatogr. Sci., 20 (1982) 91.
123. A. N. Freedman, J. Chromatogr., 190 (1980) 263.
124. G. I. Senum, J. Chromatogr., 205 (1981) 413.
125. A. N. Freedman, J. Chromatogr., 236 (1982) 11.
126. J. N. Driscoll, J. Ford, L. F. Jaramillo, and E. T. Gruber, J. Chromatogr., 158 (1978) 171.
127. M. K. Casida and K. C. Casida, J. Chromatogr., 200 (1980) 35.
128. M. L. Langhorst, J. Chromatogr. Sci., 19 (1981) 98.
129. W. A. Aue and S. Kapila, J. Chromatogr. Sci., 11 (1973) 225.
130. E. D. Pellizzari, J. Chromatogr., 98 (1974) 323.
131. A. Zlatkis and D. C. Fenimore, Rev. Anal. Chem., 2 (1975) 317.
132. S. O. Farwell and R. A. Rasmussen, J. Chromatogr. Sci., 14 (1976) 224.
133. S. O. Farwell, D. R. Gage, and R. A. Kagel, J. Chromatogr. Sci., 19 (1981) 358.
134. A. Zlatkis and C. F. Poole (Eds.), "Electron Capture. Theory and Practice in Chromatography", Elsevier, Amsterdam, 1981.
135. C. F. Poole, J. High Resolut. Chromatogr. Chromatogr. Commun., 5 (1982) 454.
136. G. R. Shoemake, D. C. Fenimore, and A. Zlatkis, J. Gas Chromatogr., 3 (1965) 285.
137. D. J. Dwight, E. A. Lorch, and J. E. Lovelock, J. Chromatogr., 116 (1976) 257.
138. W. E. Wentworth, A. Tishbee, C. F. Batten, and A. Zlatkis, J. Chromatogr., 112 (1975) 229.
139. S. Kapila, D. J. Bornhop, S. E. Manahan, and G. L. Nickell, J. Chromatogr., 259 (1983) 205.
140. A. Neukermans, W. Kruger, and D. McManigill, J. Chromatogr., 235 (1982) 1.
141. J. E. Lovelock, Anal. Chem., 33 (1961) 162.
142. J. E. Lovelock, Anal. Chem., 35 (1963) 474.
143. P. Devaux and G. Guiochon, Chromatographia, 2 (1969) 151.
144. P. G. Simmonds, D. C. Fenimore, B. C. Pettit, J. E. Lovelock and A. Zlatkis, Anal. Chem., 39 (1967) 1428.
145. D. C. Fenimore, P. R. Loy, and A. Zlatkis, Anal. Chem., 39 (1971) 1972.
146. E. D. Pellizzari, J. Chromatogr., 92 (1974) 299.
147. C. H. Hartmann, Anal. Chem., 45 (1973) 733.
148. P. L. Patterson, J. Chromatogr., 134 (1977) 25.
149. F. J. Yang and S. P. Cram, J. High Resolut. Chromatogr. Chromatogr. Commun., 2 (1979) 487.
150. J. Connor, J. Chromatogr., 210 (1981) 193.
151. A. P. J. M. de Jong, J. High Resolut. Chromatogr. Chromatogr. Commun., 4 (1982) 213.
152. W. E. Wentworth, E. Chen, and J. E. Lovelock, J. Phys. Chem., 70 (1966) 445.
153. D. C. Fenimore, A. Zlatkis, and W. E. Wentworth, Anal. Chem., 40 (1968) 1594.
154. D. C. Fenimore and C. M. Davis, J. Chromatogr. Sci., 8 (1970) 519.

155. R. J. Maggs, P. L. Joynes, A. J. Davies., and J. E. Lovelock, Anal. Chem., 43 (1971) 1966.
156. J. E. Lovelock and A. J. Watson, J. Chromatogr., 158 (1978) 123.
157. J. J. Sullivan and C. A. Burgett, Chromatographia, 8 (1975) 176.
158. E. P. Grimsrud and W. B. Knighton, Anal. Chem., 54 (1982) 565.
159. C. F. Poole, Chem. & Ind. (London), (1976) 479.
160. J. Vessman, J. Chromatogr., 184 (1980) 313.
161. C. A. Clemons and A. P. Altshuller, Anal. Chem., 38 (1966) 133.
162. D. A. Miller and E. P. Grimsrud, Anal. Chem., 51 (1979) 851.
163. W. L. Zielinski, L. Fishbein, and R. O. Thomas, J. Chromatogr., 30 (1967) 77.
164. Y. Hattori, Y. Kuge, S. Nakagawa, Bull. Chem. Soc. Jpn., 51 (1978) 2249.
165. E. C. M. Chen and W. E. Wentworth, J. Chromatogr., 217 (1981) 151.
166. E. C. M. CHen and W. E. Wentworth, J. Chromatogr., 68 (1972) 302.
167. C. F. Poole, J. Chromatogr., 118 (1976) 280.
168. W. E. Wentworth and E. Chen, J. Gas Chromatogr., 5 (1967) 170.
169. J. E. Lovelock, J. Chromatogr., 99 (1974) 3.
170. J. Connor, J. Chromatogr., 200 (1980) 15.
171. J. Rosiek, I. Sliwka, and J. Lasa, J. Chromatogr., 137 (1977) 245.
172. P. L. Golby, E. P. Grimsrud, and S. W. Warden, Anal. Chem., 52 (1980) 473.
173. F. Pacholec and C. F. Poole, Anal. Chem., 54 (1982) 1019.
174. K. Ballschmiter and M. Zell, Z. Anal. Chem., 293 (1978) 193.
175. J. E. Lovelock, J. R. Maggs, and E. R. Adlard, Anal. Chem., 43 (1971) 1962.
176. E. P. Grimsrud and S. W. Warden, Anal. Chem., 52 (1980) 1842.
177. P. G. Simmonds, J. Chromatogr., 166 (1978) 593.
178. E. P. Grimsrud and R. G. Stebbins, J. Chromatogr., 155 (1978) 19.
179. M. P. Phillips, R. E. Sievers, P. D. Golden, W. C. Kuster, and F. C. Fehnsenfeld, Anal. Chem., 51 (1979) 1819.
180. E. P. Grimsrud and D. A. Miller, Anal. Chem., 50 (1978) 1141.
181. D. A. Miller, K. Skogerboe, and E. P. Grimsrud, Anal. Chem., 53 (1981) 464.
182. J. A. Campbell, E. P. Grimsrud, and L. R. Hageman, Anal. Chem., 55 (1983) 1335.
183. M. A. Wizner, S. Singhawangcha, R. M. Barkley, and R. E. Sievers, J. Chromatogr., 239 (1982) 145.
184. R. S. Brazell, F. F. Andrawes, and E. K. Gibson, J. High Resolut. Chromatogr. Chromatogr. Commun., 4 (1981) 188.
185. F. F. Andrawes, T. B. Byers, and E. K. Gibson, Anal. Chem., 53 (1981) 1544.
186. F. F. Andrawes, R. S. Brazell, and E. K. Gibson, Anal. Chem., 52 (1980) 89.
187. A. E. Lawson and J. M. Miller, J. Gas Chromatogr., 4 (1966) 273.
188. J. Johns and A. L. Stapp, J. Chromatogr. Sci., 11 (1973) 234.
189. D. M. Rosie and E. F. Barry, J. Chromatogr. Sci., 11 (1973) 237.
190. M. G. Proske, M. Bender, H. Schirrmeister, and G. Bottcher, Chromatographia, 11 (1978) 715.
191. P. E. Pecsar, R. B. DeLew, and K. R. Iwao, Anal. Chem., 45 (1973) 2191.
192. C. H. Lochmuller, B. M. Gordon, A. E. Lawson, and R. J. Mathieu, J. Chromatogr. Sci., 16 (1978) 523.
193. C. Roy, G. R. Bellemare, and E. Chornet, J. Chromatogr., 197 (1980) 121.
194. R. T. Witterbroad, Chromatographia, 5 (1972) 454.
195. G. Wells and R. Simon, J. Chromatogr., 256 (1983) 1.
196. D. M. Rosie and R. L. Grob, Anal. Chem., 29 (1957) 1263.
197. J. W. Carson, G. Lege, and R. Gilbertson, J. Chromatogr. Sci., 16 (1978) 507.
198. J. Sevcik, Chromatogrpahia, 4 (1971) 195.
199. E. J. Sowinski and I. H. Suffet, J. Chromatogr. Sci., 9 (1971) 632.
200. R. Ross and T. Shafik, J. Chromatogr. Sci., 11 (1973) 46.

201. W. A. Aue and C. G. Flinn, J. Chromatogr., 142 (1977) 145.
202. W. A. Aue and C. G. Flinn, Anal. Chem., 52 (1980) 1537.
203. C. G. Flinn and W. A. Aue, J. Chromatogr. Sci., 18 (1980) 136.
204. C. G. Flinn and W. A. Aue, J. Chromatogr., 153 (1978) 49.
205. W. A. Aue and C. G. Flinn, J. Chromatogr., 158 (1978) 161.
206. C. G. Flinn and W. A. Aue, J. Chromatogr., 186 (1979) 136.
207. M. L. Selucky, Chromatographia, 4 (1971) 425.
208. T. J. Cardwell and P. J. Marriott, J.Chromatogr. Sci., 20 (1982) 83.
209. M. Maruyama and M. Kakemoto, J. Chromatogr. Sci., 16 (1978) 1.
210. J. Sevcik and N. T. P. Thao, Chromatographia, 8 (1975) 559.
211. M. Dressler, J. Chromatogr., 262 (1983) 77.
212. W. E. Dole and C. C. Hughes, J. Gas Chromatogr., 6 (1968) 603.
213. R. Pigliucci, W. Averill, J. E. Purcell, and L. S. Ettre, Chromatographia, 8 (1975) 175.
214. W. P. Cochrane and R. Greenhalgh, Chromatographia, 9 (1976) 255.
215. M. C. Bowman and M. Beroza, Anal. Chem., 40 (1968) 1448.
216. C. A. Burgett and L. G. Green, J. Chromatogr. Sci., 12 (1974) 556.
217. R. Greenhalgh and M. Wilson, J. Chromatogr., 128 (1978) 157.
218. S.-A. Fredriksson and A. Cedergren, Anal. Chem., 53 (1981) 614.
219. P. L. Patterson, R. L. Howe, and A. Abushumays, Anal. Chem., 50 (1978) 339.
220. P. L. Patterson, Chromatographia, 16 (1982) 107.
221. P. L. Patterson , Anal. Chem., 50 (1978) 345.
222. C. R. Vogt and S. Kapila, J. Chromatogr. Sci., 17 (1979) 546.
223. C. H. Burnett, D. F. Adams, and S. O. Farwell, J. Chromatogr. Sci., 16 (1978) 68.
224. S. Sass and G. A. Parker, J. Chromatogr., 189 (1980) 331.
225. P. J. Marriott and T. J. Cardwell, Chromatographia, 14 (1981) 279.
226. J. G. Eckhardt, M. B. Denton, and J. L. Moyers, J. Chromatogr. Sci., 13 (1975) 133.
227. C. H. Burnett, D. F. Adams, and S. O. Farwell, J. Chromatogr. Sci., 15 (1977) 230.
228. A. R. Baig, C. J. Cowper, and P. A. Gibbons, Chromatographia, 16 (1982) 297.
229. L. N. Zotov, G. V. Golovkin, and R. V. Golovnya, J. High Resolut. Chromatogr. Chromatogr. Commun., 4 (1981) 6.
230. D. H. Fine, F. Rufeh, D. Lieb, and D. P. Rounbehler, Anal. Chem., 47 (1973) 1188.
231. D. H. Fine and D. P. Rounbehler, J. Chromatogr., 109 (1975) 271.
232. D. H. Fine, D. P. Rounbehler, E. Sawicki, and K. Krost, Environ. Sci. Technol., 11 (1977) 577.
233. J. L. Owens and O. E. Kinast, Anal. Chem., 53 (1981) 1961.
234. K. S. Webb and T. A. Gough, J. Chromatogr., 177 (1979) 349.
235. D. P. Rounbehler, S. J. Bradley, B. C. Challis, D. H. Fine, and E. A. Walker, Chromatographia, 16 (1982) 354.
236. R. C. Hall, J. Chromatogr. Sci., 12 (1974) 152.
237. D. M. Coulson, Adv. Chromatogr., 3 (1966) 197.
238. M. L. Selucky, Chromatographia, 5 (1972) 359.
239. P. Jones and G. Nickless, J. Chromatogr., 73 (1972) 19.
240. G. G. Patchett, J. Chromatogr. Sci., 8 (1970) 153.
241. L. P. Sarna and G. R. B. Webster, J. Ass. Off. Anal. Chem., 57 (1974) 1279.
242. B. E. Pape, D. H. Rodgers, and T. C. Flynn, J. Chromatogr., 134 (1977) 1.
243. B. J. Ehrlich, R. C. Hall, R. J. Anderson, and H. G. Cox, J. Chromatogr. Sci., 19 (1981) 245.
244. S. Gluck, J. Chromatogr. Sci., 20 (1982) 103.
245. R. K. S. Goo, H. Kanai, V. Inouye, and H. Wakatsuki, Anal. Chem., 52 (1980) 1003.
246. E. M. Fredericks and G. A. Harlow, Anal. Chem., 35 (1964) 263.
247. R. L. Martin and J. A. Grant, Anal. Chem., 37 (1965) 644.
248. J. Sevcik, Chromatographia, 4 (1971) 102.

249. F. J. Yang, J. Chromatogr. Sci., 19 (1981) 523.
250. E. L. Anderson and W Bertsch, J. High Resolut. Chromatogr. Chromatogr. Commun., 1 (1978) 13.
251. R. Digliucci, W. Averill, J. E. Purcell, and L. S. Ettre, Chromatographia, 8 (1975) 165.
252. H. A. McLeod, A. G. Butterfield, D. Lewis, W. E. J. Phillips, and D. E. Coffin, Anal. Chem., 47 (1975) 674.
253. P. L. Coduti, J. Chromatogr. Sci., 14 (1976) 423.
254. T. H. Parliament and M. D. Spencer, J. Chromatogr. Sci., 19 (1981) 432.
255. V. Lopez-Avila and R. Northcutt, J. High Resolut. Chromatogr. Chromatogr. Commun., 5 (1982) 67.
256. B. V. Burger and Z. Munro, J. Chromatogr., 262 (1983) 95.
257. F. Etzweiler and N. Neuner-Jehle, Chromatographia, 6 (1973) 503.
258. M. Hrivnac, W. Frischknecht, and L. Cechova, Anal. Chem., 48 (1976) 937.
259. F. Rinderknecht and B. Wenger, J. High Resolut. Chromatogr. Chromatogr. Commun., 2 (1979) 741.
260. J. Roeraade and C. R. Enzell, J. High Resolut. Chromatogr. Chromatogr. Commun., 2 (1979) 123.
261. P. Sandra, T. Saeed, G. Redant, M. Goderfroot, M. Verstappe, and M. Verzele, J. High Resolut. Chromatogr. Chromatogr. Commun., 3 (1980) 107.
262. J. R. Hudson and S. L. Morgan, J. High Resolut. Chromatogr. Chromatogr. Commun., 4 (1981) 186.
263. V. Pretorius, P. Apps, and W. Bertsch, J. High Resolut. Chromatogr. Chromatogr. Commun., 6 (1983) 104.
264. D. W. Later, B. W. Wright, and M. L. Lee, J. High Resolut. Chromatogr. Chromatogr. Commun., 4 (1981) 406.
265. K. Grob, Chromatographia, 9 (1976) 509.
266. L. Lopez-Avila, J. High Resolut. Chromatogr. Chromatogr. Commun., 3 (180) 545.
267. A. Bjorseth and G. Eklund, J. High Resolut. Chromatogr. Chromatogr. Commun., 2 (1979) 22.
268. A. Sodergren, J. Chromatogr., 160 (1978) 271.
269. F. Poy, J. High Resolut. Chromatogr. Chromatogr. Commun., 2 (1979) 243.
270. W. Bertsch, J. High Resolut. Chromatogr. Chromatogr. Commun., 1 (1978) 85.
271. W. Bertsch, J. High Resolut. Chromatogr. Chromatogr. Commun., 1 (1978) 187.
272. W. Bertsch, J. High Resolut. Chromatogr. Chromatogr. Commun., 1 (1978) 289.
273. E. L. Anderson, M. M. Thomason, H. T. Mayfield, and W. Bertsch, J. High Resolut. Chromatogr. Chromatogr. Commun., 2 (1979) 335.
274. G. Schomburg, H. Husman, and F. Weeke, J. Chromatogr., 112 (1975) 205.
275. R. E. Kaiser and R. I. Rieder, J. High Resolut. Chromatogr. Chromatogr. Commun., 2 (1979) 416.
276. T. S. Bugs and T. W. Smuts, J. High Resolut. Chromatogr. Chromatogr. Commun., 4 (1981) 317.
277. D. R. Deans, Chromatographia, 1 (1968) 18.
278. D. R. Deans, J. Chromatogr., 203 (1981) 19.
279. D. C. Fenimore, R. E. Freeman, and P. R. Loy, Anal. Chem., 45 (1973) 2331.
280. D. E. Willis, Anal. Chem., 50 (1978) 827.
281. W. Bertsch, F. Hsu, and A. Zlatkis, Anal. Chem., 48 (1976) 928.
282. W. Bertsch, E. L. Anderson, and G. Holzer, Chromatographia, 10 (1977) 449.
283. A. Ducass, M. F. Gonnord, A. Arpino, and G. Guiochon, J. Chromatogr., 148 (1978) 321.
284. J. Sevcik and T. H. Gerner, J. High Resolut. Chromatogr. Chromatogr. Commun., 2 (1979) 436.
285. G. Schomburg, F. Weeke, F. Muller, and M. Oreans, Chromatographia, 16 (1982) 87.

212

286. R. J. Phillips, K. A. Knauss, and R. R. Freeman, J. High Resolut. Chromatogr. Chromatogr. Commun., 5 (1982) 546.
287. B. J. Hopkins and V. Pretorius, J. Chromatogr., 158 (1978) 465.
288. J. A. Settlage and W. G. Jennings, J. High Resolut. Chromatogr. Chromatogr. Commun., 3 (1980) 146.
289. H. U. Buser, R. Soder, and H. M. Widmer, J. High Resolut. Chromatogr. Chromatogr. Commun., 5 (1982) 156.
290. J. A. Rijks, J. H. M. Van Den Berg, and J. P. Piependall, J. Chromatogr., 91 (1974) 603.
291. H. Brotell, G. Rietz, S. Sandqvist, M. Berg, and H. Ehrsson, J. High Resolut. Chromatogr. Chromatogr. Commun., 5 (1982) 596.
292. J. F. K. Huber, E. Kenndler, W. Nyiry, and M. Oreans, J. Chromatogr., 247 (1982) 211.
293. H. J. Stan and D. Mrowetz, J. High Resolut. Chromatogr. Chromatogr. Commun., 6 (1983) 255.
294. W. Chinghai, H. Frank, W. Guanghua, Z. Liangmo, E. Bayer, and L. Peichang, J. Chromatogr., 262 (1983) 352.
295. H. J. Stan and D. Mrowetz, J. Chromatogr., 279 (1983) 173.
296. S. Nygren, J. High Resolut. Chromatogr. Chromatogr. Commun., 2 (1979) 319.
297. W. E. Harris and H. W. Habgood, "Programmed Temperature Gas Chromatography", Wiley, New York, 1966.
298. E. M. Sibley, C. Eon, and B. L. Karger, J. Chromatogr. Sci., 11 (1973) 309.
299. D. W. Grant and M. G. Hollis, J. Chromatogr., 158 (1978) 3.
300. L. S. Ettre, L. Mazor, and J. Takacs, Adv. in Chromatogr., 8 (1969) 271.
301. S. Nygren, J. Chromatogr., 142 (1977) 109.

Chapter 4

THE COLUMN IN LIQUID CHROMATOGRAPHY

4.1 <u>Introduction</u>

The expression "liquid chromatography" tells one little about the technique beyond the nature of the mobile phase and the fact that it is a separation method. The name represents a lexicon of separation techniques, of which the most important are size-exclusion, liquid-solid, liquid-liquid, and ion-exchange chromatography. Separations based on molecular size are referred to as size-exclusion chromatography. In this case, the stationary phase is an inert, porous particle or organic gel of controlled pore size. Separation is effected by the sample's ability to diffuse into and out of the pore system. From an historical viewpoint, controlled porosity gels were used for molecular size separations, thus giving rise to the names gel permeation chromatography and gel filtration chromatography. The latter name is used for those molecular size separations employing an aqueous mobile phase. Although these names are still found in the recent literature, the use of porous particles rather than gels is now common, and size-exclusion chromatography is the preferred name. Separations based on the competitive adsorption of the sample and mobile phase for active sites on a solid adsorbent are referred to as liquid-solid chromatography. When the stationary phase is a liquid bonded to a solid support and separations are effected by partition of the sample between the mobile and stationary phases, the technique is called liquid-liquid, bonded-phase, or simply liquid chromatography. For many separations the actual mechanism is not clearly defined and may involve either adsorption, partition, or a combination of both. Conventional chromatographic practice employed a polar stationary phase and a nonpolar mobile phase. In high performance liquid chromatography (HPLC) it has, however, become common practice to use a nonpolar chemically-bonded stationary phase with an aqueous mobile phase containing various proportions of a miscible organic modifier. This system is known as reversed-phase chromatography to distinguish it from the normal or conventional operating mode. Reversed-phase chromatography is the method most often used in HPLC for the separation of both polar and nonpolar organic molecules. Separations based on the interaction between a stationary phase containing immobilized charged groups and ionic or readily ionized molecules in the mobile phase is termed ion-exchange chromatography. An alternative method of separating ionic molecules is ion-pair chromatography. Here, conventional liquid-solid or reversed-phase chromatography is used to separate the sample as the lipophilic complex formed between the sample and a counterion of opposite

charge. The characteristics of the above chromatographic methods are described
in detail later in this chapter. Ion-pair chromatography is discussed in
Chapter 7.14.5.

4.2 Selection of the Separation Method

To begin a chromatographic analysis by HPLC, the first decision to be made
is the selection of the most appropriate column type. The analyst requires some
knowledge of the sample's physical characteristics, as well as the nature of the
information desired from the separation, to make this preliminary selection.
For example, if the information sought is the molecular weight distribution of a
sample, then size-exclusion chromatography would be an obvious choice.
Alternatively, if one wanted to know the number of components present in a low
molecular weight, non-ionic, organic sample, then either liquid-solid,
bonded-phase liquid-liquid, or reversed-phase chromatography would be
preferred. The decision of which method to try first could be based on such
physical properties as the sample's relative solubility in polar and nonpolar
solvents. In many cases, a full analysis may be obtained by more than one
method if the experimental conditions are optimized and the sample is not too
complex. For a complex sample, no single method may be completely adequate for
the separation and a combination of techniques might be required. In this case,
a preliminary sample fractionation might be carried out by size-exclusion
chromatography and the components of similar molecular size further resolved by
liquid-solid or one of the bonded-phase chromatographic methods.

Past experience with different separation methods can also aid in column
selection. Liquid-solid chromatography is usually the preferred method for
separating mixtures of low molecular weight, lipophilic, geometric isomers. The
constituents of a homologous series are generally best separated by
reversed-phase chromatography. The technique of ion suppression is preferred
for the separation of weak acids or bases. The pH of the mobile phase is first
adjusted to eliminate sample ionization and the sample is then separated by the
same methods used for non-ionic samples. Ion suppression cannot, however, be
used for the separation of strong bases due to the poor stability and
appreciable solubility of silica-based packing materials at the high pH required
to suppress ionization. Strong acids and bases are usually separated by
ion-pair or ion-exchange chromatography. Ion-pair chromatography is the most
versatile of the two techniques, and does not require the purchase of special
columns. The ion-pair complexes are separated by the same column techniques
used for non-ionic organic molecules.

As an aid to the selection of the best or first-choice chromatographic
method for various sample types a flow diagram is given in Figure 4.1.

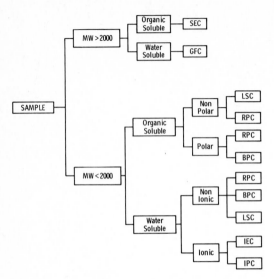

Figure 4.1   Flow diagram of column selection based on sample type. MW =
            molecular weight; SEC = size-exclusion chromatography; GFC = gel
            filtration chromatography; LSC = liquid-solid chromatography;
            RPC = reversed-phase chromatography; BPC = bonded-phase
            chromatography; IEC = ion-exchange chromatography; and IPC =
            ion-pair chromatography.

The selection is based on the knowledge of the molecular weight of the sample,
its relative solubility in water or a nonpolar organic solvent and whether or
not the sample is ionic. This information would generally be known in advance
or could be readily established with a few simple experiments. Normally, the
analyst combines the information concerning the properties of the sample with
the information desired from the separation to select the most appropriate
chromatographic method. The selection of the molecular weight cut-off of about
2000 is quite arbitrary and is based on experimental convenience. A limited
number of size-exclusion packings for the separation of molecules with a
molecular weight below 2000 are available. Likewise, silica gel and low loaded
bonded-phase column packings of wide pore diameter have been developed for the
separation of large molecular weight molecules by adsorption or partition
chromatography.

4.3 Column Packing Materials for HPLC

The high performance packings used in modern liquid chromatography are
comprised of small, rigid particles having a narrow particle size distribution.
There are three general types of packing materials available, Figure 4.2,
although the most commerically important are the totally porous microparticulate
packings. Rigid, porous, polymeric beads prepared from polystyrene crosslinked

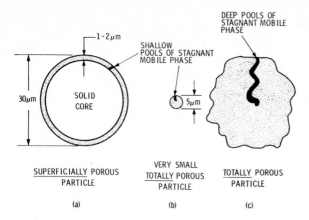

Figure 4.2 Types of particles for HPLC column preparation.

with divinylbenzene are used for ion-exchange and size-exclusion chromatography. Although of historical importance, the porous polymeric bead packings have been replaced by more efficient and mechanically stable porous silica-based packings for many applications. Pellicular packings were originally prepared to replace the porous polymer beads routinely used at that time for high performance ion-exchange chromatography. They consisted of a glass bead core around which a thin, porous polymeric film was attached [1]. The rigid core of these packings permitted operation at high pressures, but the packings produced less than desirable column efficiencies, due mainly to poor stationary phase mass transfer properties. Mass transfer was improved by creating a "superficially porous" layer around the bead core. Known as porous layer beads, they had particle diameters of 30 to 55 micrometers and consisted of a glass bead core to which was fused an intermediate porous silica or alumina layer 1-3 micrometers thick. They can be used without modification in liquid-solid chromatography, or in partition chromatography as a support for a mechanically-held liquid phase. Bonded phases were prepared by reacting a suitably substituted organosilane with the porous silica layer. The porous layer beads have a very small pore volume, providing rapid solute diffusion in and out of the stationary phase.

The importance of porous layer beads in HPLC declined when totally porous particles in narrow particle size ranges with diameters below 10 micrometers became available. Compared with the porous layer beads, totally porous silica particles can provide increases of up to an order of magnitude in column efficiency, sample capacity, and speed of analysis. The chromatographic properties of the two types of column packings are compared in Table 4.1 [2]. For particles having identical diameters, the porous layer beads would provide

more rapid mass transfer and less band broadening than totally porous
particles. However, by reducing the particle size of the totally porous
particles, the pore depth decreases proportionately and column efficiency
increases. Technical difficulties prevent the production of porous layer beads
in the same size ranges available for porous particles. Thus, porous particles
of 3 to 10 micrometers diameter are the dominant column packings used in modern
analytical HPLC.

TABLE 4.1

PROPERTIES OF HPLC COLUMN PACKINGS

Property	Porous Layer Beads	Microporous Particles
Average Particle size, micrometers	30–40	5–10
Best HETP values, mm	0.2–0.4	0.01–0.03
Typical column lengths, cm	50–100	10–30
Typical column diameters, mm	2	2–5
Pressure drop, p.s.i./cm	2	20
Sample capacity, mg/g	0.05–0.1	1–5
Surface area, m^2/g	10–15	400–600

The small sample capacity of the porous layer beads is due to the low ratio
of stationary phase per unit column volume. This feature can be advantageous
when very fast analyses are required or when samples with above average
retention are to be separated. The surface area and, therefore, the retentive
capacity of the porous particles are much higher than the porous layer beads.
Porous layer beads are thus well-suited for separating compounds with average
retention properties.

4.3.1 Microporous Adsorbents for Liquid-Solid Chromatography

Silica gel is by far the most important adsorbent used in HPLC. It is also
the base material used to prepare bonded-phase packings, giving it a preeminent
position in modern column technology. Alumina column packings are also
commercially available but account for only a few percent of all published
separations using liquid-solid chromatography [3]. Alumina may have some
selectivity advantages over silica gel for the separation of unsaturated
hydrocarbons and halogen-containing compounds. It may also be useful for the
separation of very basic compounds which are too strongly absorbed to acidic
silica gel. Additionally, alumina is more stable than silica gel at high pH
[4]. Adsorbent packings are available in two forms, irregular or spherical,

with particle diameters in the range 3 to 20 micrometers. Packings with 5 and 10 micrometer diameters and a narrow size distribution are the most common. It is not clear whether the spherical packings have any major advantages over irregular packings [5]. Both types of packings yield columns of similar efficiency and stability, although columns packed with spherical particles are slightly more permeable.

The commercial preparation of silica gel for HPLC packings is a rather complex process. It has been described in some detail by Unger [6] and Scott [7,8]. Irregular microporous silica gels are prepared by the sol-gel procedure. A hydrogel is initially prepared by the addition of sodium silicate to an acidic aqueous solution. An insoluble cake is formed, allowed to age, and eventually crushed to suitably-sized portions. Residual sodium salts are then washed away and the hydrogel is dehydrated by heating at about 120°C for several hours to form a hard, porous xerogel. After milling and sieving to the correct particle size, the xerogel is ready for chromatographic use.

Spherical particles are more difficult to manufacture. One method prepares silica hydrogel beads by emulsification of a silica sol in an immiscible organic liquid. Bead size and size distribution are dependent on such experimental conditions as stirring speed, solvent viscosity, pH, temperature, and the concentration of the silica sol. Alternative methods for preparing spherical silica particles include condensation of silica sols using spray-drying techniques and copolymerization of an organic reagent with a silica sol to form a gelatinous bead. In the latter case, the microporous spherical silica particles are obtained by burning out the organic reagent.

Apart from shape, microporous silica particles are characterized by their mean pore diameter, specific surface area, and specific pore volume. These parameters can be measured by well-established methods based on the physisorption of gases or vapors and the controlled penetration of fluids [6,9]. The specific surface area can be separated into two components: the surface area within the pores and the external surface area of the particle. The pore surface area is several orders of magnitude larger than the external surface area and, in general, the larger the surface area of the packing material, the smaller will be the pore diameter. Commercial porous silica packings for HPLC have surface areas ranging from 100 to 860 m^2/g with an average value around 400 m^2/g, a large pore volume of 0.7 to 1.2 ml/g and a moderate mean pore diameter of 50 to 250 Å [10,11]. One should keep in mind that only the specific pore volume, defined as the amount of liquid that fills the total volume of the pores per gram of adsorbent, has any real physical significance. The specific surface area and pore diameter are calculated by reference to an appropriate model. The values obtained are only as accurate in absolute terms as the model can be said to be a perfect replica of the real

220

situation. For porous silica particles a globular model is usually used to estimate the pore diameter. This model assumes that the porous system can be described by a regular arrangement of densely packed spheres. The pore system is formed by the crevices between the spheres and is controlled by the size of the spheres and their packing density. The pore space can be envisioned as a matrix of alternating wide cavities and narrow constrictions. For liquid–solid HPLC, pore dimensions should ideally be large compared to the molecular dimensions of the solute to allow uninhibited solute diffusion into and out of the particles. For normal size molecules the pore size is rarely a problem, provided that the pore size distribution is not so wide that micropores (< 10–20 Å) exist. Micropores can produce size–exclusion effects or strong irreversible adsorption due to the high surface energy existing within the narrow pores. Size–exclusion chromatography of polystyrene standards can be used to estimate average pore diameters [12].

An understanding of the surface chemistry of silica is important to an interpretation of its chromatographic properties. The silica surface consists of a network of silanol groups, some of which may be hydrogen bonded to water, and siloxane groups as shown in Figure 4.3 [13]. The composition of the silica surface depends on the method of preparation and the thermal treatment. The nature of the silica surface with respect to adsorbed water is depicted in Figure 4.4 [8,14]. Heating at temperatures below 150°C removes physically

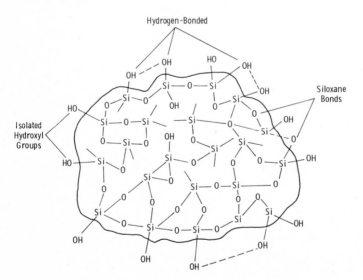

Figure 4.3   Structure of silica gel, depicting the various types of bonds and silanol groups present. (Reproduced with permission from ref. 13. Copyright Academic Press, Inc.)

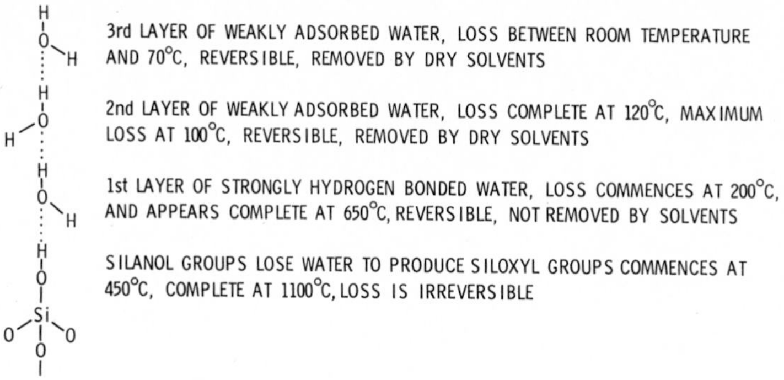

H
O
:H

3rd LAYER OF WEAKLY ADSORBED WATER, LOSS BETWEEN ROOM TEMPERATURE
AND 70°C, REVERSIBLE, REMOVED BY DRY SOLVENTS

H
O

2nd LAYER OF WEAKLY ADSORBED WATER, LOSS COMPLETE AT 120°C, MAXIMUM
LOSS AT 100°C, REVERSIBLE, REMOVED BY DRY SOLVENTS

H
O
:H

1st LAYER OF STRONGLY HYDROGEN BONDED WATER, LOSS COMMENCES AT 200°C,
AND APPEARS COMPLETE AT 650°C, REVERSIBLE, NOT REMOVED BY SOLVENTS

H
O

SILANOL GROUPS LOSE WATER TO PRODUCE SILOXYL GROUPS COMMENCES AT
450°C, COMPLETE AT 1100°C, LOSS IS IRREVERSIBLE

Si

Figure 4.4   Schematic representation of water bound to a silanol group.
(Reproduced with permission from ref. 14. Copyright Elsevier
Scientific Publishing Co.)

adsorbed water. Above 200°C the primary layer of strongly hydrogen-bonded water
is removed and the process is complete at about 650°C. Above 450°C,
dehydroxylation occurs between neighboring silanol groups, resulting in an
increased concentration of siloxane groups. At still higher temperatures the
dehydration of silanol groups is complete, forming a hydrophobic surface covered
with siloxane groups. Sintering of the silica particles starts at temperatures
in excess of 900°C. A completely dehydrated silica surface adsorbs moisture
extremely slowly. However, predried hydrated silica rapidly picks up water
vapor from the atmosphere by physical adsorption and capillary condensation.
For most commercial silica gels dried below 200°C, the surface concentration of
silanol groups is about 8.7 ± 0.2 micromoles/m^2 [9,13].

The accessible surface concentration of silanol groups is most important in
liquid-solid HPLC since adsorption of the solute to the silica surface occurs at
these sites. The absolute retention is controlled by the accessible
concentration of silanol groups per unit column volume, defined as the product
of the surface concentration of silanol groups and the surface area of adsorbent
per unit column volume. The siloxane groups give rise to weak, nonspecific
interactions resulting in poor solute selectivity.

The surface of pure silica is expected to be weakly acidic due to the
presence of silanol groups having a pK_a of about 9.5. A suspension of silica in
neutral, salt-free water should have a pH of about 5. However, it has been show
that the pH of commercial packings deviate widely from this value, Table 4.2
[9], especially for spherical packing materials. Although these pH values
cannot be directly related to the pH of the silica used in a particular column
separation employing a nonpolar eluent, it can, however, be demonstrated that
the aqueous suspension pH value correlates well with the retention behavior of
polar and nonpolar solutes. For polar solutes the surface pH affects such

chromatographic parameters as relative retention, peak shape, and column efficiency. It was observed that "acidic silicas" were more efficient at eluting neutral and acidic solutes while basic nitrogen-containing compounds could not be eluted as measurable peaks within a reasonable time [9]. The opposite situation exists with "basic silicas". Thus, there is no ideal silica for the simultaneous separation of both acidic and basic samples. The tailing of polar compounds on silica gel can often be controlled by coating the silica packing with a buffer salt [15,16]. This can be done in a batch process prior to column packing or by an in situ coating procedure for packed columns.

TABLE 4.2

PHYSICAL PROPERTIES OF SOME COMMERCIALLY AVAILABLE SILICA GELS

Manufacturer's Name	Specific Surface Area	Average Pore Diameter (nm)	Specific Pore Volume ml/g	pH of a 1% w/w Aqueous Suspension	Particle Type
Hypersil	170	11.5	0.7	9.0	Spherical
LiChrosorb Si 100	320	11.1	1.2	7.0	Irregular
Porasil	350	10.0	1.1	7.2	Irregular
Spherisorb	190	8.1	0.6	9.5	Spherical
Zorbax BPSil	300	5.6	0.5	3.9	Spherical

Silica gel is only slightly water soluble at neutral pH, but its solubility increases exponentially above pH 8 due to formation of the silicate anion. Aqueous solubility can be controlled by using a silica gel saturator column placed before the injector [9]. Normally, it is not recommended that silica-based packings be used at pHs greater than 8.0 and less than 1.0.

4.3.2 Bonded-Phase Column Packings

Prior to the development of bonded phases for liquid-liquid chromatography, silica gel columns coated with a physically adsorbed liquid stationary phase were widely used. Reproducible columns were easy to prepare and could be renewed in service by in situ coating techniques. These columns could also tolerate relatively large sample loads without loss in efficiency. However, their advantages were more than offset by disadvantages associated with stripping of the stationary phase from the column. A precolumn was necessary to saturate the mobile phase with stationary phase, mobile phase selection was limited to those solvents with low stationary phase solubility, and high flow rates and gradient elution techniques were not possible [17,18]. Columns of this type are rarely used today.

Several approaches have been made to chemically bond the liquid phase to its mechanical support and thereby eliminate those problems associated with the stripping of stationary phase from the column. Early attempts were only partially successful [19,20]. The "estersils" or silicate esters were among the first bonded phases to be evaluated [21]. They were prepared by reacting an alcohol or a phenylisocyanate with a porous silica packing, accompanied by the continuous removal of water. The surface of these monomolecular-layered particles may be pictured as a forest of organic groups standing on end and hence, the term "brushes" is sometimes used to describe them.

$$-\overset{\mid}{\underset{\mid}{Si}}-OH \ + \ HO-R \qquad -\overset{\mid}{\underset{\mid}{Si}}-O-R \ + \ H_2O$$

| silica | alcohol | estersil |
| gel | | packing |

Because of their simple monomeric structure, the estersil phases exhibit faster mass transfer kinetics than are obtained by coating a bulk liquid onto an inert support. The selectivity of the phase is controlled by the nature of the substituents attached to the alkyl chain. For example, nonpolar phases were prepared using octanol as the esterifying reagent, while 3-hydroxypropionitrile was used for polar phases. Unfortunately, the estersil phases had extremely poor hydrolytic stability and could not be used with mobile phases or samples containing traces of water, base, or acid.

Bonded phases with an $\equiv$Si-R or $\equiv$Si-NH-R structure are more hydrolytically stable than the estersils. Their preparation may involve several steps. The silica gel is usually dried to remove physically adsorbed water and then refluxed with a reagent such as thionyl chloride to convert surface silanol groups to the more reactive $\equiv$Si-Cl groups. The organic group is then attached to the silica surface by a Grignard, organolithium, Wurtz, or Friedel-Crafts reaction [22]. The $\equiv$Si-NH-R bonded phases are prepared by reacting the $\equiv$Si-Cl groups with substituted amines. The structure of these phases is similar to the estersils and they also exhibit fast mass transfer kinetic properties. The preparation of the above bonded phases is not always straightforward as reaction conditions may be difficult to optimize, the concentration of the surface organic groups is relatively low, and the organometallic reagents used in the preparation often leave undesirable residues which are difficult to wash out of the final product.

Most commercially available bonded phases are of the siloxane type Si-O-Si-R [23-25]. They are prepared by reacting the surface silanol groups on silica with an organochlorosilane or organoalkoxysilane. In a typical reaction, the concentration of surface silanol groups is first maximized by acid hydrolysis. Heating in 0.1 M hydrochloric acid at 90°C for at least 24 h is usually adequate. Physically adsorbed water is removed from the hydrolyzed packing by vacuum drying at 200°C for 8-12 h. The organic groups are attached to the silica surface by refluxing with the organosilane reagent in a high boiling solvent such as toluene. When an organochlorosilane is used in the reaction, an acid acceptor catalyst such as pyridine may be added to the reaction mixture. Reaction times are usually fairly long (16-24 h). Reactive organosilanes with ethyl, octyl, hexyl, octadecyl, phenyl, chloropropyl, aminopropyl, or cyanopropyl groups are commercially available, making this a very general synthetic route for preparing bonded-phase packings having different selectivities.

Bonded-phase packings can be prepared by reaction with mono-, di-, or trichloroorganosilanes (or the corresponding alkoxysilanes) to produce products having different chromatographic properties. Monochloro-organosilanes react with only one silanol group to form a monomolecular layer of organic groups anchored by siloxane bridges to the silica surface. The most frequently used organochlorosilane reagents for surface modification have organic substituents with critical dimensions greater than the distance between the surface silanol groups. Consequently, the concentration of residual silanol groups at the surface is significant even for exhaustive silanization. The most accessible and therefore the most chromatographically active of the residual silanol groups may be removed by endcapping. This reaction is carried out after silanization by reaction with a less bulky reagent such as trimethylchlorosilane. Endcapping can dramatically affect the chromatographic performance of some molecules on bonded-phase packings. An example is shown in Figure 4.5 for the separation of some PTH-amino acid derivatives on a reversed-phase packing before and after endcapping [26]. It should be noted that even after endcapping, some unreacted silanol groups remain on the support surface. These groups are largely inaccessible to solute molecules of average size and do not influence the general chromatographic properties of the bonded phase.

Di- and trichloroorganosilanes are more chemically reactive than the monochloroorganosilanes. The reaction of these reagents with surface silanol groups can result in surface attachment via one or two siloxane bonds as shown in Figure 4.6. Unreacted Si-Cl groups are usually converted to silanol groups by hydrolysis. For steric reasons, it is doubtful that all the silica silanol groups will react with the silylating reagent. Thus, the reaction conditions may generate additional silanol groups and some surface silanol groups may remain. These are generally minimized by endcapping.

The reaction between a di- or trichloroorganosilane and silica in the presence of moisture or a protic agent results in the formation of a linear or crosslinked polymeric layer. The structure of these polymeric bonded phases is

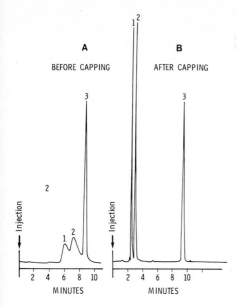

Figure 4.5 The influence of endcapping on peak shape and retention of some
PTH–amino acids using a reversed-phase separation system. Peak
identification: 1. PTH-histidine; 2. PTH-arginine; 3. PTH-valine.
(Reproduced with permission from ref. 26. Copyright Preston
Publications, Inc.)

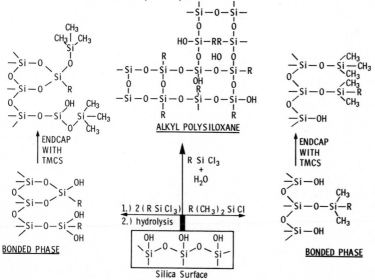

Figure 4.6 Preparation of siloxane bonded phases by surface silanization of
microporous silica.

not as well-defined as the monomeric phases. Depending on the thickness and
nature of the polymeric layer, the chromatographic performance of these bonded
phases may be poor compared to the monomeric phases. The polymer network may

swell to different extents with changes in the mobile phase and the polymer cavities may act as reservoirs for stagnant pools of mobile phase, promoting slow mass transfer and lower column efficiency.

To classify bonded-phase packings, some knowledge of the parameters controlling retention and selectivity are required. The important parameters are the type and concentration of the bonded organic modifier and the concentration of accessible unreacted silanol groups [6,27-29]. The surface concentration of silanol groups on silica is given by equation (4.1).

$$a_{OH} = \frac{10^6 \, m_{OH}}{S_{Si}} \qquad (4.1)$$

a_{OH} = number of micromoles of silanol groups per square meter of BET surface area of silica

m_{OH} = moles of surface silanol groups per gram of column material

S_{Si} = BET surface area of silica

For the preparation of bonded phases the surface of the starting material is chemically treated to maximize the concentration of silanol groups. Typical a_{OH} values fall in the range of 7-9 micromoles/m^2, corresponding to about 4-5 silanol groups per nm^2 of silica surface. The surface concentration of bonded organic modifier is given by equation (4.2)

$$a_1 = \frac{10^6 \, w}{M \, S_{Si}^*} \qquad (4.2)$$

a_1 = number of micromoles of bonded phase per m^2 of BET surface area of silica support

w = weight of bound organic modifier in grams per gram of adsorbent

M = molecular weight of bound organic modifier

S_{Si}^* = specific surface area of the silica support corrected for the weight increase upon silanization

Commercial bonded phase packings have a_1 values ranging from 2 to 4.

For monomeric bonded phases prepared with monofunctional silanizing reagents, the conversion of surface silanol groups can be obtained from equation (4.3)

$$X_{OH} = \frac{a_1}{a_{OH}} \qquad (4.3)$$

X_{OH} = conversion factor for surface silanol groups which are silanized in the preparation of the bonded phase

Expressed as a percentage, the surface coverage by the organic modifier varies

from less than 10% to a maximum of approximately 60%; the upper limit is
established by the size of the organic modifier and its ability to limit reagent
access to some of the silanol groups. The presence of some residual silanol
groups may be chromatographically desirable to allow "wetting" of polar
bonded-phase packings by the mobile phase.

4.3.3 Column Packings for Ion-Exchange Chromatography

Four types of microparticulate packings are available for high performance
ion-exchange chromatography. Historically, porous microbeads having a
polystyrene-divinylbenzene backbone substituted with ionogenic groups held a
prominent position in this area. General improvements in their synthesis over
the years has resulted in the availability of semirigid packings of 5 to 10
micrometer diameters, used extensively in autoanalyzers for amino acid, peptide,
and carbohydrate analyses. Slow diffusion of sample molecules within the
polymeric matrix generally makes these packings less efficient than alternative
materials. To improve mass transfer properties, pellicular ion-exchange
packings were introduced [1,20]. They have an impenetrable glass core with a
thin polymer film around the outside, or are superficially porous having a thin
intermediate coating of silica to which the polymer film is chemically attached
or deposited by copolymerization. The pellicular packings are available in only
relatively large particle sizes, which limits their ultimate column efficiency,
and have very low sample capacity as well. The widespread adoption of
microporous silica packings in other branches of HPLC quite naturally lead to
the development of silica based ion-exchange packings. Their preparation is
very similar to the methods discussed previously for bonded phases. Monochloro-
or monoalkoxy-substituted silanes are reacted with silica to produce a monomeric
layer of organic groups attached to the surface by siloxane bonds. The
ionogenic functional groups are usually introduced into the organic backbone in
a subsequent reaction sequence. Polymeric layers may also be prepared but these
materials are usually less efficient than the monomeric silica-based packings
[30]. However, their higher sample capacity might be preferred for large scale
separations. Controlled porosity glass column packings with attached
hydrophilic polymeric groups have been introduced for the high speed separation
of large, ionic molecules such as proteins and nucleic acids, etc. [31,32].
Samples of this type have traditionally been difficult to separate with
conventional ion-exchange materials due to irreversible sample adsorption or
denaturation. The use of these packings is likely to become more widespread in
the future as HPLC assumes a much greater role in the separation of high
molecular weight biopolymers.

The chromatographic properties of ion-exchange packings used for analytical

separations have been summarized by Majors, Table 4.3 [33]. Virtually all
commercially available materials contain a sulfonate (strong cation exchange
group), carboxylate (weak cation exchange group), tetraalkylammonium ion (strong
anion exchange group), or an amine (weak anion exchange group) as the ionogenic
functional group. Direct reaction of aromatic rings with oleum or
chlorosulfonic acid, or chloromethylation with chloromethyl ether and zinc
chloride followed by amination, are widely exploited reactions for introducing
sulfonate and tetraalkylammonium groups, respectively, into the column packing
materials (Figure 4.7).

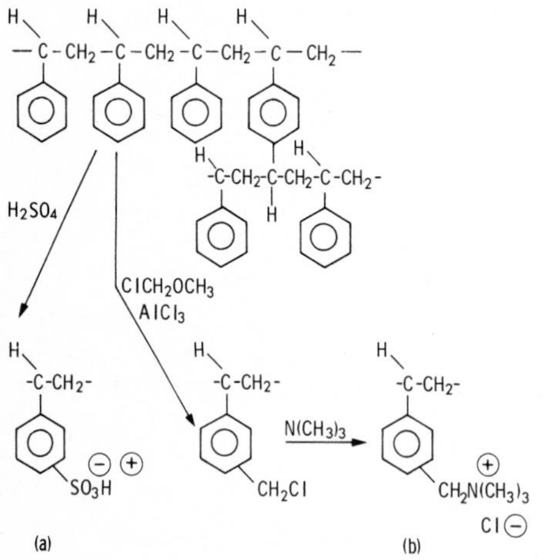

Figure 4.7 Methods for the preparation of ion-exchange resins.

Porous polystyrene-divinylbenzene microbeads containing between 6 and 32%
of crosslinking agent are commercially available. Resin beads prepared with a
low concentration of crosslinking agent are soft and deform at high pressures.
Changes in size due to swelling or shrinking may lead to changes in the
experimental conditions. These large pore materials are used to separate
biopolymers at relatively low pressures. Increasing the relative concentration
of crosslinking agent improves the mechanical stability of the resin beads.
However, bead swelling is a function of the degree of crosslinking and column
temperature as well as the ionic strength, pH, type of buffer, and concentration
of organic modifier employed as mobile phase. For some porous resin microbead
packings it may not be possible to adjust the composition of the mobile phase or
change the experimental conditions without having to repack the column.
Overall, the porous microbead resins show poor chromatographic efficiency and
for this reason are frequently operated at elevated temperatures (50-80°C) to

TABLE 4.3

QUALITATIVE COMPARISON OF HPLC ION-EXCHANGE PACKINGS

Property	Pellicular	Silica	PS–DVB Resins
Particle diameters, micrometers	30–40	5–10	7–10
Ion exchange capacity, mequiv/g	0.01–0.1	0.5–2	3–5
Resistance to pressure deformation	Excellent	Very good	Fair to poor
Shape	Spherical	Spherical or Irregular	Spherical
Pressure drop	Low	High	Highest
Efficiency	Moderate	High	Moderate
Packing Technique	Dry	Slurry	Slurry
pH Range[a]	2–12(coated) 2–7.5(bonded)	2–7.5	0–12(anion) 0–14(cation)
Regeneration Rates	Fast	Moderate	Slow

[a] An upper limit of pH 9 for bonded phases is claimed by some manufacturers.

reduce solvent viscosity and improve solute diffusion within the resin bead. Although not comparable in efficiency to silica-based ion-exchange materials, they do have some advantages over the latter. Their ion-exchange capacity, usually in the range of 3–5 mequiv/g, is 5 to 10 times greater than the silica-based packings. They may be used over a much wider pH range; bonded-phase silica packings are unstable outside the pH range 2–7.5 [34]. Microporous silica ion-exchange materials, if densely covered with bonded phase, may be used up to a pH of 10.5 if a saturator column is used ahead of the injector to saturate the mobile phase with soluble silica. In general, microbead resin columns are more robust and have longer usable lifetimes than silica-based ion-exchange columns.

Microporous silica-based bonded-phase ion-exchange packings provide higher column efficiencies, can conveniently be used at ambient temperatures, and respond more rapidly to changes in the mobile phase composition. Gradient elution presents less of a problem since the volume of the packing structure does not change during the analysis.

When efficiency is the main criteria for selection, then the silica-based packings are preferred. When capacity is the main requirement, the resin microbead packings should be selected. As an alternative to ion-exchange, ion-pair chromatography should always be considered. This is discussed in

Chapter 7.14.5.

4.3.4 Column Packings for Size-Exclusion Chromatography

The choice of column packing for size-exclusion chromatography is most important since the separation obtained is based on molecular size and controlled entirely by the pore dimensions of the porous material. Particle sizes are in the 5-20 micrometer diameter range to provide good kinetic column efficiency. This is particularly important for the separation of macromolecules which have small diffusion coefficients. Three types of packing materials are available for size-exclusion chromatography: ·semirigid organic gels, porous silica, and controlled pore glasses. Table 4.4 lists a few of the materials available and their separation ranges [10,35,36].

Semirigid organic gels prepared from polystyrene crosslinked with various amounts of divinylbenzene have been used for many years for gel permeation chromatography. These gels are not wet by aqueous solvents and are thus used exclusively with organic solvents. Their main advantage is that they can be prepared in a wide range of pore diameters to separate samples in the molecular weight range of about 10^2-10^7. Polystyrene gels with a low degree of crosslinking are the least stable to high pressures and may shrink or swell with changes in mobile phase. Once packed in a particular solvent, it may not be possible to change solvents with these columns. Increasing the degree of crosslinking produces more rigid gels. These are suitable for separating lower molecular weight samples and can tolerate a narrow range of solvent changes. Besides polystyrene, packings prepared from vinyl acetate and polyacrylamide are also available. Porous vinyl-acetate-based gels have properties similar to polystyrene but cover a somewhat smaller molecular weight range. Polyacrylamide gels can be used in aqueous solution for gel filtration chromatography.

The semirigid organic gels have been replaced in many applications by microporous silica packings. However, the organic gels are preferred for the separation of small molecules or for molecules which are adsorbed or interact strongly with silica gel. Rigid particles are packed relatively easily into homogeneous column beds and are mechanically stable over long periods. They can withstand high pressures, flow rates, and temperatures and are compatible with a wide variety of solvents, including aqueous solvents for gel filtration chromatography.

The main disadvantage of microporous silica packings is their surface activity which can cause problems in the separation of biochemically important molecules or polar polymeric samples. Low recoveries for proteins and the loss of enzyme activity are two manifestations of the problem. Surface deactivation of silica or porous glass packings by silylation reduces adsorption but in

TABLE 4.4

MICROPARTICULATE PACKINGS FOR SIZE-EXCLUSION CHROMATOGRAPHY

Commercial Name	Packing Material[a]	Average Pore Size (Å)	Molecular Weight Exclusion Limit or Range
μSpheragel	PS-DVB	50	2×10^3
		100	$(1 \text{ to } 50) \times 10^2$
		500	$(5 \text{ to } 10) \times 10^3$
		1000	$(1 \text{ to } 50) \times 10^3$
		10,000	$(1 \text{ to } 50) \times 10^4$
		100,000	$(1 \text{ to } 50) \times 10^5$
		1,000,000	$< 10^6$
μStyragel	PS-DVB	100	$< 7 \times 10^2$
		500	$(5 \text{ to } 100) \times 10^2$
		1000	$(1 \text{ to } 20) \times 10^3$
		10,000	$(1 \text{ to } 20) \times 10^4$
		100,000	$(1 \text{ to } 20) \times 10^5$
		1,0000,000	$(1 \text{ to } 20) \times 10^6$
TSK Type H	PS-DVB	40	$(0.5 \text{ to } 10) \times 10^2$
		250	$(0.5 \text{ to } 60) \times 10^2$
		1500	$(1 \text{ to } 60) \times 10^3$
		10,000	$(0.5 \text{ to } 40) \times 10^4$
		100,000	$(0.2 \text{ to } 400) \times 10^4$
LiChrospher	Silica	100	$(5 \text{ to } 8) \times 10^4$
		500	$(3 \text{ to } 6) \times 10^5$
		1000	$(6 \text{ to } 14) \times 10^5$
		4000	$(2 \text{ to } 5.8) \times 10^6$
TSK Type SW	Silica	130	$< 2 \times 10^4$
		240	$< 4 \times 10^4$
		450	$< 2 \times 10^5$
μBondagel E	Silica	125	$(0.2 \text{ to } 50) \times 10^3$
		500	$(0.5 \text{ to } 50) \times 10^4$
		1000	$(5 \text{ to } 20) \times 10^5$
CPG	Porous Glass	40	$(1 \text{ to } 8) \times 10^3$
		100	$(1 \text{ to } 30) \times 10^3$
		250	$(2.5 \text{ to } 125) \times 10^3$
		550	$(1.1 \text{ to } 35) \times 10^4$
		1500	$(1 \text{ to } 10) \times 10^5$
		2500	$(2 \text{ to } 15) \times 10^5$

[a] Available in 5 or 10 micrometer particle sizes.

aqueous media the bonded phases cause hydrophobic interactions which are detrimental to sample recovery. Column packings with low absorptivity and low hydrophobicity have been prepared by silylation with suitably substituted reagents to form a water compatible bonded phase [37,38].

The controlled pore glass packings are nearly pure silica in composition but are prepared by a different method to the microporous silica materials discussed previously. A borosilicate glass is thermally treated to cause separation of the borate and silicate anions. The material is crushed and sieved to the desired particle size and the pore network is created by chemically leaching out the borate anions. The pore size is controlled by the thermal treatment. The chromatographic properties of the controlled pore glass column packings are, not unsurprisingly, similar to the microporous silica packings. Adsorptive activity can be reduced in the same manner as described for silica packings.

4.4 Column Preparation

The decision whether to pack or purchase HPLC columns depends on the type of laboratory and the rate of column consumption. In the research laboratory where the emphasis is on fundamental studies or the preparation of novel column packings, there may be no choice but to prepare all columns. In a large analytical laboratory where the consumption of columns is substantial, it may be economically justifiable to pack all the columns used. Some laboratories prefer to pack their own columns simply to avoid problems with changes in column specification by the manufacturer or to avoid periodic shortages in supply. However, for those laboratories using only a few columns per year, it is unlikely that a column packing facility is economically justifiable. A perusal through the scientific literature indicates that this is the general case and few laboratories currently pack their own columns. The general quality of commercial columns has improved substantially over the last few years and competition for the available market has stabilized prices at reasonable levels. However, even if the manufacturer provides a test chromatogram with the column, it is necessary that the receiving laboratory repeat the test specification to check for damage or change during shipment. Since testing individual columns requires equipment and personnel, the manufacturers would like to transfer their quality control requirements to the purchaser in return for a lower price per column. This is attractive to many laboratories with restricted budgets (particularly university laboratories with low labor overheads!), but requires that the laboratory is conversant with column testing procedures.

4.4.1 The Column Blank

The choice of the column blank and its associated fittings can influence
both the performance of the column and the applications in which the column can
be used. The construction material must be able to withstand the pressure at
which the column is to be operated and be chemically resistant to a wide range
of mobile phases. For this reason, seamless polished 316 stainless steel is the
preferred material. It is mechanically strong and is inert to most solvents
with the exception of halide salts, particularly at low pH [2]. Corrosion
occurs when the halide salt containing mobile phase is in prolonged contact with
the column. Flushing the system with water at the end of the day will usually
prevent any serious problems. If corrosion is, however, envisaged as a problem
(this is rarely the case) then glass may be used instead of stainless steel.
Heavy-walled glass tubing with an upper operating pressure limit of about 600
p.s.i. is commercially available. Glass-lined stainless steel columns,
combining the smoothness and inertness of glass with the strength of stainless
steel, are also available [39].

Prior to packing, the interior wall of the empty column must be cleaned and
polished. It is particularly important that the column blank has a smooth inner
surface and that residual grease and metal fines from the tube drawing process
are removed. This is achieved by a thorough washing sequence involving aqueous
and organic solvents. Four or five solvents of different polarity are usually
sufficient. Dilute acids and detergent solutions may also be included in the
washing sequence. The air-dried tube is then polished internally by passing a
lint-free cloth attached to a nylon thread or a pipe cleaner through the column
several times. The inner wall of the polished column should be smooth and
reflective when viewed against the light, and free from any indentations and/or
burrs.

Column end fittings and connectors must be designed with a minimum dead
volume to eliminate extracolumn band broadening. Standard tube fittings, such
as those employed in gas chromatography and general laboratory plumbing, are not
suitable for column connection. Some typical low dead-volume column fittings
for HPLC are shown in Figure 4.8. These fittings may be obtained from most
chromatographic supply houses. Zero dead-volume fittings are also available and
provide a minimum of dead volume by eliminating the space between the column
ends within the fitting.

The packing material is retained and protected from the intrusion of
particle matter by porous frits or screens. These may be either incorporated
into the column ends themselves or made an integral part of the column end
fittings. The frits are made from porous stainless steel, Hastaloy, or PTFE and
have an average pore diameter less than the particle size of the column packing

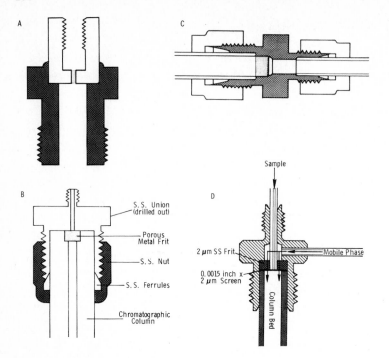

Figure 4.8 Column end fittings and connectors for HPLC. A, low dead-volume
            column end fitting; B, standard column terminator; C, standard
            reducing union; D, coaxial flow column end fitting.

(typically 2 micrometers).

     The most common size for analytical columns packed with 5 to 10 micrometer
diameter materials is 25 to 30 cm in length with an internal diameter of 4.0 to
4.6 mm. Longer columns with a narrower internal diameter (e.g., 50 cm x 2.1 mm
I.D.) are used to pack porous layer beads with average particle diameters
greater than 20 micrometers. With the introduction of column packings having 3
and 5 micrometer diameters there has been a trend towards the use of shorter
columns of 5-7.5 or 10-15 cm with 4.0-4.6 or 8.0 mm internal diameters,
respectively [40]. Narrow bore columns are favored for trace analysis as they
consume less solvent for a separation and provide less peak dilution for a
component at the detector. They also require less packing material to prepare
the column. Wider bore columns can separate larger sample sizes without loss in
efficiency and are operated at higher flow rates to minimize the influence of
extracolumn band broadening. The packing of columns longer than mentioned above
usually results in columns of lower overall efficiency. When longer columns are
needed for a particular separation, superior results are obtained by the series
coupling of standard columns with short lengths of 0.25 mm I.D. stainless steel
capillary tubing [41].

4.4.2 Selection of Column Packing Method

The method chosen for column packing depends mainly on the mechanical
strength of the packing and its particle size and particle size distribution.
Rigid particles of greater than 20 micrometer diameter can be dry packed
[18,42]. Particles of smaller diameter have a high surface energy to mass ratio
and tend to form aggregates when dry packed. This results in a nonuniform
packing with widely varying flow velocities along the column and poor overall
column efficiency. For particles with diameters less than 20 micrometers,
slurry packing techniques are used. The main requirements of the slurry packing
solvent are that it must thoroughly wet the packing and provide adequate
dispersion of the packing material. Many common solvents and solvent mixtures
have been used for this purpose. No single solvent system is suitable for all
column packing operations due to the diverse chemical nature of the surfaces of
the column packings used. Bonded-phase materials with hydrocarbon-like surfaces
become highly charged when packed under conditions of rapid solvent flow. The
charge repulsion caused by the build-up of static charges results in the
formation of an open packing that collapses when the flow is stopped. This can
be remedied by adding a trace amount of an electrolyte to the slurry solvent.
With particle diameters of about 10 micrometers, balanced-density slurry
techniques are most commonly used. Balanced-density slurry solvents are usually
mixtures of such solvents as tetraboromethane, tetrachloroethylene, and methanol
mixed in various proportions to match the density of the column packing
material. Particle segregation by sedimentation decreases as the density of the
slurry solvent approaches that of the packing material. Consequently, the
slurry solvent mixture must be tailored to the properties of the individual
packing material and no universal recipe for all packings exists [43-45]. The
optimum slurry solvent combination is easily established by trial and error.
After ultrasonic agitation, the balanced-density slurry should be stable at room
temperature in a draft-free environment for at least 30 minutes without signs of
sedimentation. Particle sedimentation towards the bottom of the flask indicates
that the concentration of the denser slurry solvent should be increased. The
opposite is true when sedimentation occurs against the influence of gravity.
Some suitable solvents for preparing balanced-density solvent mixtures are given
in Table 4.5. The brominated or iodated alkanes are somewhat expensive, highly
toxic, and may react with some packing materials. Iodated solvents may
chemically react with amino-containing phases and must be avoided in this case.
Packing materials with tight particle size distributions and average
particle diameters of 5 micrometers or less sediment very slowly with good
dispersive properties and therefore a balanced-density slurry is rarely required
for their packing [46-51]. A slurry solvent mixture containing 1.0% of a

non-ionic surfactant has been found effective for packing a wide range of 5 micrometer bonded-phase materials [51].

TABLE 4.5

PROPERTIES OF SOME SLURRY-PACKING SOLVENTS

Solvent	Density g/ml	Viscosity cP, 20°C
Diiodomethane	3.3	2.9
1,1,2,2-Tetrabromomethane	3.0	---
Dibromomethane	2.5	1.0
Iodomethane	2.3	0.5
Tetrachloroethylene	1.6	0.9
Carbon tetrachloride	1.6	1.0
Chloroform	1.5	0.6
Trichloroethylene	1.5	0.6
Bromoethane	1.5	0.4
Dichloromethane	1.3	0.4
Ethylene glycol	1.11	1.7
Water	1.0	1.0
Tetrahydrofuran	0.9	0.5
n-Butanol	0.8	3.0
n-Propanol	0.8	2.3
Methanol	0.8	0.6
Cyclohexane	0.8	1.0
n-Heptane	0.7	0.4
Isooctane	0.7	0.5

As particles sediment slowly in solvents of high viscosity, efficient columns can be packed with viscous solvents as the slurry medium [52]. However, solvents of high viscosity increase the resistance to flow through the column so that the packing procedure is slow and requires very high pressures. Consequently, this method of column packing has not been widely used.

Review articles devoted to column packing procedures are available [2,18,30]. Knox has summarized the critical points for successful slurry packing of columns, Table 4.6 [53].

TABLE 4.6

CRITICAL POINTS IN SLURRY PACKING

1. The particles must not sediment too fast during the procedure.

2. The particles must not agglomerate.

3. The particles must hit the accumulating bed at a high impact velocity.

4. Each particle should have time to settle in before it is buried by other particles landing on top.

5. The liquid used to support the slurry must be easily washed out of the packing and must not react with it.

4.4.3 Dry-Packing Column Procedures

Rigid particles with diameters > 20 micrometers can be dry-packed efficiently with relatively simple apparatus using the tap-fill method [18]. The empty column is held vertically with an end fitting at its lower end and a small increment of packing, equivalent to about 3-5 mm of column bed, is added from a funnel. The packing is consolidated by tapping firmly on a hard surface 2-3 times a second (80-100 times per packing increment) while lightly rapping on the side of the column at the approximate packing level and rotating the column slowly. Further increments of packing are added and the process is repeated until the column is filled. Verticle tapping is then continued for 3-5 minutes and the inlet fitting attached without disturbing the column packing. The packed column is then attached to the liquid chromatograph and equilibrated with the mobile phase until no further bubbles apppear at the column outlet and the column pressure and flow rate have stabilized.

During a recent study of dry-packing techniques for an irregular silica packing it was found that vertical tapping during the packing process was detrimental to column performance [42]. The best results were obtained when incremental addition, lateral tapping at the level of the column packing and rotation of the column were employed. Incremental addition of the column packing, followed by consolidation of the packed bed, gave better results than bulk filling. The column characteristics such as plate height, permeability, etc., can be reproduced to ± 10% for the same packing material using either manual or automatic dry-packing procedures.

Lateral tapping is performed with a hard metal tool. Vibration devices, similar to those used to pack GC columns, should not be used as they tend to produce particle sizing across the diameter of the column. This results in a heterogeneous bed with large particles near the wall and the smaller particles at the center.

4.4.4 Down-Fill Slurry Column Packing

Slurry packing techniques are required for the preparation of efficient columns with rigid particles of < 20 micrometer diameter. The same general packing apparatus, Figure 4.9, can be used to pack columns by the balanced-density slurry, nonbalanced-density slurry, or the viscous slurry method.

Prior to the preparation of the slurry, some preliminary treatment of the packing material is required. Silica gel packings are heated in an oven to remove physically adsorbed water which tends to cause particle sedimentation during packing. The column packings are suspended in a solvent to remove fines or agglomerates. Typical slurry concentrations are 1–30% w/w with 5–15% w/w considered optimum. The slurry is degassed in an ultrasonic bath prior to loading it into the packing reservoir.

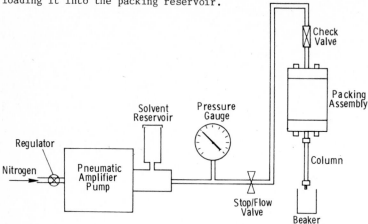

Figure 4.9   Down–fill slurry packing apparatus. (Reproduced with permission from ref. 48. Copyright Elsevier Scientific Publishing Co.)

Using Figure 4.9 as a guide, the packing procedure using the down–fill techniques will be described. At the lower end of the precleaned column blank a retaining porous frit or screen and a standard Swagelok fitting are attached. The Swagelok fitting is used to allow maximum solvent flow through the column during packing and must be replaced with a zero dead–volume fitting before installing the column in the chromatograph. A 3–5 cm length of connector tubing, which is affixed to the base of the packing reservoir, is attached to the top end of the column. The connector tube behaves as a mini–reservoir and guides the slurry into the column blank and also ensures that the column is completely filled at the end of the packing procedure. The internal diameter of the connector tube should be the same as the column blank to ensure a rapid and smooth passage of the slurry into the column without subjecting the packing to undue sheer forces. The column blank and connector tube are filled with either the slurry solvent itself or a solvent of higher density and degassed by tapping the side of the column with a metal rod. A stopcock is attached to the lower end of the column to retain the solvent. The slurry is added to the reservoir and the connecting tubing is filled with solvent back to the stop–flow valve. This solvent may be the same or different than the slurry solvent. The pressure used for the packing operation is then established up to the stop–flow valve. The column is packed at a pressure of 5000–12,000 p.s.i. using a gas–driven pneumatic amplifier pump. The packing process is rapid; with the stopcock open, the stop–flow valve is also opened and a constant pressure is maintained while about 100 ml of the packing solvent is pumped through the column. The stop–flow valve is then closed and the pressure allowed to bleed through the column until the solvent flow ceases. The performance of the column packing is often improved by bed consolidation. The packing apparatus is pressurized to its original value, the stop-flow valve is opened, and a further 50-100 ml of solvent is pumped through the column. The bed consolidation process is repeated several times until the column permeability is constant. That is, at a constant pressure, the column flow rate does not change after repressurization. The column is then removed from the packing apparatus, excess packing is brushed from the column top with a razor blade, and a screen or frit and a zero dead-volume fitting are attached. Prior to testing and use, the column may require further conditioning to wash out the packing solvent or to control the activity of adsorbent packings.

4.4.5 Up-Fill Slurry Column Packing

The up-fill slurry packing method is performed with an apparatus similar to that in Figure 4.10. It is the only packing method which has been used successfully to pack columns longer than 25 cm with 5 micrometer particles. The head of the packing reservoir can be modified to allow six columns to be packed simultaneously [54].

The principle of the packing method is based on the upward displacement of

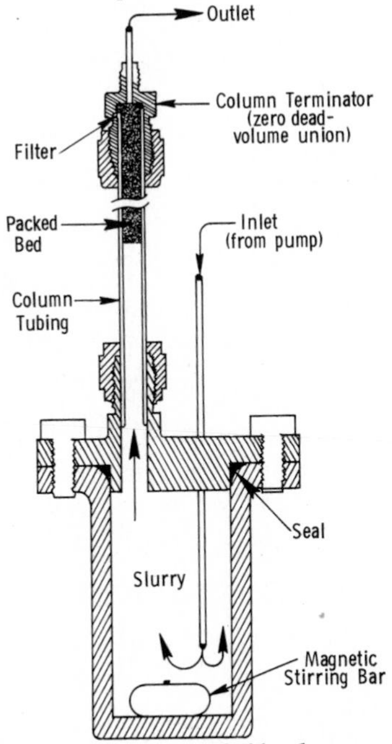

Figure 4.10 Up-fill slurry packing apparatus. (Reproduced with permission from Micromeritics Instruments Corp.)

a stable slurry into the column blank from a reservoir. The contents of this reservoir are continuously diluted by the slurry solvent being pumped into it. Dilute slurries (1-10% w/w) and continuous stirring are used to minimize the possibility of agglomeration of the packing slurry in the reservoir. This becomes important with packings having diameters greater than 5 micrometers. The viscosity of the slurry solvent does not influence the efficiency of the packed column but does, however, control the packing rate. Therefore, slurry solvents of low viscosity are preferred. The controlling factor for packing columns by the up-fill method is that the velocity of both the particles striking the bed and the solvent transporting them must be high enough to prevent the particles from falling back into the reservoir. If this condition is met, each particle in a dilute slurry is not influenced by the other particles and packs randomly to form an even, densely-packed bed.

To pack a column, the reservoir in Figure 4.10 is charged with the packing slurry. The slurry is stirred constantly to maintain a homogeneous suspension. With the lid of the reservoir secured and the outlet valve open, the slurry solvent is pumped gently into the attached column until the first drop of slurry appears at the column exit. This purges trapped air from the system. The outlet valve is closed, the pressure used to pack the column, (2000-7500 p.s.i.) is established, and the valve is reopened to force the slurry into the column. The packing process is complete when the flow rate through the column has stabilized. The pump and magnetic stirrer are stopped, the pressure is allowed to bleed away, and the column is disconnected from the reservoir. A zero dead-volume fitting and frit are attached to the column inlet end and the column is ready for equilibration and testing.

4.4.6 Packing Columns with Semirigid Particles

Packing procedures for semirigid particles are similar to those used for rigid particles, except that the column packing pressure is limited by the lower mechanical strength of the packing. Ion-exchange resins with a polymeric backbone are usually packed by the salt-balanced-density slurry method. Organic resins must be swollen in the slurry solvent prior to packing. The resin is packed into the column at a pressure dependent on the mechanical strength of the swollen resin beads, usually < 5000 p.s.i. for even the strongest beads. Semirigid organic gels are usually packed by the balanced-density slurry packing technqiue after first being swollen for several hours in the slurry medium. These gels are normally packed at pressures less than 3000 p.s.i. The organic gels have a low density compared to silica and require lower density solvents for slurry preparation.

Columns of soft gels cannot be packed from dry particles, nor can the high pressure slurry-packing process be used since the gel structure collapses even under relatively low pressures. These soft gels are not useful for HPLC.

4.5 Column Testing and Evaluation

A new column, whether purchased or packed in the laboratory, should always be tested before use to ensure that it meets the proper specifications for its intended use. The test chromatogram also provides a reference of column performance which may be used later to assess column changes. Repeating the test chromatogram at periodic intervals should be a part of good laboratory practice so that the status of the column is known. Changes in such primary column parameters as the capacity factor, plate count, separation factor, and pressure drop are all associated with changes in the chromatographic system and may be reversible. A routine column testing procedure will be described shortly that utilizes the parameters obtained from the test chromatogram. The results obtained are similar to those provided with pretested commercial columns. The parameters used to describe column performance are summarized in Table 4.7. A complete evaluation of column performance is more complex and is far too detailed and time consuming for use in most analytical laboratories. Bristow and Knox have summarized the parameters required to completely define column performance, Table 4.8 [55].

TABLE 4.7

PARAMETERS FOR ROUTINE COLUMN TESTING

- Plate count for several solutes (e.g., k = 0,3,10)
- Retention (e.g., k values) for appropriate solutes
- Relative retention (values) for unlike solutes
- Peak symmetry (skew or asymmetry factor)
- Column pressure drop or permeability measured at a given flow rate, mobile phase, and temperature
- Concentration of bonded organic phase

4.5.1 Routine Column Test Methods

No single set of conditions exists for testing the performance of all column types. The chemical nature of the column packing dictates the choice of test solutes and the mobile phase. Some examples of suitable test mixtures for evaluating the performance of different column types are given in Table 4.9. Other practical considerations in designing a test system are given in Table 4.10.

TABLE 4.8

PARAMETERS FOR COMPLETE COLUMN EVALUATION

Operating Conditions:	Column temperature Packing designation Packing method Mobile phase composition Test sample composition Detection method Injection method
Properties of the Mobile Phase and Solute(s):	Mobile phase viscosity Diffusion coefficient of the solute(s) in the mobile phase
Geometrical Parameters:	Column bed length Column bore Particle size
Chromatographic Parameters:	Volume of injected sample Peak retention times or retention volumes for the unretained and retained solute(s) Peak width at half-height of solutes on a time or retention volume basis Column pressure drop Volume flow rate of mobile phase
Chromatographic Properties:	Reduced plate height Reduced linear mobile phase velocity Column flow resistance
Subsidiary Properties:	Column capacity ratio Knox-Parcher ratio Total porosity

Having chosen the test sample and mobile phase composition, the chromatogram is run, usually at a fairly fast chart speed to reduce errors associated with the measurement of peak widths, etc. The parameters calculated from the chromatogram are the retention volume and capacity factor of each component, the plate count for the unretained peak and at least one of the retained peaks, the peak asymmetry factor for each component, and the separation factor for at least one pair of solutes. The pressure drop for the column at the optimum test flow rate should also be noted. This data is then used to determine two types of performance criteria. These are kinetic parameters, which indicate how well the column is physically packed, and thermodynamic parameters, which indicate whether the column packing material meets the manufacturer's specifications. Examples of such thermodynamic parameters are whether the percentage of bonded phase is correctly stated, and whether this particular column will perform a separation similar to a column of the same type used previously in the laboratory.

TABLE 4.9

TEST MIXTURES FOR EVALUATING COLUMN PERFORMANCE

Test Mixture	Mobile Phase
a) Adsorbents (and some polar bonded phases)	
Carbon tetrachloride/benzene/naphthalene	Heptane (40 ml/h)
Pentane/toluene/nitrobenzene/acetophenone/ 2,6-dinitrotoluene/1,3,5-trinitrobenzene	Hexane-methanol (99.5:0.5)
Benzene/benzyl alcohol/dimethylaniline	- - - - -
Naphthalene/m-dinitrobenzene/o-nitroaniline	Hexane-methylene chloride-2-propanol (89.5:10:0.5) (20 ml/h)
Toluene/phenanthrene/nitrobenzene	Hexane-acetonitrile (99:1)
m-Xylene/nitrobenzene	Hexane-acetonitrile (99:1)
b) Reversed-Phase	
Benzene/cumene/tetramethylbenzene	- - - - -
Methanol/acetone/phenol/p-cresol/ 2,5-xylenol/anisole/phenetole	Methanol-water (60:40)
Benzene/naphthalene/biphenyl	Methanol-water
Resorcinol/naphthalene/anthracene	Acetonitrile-water (55:45)
Toluene/benzene/acetophenone	Methanol-water
Anthraquinone/2-methylanthraquinone/ 2-ethylanthraquinone	Methanol-water (70:30)
c) Ion-exchange	
i) Anion exchangers Cytidine-5-monophosphate/ uridine-5-monophosphate/ guanosine-5-monophosphate	0.05 M KH_2PO_4 pH 3.35 (40 ml/h)
ii) Cation exchangers Uracil/guanine/cytosine	0.2 M $NH_4H_2PO_4$ pH 3.5 (40 ml/h)
d) Size Exclusion[a]	
Toluene Acetone Ethylene glycol	Tetrahydrofuran Water

[a] To indicate plate count. Size calibration is treated elsewhere

TABLE 4.10

DESIRABLE PROPERTIES OF TEST SOLUTES AND MOBILE PHASES FOR COLUMN TESTING

- Test solutes should be of low molecular weight to ensure rapid diffusion and easy access to the packing pore structure.

- Test mixture should contain components that correctly characterize the column in terms of both kinetic and thermodynamic performance.

- A value for the column dead volume is required in most calculations. It is convenient to have one component of the test mixture as an unretained solute.

- At least two components of the test mixture should have k values between 2 and 10.

- All measurements should be made with a mobile phase of simple composition, low viscosity, and under isocratic conditions.

- The sample volume and/or amount of test sample should not overload the column.

- It is convenient to use test solutes with a strong absorbance at 254 nm. A solute giving 50-100% FSD with a detector setting of 0.05 or 1.0 AUFS is convenient.

The method used to calculate the peak asymmetry factor is shown in Figure 4.11. It follows the recommendations of Kirkland [56] but is not universally adopted by all column suppliers. In this discussion, the peak asymmetry factor is measured at 10% of the peak height. Some column suppliers use the baseline measurement to specify peak asymmetry, leading to larger limiting values than those given here. Peak asymmetry, especially of the unretained peak or of weakly-retained peaks (k < 3), is typical of poorly packed columns (if instrumental contributions can be excluded). If only the unretained peak (k < 1) is asymmetric and/or there is a significant difference (> 15%) between the plate counts for the unretained and retained peaks, then extracolumn effects are implicated and should be investigated prior to repeating the test. As a general guide, columns yielding peak asymmetry factors greater than 1.2 are of poor quality and those with values greater than 1.6 should be discarded.

Some typical minimum values for the plate counts of commercially available columns are given in Table 4.11. The number of theoretical plates obtained from a particular column type will correlate primarily with the average particle size of the packing. It is also influenced by the column flow rate at which the measurements are made and by the composition of the mobile phase. Also, the calculation of the plate count will produce values having a large positive deviation from the true result when applied to asymmetric peaks. Lower than average values for the plate count indicate a poorly packed column.

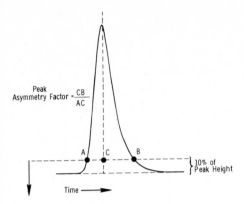

Figure 4.11   Calculation of the peak asymmetry factor.

TABLE 4.11

EXPECTED EFFICIENCY OF COMMERCIALLY AVAILABLE PACKED COLUMNS

Type	Minimum Efficiency (plates/meter)
a) Adsorbent	
Pellicular (35-45 micrometers)	2000-5000
Porous (10 micrometers)	24,000
(5 micrometers)	40,000
b) Reversed-Phase	
Pellicular (35 micrometers)	1000
Porous (10 micrometers)	12,000-20,000
(5 micrometers)	35,000-45,000
(3 micrometers	80,000-100,000
c) Ion-Exchange	
Pellicular (35 micrometers)	1000
Porous (10 micrometers) Anion	10,000
Cation	15,000
d) Size-Exclusion	
Porous Silica (10 micrometers)	15,000-20,000
Semirigid Organic Gels (10 micrometers)	9000-12,000

Under the test conditions, large changes in k and α values for the test solutes compared to those found previously for similar columns indicate a change in the chemical nature of the packing material. The separation characteristics of the column will therefore be different from notionally similar columns used previously. Significant changes in the column pressure drop between comparable columns measured under the same test conditions indicate that either the column or its fittings are partially blocked. If cleaning the fittings does not restore the pressure drop to the normal range the column should be discarded.

For the comparison of columns of a similar type, it is essential that the experimental conditions for the test chromatogram be faithfully reproduced and sufficient time is allowed for column equilibration before starting the test. The expected changes in column performance parameters due to changes in the experimental conditions are summarized in Table 4.12. Some typical values for column performance parameters and their acceptable ranges are given in Table 4.13 for a reversed-phase column obtained from a·commercial source.

4.5.2 The Use of Reduced Parameters for Column Testing

The use of reduced parameters (h, v, Θ,) rather than absolute parameters (HETP, u, K_o) enables the performance of different column types to be compared, even if the operating conditions are not identical. Columns packed with materials of different sizes, eluted with solvents of different viscosities, and tested with solutes having different diffusion coefficients may be compared directly using the parameters defined below [17,55,57].

TABLE 4.12

EXPERIMENTAL PARAMETERS AFFECTING COLUMN EFFICIENCY

Parameter	Change in Efficiency (n)
Linear velocity,	Lower velocities give higher values of n in the velocity range above 0.05 cm/sec
Particle size, d_p	Smaller particle size gives higher values of n
Column length, L	n is proportional to L
Mobile phase viscosity,	Lower values give high values of n
Temperature, T	High values reduce viscosity and give higher values of n
Capacity factor, k	Low k ($<$ 2) give lower values of n; for high k values ($>$ 2) n is little influenced
Extracolumn volumes	n is decreased due to band broadening contributions to peak width
Sample amount and size	Large amounts (mg) or large volumes (several hundred microliters) will decrease n

The reduced plate height is the number of particles to the theoretical plate and is calculated from equation (4.4)

TABLE 4.13

TEST CONDITIONS AND SPECIFICATIONS FOR A 30 cm x 4.0 mm I.D. REVERSED-PHASE
COLUMN
Particle size 10 micrometers

Test Conditions

Sample: Resorcinol/naphthalene/anthracene
Detector: UV,   = 254 nm, 0.16 AUFS
Sample Size: 2.0 microliters, concentration unknown
Temperature: Ambient
Mobile Phase: Acetonitrile-water (55:45) flow rate 2.0 ml/min
Chart Speed: 1.0 cm/min

Test Sample Retention

Compound	Retention Time(min)	Retention Volume(ml)
Resorcinol	1.3	2.7
Naphthalene	6.6	13.2
Anthracene	13.9	27.7

Column Performance

	Test Results	Specification
Retention volume (resorcinol)	2.7	1.8-3.4
Retention volume (anthracene)	27.7	19.0-31.0
Asymmetry factor (anthracene)	1.26	0.85-1.80
Capacity factor (anthracene)	9.40	
Number of plates per column (anthracene)	5437	>5000
Number of plater per meter	18,122	
HEPT (mm)	0.055	
Selectivity factor (anthracene/naphthalene)	2.4	1.9-3.0

$$h = \frac{HETP}{d_p} = \frac{1}{5.54}\left[\frac{L}{d_p}\right]\left[\frac{w_h}{t_R}\right]^2 \qquad (4.4)$$

HETP = height equivalent to a theoretical plate

L   = column length

d_p = particle diameter

w_h = peak width at half height

t_R = retention time of a retained solute

The reduced velocity measures the rate of flow of the mobile phase relative to
the rate of diffusion of the solute over one particle diameter and is calculated
from equation (4.5)

$$v = \frac{ud_p}{D_m} = \frac{Ld_p}{t_m D_m} \qquad (4.5)$$

v = reduced linear velocity

u = linear velocity

t_m = retention time of an unretained peak

D_m = solute diffusion coefficient (m^2/s)

It is unlikely that an accurate value of D_m will be known for the test solute in the mobile phase. An approximate value can be calculated from the Wilke–Chang equation (4.6) [58].

$$D_m = 7.4 \times 10^{-12} \; \frac{(\psi M)^{1/2} T}{V_s \eta} \tag{4.6}$$

ψ = solvent-dependent constant
M = relative molecular mass of the solvent
T = absolute temperature
η = solvent viscosity
V_s = molar volume of the solvent

Typical values for low molecular weight solutes fall in the range $0.5\text{–}3.5 \times 10^{-9}$ m^2/s. The higher value is typical for organic solvents of low viscosity such as hexane, and the lower value for polar aqueous solvents. An educated guess at the value for D_m is adequate for column testing, but of course this means that the value for v will contain some error and should not be considered as an absolute value.

The column flow resistance, $\ominus$, is a dimensionless parameter which provides a measure of the resistance to flow of the mobile phase and takes into account the influences of column length, particle diameter, and eluent viscosity. It can be calculated from equation (4.7).

$$\ominus = \frac{d_p^2}{K_o} = \frac{\Delta P d_p^2 t_m}{\eta L^2} \tag{4.7}$$

ΔP = column pressure drop

According to chromatographic theory, the reduced plate height is related to the reduced linear velocity by equation (4.8) [53].

$$h = \frac{B}{v} + A v^{0.33} + C v \tag{4.8}$$

| contribution from axial diffusion | contribution from flow in interparticle space | contribution from slow equilibrium between mobile and stationary phases |

The constant B reflects the geometry of the eluent in the column and the extent to which diffusion of the solute is hindered by the presence of the packing. For columns of acceptable performance, B should be less than 4 and an approximate value of 2 is desired in practice. The constant A is a measure of how well the column is packed. A well-packed column will have a value of A between 0.5 and 1.0 while a value greater than 2 indicates a poor column. The value of C reflects the efficiency of mass transfer between the stationary and mobile phases. At high reduced mobile phase velocities the C term dominates the reduced plate height value, and therefore column efficiency. C is close to zero for pellicular packings but has a greater value for porous packings. A value for C of 0.05 is reasonable for the latter.

The constants A, B and C can be calculated by curve fitting a plot of h against **v**. Their values depend only on how well the column has been packed and, for "good" columns, will have constant minimum values that are independent of particle size. Consequently, a plot of h against **v** for any "good" column should fall on the same general curve, dependent only on whether the packing is porous or pellicular in nature. These "idealized" curves for "good" columns are shown in Figure 4.12. The idealized curves are described by equations (4.9) and (4.10).

$$h = \frac{2}{v} + v^{0.33} + 0.003v \qquad \text{(pellicular)} \qquad (4.9)$$

$$h = \frac{2}{v} + v^{0.33} + 0.05v \qquad \text{(porous)} \qquad (4.10)$$

In agreement with theory, the experimental data for "good" columns fall on top

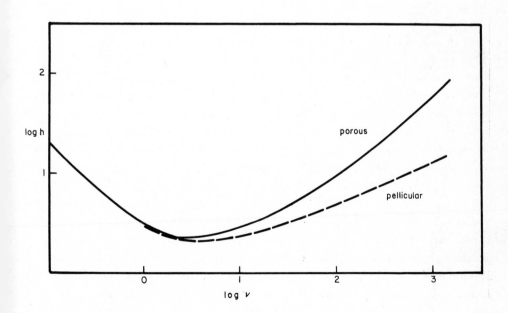

Figure 4.12   A plot of reduced plate height against reduced mobile phase velocity for a good column packed with pellicular or porous packings.

of one of the two "idealized" curves, whereas the experimental data for "poor" columns lie above the appropriate curve. The deviation of the experimental from the theoretical curve can be used to diagnose the reason for the poor column performance of the former. If the minimum h value is less than 3 for a reduced mobile phase velocity of $3 < v < 10$, then the column is well-packed. If only a few data points have been determined, the value of h should not exceed 3 or 4 at v about 5 and 10 to 20 at $v = 100$ for a "good" column. Returning to the idealized curves in Figure 4.12, if the value of h lies above the appropriate minimum and the experimental curve is flat in this region, then the column is poorly packed. Higher experimental values of h than predicted with values of $v > 20$ indicate a poor quality column packing material.

The experimental value for $\ominus$ provides information on how the chromatographic system as a whole is performing. Typical values for porous packings are in the range 250-650. A very high value, for example ten times the normal value, is indicative of a partial blockage in the chromatographic system. This should be diagnosed and rectified before proceeding further with the column tests.

4.5.3 Quality Control of Column Packing Materials

Many laboratories are concerned about the batch-to-batch reproducibility of column packing materials since this will obviously influence column reproducibility. Variations can be expected between lots from a single manufacturer and even larger variations between different manufacturers of the same notional material. The general variability in the performance of column packing materials is greatest for the chemically-bonded stationary phases. This arises because all chemically-bonded phases sold by the same general description are in fact not chemically identical. For example, there are at least three kinds of octadecylsilyl reversed-phase column packing materials. Depending on the method of preparation, monomeric, polymeric, and either of the previous two types having residual silanol groups endcapped with low molecular weight alkylsilyl groups are commercially available. The primary structure of the silica gel backbone will influence the surface area of the bonded packing, the concentraion of bonded hydrocarbon groups, and the ability of the mobile phase to wet the column [59]. There is thus a wide variation in what is sold as an octadecyl reversed-phase packing. The large variability in separtion properties of reversed-phase materials from different sources has been demonstrated for the separation of polycyclic aromatic hydrocarbons [60-65], steroids [64], and aromatic compounds [65,66].

There are a few studies published on the lot-to-lot variation in the separation properties of the column packing materials in current use [67,78].

This variation may be quite large, especially for reversed-phase packings. Only a fraction of this variation can be assigned to the reproducibility of the experimental test conditions, Table 4.14. This data was based on 21 lots of a reversed-phase column packing material from one manufacturer over a 1.5 year period. Variations in column-to-column parameters for the same lot of packing material were accounted for almost entirely by the variation in the mobile phase composition and column temperature between tests. The error in the composition of the organic modifier present in the mobile phase was estimated as about ± 0.3% and the laboratory temperature as ± 1°C RSD.

TABLE 4.14

THE VARIATION OF TEST PARAMETERS FOR THE QUALITY CONTROL EVALUATION OF A REVERSED-PHASE PACKING

Test Compound/ Pair of Compounds	Capacity Factor (k)	Separation Factor (a)	Temperature Coefficient of k or a (24–45°C)	Change in k or a with Per- cent Organic Modifier
Biphenyl	1.82 ± 6.7		-1.2%/°C	-4.1%
Anthracene	3.99 ± 5%		-2.1%/°C	-7%
Androstenedione	8.64 ± 9%		-3.1%/°C	-10%
Biphenyl/ Naphthalene		1.45 ± 0.01	-0.001/°C	-0.012
Anthracene/ Phenanthrene		1.32 ± 0.06	-0.007/°C	-0.003
Androstenedione/ Testosterone		1.22 ± 0.03	0.002/°C	-0.0018

The control ranges established for the 21 lots evaluated over the test period are given in Table 4.15. For the separation factor of anthracene/phenanthrene, the quality control range of acceptable lots was established as 1.20-1.44. In resolution terms, for two columns of equal efficiency prepared from packings passing the quality control test, the resolution of anthracene/phenanthrene could differ by as much as 125%. For androstenedione/testosterone, resolution at constant column efficiency differed by as much as 75% between acceptable lots. For the same column materials, the resolution between naphthalene/biphenyl showed only a small variation between lots, indicating that the choice of test compounds, and therefore the sample to be separated, influence column reproducibility. For perspective, when nine reversed-phase materials from different manufacturers were evaluated at constant column efficiency, the variation in resolution for the test compounds used above was approximately 300%.

TABLE 4.15

QUALITY CONTROL RANGES FOR REVERSED-PHASE PACKING MATERIAL

(21 lots sampled)

Parameter	Quality Control Ranges		
	Low	High	Mean
Anthracene (k)	3.59	4.39	3.99
Anthracene/Phenanthrene (α)	1.20	1.44	1.32
Androstenedione (k)	7.10	10.18	8.64
Androstenedione/Testosterone (α)	1.16	1.28	1.22

4.5.4 Trace Metal Activity of Reversed-Phase Packings

The presence of residual trace metal ions in reversed-phase packings can dramatically lower the column efficiency with some polar molecules, especially those with acidic or basic centers [69]. Although there is no standard test for measuring this activity in column packings, Verzele has suggested a novel, qualitative test based on the separation of the hop bitter acids that are easily isolated from beer. He has shown that most commercial column packings contain a more than desirable concentration of trace metal ions. These can be removed by a series of extractions with refluxing hydrochloric acid. Whenever severe tailing is observed with acidic samples, the addition of 0.5% v/v phosphoric acid to the mobile phase is recommended. Certain metal complexing reagents such as EDTA may also be effective in eliminating metal/sample interactions when added to the mobile phase.

4.5.5 Measurement of the Column Dead Volume

The elution of a solute from different columns is not correctly described by the elution volume, V_R, since V_R varies with the inter- and intraparticle volumes associated with the column packing, as well as with differences in tube dimensions, guard columns, and other hardware placed between the points of injection and detection. To accurately describe the retention of a solute on a column we define a term called the capacity factor, k, equation (4.11)

$$k = \frac{V_R - V_m}{V_m} \qquad (4.11)$$

V_R = solute retention volume

V_m = column dead volume

A correct measure of the column dead volume, V_m, is essential for the accurate calculation of k, from which many other chromatographic relationships are derived. The column dead volume is obtained experimentally by measuring the retention volume of an unretained solute. It is the sum of the interparticle and intraparticle fluid volumes and the extracolumn dead volume. In a well-designed instrument the extracolumn dead volume due to the injector, connector tubing, column fittings, and detector will be very small.

The choice of test solute is important for measuring V_m [70-76]. It is essential that the test substance is not excluded from the intraparticle volume, otherwise the value obtained will be less than the real value. For adsorption columns, benzene is used as the test solute and methanol as the mobile phase. For reversed-phase columns the choice of test solute presents some problems. First of all, most organic test compounds show some partition with the stationary phase and lead to a value of V_m which is too high. Many ionic compounds in unbuffered eluents are excluded from the intraparticle volume by an ion repulsion mechanism, leading to a value for V_m which is too small. Wells and Clark, and Jinno have compared of the many methods used to measure V_m. For reversed-phase columns they recommend the use of sodium nitrate or nitrite at a concentration greater than 3×10^{-6} moles injected onto the column in an unbuffered mobile phase [72,75]. Other workers have shown that a 0.01% solution of potassium bromide with a detector operating at 200 nm is a suitable alternative to sodium nitrate [76]. Any test solute for measuring V_m must meet certain requirements: the value for V_m should be independent of the column flow rate, the mobile phase composition, and the sample concentration injected. Test solutes not meeting these requirements will provide erroneous values for V_m.

4.6 Radial Compression Columns

Even in well-packed columns, solute dispersion occurs to a greater extent in the region of the column wall than at the column core. This phenomena, known as the "wall effect", is due to differences in the packing density of particles close to the wall and those at the column center. This leads to inhomogeneous flow patterns across the diameter of the column and results in a lower overall column efficiency. Radial compression columns were designed to minimize both column void volumes and the wall effect. Compared to standard packed columns they have two novel features. A series of distribution plates at the entrance and exit of each column prevents irregular flow patterns from forming in the direction of sample migration. Secondly, and more importantly, the columns are prepared from heavy wall polyethylene cartridges (packed with normal chromatographic packings) that are uniformly radially compressed by hydraulic pressure in a purpose-built hydraulic press, Figure 4.13. The column wall is forced to deform and mold to the shape of the internal packing structure. This stabilizes the packing structure and reduces the number of channels available to the mobile phase at the wall/packing interface. Radial compression columns are

254

characterized by high permeability, low operating pressures, and high efficiency over a wide range of flow rates [77]. Compared to a conventional stainless steel column packed with the same material, an increase in efficiency approaching a factor of two might be expected. On an equivalent length basis, the pressure drop across the radial compression column will be greater at a fixed flow rate due to its higher packing density.

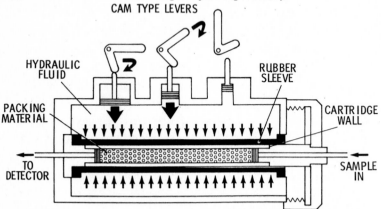

Figure 4.13 Hydraulic press for radial compression column chromatography. (Reproduced with permission from Waters Associates).

Radial compression columns were first described for use in preparative HPLC with column dimensions of 30 x 5.7 cm and packed with particles of 75 micrometers average diameter [78,79]. These columns were used to separate gram quantities of material, often at flow rates exceeding 100 ml/min. Analytical columns are available in 10 cm lengths with 5 or 8 mm internal diameters and packed with either 5 or 10 micrometer particulate packings. These have been shown to provide good separations of peptides [80] and low molecular weight serum constituents [81]. Their area of application is not limited to the biomedical field; they can be used to separate any sample which could be separated on a rigid stainless steel column.

4.7 Guard Columns

A guard column is a short column that is placed between the injector and analytical column to protect the latter from damage or loss of efficiency due to the presence of particulate matter or strongly adsorbed material in the samples to be analyzed [82]. It also functions as a saturator column to prevent dissolution of the stationary phase in the analytical column. Consequently, the guard column is usually packed with the same stationary phase as the analytical column and is discarded at intervals dictated by the contamination level of the sample. To maintain an adequate capacity for sample impurities without introducing excessive peak dispersion, the volume ratio of the guard column to that of the analytical column should be in the range 1:15 to 1:25.

The guard column is incorporated into the chromatographic system for practical convenience but represents a further component in which peak dispersion and loss of chromatographic efficiency can occur. In addition to the particle diameter and the packing density of the stationary phase, the choice of fittings and connecting devices is also important. Experimental observations have shown that the connection to the analytical column should be made with a low dead-volume fitting rather than a zero dead-volume fitting, and that short lengths of connecting tubing (0.15 to 0.60 mm I.D.) may be used without significantly diminishing the efficiency of the chromatographic system [83]. A well-designed guard column should not increase sample dispersion by more than 5-10%, a worthwhile sacrifice for the extended life of the analytical column [84]. When a greater loss of column efficiency can be tolerated, guard columns packed with microporous or pellicular packings > 20 micrometer in diameter may be used. Such columns have a cost and convenience advantage as they may be packed in the laboratory using the tap and fill method. Guard columns containing packings of smaller particle diameters must be slurry packed.

4.8 Mobile Phase Selection in HPLC

Retention in liquid chromatography depends on the strength of the solute's interaction with both the mobile and stationary phases. In gas chromatography, the mobile phase does not contribute to the selectivity of the separation system and, consequently, changing the stationary phase (i.e., column) is the principal method of adjusting this parameter. In liquid chromatography the opposite approach is generally adopted. An intelligent selection of the type of stationary phase for the separation is made and selectivity is adjusted by modifying the mobile phase. The selection of the mobile phase for a particular separation is thus a very important consideration in HPLC.

For liquid chromatography some fairly broad generalizations can be made about the selection of certain preferred solvents from the large number available. A suitable solvent will preferably have low viscosity, be compatible with the detection system, be readily available in a pure form, and, if possible, have low flammability and toxicity. Few solvents meet all of these requirements and a compromise must be made. Since detection in HPLC occurs on-line, compatibility with the detection system employed is very important although, of course, not related to the function of the solvent as a mobile phase. Solvents of low viscosity promote good solute diffusivity which improves the chromatographic performance of the column. Many of the common solvents used in HPLC are flammable or have some adverse health effects. Consequently, it is generally recommended that HPLC instrumentation be used in a well-ventilated laboratory. Also, in practice any solvent used must .be able to completely dissolve the sample without reacting with it chemically.

A short list of useful solvents for HPLC is given in Table 4.16. More extensive compilations are available [18,85-87]. Special grades of solvents are available for HPLC; the majority have been carefully purified to remove UV-absorbing impurities and particulate matter. Some solvents contain a preservative or antioxidant which can influence their chromatographic or detection compatibility properties. These may have to be removed prior to use. Suitable methods for the elimination of these and other potential problems which can arise from using commercial solvents are discussed elsewhere [2,82]. A chromatographic test for the purity of water, and any binary mixture of water with a miscible organic solvent, has been described by Bristol [88]. Using reversed-phase chromatography, the trace-enrichment technique (Chapter 7.6.6), and elution by forward and reverse gradients, it is possible to establish the level of solvent impurities in either the water or the organic component of a binary mixture which interferes in detector operation. In general, solvent impurities cause a drift in the detector baseline and diminished sensitivity under isocratic conditions, and large baseline fluctuations and spurious interfering peaks when gradient elution is used. If the impurities are not eluted from the column they will accumulate in the stationary phase, resulting in a change in solute retention.

The solvent properties of greatest chromatographic interest are strength and selectivity. For non-electrolytic solvents, solvent strength is synonymous with "polarity" in normal phase liquid-solid and liquid-liquid chromatography. Strong or polar solvents are characterized by their ability to dissolve polar samples and to elute solutes with low capacity factors in normal phase chromatography. In reversed-phase chromatography there is an inverse relationship between solvent strength and chromatographic elution power. For example, water is a strong solvent but has little elution power in reversed-phase chromatography and must, therefore, be mixed with a less polar organic modifier to elute a solute with a low capacity factor value.

Snyder has described solvent selectivity as "the ability of a given solvent to selectively dissolve one compound as opposed to another, where the polarities of the two compounds are not obviously different" [89]. The general strategy in chromatographic separations is to first adjust the solvent strength to maintain sample capacity factor values in the range 1-10 and then, while holding the solvent strength constant, modify the mobile phase to alter its selectivity until the required sample resolution is obtained. Thus, a general scheme that provided a means of classifying solvents according to their selectivity and strength would be useful.

The principal intermolecular forces responsible for describing the

TABLE 4.16

PROPERTIES OF SOME COMMON SOLVENTS USED IN HPLC

Solvent	UV Cut-off (nm)	Refractive Index	Boiling Point (°C)	Viscosity (cP, 25°C)	Solvent Polarity Parameter (P')	Solvent Strength Parameter (ϵ°)	Selectivity Group
Isooctane	197	1.389	99	0.47	0.1	0.01	— —
n-Hexane	190	1.372	69	0.30	0.1	0.01	— —
Methyl t-butyl ether	210	1.369	56	0.27	2.5	0.35	I
Benzene	278	1.501	81	0.65	2.7	0.32	VII
Methylene chloride	233	1.421	40	0.41	3.1	0.42	V
n-Propanol	240	1.385	97	1.9	4.0	0.82	II
Tetrahydrofuran	212	1.405	66	0.46	4.0	0.82	II
Ethyl acetate	256	1.370	77	0.43	4.4	0.58	VIa
Chloroform	245	1.443	61	0.53	4.1	0.40	VIII
Dioxane	215	1.420	101	1.2	4.8	0.56	VIa
Acetone	330	1.356	56	0.3	5.1	0.56	VIa
Ethanol	210	1.359	78	1.08	4.3	0.88	II
Acetic acid		1.370	118	1.1	6.0	Large	IV
Acetonitrile	190	1.341	82	0.34	5.8	0.65	VIb
Methanol	205	1.326	65	0.54	5.1	0.95	II
Water		1.333	100	0.89	10.2	Very Large	VIII

properties of non-electrolytic solvent are dispersion, dipole, and hydrogen bonding interactions [18,90-92]. Dispersive interactions result from the electronic asymmetry that exists in any molecule at any given instant. Dispersive forces are weak, nonselective, and therefore not of great chromatographic significance. However, they are the only intermolecular forces that exist in hydrocarbon solvents. Polar molecules possessing a permanent dipole moment are capable of both dipole induction and orientation interactions. These dipole interactions usually occur between the functional groups of molecules and are strongest for solvents having large dipole moments. Hydrogen bonding interactions occur between molecules which can behave as proton donors and acceptors. Strong donor solvents preferentially interact with and dissolve strong acceptor sample compounds, and vice versa.

Snyder [89,92] has described a scheme for classifying common solvents according to their polarity or chromatographic strength (P' values), and according to their selectivity or relative ability to engage in hydrogen bonding or dipole interactions. Various common solvents were classified into eight groups (I, II,....,VIII) showing significantly different selectivities. The P' values and selectivity group classifications for some solvents commonly used in liquid chromatography are given in Table 4.16.

The values of P' and selectivity factors are calculated from the experimentally-derived solute polarity distribution coefficients, K_g'', for the test solutes ethanol, dioxane, and nitromethane. The solute distribution coefficients are corrected for effects due to solute molecular size, solute/solvent dispersion interactions, and solute/solvent induction due to solvent polarizability. The resultant parameters P' and solvent selectivity (X_e, X_d, and X_n) should reflect only the selective interaction properties of the solvent. The test solutes ethanol, dioxane, and nitromethane are used to measure the strengths of solvent proton acceptor, proton donor, and strong dipole interactions, respectively. The polarity index P' is calculated as the sum of $\log K_g''$ values from the three solutes and the selectivity parameters X_e, X_d, and X_n are calculated as the ratio $\log K_g''/P'$ for each solute.

$$P' = \log (K_g'')_{ethanol} + \log (K_g'')_{dioxane} + \log (K_g'')_{nitromethane}$$

$$X_e = \frac{\log (K_g'')_{ethanol}}{P'}$$

$$X_d = \frac{\log (K_g'')_{dioxane}}{P'}$$

$$X_n = \frac{\log (K_g'')_{nitromethane}}{P'}$$

Solvents of similar selectivity can be grouped together by comparing their X_e, X_d, and X_n values. Arranging the experimental data in diagramatic form, Figure 4.14, illustrates that the many available solvents can be grouped into eight classes with distinctly different selectivities. Choosing a second solvent in the same selectivity group is not likely to significantly change the chromatographic separation obtained. However, choosing another solvent with similar strength from a different selectivity group can alter the separation. The solvent classification scheme is most valuable for identifying solvents with different chromatographic selectivities. Some typical examples of selective solvents are given in Table 4.17. The underlined solvents are those generally preferred for HPLC.

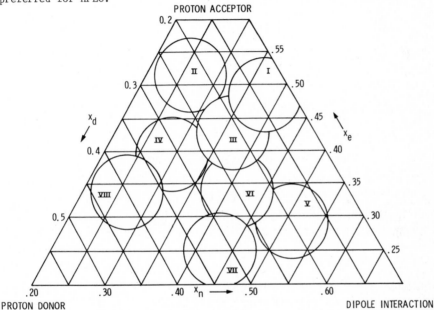

Figure 4.14 Selectivity triangle for solvents. (Reproduced with permission from ref. 92. Copyright Preston Publications, Inc.)

It is not unusual, and indeed in many instances it is preferable, to use mixtures of solvents as the mobile phase rather than a single pure solvent. In the case of binary solvents, mixing a strength-adjusting solvent with various volume fractions of a strong solvent enables the complete polarity or solvent strength range between the extremes represented by the pure solvents themselves to be covered. The strength-adjusting solvent is usually a non-selective solvent such as water for reversed-phase chromatography and hexane for normal phase applications. The solvent strength of a binary solvent mixture is the arithmetic average of the solvent strength weighting factors, Table 4.18, adjusted according to the volume fraction of each solvent. For normal phase chromatography, the solvent strength weighting factor, S_i, is the same as the polarity index, P'. In reversed-phase chromatography a different set of experimentally determined solvent weighting factors are used [93]. The solvent strength for any solvent mixture can be calculated from equation (4.12).

TABLE 4.17

CLASSIFICATION OF SOLVENT SELECTIVITY

Group Designation	Solvents
I	Aliphatic ethers, methyl t-butyl ether, tetramethylguanidine, hexamethyl phosphoric acid amide (trialkyl amines)
II	Aliphatic alcohols, methanol
III	Pyridine derivatives, tetrahydrofuran, amides (except formamide), glycol ethers, sulfoxides
VI	Glycols, benzyl alcohol, acetic acid, formamide
V	Methylene chloride, ethylene chloride
VI	a) Tricresyl phosphate, aliphatic ketones and esters, polyesters, dioxane b) Sulfones, nitriles, acetonitrile, propylene carbonate
VII	Aromatic hydrocarbons, toluene, halosubstituted aromatic hydrocarbons, nitro compounds, aromatic ethers
VIII	Fluoroalcohols, m-cresol, water, (chloroform)

TABLE 4.18

SOLVENT CHARACTERISTICS OF MIXED SOLVENTS

Solvent	Selectivity Group	Solvent Strength Weighting Factor (S_i)	
		Reversed-Phase[a]	Normal Phase
Methanol	II	2.6	5.1
Acetonitrile	VI	3.2	5.8
Tetrahydrofuran	III	4.5	4.0
Water	---	0	10.2
Chloroform	VIII		4.1
Methylene chloride	V		3.1
Methyl t-butyl ether	I		ca.2.5
Ethyl ether	I		2.8
Hexane	---		0

[a] Approximate values for some other common solvents are acetone (3.4), dioxane (3.5), ethanol (3.6), and isopropanol (4.2).

$$S_T = \Sigma_i \; S_i \; \Theta_i \qquad\qquad\qquad (4.12)$$

S_T = total solvent strength of the mixture

S_i = solvent strength weighting factor

Θ_i = volume fraction of solvent in the mixture

For the binary solvent mixture containing 60% methanol and 40% water, the mixed

solvent strength for reversed-phase chromatography is calculated as follows

$$S_T = S_{CH_3OH} \; \Theta_{CH_3OH} + S_{H_2O} \; \Theta_{H_2O}$$
$$S_T = (2.6) \; (0.6) + (0) \; (0.4)$$
$$S_T = 1.56$$

Solvent mixtures of identical strength but different selectivity are sometimes
described as iso-eluotropic solvents [94]. Suitable binary solvent mixtures can
be selected from the solvent selectivity triangle discussed previously. To
maximize the differences in selectivity, solvents are selected from different
selectivity group designations which lie close to the triangle apices, Figure
4.15. For reversed-phase chromatography a suitable selection might be methanol,
acetonitrile, and tetrahydrofuran mixed with water to control solvent strength.

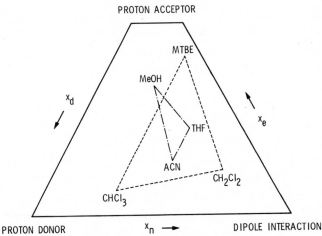

Figure 4.15   Selectivity triangle for preferred solvents in reversed-phase
(- - -) and normal phase (———) chromatography.

If, as in the methanol-water (60:40) example used above, the solvent strength
was about optimum for the separation and one wished to change solvent
selectivity to adjust resolution, then the volume fraction of the new solvent
required to give an iso-eluotropic mixture could be calculated from equation
(4.12). Using acetonitrile as an example and a value of S_T = 1.56, we have

262

$$1.56 = S_{CH_3CN} \Theta_{CH_3CN} + S_{H_2O} \Theta_{H_2O}$$
$$1.56 = (3.2) \Theta_{CH_3CN} + (0) \Theta_{H_2O}$$

and

$$\Theta_{CH_3CN} = \frac{1.56}{3.20} = 0.49 \text{ or } 49\%$$

Thus a mixture of acetonitrile-water (49:51) has similar solvent strength to a mixture of methanol-water (60:40). Likewise, it can be shown that a mixture of tetrahydrofuran-water (35:65) is iso-eluotropic with the two previous binary solvent mixtures.

Binary solvent mixtures provide a simple means for controlling solvent strength but only limited opportunities for controlling solvent selectivity. With ternary and quaternary solvent mixtures it is possible to fine tune solvent selectivity while maintaining a constant solvent strength [94-96]. In addition, there are only a very small number of organic modifiers that can be used as binary mixtures with water. Many more mobile phases of the same solvent strength can be prepared if ternary and quaternary solvent systems are used. An example of mobile phase selectivity optimization using a ternary solvent mixture is shown in Figure 4.16 [94]. The separation in Figure 4.16A was obtained with the binary mobile phase methanol-water (50:50). The solvent strength of this mixture is about correct for the separation based on the sample capacity factor values, but components 1 and 2, benzyl alcohol and phenol, are not resolved. This indicates that a change in solvent selectivity is required. The separation obtained with the iso-eluotropic binary solvent mixture tetrahydrofuran-water (32:68), Figure 4.16B, shows a change in component resolution with an analysis time similar to the separation in Figure 4.16A. Sample resolution is not complete here but after adjusting the mobile phase selectivity, phenol and benzyl alcohol are well separated. In the second solvent system, phenol and 3-phenylpropanol are unresolved. Mobile phase strength and selectivity are optimized in the ternary solvent system methanol-tetrahydrofuran-water (10:25:65), Figure 4.16C, in which all the sample components are adequately resolved.

In many instances, selecting a mobile phase is still a trial and error procedure using the guidelines discussed above. The Snyder solvent selectivity triangle concept can be combined with a mixture-design statistical technique to define the optimum mobile phase composition for a particular separation [97]. A feature of this mixture-design technique is that it leads to the selection of a quaternary mobile phase system for most separations. The selection process can be controlled by a microprocessor in an interactve way if the solvent delivery system can pump four solvents simultaneously. Method development proceeds in

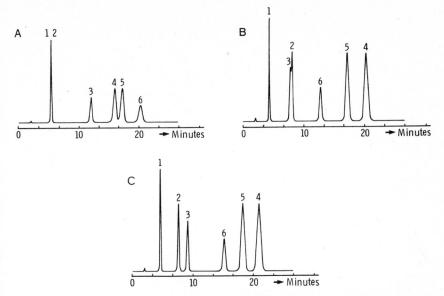

Figure 4.16   An example of the use of ternary solvents to control mobile phase
strength and selectivity in reversed-phase chromatography.
A, methanol-water (50:50); B, tetrahydrofuran-water (32:68);
C, methanol-tetrahydrofuran-water (35:10:55). Peak identification:
1, benzyl alcohol; 2, phenol; 3, 3-phenylpropanol;
4, 2,4-dimethylphenol; 5, benzene; 6, diethylphthalate.
(Reproduced with permission from ref. 94.   Copyright Elsevier
Scientific Publishing Co.)

the following manner.   An initial binary solvent system with the correct solvent

strength for the separation is selected.   This binary solvent system forms one

apex of the selectivity triangle and defines the solvent strength used to

calculate the composition of the mobile phase at the other apices of the

triangle.   Thus each of the three initial chromatograms is obtained with binary

solvents of equivalent solvent strength containing a common strength-adjusting

solvent and three selectivity-adjusting solvents from different solvent

selectivity groups.   The area bound by the sides of the selectivity triangle

defines the selectivity space in which the optimum mobile phase will be found.

Further optimization is achieved by mixing each of the binary solvents in turn

to generate three ternary mobile phase systems forming the apices of a new

solvent selectivity triangle.   Visual comparison of the six chromatogrms will

normally allow the optimum solvent composition for the separation, the seventh

chromatogram, to be predicted.   The advantage of the above approach is that it

can be used to achieve a prechosen level of resolution for all the components of

a complex mixture or, alternatively, a single pair or several specific pairs of

compounds within the mixture using a precisely defined and limited experimental

framework.   In theory, only seven definitive isocratic experiments are needed to

adequately evaluate the entire selectivity space of the solvent selectivity
triangle statistically.

The major problem experienced in optimizing a chromatographic system is the
difficulty in distinguishing a good separation from a poor one. A number of
methods for measuring the performance of a multicomponent system have been
proposed, most of which involve the linear combination of the measurements for
particular pairs to give a single value related to separation performance under
a single set of conditions. One such method is the chromatographic optimization
function (COF) given by equation (4.13)

$$COF = \sum_{i=1}^{K} A_i \ln\frac{R_i}{R_{id}} + B(t_x - t_A) \qquad (4.13)$$

A_i = weighting factor used to indicate which peaks are more important to
separate than others
R_i = resolution of the ith pair
R_{id} = desired resolution of the ith pair
B = arbitrary weighting factor
t_x = maximum acceptable analysis time
t_A = experimental time
K = number of adjacent pairs of peaks

The COF reduces data from each chromatogram to a single number convenient for
mathematical interpretation. Chromatograms with good peak resolution have COF
values approaching zero or a small negative value and those with poor separation
result in large negative values. The COF value is satisfactory if all
components have the same relative order of retention in all solvents. It is not
uncommon for peak crossover to occur with changes in mobile phase selectivity
and this change may not be reflected by the COF value. Also, it is difficult
for the chromatographer to easily relate the COF value to chromatographic terms
which are familiar to him. The visual impact of the separation process is
lost. A more recent approach makes use of overlapping resolution maps (ORM) in
which a resolution contour map is generated for each pair of peaks in all
solvent compositions within a selected solvent triangle [97]. Any area of the
solvent triangle that has a resolution exceeding the desired minimum value
represents a choice of solvents which can separate that particular pair. By
overlapping acceptable regions of separation for all sample component pairs,
areas corresponding to particular solvent mixtures which will achieve the
desired resolution for all component pairs can be identified. Methods using
overlapping resolution maps provide a visual approach to mobile phase
optimization, assuming the availability of computer and graphics equipment for

data handling.

For liquid-solid chromatography using adsorbents such as silica gel and alumina, the solvent strength parameter, $\varepsilon°$, is a useful measure of solvent eluting power. The empirical arrangement of solvents in order of increasing elution strength is called an eluotropic series. For liquid-solid chromatography, the series was established by determining the retention volumes with various solvents for a particular substance on a given adsorbent. The more "polar" the solvent, the smaller the retention volume. All values of $\varepsilon°$ are referenced to the value for n-pentane, which is taken to be zero. Some typical values for the solvent strength parameter ($\varepsilon°$) are given in Table 4.16. Details of this parameter are covered in the section dealing with liquid-solid chromatography.

4.9 Gradient Elution in Liquid Chromatography

So far we have discussed elution in liquid chromatography under isocratic conditions; that is, with a mobile phase whose solvent strength remains constant throughout the separation. This is the favored method of separation in liquid chromatography and has many advantages in terms of convenience, simplicity, and reproducibility of the chromatographic data. However, isocratic elution is ineffective for the separation of samples containing components of widely different relative retention. For samples of this type isocratic elution produces poor resolution of early eluting peaks and inconveniently long retention times and broad peaks for late eluting components. This situation is frequently referred to as "the general elution problem". One solution to this problem is gradient elution liquid chromatography in which the composition of the mobile phase is varied throughout the separation so as to provide a continual increase in solvent strength and thereby a more convenient elution time and sharper peaks for all sample components [98]. Improvements are seen in peak shape, resolution, detection sensitivity, and analysis time.

In this section a phenomenological approach to the description of solvent gradients will be adopted. In particular, the linear-solvent-strength gradient will be emphasized since this represents the simplest gradient type to analyze and fairly closely approximates the conditions most frequently used in the laboratory [93-99]. Mathematical models for the gradient elution process can get rather complex and will not be treated here. As well as the linear-solvent-strength model [93-99], more complex approaches have been described by Jandera and Churacek [100-104] and Schoenmakers, Billiet, and De Galan [105-107]. General reviews of gradient elution liquid chromatography are also available [93,99,108-110].

Usually in gradient elution liquid chromatography a binary mixture of solvents A and B is used in which the volume fraction of the stronger eluting solvent B is progressively increased with time. The shape of the solvent gradients used tend to conform to one of the idealized shapes shown in Figure 4.17. Simple gradients usually suffice to solve most problems and are easier to interpret chromatographically. Complex gradients can be constructed by combining several gradient segments represented by the shapes in Figure 4.17 to form the complete gradient program. As well as gradient segments, isocratic segments can be included in the program. These are most frequently used at the start or end of the program but can be used at any point in the gradient sequence to improve sample resolution. Step-function gradients can also be used but these have few advantages over continuous gradients [103].

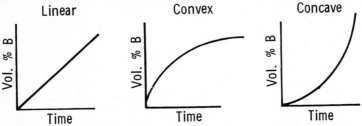

Figure 4.17   Idealized shapes for mobile phase gradients.

In a linear-solvent-strength gradient, the logarithm of the capacity factor for each sample component, k_i, is assumed to decrease linearly with time according to equation (4.14).

$$\log k_i = \log k_o - b \left[\frac{t}{t_m} \right] \tag{4.14}$$

k_i = capacity factor for the ith component
k_o = value of k_i determined isocratically in the starting solvent A
b = gradient steepness parameter
t = time after start of gradient and sample injection
t_m = column dead time

This equation predicts that the average effective value of k during migration is the same for every band and the terminal k value at the time of elution from the column will also be identical for each band. Consequently a linear-solvent-strength gradient should result in equal resolution of every band and equal bandwidth for all components. These properties are apparent in the separation shown in Figure 4.18 [106]. If equation (4.14) were strictly true then a plot of $\log k_i$ against t would be a straight line. This is found to be a good approximation for the range most widely used in liquid chromatography (1 < k_i < 10). The gradient steepness parameter, b, should ideally remain constant

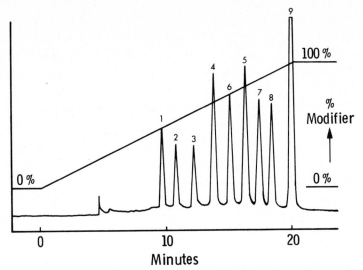

Figure 4.18   Example of an optimum linear-solvent-strength gradient.   Peak
           identification: 1, benzyl alcohol; 2, 2-phenylethanol;
           3, o-cresol; 4, nitrobenzene; 5, diethyl o-phthalate;
           6, benzophenone; 7, naphthalene; 8, biphenyl; 9, anthracene.
           (Reproduced with permission from ref. 106. Copyright Elsevier
           Scientific Publishing Co.)

throughout the solvent program and hence have the same value for all sample
components, a condition which is never exactly possible under normal operating
conditions.   There are certain cases where b does in fact increase regularly
with solute retention.   Examples of such cases are homologous series,
benzologous series (e.g., polycyclic aromatic hydrocarbons), mixtures of
oligomers, samples comprising a parent compound with varying degrees of
functional-group substitution, and samples containing similar compounds that
bear different net charges (ion-exchange chromatography).   Here a
linear-solvent-strength gradient does not represent an optimum situation since b
increases successively for later eluting bands, and therefore a convex gradient
should be employed to provide constant spacing and bandwidths for the various
sample components.   In general, the linear-solvent-strength gradient is the
preferred separation method in gradient elution liquid chromatography, except as
noted above.

     A suitable gradient for a separation can be described by three
characteristics: the initial and final mobile phase compositions, the gradient
shape, and the gradient steepness.   Solvent selection is as important in
gradient elution as in isocratic analysis.   Rules and observations similar to
those discussed previously can be applied.   Binary solvent mixtures are most
frequently employed and an initial solvent A that is too weak or a final solvent
B that is too strong will lead to a degradation in the quality of the

separation. The choice of solvent A has an influence on the separation of the initial bands in the chromatogram but very little influence on the separation of the later eluting bands. The initial solvent may be a pure solvent or a binary solvent mixture of A and B adjusted in strength to provide optimum resolution for the early eluting bands in the chromatogram with minimum time delay before the appearance of the first peak. The strength of the final solvent B influences the selectivity of the separation (i.e., relative peak position) and the retention time and peak shape of later eluting bands. When later bands elute long after the end of the gradient program a stronger B solvent is needed. If B is too weak, the analysis time may become inconveniently long and later eluting bands broadened and difficult to detect. Within these solvent constraints, the gradient program is selected and optimized.

TABLE 4.19

SELECTION OF OPTIMAL LINEAR-SOLVENT-STRENGTH GRADIENTS IN LIQUID CHROMATOGRAPHY

Column Type	Gradient Shape	Range or Rate
Reversed-Phase BPC	Linear increase in %B (Organic modifier)	5-7% B/t_m
Normal-Phase BPC	Vary composition of binary mobile phase linearly to give change in polarity equal to 0.5 P'/t_m	12-15% B/t_m
Liquid-Solid	Vary composition of binary mobile phase in a concave fashion to give a change in solvent strength equal to 0.02 $\varepsilon°/t_m$	$(10-18)t_m$ for a 0-100% gradient[a]
Ion-Exchange	Increase salt concentration by 1.6 fold per t_m in a concave gradient	
	Increase pH by 0.2 units/t_m, linear gradient (cation exchange)	pH varied from 2-6 in $20t_m$
	Decrease pH by 0.2 units/t_m, linear gradient (anion exchange)	pH varied from 8-2 in $30t_m$

[a] Gradients from 5-95% B are generally preferred from a practical point of view

Some general rules for predicting the optimum gradient for linear-solvent-strength separations are given in Table 4.19. Note that the gradient steepness is dependent on the dead volume of the column, t_m, and that a linear-solvent-strength gradient in liquid-solid chromatography or a salt gradient in ion-exchange chromatography has a concave shape. The generalizations in Table 4.19 can be predicted by theory and have been validated experimentally [99]. This table provides a starting point for gradient selection but some fine tuning will probably be necessary for optimization.

A convex gradient provides a faster increase in the percent volume of B at the beginning of the separation than a linear gradient. Where a linear gradient was the optimal profile, a convex gradient leads to elution of bands with a lower average capacity factor and a shorter total analysis time. As a result, early eluting bands appear sharpened and bunched together with a loss in resolution and later eluting bands appear wider and better resolved. A concave gradient produces an effect opposite to the convex gradient with wider, better resolved bands at the start of the chromatogram and sharper, less well resolved bands appearing at the end. Whichever gradient shape is selected for the separation, the ideal situation is reached when all bands are of equal width across the chromatogram.

Gradient elution can be used as a method for calculating the composition of the mobile phase required to elute a substance with a certain capacity factor value under isocratic conditions. Thus gradient elution analysis is a useful and rapid method for "scouting" the optimal conditions for an isocratic separation of a sample using the same chromatographic system [99,103]. In gradient elution chromatography, the composition of the mobile phase entering the column as a function of time is a known parameter and determines the relationship of the sample capacity factors with time, conforming to equation (4.14). The mobile phase composition at the column exit lags behind that at the column inlet by the time required to traverse the length of the column, equivalent to t_m. Consequently, the capacity factor of a substance at the time of elution k_f is defined by a modified form of equation (4.14), equation (4.15).

$$\log k_f = \log k_o - b \left[\frac{t_g - t_m}{t_m} \right] \tag{4.15}$$

By eliminating k_o between equations (4.14) and (4.15), we have

$$\log k_i = -\log k_f + b \left[\frac{t_g - t - t_m}{t_m} \right] \tag{4.16}$$

t_g = time from sample injection to elution of the band maximum in gradient elution

t = time after start of gradient and sample injection

t_m = column dead time

For bands that are fairly well retained by the column in the gradient program, k_f is approximately equal to $1/(2.3)b$ and substituting this value into equation (4.16) enables one to describe k_i in terms of b, t_g, t, and t_m. The values of t_g and t_m are determined by the experiment and values for b can usually be estimated. Thus k_i can be determined as a function of time, t, and since the mobile phase composition as a function of time is known from the gradient program, k_i as a function of mobile phase composition is also known. For typical values of b (0.2-0.4) Table 4.20 can be used to calculate Θ_B, the percentage volume of the stronger solvent B in a binary mobile phase, that will provide a predetermined value of k_i under isocratic elution conditions [99]. For example, if a value of k = 3 is desired in the isocratic separation, the value of Θ_B (from the gradient device) should be observed at a time t = $t_g - 2t_m$. That value of Θ_B would then give a value of about 3 for k in the isocratic separation.

To date ternary gradients have been rarely used in gradient elution chromatography [104]. With ternary gradients, it should be possible to adjust retention and selectivity independently to a certain extent and to change both selectivity and retention with time much more efficiently then when using binary gradients. The simplest type of ternary gradient system involves adding an equal volume percent of a third solvent C to both initial solvents A and B which are then mixed as a binary gradient. The ternary gradient formed then contains variable amounts of weak solvent A and strong solvent B with a fixed amount of a selectivity-adjusting solvent C. Continuously variable gradients of solvents B and C in A have not been investigated thoroughly.

A novel method, sequential isocratic step liquid chromatography, has been described the either scouting for preliminary conditions for an isocratic separation with a sample of unknown composition or for determining the polarity range and complexity of an unknown sample [111]. The method is comprised of a series of short isocratic separations of a sample in a sequence of steps going from 100, 80, 60, 40, 20, and 0 percent volume of strong eluting solvent B in weak solvent A. The time of each isocratic segment is established to cover the range k = 0-10 and is equal to 10 times the column dead time. At high flow rates (e.g., 5 ml/min), the total time for the complete stepwise analysis can be

TABLE 4.20

THE RELATIONSHIP BETWEEN k AND t FOR DIFFERENT VALUES OF THE GRADIENT
STEEPNESS FACTOR b

Time $t =$	Value of k for Θ_B at the column inlet	
	b = 0.02	b = 0.4
t_g	1.4	0.4
$t_g - t_m$	2.2	1.1
$t_g - 2t_m$	3.4	2.7
$t_g - 3t_m$	5.4	6.8

kept short (less than 1 hour). A further advantage is that the whole process
can be automated so that unattended operation is possible and, by adroit
selection of solvents, near-universal wavelength detection at 190 nm can be
used. A visual comparison of the individual segments provides a good starting
point to select the solvent strength required for an isocratic separation or to
decide the initial and final solvent strengths for gradient elution.

4.10 Liquid-Solid Chromatography

Separations in liquid-solid chromatography result largely from the
interaction of sample polar functional groups with discrete adsorption sites on
the stationary phase. In this respect, it differs greatly from reversed-phase
chromatography where sample retention is controlled by the lipophilicity of the
sample. The strength of these polar interactions (see Table 4.21) is
responsible for the selectivity of the separation in liquid-solid
chromatography. Thus liquid-solid chromatography is generally considered to be
suitable for the separation of non-ionic molecules which are soluble in organic
solvents. Very polar compounds, for example compounds with high solubility in
water and low solubility in organic solvents, interact very strongly with the
adsorbent surface and result in peaks of poor symmetry and poor efficiency. It
can be seen from Figure 4.19 that liquid-solid chromatography is suitable for
the separation of moderately polar compounds [112].

There are some sample types for which liquid-solid chromatography shows a
unique separating ability. It is generally the method of choice for separating
geometric isomers and for class separations [113,114]. The ability of
liquid-solid chromatography to separate geometric isomers has been attributed to
a lock-key type steric fitting of solute molecules with the discrete adsorption
sites on the silica surface. An example is shown in Figure 4.20 for

TABLE 4.21

CLASSIFICATION OF FUNCTIONAL GROUP ADSORPTION STRENGTHS

Relative Sample Adsorption	Compound Type
Nonadsorbed	Aliphatics
Weakly adsorbed	Alkenes, mercaptans, sulfides, aromatics and halogenated aromatics with one or two rings
Moderately adsorbed	Polynuclear aromatics, ethers, nitriles, nitro compounds, and most carbonyls
Strongly adsorbed	Alcohols, phenols, amines, amides, imides, sulfoxides, acids
General trends	1) F < Cl < Br < I 2) Internal hydrogen bonding between functional groups diminishes retention 3) Bulky alkyl groups adjacent to the polar functional group diminish retention 4) cis compounds are retained more strongly than trans 5) Equitorial groups in cyclohexane derivatives (and steroids) are more strongly retained than axial derivatives

the separation of the four geometric isomers of biliverdin IX dimethyl esters [115]. The unique ability of adsorbents to differentiate solutes based on differences in their polar functional groups leads to their use in class separations. The chromatogram in Figure 4.21 shows a separation of a mixture of aniline derivatives and aromatic alcohols on silica gel. Note that the two different functional group types of the sample are well-resolved. Sample ordering by functional group type, number of functional groups, and polarity is the basis for class separations of complex mixtures where the isolation or measurement of the contribution of compound types to the overall sample is required rather than a within-group separation. An example is the determination of the sterol and triglyceride fractions of a lipid sample [116]. Consider also Figure 4.22, which is the chromatogram obtained for the semipreparative fractionation of the soluble organic fraction from a diesel engine particulate exhaust sample [117,118]. Each peak contains components of a similar type which can be collected and separated in a subsequent chromatographic analysis. Here liquid-solid chromatography has a distinct advantage over reversed-phase chromatography, which provides a separation based on sample hydrophobicity and is better reserved for "within-class" separations. Combining the two separation methods provides a powerful tool for complex mixture analysis.

Lipophilic amines, difficult to separate on bonded phase columns due to

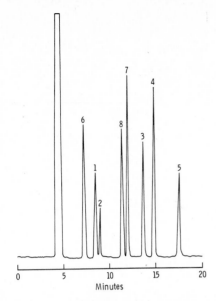

Figure 4.19   Separation of linuron and its metabolites on µPorasil by gradient
elution from hexane-2% (propanol-25% NH$_4$OH) to hexane-
35% (propanol-25% NH$_4$OH) over 30 min. Flow rate 1.5 ml/min.
Peak identification: 1, linuron; 2, 3,4-dichloroaniline;
4, 3-(3,4-dichlorophenyl)-1-methylurea; 5, 3,4-dichlorophenylurea;
6, methyl-N-(3,4-dichlorophenyl)carbamate; 7, 3,4-dichloro-
acetanilide; 8, 3-(3,4-dichlorophenyl)-1-ethylurea. (Reproduced
with permission from ref. 112. Copyright Preston Publications,
Inc.)

poor selectivity and peak tailing, may be separated with good peak symmetry on
silica gel [119]. The mobile phase is usually a mixture of water and an organic
solvent containing an inorganic or organic base as a selectivity modifier. The
separation mechanism in this example is believed to be controlled by
electrostatic adsorption forces.

The mechanism by which retention occurs in liquid-solid chromatography is
not understood with absolute certainty [120]. Two models of the adsorption
process will be discussed here: the competition model developed by Snyder and
Soczewinski [121-125] and the solvent interaction model proposed by Scott and
Kucera [8,126,127]. The discussion will be from a mainly phenomenological point
of view and directed towards understanding the influence of the mobile phase on
selectivity and retention in liquid-solid chromatography.

For nonpolar and moderately polar mobile phases which interact with the
adsorbent surface largely by dispersive and weak dipole interactions, the
competition model assumes that the entire adsorbent surface is covered by a
monolayer of solute and mobile phase molecules. Under normal chromatographic
conditions, the concentration of sample molecules will be small and the adsorbed
monolayer will consist mainly of mobile phase molecules. Retention of a solute

274

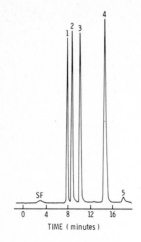

TIME (minutes)

Figure 4.20 Isocratic separtion of biliverdin IX isomers by liquid-solid
chromatography. Stationary phase Zorbax Sil-850, mobile phase
methylene chloride-methanol-water (99:0.9:0.1), flow rate 1.5
ml/min. Peak identification: Biliverdin IX β, $R_1 = R_4 = R_5$
$= R_6 = H$, $R_2 = R_3 = A$, $R_6 = R_8 = B$; Biliverdin IX α, $R_1 = R_3 = R_6$
$= R_7 = H$, $R_4 = R_5 = A$, $R_2 = R_8 = B$; 3: Biliverdin IX γ, $R_2 = R_3$
$= R_5 = R_7 = H$, $R_1 = R_8 = A$, $R_4 = R_6 = B$; 4: Biliverdin IX δ,
$R_1 = R_3 = R_5 = R_8 = H$, $R_6 = R_7 = A$, $R_2 = R_4 = B$; $A= -(CH_2)_2CO_2CH_3$;
$B= -CH=CH_2$. (Reproduced with permission from ref. 115.
Copyright Academic Press, Inc.)

molecule then occurs by displacing a roughly equivalent volume of mobile phase
molecules from the monolayer to make the surface accessible to the adsorbed
solute molecule, as depicted in Figure 4.23. The model also assumes that
solution energy terms involving the interaction between solute and mobile phase
molecules are cancelled by similar interactions in the adsorbed phase. Under
the above set of circumstances, solute retention can be represented by the
equilibrium, equation (4.18),

$$X_n + nB_a \rightleftharpoons X_a + nB_n \qquad (4.18)$$

X_n = solute molecule in the mobile phase

B_a = mobile phase molecule forming the adsorbent monolayer

n = number of molecules of B displaced by adsorption of X

X_a = adsorbed solute molecule (now part of the adsorbent monolayer)

B_n = n molecules of B in the mobile phase

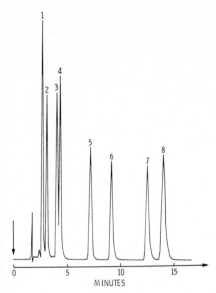

Figure 4.21   Separation of a mixture of aniline derivatives and aromatic
              alcohols on LiChrosorb Si 60, d_p = 5 μm. Mobile phase
              isooctane-dichloromethane-isopropanol (900:90:10, v/v/v), flow
              rate 1.5 ml/min. Peak identification: 1, 2,6-diethylaniline;
              2, 2-methyl-6-ethylaniline; 3, o-isopropylaniline; 4, o-ethyl-
              aniline; 5, 2-phenyl-2-propanol; 6, α-methylbenzyl alcohol;
              7, benzyl alcohol; 8, cinnamyl alcohol.

while solute retention can be envisaged as a continuous competitive displacement
process between solute and mobile phase molecules on the adsorbent surface. The
net retention will depend on the relative interaction energy of the solute and
mobile phase molecules with the adsorbent surface. The model leads to two
fundamental equations describing solute retention and solvent strength which
have experimental validation. The variation of solute retention as a function
of solvent strength is given by equation (4.19)

$$\log \frac{k_1}{k_2} = a' \, A_s \, (\varepsilon_2 - \varepsilon_1) \qquad\qquad (4.19)$$

k_1 = solute capacity factor in mobile phase 1
k_2 = solute capacity factor in mobile phase 2
a' = adsorbent activity parameter
A_s = cross-sectional area of the molecule X
ε_1 = solvent strength ($\varepsilon°$-value) of mobile phase 1
ε_2 = solvent strength ($\varepsilon°$-value) of mobile phase 2

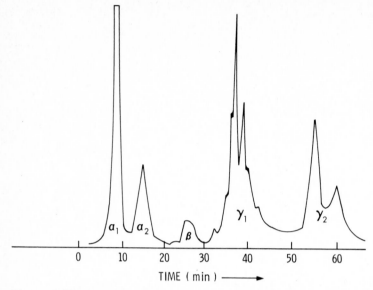

TIME (min) ⟶

Figure 4.22 Semi-preparative class separation of a diesel engine exhaust
sample. Column 25 cm x 7.9 mm, 10 μm μPorasil. Solvent gradient hexane to 5% methylene chloride over 5 min., linear gradient
to 100% methylene chloride over 25 min., isocratic for 10 min.,
linear gradient to 100% acetonitrile over 10 min., isocratic for
5 min., step change to tetrahydrofuran for 10 min., step reverse
to acetonitrile for 5 min. The α_1 and α_2 fractions
contain 2-4 and 4-6 ring PAHs and saturated aliphatics; β fraction
contains 6-8 ring PAHs, hydroxy- and nitro-PAHs with 2 rings; the
γ_1 fraction contains hydroxy-PAHs (3-4 rings), nitro-PAHs
(2-6 rings), PAH quinones (3-5 rings), and PAH ketones (2-3
rings); and the γ_2 fraction contains PAH quinones (3-5 rings),
hydroxy-PAHs (5-7 rings), PAH ketones (3-4 rings) and dinitro-PAHs
(3 rings). (Reproduced with permission from ref. 117. Copyright
Elsevier Scientific Publishing Co.)

and the solvent strength of a binary mobile phase, ε_{AB} is given by
equation (4.20) [128].

$$\varepsilon_{AB} = \varepsilon_A + \log\left[\frac{N_B \, 10^{a'n_b(\varepsilon_B - \varepsilon_A)} + 1 - N_B}{a'n_b}\right] \qquad (4.20)$$

ε_{AB} = solvent strength of the binary mobile phase AB

ε_A = solvent strength of the weaker solvent A

ε_B = solvent strength of the stronger solvent B

N_B = mole fraction of the stronger solvent B in the mobile phase

n_b = molecular area of solvent molecule B

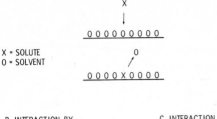

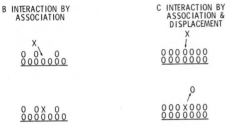

Figure 4.23 Different mechanisms proposed for the interaction of solute and
mobile phase molecules with a silica surface.

The parameter a', the adsorbent activity function, is related to the ability of
the adsorbent to interact with adjacent molecules of solute or solvent and can
be treated as a fixed numerical constant for the purpose of equations (4.19) and
(4.20). Equation (4.20) is applicable to weak non-hydrogen bonding solvents in
which $\varepsilon_B < 0.3$. As the polarity of the solvent (or solute) increases, equation
(4.20) is no longer valid since two of the premises of the competition model are
no longer true. With polar molecules, solute-solvent interactions are specific
in nature and not necessarily identical for the molecules X and B. These
solute-solvent interactions may be different in the adsorbed monolayer and bulk
mobile phase and can no longer be considered equal and self-cancelling.
Secondly, the premise of an adsorbent surface which is effectively homogeneous
and continuous with regard to solvent adsorption is no longer true. For weak
solvents there is little preference for adsorption on any given part of the
surface and little tendency for weakly retained molecules to localize.
Molecules with strong dipoles will attempt to adsorb directly on top of a
preferred site on the adsorbent surface [129,130]. For silica gel, the
adsorption sites preferred by polar molecules are associated with the position
of the silanol groups. The silica surface is thus comprised of active sites
where localization of polar molecules can occur and other regions where
preference for localization is not expected. When both solute and solvent
molecules compete for localized adsorption on the same surface site, the
relative adsorption of the solute will be decreased compared to the case of a
non-localized solvent.

An important consequence of surface localization of polar molecules is that the solvent strength of the mobile phase will vary continuously with the mole fraction of polar solvent in the mobile phase and in the adsorbed monolayer. Experiment shows that localization of the strong solvent B in a binary mobile phase AB will occur provided that the mole fraction of B in the adsorbed monolayer, Θ_B, is less than 0.75. Once most of the surface is covered by localized B molecules, the remaining spaces on the surface are not able to accommodate B molecules in the configuration and positioning required for the localizing effect. For larger volumes of Θ_B the coverage of the remaining surface by B molecules must involve delocalized adsorption. Localized molecules are held much more strongly to the adsorbent surface than are delocalized molecules. The net result is that values for ε_B progressively decline with an increase in Θ_B. If the value of ε_B is ε_B' for $\Theta = 0$, and ε_B'' for $\Theta = 1$, then a localization function $\%_{1c}$ can be defined as

$$\%_{1c} = (\varepsilon_B - \varepsilon_B'')/(\varepsilon_B' - \varepsilon_B'') \tag{4.21}$$

ε_B = solvent strength ($\varepsilon°$) for pure solvent B

ε_B'' = value of ε_B as $\Theta_B \longrightarrow 1$; for delocalized solvent B

ε_B' = value of ε_B as $\Theta_B \longrightarrow 0$; for localized solvent B

When the function $\%_{1c}$ is plotted against Θ_B, the curve generated can be fitted empirically to equation (4.22)

$$\%_{1c} = (1 - \Theta_B)\left[\frac{1}{(1 - 0.94\Theta_B)} - 14.5\Theta_B^9\right] \tag{4.22}$$

Combining equations (4.21) and (4.22) allows the calculation of ε_B for any mole fraction of solvent B in the mobile phase and the value of ε_{AB} to be calculated from equation (4.20). The calculation is rather involved and reference [128] should be consulted for details. This approach is quite successful for predicting the solvent strength of both binary [128] and ternary and quaternary [129] mobile phases in liquid-solid chromatography. Possible exceptions are those mobile phases containing a strong solvent capable of self hydrogen bonding, such as alcohols.

The solvent interaction model differs from the competition model by proposing the formation of solvent bilayers adsorbed onto the adsorbent surface. The composition and extent of bilayer formation depends on the concentration of polar solvent in the mobile phase. Solute retention occurs by interaction (displacement or association) of the solute with the second layer of adsorbed mobile phase molecules as shown in Figure 4.23.

For weak solvents interacting with the adsorbent surface largely by dispersive forces, the solvent interaction model predicts results similar to the competition model. The adsorbent surface is covered by a monolayer of mobile phase molecules and solute retention is caused by displacement of solvent molecules by the solute, which can then interact directly with the adsorbent surface. For a binary solvent AB in which the concentration of the stronger solvent B is low (e.g., 0.5%), the adsorbent surface is covered with a monolayer of B molecules hydrogen bonded to the hydrated silanol groups. As the concentration of B molecules increases, polar interactions between the solvent molecules of the second layer and the hydrogen bonded solvent of the primary layer form a bilayer. When the mobile phase contains low concentrations of the polar modifier B, the second layer of solvent molecules is incomplete, and solute retention can be explained by association of the solute within this layer and without the need for displacement (see Figure 4.23). At high concentrations of B, the second layer of the hydrogen bonded B molecules is complete and retention occurs by solute displacement of B molecules from this layer. The solute molecule is now associated with the primary layer of adsorbed solvent molecules but does not displace it. For solutes with high capacity factors (k > 20), indicating that the polarity of the solute is equal to or greater than that of the B molecules, displacement of the primary solvent layer is possible by direct interaction of the solute with the adsorbent surface. The solvent interaction model thus provides a variety of mechanisms for solute retention that depend mainly on the relative polarity of the solute and B solvent molecules and on the concentration of B molecules in the mobile phase. Note also that in this model the active centers for the adsorption of polar molecules are the silanol groups which are assumed to be hydrated. Evidence for this assumption was discussed earlier in this chapter.

The two models discussed above provide a quantitative description of solvent strength in liquid-solid chromatography. The calculations are rather involved and a more empirical approach can be justified for routine purposes or for the separation of simple mixtures. Solvent strength and selectivity are conveniently controlled by using binary or higher order solvent mixtures. The change in solvent strength as a function of the volume percent of the more polar component is not a linear function. Some representative examples of these curves are shown in Figure 4.24. At low concentrations of the polar solvent, small increases in concentration produce large increases in solvent strength; at the other extreme, relatively large changes in the concentration of the polar solvent affect the solvent strength of the mobile phase to a lesser extent. Once the optimal solvent strength of the mobile phase has been determined for a separation, the resolution of the sample is improved by changing solvent selectivity at a constant solvent strength. Graphical methods have been developed for obtaining the percent volume composition of mobile phases having

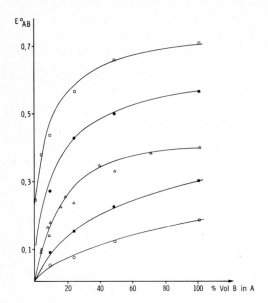

Figure 4.24 Solvent strengths of eluent mixtures on alumina. (o) pentane-
carbon tetrachlorides, (■) pentane-n-propyl chloride, (Δ) pentane-
dichloromethane, (●) pentane-acetone, and (□) pentane-pyridine.

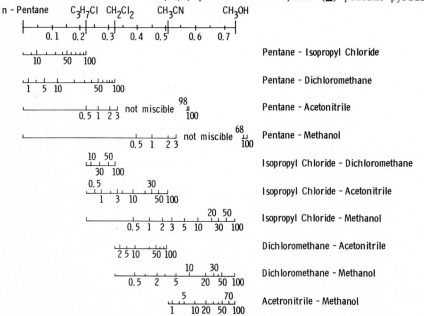

Figure 4.25 Solvent strengths of eluent mixtures for liquid-solid
chromatography. (Reproduced with permission from ref. 131.
Copyright Springer-Verlag).

similar solvent strength but different solvent constituents and therefore different selectivity [114]. Figure 4.25 depicts one of these graphical selection schemes [131].

Retention and selectivity in liquid-solid chromatography are dramatically influenced by the presence of even low concentrations of polar additives in the mobile phase. These additives, known as moderators or modulators, exert an overriding influence on the separation properties of the chromatographic system. The most ubiquitous example is water. The influence of this moderator is most pronounced when the mobile phase is nonpolar. Even though the solubility of water in n-heptane at ambient temperatures is only 0.01% v/v at saturation, a change of a few parts per million in its concentration is sufficienct to greatly vary sample retention. The controlled addition of a moderator to a chromatographic system does, however, provide certain advantages which outweigh the experimental difficulties encountered in controlling the moisture content of low polarity solvents.

The moderator is preferentially adsorbed from the mobile phase by the most active sites on the surface of the solid stationary phase. The net effect is that the initially heterogeneous energetic surface becomes more homogeneous in terms of its adsorptive interactions, resulting in less variation of sample retention from run-to-run and a substantial increase in sample capacity due to an improvement in the linearity of the adsorption isotherm. Higher column efficiencies, reduced band tailing, and a diminished tendency for sample decomposition may also be observed in some cases [132].

Solvents which correspond to the same hydration level as the adsorbent are called isohydric solvents [133]. They provide rapid column equilibration after adjusting the mobile phase since the column and new solvent are already roughly equilibrated with respect to water. When changing between mobile phases containing different concentrations of water, column equilibration is a slow process and several hours of adsorbent conditioning are usually required before constant k values are obtained. The situation can be even more complex with gradient elution regardless of whether or not isohydric solvents are used. If both solvents A and B are isohydric at the start of the gradient program, intermediate compositions of A and B are usually nonisohydric, leading to the possible uptake of water by the column. This adversely affects the separation and leads to irreproducible separations and long column regeneration times. The use of a 100% saturated solvent is undesirable because such liquid-solid chromatographic systems are often unstable. Under these conditions the pores of the adsorbent apparently fill gradually with water, leading to changes in retention with time and possibly also to a change in the retention mechanism as liquid-liquid partition effects become more important. When silica is the adsorbent, 50% saturation of the mobile phase has been recommended for stable chromatographic conditions [131-133]. Solvents with 50% water saturation can be prepared by mixing dry solvent with a 100% saturated solvent or, preferably, by using a moisture control system [134]. The latter consists of a water-coated, thermostatted adsorbent column through which the mobile phase is recycled for the time required to reach the desired degree of saturation.

A column which has been deactivated with water may no longer show adequate separation properties. Restoring the activity of the column by pumping a large volume of dry mobile phase through the column is a slow and expensive process. Alternatively, reactivation can be accomplished chemically using the acid-catalyzed reaction between water and 2,2-dimethoxypropane, the products of which, acetone and methanol, are easily eluted from the column [135].

In addition to water, virtually any organic polar modifier may be used to control solute retention in liquid-solid chromatography. Alcohols, acetonitrile, tetrahydrofuran, and ethyl acetate in volumes of less than one percent can be incorporated into nonpolar mobile phases to control adsorbent activity. In general, column efficiency declines for alcohol-moderated eluents compared to water-moderated eluent systems. Many of the problems discussed above for water-moderated eluents are true for organic-moderated eluents as well.

4.11 Polar Bonded Phases

Polar bonded phases containing diol, cyano, diethylamino, amino, or diamino functional groups are commercially available. Representative structures are shown in Table 4.22. The diol- and diethylamino-substituted stationary phases are used in size exclusion and ion-exchange chromatography, respectively, and will be dealt with later. The alkylnitrile- and alkylamine-substituted stationary phases, when used with a mobile phase of low polarity, behave in a manner similar to the solid adsorbents discussed in the previous section. That is, the retention of the sample increases with solute polarity and increasing the polarity of the mobile phase reduces the retention of all solutes. The polar bonded-phase packings are generally less retentive than adsorbent packings but are relatively free from the problems of chemisorption, tailing, and catalytic activity associated with silica and alumina. The bonded-phase packings respond rapidly to changes in mobile phase composition and can be used in gradient elution analyses. Adsorbent packings respond slowly to changes in mobile phase composition due to slow changes in surface hydration, making gradient elution analysis difficult. Because of the above advantages, the polar bonded-phase packings have been proposed as alternatives to microporous adsorbents for separating the same sample types [13]. The alkylnitrile-substituted phase is of intermediate polarity and is less retentive than silica gel but displays similar selectivity. It provides good selectivity for the separation of double bond isomers and ring compounds differing in either the position or number of double bonds [115]. With aqueous mobile phases the alkylnitrile-substituted stationary phases have been used for the separation of saccharides that are poorly retained on reversed-phase columns. The alkylamine-substituted phases provide a separation mechanism complementary to either silica gel or alkylnitrile-substituted phases. The amino function imparts strong hydrogen bonding properties to the stationary phase as well as acid or base properties, depending on the nature of the solute. The aminoalkyl-substituted stationary phase has been used for the class separation of polycyclic aromatic hydrocarbons [136,137]. Retention is based primarily on change transfer interactions between the aromatic π-electrons of the polycyclic aromatic hydrocarbons and the polar amino groups of the stationary phase. Samples are fractionated into peaks containing those components with the same number of rings. Retention increases incrementally with increasing ring number, but is scarcely influenced by the presence of alkyl ring substituents. In contrast, reversed-phase separations show poor separation between alkyl-substituted polycyclic aromatic hydrocarbons and polycyclic aromatic hydrocarbons of higher ring number.

TABLE 4.23

STRUCTURES OF POLAR BONDED PHASES

Polar Functional Group	Structure	Applications
Diol	$-(CH_2)_3OCH_2CH(OH)CH_2(OH)$	Surface modifying groups for silica packings used in size-exclusion chromatography
Cyano	$-(CH_2)_3CN$	Partition or adsorption chromatography
Amino	$-(CH_2)_nNH_2$ n = 3 or 5	Adsorption, partition, or ion-exchange chromatography
Dimethylamino	$-(CH_2)_3N(CH_3)_2$	Ion-exchange chromatography
Diamino	$-(CH_2)_2NH(CH_2)_2NH_2$	Adsorption or ion-exchange chromatography

The practice of bonded-phase chromatography is similar to that described for liquid-solid chromatography. A polar solvent modifier, such as isopropanol at the 0.5-1.0% v/v level, is used in nonpolar solvents to improve peak symmetry and retention time reproducibility. It is believed that the polar modifier solvates the polar groups of the stationary phase, leading to an improvement in mass transfer properties. For the separation of carboxylic acids or phenols, either glacial acetic acid or phosphoric acid is used at low levels as a tailing inhibitor. Likewise, propylamine is a suitable modifier for the separation of bases.

Certain specific problems arise with the use of alkylamine-substituted stationary phases. Since amines are readily oxidized, degassing the mobile phase and avoiding solvents which may contain peroxides (e.g., diethyl ether, tetrahydrofuran) are recommended. Samples or impurities in the mobile phase containing ketone or aldehyde groups may react chemically with the amine group of the stationary phase, forming a Schiff's base complex [115]. This reaction will alter the separation properties of the column. The column may be regenerated by flushing with a large volume of acidified water [138].

4.12 Reversed-Phase Chromatography

Reversed-phase chromatography, so-called because it represented a reversal of the historical, or "normal" practice of column chromatography, is characterized by using a polar mobile phase in conjunction with a nonpolar stationary phase. In terms of current popularity, this mode of operation may be considered the "normal mode", as it is estimated that somewhere between 75-90% of all HPLC separations are carrried out in the reversed-phase mode. The reasons for this include the simplicity, versatility, and scope of the reversed-phase method, Table 4.23 [139]. The hydrocarbon-like stationary phases equilibrate rapidly with changes in mobile phase composition and are therefore eminantly suitable for use with gradient elution. Retention in reversed-phase chromatography occurs by non-specific hydrophobic interactions of the solute with the stationary phase. The near universal application of reversed-phase chromatography stems from the fact that virtually all organic molecules have hydrophobic regions in their structure and are capable of interacting with the stationary phase. Reversed-phase chromatography is thus ideally suited to separating the components of a homologous or oligomous series. Figure 4.26 illustrates the separation of a series of phenylene oxide oligomers; retention is based on an increase in hydrophobicity with oligomer molecular weight [140]. Within a homologous series, the logarithm of the capacity factor is generally a linear function of the carbon number. Branched chain compounds are generally retained to a lesser extent than their straight chain analogs and unsaturated compounds are eluted before the corresponding saturated analogs. Since the mobile phase in reversed-phase chromatography is polar and generally contains water, the method is ideally suited to the separation of polar molecules which are either insoluble in organic solvents or bind too strongly to solid

TABLE 4.23

CONSIDERATIONS FOR THE WIDESPREAD USE OF REVERSED-PHASE LIQUID CHROMATOGRAPHY

Advantages	Limitations
1. It can separate a broad spectrum of non-ionic, ionizable, and ionic compounds.	1. With silica-based column packings the usuable pH range is limited to pH 2-7.5.
2. There is great flexibility for separating ionic and ionizable compounds through selective equilibria such as ion pair, ion suppression, and ligand exchange.	2. If unreacted silanol groups are present on the silica surface, adsorption of the solutes to the silica surface can occur, resulting in poor peak shapes.
3. The system is relatively simple to operate.	3. The retention mechanism is more complex than in other forms of chromatography and a better understanding is needed to control it.
4. Since the stationary phases are chemically bonded, columns are stable and separations are reproducible.	
5. Because of the weak surface energies of bonded phases, analyses are rapid and re-equilibration times short.	
6. By using injection solvents significantly weaker than the mobile phase, selected solutes can be preconcentrated at the head of the column. A method of "trace enrichment".	
7. Reversed-phase chromatography can be used for determining various physiochemical properties such as hydrophobicity, dissociation constants, and complexation constants.	

adsorbents for normal elution. Many samples of a biological origin fall into this category. Figure 4.27 shows the separation of some polar constituents of biological fluids in which the more polar components are eluted at the start of the chromatogram and the less polar constituents later [141]. In addition, compounds of substantially different polarity can be separated in the same run using gradient elution, within a convenient time frame.

Retention in reversed-phase chromatography is a function of sample hydrophobicity whereas the selectivity of the separation results almost entirely from specific interactions of the solute with the mobile phase [142]. Generally, the selectivity may be conveniently adjusted by changing the type of organic modifier in the mobile phase. For ionic or ionizable solutes, pH buffers, which suppress ionization, or ion-pairing reagents, used to form lipophilic complexes, increase the degree of solute transfer to the stationary phase and may be used to control selectivity [143]. Metal-ligand complexes and chiral reagents can be added to the mobile phase to separate optically active

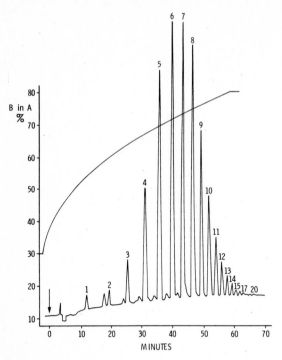

Figure 4.26   Reversed-phase separation of poly-(2,6-diphenyl-p-phenylene oxide) oligomers.   Bonded phase C$_{18}$ column, concave gradient of 30 to 80% solvent B in A over 60 min.   Solvent A, water-dioxane (1:1) and solvent B, water-dioxane (1:99). Peak identification: 1, monomer; 2, dimer, etc.   (Reproduced with permission from ref. 140.   Copyright Elsevier Scientific Publishing Co.)

isomers.   The general use of mobile phase additives to control secondary chemical equilibria in the mobile phase and its effect on sample separation is discussed in Chapter 7.   Reversed-phase chromatography is also gaining increasing attention as a method for separating large biological molecules such as proteins and enzymes [144-148].   Sample retention is controlled by the hydrophobicity of the molecule and provides a complementary separation mechanism to ion-exchange.   Much work remains to be done in this area to prepare packings tailor-made to the particular problems encountered.

The details of the mechanism governing retention in reversed-phase chromatography using chemically-bonded hydrocarbonaceous phases is not completely understood [17,27,149].   The solvophobic theory, discussed below, provides a semiquantitative explanation of solute retention but is mathematically complex.   Also, unlike the models used to explain retention in other chromatographic systems, the theory does not provide a simple picturesque description of the physical processes involved.   Its origins are couched in the thermodynamic properties of ordered liquid systems which do not naturally lend

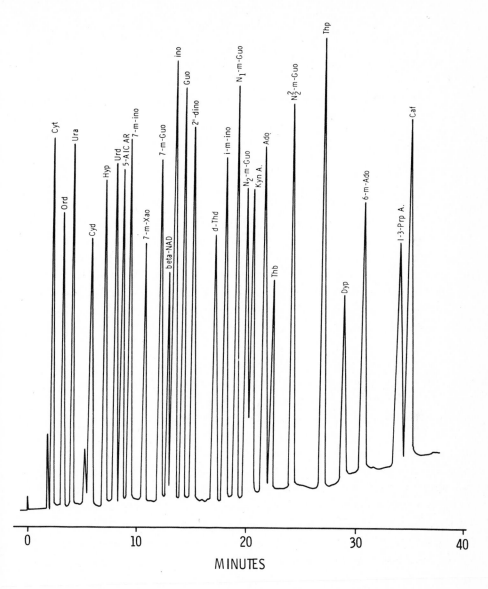

Figure 4.27 Separation of nucleosides, bases, nucleotides, aromatic amino acids, and some of their metabolites. Column: Bondapak C$_{18}$, (10 micrometers particle size); eluants: low strength, 0.02 M KH$_2$PO$_4$, pH 5.6; high strength, 60% methanol; gradient slope 0.69%/min, linear; temperature, ambient; flow: 1.5 ml/min. (Reproduced with permission from ref. 141. Copyright Academic Press, Inc.)

themselves to a simple visual concept. Before describing the solvophobic theory proper, albeit in a simplified form, the boundary conditions existing within the chromatographic system must be considered since they affect retention.

Solute retention in reversed-phase chromatography could proceed either via partitioning between the hydrocarbonaceous surface layer of the nonpolar stationary phase and the mobile phase or by adsorption of the solute to the nonpolar portion of the stationary phase. In this context, the partitioning mechanism would seem unlikely since the hydrocabronaceous layer is only a monolayer thick and lacks the favorable properties of a bulk liquid for solubilizing the solutes. However, as suggested by Scott and Kucera [66], the less polar solvent component of the mobile phase could accumulate near the apolar surface of the stationary phase, forming an essentially stagnant layer of mobile phase rich in the less polar solvent. As a result, the solute could partition between this layer and the bulk mobile phase without directly interacting with the stationary phase proper. The balance of evidence favors the adsorption mechanism either with the stationary phase surface itself or by interaction with ordered solvent molecule layers at the stationary phase surface [150]. To provide a simple view of the solvophobic theory we will assume that solute retention occurs by adsorption of the solute to the stationary phase, without further defining the nature of the stationary phase.

The solvophobic theory assumes that aqueous mobile phases are highly structured due to the tendency of water molecules to self-associate by hydrogen bonding and that this structuring is perturbed by the presence of nonpolar solute molecules. As a consequence of the very high cohesive energy of the solvent, the less polar solutes are literally "squeezed out" of the mobile phase and are bound to the hydrocarbon portion of the stationary phase. In this instance, the driving force for solute retention is not the familiar mechanism used to explain retention in other chromatographic systems, the favorable affinity of the solute for the stationary phase, but rather the effect of the solvent forcing the solute to the hydrocarbonaceous layer. Thermodynamically, this driving force can be expressed as the increased entropy of water in the mobile phase accompanying the transfer of a solute from the mobile phase to the nonpolar stationary phase. If the solute contains polar groups then the dipolar or hydrogen bonding interaction of these groups with the mobile phase will oppose the solute transfer mechanism. The differences in interactive energies between nonpolar and polar solutes with the mobile phase and the differences in hydrophobic solute molecular surface areas are responsible for the functional group selectivity observed in reversed-phase chromatography [151].

Hydrophobic selectivity in reversed-phase chromatography arises as a consequence of differences in the nonpolar surface areas of different solutes. For this reason, reversed-phase chromatography is the preferred technique for separating homologous samples. Selectivity for polar and ionic molecules can be manipulated by secondary chemical equilibria. These equilibria are affected by changes in the mobile phase composition or by the addition of buffers or reagents (e.g., ion-pairing or metal-ligand complexing reagents) to the mobile phase. The solvophobic theory can be extended to include the effect of secondary chemical equilibria and at least provide a semiquantitative explanation of retention with mobile phases of widely different compositions [23,27]. The previous discussion assumed that retention could be explained entirely by solvophobic interactions. In practice, solute retention is often complicated by solute interactions with the residual silanol groups on the stationary phase surface, thus both solvophobic and silanophilic binding must be considered [152].

The theory states that the free energy change of the binding process is equal to the difference between the "solvophobic effect", arising from bringing the complex into solution, and placing the individual components into the solvent [23]. Using Figure 4.28 as a guide, the energy required to bring each component from a hypothetical gas phase into the solvent is the sum of two terms. ΔG_c is the energy required to prepare a solvent cavity of suitable size and shape to accommodate the solute. The quantity ΔG_{INT} accounts for the interaction energy between the solute and the surrounding solvent, corrected for the change in the free volume of the solvent when the cavity is filled. Summing the individual energy terms involved in this process leads to equation (4.23) which relates retention of a solute to factors which depend only on the mobile phase composition. The derivation of equation (4.23) is rather complex and references [23] and [27] should be consulted for details.

$$\ln k = \Theta + \frac{1}{RT}\left[\Delta A(N\gamma + a) + NA_s\gamma(X^e-1) + W-\frac{\Delta Z}{\varepsilon}\right] + \ln\frac{RT}{P_o V} \qquad (4.23)$$

k	=	solute capacity factor
Θ	=	volume ratio of stationary and mobile phases
R	=	gas constant
T	=	temperature
ΔA	=	contact area (the difference between the surface areas of the solute and ligand and that of the complex)
N	=	Avogadro's number
γ	=	surface tension
W and a	=	solvent dependent constants arising from the van der Waals contribution to the binding energy
V	=	mole volume of solvent
ε	=	dielectric constant
ΔZ	=	contribution of electrostatic interactions to the binding process
A_s	=	surface area of the solvent molecule
X^e	=	factor which adjusts the macroscopic surface tension to molecular dimensions
P_o	=	1 atmosphere

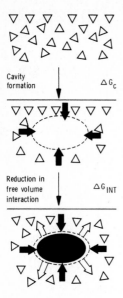

Figure 4.28 Two step process of solute solvation according to the solvophobic
theory. In the first step, a cavity of suitable size and shape is
created in the solvent for the incoming solute molecule; in the
second step, the solute enters the cavity and interacts with the
surrounding solvent. (Reproduced with permission from ref. 23.
Copyright Preston Publications, Inc.)

At first glance equation (4.23) might indicate that the relationship between
retention and the properties of the mobile phase is very complex. Both
experiments and theoretical calculations indicate that only the terms involving
surface tension, γ and X^e, and the dielectric constant, ε, are important. For
non-ionized molecules those terms involving the dielectric constant can be
neglected. In agreement with experiment, equation (4.23) predicts that the
capacity factor will decrease with a decrease in surface tension and that an
approximately linear relationship exists between the logarithm of the capacity
factor and the volume percent of organic modifier in the aqueous mobile phase.
It also predicts an increased solute capacity factor for neutral solutes upon
the addition of salts to the mobile phase and a reduction in the capacity factor
for solutes which ionize, in agreement with experimentally observed phenomena.

4.13 Size-Exclusion Chromatography

Size-exclusion chromatography is a liquid chromatographic technique which
separates molecules according to size. The retention mechanism is unique in
that solute distribution between phases is established by entropy rather than
enthalpy differences. Ideally no interaction occurs between the sample and
support. The support simply acts as a porous matrix containing mobile phase and

the separation is effected by the ability of the sample molecules to gain access to the "stagnant mobile phase" via diffusion through the support pore network. Thus the size and distribution of support pores primarily determine retention and resolution. Unlike other HPLC methods, the mobile phase in size-exclusion chromatography is not varied to control resolution. The mobile phase is chosen to be a good solvent for the sample and, since macromolecules have low diffusion coefficients, to have a low viscosity at the temperature of the separation to ensure a high column plate count.

Size-exclusion chromatography is the preferred method for separating high molecular weight components (MW > 2000). It is also a powerful exploratory method for the separation of unknown samples since it quickly and conveniently provides an overall view of sample composition within a predictable time. Virtually no method development is needed when the mobile phase is a good solvent for the sample. All sample components elute before t_m, the column dead time, which can be predicted in advance. Since sample retention is low the bands elute as sharp peaks, which aids detection. Because of low sample retention, the peak capacity of the method is small and only a few separated bands can be accommodated within the separation. This is shown in Figure 4.29 for the separation of a mixture of proteins [18]. Consequently, size-exclusion chromatography is not the method of choice for resolving multicomponent samples into individual peaks. However, it is an excellent method for obtaining a preliminary sample fractionation for further analysis by other techniques. Freedom from sample loss or reaction and the absence of column deactivation problems are benefits in this respect.

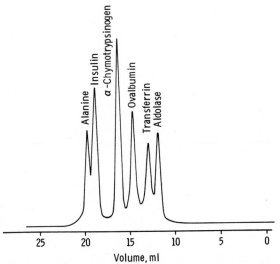

Figure 4.29   Separation of proteins on a small particle gel column. Column 60 x 0.75 cm, TSL-Gel 2000 SW; mobile phase, 0.01 M phosphate buffer (pH 6.5) with 0.2 M sodium sulfate; flow rate, 0.3 ml/min.

Size-exclusion chromatography is principally used to separate samples of different molecular size and to obtain molecular weight and molecular weight distribution information for polymers, Table 4.24. It can not be used to separate compounds of similar size; other methods must be used in this case. Polymers containing narrow molecular weight distributions may be separated as narrow bands, Figure 4.30 [153]. Broad molecular weight polymers are eluted as a single broad band with a profile characteristic of the molecular weight distribution of the sample, Figure 4.31 [18]. Size-exclusion columns with pore diameters of 60 Å or 100 Å are now available and can be used to separate molecules with molecular weights below 1000 [154]. In terms of efficiency and separating power these columns cannot compete with conventional separation columns. For simple mixtures containing components of different sizes or for the analysis of small molecules in a high molecular weight matrix, size-exclusion columns can, however, provide a very rapid and sensitive analytical method. Because of these benefits, they are likely to be more widely used when the nature of the sample allows separation by size.

TABLE 4.24

PRINCIPAL AREAS OF APPLICATION FOR SIZE-EXCLUSION CHROMATOGRAPHY

A. Gel Filtration Chromatography

Used for the separation of water-soluble macromolecules often of biochemical origin. The following information may be obtained:
1. Molecular fractions for characterization or further use
2. To estimate molecular weights using calibration standards
3. To serve as a method for desalting or buffer exchange (mainly in conventional GFC)
4. To estimate molecular association constants
   a. complexes of small molecules with macromolecules
   b. macromolecular aggregation

B. Gel Permeation Chromatography

Normally used as an analytical technique for separating samples soluble in organic solvents. The following information may be obtained:
1. Small molecules may be separated by differences in size
2. To obtain molecular weight averages or the molecular weight distribution of polymers
3. To prepare molecular weight fractions for further use

In size-exclusion chromatography separation results from the distribution of the sample between the mobile phase inside and outside the pores of the column packing. There is no sample retention in the sense that there is no direct interaction with the stationary phase. Indeed, a separation is complete in size-exclusion chromatography when a volume of mobile phase equivalent to the void volume of the column has passed through it. In other liquid

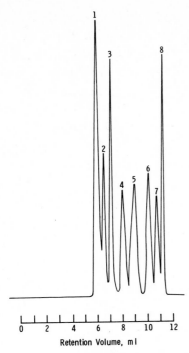

Retention Volume, ml

Figure 4.30 Separation of narrow molecular weight polystyrene standards on a μ-Bondagel column combination, 125, 300, 500, and 1000 Å; mobile phase methylene chloride, flow rate 0.5 ml/min. Polystyrene standards 1, 2,145,000; 2, 411,000; 3, 170,000; 4, 51,000; 5, 20,000; 6, 4000; 7, 600; and 8, benzene. (Reproduced with permission from ref. 153. Copyright Preston Publications, Inc.)

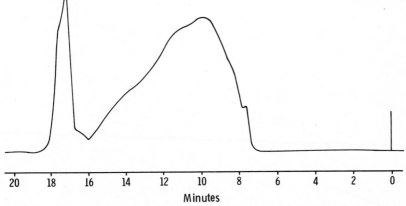

Minutes

Figure 4.31 Molecular wieght distribution of a carboxymethyl-cellulose sample. Column combination of LiChrospher SI-100 and 500; mobile phase 0.5 M aqueous sodium acetate, pH 6, flow rate 0.5 ml/min.

chromatographic techniques, retention is measured as the difference between the elution volume and the void volume, and is always greater than the void volume.

Because of the fundamental difference in elution, the general methods of describing retention and resolution must be modified for size-exclusion chromatography. In particular, the capacity factor k and selectivity factor have no real meaning in size-exclusion chromatography. All peaks elute with a negative value of k, using conventional terminology. Retention in size-exclusion chromatography is better described by the distribution coefficient, K_D, which is related to the experimental parameters by equation (4.24).

$$V_e = V_o + K_D V_i \qquad (4.24)$$

V_e = elution volume of the solute
V_o = column void volume (mainly the interstitial liquid volume
    between the packing particles)
K_D = solute distribution coefficient
V_i = internal pore volume

To avoid confusion in terminology, it should be remembered that the void volume, also known as the dead volume, is comprised of the interstitial particle volume and pore volume in most liquid chromatographic techniques; in size-exclusion chromatography it is used to describe the interstitial particle volume only. The solute distribution coefficient represents the ratio of the average solute concentration in the pores of the stationary phase support to that outside the support. The stationary phase is thus the stagnant mobile phase pool contained in the pores of the column packing. The solute concentration inside the pores decreases with increasing solute size, accounted for by the fact that the entire pore volume is not accessible to large solutes. K_D is constrained to values between 0 and 1, representing the extremes of complete exclusion and permeability of the pore volume to the solute.

The conceptual idea of a theoretical plate can be used in size-exclusion chromatography to measure column efficiency and to compare the performance of packed columns. It is usually measured with small molecules, for example toluene, acetone, or benzyl alcohol, which can explore all the pores of the packing and thus have K_D values of unity. Plate counts measured in this way produce HETP values which are lower than the actual values measured with polymers or proteins partially excluded from some of the pore volume. With biopolymers having a defined structure it is legitimate to talk about an HETP value; however, most synthetic polymers are not single substances but mixtures with a mean molecular size distribution. In this case peak dispersion is comprised of chromatographic dispersion and the molecular dispersion of the solute. Under these circumstances the HETP value is an inappropriate measure of column efficiency. In size-exclusion chromatography the HETP value is due almost entirely to the contributions from stagnant mobile phase dispersion and interparticle mobile phase mass transfer. Contributions from longitudinal diffusion are insignificant since the large polymeric molecules encountered in size-exclusion chromatography have small diffusion coefficients.

Peak-to-peak resolution in size-exclusion chromatography can be calculated by the ratio of peak separation at the peak maxima to the sum of the baseline peaks widths. This general definition of resolution is less useful in size-exclusion chromatography where a measure of the ability of the column to separate solutes of different molecular weight is required. For this purpose, we define a new term, the specific resolution factor, which relates peak resolution to sample molecular weight [155]. It is assumed that all measurements are made within the linear region of the molecular weight calibration curve.

$$R_{sp} = R_s \frac{1}{\log (M_1/M_2)} \tag{4.25}$$

R_{sp} = specific resolution factor
R_s = chromatographic resolution
M_1/M_2 = molecular weight ratio of two standards

If we accept a resolution of unity ($R_s = 1$) as a desirable separation factor (this would correspond to 2% peak overlap for peaks of equal height), then the ratio M_1/M_2 can be defined as the minimum molecular weight ratio, R_m, equation (4.26).

$$R_{sp} = \frac{1}{\log R_m} \tag{4.26}$$

R_m = minimum molecular weight ratio (M_1/M_2) having $R_s = 1.0$

The minimum molecular weight ratio is a useful parameter for comparing column performance and can be used to relate resolution to the basic properties of the chromatographic system, equation (4.27).

$$\log R_m = \frac{4 \, m \, V_e}{V_i \sqrt{n}} \tag{4.27}$$

m = slope of the linear region of the molecular weight calibration curve
V_e = elution volume of standard M_1
V_i = pore volume
n = number of theoretical plates

Equation (4.27) indicates that in order to obtain increased resolving power (corresponding to a minimum R_m value) for a particular column, the column efficiency and the internal pore volume of the packing must be maximized and the slope of the calibration curve minimized. Some of the experimental variables which influence resolution in size-exclusion chromatography are summarized in Table 4.25. Theoretical details are discussed in references [6,33,36,155-159].

TABLE 4.25

FACTORS INFLUENCING RESOLUTION IN SIZE-EXCLUSION CHROMATOGRAPHY

A. General Features

   1. Inherent resolving power of the support material is governed by particle size, pore size distribution, and pore volume.

   2. The dimensions of the column and packing density of the support.

   3. Column operating conditions: mobile phase velocity, temperature, viscosity, solute diffusion coefficient, and sample amount.

B. Factors Improving Resolution

   1. Small particles with a narrow particle size distribution

   2. Particles with a narrow pore size distribution

   3. Particles with a large pore volume

   4. Low mobile phase velocities

   5. Increased column temperatures

C. Factors Diminishing Resolution

   1. The opposite of those factors mentioned in B.

   2. Increased sample molecular weight (due to a decrease in diffusion coefficients).

   3. Mobile phases of high viscosity.

    The above discussion has treated resolution in terms of the separation obtained between two peaks eluting close together. This is a realistic picture for the separation of proteins of a definite molecular weight and a reasonable approximation for oligomers with a narrow molecular weight distribution. Note that under these circumstances resolution can be predicted from equation (4.27) by measuring the calibration curve slope (m), the support pore volume (V_i), and the number of theoretical plates (n). Size-exclusion chromatography is often used to fingerprint polymers of a broad molecular weight distribution. Here the peak separation model is not a useful description of the separation, which may show a continuous distribution of the sample over most of the elution range

without any peak-to-peak resolution. However, the calculation of the column
parameters discussed above is a useful measure of column quality or
effectiveness if not a good description of the sample separation.

In size-exclusion chromatography there should ideally be no interaction
between the sample and the stationary phase support. All solutes should elute
in the range $0 < K_D < 1$. A value of $K_D > 1$ for a non-excluded, small molecule
probe is indicative of an active support [157]. For the bonded phase silica
supports used in aqueous size-exclusion chromatography, three different solute
support interaction mechanisms are possible, depending on the nature of the
solute and the pH and ionic strength of the mobile phase [33,156,160,161]. The
support may show cationic or anionic ion-exchange properties. Ionic sites on
the support may also cause ion exclusion, the process whereby a charged solute
is prevented from exploring the pore volume due to ion repulsion.
Alternatively, the support may show reversed-phase properties and interact with
the sample by a hydrophobic mechanism. The strength of these support
interactions may be estimated from differences in the elution volumes obtained
experimentally from those predicted theoretically for a series of macromolecular
and small molecule test probes [157]. The details of these tests for
biopolymers are summarized in Table 4.26. All supports show some ionic or
solvophobic interactions which can often be minimized by controlling the pH and
ionic strength of the mobile phase. For proteins an ionic strength between 0.1
and 0.6 M is considered optimum. A low ionic strength favors ionic interactions
whereas a high ionic strength leads to solvophobic interactions of the proteins
with the support.

The information sought from an analysis by size-exclusion chromatography is
the sample molecular weight or molecular weight distribution. The information
yielded by the experiment is a separation based on molecular dimensions - size
and shape - a parameter which is not necessarily synonomous with molecular
weight. Even monodisperse samples of the same molecular weight do not
necessarily have the same molecular size. The size of a molecule in solution is
a function of sample/solvent interactions and depends on the nature and strength
of intramolecular forces. Biopolymers adopt different conformations, and
therefore different molecular sizes, when dissolved in various solvents. Some
solutes are hydrated or otherwise associated with small molecules in solution;
such interactions change their hydrodynamic size and separation
characteristics. For monodisperse samples careful column calibration with
standards of similar structural properties is required to obtain molecular
weight information. For polydisperse samples there is no absolute, well-defined
value for the molecular weight. Many synthetic polymers are polydisperse
systems characterized by some distribution of species of different absolute

TABLE 4.26

METHOD FOR THE CHARACTERIZATION OF COLUMNS USED TO SEPARATE WATER SOLUBLE
BIOPOLYMERS

Macromolecular Probes

Proteins are recommended as probes as they are monodisperse. Information
that can be obtained about the column includes:
1. The column void volume V_o. Calf thymus DNA is totally excluded
   from all packings with a pore diameter $< 10,000$ nm.
2. Column Calibration. Proteins that cover a wide molecular weight
   range are available (see Figure 4.35).
3. Surface Adsorption. Underivatized surface silanol groups have a
   slight negative charge. The positively charged protein lysozyme can
   be used to determine the concentration of reactive silanol groups.
4. Retention of Biological Activity. In the case of enzymes, it is
   possible to recover all the sample mass applied to the column and lose
   a large portion of their catalytic activity through solute-support
   interactions. Trypsin is sensitive to this type of interaction on
   silica-based supports and can be used to estimate column performance
   for the recovery of biological activity.

Small Molecule Probes

Small molecule probes can be used to identify conditions under which
nonsize-exclusion chromatography occurs. All probes not associating with
the column would have a K_D value of 1.0. A deviation from this value
indicates an interaction with the support due to ionic or solvophobic
interactions. A mobile phase of low ionic strength favors ionic
interactions with the support while a mobile phase of high ionic strength
favors solvophobic interactions. For column evaluations an aqueous mobile
phase of phosphate buffer pH 7.05 and ionic strength 0.026 is suitable.
1. Cationic-exchange interactions. Arginine and lysine are suitable
   probes.
2. Anionic-exchange interactions. Glutamic acid, oxalic acid, and citric
   acid are suitable probes.
3. Solvophobic effect. Neutral, water soluble molecules such as benzyl
   alcohol or phenylethanol are suitable probes.

---

molecular weight around an average or central value. The molecular weight of a
polydisperse sample must be described by a statistical function such as the
number-average molecular weight ($\overline{M}_n$) or the weight-average molecular weight
($\overline{M}_w$). The calculation of these values from the chromatogram is illustrated in
Figure 4.32. The calculations can be performed manually or, if suitable data
handling equipment is available, automatically.

The most critical decision in size-exclusion chromatography is column
selection. This is also a fairly simple decision since the molecular weight
separating range of the column is defined by the pore size and pore size
distribution of the packing. The molecular size separating range of a packing
is described by the shape of its calibration plot. Some representative
calibration plots for LiChrospher microporous silica packings are shown in
Figure 4.33. The flat central portion of the curve, the fractionation range,

Term Describing Molecular Weight	Symbol	Formula for Calculation	Formula for Calculation from the Chromatogram
Number-Average molecular weight	$\overline{M}_n$	$\dfrac{\Sigma N_i M_i}{\Sigma N_i} = \dfrac{\Sigma W_i}{\Sigma(W_i/M_i)}$	$\dfrac{\Sigma_{i=1}^{N} h_i}{\Sigma_{i=1}^{N} (h_i M_i)}$
Weight-Average molecular weight	$\overline{M}_w$	$\dfrac{\Sigma N_i M_i^2}{\Sigma N_i M_i} = \dfrac{\Sigma W_i M_i}{\Sigma W_i}$	$\dfrac{\Sigma_{i=1}^{N} (h_i M_i)}{\Sigma_{i=1}^{N} h_i}$
Z-Average molecular weight	$\overline{M}_z$	$\dfrac{\Sigma N_i M_i^3}{\Sigma N_i M_i^2}$	
Polydispersity INDEX	I	$\dfrac{\overline{M}_w}{\overline{M}_n}$	

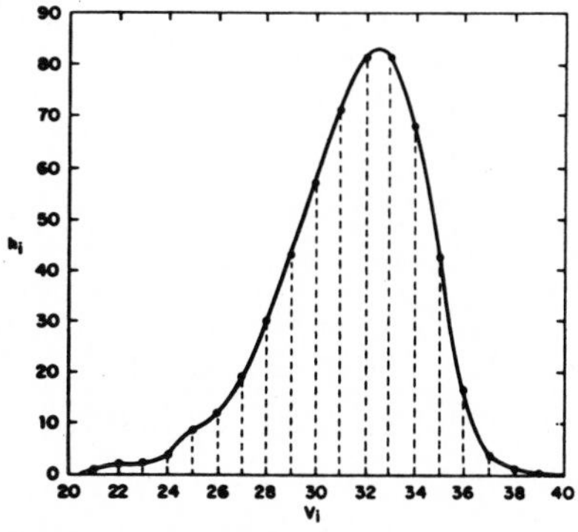

1 to calculate $\overline{M}_n$ or $\overline{M}_w$, manually digitize chromatogram (see insert). Choose V_i to be equal volume increments such that i>8 an preferably >20

2 measure the height h_i corresponding to each segment V_i

3 calculate M_i for each value of V_i from a molecular weight calibration curve generated separately

4 Use equations above to calculate $\overline{M}_n$ and $\overline{M}_w$

NOTES
a) N_i and W_i are the number and weight of molecules having a molecular weight M_i
b) $\overline{M}_n = \overline{M}_w$ for a monodisperse system
c) $\overline{M}_w > \overline{M}_n$ for a polydisperse system
d) I is a measure of the breadth of the polymer molecular weight distribution

Figure 4.32 Manual methods for determining the molecular weight of polydisperse samples.

300

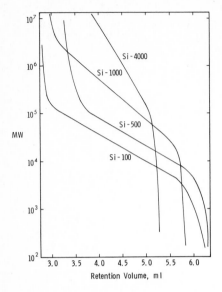

Figure 4.33 Calibration curves for LiChrospher columns. Calibration standards
polystyrene and mobile phase tetrahydrofuran.

represents the usable molecular weight separating range of the packing. The
sharp breaks in the calibration plot that occur at either end of the
fractionation range correspond to the molecular weight regions in which the
sample is either totally excluded or free to explore the total pore volume of
the packing. In these regions the separation properties of the packing are
poor. To separate two components of different molecular size, a column packing
for which the two components elute in the middle of the fractionation range
should be selected. All other things being equal, the column with the smallest
gradient for the fractionation range will provide the highest resolution. Once
the optimum column has been selected, resolution can be increased by coupling
two or more columns of the same pore size to increase the chromatographic
efficiency of the system. In this case the range of molecular weight separation
remains the same, but the volume in which the separation is made is increased by
a factor equal to the increase in column length.

    For separating samples with a relatively narrow molecular weight range   (<
500), packings with a single pore size are normally used. To separate samples
with a broad molecular weight distribution, columns containing packings with a
wide pore range are used. This is achieved by coupling columns having
nonoverlapping but closely adjacent fractionation ranges [162]. A wide linear
molecular weight calibration range is obtained with a bimodal pore size
configuration. Columns containing only two discrete pore sizes and a narrow
size distribution, differing by approximately a factor of ten in pore size and

of approximately equal pore volume, are coupled for this purpose. The linear separation range covered by two commercial column packing materials is shown in Figure 4.34. The linear portion of the calibration curve can be described by equation (4.28).

$$V_R = C_1 - C_2 \log (MW) \qquad\qquad (4.28)$$

V_R = elution volume

C_1 = intercept of the linear portion of the molecular weight
   calibration curve

C_2 = slope of the linear portion of the molecular weight calibration
   curve in ml per decade of molecular weight

MW = molecular weight

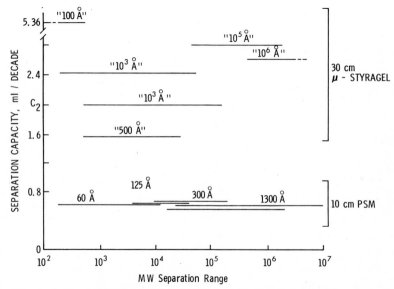

Figure 4.34 Selection of bimodal pore size column packing for maximum range and linearity. (Reproduced with permission from ref. 162. Copyright Elsevier Scientific Publishing Co.)

Consider the μStyragel columns described in Figure 4.34. The pore volumes of the different packings can be estimated by the C_2 terms which are additive for coupled columns. To obtain a separation system with a linear range of 5×10^2 to 1×10^6, two 500 Å columns (yielding a slope C_2 of $2 \times 1.6 = 3.2$) must be coupled with one column of 10^5 Å (C_2 ca. 3). From Figure 4.34, it can also be seen that selecting porous silica microsphere columns for bimodal operation is simpler and more accurate because the internal pore volumes and the fractionation ranges are nearly uniform for all pore sizes. By coupling microporous silica columns of 60 Å and 750 Å pore sizes, a linear molecular weight separation range greater than four orders of magnitude can be obtained,

Figure 4.35. The bimodal pore size configuration provides better resolution and
a wider linear calibration range than can be obtained by the series coupling of
several column types with overlapping molecular weight ranges.

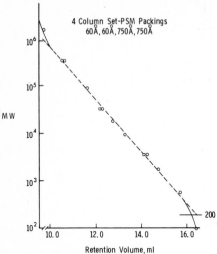

Figure 4.35   Broad-range, linear calibration for polystyrene standards using
               the bimodal pore size selection principle. (Reproduced with
               permission from ref. 162. Copyright Elsevier Scientific
               Publishing Co.)

The first step in extracting molecular weight information from a
size-exclusion chromatogram is to establish a calibration curve relating the
retention volume (or K_D) to the molecular weight of the polymer sample.
Molecular weight, rather than the size of the polymer molecule, is used as the
calibration parameter since the former is independent of experimental
conditions. True or approximately monodisperse samples may be eluted as
separate peaks in the chromatogram whereas most synthetic polymers are
polydisperse and elute as a broad band with a profile characteristic of the
molecular weight distribution of the polymer. For monodisperse samples an
absolute molecular weight value is obtained while a statistical average
molecular weight parameter is calculated for polydisperse samples. For
monodisperse samples the primary method of calibration is the peak position
calibration method. A series of different molecular weight standards are
chromatographed under constant experimental conditions and their elution volumes
measured. A plot of molecular weight against elution volume constitutes the
calibration plot. Figure 4.36 is an example of such a plot for a series of
protein standards. The peak position calibration method is limited in scope by
the availability of suitable standards. Apart from polystyrenes, very few other
narrow molecular weight polymer standards are available for characterizing
synthetic polymers. The accuracy of the peak positon calibration method depends

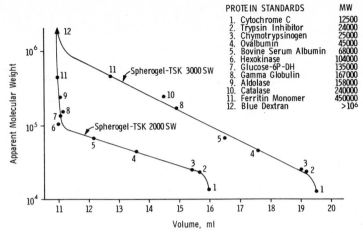

PROTEIN STANDARDS	MW
1. Cytochrome C	12500
2. Trypsin Inhibitor	24000
3. Chymotrypsinogen	25000
4. Ovalbumin	45000
5. Bovine Serum Albumin	68000
6. Hexokinase	104000
7. Glucose-6P-DH	135000
8. Gamma Globulin	167000
9. Aldolase	158000
10. Catalase	240000
11. Ferritin Monomer	450000
12. Blue Dextran	>10^6

Figure 4.36   Protein calibration curves for Spherogel TSK–SW 2000 and SW 3000
columns. Mobile phase phosphate buffer 0.2 M, pH 6.8, flow rate
1.0 ml/min.

on whether the solution conformations of the standards and samples are similar,
the condition required for the molecular weight and molecular size to be
uniquely correlated. Large errors in estimating the sample molecular weight can
result when a calibration prepared from narrow molecular weight standards of one
polymer is used to characterize polymers of other types. For this reason,
various attempts have been made to determine a universal calibration parameter
to characterize the effective dimensions or sizes of macromolecules and various
polymers. Macromolecules have solution conformations which may be approximately
described by random coil, rigid sphere, or rigid rod models [163]. The best
approximation of molecular size for the above models is given by the radius of
gyration. This parameter provides the common denominator for comparing solute
molecular sizes for molecules of different chemical structure and shapes. The
radius of gyration is not independent of structural conformation changes and
their effect on the separation mechanism. The hydrodynamic volume of a polymer
is also a size characteristic parameter which may be used for calibration. The
hydrodynamic volume is equal to the product of the intrinsic velocity and the
molecular weight of the polymer. The intrinsic viscosity of the polymer is an
experimental quantity derived from the measured viscosity of the polymer in
solution or calculated using Mark–Houwink constants [36,163]. A plot of the
logarithm of the hydrodynamic volume against elution volume provides a
calibration curve that is approximately valid for all polymers. This approach
provides a universal calibration curve for virtually all polymers, independent
of shape. The universal calibration method is conceptually sound, but its use
is still rather limited. The accuracy and the precision of the method have not
yet been fully evaluated.

Various computer-assisted calibration methods are available for using broad molecular weight distribution standards for calibration purposes. The details are complex and are reviewed in references [36,163,164].

From the above discussion, it is apparent that establishing an accurate calibration curve requires a considerable amount of effort and may be limited by the availability of suitable standards. Sample molecular weight distributions may be evaluated quantitatively by molecular weight measurements using light scattering or viscometry by either continuous monitoring of the column eluent or in a discontinuous way. Average molecular weight distributions can be obtained directly as the sample elutes from the column using low angle laser light scattering (LALLS). The column eluent first flows through a light-scattering detector and then through a refractometer which serves as a concentration-sensitive detector. The output of the light-scattering detector is proportional to the product of the molecular weight (MW) of the eluting sample and its concentration. By examining the outputs from both detectors, average molecular weights and molecular weight distributions can be obtained with a data processor [35,164]. Although this is a very elegant approach towards calibration, the general use of the LALLS detector is limited by its high purchase price.

4.14   Ion-Exchange Chromatography

Ion-exchange chromatography is a flexible technique used mainly for the separation of ionic or easily ionizable species. The stationary phase is characterized by the presence of charged centers bearing exchangeable counterions. Both cations, Figure 4.37, and anions, Figure 4.38, can be separated by selection of the appropriate ion-exchange medium [165,166]. Ion-exchange finds application in virtually all branches of chemistry. In clinical laboratories it is used routinely to profile biological fluids, Figure 4.39, and for diagnosing various metabolic disorders [167]. Ion-exchange has long been used as the separation mechanism in automated amino acid and carbohydrate analyzers. Figure 4.40 is an example of a separation of a mixture of basic amino acids and polyamines obtained with an automated amino acid analyzer [168]. More recently, ion-exchange chromatography has been used to separate a wide range of biological macromolecules using special wide pore, low capacity packings designed for this purpose [31,144,169]. Ion-exchange may also be used to separate neutral molecules as their charged bisulfite or borate complexes, Figure 4.41, and certain cations as their negatively-charged complexes (e.g., $FeCl_4^-$). In the case of the borate complexes, carbohydrate compounds having vicinal diol groups can form stable charged adducts that can be resolved by anion exchange chromatography [170]. Ligand-exchange chromatography

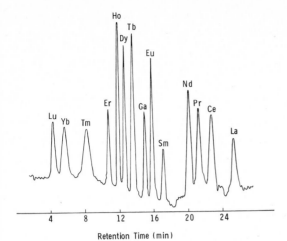

Figure 4.37 Separation of the lanthanides on a 10 cm x 4 mm I.D. strong
cation-exchange resin Ammex A-5, particle size 13 μm. Linear
gradient, 20 min, 0.018 to 0.070 mol/1 hydroxybutyric acid. The
lanthanides were detected at 600 nm using Arsenazo I and a postcolumn reaction detector. (Reproduced with permission from ref.
165. Copyright American Chemical Society).

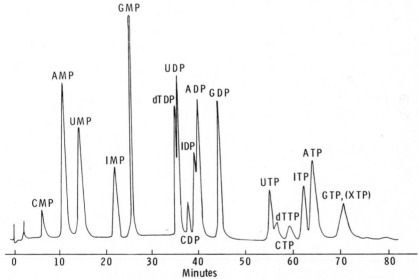

Figure 4.38 Separation of mono-, di-, and triphosphate nucleotides on a 25 cm
x 4.6 mm I.D. PARTISIL-IOSAX Column. Linear gradient 0-100% in
45 min of 0.007 F KH_2PO_4, pH 4.0 to 0.25 F KH_2PO_4-0.05 F KCl,
pH 4.5. Flow rate 1.5 ml/min. (Reproduced with permission from
ref. 166. Copyright Whatman, Inc.)

has been used with cation exchangers (in the nickel or copper form) for the
separation of amino acids and other bases (see Chapter 7.19 for details).

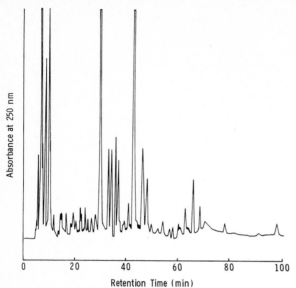

Figure 4.39 Separation of UV-absorbing anions in urine on a 25 cm x 4.6 mm
I.D. Hitachi Gel 3013-N column using a complex stepwise buffer
gradient from water to 0.3 mol/1 NH$_4$Cl-0.05 mol/1 NH$_4$H$_2$PO$_4$
-30% v/v CH$_3$CN, pH 4.8. Column temperature was also raised
in increments during the run. (Reproduced with permission from
ref. 167. Copyright Elsevier Scienitific Publishing Co.)

$$RM-L \; + \; X \rightleftharpoons RM-X \; + \; L$$

RM = metal/ion-exchange ion-pair
L = mobile phase ligand which can form a complex with the metal M
X = sample ligand

Ion-exchange packings may also be used to separate neutral and charged species
by mechanisms not involving ion-exchange. Oligosaccharides and related
materials can be separated by a partition mechanism on ion-exchange columns when
water-alcohol mobile phases are employed [171]. At equilibrium, the water
concentration in the ion-exchange resin is greater than that in the surrounding
solution. Under these circumstances, the ion-exchange resin functions as a
solid support and the sugars are separated by partition between the water-rich
resin phase and the eluent.

Ion-exclusion may be used to separate charged from uncharged species and
also charged species from one another on the basis of their degree of Donnan
exclusion from the resin pore volume. An ion-exchange packing having the same
charge as the sample ions is selected for this purpose. Retention is dependent
on the degree of sample access to the packing pore volume. An example of this
mechanism is the separation of organic acids with a cation-exchange packing in
the hydrogen form [31,172]. Strong acids are completely excluded and elute
early, weak acids are only partially excluded and have intermediate retention

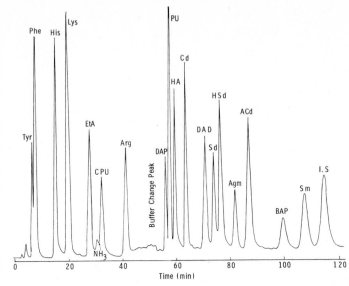

Figure 4.40 Separation of basic amino acids and polyamines using an amino
acid analyzer. Column was 11.5 x 0.45 cm I.D. cation-exchanger,
Durrum DC-4A. The first buffer, 0.20 M sodium citrate dihydrate
-0.3 M sodium chloride-4% ethanol (pH 5.6) for the first 44
min and then 0.20 M sodium citrate dihydrate-2.50 M sodium
chloride-6% ethanol (pH 5.65). Two temperatures were used:
61° C during the first 48 min and 78°C thereafter. EtA,
ethanolamine; CPU, carbamylputrescine; DAP, 1,3-diaminopropane;
PU, putrescine; HA, histamine; Cd, cadaverine; Sd, spermidine;
Agm, agmatine; Sm, spermine; and I.S., internal standard.
(Reproduced with permission from ref. 168. Copyright Elsevier
Scientific Publishing Co.)

values, and neutral molecules are not influenced by the Donnan membrane
potential and can explore the total pore volume.

Ion-exchange packings are characterized by the presence of charge-bearing
functional groups. From a global point of view, sample retention could be
envisioned as a simple exchange between the sample ions and those counterions
originally attached to the charge-bearing functional groups. However, this
simple picture is a poor representation of the actual retention process.
Retention in ion-exchange chromatography is known to depend on factors other
than coulombic interactions. With organic ions, for example, hydrophobic
interactions between the sample and the non-ionic regions of the support are
important. In ion-exchange chromatography the mobile phase is often of high
ionic strength; this favors hydrophobic interactions by "salting-out" the
sample. Indeed, from a qualitative point of view, the retention of organic ions
probably proceeds by a hydrophobic reversed-phase interaction with the support
followed by diffusion of the sample ion to the fixed charge center where an
ionic interaction occurs.

308

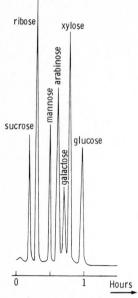

Figure 4.41 Separation of a carbohydrate mixture as borate complexes on a 19.0 x 0.6 cm I.D. column of DA-Z4 anion-exchange resin. Mobile phase: boric acid buffer 0.4 mol/1, pH 9.2, flow rate 1.3 ml/min, and temperature 60°C. (Reproduced with permission from ref. 170. Copyright Elsevier Scientific Publishing Co.)

Size-exclusion and ion-exclusion are two other retention mechanisms which may have to be considered under certain circumstances. Size-exclusion chromatography was described in the previous section and is important for explaining the retention of large ionic molecules which, because of their size, are prevented from exploring the total pore volume of the packing where the highest concentration of ion-exchange centers are located. Also, since Donnan membrane potentials are developed on charge-carrying supports, ions having the same charge as the support may be excluded from the pore volume by ion repulsion. This mechanism is important for separating charged from neutral molecules since only the charged molecules are excluded from the pore volume. The net result of size-exclusion and ion-exclusion processes is the elution of the sample in an eluent volume which is less than the total column void volume. Sample retention in conventional ion-exchange resins having comparatively large particle diameters has been studied extensively [173-177]. Both sample retention and column efficiency are influenced by diffusion-controlled processes, of which the following steps are considered the most important:

- Diffusion of ions through the liquid film surrounding the resin bead
- Diffusion of ions within the resin particle to the exchange sites
- The actual exchange of one counterion for another
- Diffusion of the exchanged ions to the surface of the resin bead
- Diffusion of the exchanged ions through the liquid film surrounding the resin bead into the bulk solution

Slow diffusion of the sample ions within the resin beads contributes significantly to poor column performance. Reducing the particle size to less than 10 micrometers in diameter compensates for the poor mass transfer kinetics exhibited by conventional resin beads by reducing the length of intraparticulate channels.

Because the column packings used in ion-exchange contain charged functional groups, an unequal distribution of mobile phase ions inside and outside the resin bead develops in accordance with the Donnan membrane effect. The ion-exchange bead behaves as a concentrated electrolyte solution in which the resin charges are fixed, whereas the counterions are free to move. The contact surface between the resin bead and the mobile phase can be envisioned as a semipermeable membrane. When equilibrium is attained between the external and internal solution and one side of the membrane contains a nondiffusible ion, then a combination of the Donnan membrane effect and the need to preserve overall electrical neutrality results in a greater concentration of free electrolyte within the bead. Diffusion of sample ions and counterions across the Donnan membrane barrier is often the rate-controlling process in ion-exchange chromatography.

The retention mechanism in ion-exchange chromatography can also be described by the law of mass action as shown in the example below:

$$n\ RSO_3^- X^+ + Y^{n+} \rightleftharpoons (RSO_3^-)_n Y^{n+} + nX^+$$

$$K_q = \frac{[(RSO_3^-)_n Y^{n+}][X^+]^n}{[RSO_3^- X^+]^n [Y^{n+}]}$$

K_q = equilibrium quotient

X^+ = stationary phase counterion

Y^{n+} = sample cation of valence charge n

n = an integer equal to the valence charge of the sample cation

R = polymer matrix

Since the activities of ions in the resin phase cannot be evaluated accurately, the equilibrium quotient, K_q, is not a true thermodynamic constant. This

quotient is influenced by many experimental values and is therefore of limited
use in predicting retention. It is, however, a good measure of the resin's
preference for one particular counterion relative to another. Selectivity
series have been established for many commmon counterions.

$$Li^+ < H^+ < Na^+ < NH_4^+ < K^+ < Cs^+ < Ag^+ \ Cu^{2+} < Cd^{2+} < Ni^{2+} < Ca^{2+} < Sr^{2+} < Pb^{2+}$$
$$< Ba^{2+}$$

$$F^- < OH^- < acetate < formate < Cl^- < SCN^- < Br^- < I^- < NO_3^- < SO_4^{2-} < citrate$$

The absolute order depends on the individual ion-exchanger but deviations from
the above order are usually only slight for different cation and anion
exchangers. For weak-acid resins, H^+ is preferred over any common cation, while
weak-base resins prefer OH^- over any of the common anions.

Once a selection of the column type has been made, sample resolution is
optimized by adjusting the ionic strength, pH, temperature, flow rate, and
concentration of buffer or organic modifier in the mobile phase [178]. The
influence of these parameters on solute retention is summarized in Table
4.27. Increasing the concentration of counterions in the mobile phase by either
increasing the buffer concentration or by the addition of a neutral salt
provides stronger competition between the sample and counterions for the
exchangeable ionic centers and generally reduces retention. As the ionic
strength of the mobile phase has the greatest influence on solvent strength,
this parameter is usually optimized first to obtain sample capacity factor
values between 2-10. Selectivity is adjusted by varying the pH of the mobile
phase. Changing the pH modifies both the character of the ion-exchange media
and the acid/base equilibria and ionization of the sample. A pH gradient can
also be used to control solvent strength. Slow column equilibration can,
however, cause problems if the pH ranges over several pH units. Under these
conditions the separation may not be reproducible. Ionic strength gradients are
usually used for controlling retention whereas pH gradients of a narrow range
are used to control selectivity. The operating pH range for a separation can be
estimated from the pK_a values of the sample components. When sample pK_a values
are unknown, approximate values can be estmiated by considering the molecular
structure and the number and type of functional groups present, Table
4.28. Sample pK_a and mobile phase pH values are related by the
Henderson-Hasselbalch equation (4.29).

$$pH = pK_a \pm \log \frac{[\text{ionized}]}{[\text{non-ionized}]} \tag{4.29}$$

TABLE 4.27

FACTORS INFLUENCING RETENTION IN ION-EXCHANGE CHROMATOGRAPHY

Mobile Phase Parameter	Influence on Mobile Phase Properties	Effect on Sample Retention
Ionic Strength	Solvent Strength	Solvent strength generally increases with an increase in ionic strength. Selectivity is little affected by ionic strength except for samples containing solutes with different valence charges. Nature of mobile phase counterion, K_q value, controls the strength of the interaction with the stationary phase.
pH	Solvent Strength	Retention increases in cation-exchange and decreases in anion-exchange chromatography with an increase in pH.
	Solvent Selectivity	Small changes in pH can have a large influence on separation selectivty.
Temperature	Efficiency	Elevated temperatures increase the rate of solute exchange between the stationary and mobile phases and also lower the viscosity of the mobile phase.
Flow Rate	Efficiency	Flow rates may be slightly lower than in other HPLC methods to maximize resolution and improve mass transfer kinetics.
Buffer Salt	Solvent Strength and Selectivity	Solvent strength and selectivity are influenced by the nature of the counterion i.e., its K_q value. A change in buffer salt may also change the mobile phase pH.
Organic Modifier	Solvent Strength	Solvent strength generally increases with the volume percent of organic modifier. Its effect is most important when hydrophobic mechanisms contribute significantly to retention. In this case, changing the organic modifier can be used to adjust solvent selectivity as normally practiced in reversed-phase chromatography.
	Efficiency	Lowers mobile phase viscosity and improves solute mass transfer kinetics.

Only ionized solutes are retained by an ion-exchange mechanism and, as a general rule of thumb, equation (4.29) predicts that the optimum buffer pH should be 1 or 2 pH units below the pK_a of bases and 1 or 2 pH units above the pK_a of acids. When choosing a buffer salt two criteria must be met. The buffer must be able to establish the operating pH for the separation and the exchangeable buffer counterion must provide the desired solvent strength. Some common pH buffers and their usable ranges are summarized in Table 4.29. The solvent strength of the buffer can be estimated from the K_q value for the counterion. Typical buffer concentrations are 0.001-0.5 M.

TABLE 4.28

pK_a VALUES OF REPRESENTATIVE ORGANIC COMPOUNDS

Compound Type	Approximate pK_a Value
Amides	-2 - 1
Pyrroles	0.3
Thiazoles	1 - 3
Amino acids	2 - 4 and 9 - 12
Carboxylic acids	5
Aromatic amines	5 - 7
Aliphatic amines	9.5 - 11.0
Phenols	10

TABLE 4.29

BUFFER SALTS WIDELY USED IN ION-EXCHANGE CHROMATOGRAPHY

Buffer Salt	Useful pH Range
Citric acid	2.0 - 6.0
Ammonium phosphate	2.2 - 6.5
Potassium hydrogen phthalate	2.2 - 6.5
Disodium hydrogen citrate	2.6 - 6.5
Sodium formate	3.0 - 4.4
Sodium acetate	4.2 - 5.4
Triethanolamine	6.7 - 8.7
Sodium borate	8.0 - 9.8
Sodium perchlorate	8.0 - 9.8
Sodium nitrate	8.0 - 10.0
Ammonia	8.2 - 10.2
Ammonium acetate	8.6 - 9.8
Sodium dihydrogen phosphate	2.0 - 6.0
	8.0 - 12.0
Potassium dihydrogen phosphate	2.0 - 8.0
	9.0 - 13.0

Ion-exchange columns are generally less efficient than other column types used in HPLC. To improve solute diffusion and mass transfer, ion-exchange columns are often operated at elevated temperatures. As well as increasing column efficiency, an increase in column temperature usually results in low capacity factor values. Small changes in column temperature often result in large changes in separation selectivty, particularly for structurally dissimilar compound types. Ion-exchange columns are also often operated at low mobile phase flow rates to maximize column efficiency. For most separations the mobile phase is usually completely aqueous. Water-miscible organic solvents may be added to the mobile phase to increase column efficiency and to control solvent strength. The presence of an organic modifier in the mobile phase has most influence on the separation when solute retention is at least partly controlled by a reversed-phase mechanism. When this is the case, solvent strength and selectivity can be adjusted as described for reversed-phase chromatography.

4.15 Ion Chromatography

Ion chromatography is a recent development which has found widespread application for the analysis of inorganic and organic ions with pK_a values less than 7. It combines the techniques of ion-exchange separation, eluent suppression, and conductivity detection for the quantitative determination of such ions as Li^+, Na^+, K^+, NH_4^+, Ca^{2+}, Mg^{2+}, F^-, Cl^-, Br^-, I^-, SO_4^{2-}, NO_2^-, NO_3^-, PO_4^{2-}, alkylamines, organic acids, etc. [179-184]. Its explosive growth is due in part to the difficulty of determining these ions by other methods; it has replaced many tedious wet chemical analyses with a simple automated instrument that can analyze multiple ions in the same sample injection. Examples of ion chromatographic separations are shown in Figure 4.42 for nine common anions and Figure 4.43 for the alkali earth elements [185].

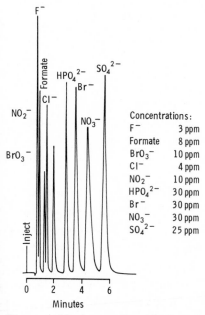

Figure 4.42   Separation of nine common ions by ion chromatography using a hollow fiber suppressor and conductivity detection. Mobile phase was 2.8 mM $NaHCO_3$-2.2 mM Na_2CO_3 and flow rate 2.0 ml/min.

The instrumentation used in ion chromatography does not vary significantly from that used in HPLC. The mechanical strength of the column packings limits pressures to about 2000 p.s.i. The column packings are styrene-divinylbenzene bead-type resins that are surface functionalized to give low ion-exchange

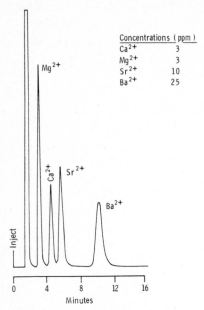

Figure 4.43 Separation of alkali earth elements. Mobile phase 2.5 mM
m-phenylenediamine dihydrochloride-2.5 mM HNO$_3$, flow rate
3.5 ml/min. The separator column was a surface-sulfonated
35-55 μm macroporous resin with a conventional suppressor column
and conductivity detector.

capacities of 0.001 to 0.05 mequiv/g [185,186]. These resins have good
structural rigidity, allowing operation at high flow rates with only moderate
back pressure. The thin skin of surface ion-exchange groups provides short
diffusion paths and higher column efficiencies than conventional resins. These
resin beads are stable over the pH range 1-14. The limited hydrolytic stability
of silica-based packings curtails their use in ion chromatography compared to
their dominant position in the modern practice of HPLC. For anion separations,
a special packing which has an outer layer of fine (0.1 to 0.5 micrometer),
aminated latex particles agglomerated to the surface of a surface-sulfonated
resin bead is frequently used [187,188]. The latex layer is strongly attached
to the surface-sulfonated core by a combination of electrostatic and van der
Waals forces. The latex layer is not displaced from the substrate by the
eluents or conditions normally used in ion chromatography. The thinness of the
exchange layer and the Donnan exclusion effect of the intermediate sulfonated
layer provide excellent sample mass transfer properties by ensuring that sample
penetration is confined to the outer latex layer. In general, the efficiency of
the columns used in ion chromatography is limited by the large-sized particles
and broad particle size distributions of the resin packings. Resin beads are
currently available in the ranges 20-30, 37-74, 44-57 micrometers.

The principal problem in the devlopment of ion chromatography was not one of column technology, but rather one of providing a suitable on-line detection system. Most common ions cannot be detected photometrically and a method was needed which could detect the separated ions in the background of a highly conducting eluent. Since conductivity is a universal property of ions in solution and can be simply related to ion concentration, it was considered a desirable detection method provided that the contribution from the eluent background could be eliminated. The introduction of eluent suppressor columns for this purpose led to the general acceptance of ion chromatography [189]. The ion chromatograph consists of two ion-exchange columns in series followed by a flow-through electrical conductivity detector. The first column separates the ions in the injected sample while the second column suppresses the conductance of the electrolyte in the eluent, and thereby enhances the conductance of the separated ions since these are converted to the highly conducting acid or base form. By way of example, consider the separation of a mixture of Li^+, Na^+, and K^+ on a cation-exchange column with dilute hydrochloric acid as the mobile phase. The three cations are separated on the first column according to their strength of interaction with the column packing and leave the column in the form of their chloride salts dissolved in the mobile phase. As the eluent enters the suppressor column two reactions occur as shown below:

$$HCl + Resin^+ OH^- \longrightarrow Resin^+ Cl^- + H_2O$$

$$MCl + Resin^+ OH^- \longrightarrow Resin^+ Cl^- + MOH$$

$$M = Li^+, Na^+, K^+$$

The eluent leaving the suppressor column contains the sample ions as their highly conducting hydroxide salts in a weakly conducting background of water. Anion suppression occurs in an analogous manner except that the conducting species is now an acid in a background eluent of water. The suppressor column is packed with an ion-exchange packing opposite in type to the separator column and of higher capacity since it must neutralize a large volume of column eluent. The column packing in the suppressor column should have a low surface area (microporous resin), a small bead diameter to minimize band dispersion, a moderate degree of crosslinking to minimize adsorption and swelling effects, and a minimal total volume to minimize the effects of dispersion and the retention of weak electrolytes [185,189]. As far as sample dispersion is concerned, the void volume of the suppressor column is less than the total column void volume since the Donnan exclusion phenomena prevents entry of the sample ions into the pore volume. Only the interstitial particle volume need normally be considered. The low concentration eluents used to separate the ions on the separator column allow a large number of samples to be analyzed before the suppressor column is completely expended. Typically, 40 to 60 anion samples and 20 to 30 cation samples can be run before the suppressor resin beds require regeneration. Some instruments contain two suppressor columns in parallel so that one column can be regenerated while the other is in use.

The suppressor column is included in the chromatographic system by necessity rather than desire. As it has several deleterious effects on the quality of the separation, a natural progression in the further development of ion chromatography was the search for nonsuppressor-based technologies. The principal disadvantages of suppressor columns are the need to periodically replace or regenerate the suppressor column; the varying elution times for weak acid anions or weak base cations due to ion-exclusion effects in the unexhausted portion of the suppressor column; the apparent reaction of some ions such as nitrite with the unexhausted portion of the suppressor column, resulting in a varying response depending on the percentage exhaustion of the suppressor column; and, interference in the baseline of the chromatogram by a negative peak characteristic of the eluent, which varies in elution time with the degree of exhaustion of the suppressor column. Finally, there is some band spreading in the suppressor column which diminishes the efficiency of the separator column and reduces detection sensitivity. Many of the problems encountered with conventional suppressor columns can be eliminated by using bundles of empty or packed ion-exchange hollow fibers for eluent suppression [190-192]. The hollow polyethylene fibers (6 ft x 0.3 mm I.D.) are surface sulfonated to allow cations to permeate the fiber wall into the regenerating solution and hydrogen ions to pass into the column eluent. The sample anions in the column eluent do not permeate the fiber wall because of Donnan exclusion forces. The main advantage of the hollow fiber ion-exchange suppressor is that it allows continuous operation of the ion chromatograph without varying interferences from base line discontinuities, ion-exclusion effects, or chemical reactions. Continuous operation is made possible by constantly renewing the regenerant solution in which the hollow fibers are bathed. The main disadvantage is that the hollow fiber ion-exchange suppressors have approximately 2-5 times the void volume of conventional suppressor columns. This obviously leads to some loss of separation efficiency, particularly for ions eluting early in the chromatogram. Other limitations include restrictions on useable column flow rates to ensure complete suppression, the need for an excess of the exchangeable counterion in the regeneration solution, and the possible "leakage" of the regeneration solution into the eluent stream.

Common eluents in suppressor ion chromatography are dilute solutions of mineral acids or phenylenediamine salts for cation separations and sodium bicarbonate/sodium carbonate buffers for anion separations. These eluents are too highly conducting to be used without a suppressor column for conductivity

detection. Fritz and co-workers have shown that if separator columns of very
low capacity are used in conjunction with an eluent of high affinity for the
separator resin but of low conductivity, then a suppressor column is not
required [193-195]. A specially designed conductivity detector with some
tolerance for background conductance was needed, however. The single most
important requirement of the eluent selected for nonsuppressed ion
chromatography is that the background conductivity of the eluent must be low,
generally in the 75-200 μohm range. The salts of aromatic acids such as benzoic
and phthalic acids, in low concentration are suitable eluting agents. As the
conductance of both H^+ and OH^- ions are unusually high, the pH of the eluent
must be carefully adjusted to allow ionization of the mobile phase counterion
without increasing the conductance of the eluent to an undesirable level. For
example, sodium phthalate solutions may be used in the pH range 4.0-6.5 without
swamping the detector response. Since the detector baseline noise is primarily
a function of background eluent conductivity there is some loss in the signal to
noise ratio of the sample ion in the nonsuppressed mode. There is also a loss
of sensitivity due to the "matrix effect" [185]. As the protonated solute ion
is being measured in a more conductive background, there is an inherent loss of
sensitivity. In general, nonsuppressed ion chromatography is at least an order
of magnitude less sensitive than suppressed ion chromatography and has a more
restricted linear operating range.

An alternative approach to conductivity monitoring using a photometric
detector has been described by Small et al. [196-198]. It is based on the use
of vacancy chromatography to determine sample ions lacking absorption in the
UV-visible range. An eluent containing a UV-absorbing counterion such as sodium
phthalate, sodium sulfobenzoate, sodium trimesate, sodium iodide, or copper
nitrate is used for the separation. Under equilibrium conditions a constant
signal due to absorbance by the eluent is determined by the detector. When a
sample ion elutes from the column, the concentration of UV-absorbing counterion
in the eluent peak volume is depleted by the ion-exchange process on the column,
and the detector registers a negative deflection in proportion to the reduction
of absorbing species in the eluent. Sample ions with pK_a values greater than 7
can be analyzed by this technique since the detection is no longer dependent on
the native conductance of the sample ion. In the future both higher efficiency
separator columns and new methods of nonsuppressed detection can be expected to
appear for ion chromatography.

4.16 Microcolumns in Liquid Chromatography

There were three incentives for reducing the internal diameters of the
columns used in liquid chromatography. Analogous to capillary column gas

chromatography, it was hoped that microcolumns would yield the very high column efficiencies required for the resolution of complex mixtures. Since these microcolumns are operated at flow rates of a few microliters/min, they are extremely economical in solvent consumption. These low column flow rates also offer new possibilities for sensitive detection, provided, of course, that detection volumes can be sufficiently miniaturized for this purpose.

Three types of microcolumns are currently in use. Microbore columns are similar in construction to conventional packed columns except that the column bore is reduced to about 1 mm and the mobile phase flow rate is about 30-40 microliters/min. Packed microcapillaries are a hybrid of microbore packed columns and open microtubular columns. They have a column bore of 70 micrometers or less and are loosely packed with particles having diameters of from 5 to 30 micrometers. During the drawing process, some of the particles become mechanically embedded in the column wall and result in the characteristic open bed, zig-zag structure of these columns. Mobile phase flow rates are on the order of 1 microliter/min. Open microtubular columns are the equivalent of the capillary column in gas chromatography. Ideally they have column bores of 10-30 micrometers and contain a liquid phase or an adsorbent either coated on or chemically attached to the column wall [199]. Flow rates may be in the sub-microliter range for operation at maximum efficiency. The three column types are shown in cross-section in Figure 4.44. The use of microcolumns in liquid chromatography has been reviewed by Novotny [200].

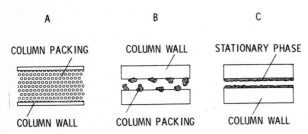

Figure 4.44   Cross-sectional view of microtubular columns. A, microbore column; B, packed microcapillary column; C, open microtubular column. (Reproduced with permission from ref. 200. Copyright American Chemical Society).

The theory of conventional packed columns in HPLC is adequate to describe the processes taking place in packed microbore columns and, to some extent, in packed microcapillary columns [55]. Some disagreement with theory for the latter results from changes in flow dynamics and mass transfer phenomena resulting from the open zig-zag packing structure [201]. The approach to the development of a theory for the kinetic optimization of open microtubular columns is similar to that for packed columns except that those terms containing the particle diameter are replaced by similar terms containing the diameter of the column bore [202-204]. The flow resistance parameter is given exactly by Poiseuille's equation and is equal to 32 for an open microtubular column. Likewise, the reduced plate height for a straight open tube is derived from the Golay equation and is given by equation (4.30).

$$h = \frac{2}{v} + \left[\frac{1 + 6k + 11k^2}{96 (1 + k)^2}\right] v + \left[\frac{2 k d_f^2 D_m}{3 (1 + k)^2 d_c^2 D_s}\right] v \qquad (4.30)$$

h = reduced plate height
v = reduced linear velocity of the eluent
k = capacity factor
d_f = stationary phase film thickness
d_c = diameter of the column bore
D_m = diffusion coefficient of the solute in the mobile phase
D_s = diffusion coefficient of the solute in the stationary phase

The last term in equation (4.30) represents slow equilibrium in the stationary phase. To a first approximation it can be ignored in HPLC, leading to the following approximations for h_{min} and v_{opt}:

$$h_{min} = \frac{1}{(1 + k)} \left[\frac{1 + 6k + 11k^2}{12}\right]^{1/2} \qquad (4.31)$$

$$v_{opt} = (1 + k) \left[\frac{192}{1 + 6k + 11k^2}\right]^{1/2} \qquad (4.32)$$

In order to achieve a particular value of h, which implies a particular value of , it is necessary to use the correct column bore. This diameter is given by equation (4.33).

$$d_c = \left[\frac{n \, h \, v \, \Theta \, \eta \, D_m}{\Delta p}\right]^{1/2} \qquad (4.33)$$

n = number of plates to which the column is equivalent
η = eluent viscosity
Δp = pressure drop across the column

As an example, consider a solute with a value of k = 3; thus h_{min} = 0.8 and v_{opt} = 5. Substituting these values into equation (4.33) and assuming approximate values for η and D_m, and practical values for n and Δp, one can calculate values for d_c. However, a column cannot be operated independently of a chromatograph and therefore it is necessary to consider solute dispersion effects in the injector/detector system when evaluating column performance. Whether a given column can be operated in practice with a given injector/detector system will depend upon the relationship between the volume of eluent within which the peak is eluted and the volumetric dispersion introduced by the detector. To avoid significantly degrading the column efficiency, the detector volume must be limited to less than the volume represented by a single standard deviation of an eluted peak. For an open microtubular column, the standard deviation of an eluted peak is given by equation (4.34).

$$\sigma_v = \frac{1}{\sqrt{n}} \left[\frac{\pi d_c^2 L}{4} \right] (1 + k) = \sqrt{n} \left[\frac{\pi d_c^3 h}{4} \right] (1 + k) \qquad (4.34)$$

L   = column length
σ_v = standard deviation of an eluted peak

The equations given above can predict the optimum conditions (i.e., h_{min} = 0.8) required to operate an open microtubular column at different values of Δp and various values of n. Examples of such calculations are shown in Table 4.30 [202]. From Table 4.30 it can be concluded that the operation of open microtubular columns near their optimum theoretical efficiency will be feasible only for columns with exceptionally high values of n. The principal limitation is the low value of σ_v° (special case of the standard deviation of an unretained peak at k = 0), which is difficult to attain in practice with current instrument technology.

Although practical considerations may prevent open microtubular columns from being operated under optimum conditions, it may be possible to use columns of wider bore operated at higher reduced velocities. In this way, efficiencies higher than conventional packed columns may be obtained, even if the latter are operated at their optimum value. To compare the efficiency of the various packed and open tubular column types under different conditions, we can use the separation impedance. The separation impedance is defined by equation (4.35) [205].

$$E = \left[\frac{t_m}{n} \right] \left[\frac{\Delta p}{n} \right] \left[\frac{1}{\eta} \right] = h^2 \Theta \qquad (4.35)$$

E   = separation impedance
t_m = elution time of an unretained solute

TABLE 4.30

CAPILLARY COLUMNS OPERATED UNDER OPTIMAL CONDITIONS

$h = 0.8; \Theta = 32; \eta = 1.0 \times 10^{-3}$ N.s/m^2; $D_m = 10^{-9}$ m^2/s

n	t_m (sec)	d_c (m)	L (m)	σ_v° (µl)
Δp = 220 bar ca. 3000 p.s.i.				
10,000	0.1	0.25	0.002	1×10^{-9}
30,000	0.9	0.43	0.01	0.8×10^{-8}
100,000	10.2	0.8	0.063	1×10^{-7}
300,000	90	1.37	0.33	0.8×10^{-6}
1,000,000	1000	2.5	2.0	1×10^{-5}
3,000,000	9000	4.3	10	0.8×10^{-4}
10,000,000	10,000	8	63	1×10^{-3}
Δp = 20 bar ca. 300 p.s.i.				
10,000	1.0	0.8	0.006	3.1×10^{-8}
30,000	9	1.4	0.033	2.8×10^{-7}
100,000	100	2.5	0.2	3.1×10^{-6}
300,000	900	4.3	1.0	2.8×10^{-5}
1,000,000	10,00	8	6.4	3.1×10^{-4}
3,000,000	90,000	25	33	2.8×10^{-3}

The separation impedance is a dimensionless quantity and will be smallest for
columns operating at their optimum efficiency. Optimum values for columns of
different types are given in Table 4.31. For given values of n and Δp, and
assuming E_{min} can be achieved, then the open microtubular column would be
capable of producing a separation 100 times faster than the microbore column.
The minimum value of E for the open microtubular columns cannot be obtained in
practice for reasons discussed above. For a packed microbore column with a
column bore of 1 mm the value for σ_v° will lie between 0.2-20 microliters. As
detectors having volumes approaching 0.1 microliters can be made, microbore
packed columns operated under conditions providing E_{min} values are entirely
feasible.

Returning now to open microtubular columns, the limiting feature of their
operation is clearly the dispersion in the injector/detector systems. In turn,
minimum values for this dispersion can be used to calculate the separation
impedance under operating conditions which might be possible in practice. Under
these circumstances, theory predicts that there is always a value of n above

TABLE 4.31

OPTIMUM VALUES FOR h, v, ⊖, and E

Column Type	h_{min}	v_{opt}	⊖	E_{min}
Microbore packed column's	2	2-5	500	2000
Microcapillary packed column's	2	5	150	600
Open microtubular columns (k = 3)	0.8	5	32	20

which the open microtubular column will provide performance superior to
conventional packed columns [202,204,205]. The open microtubular column will
show superiority over packed columns operated under optimum conditions when the
plate count is in excess of about 500,000 with a detector volume of 100 nl. If
the detector volume could be reduced to 10 nl then the superiority of the open
microtubular column would occur at 200,000 theoretical plates, and about 70,000
plates at 1 nl. Thus, in practice, open microtubular columns are likely to be
useful for analyses demanding very high plate numbers since they will prove
superior to packed columns in terms of speed. Their potential for much higher
performance than packed columns can, however, only be realized by a great
reduction in the effective volume of the injection and detection devices. An
ultimate aim should be a dispersion from these sources of around 1 nl or less.
For example, with a detector/injector dispersion of 1 nl and an open
microtubular column of 10 micrometer bore generating 1 million plates,
separations could be performed 27 times faster than with a packed column [205].

4.16.1 Instrumental Requirements for Microcolumn HPLC

Columns of very high efficiency and columns of small cross-sectional area
produce very narrow peaks in the sense that the total volume of mobile phase
contained in them may be in the sub-microliter range. If such peaks pass
through a detector connecting tube and cell of significant volume relative to
that of the peaks, then the peaks will be broadened. Such broadening is due to
both the dispersion resulting from the parabolic velocity profile of the liquid
through the tubes and cell and to the logarithmic dilution effect resulting from
the finite volume of the cell [17,206]. Extracolumn dispersion can arise from
the finite volume of sample being employed, dispersion in the connecting tubes
between the injector and column and the column and detector, dispersion in the
scintered frits which contain the column packing, and finally, the detector cell
itself [207]. In Table 4.32 the values calculated for the volume of mobile
phase equivalent to one standard deviation for a peak eluting at capacity

factors of k = 0 and k = 5 on a conventional wide bore and a microbore packed column operating at optimum efficiency are compared [207]. Ideally, the minimum acceptable value of instrumental dispersion would be somewhat less than $\sigma_v°/2$. For the microbore column this means a value less than 1.0 microliter is desirable. For open microtubular columns a similar calculation indicates that a value for dispersion less than 1-10 nl is required.

TABLE 4.32

STANDARD DEVIATION OF MOBILE PHASE FOR PEAKS ELUTED FROM A CONVENTIONAL 25 cm x 4.2 mm I.D. COLUMN WITH 3.0 ml DEAD VOLUME AND A 100 cm x 1.0 mm I.D. MICROBORE COLUMN WITH A DEAD VOLUME OF 0.7 ml

Particle Diameter (μm)	Limiting HETP (μm)	Maximum Efficiency (plate number)	Standard Deviation (μl)	
			k = 0	k = 5
Conventional Wide Bore Column				
35	70	3,570	50	301
20	40	6,250	38	228
10	20	12,500	27	161
5	10	25,000	19	114
Microbore Column				
35	70	14,285	5.86	35.1
20	40	25,000	4.43	26.6
10	20	50,000	3.13	18.8
5	10	100,000	2.21	13.3

The operation of microbore columns near their optimum efficiency is possible by modifying liquid chromatographs designed for use with wide bore packed columns [206-210]. The injection valve should be miniaturized to handle a sample volume of 0.2 or 0.5 microliters. Valco makes a suitable valve that contains the sample within the grooves of the valve's spigot [206]. This valve has a maximum operating pressure of 7000 p.s.i. To minimize connecting volumes, the column is attached directly to the valve's housing. A stainless steel frit is situated between the end of the column and the valve's seat, the assembly being retained by the sample valve union, Figure 4.45. The connection dead volume is limited to the volume contained in the sintered frit and the small aperture in the valve wall between the spigot and the frit. To make column-to-column connections, a modified Swagelok union can be used, Figure 4.45. The columns are butted together to eliminate any dead volume in the connection. The microbore column is connected directly to the detector cell by a screw-thread, thus virtually eliminating the necessity for connecting tubes. The modified detector cell can be fitted into the light path of a standard commercial UV detector and has a dead volume of about 1.0 microliter [208].

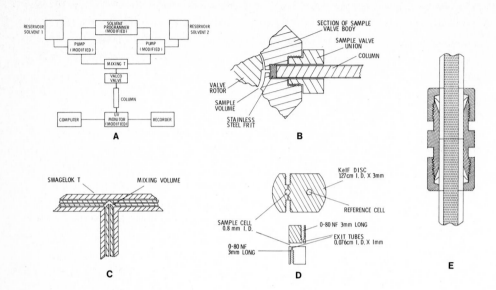

Figure 4.45   Instrumentation used in microbore HPLC. A, block diagram of
equipment; B, injection valve cross section showing column
connection; C, microvolume mixing–T; D, detector cell in cross
section ; E, device for coupling columns.

The associated time constant of the photosensor and the electronics of currently
available commercial UV detectors may, however, be inadequate for use with
microbore columns, particularly at high flow rates. Typical detector time
constants on the order of a few seconds will not provide a faithful record of
the chromatogram; a time constant of about 40 milliseconds is required. Scott
has explained how such time constants may be achieved by modification of the
detector electronics and recording system of a conventional liquid chromatograph
[208]. Microbore HPLC is performed at flow rates about 100–fold less than
conventional wide bore columns, thus necessitating modification of the gradient

former and pump. The frequency generator of a Waters 6000 M pump was replaced
by a wide range function generator to provide variable flow rates in the range
1.0 microliter/min to 1.0 ml/min. Scott and Kucera have also designed a
mixing-T with a volume of 2 microliters for use with a modified solvent
programmer, enabling mobile phase gradients to be used with microbore columns
[211]. More recently, the construction of a microvolume dynamic mixer with
superior performance has been described [212]. Progress in microbore HPLC
technology has been reviewed by Scott [17,213] and Kucera [214].

Instrumental dispersion in open or packed microtubular liquid
chromatography is a much more severe restraint than is true of microbore
columns. Instrumental dispersion of 1-10 nl is desired, but this requirement
cannot be met by the equipment currently available. Early attempts to attain
zero dead volumes adopted the procedures common to capillary column gas
chromatography by employing injection splitters and an additional flow of liquid
at the column outlet [200]. For packed microcapillaries, typical conditions are
an injection split ratio of 1:1000 with a column flow rate of about 10
microliters/min for a column of 70 micrometers internal diameter and a detector
make-up liquid flow rate of 20-50 microliters/min [215,216]. Injection
splitting allows the use of large sample volumes (1-10 microliters) but sample
detection can become a problem with the large split ratios required. In the
injection splitting device described by Novotny [217], a sample plug is
introduced with a conventional valve injector, and a heartcut fraction of the
sample is diverted onto the microcolumn using sequentially-timed valves. By
varying the injection time and the flow rate injection volumes from 1 nl to 1
microliter or more can be selected. Other injection methods used with
microtubular columns attempt to eliminate dead volumes by loading the sample
directly onto the column itself. The stainless steel tube [218], the in-column
method [219], and the static splitter [204] have been used to inject sample
volumes of 0.01 to 0.3 microliters. In the stainless steel tube method, the
sample, in a narrow bore capillary tube, is pushed through a PTFE connector
until the sample is in contact with the microtubular column. The mobile phase
is then pumped through the complete assembly, Figure 4.46. The in-column method
uses the microtubular column itself as the injection tube. A very small volume
of sample is sucked into the column either by a micro-feeder or by thermal
contraction (after preheating the column end and dipping it into the sample
solution). The last two methods require a great deal of manual skill, and are
generally imprecise.

Since microtubular columns have little flow resistance at the low mobile
phase flow rates typically used, an air-tight 50-250 microliter syringe
controlled by a micro-feeder can be used as a pump [220]. A micro-feeder
consists of a small synchronous motor, gears, and a screw-thread to advance the

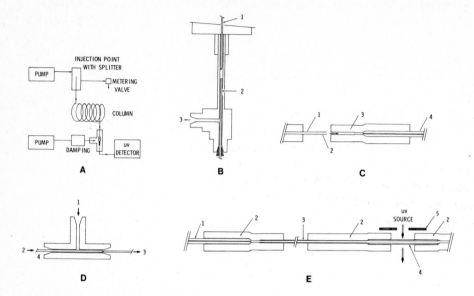

Figure 4.46 Instrumentation used in microtubular HPLC. A, block diagram of
equipment using an injection splitter; B, injection splitter:
1. syringe needle, 2. microtubular column, 3. discard flow;
C, stainless steel tube injection method: 1. stainless steel
capillary tube, 2. sample, 3. PTFE tube, 4. microtubular column;
D, make-up liquid mixing-T: 1. make-up liquid, 2. column eluent,
3. flow to detector, 4. mixing-T; E, microvolume UV detector:
1. microtubular column, 2. PTFE support sleeve, 3. stainless
steel connecting capillary, 4. quartz capillary tube (detector
cell), 5. slits.

syringe plunger at a constant rate. Commercially available syringe pumps that
have been modified to provide low flow rates are also suitable for use with
microtubular columns.

Detection remains one of the principal problems in microtubular HPLC
[204]. As discussed previously, low dispersion and high sensitivity must be
obtained simultaneously. A UV detector with a total dead volume of 1.0
microliter for the connecting parts and 0.1 microliter for the detector cell has
been described, Figure 4.46 [220]. This detector can be used without an

additional flow of make-up liquid and has sensitivity similar to conventional UV detectors. On-column detection using the microtubular column itself as the detector cell eliminates problems with dead volumes in connectors and is now the method of choice in microtubular HPLC [204,221]. Other detection devices with low detector cell disperion volumes have recently been described. These include electrochemical detectors [222-224], infrared monitors [225], post-column reaction detectors [226], flame ionization detectors [227], thermionic detectors [228], flame photometric detectors [229], and fluorescence detectors [204]. A method for generating gradients with microtubular columns has also been discussed [230].

4.16.2 Preparation of Microcolumns

Microbore columns of 1.0 mm internal diameter and up to 1.0 m in length are slurry packed by methods similar to those used for wide bore columns. A balanced density slurry of low viscosity is preferred for silica gel while a viscous medium (e.g., methanol-glycerol, 3:1) is generally used for reversed-phase packings. A 1.0 m column of 1.0 mm internal diameter requires about 0.5 g of packing and should be packed rapidly and at high pressure for maximum efficiency [206]. If the columns are coiled either before or after packing, then the minimum coil radius which can be formed without loss of efficiency was found to be 12 cm. Overloading occurs for packed microbore columns with sample amounts in excess of 10-20 micrograms.

A distinctive feature of microbore columns is their capability of producing extremely high intrinsic efficiencies; they can be connected together in series to provide column efficiencies that increase linearly with column length. If wide bore columns are joined together they soon reach a limiting efficiency above which further addition of columns does not increase the overall efficiency. This is clearly illustrated in Table 4.33 [231,232]. There appears to be about a 60-70% efficiency loss in coupling the wide bore columns. Two reasons have been proposed to explain this non-linearity between efficiency and column length for wide bore columns [206,232]. The first explanation takes into account the heat generated by the high pressure drop that develops across long columns at normal mobile phase flow rates. Since liquids have significant viscosity, a considerable' amount of work must be done to force the liquid through the long column. This work is converted into heat and thus the temperature of both the mobile phase and stationary phase rises. It is believed that this temperature rise affects the transfer properties of the solute between the two phases and results in peak dispersion and possibly peak asymmetry as well. The second explanation involves the heterogeneous permeability of the column across its radius that results in channeling throughout the stationary phase and allows the mobile phase to flow more rapidly through one part of the column. The heterogeneity of the column permeability causes an exaggerated multipath term which will increase in importance as the column length increases, particularly if the column length is made up of a series of individual columns. Whichever explanation is correct, the microbore column obviates both these problems. If we consider the thermal effect, the heat generated is more easily conducted away from the column due to its smaller dimensions. At the low flow rates involved, a few microliters/min, the heat generated is relatively small even though the pressure drop may be quite high. In the second case, since the packing has a much lower cross-sectional area, the likelihood of a homogeneous permeability is much greater and there will be less dispersion resulting from the multipath effect.

TABLE 4.33

COMPARISON OF COLUMN CONCATENATION FOR COLUMNS OF DIFFERENT DIAMETERS

Column	Number of Connected Columns	Average Plate Count	Column Pressure Drop (p.s.i.)
50 cm x 1.0 mm I.D.	1	15,400	730
	3	47,420	2260
	6	92,800	4530
	9	139,060	6800
25 cm x 4.6 mm I.D.	1	8350	370
	2	11,830	810
	4	16,690	1540
	6	19,500	2355

From the previous discussion, it may appear as if an infinite number of theoretical plates could be obtained by simply coupling a large number of microbore columns together. However, there is a finite pressure limit at which both the instrument and the column packing material can operate. This pressure limit establishes a maximum value for the number of theoretical plates attainable. With current state-of-the-art instrumentation, this efficiency limit is about one million theoretical plates using 8 micrometer particles and an 18 m long column operated at its optimum reduced velocity [206]. An example of a high efficiency separation using coupled microbore columns is shown in Figure 4.47 for the UV-absorbing components in a dwarf pine essential oil sample [206].

Packed microcapillary columns are prepared using a conventional capillary column glass drawing machine. A glass column blank of 0.25 mm internal diameter and 5.5 mm outer diameter is packed with a microparticulate adsorbent such as alumina or silica gel and is drawn out to capillaries with an internal diameter

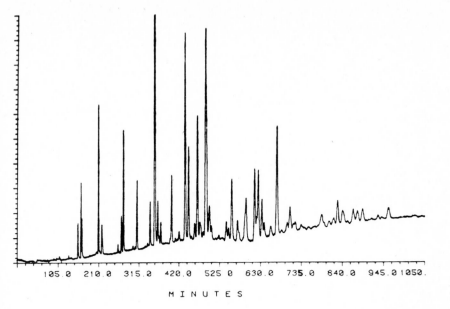

MINUTES

Figure 4.47   Chromatogram of dwarf pine essential oil on a 4.5 m x 1 mm I.D.
ODS reversed-phase microbore column. Mobile phase methanol-
water (85:15) at a flow rate of 30 µl/min. (Reproduced with
permission from ref. 206. Copyright Elsevier Scientific
Publishing Co.)

of 30-100 micrometers and a wall thickness of about 250-300 micrometers
[201,233,234]. Under these circumstances, the particles are partially welded to
the capillary wall, thus giving the columns adequate mechanical stability at
high operating pressures. Although the surface chemistry of the adsorbent
undergoes certain changes due to the temperature necessary to melt the glass,
alumina and certain kinds of silica gel can survive this process and produce
columns suitable for use in the adsorption mode, or, after chemical
modification, as reversed-phase materials. The latter method is necessary for
the preparation of bonded phases since it is unlikely that any organic surface
groups would remain intact at the temperatures used for column drawing. The
bonded stationary phases are prepared by methods similar to those used in
capillary column GC. Organo-substituted silanes of various structures are
employed to prepare polar and nonpolar stationary phases [234,235]. An example
of a separation using a chemically-modified packed microcapillary column is
shown in Figure 4.48A.

The current practice of packed microcapillary HPLC is a compromise between
column efficiency and available instrumentation. Columns with internal
diameters between 50 to 100 micrometers packed with particles around 30

micrometers have been generally used to minimize operating difficulties whereas columns of 30–50 micrometer internal diameter packed with finer particles are needed to maximize efficiency.

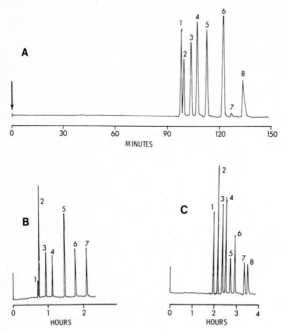

Figure 4.48 A. Separation of aromatic hydrocarbons on a 27 m microcapillary packed with acidic alumina modified with octadecylsilane. Mobile phase was acetonitrile–water (65:35), linear velocity 0.5 cm/s at a column temperature of 50°C. 1. toluene; 2. naphthalene; 3. fluorene; 4. anthracene; 5. pyrene; 6. chrysene; 7. impurity; 8. benzo[e]pyrene.

B. Separation of aromatic amines. Column: 23 m x 33 μm I.D., coated with β,β'-oxydipropionitrile (OPN); mobile phase hexane saturated with OPN; inlet pressure 20 atm.; UV detection, 235 nm. 1. isooctane; 2. N,N-dimethylaniline; 3. N-phenyl-α-naphthylamine; 4. N-phenyl-β-naphthylamine; 5. aniline; 6. α-naphthylamine; 7. β-naphthylamine.

C. Separation of aromatic hydrocarbons on a 21 m x 43 μm I.D. open microtubular column with a chemically bonded octadecyl-silane phase. Mobile phase is acetonitrile–water (40:60) and flow rate 0.28 μl/min; 1. benzene; 2. naphthalene; 3. biphenyl; 4. fluorene; 5. phenanthrene; 6. anthracene; 7. fluoranthene; 8. pyrene.

Related to packed microtubular columns are the narrow-bore microparticle-packed fused silica columns. These have internal diameters < 500 micrometers and are packed with conventional packing materials having particle diameters < 10 micrometers [221,236–240]. The flexible fused silica columns are identical to those used in gas chromatography. From the chromatographic point of view, the fused silica columns offer the advantages of ease of handling; high pressure capability; a smooth, relatively metal-free, chemically inert surface; and the convenience of being able to easily coil long column lengths. A unique advantage of these columns is that since they are made from silica which is optically transparent in the UV-region, the column itself can act as the detector cell. This minimizes the effective detector dead volume to the volume fixed by the geometry of the column and the dimensions of the excitation slit. Although these columns do not approach the theoretical efficiency of open tubular and packed capillary columns discussed above, they do possess their own significant advantages, including very high efficiency, adequate sample capacity, and very low flow rates. Their performance characteristics and instrumental requirements are intermediate between those of microbore and open tubular columns.

Open microtubular columns are prepared by drawing glass capillaries to appropriate sizes and then chemically modifying the inner wall to create an adsorbent surface [199,204,219,241-243]. This thin adsorbent layer can then be either coated with a liquid stationary phase or reacted directly with an organically substituted silane to prepare bonded phases. A thin layer of silica gel can be created by etching the inner wall of the drawn capillary with a sodium hydroxide solution. Experiment shows that the optimum sodium hydroxide concentration, the reaction temperature, and reaction time must be established by trial and error and vary with the internal diameter and composition of the glass capillary [219,237,243]. The adsorbent layer can be used directly or may be modified in a variety of ways using the methods common to the preparation of capillary columns for GC. The dynamic coating method can be used to coat the etched microtubular columns with such stationary phases as ethylene cyanohydrin, β,β'-oxydipropionitrile, or 1,2,3-tris(2-cyanoethoxy)propane. After heating in a stream of nitrogen to remove residual solvents, a thin film of liquid remains coated on the etched wall. Since the liquid phase is held by adsorptive forces, the mobile phase employed must be saturated with the stationary phase to prevent stripping of the stationary phase from the column. An example of the performance of such a column is shown in Figure 4.48B for the separation of a mixture of aromatic amines. Chemically-bonded phases are prepared by dynamically coating the etched capillary with a suitable reagent such as an organically-substituted silane followed by a drying and reaction sequence. The latter is usually accomplished by heating the column to an elevated temperature for several hours [241]. A variety of nonpolar and polar stationary phases, including ion-exchange columns, were prepared in this way [218,219,242]. An example of the performance of a chemically-bonded open microtubular column is shown in Figure 4.48C for the separation of a mixture of aromatic hydrocarbons. Instrumental constraints prevent the maximum theoretical efficiency of open microtubular columns from being achieved. Theory predicts that columns in the range of 1-3 micrometer internal diameter and an instrumental dead volume no greater than about 1-10 nl would be optimal. The current practice of open microtubular HPLC using columns with internal diameters in the range of 15-75 micrometers is a practical compromise between column efficiency and instrumental constraints.

4.17 High Speed Liquid Chromatography

The previous discussion dealt with the use of microcolumns of low cross-sectional area for the attainment of very high separation efficiencies. Columns with very high theoretical plate counts were described, but much of this efficiency was obtained at the expense of analysis time. This compromise remains essential when the emphasis is placed on maximizing resolution with complex samples. An alternative separation philosophy, suitable for less complex mixtures, is to minimize analysis time. Scott used a short packed microbore column, 25 cm long, to separate a seven component mixture in under 30 seconds with a mobile phase velocity of 8 cm/s [208]. Separation speed was limited by the column back pressure at high mobile phase velocities and instrumental consideratinos such as the extracolumn dead volume and the time constraints of the detector and recording systems. Although instrumental constraints present less of a problem when short, high efficiency, wide bore columns are used, modifications to a conventional liquid chromatograph are still required for high speed separations. The volume of the sample flow path and the response times of the detector and recorder systems must all be reduced. A commercially available high speed liquid chroamtography fitted with a modified injector, connecting tubes, detector, and gradient mixer has been designed to take advantage of the possibility of performing very fast separations on short, wide bore columns packed with particles of 3 or 5 micrometer diameter [244-246]. Compared to a conventional liquid chromatograph, the instrumental band broadening was reduced by about 80% and the response of the detction system was about 10 times faster. Instrumental band broadening was minimized by using a smaller injection volume than normally used (less than 6 microliters), and a detector with a 2.4 microliter flow cell [247].

Column seelction is very important in high speed liquid chromatography. Small particles of 3 or 5 micrometer diameter are used to generate high efficiencies per column. The column length is kept short, 5 to 12.5 cm, to decrease analysis times and to minimize the column pressure drop developed at high mobile phase flow rates. The highest efficiencies are obtained with the 3 micrometer particle column. The 5 micrometer particle column permits operation at a substantially lower back pressure and doubles the surface area of the

stationary phase, allowing a higher sample loading without loss of efficiency.
Typical flow rates with these column are 1-5 ml/min corresponding to a linear
velocity of 0.5 to 0.9 cm/s. However, since the analysis time is significantly
reduced, the actual consumption of mobile phase needed for an anaysis is
generally less than that required for a conventional separation. Figure 4.49
shows a typical example of a high speed liquid chromatogram in which a seven
component test mixture is separated in about 90 seconds. Under favorable
conditions as many as 352 theoretical plates per second for bonded phases and
444 theoretical plates per second for silica gel particles, both of 3 micrometer
diameter, have been obtained [244,245]. This compares favorably with values of
150-250 theoretical plates per second for open tubular capillary columns,
100-140 theoretical plates per second for SCOT columns, and 20 theoretical
plates per second for packed columns in gas chromatography. However, the total
number of theoretical plates available for a separation is of course limited by
the maximum operating pressure of the column, which is a more rigid constraint
in liquid than in gas chromatography.

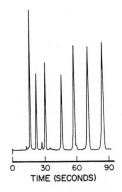

0   30   60   90
TIME (SECONDS)

Figure 4.49   Analysis of a test mixture on a reversed-phase column   Column:
100 x 4.6 mm I.D., C_{18}-bonded phase packing, 3 μm particles.
Mobile phase acetonitrile-water (65:35) at 4.4 ml/min, temperature
40°C. Peak identification in order of elution: uracil, phenol,
nitrobenzene, toluene, ethylbenzene, isopropylbenzene, and
t-butylbenzene. (Reproduced with permission from ref. 244.
Copyright Friedr. Vieweg & Sohn).

In addition to high speed separations, the short packed columns can be
coupled to provide higher resolution while still providing reasonable analysis
times. An example of a chromatogram obtained in this way is shown in Figure
4.50. With 3 micrometer particles, pressure limitations at high flow rates and
the non-additivity of the efficiency of coupled wide bore columns discussed
earlier are an overall limitation to the maximum obtainable efficiency. In
terms of analysis speed, the pressure drop, instrument band broadening, and

334

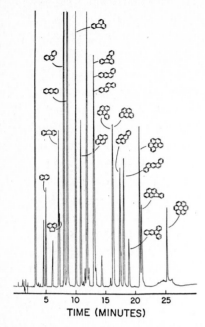

5   10   15   20   25
TIME (MINUTES)

Figure 4.50   Analysis of polycyclic aromatic hydrocarbons. Column: two series
coupled 100 x 4.6 mm I.D. C$_{18}$-bonded pahse, 3 m packing
columns. Mobile phase: acetonitrile-water, linear gradient from
65 to 90 % in 20 minutes at 1.8 ml/min, UV detection at 254 nm.
(Reproduced with permission from ref. 244. Copyright Friedr.
Vieweg & Sohn).

radial temperature gradients represent the current limits on this interesting
development in the practice of liquid chromatography. Even with these
constraints, high speed liquid chromatography can provide very fast separaions
of simple mixtures with a considerble savings in time and solvent consumption.
In addition, detection limits are improved almost fivefold using high speed
liquid chromatography compared to those observed under normal operating
conditions with a 10 micrometer particle packed column.

4.18 Multidimensional Liquid Chromatography

Gradient elution analysis is frequently used for separating sample
components with a wide range of capacity factor values. This situation,
referred to as the general elution problem, can also be solved by
multidimensional techniques. Both methods lead to an optimization of resolution
in a minimum analysis time.

In multidimensional analysis the sample is separated by switching between
two or more columns possessing complementary separation characteristics. To
enhance the separation or to increase sample throughput, the mobile phase and

its direction may also be changed. Multidimensional chromatography has been variously called column switching, coupled column chromatography, recycle chromatography, mode sequencing, stationary phase programming, selectivity switching, and column backflushing. Some of these terms describe operations frequently used in multidimensional chromatography, several of which may be combined in a particular separation [248,249]. Multidimensional liquid chromatographic techniques are now widely used for sample enrichment and sample cleanup procedures prior to analysis by gas or liquid chromatography. The general requirements for this use are described in Chapter 7.6.6.

Multidimensional liquid chromatography can be performed either on- or off-line [248]. The off-line approach is simple but time consuming. Fractions are collected manually or with a fraction collector as the effluent leaves the detector from the primary column and then reinjected, perhaps after concentration, into a second column. The off-line approach has several advantages in addition to simplicity: the sample can be easily concentrated by either evaporation or extraction and the mobile phases used in each column need not be mutually compatible. Some disadvantages are the time constraints arising from the range of manual procedures involved and the possibility of sample loss or contamination during handling. On-line techniques provide the advantages of automation by using pneumatic or electronically-controlled switching valves to divert the column effluent from the primary column into the secondary column or columns. Automation improves reliability, sample throughput, and shortens analysis time, as well as minimizing sample loss or change since the analysis is performed in a closed-loop system. The main limitation is that the mobile phase system used in the coupled columns must be compatible in both miscibility and solvent strength. This requirement arises since the eluent from the first column is the injection solvent for the second column; consequently, not all column types are mutually compatible. When the second column is also used for sample concentration, the constraints on mobile phase selection become particularly severe. The relative merits of off- and on-line multidimensional liquid chromatography are summarized in Table 4.34.

As the procedures used in off-line multidimensional liquid chromatography are quite straightforward, further discussion will be limited to automated methods. Some comments concerning nomenclature are pertinent at this point. With reference to Figure 4.51, a typical chromatographic peak could be sliced in different ways to provide the fractions of most interest [248]. The leading edge, peak centrum, or tailing edge could be switched to the second column to provide "endcut", "heartcut", or "frontcut" fractions, respectively. The entire peak or any of the above cuts could be returned to the head of the column and

TABLE 4.34

COMPARISON OF OFF- AND ON-LINE MULTIDIMENSIONAL LIQUID`CHROMATOGRAPHIC TECHNIQUES

Off-Line Multidimensional Techniques

Advantages	Disadvantages
A. Easy to carry out by collection of column effluent	A. Difficult to automate; more cumbersome and inconvenient
B. Can concentrate trace solutes from large volumes	B. Greater chance of sample loss (e.g., adsorption, evaporation, oxidation)
C. Can work with two LC modes that use incompatible solvents	C. More time consuming
	D. More difficult to quantitate and reproduce

On-Line Multidimensional Techniques

Advantages	Disadvantages
A. Easy to automate, especially with modern chromatographs	A. Requires more complex hydraulics (or pneumatics), switching valve(s), more expensive
B. Less chance of sample loss since experiment carried out in closed system	B. Difficult to handle trace compounds since very dilute and in large volume; can compensate for this by on-line concentration methods
C. Can configure switching system which best suits needs (e.g., backflush, heartcutting, on-column concentration)	C. Solvents from primary and secondary modes must be compatible, both from miscibility and strength requirements
D. Decreased total analysis time	
E. More reproducible	
F. Can increase sample throughput	

---

reinjected for "recycle chromatography". If the sample elutes from the first column in a solvent which is a weak solvent for the second column, then a large volume of mobile phase can be passed through the second column to concentrate the sample at the head of this column. This is an example of "on-column concentration". After the components of interest have been switched to the second column, the direction of mobile phase flow to the first column can be reversed to rapidly elute sample components with a high column retention. This is an example of "column backflushing". In all the above techniques whole areas of the chromatogram as well as single peaks or fractions of a single peak may be

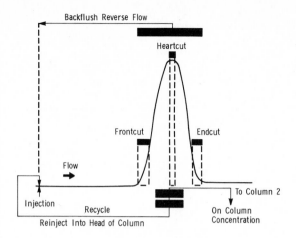

Figure 4.51 Single peak representation of the nomenclature used in multidimensional chromatography. The black rectangles refer to volume
increments. (Reproduced with permission from ref. 248. Copyright
Preston Publications, Inc.)

switched.

A standard high performance liquid chromatograph, with the addition of a
switching valve, is used for multidimensional chromatography. This valve may be
a simple, manually operated six or ten-port valve or it may be automatically
controlled either via a fixed interval timer or coupled to the operation of an
in-line detector and controlled by a microprocessor [250-252]. Depending on the
complexity of the column switching network, several switching valves may be
needed. Each valve must be capable of high pressure operation without
deterioration and provide a low dead volume sample path so as not to
significantly broaden the peaks that pass through it. The valve may also
contain a loop of fixed volume for trapping the switched peak prior to injection
on the second column or it may simply be used as a switch to divert flow between
columns. An example of a two column multidimensional liquid chromatograph using
a six-port switching valve, intermediate trap and injection valve is shown
schematically in Figure 4.52 [253]. Column one is a size-exclusion column and
column two a reversed-phase column. This column combination has found many
practical applications due to the complementary separation mechanisms of the two
column types and their mobile phase compatibility [253,254]. The detector
positioned between the two columns is not an essential component. For complex
chromatograms the precise location of the peak or peaks to be switched is very
important for quantitative analysis since the separation of the switched peaks
on the second column must not be compromised by switching additional impurity
peaks. The intermediate detector thus conveniently locates the position of the
peak to be switched and minimizes sample loss and contamination. When the

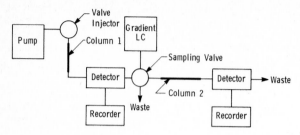

Figure 4.52   Schematic of an on-line multidimensional two column liquid
chromatograph.   (Reproduced with permission from ref. 253.
Copyright Elsevier Scientific Publishing Co.)

detector function is not used to control the switching process a sequential
timer can be used to automatically switch the peaks of interest.   However,
interval timers are unable to account for variability in the elution volume of
the sample from the first column.   Figure 4.53 shows a more complex switching
arrangement using a ten-port valve for sample injection with the frontcut
diverted to the detector and the heart- and endcuts switched to a second
column.   Column switching is more readily carried out with a ten-port
multifunctional valve which is able to accommodate sample injection and solute
switching in the same device for most applications.   Several other valve
configurations for column switching using a ten-port valve are given in
reference [251].

As multidimensional chromatography is used to solve the general elution
problem there is obviously some overlap of the technique with gradient elution
analysis.   Multidimensional techniques are preferred when a very large number of
theoretical plates is required for a separation, when a large number of samples
of similar kind are to be analyzed in the shortest possible time, and for the
analysis of complex mixtures which contain only a few adjacent components of
interest [255].   For the above separation problems, unidimensional techniques
are inherently inefficient as only a small fraction of the column is actually in
use at any given time.   In other words, the number of plates generated per unit
time measured for the part of the chromatogram of interest is unnecessarily
low.   The multidimensional separation methods provide optimum efficiency for the
separation of the components of interest while simultaneously minimizing
analysis time by decreasing the time spent in separating components of the
sample which are not of particular interest.

Freeman [249] has shown that the peak capacity for a multidimensional
system is given by equation (4.36) while for a similar number of series coupled
columns the peak capacity is given by equation (4.37).

A. FILL SAMPLE LOOP

B. INJECT SAMPLE ONTO COLUMN 1 AND
FRONT CUT TO DETECTOR

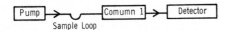

A. HEART AND ENDCUT TO COLUMN 2
TO DETECTOR

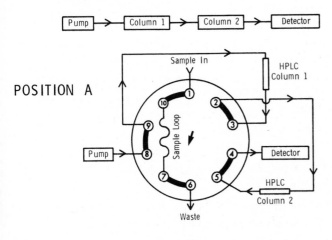

POSITION A

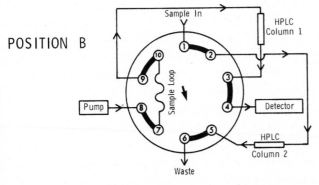

POSITION B

Figure 4.53   The use of a single ten-port switching valve for sample injection
and front- and heartcut peak diversion in a multidimensional
liquid chromatograph.   (Reproduced with permission from Valco
Instruments Co.)

$$\Theta_r = \Theta_1 \cdot \Theta_2 \cdot \Theta_3 \text{ ----} \Theta_n \qquad (4.36)$$

$$\Theta_r = \eta^{1/2}. \qquad (4.37)$$

Θ = individual column peak capacity
η = number of columns
Θ_r = resultant peak capacity for the system

Thus, if two identical columns with a peak capacity of $\Theta = 25$ are coupled in series, then the resultant peak capacity would be 35 compared to a value of 625 if the same columns were used in a multidimensional mode.

To provide a better understanding of multidimensional chromatography and its uses, two common areas of application will be described. First, consider a sample comprised of components having a wide range of distribution coefficients. If a column of high selectivity is used, components eluting at the start of the chromatogram will be well resolved but later eluting peaks with long retention times may be difficult to detect [256]. Conversely, if the column has low selectivity the analysis time will be shortened but early eluting components will be poorly resolved. In the multidimensional mode, two columns of the same stationary phase but of different length or two columns of similar length but of different selectivity are used. The first column is short or of low selectivity and is used to separate the later eluting components of the chromatogram. The components of low retention are switched to the second column and remain there while the most retained components are separated on the first column. When these components have been separated, the least retained components are then separated on the longer or more selective second column. Compared to the normal chromatogram, the multidimensional chromatogram is reversed: the most retained components are at the beginning while the least retained are at the end. The total analysis time is reduced, there is less dead space between peaks, and as the peaks elute in a smaller volume of mobile phase, they are easier to detect. The separation may be carried out isocratically provided that the total number of sample components is not too great. This simplifies the instrumentation required, permits the use of nongradient compatible detectors, and eliminates the problem of column equilibration required in gradient elution analysis.

As a second example, consider the problem in which quantitative analysis of few components of similar retention is required in a complex mixture containing components having a wide range of capacity factor values. Assume further that the peaks of interest are buried in the center of the normal chromatogram. The time used in separating the components prior to and after the peaks of interest is wasted as it is not contributing effectively to the separation of the desired components. Using the column configuration of the previous example, the short column or the column of low selectivity is used to quickly isolate the area of the chromatogram of interest which is then switched to the second column for separation. The front- and endcuts of the chromatogram are vented to waste. If the components of the endcut are well retained they may be backflushed by reversing the direction of column flow. The components of interest and any overlapping bands from the first column are meanwhile being efficiently separated on the second column. Snyder has adapted this technique to a process he termed "boxcar chromatography" [255]. In boxcar chromatography samples are switched from column one to column two before the previous sample has migrated the complete length of the second column. In this way the separating power of the second column is fully utilized since it is continuously resolving samples separated only by the minimum time required to ensure that inter-sample mixing does not occur. This method is capable of a very high sample throughput and can be operated automatically for routine analysis.

Multidimensional chromatography can be used to perform separations requiring very large plate numbers by operating in the recycle mode. Recycle chromatography increases the effective length of the column by returning the eluted sample to the head of the column for further separation. There are some attendant advantages to this approach compared to the alternative of coupling columns in series. For coupled columns sample resolution increases linearly with the square root of the column length but the column pressure drop increases linearly with the length. Thus practical operating constraints essentially limit the number of columns that may be coupled together and therefore restricts the total number of theoretical plates that can be generated in a reasonable time. Also, the total number of theoretical plates obtained for conventional packed columns is usually somewhat less than the sum of the theoretical plates for the individual columns connected. In the recycle mode the pressure constraint limiting the theoretical plate count is eliminated because the column length is constant; thus large numbers of theoretical plates can be obtained by increasing the number of sample passes through the column. The limiting factor in recycle chromatography is established by remixing, which occurs when the fastest moving peak of the sample eventually overtakes the slowest moving component after a certain number of cycles. Generally, some facility within the recycle chromatograph is made for drawing off either completely separated or unwanted components, permitting additional cycles. A further limitation of performance in the recycle mode is the extracolumn band broadening occurring in the pump, valves, connecting tubes, and detector at each cycle in the detector-to-injector transfer. The influence of column and extracolumn effects on the maximum efficiency realizable in recycle chromatography has been treated theoretically [257,258]. Snyder has shown that plate counts as high as 5-10 million are possible in the recycle mode [255]. However, in spite of the above possibilities recycle chromatography is rarely used in practice. Most applications have been made for size-exclusion and preparative liquid chromatography. In size-exclusion chromatography it is used to improve column efficiency and resolution, which are limited by the low retention typical of size-exclusion columns. In preparative liquid chromatography it minimizes the pressure drop at the high column flow rates normally used. Because of these high flow rates extracolumn band broadening has little influence on the separation.

Three instrument designs have been described for recycle chromatography. In its simplest form it is a closed-loop system comprised of an injector, column, optical detector, draw-off valve, and pump [259-261]. For analytical work the main disadvantage of this system is the relatively large extracolumn dead volume, particularly in the pump head. This is minimized by designs using the alternate pumping principle [258,262,263]. Here the pump is eliminated from the sample path during recycle by using two identical separation columns in parallel interconnected by a six-port valve. Switching the valve directs the sample alternately through columns one and two without passing it back through the injector and pump head. More recently a circulation valve has been described for the automatic continuous injection of all the eluent from the column outlet back into the column inlet [264]. The advantage of a circulation valve is that it theoretically should not introduce any extracolumn band broadening into the system.

4.19 Temperature and Flow Programming in Liquid Chromatography

Temperature affects sample solubility, solute diffusion, and mobile phase viscosity in liquid chromatography [265,266]. The solute diffusion coefficient tends to increase while the mobile phase viscosity decreases with increasing temperature, producing a favorable influence on the column efficiency but usually a less noticeable influence on the selectivity factor. Increasing the temperature usually reduces retention, so that changes in sample resolution as a function of column temperature are fairly modest in many cases. A notable exception is in ion-exchange chromatography where temperature can dramatically influence selectivity. At the other extreme, selectivity in size-exclusion chromatography is essentially independent of temperature although resolution is often found to increase with increasing temperature due to an improvement in column efficiency. This improvement is attributed to favorable changes in solute diffusivity and mobile phase viscosity, and to a reduction of non-size-exclusion temperature dependent interactions. Some samples, for example biological polymers, are unstable at elevated temperatures, which effectively limits the temperature range over which they may be separated. The effect of temperature on retention in normal and reversed-phase chromatography

is largely determined by the enthalpy of the solute interaction with the
stationary phase. The enthalpy contribution is evaluated from the slope of
plots of log k vs. 1/T (K), called van't Hoff plots [267-272]. If an exact
linear dependence of $\Delta H°$ on log k is observed then the solute retention order
will be temperature independent for most systems. Also, for compounds with
different capacity factor values at a particular temperature, the resolution and
selectivity factor will decrease with increasing temperature [269,270]. There
are several examples of improved resolution resulting from a decrease in
temperature, including the use of subambient temperatures, but this is not
always a wise approach to separation optimization because column efficiency
declines and operating pressures and retention increase [265,267-269,271]. For
certain solutes the van't Hoff plots are nonlinear and may have sharp
discontinuities in their gradients. Causes for this include retention by a
mixed mechanism, a change in solute conformation which influences binding, or
the existance of more than one form of the solute having different retention
characteristics. Under these conditions the order of retention, the selectivity
factor, and resolution may all vary with temperature. For these systems an
optimum temperature may exist for the separation. Most separations are
performed at ambient temperatures because it is a reasonably efficient
temperature for small molecules and it is experimentally inconvenient to heat
columns, mobile phases, etc. Therefore, temperature is often the last parameter
to be optimized in normal and reversed-phase chromatography and is used mainly
for fine tuning the separation.

Retention is not controlled entirely by enthalpy but is also influenced by
changes in entropy which accompany the interaction of the solute with the
stationary phase. The process termed enthalpy-entropy compensation leads to two
generally useful equations which can be used to test the similarity of the
interaction mechanism of different solutes [268,272,273]. The capacity factor
is related to the enthalpy and entropy of the binding interaction by equation
(4.38).

$$\log k = \frac{-\Delta H°}{2.3RT} + \frac{\Delta S°}{2.3R} + \log \beta \qquad (4.38)$$

k $\quad$ = solute capacity factor

$\Delta H°$ = standard enthalpy of the transfer of a solute from the mobile
$\qquad$ phase to the stationary phase

$\Delta S°$ = associated change in standard entropy

T $\quad$ = column temperature (K)

R $\quad$ = gas constant

β $\quad$ = phase ratio of column

When $\Delta H°$ is independent of temperature, equation (4.38) is similar to the well-known van't Hoff equation. The slope of the enthalpy-entropy plot is called the compensation temperature and is calculated from equation (4.39)

$$\log k_T = \frac{-\Delta H°}{2.3R} \left[\frac{1}{T} - \frac{1}{T_c} \right] - \frac{\Delta G°}{2.3RT_c} + \log \beta \qquad (4.39)$$

k_T = capacity factor at temperature T

T_c = compensation temperature

$\Delta G°$ = Gibbs free energy change for the solute/stationary phase interaction

A similarity in values for the compensation temperature suggests that the solutes are retained by essentially identical interaction mechanisms and thus it is a useful tool for comparing retention mechanisms in different chromatographic systems.

Temperature programming is not widely used in liquid chromatography [274-278]. As the principal influence of increasing temperature is to reduce retention, temperature programming is used to improve front end resolution and to reduce retention of later eluting peaks. The maximum permissible operating temperature is about 20°C below the boiling point of the mobile phase. In most cases this means that the range of retention which can be controlled by changing temperature is less than the range available by adjusting solvent strength. With adsorbent stationary phases stripping of modifier from the column is also a problem as the temperature increases [269,274,275]. To restore column equilibrium, i.e., to obtain identical capacity factor values before and after programming, a long time is required. The detection sensitivity of later eluting peaks in temperature program analysis is often enhanced as they elute in a smaller volume of mobile phase compared to isothermal analysis. This situation is opposite to flow programming where the sensitivity of later eluting peaks is diminished by volumetric dilution.

Temperature programming can be used with spectrophotometric detectors but not with the temperature sensitive refractive index detector. A backend restrictor is required to prevent bubble formation at temperatures approaching the maximum allowable temperature established by the vapor pressure of the mobile phase. As column packings are fairly poor heat conductors, a liquid thermostat is required to adjust temperature and the mobile phase should be preheated to the column temperature. Failure to do so can result in the formation of asymmetric peaks [279]. Air thermostats are adequate for isothermal operation where stable temperature control of ± 0.2 - 1.0°C is considered sufficient to obtain reproducible retention data [280].

The results of flow programming resemble those of temperature programming;

it improves front end resolution and reduces the retention time of later eluting peaks [275,281]. However, flow is easier to change than temperature since it is controlled by the operating pressure of the pump, which can be changed rapidly to generate step flow gradients or varied continuously in a linear or exponential manner. Generally, a linear increase in inlet pressure and, hence, in flow velocity, results in a linear decrease in retention time, often with a decrease in resolution for later eluting peaks. Changing the mobile phase velocity within sensible ranges has only a small effect on column efficiency and selectivity. It is thus a less powerful way of changing separation characteristics than gradient elution. Disadvantages of flow programming are the limited range of pressures available for changing flow and the volumetric dilution of later eluting peaks, which reduces sensivitity. Advantages lie in the fact that the separation conditions are easily returned to the initial conditions by depressurizing to the original pressure and that it may be used with both refractive index and spectrophotometric detectors. Differential refractometer detectors are likely to exhibit baseline drift as the flow velocity changes.

4.20 References

1. I. Halasz and C. Horvath, Anal. Chem., 36 (1964) 1179.
2. R. E. Majors, J. Assoc. Off. Anal. Chem., 60 (1977) 186.
3. J. M. Bather and R. A. C. Gray, J. Chromatogr., 156 (1978) 21.
4. C. Laurent, H. A. H. Billiet, and L. de Galan, Chromatographia, 17 (1983) 253.
5. R. Ohmacht and I. Halasz, Chromatographia, 14 (1981) 216.
6. K. K. Unger, "Porous Silica", Elsevier, Amsterdam, 1979.
7. R. P. W. Scott, J. Chromatogr. Sci., 18 (1980) 297.
8. R. P. W. Scott, Adv. Chromatogr., 20 (1982) 167.
9. H. Engelhardt and H. Muller, J. Chromatogr., 219 (1981) 395.
10. R. E. Majors, J. Chromatogr. Sci., 18 (1980) 488.
11. H. Engelhardt and H. Elgass in C. Horvath (Ed.), "High-Performance Liquid Chromatography, Advances and Perspectives", Academic Press, New York, Vol. 2, 1980, p 57.
12. I. Halasz and K. Martin, Agnew. Chem., 90 (1978) 954.
13. R. E. Majors in C. Horvath (Ed.), "High-Performance Liquid Chromatography, Advances and Perspectives", Academic Press, New York, Vol. 1, 1980, p 75.
14. R. P. W. Scott and S. Trainman, J. Chromatogr., 196 (1980) 193.
15. R. Schwarzenbach, J. Liq. Chromatogr., 2 (1979) 205.
16. R. Schwarzenbach, J. Chromatogr., 202 (1980) 397.
17. C. F. Simpson (Ed.), "Techniques in Liquid Chromatography", Wiley, New York, 1982.
18. L. R. Snyder and J. J. Kirkland, "Introduction to Modern Liquid Chromatography", 2nd Edn., Wiley, New York, 1979.
19. V. Rehak and E. Smolkova, Chromatographia, 9 (1976) 219.
20. E. Grushka (Ed.), "Bonded Stationary Phases in Chromatography", Ann Arbor Science, Ann Arbor, 1974.
21. I. Halasz and I. Sebastian, Agnew. Chem. (Int. Ed. Engl.), 8 (1969) 453.
22. D. C. Locke, J. J. Shermud, and B. Banner, Anal. Chem., 44 (1972) 90.
23. C. Horvath and W. Melander, J. Chromatogr. Sci., 15 (1977) 393.
24. E. Grushka and E. J. Kikta, Anal. Chem., 49 (1977) 1005A.
25. H. Engelhardt and G. Ahr, Chromatographia, 14 (1981) 129.

26. N. H. C. Cooke and K. Olsen, J. Chromatogr. Sci., 18 (1980) 512.
27. W. R. Melander and C. Horvath in C. Horvath (Ed.), "High-Performance Liquid Chromatography, Advances and Perspectives", Academic Press, New York, Vol. 2., 1980, p 114.
28. J. L. M. van de Veene, J. P. M. Rindt, G. J. M. M. Coenen, and C. A. M. G. Cramers, Chromatographia, 13 (1980) 11.
29. M. Verzele, J. Lammens, and M. Van Roelenbosch, J. Chromatogr., 178 (1979) 463.
30. M. Caude and R. Rosset, J. Chromatogr. Sci., 15 (1977) 405.
31. F. E. Regnier, K. M. Godding, and S.-H. Chang in D. Hercules (Ed.), "Contemporary Topics in Analytical Clinical Chemistry", Plenum, New York, Vol. 1, 1977, p 1.
32. W. Kopaciewicz, M. A. Rounds, J. Fausnaugh, and F. R. Regnier, J. Chromatogr., 266 (1983) 3.
33. R. E. Majors, J. Chromatogr. Sci., 15 (1977) 334.
34. R. Woods, L. Cummings, and T. Jupille, J. Chromatogr. Sci., 18 (1980) 551.
35. H. G. Barth, J. Chromatogr. Sci., 18 (1980) 409.
36. W. W. Yau, J. J. Kirkland, and D. D. Bly, "Modern Size-Exclusion Liquid Chromatography", Wiley, New York, 1979.
37. D. E. Schmidt, R. W. Giese, D. Conron, and B. L. Karger, Anal. Chem., 52 (1980) 177.
38. P. Roumeliotis and K. K. Unger, J. Chromatogr., 218 (1981) 535.
39. G. Vigh, E. Gemes, and I. Inczedy, J. Chromatogr., 147 (1978) 59.
40. R. W. Stout, J. J. De Stefano, and L. R. Snyder, J. Chromatogr., 261 (1983) 189.
41. I. Halasz and G. Maldener, Anal. Chem., 55 (1983) 1842.
42. R. D. Davies, J. High Resolut. Chromatogr. Chromatogr. Commun., 4 (1981) 270.
43. R. E. Majors, Anal. Chem., 44 (1972) 1722.
44. H. R. Linder, H. P. Keller, and R. W. Frei, J. Chromatogr. Sci., 14 (1976) 234.
45. J. J. Kirkland, Chromatographia, 8 (1975) 661.
46. J. J. Kirkland, J. Chromatogr. Sci., 10 (1972) 593.
47. Y. Yamauchi and J. Kumanotani, J. Chromatogr., 210 (1981) 512.
48. T. J. N. Webber and E. H. McKerrel, J .Chromatogr., 122 (1976) 243.
49. P. A. Bristow, P. N. Brittain, C. M. Riley, and B. F. Williamson, J. Chromatogr., 131 (1977) 57.
50. M. Broquaire, J. Chromatogr., 170 (1979) 43.
51. K. Kawata, M. Uebori, and Y. Yamazaki, J. Chromatogr., 211 (1981) 378.
52. J. Asshauer and I. Halasz, J. Chromatogr. Sci., 12 (1974) 139.
53. J. H. Knox, J. Chromatogr. Sci., 15 (1977) 353.
54. H. P. Keller, F. Erni, H. R. Lindner, and R. W. Frei, Anal. Chem., 49 (1977) 1958.
55. P. A. Bristow and J. H. Knox, Chromatographia, 10 (1977) 279.
56. J. J. Kirkland, W. W. Yau, H. J. Stoklosa, and C. H. Dilks, J. Chromatogr. Sci., 15 (1977) 303.
57. J.-C. Chen and S. G. Weber, J. Chromatogr., 248 (1982) 434.
58. C. R. Wilke and P. Chang, Amer. Inst. Chem. Engr. J., 1 (1955) 264.
59. N. Tanaka, H. Goodell, and B. L. Karger, J. Chromatogr., 158 (1978) 233.
60. R. D. Smillie, D. T. Wang, and O. Merez, J. Environ. Sci. Health, A13 (1978) 47.
61. K. Ogan and E. Katz, J. Chromatogr., 188 (1980) 115.
62. K. Ogan, E. Katz, and W. Slavin, Anal. Chem., 51 (1979) 1315.
63. A. L. Colmsjo and J. C. MacDonald, Chromatographia, 13 (1980) 350.
64. E. C. Nice and M. J. O'Hare, J Chromatogr., 166 (1978) 263.
65. F. M. Rabel, E. T. Butts, M. L. Hellman, and D. J. Popovich, J. Liq. Chromatogr., 1 (1978) 631.
66. R. P. W. Scott and P. Kucera, J. Chromatogr., 142 (1977) 213.
67. J. G. Atwood and J. C. Goldstein, J. Chromatogr. Sci., 18 (1980) 650.
68. A. P. Goldberg, Anal. Chem., 54 (1982) 342.

69. M. Verzele and C. Dewaele, J. Chromatogr., 217 (1981) 377.
70. H. A. H. Billiet, J. P. J. van Dalen, P. J. Schoenmakers, and
     L. de Galan, Anal. Chem., 55 (1983) 847.
71. K. Jinno, N. Ozaki, and T. Sato, Chromatographia, 17 (1983) 341.
72. K. Jinno, Chromatographia, 17 (1983) 367.
73. O. A. G. J. van der Houwen, J. A. A. van der Linden, and
     A. W. M. Indemans, J. Liq. Chromatogr., 5 (1982) 2321.
74. E. Grushka, H. Colin, and G. Guiochon, J. Chromatogr., 248 (1982) 325.
75. M. J. Wells and C. R. Clark, Anal. Chem., 53 (1981) 1341.
76. D. E. Berendsen, P. Schoenmakers, and L. de Galan, J. Liq. Chromatogr.,
     3 (1980) 1669.
77. J. S. Landy, J. L. Ward, and J. G. Dorsey, J. Chromatogr. Sci., 21
     (1983) 49.
78. J. N. Little, R. L. Cotter, J. A. Pendergast, and P. D. McDonald,
     J. Chromatogr., 126 (1976) 439.
79. C. A. Bishop, D. R. K. Harding, L. J. Meyer, W. S. Hancock, and
     M. T. W. Hearn, J. Chromatogr., 192 (1980) 222.
80. W. S. Hancock, J. D. Capara, W. A. Bradley, and J. T. Sparrow,
     J. Chromatogr., 206 (1981) 59.
81. S. P. Assenza and P. R. Brown, J. Liq. Chromatogr., 3 (1980) 41.
82. F. M. Rabel, J. Chromatogr. Sci., 18 (1980) 394.
83. G. D. Reed and C. R. Loscombe, Chromatographia, 15 (1982) 15.
84. E. Lundanes, J. Dohl, and T. Greibrokk, J. Chromatogr. Sci., 21 (1983)
     235.
85. J. T. Przybytek (Ed.), "High Purity Solvent Guide", Burdick and Jackson
     Laboratories, Muskegon, 1980.
86. L. R. Snyder in E. S. Perry and A. Weissberger (Eds.), "Separation and
     Purification", John Wiley, New York, 3rd Edn., 1978, p 25.
87. R. P. W. Scott, "Contemporary Liquid Chromatography", John Wiley, New
     York, 1976, p 240.
88. D. W. Bristol, J. Chromatogr., 188 (1980) 193.
89. L. R. Snyder, J. Chromatogr., 92 (1974) 223.
90. B. L. Karger, L. R. Snyder, and C. Eon, J. Chromatogr., 125 (1976) 71.
91. R. P. W. Scott, J. Chromatogr., 122 (1976) 35.
92. L. R. Snyder, J. Chromatogr. Sci., 16 (1978) 223.
93. L. R. Snyder, J. W. Dolan, and J. R. Gant, J. Chromatogr., 165 (1979) 3.
94. P. J. Schoenmakers, H. A. H. Billiet, and L. de Galan, J. Chromatogr.,
     218 (1981) 259.
95. S. P. Bakalyar, R. McIlwrick, and E. Roggendorf, J. Chromatogr., 142
     (1977) 353.
96. E. Roggendorf and R. Spatz, J. Chromatogr., 204 (1981) 263.
97. J. L. Glajch, J. J. Kirkland, K. M. Squire, and J. M. Minor,
     J. Chromatogr., 199 (1980) 57.
98. J. J. Kirkland and J. L. Glajch, J. Chromatogr., 255 (1983) 27.
99. L. R. Snyder in C. Horvath (Ed.), "High Performance Liquid
     Chromatography, Advances and Perspectives", Academic Press, New York,
     Vol. 1, 1980, p. 207.
100. J. Jandera and J. Churacek, J. Chromatogr., 192 (1980) 1.
101. J. Jandera and J. Churacek, J. Chromatogr., 192 (1980) 19.
102. J. Jandera, M. Janderova, and J. Churacek, J. Chromatogr., 148 (1978) 79.
103. J. Jandera and J. Churacek, J. Chromatogr., 170 (1979) 1.
104. J. Jandera, J. Churacek, and H. Colin, J. Chromatogr., 214 (1981) 35.
105. P. J. Schoenmaker, H. A. H. Billiet, R. Tijssen, and L. de Galan,
     J. Chromatogr., 149 (1978) 519.
106. P. J. Schoenmakers, H. A. H. Billiet, and L. de Galan, J. Chromatogr.,
     185 (1979) 179.
107. P. J. Schoenmakers, H. A. H. Billiet, and L. de Galan, J. Chromatogr.,
     205 (1981) 13.
108. C. Liteaunu and S. Gocan, "Gradient Elution Chromatography", Wiley,
     New York, 1974.
109. P. Jandera and J. Churacek, Adv. Chromatogr., 19 (1981) 125.

348

110. L. R. Snyder, M. A. Stadalius, and M. A. Quarry, Anal. Chem., 55 (1983) 1412A.
111. V. V. Berry, J. Chromatogr., 119 (1980) 219.
112. G. Glad, T. Popoff, and O. Theander, J. Chromatogr. Sci., 16 (1976) 118.
113. D. L. Saunders, J. Chromatogr. Sci., 15 (1977) 372.
114. S. R. Abbott, J. Chromatogr. Sci., 18 (1980) 540.
115. R. D. Rasmussen, W. H. Yokoyama, S. G. Blumenthal, D. E. Bergstrom, and B. H. Ruebner, Anal. Biochem., 101 (1980) 66.
116. K. Aitzemuller, J. Chromatogr., 113 (1975) 231.
117. S. P. Levine and L. M. Skewes, J. Chromatogr., 235 (1982) 532.
118. D. Schuetzle, T. L. Riley, T. J. Prater, T. M. Harvey, and D. F. Hunt, Anal. Chem., 54 (1982) 265.
119. B. A. Bidlingmeyer, J. K. Del Rios, and J. Korpi, Anal. Chem., 54 (1982) 442.
120. Y. I. Yashin, J. Chromatogr., 251 (1982) 269.
121. L. R. Snyder, "Principles of Adsorption Chromatography", Dekker, New York, 1968.
122. L. R. Snyder, Anal. Chem., 46 (1974) 1384.
123. L. R. Snyder and H. Poppe, J. Chromatogr., 184 (1980) 363.
124. L. R. Snyder, Liq. Chromatogr. HPLC Magzn., 1 (1983) 478.
125. L. R. Snyder, J. Chromatogr., 255 (1983) 3.
126. R. P. W. Scott and P. Kucera, J. Chromatogr., 171 (1979) 37.
127. R. P. W. Scott and P. Kucera, J. Chromatogr., 149 (1978) 93.
128. L. R. Snyder and J. L. Glajch, J. Chromatogr., 214 (1981) 1.
129. J. H. Glajch and L. R. Snyder, J. Chromatogr., 214 (1981) 21.
130. L. R. Snyder, J. L. Glajch, and J. J. Kirkland, 218 (1981) 299.
131. H. Engelhardt, "High Performance Liquid Chromatography", Springer-Verlag, Berlin, 1979, p 115.
132. H. Engelhardt, J. Chromatogr. Sci., 15 (1977) 380.
133. J.-P. Thomas, A. P. Brun, and J. P. Bounine, J. Chromatogr., 172 (1979) 107.
134. H. Engelhardt and W. Bohme, J. Chromatogr., 133 (1977) 67.
135. R. A. Bredeweg, L. D. Rothman, and C. D. Pfeiffer, Anal. Chem., 51 (1979) 2061.
136. S. A. Wise, S. N. Chesler, H. S. Hertz, L. P. Hilpert, and W. E. May, Anal. Chem., 49 (1977) 2306.
137. J. Chmielowiec and A. E. George, Anal. Chem., 52 (1980) 1154.
138. D. Karlesky, D. C. Shelley, and I. Warner, Anal. Chem., 53 (1981) 2146.
139. A. M. Krstulovic and P. R. Brown, "Reversed-Phase High Performance Liquid Chromatograhy: Theory, Practice and Biomedical Applications", Wiley, New York, 1982.
140. F. P. B. van der Maeden, M. E. F. Piemond, and P. C. G. M. Janssen, J. Chromatogr., 149 (1978) 539.
141. P. R. Brown and A. M. Krstulovic, Anal. Biochem., 99 (1979) 1.
142. P. Jandera, H. Colin, and G. Guiochon, Anal. Chem., 54 (1982) 435.
143. M. Otto and W. Wegscheider, J. Chromatogr., 258 (1983) 11.
144. F. E. Regnier and K. M. Gooding, Anal. Biochem., 103 (1980) 1.
145. M. T. W. Hearn, Adv. Chromatogr., 20 (1982) 1.
146. F. E. Regnier and J. Fausnaugh, Liq. Chromatogr. and HPLC Magzn., 1 (1983) 402.
147. M. T. W. Hearn and B. Grego, J. Chromatogr., 266 (1983) 75.
148. M. T. W. Hearn and B. Grego, J. Chromatogr., 203 (1981) 349.
149. C. Horvath, Liq. Chromatogr. and HPLC Magzn., 1 (1983) 552.
150. H. Colin and G. Guiochon, J. Chromatogr., 158 (1978) 183.
151. B. L. Karger, J. R. Gant, A. Hartkopf, and P. H. Weiner, J. Chromatogr., 128 (1976) 65.
152. P. C. Sadek and P. W. Carr, J. Chromatogr. Sci., 21 (1983) 314.
153. R. V. Vivileechia, B. G. Lightbody, N. Z. Thinot, and H. M. Quinn, J. Chromatogr. Sci., 15 (1977) 424.
154. A. Krishen, J. Chromatogr. Sci., 15 (1977) 434.
155. D. D. Bly, J. Polym. Sci. Part (C), 21 (1968) 13.

349

156. W. W. Yau, J. J. Kirkland, D. D. Bly, and H. J. Stoklosa, J. Chromatogr., 125 (1976) 219.
157. E. Pfannkock, K. C. Lu, F. E. Regnier, and J. G. Barth, J. Chromatogr. Sci., 18 (1980) 430.
158. J. Cazes (Ed.), "Liquid Chromatography of Polymers and Related Materials", Dekker, New York, Vol. 1 (1977), Vol. 2 (1980), Vol 3 (1981).
159. B. W. Hatt in C. E. H. Knapman (Ed.), "Developments in Chromatography", Applied Science Publisher, London, Vol. 1, 1978, p 157.
160. S. N. E. Omorodion, A. E. Hamielec, and J. L. Brash in T. Provodor (Ed.), "Size-Exclusion Chromatography", American Chemical Society, Washington, D.C., 1980, p 267.
161. A. R. Cooper and D. S. van Derveer, J. Liq. Chromatogr., 1 (1978) 693.
162. W. W. Yau, C. R. Ginnard, and J. J. Kirkland, J. Chromatogr., 149 (1978) 465.
163. W. W. Yau and D. D. Bly in T. Provdor (Ed.), "Size-Exclusion Chromatography", American Chemical Society, Washington, D.C., 1980, p. 197.
164. J. Janca, Adv. Chromatogr., 19 (1981) 37.
165. S. Elchuk and R. M. Cassidy, Anal. Chem., 51 (1979) 1434.
166. Whatman Inc., Ion-Exchange Application Brochure, 1979.
167. H. Miyagi, J. Miura, Y. Takata, S. Kamitake, S. Ganno, and Y. Yamagata, J. Chromatogr., 239 (1982) 733.
168. R. C. Adlakha and V. R. Villanueva, J. Chromatogr., 187 (1980) 442.
169. D. N. Vacik and E. C. Toren, J. Chroamtogr., 228 (1982) 1.
170. W. Voelter and H. Bauer, J. Chromatogr., 126 (1976) 693.
171. O. Samuelson, Adv. Chromatogr., 16 (1978) 113.
172. K. Tanaka and T. Shizuka, J. Chromatogr., 174 (1979) 157.
173. F. Helfferich, "Ion Exchange", McGraw-Hill, New York, 1962.
174. J. A. Marinsky (Ed.), "Ion Exchange, A Series of Advances", Marcel Dekker, New York, Vol. 1, 1966.
175. C. J. O. R. Morris and P. Morris, "Separation Methods in Biochemistry", Wiley, New York, 2nd Edn., 1976, p 249.
176. J. X. Khym, "Analytical Ion-Exchange Procedures in Chemistry and Biology", Prentice-Hall, Englewood Cliffs, 1974.
177. P. R. Brown and A. M. Krstulovic in E. S. Perry and A. Weissberger (Eds.), "Separation and Purification Methods", Wiley, New York, 3rd Edn., 1978, p 199.
178. F. M. Rabel, Adv. Chromatogr., 17 (1979) 53.
179. E. Sawicki, J. D. Mulik, and E. Wattgenstein, (Eds.), "Ion Chromatographic Analysis of Environmental Pollutants", Ann Arbor Science, Ann Arbor, Vol. 1, 1978.
180. J. D. Mulik and E. Sawicki (Eds.), "Ion Chromatographic Analysis of Environmental Pollutants", Ann Arbor Science, Ann Arbor, Vol. 2, 1979.
181. F. C. Smith and R. C. Change, CRC Crit. Revs. Anal. Chem., 9 (1980) 197.
182. H. Small, in J. F. Lawrence (Ed.), "Trace Analysis", Academic Press, New York, Vol. 1, 1982, p 269.
183. J. S. Fritz, D. T. Gjerde, and C. Pohlandt, "Ion Chromatography", Huthig, Heidelberg, 1982.
184. H. Small, Anal. Chem., 55 (1983) 235A.
185. C. A. Pohl and E. L. Johnson, J. Chromatogr. Sci., 18 (1980) 442.
186. D. J. Gjerde and S. J. Fritz, J. Chromatogr., 176 (1979) 199.
187. T. S. Stevens and H. Small, J. Liq. Chromatogr., 1 (1978) 123.
188. T. S. Stevens and M. A. Langhorst, Anal. Chem., 54 (1982) 950.
189. H. Small, T. S. Stevens, and W. C. Bauman, Anal. Chem., 47 (1975) 1801.
190. T. S. Stevens, J. C. Davis, and H. Small, Anal. Chem., 53 (1981) 1488.
191. Y. Hanaoki, T. Murayama, S. Muramoto, T. Matsura, and A. Nanba, J. Chromatogr., 239 (1982) 537.
192. T. S. Stevens, G. L. Jewett, and R. A. Bredeweg, Anal. Chem., 54 (1982) 1206.

350

193. D. T. Gjerde, J. S. Fritz, and G. Schmuckler, J. Chromatogr., 186 (1979) 509.
194. D. T. Gjerde, J. S. Fritz, and G. Schmuckler, J. Chromatogr., 187 (1980) 35.
195. S. Matsushita, Y. Tada, N. Baba, and K. Hosako, J. Chromatogr., 259 (1983) 459.
196. H. Small and T. E. Miller, Anal. Chem., 54 (1982) 462.
197. R. A. Cochrane and D. E. Hillman, J. Chromatogr., 241 (1982) 392.
198. J. R. Larson and C. D. Pfeiffer, J. Chromatogr., 259 (1983) 519.
199. D. Ishii and T. Takeuchi, Adv. Chromatogr., 21 (1983) 131.
200. M. Novotny, Anal. Chem., 53 (1981) 1294A.
201. T. Tsuda and M. Novotny, Anal. Chem., 50 (1978) 271.
202. J. H. Knox and M. T. Gilbert, J. Chromatogr., 186 (1979) 405.
203. G. Guiochon, Anal. Chem., 53 (1981) 1318.
204. J. W. Jorgenson and E. J. Guthrie, J. Chromatogr., 255 (1983) 335.
205. J. H. Knox, J. Chromatogr. Sci., 18 (1980) 453.
206. R. P. W. Scott and P. Kucera, J. Chromatogr., 169 (1979) 51.
207. C. E. Reese and R. P. W. Scott, J. Chromatogr. Sci., 18 (1980) 479.
208. R. P. W. Scott, P. Kucera, and M. Munroe, J. Chromatogr., 186 (1979) 475.
209. T. Tsuji and R. B. Binns, J. Chromatogr., 253 (1982) 227.
210. J. Bowermaster and H. McNair, J. Chromatogr., 279 (1983) 431.
211. R. P. W. Scott and P. Kucera, J. Chromatogr., 185 (1979) 27.
212. H. E. Schwartz, B. L. Karger, and P. Kucera, Anal. Chem., 55 (1983) 1752.
213. R. P. W. Scott, J. Chromatogr. Sci., 18 (1980) 49.
214. P. Kucera, J. Chromatogr., 198 (1980) 93.
215. T. Tsuda and M. Novotny, Anal. Chem., 50 (1978) 271.
216. M. Krejci, K. Tesarik, and J. Pajurek, J. Chromatogr., 198 (1980) 17.
217. V. L. McGuffin and M. Novotny, Anal. Chem., 55 (1983) 580.
218. T. Tsuda, K. Hibi, T. Nakanishi, and D. Ishii, J. Chromatogr., 158 (1978) 227.
219. T. Tsuda, T. Tsuboi, and G. Nakagawa, J. Chromatogr., 214 (1981) 283.
220. D. Ishii, K. Asai, K. Hibi, T. Jonokuchi, and M. Nagaya, J. Chromatogr., 144 (1977) 157.
221. F. J. Yang, J. High Resolut. Chromatogr. Chromatogr. Commun., 3 (1980) 589.
222. Y. Hirata, P. T. Lin, M. Novotny, and R. M. Wightman, J. Chromatogr., 181 (1980) 287.
223. Z. Frobe, R. Richon, and W. Simon, Chromatographia, 17 (1983) 467.
224. K. Slais and D. Kourilova, J. Chromatogr., 258 (1983) 57.
225. K. Jinno and C. Fujimoto, Chromatographia, 17 (1983) 259.
226. T. Takeuchi, S. Saito, and D. Ishii, J. Chromatogr., 284 (1983) 125.
227. M. Krejci, K. Tasarik, M. Rusek, and J. Pajurek, J. Chromatogr., 218 (1981) 167.
228. V. L. McGuffin and M. Novotny, Anal. Chem., 55 (1983) 2296.
229. V. L. McGuffin and M. Novotny, Anal. Chem., 53 (1981) 946.
230. T. Takeuchi and D. Ishii, J. Chromatogr., 253 (1982) 41.
231. P. Kucera and G. Manius, J. Chromatogr., 216 (1981) 9.
232. C. Dewaele and M. Verzele, J. High Resolut. Chromatogr. Chromatogr. Commun., 3 (1980) 273.
233. T. Tsuda, I. Tanaka, and G. Nakagawa, J. Chromatogr., 239 (1982) 507.
234. V. L. McGuffin and M. Novotny, J. Chromatogr., 255 (1983) 381.
235. M. Novotny, J. Chromatogr. Sci., 18 (1980) 463.
236. T. Takeuchi and D. Ishii, J. Chromatogr., 213 (1981) 25.
237. T. Takeuchi and D. Ishii, J. Chromatogr., 238 (1982) 409.
238. F. J. Yang, J. Chromatogr., 236 (1982) 265.
239. D. Ishii and T. Takeuchi, J. Chromatogr., 255 (1983) 349.
240. J. C. Gluckman, A. Hirose, V. L. McGuffin, and M. Novotny, Chromatographia, 17 (1983) 303.
241. D. Ishii, T. Tsuda, and T. Takeuchi, J. Chromatogr., 185 (1979) 73.
242. D. Ishii and T. Takeuchi, J. Chromatogr. Sci., 18 (1980) 462.

243. T. Takeuchi and D. Ishii, J. Chromatogr., 279 (1983) 439.
244. J. L. DiCesare, M. W. Dong, and L. S. Ettre, Chromatographia, 14 (1981) 257.
245. J. L. DiCesare, M. W. Dong, and J. G. Atwood, J. Chromatogr., 217 (1981) 369.
246. M. W. Dong and J. L. DiCesare, J. Chromatogr. Sci., 20 (1982) 517.
247. E. Katz and R. P. W. Scott, J. Chromatogr., 253 (1982) 159.
248. R. E. Majors, J. Chromatogr. Sci., 18 (1980) 571.
249. D. H. Freeman, Anal. Chem., 53 (1981) 2.
250. F. W. Willmott, I. Mackenzie, and R. J. Dolphin, J. Chromatogr., 167 (1978) 31.
251. M. C. Harvey and S. D. Stearns, Amer. Lab., 14 (1982) 68.
252. C. J. Little, D. J. Tompkins, O. Stahel, R. W. Frei, and C. E. Werkhoven-Goewie, J. Chromatogr., 264 (1983) 183.
253. E. L. Johnson, R. Gloor, and R. E. Majors, J. Chromatogr., 149 (1978) 571.
254. F. Erni and R. W. Frei, J. Chromatogr., 149 (1978) 561.
255. L. R. Snyder, J. W. Dolan, and Sj. Van Der Wal, J. Chromatogr., 203 (1981) 3.
256. J. F. K. Huber, R. van der Linden, E. Ecker, and M. Oreans, J. Chromatogr., 83 (1973) 267.
257. M. Martin, F. Verillon, C. Eon, and G. Guiochon, J. Chromatogr., 125 (1976) 17.
258. P. Kucera and G. Manius, J. Chromatogr., 219 (1981) 1.
259. K. J. Bombaugh, W. A. Dark, and R. F. Levangie, J. Chromatogr. Sci., 7 (1969) 42.
260. K. J. Bombaugh and R. F. Levangie, J. Chromatogr. Sci., 8, (1970) 560.
261. K. J. Bombaugh and R. F. Levangie, Sep. Sci., 5 (1970) 751.
262. K. J. Bombaugh, J. Chromatogr., 53 (1979) 27.
263. R. A. Henry, S. H. Burne, and D. R. Hudson, J. Chromatogr. Sci., 12 (1974) 197.
264. M. Minarik, M. Popl, and J. Mostecky, J. Chromatogr. Sci., 19 (1981) 250.
265. J. S. Kowalczyk and G. Herbut, J. Chromatogr., 196 (1980) 11.
266. G. Herbut and J. S. Kowalczyk, J. High Resolut. Chromatogr. Chromatogr. Commun., 4 (1981) 27.
267. R. B. Diaso and M. E. Wilburn, J. Chromatogr. Sci., 17 (1979) 565.
268. K. Jinno and Y. Hirata, J. High Resolut. Chromatogr. Chromatogr. Commun., 5 (1982) 85.
269. W. R. Sisco and R. K. Gilpin, J .Chromatogr. Sci., 18 (1980) 41.
270. L. R. Snyder, J. Chromatogr., 179 (1979) 167.
271. R. K. Gilpin and J. A. Squires, J. Chromatogr. Sci., 19 (1981) 195.
272. W. R. Melander, B.-K. Chen, and C. Horvath, J. Chromatogr., 185 (1979) 99.
273. Gy. Vigh and Z. Varga-Puchony, J. Chromatogr., 196 (1980) 1.
274. H. Engelhardt, J. Chromatogr. Sci., 15 (1977) 380.
275. L. R. Snyder, J. Chromatogr. Sci., 8 (1970) 692.
276. J. Chmielowiec and H. Sawatzky, J. Chromatogr. Sci., 17 (1979) 245.
278. E. J. Kikta, A. E. Stange, and S. Lam, J. Chromatogr., 138 (1977) 321.
279. R. J. Perchalski and B. J. Wilder, Anal. Chem., 51 (1979) 775.
280. R. K. Gilpin and W. R. Sisco, J. Chromatogr., 194 (1980) 285.
281. H. Wiedemann, H. Engelhardt, and I. Halasz, J. Chromatogr., 91 (1974) 141.

Chapter 5

INSTRUMENTAL REQUIREMENTS FOR HIGH PERFORMANCE LIQUID CHROMATOGRAPHY

5.1 Introduction

    The equipment used in modern high performance liquid chromatography is very different from the simple gravity-fed devices which dominated the practice of liquid chromatography for most of this century. Theory and practice have indicated that pressures well above atmospheric are required to operate high efficiency columns, packed with small diameter particles, at flow rates of a few milliliters per minute. A block diagram of a suitable instrument for HPLC is shown in Figure 5.1. For isocratic operation, a solvent or solvent mixture is pressurized by a pump and delivered as a pulse-free flow to the column. In gradient elution two or more solvents are mixed in proportions such that the concentration of the stronger solvent increases with time. Depending on the method of gradient production more than one pump may be required. Between the pump and the injector there may be a series of devices which correct or monitor the pump output. Such devices include pulse dampers, mixing chambers, flow controllers and pressure transducers. Their function is to ensure that a homogeneous, pulse-free liquid flow is delivered to the column at a known pressure and volume flow rate. They may be operated either independently of the

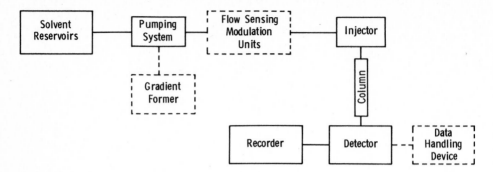

Figure 5.1 Block diagram of a high pressure liquid chromatograph. Dotted lines refer to components which are optional.

pump or in a feedback network which continuously updates the pump output. An injection device is connected to the head of the column for loading the sample by syringe or valve. A flow-through detector, connected to the end of the column, monitors the separation. The chromatogram is recorded on a strip chart recorder, a computing integrator, or similar data handling device. In succeeding sections of this chapter the above components will be described in detail.

The reader should be aware that the equipment available for HPLC covers a broad range of technological innovation and sophistication [1,2]. Not all devices work equally well nor cost similar amounts. A liquid chromatograph may be assembled from modular units designed to work independently of each other or as a single, integrated unit. The latter approach is becoming more common with the availability of inexpensive microprocessor technology. The main advantage of this approach is that the purchaser acquires an instrument of optimized performance. The disadvantages are higher cost and the increasing likelihood that, due to the integrated approach, it may not be possible to operate components independently of one another or the central processing unit.

Another trend in modern instrumentation is automation of the operation of the liquid chromatograph, as well as injection and sample processing [3,4]. Automatic injectors have long been available for unattended operation but sophisticated equipment, capable of multistep sample cleanup and extraction prior to injection, has recently become available [4,5].

5.2 Equipment Specifications

Under ideal conditions, the peak profile recorded during an HPLC separation should depend only on the operating characteristics of the column and should be independent of the instrument in which the column resides. Under less than ideal conditions, the peak profile will be broader than the column profile by an amount equivalent to the extracolumn band broadening. This broadening arises from dispersion and mixing phenomena which occur in the injector, column connecting tubes, and detector cell, as well as from electronic constraints which govern the response speed of the detector and recorder. The quality of an instrument might therefore be judged by its ability to minimize extracolumn band broadening and reproduce retention volumes. The constancy of retention volumes is primarily a function of the solvent delivery system and is discussed later. The independent factors which contribute to extracolumn band broadening can be treated as additive in their second moments or variances (σ^2), according to the relationship [6-11]:

$$\sigma^2_T = \sigma^2_{col} + \sigma^2_{inj} + \sigma^2_{tub} + \sigma^2_{det} + \sigma^2_{TC} + \sigma^2_{RC} \qquad (5.1)$$

σ^2_T = total variance measured from the chromatogram

σ^2_{col} = column variance

$\sigma^2_{inj} + \sigma^2_{tub} + \sigma^2_{det}$ = variance due to instrument volumes

$\sigma^2_{TC} + \sigma^2_{RC}$ = variance due to electronic response time functions

The assumption that these individual contributions are independent of one another may not be true in practice. For an accurate calculation of instrument variance it may be necessary to couple some of the individual contributions [12]. We will treat them separately for convenience.

Band broadening due to injection arises because the sample is introduced into the column as a finite volume. A solute zone is formed at the column head which reflects the degree of sample axial displacement during the injection time. This solute zone is generally less than the injection volume due to retention of the solute by the column packing. The elution strength of the injection solvent and the effect of solvent dilution with the column mobile phase will determine the extent of the zone displacement. This situation is too complex to permit description by a simple mathematical model. Two extreme views, plug or exponential sample injection, can be used to define the limiting cases. In plug sample injection, the sample profile is considered to be a square wave function defined by the injection volume, injection time, and a constant, λ, related to the design of the injector, equation 5.2.

$$\sigma^2_{inj} = \frac{V^2_{inj}}{\lambda^2} \qquad (5.2)$$

(for a rectangular profile, $\lambda^2 = 12$)

Plug injection assumes no mixing of the sample with the mobile phase; this is a fairly good approximation for a valve injection where the sample loop is momentarily inserted into the mobile phase stream. As valve or stop-flow injection occurs by displacement rather than by mixing, the time of injection is not important. The opposite is true of syringe injection into a flowing stream; here, injection time will directly influence the length of the solute zone and the exponential injection model is more appropriate.

For analytical purposes the maximum allowable sample volume and injection time to produce an acceptable degree of band broadening, $\theta^2 \sigma^2_{inj}$, can be calculated from the formulas in Table 5.1. The value of the numerical constant θ^2 can be selected to predict any desired set of boundary conditions. A value of $\theta^2 = 1\%$ is generally considered reasonable.

Band dispersion in open tubes is due to poor radial mass transfer of the solute resulting from the laminar nature of solvent flow. Its contribution to total system variance can be calculated using the formulas in Table 5.1. For connecting tubes of less than 5 cm in length and internal diameters of 0.5 to 0.15 mm this contribution is very small.

The influence of the detector cell volume on column performance is generally more significant than injection and connecting tube contributions in modern instruments. Depending on the design of the cell, it can either have the properties of a tube with plug flow or act as a mixing volume. In actuality, most detector cells behave in a manner somewhere between those represented by the two extreme models. If the cell volume is approximately 10% of the peak volume detected, then extracolumn band broadening from the detector will be insignificant. These conditions are met in practice by commercially available detectors having cell volumes of 6-10 microliters if they are used to detect well retained peaks from standard analytical columns operated under normal conditions. For solutes of low retention or for short, high efficiency columns packed with 3 to 5 micrometer packings, these detector volumes are too large and add significantly to instrument extracolumn band broadening [6,13]. To take full advantage of the separating power of high efficiency analytical columns, a detector cell volume of about 1.0 microliter is required.

Detection and recording devices can cause band broadening due to their response speed which is primarily a function of the time constant associated with resistance-capacitance filter networks used to diminish high frequency noise. Detector time constants of commercially available instruments are in the range 0.5 to 3.0 seconds. For high efficiency analytical columns a value of 0.1 s or smaller is desirable.

5.3 Solvent Reservoirs and Solvent Degassing

The solvent reservoir is a storage container of a material resistant to chemical attack by the mobile phase. In its simplest form a glass jug, solvent bottle, or Erlenmeyer flask with a cap and flexible hose connection to the pump

TABLE 5.1

FORMULAS FOR CALCULATING THE CONTRIBUTION OF EXTRACOLUMN BAND BROADENING TO COLUMN EFFICIENCY

Symbols used have their conventional chromatographic meaning (see Chapters 1 and 4). θ is a selectable factor to indicate the amount of band broadening that is acceptable; λ = constant governed by the design of the injector; ϵ = column permeability; ℓ = length of connecting tube; Q = column flow rate.

A. Injection

$$\sigma^2_{inj} = \frac{V^2_{inj}}{\lambda^2} = \frac{\theta^2 V^2_R}{\sqrt{n}}$$

Maximum Sample Size, $V_{inj,m}$:

$$V_{inj,m} = \frac{\theta \lambda V_R}{\sqrt{n}} = \theta\lambda \frac{\pi d_c^2}{4} \epsilon(1+k)hd_p \sqrt{n}$$

Maximum Time for Injection, $t_{inj,m}$:

$$t_{inj,m} = \frac{\theta \lambda t_R}{\sqrt{n}}$$

B. Connecting Tubes

$$\sigma^2_{tub} = \frac{\pi r^4 \ell}{24 D_m Q}$$

Limiting Conditions for Dimensions:

$$r^4 \ell < 6\theta^2 \epsilon(1+k)^2 d_c^2 nd_p^3 \frac{h^2}{v}$$

C. Detector Cell Volume

$$\sigma^2_{det} = \frac{V^2_{det}}{Q^2}$$

Maximum Allowable Detector Volume, $V_{det,m}$:

$$V_{det,m} = \frac{\theta \pi d_c^2}{4} \epsilon(1+k)hd_p \sqrt{n}$$

D. Detector Time Constant

Maximum Allowable Detector Time Constant, τ_m:

$$\tau_m = \frac{\theta t_R}{\sqrt{n}}$$

If 5% band broadening is acceptable

$$\tau_m = \frac{32 \theta V_R}{Q \sqrt{n}}$$

is adequate. The PTFE connecting hose is terminated on the solvent side with a 2-10 micrometer filter to prevent suspended particle matter from reaching the pump. In more sophisticated equipment the solvent reservoir may also be equipped for solvent degassing. Degassing is required to prevent gas bubble formation, which degrades pump and detector performance, when different solvents are mixed or the mobile phase is depressurized. This procedure can be carried out by applying a vacuum above the solvent, heating with vigorous stirring, ultrasonic treatment, or sparging with nitrogen or helium [14-16]. As helium has very low solubility in most common solvents vigorous sparging with helium followed by a slow flow of gas to prevent redissolution of air is perhaps the most convenient method, although all the above methods are fairly efficient at preventing bubble formation. A backpressure restrictor of 5-10 atmospheres at the detector exit also helps minimize the occurrence of bubble formation.

A further problem associated particularly with oxygen in the mobile phase is the oxidative degradation of samples and phases and a reduction in sensitivity and operating stability of ultraviolet, refractive index, electrochemical, and fluorescence detectors. Oxygen absorbs significantly in the low wavelength UV region, 190-260 nm, and changes in the mobile phase oxygen concentration lead to detector baseline drift, random noise, a loss of sensitivity, and a large baseline offset at high sensitivity. The above observations also apply to the fluorescence detector with the additional possibility that substance-specific oxygen quenching of the fluorescence signal may occur. Temperature fluctuations and oxygen concentration may be coupled together in reducing detector baseline stability at high sensitivity. Refluxing solvents was found to be the most efficient method of reducing oxygen concentrations while sparging with nitrogen or helium was less efficient but more convenient [16]. Particular care is needed when sparging with helium to prevent back diffusion of air into the solvent against the much lighter helium gas. Vacuum degassing and ultrasonic treatment were not very effective in reducing oxygen concentrations.

5.4 High Pressure Pumps

The pump is one of the most important components of the liquid chromatograph since its performance directly affects retention time reproducibility and detector sensitivity. Other components such as check valves, flow controllers, mixing chambers, pulse dampers, and pressure transducers may be associated with the pump's operation. Some of these components may be electronically linked to the pump to control its output or switch it off if a default value is exceeded.

In modern HPLC the pumping system must meet certain general requirements.

For analytical and semi-preparative applications where 25-50 cm long columns having 4.0 to 10 mm internal diameters, packed with 5 or 10 micrometer particles are typically used, the pump should be able to provide flow rates of 0.5 to 10 ml/min at pressures reaching 5000 p.s.i.*. The volumetric flow rate should be stable and accurate to better than a few percent over this range. A set of criteria for measuring pump performance is given in Table 5.2. In general terms, a high degree of accuracy in pump resetability and flow rate control with a minimum of pump pulsation and drift are the hallmarks of a good pump. Other considerations in pump selection might be price, chemical resistance to corrosive solvents, serviceability, ease of operation, and the time required for solvent changeover. Using the above check list virtually only the most expensive pumps come close to meeting these requirements. Many commercial systems represent a compromise in which different weighting factors are applied to the importance of these criteria.

TABLE 5.2

PUMP SPECIFICATION

Specification	Description
Pump Resetability	The ability to reset the pump to the same flow rate repeatedly.
Flow Rate Accuracy	The ability of the pump to deliver exactly the flow rate indicated by a particular setting.
Pump Pulsation or Noise	Flow changes sensed by the detector as a result of pump operations such as piston movement and check valve operation.
Short Term Precision	The accuracy of the volume output of the pump over a few minutes.
Drift	Measure of the generally continuous increase or decrease in the pump output over relatively long periods (e.g., hours)

The types of pumps used in HPLC instruments can be divided into two categories: constant pressure pumps, such as the gas displacement and pneumatic amplifier pumps, and constant volume models, such as reciprocating and syringe pumps. Some examples of the various types are shown in Figure 5.2 and their properties compared in Table 5.3 [17-19].

The simplest and least expensive of all HPLC pumps is the gas displacement pump. A gas cylinder with regulator is used to pressurize a liquid in a

* Conversion factors for pressure:
14.5 p.s.i. = 1 bar = 0.9868 atm = 1.02 kg/cm^2 = 0.1 MPa (10^5 Pascals)

TABLE 5.3

PUMP CHARACTERISTICS

Pump Type	Principle of Operation	Advantages	Disadvantages
Syringe pump	Solvent displacement by a mechanically controlled piston advancing at a constant rate in a fixed volume chamber. multishpaed gradients.	Pulseless flow. High pressure capability. Tandem pumps can generate multishpaed gradients.	Limited solvent capacity. Expensive. Liquid compressibility problems (see text).
Constant-Pressure Pumps			
Gas-Displacement	Direct gas pressure applied to a reservoir is used to drive the solvent out. Reservoir has a low gas-liquid interfacial area to minimize gas-solvent.	Inexpensive. Constant flow at a constant back pressure. Pulseless flow	Limited by available gas cylinder pressure. Gas dissolving in solvent at high pressure end causing bubbles at low pressure end. Limited solvent capacity Gradient and flow programming difficult.
Gas-Amplifier (pneumatic amplifier)	Dual chamber pump containing a large area gas piston head and small area eluent piston head. Pressure on the eluent piston is proportional to the ratio of the two piston areas. A low pressure gas source can be used to generate high liquid pressures. Valving arrangement required to rapidly refill the empty eluent chamber.	High pressure capability. Continuous operation. High flow rates. Pulse free.	Pump refill operation usually causes a spike in the detector baseline. Relatively expensive. Requires frequent maintenance.

Reciprocating Pumps

Single Head	Use small volume chambers with reciprocating pistons or diaphragms to drive the flow against a back pressure. Two check valves are synchronized with the piston (or diaphragm) drive to alternate filling and emptying of eluent from the chamber. Pulse dampening normally required.	Inexpensive. Continuous operation. Constant flow at constant backpressure. Gradient elution capability.	Pulsed flow. Cavitation, i.e., loss of prime with volatile solvents.
Multiple Heads	From two to four heads controlled by a special cam drive to minimize flow pulsation. In a dual piston pump one chamber is filling while the other is emptying. Electronic or pneumatic transducers often incorporated to ensure constancy of flow.	As for single head pumps. Pulse free flow.	Expensive. Maintenance is complex.

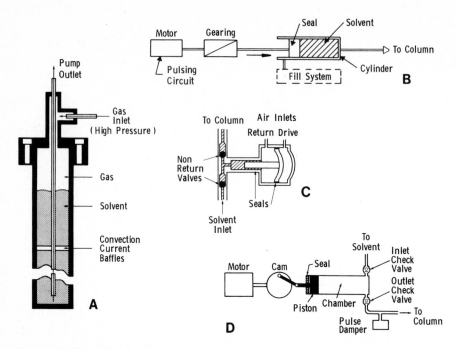

Figure 5.2 Different types of pumps used in high pressure liquid
            chromatography. A, gas displacement pump; B, syringe pump;
            C, pneumatic amplifier pump; D, reciprocating pump.

cylinder or holding coil and then force it out through a check valve. To avoid
saturating the solvent with gas the gas–liquid interface is small. The
pressurized cylinder also acts as the solvent reservoir and must be refilled
periodically. A network of valves is provided to do this conveniently [20,21].
The main advantages of the gas displacement pump are that it generates a
pulse–free output and has low maintenance costs. Disadvantages are the limited
reservoir volume and operating pressure, as well as limited capabilities for
gradient elution. The maximum operating pressure is established by the gas
cylinder pressure (ca. 2500 p.s.i.) and the constancy of flow depends on the
stability of the column back pressure.

    The pneumatic amplifier pump uses a large bore gas piston to drive a liquid
piston of smaller size. An amplification of the outlet pressure over the inlet
pressure equivalent to the ratio of the piston areas is obtained. High pressure
operation is thus possible. The pneumatic amplifier pump is equipped with a
power–return stroke that permits very rapid refilling of the empty piston
chamber with mobile phase. This pump provides essentially pulse free,
continuous pumping and can achieve relatively high volumetric flow rates. For
constant flow the column back pressure must remain constant; some pumps have an

automatic flow-feedback system that continually adjusts the air inlet pressure
to allow for fluctuations in the column back pressure.

Syringe pumps consist of a cylinder that acts as the mobile phase reservoir
and a piston, advanced by a motor connected through worm gears, that expels the
solvent. The pump output is relatively pulse free. As solvents are
compressible, a finite time is required before the volumetric flow rate is
constant at high pressures [22,23]. Most pumps are equipped with a fast pump
option to rapidly pressurize the solvent in the reservoir and minimize the delay
time when the pump is first started. During operation, stop-flow valves
maintain the pump operating pressure between injections or column
depressurization, etc. Syringe pumps require little maintenance, can operate at
high pressures and flow rates, and may be used for flow or gradient
programming. Their principal disadvantages are high cost and limited solvent
reservoir capacity (250-500 ml).

The most commonly used pumps in HPLC are of the single- or multi-head
reciprocating type. The single piston version uses an eccentric cam to drive a
piston in and out of a low volume chamber (30-1000 microliters). The piston's
movement is synchronized with the operation of two check valves which control
the direction of liquid flow during the fill and pump sequences. The volumetric
output of the pump is controlled by changing the length of the piston stroke or
the frequency of the piston movement. The flow output is delivered as a series
of pulses which must be damped to prevent interference in the detector
operation. Pulse dampers will be described later. Dual-head and triple-head
reciprocating pumps are more expensive than single-head pumps but minimize
pulsation and may be operated without a pulse damper. With a dual-head
reciprocating pump both chambers are driven by the same motor through a common
cam such that the two pistons are 180° out of phase. As one chamber is pumping
the other is refilling; thus the two flow profiles overlap, leading to partial
cancellation of the peaks and troughs in the total flow output. Likewise, with
the triple-head pump, the three pistons are 120° out of phase with one another
and the cancellation of flow pulsations is improved accordingly.

Reciprocating pumps deliver a constant flow at a fixed back pressure. At
high pressures some minor flow variability may arise due to the compressibility
of the mobile phase. Some instruments incorporate a flow controller which
provides a fixed back pressure for the pump to work against, independent of
column back pressure. The influence of pressure fluctuation, solvent
compressibility, and solvent viscosity on the volumetric output of the pump are
thereby eliminated. Reciprocating pumps can provide continuous solvent
delivery, fast solvent changeover, gradient elution compatibility, and have low
maintenance requirements.

The pump may be linked to other components which control or measure the

operating pressure, volumetric flow, and flow fluctuations in its output.
Pressure- or electronically-actuated check valves are commonly used to control
the direction of flow. The volumetric flow may be measured by a pressure
transducer which monitors the pressure drop across a calibrated flow
restrictor. This ensures that the volumetric output from the pump is constant
by operating in a feed-back loop system. Alternatively, the volumetric flow can
be measured at the detector exit using an electro-pneumatic flow controller
[24]. This device employs the bubble flow tube principle in which the flow rate
is measured electronically as the time required for a bubble introduced into the
flow stream to traverse a certain distance. For pumps that produce a pulsating
output a pulse damper is required. Pulse dampers work in a manner analogous to
that of a capacitor in an electrical circuit. The pulse damper stores energy
during the pressurizing stroke and releases it during the refill stroke of a
reciprocating pump [25]. Pulse damping has been achieved using a variety of
mechanical devices such as syringes, bellows, and coiled tubes, often in
conjunction with a fluid or gas ballast reservoir. An inexpensive pulse damper
can be prepared from a length of flattened stainless steel tubing of
significantly smaller cross-section than the pressure peak profile of the mobile
phase and a Bourdon tube pressure regulator [25]. A commercially available
pulse damper uses a flattened length of Teflon tubing immmersed in a degassed
compressible liquid. The flexibility of the tubing and the compressibility of
the fluid thus absorb any pressure fluctuations in the solvent stream. All
instruments should contain a pressure sensing device to prevent
overpressurization which can result in damage to the system. In the simplest
case this might be a fixed-pressure release valve or a Bourdon-type pressure
gauge. In more sophisticated equipment a strain-sensitive semiconductor crystal
whose resistance varies with the applied pressure is generally used [26].

5.5 Gradient Devices

Samples having a wide range of capacity factor values are not conveniently
separated using isocratic conditions. Gradient elution is frequently used in
this case. A mobile phase gradient is formed by mixing two or more solvents
either incrementally or continuously, according to some predetermined gradient
shape. The most frequently used gradients are binary solvent systems with a
linear, concave, or convex increase in the percent volume fraction of the
stronger solvent. The gradient shapes mentioned above can be described by
simple mathematical functions [27].

Linear gradient : $\theta_B = t/t_G$ (5.3)

Convex gradient : $\theta_B = 1 - (1-t/t_G)^n$ (5.4)

Concave gradient : $\theta_B = (t/t_G)^n$ (5.5)

θ_B = volume fraction of stronger eluting solvent
t = time after the gradient started
t_G = total gradient time
n = integer > 0, controls gradient steepness

If the gradient commences from a composition in which θ_B is not equal to zero, then θ_B in equation 5.3 to 5.5 is replaced by

$$\frac{(\theta_B - \theta_B{}^\circ)}{(\theta_B{}^f - \theta_B{}^\circ)} \tag{5.6}$$

$\theta_B{}^\circ$ = initial volume fraction of θ_B
$\theta_B{}^f$ = final volume fraction of θ_B

The basic requirement of a gradient device can be stated as follows:

It should have the ability to produce a reasonable choice of useful gradient profiles.

It must provide homogeneous mixing of the mobile phase before it reaches the column.

It should be able to create reproducible and accurate gradients over the full solvent mixing range (0-100% B).

It should minimize the time delay between the points at which the solvents are combined and delivered to the column.

Gradient devices are usually classified into two types, Figure 5.3, depending on whether the solvents are mixed on the low or high pressure side of the pump. Many ingeneous low-pressure mixing systems, consisting of a single pump and multiple reservoirs feeding a dilution vessel, have been described [28,29]. These systems generally do not allow rapid solvent changes and are unwieldly in operation. Modern instruments use time-proportioning electrovalves, regulated by a microprocessor or similar device, to control solvent delivery to the pump [30,31]. To obtain the highest possible accuracy the operation of the electrovalves should be synchronized with the movements of the pump system. This gradient architecture is little influenced by solvent compressibility effects and can completely eliminate errors connected with thermodynamic volume changes due to mixing of the solvents.

In high-pressure mixing devices each solvent is pumped separately in the proportions required by the gradient into a mixing chamber before being delivered to the column. Solvent compressibility and thermodynamic volume changes on mixing may influence the accuracy of the composition delivered to the column. A significant disadvantage of the high-pressure mixing architecture is the need for a separate pump for each solvent mixed.

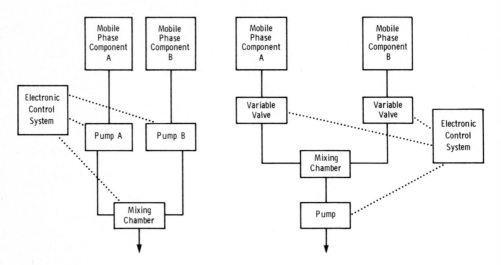

Figure 5.3 Two methods for generating binary solvent gradients.

The characteristics and pump requirements for gradient mixers were described in two useful papers [32,33]. Pneumatic or hydraulic amplifier pumps produce a variable flow output with time due to changes in the compressibility and viscosity of the mobile phase as it is mixed. Thus, unless a precise feedback flow-control device is incorporated into the pumping system, pumps of this type do not normally provide accurate gradients. The accuracy of gradients formed with syringe pumps may be strongly influenced by solvent compressibility effects when solvents of different compressibilities are mixed. Operating both pumps at a constant pressure greater than the column backpressure will generally suppress this effect. A back pressure valve is used for pressure control. In general, constant-volume reciprocating or diaphragm pumps are preferred for gradient mixing. Reciprocating pumps yield very accurate solvent mixing at volumes close to equal mixing. However, without flow-feedback control the limited range of pump speeds prevents accurate mixing of solvents at either extreme of the composition limits. Gradients containing less than 10% v/v of either solvent should normally not be mixed with reciprocating pumps.

The variability in retention volumes and peak widths in gradient elution chromatography is due primarily to the limits of precision and reproducibility of the mobile phase composition, the flow rate, and the column temperature

[33]. The most important of these are changes in the mobile phase composition and flow rate, which can arise from either column effects or instrument malfunction. The ability of the gradient former to generate a preselected gradient can be established by substituting a length of crimped capillary tubing for the analytical column and connecting this directly to the detector. If one of the solvents is UV-absorbing (e.g., acetone) and the other transparent (e.g., water, hexane) then the gradient profile can be reproduced on a strip chart recorder. Alternatively, if a conductivity detector is available, the gradient shape can be evaluated with the column in place by using water and a solution of potassium bromide as the binary mobile phase components [32]. Deviations in the actual gradient from the theoretical one indicate instrument malfunction or poor design. It should also be noted that a delay occurs between initiating the gradient program and when the new mobile phase reaches the column. This delay time depends on the flow rate and corresponds to a volume of 1-10 ml, comprising the volume elements of the pump (chamber, damper, flow controller volumes) and solvent mixing components. Ideally, this volume should be small for rapid gradient response. In practice, a compromise is made between the operational requirements of the pump and the desired gradient performance. It is important that the column dead volume and the gradient delay volume are known to accurately relate solvent composition to the elution of peaks in the gradient chromatogram. This is usually a requirement when gradient elution is used to scout for optimum isocratic separation conditions.

Assuming that the instrument gradient is perfect, substituting a column for the crimped capillary may show some deviation from the instrument gradient at the low percent volume composition end of the stronger solvent. With a reversed-phase column this is readily seen by repeating the water-acetone gradient. Deviations from the instrument gradient occur because of solvent demixing, due to preferential adsorption of the stronger eluting solvent onto the stationary phase at low mobile phase concentrations. Solvent demixing is particularly noticeable in liquid-solid chromatography where solvent localization of polar modifiers occurs readily. Solvent demixing causes large variations in the retention of early eluting compounds.

Random or systematic deviations from the preset mobile phase volume composition and flow rate are caused by imperfect functioning of the mechanical parts of pumps (plungers, valves, seals) or the electronic parts of the system. The errors observed in delivered isocratic composition, long and short term composition stability, and flow rate stability for several commercial instruments were found to be less than desirable [32]. Some instruments did not perform as well as others when solvents of different compressibility or viscosity were used to form the gradient. Deviations as large as 10% in the isocratic solvent volume composition were found for some of the pumps tested.

Dual-piston pumps with a flow-feedback system were shown to be the most reliable and accurate in generating gradients closely matching those predicted by theory.

5.6 Injection Devices

The ideal sample introduction method should reproducibly and conveniently insert a wide range of sample volumes into the column as a sharp plug without adversely affecting the efficiency of the column. These goals are met to varying extents by the following injector types: septum, septumless, stop-flow, and valve injection, Figure 5.4. Currently, microvolume sampling valves are the most frequently used devices. Automatic sample injectors are available for unattended and overnight operation.

Septum injectors, analogous to those used in gas-liquid chromatography, permit sample introduction by a high pressure syringe through a self-sealing elastomer septum. Several designs are available, including injection ports which permit sample splitting [35] and those with sliding valves to minimize contact between the septum and mobile phase [36-38]. A design employing a double septum separated by a metal spacer can be operated at pressures up to 3100 p.s.i. [38], although an upper pressure limit of 1000 - 2000 p.s.i. is more common. High pressure syringes will usually withstand pressures up to approximately 1500 p.s.i. Syringe injectors are designed so that the syringe needle just reaches the top of the column packing or, more usually, a stainless steel or PTFE screen at the head of the column. Injecting into the column packing itself may result in damage to the syringe or disrupt the column bed, leading to a loss in column efficiency. The dead volume of the injector above the column must be minimized; any poorly swept volumes into which the sample may back diffuse will result in a substantial reduction of column efficiency. Other problems with syringe injection include sample leakage (back flushing) around the syringe needle at the time of injection, septum decomposition, and the generation of ghost peaks. Septum material deposited at the top of the column may cause a loss of column permeability, efficiency, and affect retention time by partially adsorbing some sample components. Except for those designs which incorporate a sliding valve to isolate the septum from the mobile phase after injection, ghost peaks caused by adsorption of the sample to the septum or leaching of the septum material can be a problem. The reproducibility of syringe injection is rarely better than 2% and, in general, syringe injection is more troublesome than in gas chromatography.

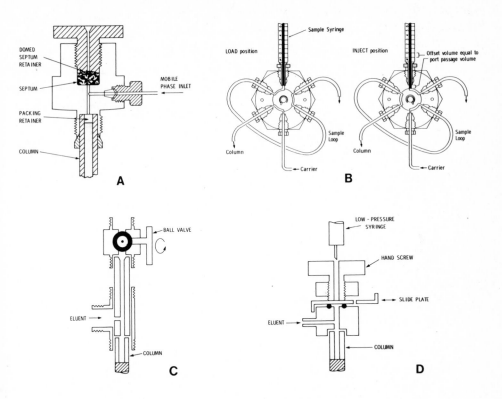

Figure 5.4   Methods of sample introduction. A, septum injector; B, valve
             injector; C and D, septumless stop-flow injection devices.

The problems associated with septum injectors can be eliminated by using
septumless injectors and stop-flow injection [39]. Usually a system of
isolation valves is used to interrupt the flow of mobile phase prior to the
injector and permit injection by syringe at approximately atmospheric pressure
through a second valve. Reversing the order in which the valves are switched

completes the injection and restores the mobile phase flow. Column efficiency is not affected by the stop-flow process since diffusion in liquids is very slow. The time required to resume full pressure after injection must be insignificant compared to the time required to sweep out the column void volume or band broadening of early eluting peaks will occur. Stop-flow injection is most easily achieved with constant-pressure pumps, although constant-volume pumps and a flow diversion valve may also be used. The introduction of air into the system during injection may disrupt the detector baseline.

Microvolume sampling valves are widely used for injection as they enable samples to be reproducibly introduced into pressurized columns without significant flow interruptions. They also provide the possibility of injection at elevated temperatures. The sample is loaded at atmospheric pressure into an external loop or groove in the valve core and introduced into the mobile phase stream by a short rotation of the valve. The volume of sample injected is normally varied by changing the volume of the sample loop or by filling a fixed volume sample loop with a combination of solvent and sample. Special variable-volume sample valves are also available [40]. Valve injections are essentially operator independent and very reproducible with injection errors of less than 0.2%.

A comparison of syringe stop-flow and valve injection indicated that a small increase in column efficiency was observed for the syringe technique [41]. A further small increase in performance was observed when the injector was modified to provide introduction of the mobile phase to the column in a curtain-flow arrangement [42]. The curtain-flow arrangement is achieved by diverting a portion of the mobile phase flow to the injection point and constraining it to flow annularly around this point, limiting sample diffusion to the column wall (see Chapter 4, Figure 4.8D). The curtain-flow technique also improves column performance slightly when valve injection is used.

With valve injection, the column efficiency is degraded slightly by the method of coupling the valve to the column [43]. The length of the connecting tube should be short, 5 cm is an acceptable and convenient length, and of small internal diameter, 0.15 or 0.50 mm. It was also found that maximum column efficiency resulted when the valve-to-column connection at the column head was made with a low dead-volume fitting and the column-to-detector connection with a zero dead-volume fitting. This is presumably due to an optimization of the flow pattern at the column injection end as the coupling arrangement does not conform to the minimum dead volume situation that might have been expected.

5.7 Detectors for Liquid Chromatography

Separations in liquid chromatography occur in a dynamic manner and therefore require detection systems which work on-line and produce an instantaneous record of the column events. Suitable detectors are generally considered to belong to either of two classes: bulk property or solute property detectors. Bulk property detectors measure the difference in some physical property of the solute in the mobile phase compared to the mobile phase alone. Examples of such physical property detectors are the refractive index, dielectric constant, and conductivity detectors. Bulk property detectors are fairly universal in application but generally have poor sensitivity and a limited dynamic range. This arises primarily from the fact that the magnitude of the detector signal does not depend solely on the properties of the solute but on the difference in properties between the solute and the mobile phase. Thus bulk property detectors are usually adversely affected by small changes in the mobile phase composition and temperature, which usually precludes the use of such techniques as flow programming or gradient elution.

Solute property detectors respond to a physical or chemical property of the solute which, ideally, is independent of the mobile phase. Examples of such detectors are spectrophotometric, fluorescence, and electrochemical detectors. Although mobile phase independence is rarely met in practice, the signal discrimination is usually sufficient to permit operation with solvent changes (e.g., flow programming, gradient elution) and to provide high sensitivity with a wide linear response range. Solute-specific detectors complement bulk property detectors as they provide high sensitivity and selectivity. Of course they do not have universal solute detection characteristics so that several detectors may be required to meet the demands of different problems.

The detection process in liquid chromatography has presented more problems than in gas chromatography. There is no equivalent to the sensitive, universal flame ionization detector for use in liquid chromatography. Indeed, the search for a liquid chromatographic detector with high sensitivity, wide linear dynamic range, good linearity, predictable response, and the capability of detecting all solutes equally seems to have run full course and this goal may never be reached. Current developments are directed towards the development of specific detectors for particular problems. Space and emphasis preclude a discussion of all the detectors available for use in HPLC and the interested reader is directed to specific reviews of detectors [44-46] and to the monograph by Scott [13]. The primary detectors -- refractive index, spectrophotometric, fluorescence and electrochemical will be described in the following sections.

5.7.1 Refractive Index Detectors

Refraction, the ability to bend light, is a property of all molecules. The range of refractive index values for most organic compounds is relatively small, providing a universal, if generally insensitive method of detection. Detection limits in favorable circumstances are on the order of 1 microgram. Refraction is also a property of the mobile phase and consequently all detectors work on the differential principle, comparing the refraction of the mobile phase in the reference cell to the mobile phase and solute in the sample cell [13,14]. Differential operation is required for detector stability as detector noise is very much influenced by the constancy of the mobile phase composition, temperature, and pressure. As small changes in the mobile phase composition cause excessive detector drift, gradient elution is not possible even with differential operation. Because the detector responds to the difference in the refractive index between the mobile phase and the mobile phase containing sample, it is normal for the detector to respond in both the positive and the negative directions for different components in a mixture. The magnitude of the detector signal is governed by the refractive index difference between the sample and the mobile phase and is thus dependent on the selection of the mobile phase. Compounds having the same refractive index as the mobile phase are not detected.

There are three types of refractive index detectors available for HPLC: the deflection-refractometer, Fresnel-refractometer and the interferometer detector. The deflection-refractometer is shown in Figure 5.5A. It senses the deflection of a light beam that passes through a triangular shaped cell (Snell's law). A beam of light is defined by a mask B, collimated by lens C, and then passed through the sample and reference cell twice, being reflected back in the direction of travel by mirror E. The light beam is detected by a position-sensitive photodetector which measures the change in angular position as the refractive index changes. Deflection-refractometers have the advantage of a wide linear range, the entire refractive index range can be covered with one prism, small detector cells can be fabricated, and the cells are relatively insensitive to air bubbles or the buildup of contaminants on the sample cell windows.

The Fresnel-refractometer uses a differential photodiode to measure the change in the amount of light reflected on passage through a glass liquid interface, Figure 5.5B. The reference cell compensates for mobile phase variations and for variations in the source output. The difference in intensity of the light reflected in the two beams is related to the refractive index difference between the sample and reference cells by Fresnel's law and has a fairly large linear range when the incident light strikes the cell near the critical angle. Small cells (3 microliters) are easily fabricated for this detector but two prisms may be required to cover the full range of refractive index values encountered for HPLC solvents. Advantages of this design are high sensitivity, compatibility with low column flow rates, and ease of cleaning.

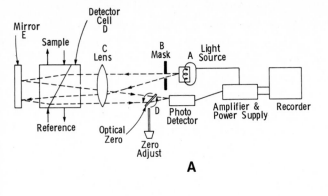

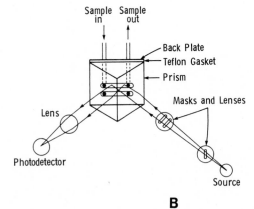

Figure 5.5 Refractive Index Detectors. A, deflection-refractometer;
B, Fresnel-refractometer.

A new type of refractometer detector uses the shearing interferometer principle. The difference between the refractive indicies of the sample and reference streams produces a difference in optical path length which is measured by the interferometer in fractions of light wavelengths. An order of magnitude increase in sensitivity compared to conventional detectors is claimed by the manufacturers.

The refractive index of a compound is a function of its molecular density and is sensitive to changes in sample concentration, solvent composition, pressure, and temperature. The background detector noise will be influenced by all these parameters as well as by the stability of the detector and recorder electronics. Some of these parameters, for example the constancy of the mobile phase composition and operating pressure, are dependent on the operating function of the chromatograph. A concentration of 1 ppm corresponds to a refractive index change of about 10^{-7} RI units, which could easily be exceeded by transient changes in the mobile phase composition. A change in pressure of

one atmosphere will cause a change of about 10^{-6} RI units. However, the largest
variations are associated with temperature fluctuations where a 1°C difference
can cause a change of 6 x 10^{-4} RI units. Commercial detectors are usually
fitted with an internal heat sink, normally a length of narrow bore tubing in
the inlet side of the cell, to stabilize the temperature of the mobile phase.
Even with dual beam operation, temperature control to ± 10^{-4}°C is desirable for
high sensitivity detection. The heat sink may contribute a dead volume of
20-100 microliters to the chromatographic system.

5.7.2 Spectrophotometric Detectors

UV-visible absorption detectors are the most widely used detectors in
liquid chromatography. As most organic compounds have some useful absorption in
the UV region of the spectrum, these detectors are fairly universal in
application, although sensitivity depends on how strongly the sample absorbs and
the availability of a transparent mobile phase at the wavelength of maximum
absorption. Table 5.4 summarizes some typical molar absorptivity values for
common functional groups. The table can be used as an aid in predicting
wavelength maxima and detector sensitivity. It also indicates that the most
universal wavelength region for detection is 180-210 nm, which in practice, is
often difficult to access due to problems in finding transparent solvents for
the separation.

The operation of spectrophotometric detectors is based on the measurement
of the absorbance of monochromatic light by the sample in accordance with the
well-known Beer-Lambert law. Most detectors provide an output in absorbance
units which is linearly related to sample concentration over a range of 10^4 to
10^5. Detection limits are in the low to subnanogram range in favorable
circumstances.

There are several types of spectrophotometric detectors commercially
available. Single wavelength detectors with a low pressure mercury source have
a high energy output at 254 nm. These detectors are simple, rugged, and
inexpensive. Using a highly regulated source, dual beam operation, and silicon
photodiode detectors, a very high signal-to-noise ratio can be achieved. Their
principal limitation is that the sample must have some absorption at the
operating wavelength of 254 nm to be detected. Multiwavelength versions of the
above detector are available, in which a medium pressure mercury lamp and a
series of filters or a phosphor screen are used to provide several selectable,
fixed wavelengths. Phosphors re-emitting 254 nm light at 280 nm are commonly
used. One detector design allows the two wavelengths, 254 and 280 nm, to be
used simultaneously. Using mercury discharge lamps and filters the wavelengths
220, 254, 280, 313, 334, and 365 nm may be selected for detection. However, the

TABLE 5.4

REPRESENTATIVE MOLAR ABSORPTIVITY VALUES FOR SOME COMMON FUNCTIONAL GROUPS

Compound Type	Chromophore	Wavelength(nm)	Molar Absorptivity
Acetylide	$-C{\equiv}C-$	175–180	6,000
Aldehyde	$-CHO$	210	1,500
		280–300	11–18
Amine	$-NH_2$	195	2,800
Azido	$C=N$	190	5,000
Azo	$-N=N$	285–400	3–25
Bromide	$-Br$	208	300
Carboxyl	$-COOH$	200–210	50–70
Disulfide	$-S-S$	194	5,500
		255	400
Ester	$-COOR$	205	50
Ether	$-O-$	185	1,000
Iodide	$-I$	260	400
Ketone	$C=O$	195	1,000
		270–285	15–30
Nitrate	$-ONO_2$	270	12
Nitrile	$-C{\equiv}N$	160	–
Nitrite	$-ONO$	220–230	1000–2000
		300–400	10
Nitro	$-NO_2$	210	strong
Nitroso	$-N=O$	302	100
Oxime	$-NOH$	190	5,000
Sulfone	$-SO_2$	180	–
Sulfoxide	$S{-}{>}O$	210	1,500
Thioether	$-S-O$	194	4,600
		215	1,600
Thioketone	$C=S$	205	strong
Thiol	$-SH$	195	1,400
Unsaturation, conjugated	$-(C=C)_3-$	260	35,000
	$-(C=C)_4-$	300	52,000
	$-(C=C)_5-$	330	118,000
Unsaturation, aliphatic	$-C=C-$	190	8,000
	$-(C=C)_2-$	210–230	21,000
Unsaturation, alicyclic	$-(C=C)_2-$	230–260	3000–8000
Miscellaneous compounds	$C=C-C{\equiv}C$	291	6,500
	$C=C-C=N$	220	23,000
	$C=C=C=O$	210–250	10,000–20,000
		300–350	weak
	$C=C-NO_2$	229	9,500
Benzene		184	46,700
		202	6,900
		255	170
Diphenyl		246	20,000

incident energy of these lines is only a fraction of the source energy at 254 nm
for a low pressure mercury source, and the signal-to-noise ratio and sensitivity
is somewhat lower than expected.

The most generally useful detector in HPLC is the continuously variable
multiple wavelength detector. Wavelengths are either selected manually in the
range of 190-900 nm or, in an automated version, the detector may be programmed
to change wavelengths during the separation and, with stop flow techniques, scan
the complete spectrum of any peak. These detectors use high energy sources,
relatively low optical resolution (i.e., wide bandpass monochromators), and
stable, low noise electronics. An example of a multiple wavelength detector is
shown in Figure 5.6. In general, these detectors produce a signal-to-noise
ratio slightly lower than the fixed wavelength detectors but are not limited in
wavelength selection.

Careful consideration must be given to the design of the detector cell as
it forms an integral part of both the chromatographic and optical systems. The
detector cell volume should be as small as possible to reduce contributions to
extracolumn band broadening. Conversely, a large illumination volume is desired
for high sensitivity. In practice, cell design is a compromise between these
two extremes. The H-cell, Z-cell, and the tapered cell are the most common
designs (Figure 5.6) [48]. Relating the design of the cell to the measurement
optics is complex and will not be discussed here [49-51].

Absorbance measurements in HPLC are influenced by refractive index effects
associated with a nonhomogeneous refractive index throughout the cell. This
arises from temperature gradients, incomplete mixing, and turbulence in the
mobile phase. Their effects cause refraction of some of the incident light to
the cell walls, where it is not detected and thus mistakenly taken to represent
sample absorption. The tapered cell shown in Figure 5.6 is designed so that,
within the refractive index range of mobile phases used in HPLC, no light is
sufficiently refracted to reach the cell wall (the liquid lens shown in the
figure is inserted to model the effect of refraction and is not a physical part
of the cell). Alternatively, for cells of a conventional design, an optical
system with aperture and field stops external to the cell is used to minimize
refraction.

It should also be noted that most spectrophotometric detectors are of the
double-beam type. This compensates for changes in the source output and permits
some background correction for those solvents having a low absorbance at the
measuring wavelength. This is not effective for gradient elution where
background correction is best done electronically.

The variable wavelength detectors can enhance the information gained about
a sample by providing spectroscopic evidence to augment retention volume data.
The coupling of two fixed wavelength detectors in series to monitor the

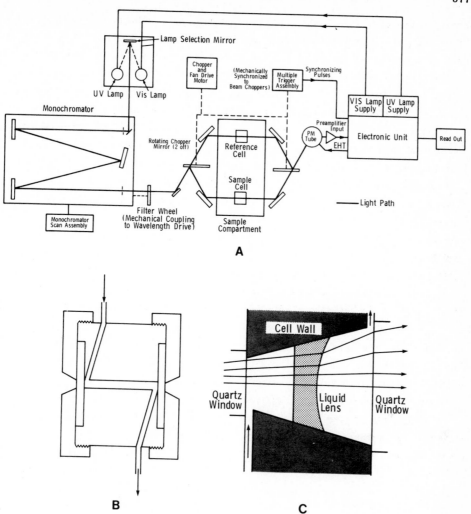

Figure 5.6 Multiple wavelength spectrophotometric detector. A, optical
diagram; B, Z-cell; C, tapered cell.

separation at two different wavelengths provides an absorbance ratio value for
each peak [52]. Baker et al. studied the separation of 101 forensic drugs by
HPLC [53]. Based on retention times alone only 9% of the drugs could be
positively identified, compared to 95% when both retention times and their
absorbance ratios at 254 and 280 nm were used. With a variable wavelength
detector the column flow can be stopped, arresting the peak in the detector,
which is then scanned manually or automatically to produce an absorbance
spectrum. This technique is known as stop-flow scanning [54-56]. The optimum
wavelength for detection and characteristic wavelength ratios for qualitative

identification are easily obtained from the spectrum. Measuring the detector peak ratios at several positions in a peak profile is a useful method of evaluating peak homogeneity and identifying the presence of unresolved contaminants. It is usually possible to measure wavelength ratios with better than 2% precision under favorable circumstances. For variable wavelength detectors set manually this may require wavelength resetability of ± 0.1 nm, a value probably not attainable in practice [57]. Precision of the measurements is also influenced by baseline location, which needs to be constantly reset as the wavelength or mobile phase is changed. It may require several minutes to perform these measurements manually but, owing to the slow diffusion of components in the mobile phase, the separation is not degraded while the flow is stopped.

In general, the absorbance spectra obtained by stop-flow techniques provides little diagnostically useful information about an unknown component unless a standard is available for comparison. The detectors used in HPLC provide low resolution spectra essentially devoid of fine structure, precluding a comparison to common spectral indicies generally recorded under high resolution conditions.

Obtaining a single spectrum by stop-flow scanning requires at least a few minutes and becomes quite tedious when more than a few peaks are sampled. Rapid-scanning multiwavelength detectors using oscillating mirrors, solid-state diode arrays or silicon-target Vidicon tubes have beed described [58-61]. With the aid of a computer or microprocessor these systems can generate a complete spectrum for a peak during its passage through the detector. The linear photodiode array detector is now commercially available and will undoubtedly grow in popularity [61].

5.7.3 Fluorescence Detectors

Only a small percentage of all inorganic and organic compounds are naturally fluorescent. However, many biologically active compounds, pharmaceutical products, and environmental contaminants are fluorescent which, combined with the very high sensitivity possible in fluorescence detection, explains its widespread use. The inherent sensitivity and selectivity of fluorescence detection has generated a great deal of interest in fluorescence labeling (precolumn derivatization) and reaction flow detectors, (postcolumn derivatization) for the trace analysis of non-fluorescent or weakly fluorescing compounds. This has furthered the interest and application of fluorescence detection in HPLC.

When a polyatomic molecule absorbs a quantum of light, an electron is promoted from the lowest vibrational level of the ground state to higher energy electronic states in accordance with the selection rules of quantum mechanics. Following this absorption of energy, the molecule relaxes to the lowest vibrational level of the first excited singlet electronic state by radiationless routes (internal conversion, vibrational relaxation, etc.), at which point a radiative transition to the ground electronic state occurs with the production of fluorescence. Fluorescent transitions result in band spectra as the electronic energy levels involved are not discrete but split by coupling with vibrational energy levels. The excitation spectrum, corrected for wavelength-dependant instrument parameters, is a replica of the absorption spectrum of the compound. For detection purposes, operation at the wavelength corresponding to the intensity maxima in the excitation and emission spectra provides maximum sensitivity. At other wavelengths the signal may be reduced in intensity but still adequate for detection purposes. This is virtually always the case when multiple component samples are separated by HPLC using fixed wavelength fluorescence detectors.

The selectivity of the fluorescence detection process arises because two wavelengths are used in the measuring process and from the fact that certain structural features are required in a molecule for fluorescence to occur. The molecular requirements for fluorescence cannot be clearly defined but many fluorescent molecules contain rigid, planar, conjugated systems. Fluorescence is a very sensitive technique since, in contrast to absorption measurements, the emitted radiation is measured against a dark background. This dark background is due to the use of filters or monochromators to isolate the emission signal from the excitation background, and is enhanced by arranging the measuring optics off-axis (usually 90°) to the excitation beam. In a well-designed instrument sensitivity is essentially limited by scattered light from the sample in solution (Rayleigh, Tyndall, and Raman) and stray light originating from the measuring cell and instrument optics. These contributions constitute the principal sources of detector noise.

For dilute solutions (< 0.05 A.U.) the measured fluorescence intensity can be related to sample concentration by equation (5.7).

$$I_f = I_o \, \theta_f \, (2.3 \, \varepsilon l c) \tag{5.7}$$

I_f = fluorescence emission intensity

I_o = excitation beam intensity

θ_f = quantum yield = $\dfrac{\text{number of photons emitted}}{\text{number of photons observed}}$

ε = molar extinction coefficient
l = path length
c = sample concentration

For fluorescence detectors used in liquid chromatography the above equation is generally linear for a concentration range of 2-3 orders of magnitude. Detection limits of 1 picogram may be obtained under favorable circumstances. Sensitivity depends on the instrument (I_o and the reduction of scattered and stray light), the sample (quantum efficiency), and the composition of the mobile phase (solvents, impurities, etc.) [62].

Detectors available for fluoroscence monitoring in HPLC differ mainly in the method used to isolate the excitation and emission wavelengths. Either filters or monochromators are used for this purpose. A block diagram of a simple filter flurorescence detector is shown in Figure 5.7 [63]. As the signal intensity is directly proportional to the source intensity, high energy line (mercury) or continuous (tungsten, deuterium, xenon) sources are used. Mercury sources are very reliable and reasonably intense, producing a series of UV line spectra superimposed on a weak continuum. Thus only certain wavelengths are available for excitation and may not overlap with the maximum excitation wavelength of the sample. For continuously variable selection of the excitation wavelength deuterium (190-400 nm), tungsten (380-700 nm), and xenon (200-850 nm) arc sources of a wide range of intensities are used. As arc sources are inherently more unstable than discharge sources a method of compensating for long term drift is required. In Figure 5.7 this is achieved by splitting off part of the excitation energy to a reference detector, electronically linked to the sample detector, which corrects for fluctuations in the source intensity. Wavelength isolation is performed using filters or diffraction grating monochromators. Cut-off filters provide low selectivity since they transmit all light above a certain wavelength. Interference filters are used to obtain a narrow bandpass (10-20 nm). Monochromators having a narrow bandpass, provide continuous wavelength selection and, with stop-flow scanning, can provide an emission spectrum for peaks of interest. The detector shown in Figure 5.7 is unusual in that it enables the emission wavelength to be varied by scanning a continuously variable interference filter past a vertical slit [63]. Cut-off filters generally provide a greater signal than is available in the narrow wavelength interval selected by a monochromator [64,65]. However, this signal enhancement is obtained at the expense of a loss in selectivity.

The comments made earlier concerning cell design for spectrophotometric detectors also pertain to fluorescence detectors. Figure 5.7 shows a diagram of a unique cell which uses a 2π steradian sample cuvette to maximize the fluorescence emission signal collected. Sample fluorescence is generated in all directions simultaneously and the high light gathering power and low cell volume of this design are good features. Detector cells with volumes of 3-5 microliters have only recently become available and many older instruments contain cells which are larger than is desirable for use with modern, high

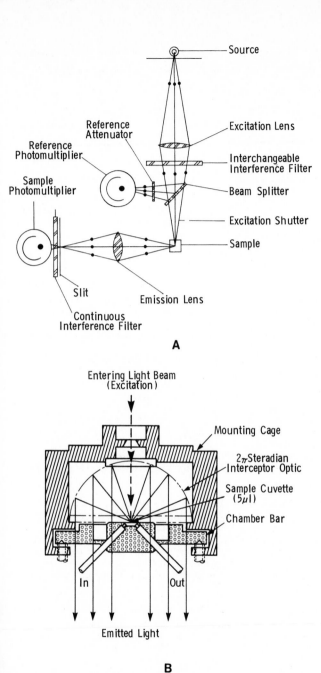

Figure 5.7 Fluorescence detector. A, optical diagram; B, 2π steradian cuvette.

efficiency HPLC columns.

It is frequently overlooked that the fluorescence signal from a sample may be dramatically affected in both wavelength and emission intensity by the mobile phase composition and even by the presence of contaminants in the mobile phase. Some of these solvent effects are summarized in Table 5.5. Under less than ideal conditions the constancy with which the pump delivers and mixes the mobile phase and the presence of contaminants in the mobile phase may influence detector sensitivity and reproducibility more than fluctuations in the detector operating systems. Fluorescence detection can be used with gradient elution and, unless the mobile phase contains a high level of fluorescent impurities, the detector baseline changes very little during the solvent program.

TABLE 5.5

THE INFLUENCE OF MOBILE PHASE COMPOSITION ON FLUORESCENCE EMISSION

Mobile Phase Parameter	Effect on Fluorescence Emission
pH	Both the emission wavelength and fluorescence intensity of ionizable aromatic compounds (i.e., those containing acidic or basic functional groups) are critically dependent upon pH and solvent hydrogen-bonding interactions.
Solvent	Intensity changes of an order of magnitude and large wavelength shifts are found for molecules which can undergo strong solvent interactions. A shift in the fluorescence spectrum to longer wavelengths is usually observed as the dielectric constant of the solvent is increased. If the solvent absorbs any of the excitation or emission energy, the sensitivity will be reduced.
Temperature	Many compounds show marked temperature dependence with increasing temperature, causing a decrease in intensity of 1-2% per °C.
Concentration	At high concentrations fluorescence emission becomes non-linear due to self-absorption by the sample itself or complete absorption of the excitation energy before it reaches the cell center. Usually occurs at >0.05A. High fluorescence intensity may overload the photomultiplier tube which returns slowly to its normal operating conditions and misrepresents the actual fluorescence signal until restabilized.
Quenching	Impurities in the mobile phase, particularly oxygen, may entirely quench the signal from low concentrations of fluorescent compounds. (see solvent degassing, discussed elsewhere).
Photodecomposition	High intensity sources may cause sample decomposition, which depends on the residence time of the sample in the detector cell.

Recently, lasers have received a great deal of attention as sources for fluorometric detection in HPLC [66]. Lasers can provide higher power intensity than conventional sources and their inherent spatial coherence and

monochromaticity allows all of the radiation to be efficiently utilized in a
small detection volume. When the cost ratio of lasers to conventional sources
become more reasonable, and the selection of operating wavelengths less limited,
then laser fluorometric detection will be very attractive for HPLC
applications.

5.7.4 Electrochemical Detectors

Electrochemical detectors include the conductivity detector for the
determination of ionic species and the amperometric, coulometric, and
polarographic detectors for the determination of compounds which can be '
electrolytically oxidized or reduced at a working electrode [67-70]. These
detectors are selective as not all organic compounds are conducting or
electrolyzable within the useable voltage range. Examples of compounds which
can be detected conveniently are phenols, mercaptans, aromatic nitro and halogen
compounds, catecholamines, heterocyclic compounds, ketones, and aldehydes. In
addition to the selectivity resulting from the electroactivity requirement, the
potential applied to the detector can be adjusted to discriminate between
different electroactive species. This enables one electroactive species to be
detected in the presence of another without interference. Other advantages of
electrochemical detectors are their reliability, simple, low volume cell design
(0.1-5 microliters), and high sensitivity. In favorable circumstances, picogram
detection limits may be obtained. Electrochemical detection requires the use of
conducting mobile phases containing inorganic salts or mixtures of water with
water-miscible organic solvents, conditions which are compatible with
reversed-phase and ion-exchange chromatography but are difficult to achieve with
other techniques. These problems can be circumvented to some extent if the
mobile phase is water miscible by adding a make-up flow of support electrolyte
at the column exit. Detector operation is critically dependent on flow rate
constancy, solution pH, ionic strength, temperature, cell geometry, the
condition of the electrode surface, and the presence of electroactive impurities
(dissolved oxygen, halides, trace metals). The background detector noise and
thus, the ultimate sensitivity of the detector, are controlled by dissolved
oxygen, ionic impurities, and the contamination of the electrode surface,
coupled with transient changes in the pump output. As the response of the
detector is critically dependent on the overall solution characteristics of the
mobile phase, gradient elution is not normally possible.

Amperometric detectors are based on thin-layer or tubular electrode cells.
Thin-layer cells are popular in practice, owing to the ease with which small
cell volumes and a variety of electrode materials can be used. Two examples of
thin-layer electrochemical cells are shown in Figure 5.8. The column eluent is

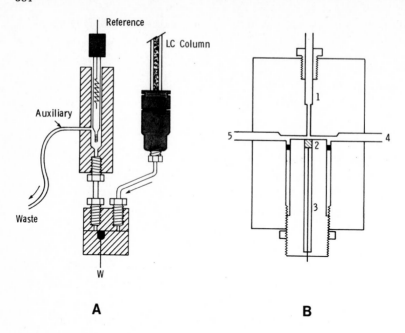

Figure 5.8 Thin-layer electrochemical amperometric detectors. A, thin-layer
detector; B, wall-jet detector.

introduced either parallel to the electrode embedded in the channel wall as in
Figure 5.8A, or perpendicular to the electrode surface followed by radial
dispersion, the so-called wall-jet detector. The wall-jet detector has high
sensitivity, can be adjusted to provide various detector cell volumes, and is
relatively free from surface adsorption problems. Carbon-paste or glassy carbon
materials are most commonly used for electrode materials [70]. The working
electrode is maintained at a fixed potential with respect to a reference
electrode, usually at or near the limiting current plateau for the compound of
interest. This may be determined from the current-potential recording of a
static sample by standard electrochemical procedures. The chromatogram is
recorded by measuring the detector cell current at a fixed potential as the
sample is eluted from the column. The background current will remain constant
as long as the mobile phase velocity and composition do not change and is
subtracted from the analytical signal. The resulting detector current is
directly proportional to the concentration of electroactive species in
accordance with Faraday's law.

When amperometric detection is used, the electrolysis of the electroactive
species is not complete, typically 1–10%. By increasing the electrode surface
area the electrolysis may reach 100%, at which point the coulometric limit is
reached. The coulometric detector is insensitive to flow rate and temperature

changes and responds in an absolute manner, eliminating the need for calibration. However, the coulometric detector is more prone to electrode contamination, presents greater design problems, and requires strict potential control over the entire working electrode surface. Although the conversion of electroactive species is much higher in the coulometric detector, the background noise is also greater than for the amperometric detector; thus both detectors exhibit similar sensitivity.

Microvolume dropping mercury polarographic detectors can be used in HPLC for the determination of electro-reducible species (the low oxidation potential of mercury precludes their use for oxidizable species). Surface contamination is rarely a problem as the electrode surface is continuously renewed. Disadvantages include high background currents due to current oscillations over the lifetime of the drop, turbulence caused by the liquid flow in the region of the drop, and the need for complex cell designs. Recent innovations in cell design and electronic damping of the signal have done much to domesticate the polarographic detector, although its general use is far less frequent then amperometric detectors. Sensitivity is similar to the amperometric detectors but its range of application is more limited.

Conductance is a fundamental property of ions in solution and exhibits a simple dependence on ion concentration. Thus the measurement of conductance is an obvious choice for the continuous and selective monitoring of ionic species in a column eluent. A typical detector cell consists of a low volume cavity or tube of insulating material, in which electrodes made of a noble metal or graphite are embedded. Generally, a constant alternating voltage is applied to the electrodes and the resistance is measured by a simple Wheatstone bridge circuit (the conductivity of a solution is equivalent to its specific resistance). The cell resistance is related to sample concentration by Ohm's law. Conductivity detectors are simple to operate and, in the absence of background electrolyte, provide high sensitivity (10^{-8} - 10^{-9} g/ml). Conducting impurities in the mobile phase limit absolute sensitivity.

5.7.5 Performance Characteristics of Liquid Chromatography Detectors

The detector performance characteristics of interest to the chromatographer are sensitivity, minimum detectability, dynamic range, response linearity, and noise characteristics. Other properties of the detection system which indicate its suitability for a particular problem are flow sensitivity and response time. Some definitions and formulas for calculating performance criteria are given in Table 5.6. It is convenient to divide chromatographic detectors into two groups: concentration sensitive devices which respond to a change of mass per unit volume (g/ml) and mass sensitive devices which respond to a change in

TABLE 5.6

PERFORMANCE CRITERIA FOR CHROMATOGRAPHIC DETECTORS

A = peak area, F = flow rate through the detector, w = sample weight, C_{max} = concentration of test substance corresponding to the upper limit of the linear range and M_{max} = mass flow rate corresponding to the upper limit of the linear range

Term	Definition	Concentration Detectors	Mass Detectors	Comments
Sensitivity	The signal output per unit mass concentration of test substance in the mobile phase.	$S = \dfrac{AF}{w}$	$S = \dfrac{A}{w}$	Of little value unless the noise level is also specified. S/N is the parameter of most interest.
Minimum Detectability	The amount of test substance that gives a detector signal equal to twice the noise.	$D = \dfrac{2N}{S}$	$D = \dfrac{2N}{S}$	Both S and N values should be determined in the same run.
Linearity	The linear range is the range of mass flow rates or concentration of test substance over which the sensitivity of the detector is constant to within ± 5.0% as determined from a linearity plot.	$L.R. = \dfrac{C_{max}}{D}$	$L.R. = \dfrac{M_{max}}{D}$	Sometimes expressed as the ratio of the upper limit of linearity obtained from the linearity plot to the minimum detectability, both measured for the same test substance.
Dynamic Range	The range of concentration or mass flow rate of the test substance over which a change in sample amount produces a change in detector signal.			The linear portion of the dynamic range is of most interest.
Noise	The amplitude of the baseline envelope which includes all random variations of detector signal of a frequency on the order of 1 cycle/min.			Refers to observed noise only. Actual noise may be larger or smaller since most instruments and recording devices employ some filtering.
Drift	This is the average slope of the noise envelope measured over a period of 0.5 h.			Usually expressed in signal units per hour.

mass per unit time (g/s). Most detectors used in liquid chromatography are of the concentration sensitive type and exhibit flow rate dependent sensitivity.

Detector noise is the ultimate limit to sensitivity. There are three characteristic types of noise (short term, long term, and drift) which may have different characteristics depending on whether it is measured under static or dynamic conditions (Figure 5.9). Static noise represents the stability of the detector when isolated from the chromatograph. Dynamic noise pertains to the normal operating conditions of the detector with a flowing eluent stream.

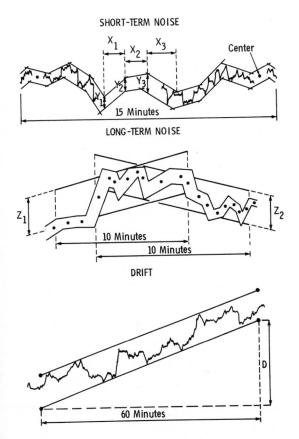

Figure 5.9 Methods for calculating short-term and long-term noise, and drift for chromatographic detectors.

Ideally, the static and dynamic noise should be very similar; the performance of the detector is otherwise being degraded by the poor performance of the solvent delivery system of the chromatograph. The noise signal is measured over a period of time on a recorder with the detector amplifer set to the maximum usable sensitivity. Short-term noise is the maximum amplitude for all random variations of the detector signal of a frequency greater than one cycle per minute. It is calculated from the recorder trace by dividing the detector output into a series of time segments less than one minute in duration and summing the vertical displacement of each segment over a fixed time interval using equation (5.8).

$$\text{short-term noise} = \sum_{R=1}^{R=N} Y_R/n \qquad (5.8)$$

(see Figure 5.9 for an explanation of symbols)

Long-term noise is the maximum detector response for all random variations of the detector signal of frequencies between 6 and 60 cycles per hour. The long-term noise is represented by the greater of Z_1 and Z_2 in Figure 5.9. The vertical distances Z_1 and Z_2 are obtained by dividing the noise signal into ten minute segments and constructing parallel lines transecting the center of gravity of the pen deflections. Long-term noise represents noise that can be mistaken for a late eluting peak. Drift is the average slope of the noise envelope measured as the vertical displacement of the pen over a period of 1 h (actually measured over 30 minutes and normalized to 1 h). For spectrophotometric detectors, the signal response is proportional to the path length of the cell and noise values are normalized to a path length of 1 cm [71,72]. Standard practices for measuring flow sensitivity and response time of spectrophotometric detectors are also detailed in references [71,72]. The dynamic range of the detector is determined from a plot of detector response or sensitivity against sample amount (mass or concentration). It represents the range of sample amount for which a change in sample size induces a descernible change in the detector signal (Figure 5.10). However, it is the linear range and not the dynamic range of the detector which is of most interest to the chromatographer. The linear range is the range of sample amount over which the response of the detector is constant to within 5%. It is usually expressed as the ratio of the highest sample amount determined from the linearity plot to the minimum detectable sample amount (Figure 5.10).

Unfortunately, most of the available detector performance criteria, emanating largely from the individual instrument companies, is not given in the above standard format. Although one can appreciate the reasoning for this it does not help the chromatographer make a sensible choice between detectors of the same kind or to compare the general properties of detectors of different types. The monograph by Scott contains the most comprehensive source of information concerning detector performance of commercially available equipment [13]. Some typical performance values for fixed wavelength spectrophotometric detectors are given in Table 5.7. Some comparative data for the most widely

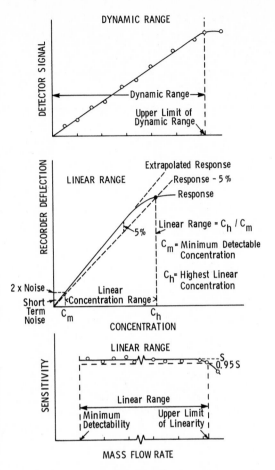

Figure 5.10   Methods for calculating the dynamic and linear response ranges
              for chromatographic detectors.

used detectors in HPLC are summarized in Table 5.8 [73].

5.8   Data Acquisition and Signal Processing

The signal for most chromatographic detectors is a voltage proportional to
the instantaneous quantity of sample eluting through the detector sensing
volume. The plot of detector output as a function of time is called the
chromatogram, the format by which the analyst judges the success of the
chromatographic experiment and the raw data from which qualitative peak
identification and sample quantitation is made. The chromatogram is usually
recorded on a potentiometric strip chart recorder, dubbed the chromatography
slave by one author [74]. The increasing sophistication of chromatographic
experiments and the need to quantify several components in a multicomponent

TABLE 5.7

TYPICAL PERFORMANCE VALUES FOR FIXED–WAVELENGTH PHOTOMETRIC DETECTORS

Measured Characteristics	Units	Typical Values
Static short–term noise per centimeter cell length	AU/cm	$(0.5 \text{ to } 1.5) \times 10^{-4}$
Dynamic short–term noise per centimeter cell length	AU/cm	$(0.5 \text{ to } 1.5) \times 10^{-4}$
Static long–term noise per centimeter cell length	AU/cm	$(1 \text{ to } 4) \times 10^{-4}$
Dynamic long–term noise per centimeter cell length	AU/cm	$(1 \text{ to } 5) \times 10^{-4}$
Static Drift	AU/h	$(5 \text{ to } 10) \times 10^{-4}$
Dynamic Drift	AU/h	$(2 \text{ to } 6) \times 10^{-4}$
Flow sensitivity	AU/min/ml	$(1 \text{ to } 5) \times 10^{-4}$
Linear range	ratio	$(5 \text{ to } 10) \times 10^{4}$
Response time	seconds	1 to 5

TABLE 5.8

TYPICAL SPECIFICATIONS FOR MOST–USED HPLC DETECTORS

Parameter (units)	UV (Absorbance)	RI (RI units)	Electrochemical (μamp)	Fluorometer	Conductivity (μMoh)
Type	Selective	General	Selective	Selective	Selective
Useful with gradients	Yes	No	No	Yes	No
Upper limit of linear dynamic range	2–3	10^{-3}	2×10^{-5}	10^{2}–10^{3}	1000
Linear range (maximum)	10^{5}	10^{4}	10^{6}	ca. 10^{3}	2×10^{4}
Sensivity at ± 1% noise, full scale	0.002	2×10^{-6}	2×10^{-9}	0.005	0.05
Sensivity to favorable sample (g/ml)	2×10^{-10}	1×10^{-7}	10^{-12}	10^{-11}	10^{-8}
Inherent flow sensitivity	No	No	Yes	No	Yes
Temperature sensitivity	Low	$10^{-4}/°C$	$1.5\%/°C$	Low	$2\%/°C$

sample has resulted in the widespread use of more sophisticated data acquisition
devices incorporating microprocessor or computer technology. A strip chart
recorder provides good visual information about the chromatogram but data
reduction by manual means is tedious for multicomponent samples. Modern
computing integrators can provide visual information in the form of a
chromatogram, as well as manipulating raw chromatographic data to report
retention values, peak parameters, relative concentrations, and column
performance parameters. The ability to normalize sample data, expand individual
sections of the chromatogram of interest, recalculate chromatographic parameters
as often as desired without rerunning the sample, and the flexible formating
available for report writing and record keeping will further the demand for
sophisticated data acquisition devices in the chromatography laboratory. Modern
instruments are now supplied with data stations as an integral part of the
instrument package or are designed to operate as part of a laboratory computer
network in which several instruments are interfaced to a host computer.

The potentiometric recorder generally used in the laboratory is a
servo-operated voltage balancing device [74]. It operates on the principle that
the input signal is continually balanced by an equal and opposite polarity
feedback signal. The small error voltage that does exist between the input and
feedback signals is applied to an error sensing servo-amplifier motor drive
system which continuously adjusts the feedback voltages and also positions the
recorder pen on the chart. The recorder may misrepresent the detector signal
due to inadequate response time, deadband and amplifier noise, and nonlinearity
effects [75,76]. The recorder response time is defined as the time required to
record a full-scale step change of signal applied to the recorder input
terminals. Values of 0.2 to 1.0 second full-scale are typical of those
recorders used in chromatographic laboratories. It should be noted that the
recorder response is not linear but S-shaped, comprising three different time
dependent events: the initial inertia and acceleration time, the full speed
slewing time, and the final deceleration time. For chromatographic purposes,
the response time to record a Gaussian function is more important than that for
a step function and this is generally taken to be approximately twice the
response time of a step signal [75]. The dead band is defined as the range
through which the signal can be varied without initiating a recorder response.
This time should be short and no greater than 10% of the response time of the
recorder to a Gaussian function. The deceleration characteristics of a
recorder's response to a step change in signal is known as damping. It is
measured as the maximum overshoot beyond the point of final rest expressed in
percent of the recorder span. Damping is provided by a resistance-capacitance
network which also acts as a filter to reject high frequency noise. The input
filter must be carefully designed as it may seriously modify the dynamic

response characteristics of the recorder. Filter networks may also be used in the detector output and, for the purpose of measuring their influence on the chromatographic signal, both filters are considered to add to extracolumn band broadening through their respective time constants. Choosing a suitable filter is often a complex problem, as it must balance the desirability of maximizing the signal while simultaneously eliminating noise originating both from the instrument and the electrical system for the laboratory [77,78].

The continuous voltage output from chromatographic detectors is not a suitable signal for computer processing. The conversion of the detector signal to a computer readable form requires an interface [79]. The interface must access the analog output of the chromatographic detector (mV), scale the output to an appropriate range (V), digitize the signal, and then transfer the data to a known location in the computer. The original input signal is transformed to a series of voltages on a binary counter and is passed to the computer as a series of binary words. The chromatogram is now in a form which is suitable for data processing.

Data processing can be performed on-the-fly or by storing the complete raw data for the chromatogram in memory for analysis at a later time [78,80]. Data processing on-the-fly provides information immediately but is limited to manipulating the data once only. It is the most appropriate method for use with microprocessor instruments of limited memory capacity. Computer systems are much more versatile and allow convenient repetitive data analysis. The raw data can also be mathematically filtered to remove noise without introducing peak skew, a process which is not possible with resistance-capacitance filters [78]. Long-term noise may also be distinguished from true peaks and eliminated from the chromatogram.

The computer can perform various mathematical procedures within its repertoire of algorithms to calculate the information desired from the smoothed raw data. These routines generally include the calculation of retention time, peak height, peak area, peak normalization, comparison with calibration data to calculate concentrations, and if desired, column parameters such as plate numbers, asymmetry factors, etc. Limitations are established simply by the number of programs stored in the computer memory. Certain features of data processing, for example peak threshold sensing, baseline correction, signal averaging, slope sensitivity, etc. are common to all computational methods. Various methods of determining these parameters were discussed by Reese [80].

The calculating power and speed of computers blind the unwary to their limitations. Every eventuality cannot be allowed for during the compilation of a computer program. Examples of peaks which are often poorly treated by computing integrators are small peaks with large peak widths, peaks on the tail of larger peaks or the solvent front, and fused peaks. Data systems are usually very reproducible but this should not be mistaken for accuracy. Data systems are most powerful when a human eye checks the graphical record against the calculated data to instigate any necessary changes.

5.9 Quantitative Analysis

The flow of information in a chromatographic experiment usually commences by qualitative peak identification based on retention time, followed by quantitation to establish the concentration of the peak in the sample. Quantitative analysis requires that a relationship between the magnitude of the detector signal and sample concentration be established. The detector signal is measured by the peak height or area from the recorder trace or taken from the print-out of a data system. Manual methods for calculating peak areas (Figure 5.11) include the product of peak height and width at half height, triangulation, trapezoidal approximation, planimetry, and cut and weigh [81]. The above methods are compared in Table 5.9. No single method is perfect and common problems include the difficulty of defining peak boundaries accurately, operator dependence on precision, and the need for a finite time to make each measurement. Those methods using triangular approximations are unreliable for measuring asymmetric and fused peaks. The trapezoidal method gives the closest approximation to the true peak area and can be used to obtain accurate area values for peaks which have substantial asymmetry [81]. Planimetry and cutting and weighing of peaks makes no assumptions about the shape of the peak profile and can be used to determine the area of skewed peaks. The proper use of a planimeter requires considerable skill and experience and, even so, obtaining accurate results requires repetitive tracings on each peak with the totals averaged. It is also obvious from Table 5.9 why computing integrators are preferred for data handling in chromatography.

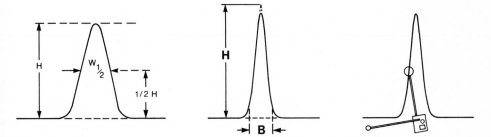

Figure 5.11 Manual methods of computing peak areas.

It is often asked whether peak height or peak area measurements provide the

TABLE 5.9

COMPARISON OF MANUAL METHODS FOR COMPUTING PEAK AREAS

Method	Formula for calculation	Area Obtained as a % of true area	Time required for measurement and calculation (min)	Comments
Peak height x width at half height	HW_h	93.9	10	Should only be applied to symmetical peaks.
Triangulation	$W_bH/2$	96.9	12	Accuracy depends on how carefully tangents are drawn. Subject to operator error.
Trapezoidal approximation	$\left[\dfrac{W_{0.15} + W_{0.85}}{2}\right]H$	100	12	Product of the average of the peak widths at 15 and 85% of the peak height multiplied by the peak height. Gives acceptable values of peak area for symmetric and asymmetric peaks.
Planimeter		100	15	Subject to operator error. Gives correct area for skewed peaks.
Cut and weigh		100	20	Depends on the accuracy of the cutting operation. Homogeneity, moisture content, and weight of paper influence precision. Can handle skewed peaks.
Computing integrator		100	ca. 0	Superior to manual methods in terms of speed and precision. Accuracy for fused peaks, rising baseline, etc., may be poor. Cannot conveniently handle negative peaks (RI detector)

best representation of the detector response in HPLC. There is no simple answer to this question as the precision and accuracy of peak height and area measurements are a function of many chromatographic variables including sample size, mobile phase composition, flow rate and column temperature [82-84]. Peak area measurements are preferred when the column flow can be controlled precisely even if the mobile phase composition shows some variability. Peak areas are relatively independent of mobile phase composition. On the other hand, if the composition can be controlled precisely but the flow rate shows some variability then peak height measurements are preferred. Peak heights are generally less dependent on flow rate constancy than peak areas. Halasz has shown that as far as mobile phase flow rate variations are concerned, it is the average short-term variations that occur while the peak is in the detector which affects quantitative precision [84]. Consequently, even if an internal standard is included with the sample, it can not be expected to improve precision due to variations of the above kind. Quantitation in gradient elution chromatography requires careful control of total flow rate when peak areas are measured and gradient composition when peak heights are measured. To test which alternative is most applicable, variation in the retention time of early eluting peaks indicates poor flow precision and variation in the retention time of late eluting peaks suggests poor precision in the mobile phase composition. Consequently, whether peak height or peak area is selected for a particular analysis depends on system performance and not necessarily on sample composition. From the practical viewpoint, peak heights are easier to measure manually than peak areas.

Rarely in HPLC is the response of the detector the same for equal amounts of different samples. Rather, detector response is related to concentration by a fixed substance-specific factor known as the response factor. Response factors must be calculated for all components in the sample to normalize their concentrations. Response factors can be calculated from a single measurement but the calibration method is preferred.

5.10 References

1. J. F. K. Huber (Ed.), "Instrumentation for High Performance Liquid Chromatography," Elsevier, Amsterdam, 1978.
2. H. M. McNair, J. Chromatogr. Sci., 16 (1978) 588.
3. F. Erni, K. Krummen, and A. Pellet, Chromatographia, 12 (1979) 399.
4. J. W. Dolan, Sj. van der Wall, S. J. Bannister, and L. R. Snyder, Clin. Chem., 26 (1980) 871.
5. Sj. van der Wal, S. J. Bannister, and L. R. Snyder, J. Chromatogr. Sci., 20 (1982) 260.
6. G. Guiochon, in C. Horvath (Ed.) "High Performance Liquid Chromatography. Advances and Perspectives", Academic Press, New York, 2 (1980) 1.
7. J. J. Kirkland, W. W. Yau, H. J. Stoklosa, and C. H. Dilks, J. Chromatogr. Sci., 15 (1977) 303.
8. R. P. W. Scott and P. Kucera, J. Chromatogr. Sci., 9 (1971) 641.

9. M. Martin, C. Eon, and G. Guiochon, J. Chromatogr., 108 (1975) 229.
10. H. Colin, M. Martin, and G. Guiochon, J. Chromatogr., 185 (1979) 79.
11. H. Poppe, Anal. Chim. Acta, 114 (1980) 59.
12. H. H. Lauer and G. P. Rozing, Chromatographia, 14 (1981) 641.
13. R. P. W. Scott, "Liquid Chromatography Detectors", Elsevier, Amsterdam, 1977.
14. V. E. Dell'Ova, M. B. Denton, and M. F. Burke, Anal. Chem., 46 (1974) 1365.
15. S. R. Bakalyar, M. P. T. Bradley, and R. Honganen, J. Chromatogr., 158 (1978) 277.
16. J. N. Brown, M. Hewins, J. H. M. Ven Der Linden, and R. J. Lynch, J. Chromatogr., 204 (1981) 115.
17. L. V. Berry and B. L. Karger, Anal. Chem., 45 (1973) 819A.
18. R. W. Yost, L. S. Ettre, and R. D. Conlon, "Practical Liquid Chromatography. An Introduction." Perkin-Elmer, Norwalk, 1980.
19. W. W. Yau, J. J. Kirkland, and D. D. Bly, "Modern Size-Exclusion Liquid Chromatography", Wiley, New York, 1979.
20. N. J. Pound, R. W. Sears, and A. G. Butterfield, Anal. Chem., 45 (1973) 1001.
21. J. L. Meek, Anal. Chem., 48 (1976) 375.
22. M. Martin, G. Blu, C. Eon, and G. Guiochon, J. Chromatogr., 112 (1975) 399.
23. P. Achener, S. R. Abbott, and R. L. Stevenson, J. Chromatogr., 130 (1977) 29.
24. K. Asai, Y.-I. Kanno, A. Nakamoto, and T. Hara, J. Chromatogr., 126 (1976) 369.
25. J. G. Nikelly and D. A. Ventura, Anal. Chem., 51 (1979) 1585.
26. A. Bylina, K. Lensniak, and S. Romanowski, J. Chromatogr., 148 (1978) 69.
27. L. R. Snyder, in C. Horvath (Ed.), "High Performance Liquid Chromatography Advances and Perspectives", Academic Press, New York, 1 (1980) 207.
28. Y. S. Kim, J. M. Kootsey, and G. M. Padilla, Anal Biochem., 78 (1977) 283.
29. C. Liteaunu and S. Gocan, "Gradient Elution Chromatography", Wiley, New York, 1974.
30. F. Erni, R. W. Frei, and W. Linder, J. Chromatogr., 125 (1976) 265.
31. P. A. Bristow, Anal. Chem., 48 (1976) 237.
32. H. A. H. Billiet, P. D. M. Keehnen, and L. De Galan, J. Chromatogr., 185 (1979) 515.
33. P. Jandera, J. Churacek, and L. Svoboda, J. Chromatogr., 192 (1980) 37.
34. A. Schmid, Chromatographia, 12 (1979) 825.
35. T. W. Smuts, D. J. Solms, F. A. Van Niekerk, and V. Pretorius, J. Chromatogr. Sci., 7 (1969) 24.
36. R. E. Jentoft and T. H. Gouw, J. Chromatogr. Sci., 8 (1970) 138.
37. H. Spaans, H. Terol, and A. Onderwater, J. Chromatogr. Sci., 14 (1976) 246.
38. B. Pearce and W. L. Thomas, Anal. Chem., 44 (1972) 1107.
39. R. M. Cassidy and R. W. Frei, Anal. Chem., 44 (1972) 2250.
40. M. Minarik, M. Popl, and J. Mostecky, J. Chromatogr., 208 (1981) 67.
41. R. J. Kelsey and C. R. Loscombe, Chromatographia, 12 (1979) 713.
42. T. J. N. Webber and E. H. McKerrell, J. Chromatogr., 122 (1976) 243.
43. N. K. Vadukul and C. R. Loscombe, Chromatographia, 14 (1981) 465.
44. P. T. Kissinger, L. J. Felice, D. J. Miner, C. R. Preddy, and R. E. Shoup, in D. M. Hercules (Ed.), "Contemporary Topics in Analytical Chemistry and Clinical Chemistry", Plenum, New York, 2 (1978) 55.
45. L. S. Ettre, J. Chromatogr. Sci., 16 (1978) 396.
46. J. L. Dicesare and L. S. Ettre, J. Chromatogr., 251 (1982) 1.
47. H. Colin, A. Jaulmes, G. Guiochon, J. Corno, and J. Simon, J. Chromatogr. Sci., 17 (1979) 485.
48. J. N. Little and G. J. Fallick, J. Chromatogr., 112 (1975) 389.
49. J. E. Stewart, Anal. Chem., 53 (1981) 1125.
50. J. E. Stewart, Appl. Optics, 20 (1981) 654.
51. J. E. Stewart, J. Chromatogr., 174 (1979) 283.
52. A. M. Krstulovic, D. M. Rosie, and P. R. Brown, Anal. Chem., 48 (1976) 1383.
53. J. K. Baker, R. E. Skelton, and C.-Y. Ma, J. Chromatogr., 168 (1979) 417.

54. A. M. Krstulovic, R. A. Hartwick, P. R. Brown, and K. Lohse, J. Chromatogr., 158 (1976) 365.
55. K. H. Falchuk and C. Hardy, Anal. Biochem., 89 (1978) 385.
56. J. W. Readman, L. Brown, and M. M. Rhead, Analyst, 106 (1981) 122.
57. R. Yost, J. Stoveken, and W. Maclean, J. Chromatogr., 134 (1977) 73.
58. M. S. Denton, T. P. De Angelis, A. M. Yacynych, W. R. Heineman, and T. W. Gilbert, Anal. Chem., 48 (1976) 20.
59. M. J. Milano, S. Lam, M. Savonis, D. B. Pautter, J. W. Pav, and E. Grushka, J. Chromatogr., 149 (1979) 599.
60. K. Saitoh and N. Suzuki, Anal. Chem., 51 (1979) 1683.
61. S. A. George and A. Maute, Chromatographia, 15 (1982) 419.
62. A. T. Rhys Williams, "Fluorescence Detection in Liquid Chromatography", Perkin-Elmer, Beaconsfield, 1980.
63. W. Slavin, A. T. Rhys Williams, and R. F. Adams, J. Chromatogr., 134 (1977) 121.
64. K. Ogan, E. Katz, and T. J. Porro, J. Chromatogr. Sci., 17 (1979) 597.
65. E. Johnson, A. Abu-Shumays, and S. R. Abbott, J. Chromatogr., 134 (1977) 107.
66. E. S. Yeung and M. J. Sepaniak, Anal. Chem., 52 (1980) 1465A.
67. B. Fleet and C. J. Little, J. Chromatogr. Sci., 12 (1974) 747.
68. D. G. Swartzfager, Anal. Chem., 48 (1976) 2189.
69. P. T. Kissinger, Anal. Chem., 49 (1977) 447A.
70. R. J. Rucki, Talanta, 27 (1980) 147.
71. T. Wolf, G. T. Fritz, and L. R. Palmer, J. Chromatogr. Sci., 19 (1981) 387.
72. American Society for Testing Materials, "Standard Practice for Testing Fixed-Wavelength Photometric Detectors used in Liquid Chromatography", Annual Book of ASTM Standards Part 42, E685, Philadelphia, Pennsylvania, 1980.
73. L. R. Snyder and J. J. Kirkland, "Introduction to Modern Liquid Chromatography", Wiley, New York, 1979.
74. R. B. Bonsall, J. Gas Chromatogr., 2 (1964) 277.
75. I. G. McWilliam and H. C. Bolton, Anal. Chem., 41 (1969) 1762.
76. P. C. Kelley and W. E. Harris, Anal. Chem., 43 (1971) 1170.
77. H. C. Smit and H. L. Walg, Chromatographia, 8 (1975) 311.
78. W. Kipiniak, J. Chromatogr. Sci., 19 (1981) 311.
79. C. E. Reese, J. Chromatogr. Sci., 18 (1980) 201.
80. C. E. Reese, J. Chromatogr. Sci., 18 (1980) 249.
81. M. F. Delaney, Analyst, 107 (1982) 606.
82. S. R. Bakalyar and R. A. Henry, J. Chromatogr., 126 (1976) 327.
83. R. P. W. Scott and C. E. Reese, J. Chromatogr., 138 (1977) 283.
84. I. Halasz and P. Vogtel, J. Chromatogr., 142 (1977) 241.

Chapter 6

PREPARATIVE-SCALE CHROMATOGRAPHY

6.1 Introduction

There is no clear concensus as to the meaning of the term preparative
chromatography. Clearly the isolation of quantities of purified materials by
chromatographic means is involved but, depending on the intended use, this may
cover a very wide mass range. Possible uses include isolation of materials for
structural elucidation, for biological or sensory evaluation, for organic
synthesis, or for commercial applications. The scale of the operation thus
includes laboratory, pilot plant, and full-scale manufacturing operations. A
narrow view of preparative-scale chromatography is adequate for our purpose. We
will concentrate on the use of preparative-scale chromatography for the
isolation of milligram to gram quantities; that is, to those procedures normally
practiced in an analytical laboratory where larger sample sizes are rarely
encountered. Operating chromatographic equipment at the other extreme, the
separation of kilogram quantities per hour, presents its own intriguing problems
of process engineering and is discussed in references [1-5]. The techniques
discussed in this chapter include gas chromatography, liquid chromatography, and
countercurrent chromatography using either commercially available analytical or
purpose-built preparative chromatographic equipment.

6.2 Gas Chromatography

Early in the development of gas chromatography considerable attention was
given to its use as a preparative as well as an analytical technique. Today
preparative gas chromatography attracts little attention due to a combination of
maturity and obsolescence. Except for the isolation of volatile compounds it
has largely been replaced by preparative liquid chromatography. The sensitivity
of on-the-fly combined gas chromatographic/spectroscopic instruments has
increased to the point where preparative isolation of samples is no longer

necessary prior to structural elucidation. Gas chromatography–mass spectrometry and gas chromatography–Fourier transform infrared spectrometry are performed routinely in many laboratories without sample isolation; these on–the–fly processes are much more arduous to achieve with liquid chromatographic equipment thus making sample isolation more common with that technique. For identification by nuclear magnetic resonance spectrometry on–line coupling to a gas chromatograph is not practical because of sample size and time constraints; in this instance isolation of microgram and preferably milligram quantities of materials is usually needed. As well as a diminishing need to isolate samples for spectroscopic identification, the poorer chromatographic performance of preparative compared to analytical columns has contributed to the decline in popularity of preparative gas chromatography.

The primary objective in preparative–scale gas chromatography is to obtain a high sample throughput. An inivatable result of this goal is that either resolution or analysis time, or both, must be compromised. The primary method of increasing the sample capacity of the column is to increase its size (i.e., the quantity of stationary phase available for the separation). Two solutions are suggested: increase the column diameter at constant length or increase the length at constant diameter. For simple separations a short, wide column is usually used, for example a column of 1–3 m x 6–10 cm I.D. For more complex separations higher column efficiencies are required and long, narrow columns, for example 10–30 m x 0.5–1.5 cm I.D., are used. It is not possible to coil wide bore columns so, unless a purpose designed preparative gas chromatograph is available, the choice may be limited to long, narrow, coiled columns. Other critical instrument considerations include the injector, detector, and method of sample collection. The maximum allowable compound size (MACS) is defined in Figure 6.1. Compared to analytical operations the sample size is much greater and is generally increased to the maximum allowed. This maximum is limited by the requirement that a minimum resolution be maintained between critical sample component pairs. Injection volumes in the range of 0.1 to 10.0 ml are common, and therefore, normal injection techniques are not applicable. Injection by syringe is possible for smaller sample sizes provided that a slow injection rate is employed. For larger sample volumes an automated injector is used. Several types have been described employing pneumatic transfer from a reservoir through a capillary restrictor, a pneumatic piston pump, or by a syringe pump. The injection process is thus controlled by time, a necessity due to the limited thermal capacity of injection block heaters and their inability to flash vaporize large sample volumes. Small sample sizes (e.g., 100 microliters) may be readily vaporized by slow injection if sufficiently volatile. Larger samples usually require an evaporation device between the injector and column. This is often a tube heated separately from the column oven and packed with glass beads

1. Construct a plot of preparative column efficiency (n) as a function of sample size

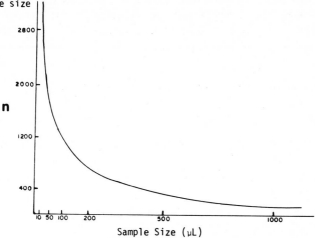

2. The required column performance to effect a given separation can be expressed as

$$n = 16R^2 \left[\frac{\alpha}{\alpha-1}\right]^2 \left[\frac{k+1}{k}\right]^2$$

3. From chromatogram the relative retention value for the least resolved pair of interest may be computed as a ratio always greater than unity (α). In practice a value of $\alpha > 1.1$ is required.

4. The relative retention ratio between the first of these peaks and an unretained peak provides a value for k. k should be between 3 and 10.

5. A value of R = 1.0 is adequate for preparative purposes so n can be calculated for the particular separation.

6. The calculated value for n enables the maximum sample size to be obtained from a plot similar to (1).

7. The relationships in (1) to (6) apply to a single component of the least resolved pair. The relative % of the peak in the largest amount is used to calculate the MACS value according to

$$MACS = \frac{100\ S_{(n)}}{x}$$

$S_{(n)}$ = maximum sample size at which the column provides a required efficiency n

x = relative percent of the largest component of the most difficult to resolve pair

Figure 6.1 Procedure for calculating the maximum allowable compound size (MACS) in preparative chromatography.

or metal spheres to increase the thermal capacity while reducing dead volume effects. The evaporation unit is usually maintained at a temperature 50°C above the boiling point of the least volatile sample component. The sample should be transferred to the column as a rectangular plug diluted in the minimum volume of carrier gas; if it is not completely vaporized then a homogeneous distribution of sample is not obtained over the cross-sectional area of the column and excessive band broadening may result. Vaporization problems may be circumvented by direct, slow, on-column injection without carrier gas flow. This is often the most practical solution for injecting large sample volumes into analytical instruments. The principal disadvantage is the possibility of stripping the stationary phase from the head of the column. The magnitude of the problem depends on the sample size, the relative solubility of the liquid phase in the sample, and the liquid phase loading. Column stripping results in diminished efficiency due to the combination of sample interactions with bare support and the irregular distribution of liquid phase produced throughout the column. For the on-column injection of both vaporized and liquid samples a substantial pressure change may result from sample vaporization. Sample backflush into the carrier gas lines can be a problem and is normally solved by positioning a backflush or needle valve in the carrier gas line.

Most separations are usually carried out isothermally. Temperature programming is not possible with wide bore columns due to the uneven radial temperature gradients generated across the column. Diatomaceous earths are in fact fairly good insulators. Temperature programming is only possible with long, narrow columns and then generally at slow rates. A precolumn switching system can be used to pass the volatile components of the sample into the main preparative column and then, using a Deans' switching system, backflush the heavy ends to a trap and thus increase the sample throughput when the heavy ends are not of interest [6]. Detection of the separated sample is usually performed with a thermal conductivity or flame ionization detector. The ideal detector would provide a stable baseline insensitive to changes in temperature, pressure, and carrier gas flow; high sensitivity; wide linear response range; and be easy to clean. These properties are most completely embodied in the flame ionization detector (FID). As the FID is destructive, it is connected to the column via a splitter so that only a few percent of the total column flow is diverted to the detector. Although the thermal conductivity detector is a nondestructive detector it, too, is usually also operated with a splitter as the filament lifetime may be shortened due to contamination by the high mass flow of material leaving the column. With nitrogen carrier gas the thermal conductivity detector may produce negative peaks, making interpretation of the chromatogram difficult.

The final operation, unique to preparative gas chromatography, is sample

collection of the vapors in the carrier gas effluent. This process may be
performed automatically or manually, and is sequenced by a signal from the
detector. In a commercial preparative scale gas chromatograph injection and
sample collection are automated for unattended operation. When using an
analytical instument both processes will normally be manual. The exit from the
detector splitter is usually led out through a side wall of the oven and
maintained at a sufficiently high temperature to avoid premature condensation of
the sample. This may involve the use of an auxiliary heating supply for the
transfer line. There is no single solution to the problem of trapping the
sample after it emerges from the transfer line. The literature is resplendent
with numerous methods involving different degrees of ingenuity of which packed
and unpacked cold traps, solution and entrainment traps, total effluent and
adsorption traps, Volman traps, and electrostatic precipitators are among the
more popular [7-10]. Some representative examples are shown in Figure 6.2. The
minimum requirement for efficient sample collection is that the partial pressure
of the solute entering the trap must be greater than its equilibrium vapor
pressure with the condensed phase at the trap temperature. The equilibrium
vapor pressure of the solute may be reduced by lowering the temperature, by
dissolving the solute in a solvent, or by adsorbing it onto a material of high
surface area. However, by far the most common approach is to reduce the
temperature of the effluent, either in a trap filled with a material such as
potassium bromide or column packing, or in a simple open tube trap. In practice
the residence time of the solute in the trap may be too short to obtain
equilibrium and the solute tends to pass directly through the trap, either as a
supersaturated vapor or as a fog composed of small liquid droplets. The
cold-trapping of samples with boiling points greater than about 150°C is almost
inevitably accompanied by the formation of aerosols which are then swept through
the trap. This can be minimized by cooling the effluent through a gradual
temperature gradient rather than subjecting it to an abrupt temperature change.
Also, turbulent flow within the trap facilitates the collision of liquid
droplets with the wall, thereby inducing their retention. The Volman trap
consists of a double-walled vessel within which turbulence is created by
maintaining the walls at different temperatures. The Volman trap is
particularly useful for recovering gram-scale quantities. Electrostatic
precipitators are also used to trap large-scale fractions. As aerosols contain
large numbers of charged droplets, the trapping efficiency can be improved by
passing the cooled effluent through a large electrostatic field. If a readily
condensable gas such as argon or carbon dioxide is used as the carrier gas then
the total effluent may be condensed, thereby entraining the small amount of
solute in a much larger quantity of carrier gas. After condensation the sample
and condensed carrier gas are separated by selective evaporation. The

efficiency of the total effluent trap can be very high but the method is inconvenient and difficult in practice. For occasional sample trapping a simple packed or empty glass U-tube cooled in a dry ice-acetone or liquid nitrogen coolant usually suffices. For further details on all aspects of preparative-scale gas chromatography several reviews are available [11-14].

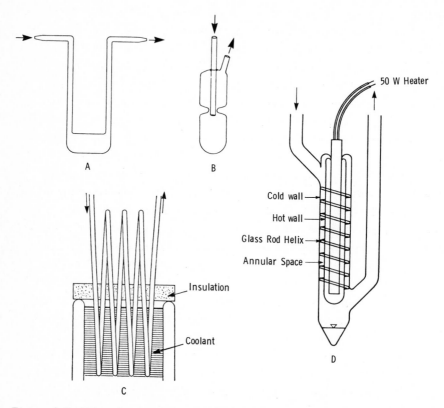

Figure 6.2 Traps for sample recovery from a gas chromatographic effluent. A, U-tube; B, simple trap; C, multiple temperature gradient trap; D, Volman trap.

Instrumental constraints are established by the equipment available; the analyst, however, has control of the column selected. Analytical separations are usually performed initially to indicate the degree of difficulty of the separation. The stationary phase is then selected from a limited number of thermally stable phases. Preparative columns are expensive to prepare and generally only a few column types are available. The support material is coarse with a narrow particle size distribution, 35-40 mesh. Sample throughput rather than efficiency is the overriding consideration. High carrier gas flow rates, in the range 100-1000 ml/min depending on the column diameter, are used, so that a coarse column packing is needed to permit operation at a reasonable pressure differential. As the stationary phase loading is greater and the column length may be longer than those used in analytical separations, operation at higher temperatures is also common. This can be a problem with thermally labile compounds and dictates the use of high quality phases and supports for column preparation. Further considerations for optimizing column conditions for a particular separation are summarized in Table 6.1 [15,16]. The selection of the column dimensions can be made with reference to Table 6.2 [14]. Packing preparative columns reproducibly and with reasonable efficiency is, at best, never easy and increases in difficulty as the column diameter is increased [13,14,17-20]. The inferior chromatographic performance of preparative columns is due to the radial unevenness of the packing structure. During the packing of large diameter columns the solid support particles segregate preferentially according to their size, with the larger particles being closer to the wall. In addition, the packing in this region is less dense, due to the physical constraint of the wall. This leads to uneven, or skew, flow profiles for the carrier gas and sample through the packed bed, inducing band broadening. To overcome this problem Verzele suggested using the shake, turn, and pressure (STP) packing method [17]. The column is shaken in the radial direction and is rotated along its long axis while being packed and periodically pressurized. Verzele also cautions that columns should be packed to a medium density with small additions of packing. The packing process, therefore, requires several hours for columns of average dimensions. Reese has described a vacuum packing method with vertical tamping which is less time consuming and produces columns with about 1800 plates per meter [20]. However, to obtain consistent results some preliminary practice and training is needed. As column permeability is an important consideration in preparative gas chromatography, the objective of any packing procedure can be defeated by the generation of fines; therefore robust packings and gentle handling are essential to obtain success. As preparative columns may contain from hundreds of grams to kilograms of packing material, failure in the above respect is failure on a grand and expensive scale. Columns with diameters less than approximately 1.5 cm are easier to pack by conventional means and for occasional preparative-scale applications long, narrow columns are recommended, if only because the initial success rate in preparation is much higher. Glass or metal columns with diameters of 1.0 cm or less can be conveniently coiled and fitted into the ovens of standard analytical gas chromatographs. Ovalized column configurations are preferred to circular coils [13,17].

TABLE 6.1

RECOMMENDATIONS FOR SELECTING OPTIMUM CONDITIONS FOR PREPARATIVE-SCALE GAS CHROMATOGRAPHY

Parameter	B. Roz et al. [15]	J. R. Conder [16]
Carrier gas	Use helium, or better hydrogen, at a flow rate two to three times larger than that corresponding to the maximum efficiency at very low sample size.	Use helium, hydrogen, or nitrogen at a velocity of 10-15 cm/s in production GC. The highest velocity permitted by the influence of carrier gas viscosity and column length on operating pressure in preparative GC.
Temperature	Select a temperature and column outlet pressure so that the Valentin equation is fulfilled. (Temperature equals boiling point of main compound at average column pressure).	Set an operating temperature equal to the mean solute boiling point at a pressure between the inlet pressure and the inlet pressure divided by the optimum solute mole fraction.
Stationary Phase	Phase should be able to withstand the temperature at which the Valentin condition is fulfilled without serious degradation.	Choose the stationary phase at least as much for its upper temperature limit as for its selectivity, especially if the feed mixture to be separated is only binary.
Support	As for Conder	Use coarse, particulate packing of narrow mesh range, e.g., 35-40 mesh.
Phase Loading	The amount of stationary phase in the column is not very critical since the retention time and sample size are both proportional to the coating ratio of the support. This ratio should be small enough so that no flooding of the column occurs at the inlet; in practice it means that the phase ratio should not exceed about 20% w/w on Chromosorb P and 4% w/w on Chromosorb G.	Not very critical. Convenient loadings are 15-25% w/w on pink and white diatomite supports and 5-7% w/w on high density diatomite.
Column Length	No optimum column length. Column lengths of 1-4 m are convenient. Longer columns may be required if $\alpha < 1.1$.	Depends on the plate height, injection time, and the number of plates required for the separation.

Sample Size	Inject important sample size so that the maximum partial pressure of the sample in the carrier gas at the column inlet is smaller than 30% of the total pressure, while the width of the injection band is large enough to allow the overlapping of the elution bands of the compounds to be isolated.	Inject the solute at a partial pressure which will give a liquid phase mole fraction = 0.2–0.5. Use wide feed bands, typically measured in minutes or fractions of a minute rather than seconds.
Mode	Eliminate compounds or impurities with long retention time either by distillation or by pre-column fractionation using GC. Superimpose injections so that the first impurities of one injection are eluted in the same time as the last impurities of the previous injection.	Inject successive samples so that their respective first and last components overlap at the column exit. Instead of aiming for complete resolution, cut the overlapping fraction with a recovery ratio of about 60% and recycle it.

TABLE 6.2

PREPARATIVE COLUMN SELECTION

Separation conditions	Sample component	Column type	Collection rate[a]	Component resolution[a]	Sample capacity[a]
Easy	Major	Long, narrow	Fair (about 5 ml/h)	Excellent	Good
		Short, wide	Excellent (about 50 ml/h)	Excellent	Excellent
	Trace	Long, narrow	Too slow	Excellent	Too low
		Short, wide	Excellent	Excellent	Excellent
Difficult	Major	Long, narrow	Fair	Excellent	Good
		Short, wide	Fair	Poor unless multiple pass used	Good
	Trace	Long, narrow	Fair	Excellent	Poor
		Short, wide	Fair	Good with multiple pass	Excellent
Wide-boiling range mixture		Long, narrow	Fair (with programming)	Good	Fair
		Short, wide	Slow	Poor	Good

[a] "Fair" to "poor" may indicate the only reasonable solution to some sample problems.

6.3 Low- and Medium-Pressure Liquid Chromatography

From its inception liquid chromatography has been used as a preparative technique [21,22]. Today liquid chromatography is widely used in analysis for sample cleanup (see Chapter 7.4), in organic synthesis and biochemistry for isolating quantities of material sufficient for identification, for purification of bulk samples, and in industrial applications for the preparation of commercial products on a much larger scale than that associated with laboratory operations. It owes its popularity to simplicity, low cost, and the ready availability of materials. No longer limited to adsorption chromatography, the same simple experimental design can be used for preparative-scale separations by size-exclusion, ion-exchange, liquid-liquid partition, and affinity chromatography [23-25]. However, liquid-solid chromatography remains the dominant technique for separating neutral organic compounds in the laboratory and our discussion will be mainly focussed on this technique.

Adsorbents available for liquid-solid chromatography include silica gel, alumina, cellulose, charcoal, florisil, polyamide, and magnesium oxide. Silica and alumina find most use while the other adsorbents are reserved for those special applications for which the former are unsuitable. This includes many biochemical separations of charged molecules that bind irreversibly to adsorbents of high activity.

Silica and alumina of suitable particle diameter (range 40-500 micrometers)

are dried prior to use, 110°C for a few hours for silica and several hours at
400°C for alumina. The dried adsorbent is cooled to room temperature and its
activity adjusted to the desired value by the addition of various amounts of
water (0–25% w/w). The addition of water is used to reduce the adsorbent
activity (see Chapter 4.10) and may be required to improve the separation and
mass recovery of polar compounds and to increase the column capacity. Acid
washing to remove trace metal impurities may be required prior to separation of
some sensitive compounds. For classical column chromatography the amount of
stationary phase and the column dimensions are established by the nature and
amount of sample to be separated [26]. A preliminary screening by thin-layer
chromatography is used to decide the complexity or difficulty of the separation
and to establish the optimum mobile phase for simple separations. A mobile
phase producing an R_F value of approximately 0.3 to 0.4 is suitable for simple
separations. In these cases a sample loading of about 10% w/w of stationary
phase may be used; for more demanding separations 3% w/w is more common.
Likewise the column length/diameter ratio is established by the same criteria,
10:1 for simple and 30:1 for difficult separations.

Two methods are commonly used for column filling. With the column in the
vertical position and a glass wool plug or porous frit in the bottom end, the
column is filled to about 60% of its height with the solvent to be used in the
separation. The sorbent is added to the column in small increments from a
filter funnel. During the addition of sorbent the solvent is allowed to run out
the bottom of the column at a rate not exceeding the rate of sorbent addition.
During the packing process the column is gently shaken or tapped to dislodge any
air bubbles trapped in the bed as it forms. It is very difficult to pack large
columns by this method; it is also time consuming as the rate of sorbent
addition must of necessity be slow. Alternatively, the sorbent can be mixed
with the mobile phase to form a slurry, and the whole poured rapidly into the
column, shaking as needed to dislodge air bubbles. In either case, the column
bed should be homogeneous and free of channels, otherwise poor results will be
obtained. Once packed the column bed should not be allowed to run dry.

Sampling is an important consideration which, for classical column
chromatography, is usually performed manually. The solvent used to pack the
column is drained off until its height is level with that of the bed. The
sample, usually dissolved in a small volume of the same solvent, is added slowly
and evenly to the top of the column bed. The solvent is drained to the top of
the column bed and the process repeated again with a second volume of solvent to
wash down the column walls, etc. The sample band should be sharp and even at
the top of the column. Sand or filter paper discs are sometimes added to the
top of the column to aid the sampling process and to prevent disruption of the
column bed. With care these should not be necessary. Some samples cannot be

dissolved in the same solvent used to pack the column. In this case the sample is usually dissolved in a more polar solvent and added to that amount of column packing material which gives a free flowing powder after solvent removal. The sample is then placed on top of the column bed in a manner similar to that used to pack the column. Alternatively, the sample may be filtered through a filter cake of sorbent supported on a Buchner funnel. The insoluble portions of the sample remain on the sorbent and the components of interest are eluted out by the passage of solvent, usually under vacuum. Excess solvent is removed and the sample is now sufficiently soluble to be handled in the usual way. This method is particularly useful for removing very polar fractions from the sample when these are not of importance to the analysis.

The sample is usually separated by elution chromatography, collecting suitable-sized fractions either manually or with a fraction collector. A single development with 4 to 5 column volumes of mobile phase at a flow rate of 0.5-5.0 ml/min, depending on the column dimensions, usually suffices for simple separations. The actual column flow rate depends on the permeability of the column bed. Thus flow rate is not a critical parameter unless the column exit is blocked and becomes the controlling factor. Of course, low flow rates add to the time required for what is already a slow technique, and it may be worth repacking the column and starting the separation again. For the separation of complex samples gradient elution conditions are used. Although numerous methods of generating continuous gradients exist [23], the preferred method is the use of a stepwise gradient generated by changing the polarity of the solvent in incremental steps. Suitable solvent combinations were discussed in Chapter 4.10 for liquid-solid chromatography. At the end of the separation the individual fractions are spotted on a plate for thin-layer chromatography (TLC) and grouped together according to their composition. Fractions containing several components can be separated on a second column to improve the overall yield of sample if required.

The procedures described above represent the operations performed in classical column chromatography. Still widely used and trusted by many, the technique is slow and demands the constant attention of the operator. In the last few years several variations have been described to either automate the method, increase its resolution, or improve sample throughput. Of course one of the principal methods of achieving all these goals is to use preparative-scale high performance liquid chromatogaphy, described in the next section. Before proceeding to this method we will review some of the less elaborate and less costly alternatives which may be suitable for the solution of many problems encountered in the laboratory.

One of the earliest attempts to improve the efficiency of classical column chromatography was dry column chromatography [27-30]. There are several versions of this technique but basically a length of plastic tube, usually thin-walled nylon or PTFE is dry-packed with TLC grade sorbent to the desired column length. The columns are disposable and, after packing, resemble a sausage tied at the lower end, leaving a small hole for air or solvent to escape. After addition of sample the column is eluted with sufficient solvent to reach the lower end of the bed. The sample is not recovered by elution but rather the column is cut into sections represented by the visualized bands and each section is then treated with solvent to remove the sample. Obviously of limited capacity, the method provides higher resolution than classical column techniques due to the use of sorbents of small particle size. It is well-suited to the recovery of small quantities of material, amounts similar to those obtained by preparative TLC.

Sorbents with diameters in the range 40-63 micrometers represent a good compromise between column efficiency and resolution if high pressure instrumentation is not used. In order to generate reasonable flow rates either a vacuum at the column base or over-pressure at the column head is required. For simple separations a vacuum filtration apparatus with a sintered-glass filter funnel containing a bed of sorbent may suffice [31,32]. The sample is eluted by vacuum suction with solvents of increasing polarity, starting with hexane. Resolution comparable to gravity-fed columns was claimed in a much shorter time. The major drawbacks to this very simple system were channeling caused by the necessary intermittent breaking of the vacuum, uneven sample application, and limited resolution due to shortness of the column. These problems were solved using the apparatus shown in Figure 6.3 [33]. This allows the column to remain under vacuum (1-10 mmHg) at all times and uses a precolumn section of celite (a diatomaceous earth of low surface activity) to provide an even distribution of sample over the cross-sectional area of the column. The columns are dry-packed with 10-40 micrometer diameter silica and consolidated by opening and closing the valve at the column head to the atmosphere. During operation solvent changes can be made very simply as the column head is maintained at atmospheric pressure. Vacuum liquid chromatography provides resolution comparable to TLC separations while using simple apparatus.

Instead of vacuum, a slight gas over-pressure can be used to increase sample throughput [34]. This method is often called flash chromatography. The only modification required to the classical technique is a flow control valve at the head of the column that is connected to a pressure-regulated gas supply. The columns are dry-packed with sorbent (40-63 micrometer diameter) with vertical tamping and then conditioned with solvent to dislodge all air pockets prior to use. As the head pressure is quite low, the bed height is restricted to less than 30 cm usually, with 15 cm commonly used. The choice of column dimensions for a particular separation can be deduced from the considerations

given in Table 6.3. Typical solvent flow rates are on the order of 5 cm/min.
Flash chromatography provides a rapid and inexpensive general method for the
preparative-scale separation of mixtures requiring only moderate resolution.

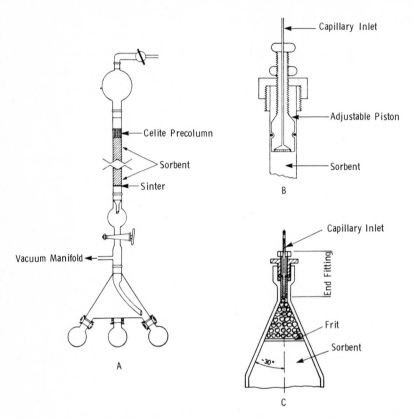

Figure 6.3 Special devices used in modern liquid chromatography. A, an
apparatus suitable for vacuum liquid chromatography; B, modified
inlet for medium-pressure liquid chromatography with piston fitting
and sample distribution devices; C, modified inlet providing an
alternative method of radial sample distribution to B.

A popular variation on the classical column technique is medium pressure
liquid chromatography [35-38]. This represents a mid-point in the transition
between the classical and high performance chromatographic methods. By using
glass and polymeric materials for construction of the apparatus and column
packings with diameters larger than 20 micrometers the cost of the system is
kept comparatively low while efficiency and sample throughput are much higher
than the classical technique. The construction materials usually limit column
pressures to less than 120 p.s.i. These pressures and the high flow rates used
with large diameter columns can be generated by inexpensive flow-metering

pumps. All connections between the pump, injector, column, and detector are made with Teflon tubes or similar materials having specially-designed plastic end fittings that make the assembly of the equipment straightforward. The operation of the medium-pressure liquid chromatograph is similar to that of a high-pressure liquid chromatograph and requires no special description. All necessary components are commerically available. A fraction collector is normally used instead of or in conjunction with a flow-through detector for sample collection. In this respect detectors of low sensitivity are required. The refractive index detector or a UV detector with a cell of short path length are frequently used.

TABLE 6.3

GENERAL CONDITIONS FOR PREPARATIVE-SCALE SEPARATIONS BY FLASH CHROMATOGRAPHY

Mobile phase selection: (1) Choose a low viscosity solvent system which separates the mixture and moves the desired component to an R_F of $\simeq 0.35$; (2) If several compounds are to be separated which run close together on TLC, adjust the solvent strength to put their midpoint at an R_F value of $\simeq 0.35$; (3) If compounds are well separated, choose a mobile phase which provides an R_F value of $\simeq 0.35$ for the least retained component.

Column Diameter (cm)	Volume of eluent[a] (ml)	Sample loading for a particular TLC resolution (mg) $\Delta R_F \geq 0.2$	$\Delta R_F \geq 0.1$	Typical fraction size (ml)
1	100	100	40	5
2	200	400	160	10
3	400	900	360	20
4	1600	1600	600	30
5	2500	2500	1000	50

[a] Typical volume of solvent required for column packing and sample separation.

Column sizes used in medium-pressure liquid chromatography are varied. The length of the column is limited by the operating pressure available and is not usually greater than one meter, with 25-60 cm commonly employed. The column diameter is established by the sample size and is usually 1-20 cm. The wider bore columns contain several kilograms of packing material and can handle samples of 15 g or more. Column packings are usually 40-63 micrometers in diameter and include reversed-phase as well as adsorbent materials [38]. The design of column end fittings is critical to avoid loss of sample resolution [36,39]. In one design, of which there are several commercially available variations, the column is a straight open-ended glass tube with a screwed end-cap attached (Figure 6.3B). The end fitting serves to locate an adjustable plunger that can be screwed in or out and locked into place with a nut. The plunger is sealed in the column by either an O-ring or an expansion mechanism. It is attached to a porous plastic disc which is in contact with the stationary phase bed and distributes the incoming liquid over the entire cross-section of the column. An identical plunger assembly supports the bed at the lower end of the column, so that the stationary phase is confined between two rigid plane parallel surfaces without dead space. In an alternative design, shown in Figure 6.3C, the column inlet is constricted to a narrow tube containing a plastic end fitting. At the injection end a frit is fitted into the cone of the column head. The space above the frit is filled with glass beads (100–500 micrometers). This simple device ensures optimum distribution, and results in a narrow, homogeneous band of sample applied over the entire cross-section of the column. The bottom of the column has a conical base similar to that at the top but without the fritted disc. The columns are usually dry-packed either manually or mechanically, details are given in references [26,36,37]. Prior to use the columns are conditioned by passing 3 to 5 column volumes of solvent through the bed and any voids which form are then filled. The sample is usually injected on-stream using a valve injector or, for very large sample volumes, through the pump. Various studies have shown that a higher column efficiency is usually attained when the sample is added to the column as a dilute solution rather than in concentrated form [23,26,36]. An attractive feature of the reversed-phase columns is that solvent equilibration is rapid so gradient elution and column reuse are possible. Medium-pressure liquid chromatography is a useful compromise between efficiency and cost for preparative-scale liquid chromatography. Requiring only simple equipment, it has rapidly become a routine technique in organic synthesis and biochemistry laboratories.

6.4 High-Pressure Liquid Chromatography

When choosing the best preparative-scale liquid chromatographic method for conditions requiring higher resolution than those discussed previously, it is necessary to first consider the intended use. For example, 1–10 mg of a pure sample would suffice for identification purposes, while 1–10 g might be the minimum amount of an intermediate useful for an organic synthesis. The amount of sample required dictates the size of the column and the operating conditions necessary for the separation, Table 6.4 [26]. This leads to two distinct possibilities for method development. Wide bore columns containing packings similar to those used in analytical columns can be used to obtain samples in the 1–100 mg range. The same columns or still larger, higher capacity columns packed with coarser particles and operated in a mass overload condition can separate 1–50 g sample amounts. The former method is the easier as it represents a simple scale-up of the analytical procedure and can probably be performed using an analytical instrument with very minor modifications. Large diameter columns operated in a mass overload condition require additional operating considerations and perhaps extensive instrument redesign. If more

than a few separations are to be carried out in this mass range then the
purchase of a purpose built preparative-scale instrument should be considered.
We will consider the two operating modes separately, although they share the
same goal, to maximize sample throughput per unit time, and differ only in
scale. In analytical high performance liquid chromatography (HPLC) speed and
resolution are the desired goals for which sample capacity is normally
compromised. By contrast, in preparative HPLC capacity (sample throughput per
unit time) is the desired goal and to maximize this parameter some of the
separation, or resolution, is sacrificed. For general reviews of
preparative-scale HPLC see references [26,40–43].

TABLE 6.4

MAXIMUM ALLOWABLE COMPOUND SIZE WITH RESPECT TO STATIONARY PHASE LOADING AND
COLUMN DIMENSIONS

Preparative column type	Application	Required amount of pure sample (mg)	Stationary phase (g)	Column I.D. (mm)	MACS[a] (mg)
Analytical	Mass spectrometry	0.001	0.2–3.2	1–5	0.2–3.2
Analytical	IR	0.1	0.2–3.2	1–5	0.2–3.2
Wide bore analytical	NMR	0.1–10	3–12	6–11	3–12
Wide bore analytical	Elemental analysis	1–25	7–25	6–11	7–25
Long narrow	Synthesis	100–1000	25–100	10–30	20–1000
Short thick	Large scale	10^3–10^5	10^2–10^4	20–100	10^3–10^5
Industrial	Commercial	10^4–10^6	10^3–10^5	10^2–10^3	10^4–10^6

[a] These values of MACS can vary and are controlled by the separation factor, a.

The scale-up approach to preparative HPLC is quite straightforward and is
the approach likely to be taken in an analytical laboratory that requires
sufficient material for identification purposes or to purify an analytical
standard on an occasional basis. Separations are carried out in the linear
region of the sorbent isotherm, which provides an upper sample mass limit of
about 0.1–1.0 mg/g of sorbent. Increasing the column diameter at constant
column length increases the weight of packing, and therefore the sample capacity
of the column, without dramatically influencing the resolution obtained if the
mobile phase linear velocity remains constant [44,45]. The loading capacity of
a column will increase with the square of the radius or directly with the column
cross-sectional area. A similar relationship holds for the mobile phase flow
rate, which must be increased linearly with the cross-sectional column area to

maintain a constant mobile phase linear velocity. Thus, to scale-up an analytical separation a larger diameter column packed with the same material and operated at the same linear velocity as the analytical column is used. The characteristics of three columns of different diameters operated at the same linear velocity are summarized in Table 6.5. Although preparative and analytical columns differ in length, the salient operating features can be seen. The analytical column with an internal diameter of 4.6 mm is adequate to isolate a few milligrams of material. The preparative column with an internal diameter of 22 mm can be used to isolate approximately 100 mg or so of sample, which is more than adequate for identification purposes. The penalty paid for the increased sample throughput is the increased mobile phase flow rate and higher column operating pressure, creating greater demands on the instrumentation used. A practical example is shown in Figure 6.4 for the scaled-up preparative separation of a mixture of bilirubin isomers [46].

TABLE 6.5

PHYSICAL PROPERTIES AND OPERATING CHARACTERISTICS OF COLUMNS OF DIFFERENT SIZE

Parameter	Column type		
	Analytical	Semipreparative	Preparative
Length (cm)	25	50	50
Internal diameter (mm)	4.5	9.4	22
Particle size (m)	10	10	10
Packing weight (g)	2.5	20	120
Flow rate (ml/min)	1	4	20
Pressure (p.s.i.)	200	300	800
Void volume (ml)	2.97	31.8	129
Maximum allowable sample size (mg)	0.25-2.5	2-20	12-120

At this point it is worth considering the demands made on the instrumentation for operation with wide bore columns and, in particular, the adaptation of analytical instruments for this purpose [42,48]. The pumping requirements differ from those in analytical HPLC as the ability to generate high flow rates at moderate back pressure is crucial to the efficient operation of wide bore columns. A flow rate maximum of 100 ml/min with a pressure limit of 3000 p.s.i. is considered adequate. Few analytical reciprocating piston-type pumps are capable of reaching this volume delivery rate; some are capable of operating at 30-60 ml/min, which suffices in many instances; still others have a maximum delivery rate below 10 ml/min, barely adequate for all but semipreparative use. As detectors are operated at low sensitivity in preparative-scale separations a relatively inexpensive HPLC pump is often adequate and can be used instead of the analytical pump. As flow rates are high the solvent reservoir should also be sufficiently large, at least a few liters, to permit reasonable operating times without the need for replenishing.

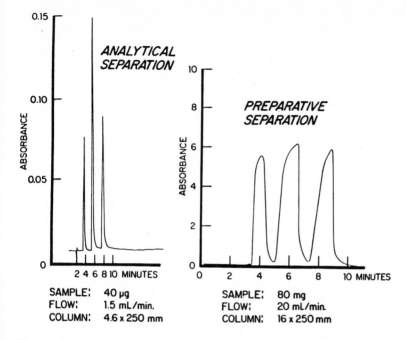

Figure 6.4   Preparative-scale separation of bilirubin isomers by HPLC. The
analytical separation was optimized for   at minimum k and then the
sample size scaled-up to that allowed by the larger amount of
packing in the preparative column. (Reproduced with permission from
ref. 46. Copyright Perkin-Elmer Corporation).

The mode of injection differs appreciably from analytical methods where the
object is to create discrete narrow bands by point or narrow zone injection of
low sample amounts. Point injection of a large sample (mass or volume) in
preparative liquid chromatography would result in a local overloading of the
column packing with a deleterious effect on column performance. To minimize
this effect the sample should ideally be applied evenly over the entire column
cross-sectional area as a dilute solution. A column with a glass-bead-filled
conical head or with a distribution or diffuser plate can be used with a
standard valve injector for this purpose. The valve must be fitted with a loop
of sufficient volume. However, for large injections, sample volumes of several
milliliters, it is easier to pump the sample onto the column with the eluent
pump or, preferably, a dedicated injection pump. On-column syringe injection
under stop-flow conditions is also possible but inconvenient and experimentally
difficult.

Column packing techniques were discussed in Chapter 4.4. Columns may be
slurry packed or dry-packed, depending on the particle diameter; various methods
for packing preparative columns have been surveyed and evaluated in references
[49,50]. Slurry packing large-diameter columns is more difficult than

small-diameter columns due to the need to keep large quantities of slurry in suspension and the high volume flow rates required to rapidly consolidate the column bed. Columns which are poorly packed may give rise to peak doubling, caused by the formation of a high permeability zone near the wall [50]. The cost of sorbents is another consideration. Coarser particles are less expensive, can be dry packed, and may provide adequate resolution for all but the most demanding applications. Macroporous polystyrene-divinylbenzene resins have been suggested as an inexpensive alternative to reversed-phase materials [52]. Stable over the entire pH range, they are ideally suited to biochemical applications such as the separations of amino acids and peptides. However, they have low maximum operating pressure limits.

Detection requirements in preparative-scale chromatography also differ from analytical operations where detectors are selected for their sensitivity. Sensitivity is not of overriding importance in preparative-scale chromatography; the ability to accommodate large column flow rates and a wide linear response range are more useful. The sensitivity of the refractive index detector is usually quite adequate for preparative work but the flow capacity of many of the analytical detectors of this type is limited. The common analytical UV detectors have a limited dynamic range and are easily overloaded when running preparative-size samples. Variable wavelength detectors can be detuned from the absorption maxima of the sample components to increase the dynamic range of the detector at the higher concentration end. Some analytical detectors have interchangeable flow cells which allow replacement of the analytical cell with one of shorter path length to reduce sensitivity. Preparative flow cells usually have wider bore inlet and outlet connections to more readily accommodate high mobile phase flow rates. In analytical instruments capillary tubing of 0.15-0.25 mm internal diameter is used in column connections to reduce dispersion. This is too restrictive at preparative-scale flow rates and should be replaced with capillary tubing of 0.4-0.6 mm I.D. To avoid damage to the flow cell it may be advisable to operate analytical detectors with a low dead-volume flow splitter so that only a few percent of the total column eluent passes throught the detector. In this case it may also be necessary to provide a flow restrictor in the line that does not lead to the detector to balance the flow resistance. Finally, if only one or two fractions are to be collected and the number of runs is low, then the sample can be collected manually in suitably-sized containers; a fraction collector is otherwise required. Various models varying from simple time- or volume-based programmable collectors to more sophisticated peak detector threshold-sensing fraction collectors are commercially available. If an anlytical instrument is to be used for more than an occasional preparative separation then automation of the whole separation process is worth considering [54-56]. This may be possible within the existing capabilities of many of the more sophisticated microprocessor-based analytical instruments.

So far we have considered preparative-scale chromatography when the column is not overloaded and the desired amount of sample has been isolated by simply scaling-up the dimensions of the column and adjusting the operating conditions accordingly. The theory pertaining to analytical columns is applicable and only minor operating modifications are required to ensure success. As there is a limit to the size of columns that can be prepared with the same efficiency as analytical columns, the only pratical way to increase sample throughput is to overload the column. There is no adequate theory for columns operating in a mass overload condition and the success of this approach must be judged on empirical grounds. As it represents the most cost effective means of obtaining bulk sample quantities and is widely used in many laboratories there is no need to fear or avoid operating above the column capacity limit. A column is considered to be in an overload condition when solute capacity factor values change by more than 10% as sample weight is increased. A column might be overloaded either by increasing the sample concentration while maintaining a constant injection volume (mass overload) or by increasing the injection volume while maintaining a constant sample concentration (volume overload) [45,57-59]. As both of these situations have different consequences we will consider then separately.

The column overload volume has been defined by Scott [57]. To a first approximation it can be described by equation (6.1).

$$V_L = V_o (a - 1)k \tag{6.1}$$

V_L = maximum volume overload
V_o = column dead volume
a = ratio of capacity factors for the separation of the most difficult peak pair
k = capacity factor value for the first peak of the peak pair used to define a

As the injection volume is increased the retention distance of the peak front remains constant while the peak width increases linearly with injection volume. The spread of the peaks is the same for each solute and is independent of the nature of the solute or its capacity factor value. The spread of the peaks toward greater retention times is characteristic of volume overload and is, in fact, the exact opposite of the situation observed for mass overload. Volume overload distorts the normal elution process of development toward frontal analysis and, when elution development is carried out with progressively increasing sample volumes, the distorted elution development culminates in frontal analysis. For a fixed sample concentration the highest sample throughput is obtained by using an injection size equal to V_L. This assumes that the sample concentration is not so high that the separation process is dominated by mass overload conditions.

A further consideration is the selection of the injection solvent. Ideally the injection solvent should be the same as the mobile phase chosen for eluting the sample or, if gradient elution is to be used, the initial solvent used in the solvent program. If this is not possible then the next most appropriate choices, in order or preference, are (1) a different solvent of similar strength but of different selectivity, (2) a weaker solvent of large volume, or (3) a small amount of a more polar solvent. Selection of the injection solvent is less of a problem in liquid-solid chromatography than in reversed-phase chromatography where the aqueous solubility of many organic compounds is rather low.

Column mass overload is too complex to treat theoretically. However, mass overload influences the separation process in basically three ways. First of all, there is a dispersive effect resulting from the limited capacity of the stationary phase. The sample will spread along the column, carried by the mobile phase, until it contacts sufficient sorbent to permit it to remain on the stationary phase surface under equilibrium conditions. Secondly, with a massive sample charge the sorbent is essentially deactivated while the overloaded solute is in the column; all solutes are thus eluted with reduced retention times. Finally, the high concentration of solute on the sorbent surface that results from mass overload will result in non-linear adsorption isotherms and the eluted peaks will exhibit pronounced tails. Figure 6.5 is an example of a separation performed under mass overload conditions. The dotted line represents the analytical separation used to optimize the separation conditions (i.e., maximize a). Trial and error must be used to establish the maximum allowable sample size in this case.

Multidimensional chromatographic techniques, described more fully in Chapter 4.18, can also be used to increase the number of theoretical plates available, reduce analysis times, or increase the sample capacity in preparative liquid chromatography. Figure 6.6 is an example of the use of recycle chromatography to separate two closely eluting components after heartcutting the original sample. The two components are completely separated after eight passes through the column. Peak shaving the ends of the fused peaks in cycle 4 could be used to collect pure fractions of both peaks. This saves time when it is not necessary to collect the entire sample. An alternative method would be to use a precolumn to fractionate the sample so that only the shaded peaks are passed to the main column. The front and back ends of the chromatogram could be vented to

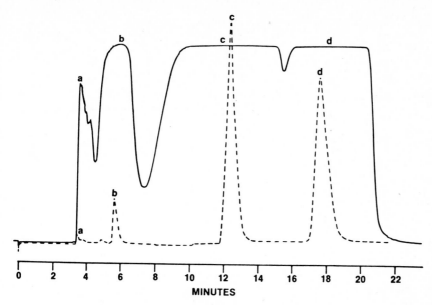

Figure 6.5 Separation of alkylbenzyl alcohol isomers by reversed–phase
semipreparative HPLC in a mass overload condition. The dotted line
is the equivalent analytical separation on the same column.
(Copyright Whatman Inc.)

waste. The separation conditions on the main column could then be optimized to
provide the desired separation in the minimum number of column cycles. Gradient
elution in preparative liquid–solid chromatography is often difficult due to
slow column equilibration and the large solvent volumes required to return the
column to its original operating conditions. For the above reasons
multidimensional chromatographic techniques are very compatible with the goals
of preparative liquid chromatography.

New column technology has been incorporated into two commerically available
preparative liquid chromatographs for separating mixtures in the 1 to 50 g
range. Radial compression columns are used by Waters Associates (see also
Chapter 4.6) [60,61]. A flexible–walled plastic cartridge, 30 x 5.7 cm, is
dry-packed with silica gel of 75 micrometers average particle diameter. This
cartridge is placed into a stainless steel pressure vessel and the ends of the
cartridge are mechanically sealed to the inlet and outlet at the ends of the
pressure vessel. Nitrogen at 500 p.s.i. is introduced into the space between
the inside wall of the pressure vessel and the outside wall of the cartridge.
The column is compressed radially, consolidating the bed, eliminating voids,
and/or the wall effect. Provided that a pressure differential of about 200
p.s.i. is maintained between the column wall and the column inlet, higher
efficiency than would otherwise be expected is obtained for these columns. Two

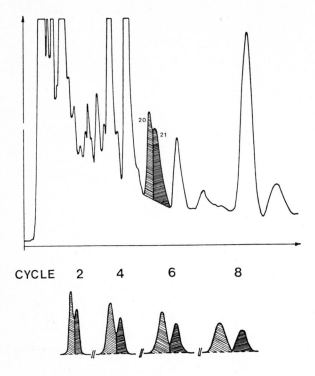

CYCLE   2   4   6   8

Figure 6.6   Preparative-scale separation of a plasma sample containing two drug
             metabolites marked 20 and 21. By heartcutting and recycle
             chromatography peaks 20 and 21 were isolated in a pure form.

cartridges can be operated in series to improve the efficiency and increase the
sample capacity. The instrument also allows the use of recycle and peak shaving
techniques. Typical eluent flow rates are on the order of 50-500 ml/min.

Jobin-Yvon have introduced a column piston device for preparing axially-
compressed columns [62-64]. This device has the advantage that both column
length and packing type can be readily changed by the operator. The column
consists of a stainless steel cylinder, 70 cm x 1.8 cm I.D. or 100 cm x 8 cm
I.D., containing a hydraulic piston at its base. The piston head is porous so
that the column eluent can be run off and is seated with a Teflon O-ring seal to
make a tight fit with the column walls. A removable injection head allows easy
column filling. A slurry of the column packing material (e.g., 20-50 micrometer
diameter silica gel) is poured into the column, the injection head is bolted
down, and the piston is pressurized to 15-150 p.s.i. The movement of the piston
consolidates the column bed and excess slurry solvent is filtered off through
the injection head. The column is then ready for use in the normal way. After
the separation, the injection head is unbolted and the packing pushed out by the
piston. It is thus not necessary to elute all components through the column;

well-retained components can be recovered or disposed of by sectioning the column bed. With the larger column, samples of 50 to 100 g can be separated per injection. Column length, capacity, selectivity, and resolution are controlled by the nature and amount of the column packing material added to the cylinder. This approach is obviously very flexible and cost effective as the column can be varied to meet the needs of each individual analysis.

To separate mixtures containing substances of widely differing polarities, a technique called flip-flop chromatography has been introduced [65]. The sample is deposited in the pores of a support such as silica gel. Depending on the pore volume of the support, up to 40% w/w of sample may be added to give a free flowing powder. The sample is then packed into a column and placed in series with two stripping columns. The stripping columns contain packings of opposite polarity, for example silica gel and reversed-phase materials, and are placed at opposite ends of the sample column. The sample column is extracted by a combination of four or more polar and nonpolar solvents. These are applied in order of alternating polarity, beginning with the most extreme, so that the tail ends of the polarity distribution of the sample are successively extracted, leaving behind material with a more restricted polarity range. The stripping columns prevent the removal of substances whose polarity is very different from that of the extracting solvent. To take full advantage of the fractionation process the column train should be used with a high-pressure liquid chromatograph. Compared to stepwise extraction using the same solvents, flip-flop chromatography provides a higher extraction efficiency and less cross-contamination of neighboring fractions.

Horvath has recently suggested that displacement chromatography be used for preparative-scale separations to ovecome the poor utilization of the stationary and mobile phases that results from the low sample concentrations permissible under normal operating conditions in elution chromatography [66-68]. In displacement chromatography, the sample is first loaded onto the column in a carrier solvent that is a weak eluent for the sample. The carrier solvent is then replaced by a solution of a displacer substance, which has a higher affinity for the stationary phase than any of the sample components and is present in sufficiently high concentration to cause complete sample displacement. If the column is sufficiently long, successive displacements of the sample components by the displacer and by each other result in a fully developed "displacement train". This train consists of adjacent square-wave concentration pulses of the individual sample components in order of increasing affinity for the stationary phase. Optimizing the above situation from the theoretical point of view is very complex but, fortuitously, appropriate conditions can be found fairly quickly using TLC [68]. Using regular analytical reversed-phase HPLC columns samples of up to several milligrams per injection of steriods and antibiotics have been separated. A disadvantage of the procedure is that after sample recovery, the displacer must be stripped from the column and the column regenerated to its original state before a second sample can be separated.

6.5 Countercurrent Chromatography

It has long been recognized that substances can be separated by their ability to distribute themselves between two immiscible liquids. The concentration of solute in either phase depends on the distribution function, a measure of the selectivity of the separation process. To effect a separation in a single distribution, or a few distributions, as might conveniently be carried out with a separatory funnel, a large difference in partition functions is required. This is obviously a restrictive criteria for all but the simplest separations. The efficiency of the separation process can be improved in two ways: by modifying the distribution coefficient or by increasing the number of equilibrium steps (extractions). Actually, in all but a few instances, both are necessary to fractionate a moderately complex mixture. This procedure would be tedious manually and automation is essential. The countercurrent distribution apparatus allows one thousand or more equilibration steps between two immiscible liquids of fixed volume and different densities [23,24,69,70]. Mixing, separation, and phase transfer between stages are automated using an electrical timer and a motor-driven shaft, on which the individual stages are mounted in series. The capacity of the apparatus for preparative work was improved by the provision of continuous sample feed. However, certain disadvantages remained, such as the complexity and physical size of the apparatus, problems with emulsion formation, the volume of solvent required was fixed by the size of the apparatus and was often excessive, and long separation times were required. Although widely used in the 1940's and 1950's, few of these countercurrent distribution machines are in use today, due to their lack of speed and efficiency. They have one basic advantage, that of the absence of a solid stationary phase on which irreversible sorption may occur. Thus they are still occasionally used for the separation of polar compounds, including natural products, polypeptides, and other macromolecules.

A renaissance of interest in countercurrent processes has occurred in the last few years due to the development of small laboratory-scale units which work in a continuous manner [72,73]. In droplet countercurrent chromatography a number of vertical, long, narrow glass tubes (e.g., 40 cm x 2 mm I.D.) are connected in series by capillary PTFE tubes [74,75]. The vertical tubes are filled with stationary phase and the mobile phase is forced to flow through the tubes as droplets without any turbulence. The sample, dissolved in a mixture of stationary and mobile phases, is injected into the apparatus and separation is achieved by multiple partition between the droplets and the stationary phase as the mobile droplets pass through the columns. Mobile phase is continuously pumped through the apparatus at the rate of 10-15 ml/h and the eluates are collected in 1-3 ml fractions. The main disadvantage of droplet countercurrent chromatography is the need to limit the choice of solvents to those which form droplets in the stationary phase [77]. A rapid screening method using TLC has been developed for solvent selection [78]. Nearly all systems used to date contain water; water-methanol-chloroform mixtures are particularly popular. Droplet countercurrent chromatography is simple and reproducible, the consumption of solvent is low, sample amounts from milligrams to several grams can be easily handled, and crude samples can be used without preliminary fractionation (no irreversible adsorption). Typical separation times are on the order of several hours.

Several commercially available countercurrent chromatographs use rotation or centrifugal forces to maintain a hydrodynamic equilibrium between a stationary phase in a series of interconnected coiled columns while mobile phase is pumped through them. In the coil planet centrifuge, the separation columns, arranged on the rotor of a centrifuge with their longitudinal axis parallel to the direction of the centrifugal force, are connected to one another by fine tubes [79,80]. The separation columns are filled with stationary phase prior to the start of the experiment. While the rotor of the centrifuge is in motion the mobile liquid phase is pumped into the separation columns and bubbles up through the stationary phase by the action of centrifugal force. The horizontal flow-through coil planet centrifuge is in fact two instruments in one, containing an analytical section counterbalanced by a planetary gear-driven preparative section [81-85]. The coiled column is rotated perpendicularly to the centrifugal field with simultaneous rotation around its own axis. The actual path described by the column and the forces experienced by the phases are fairly complex and will not be discussed here [81]. Its main advantage is that it eliminates the need for a rotating seal, which is often the weak component of other centrifugal countercurrent chromatographs. Separations of up to 100 milligrams have been demonstrated with this apparatus. For the separation of gram quantities the slowly rotating coil countercurrent chromatograph is recommended [86-89]. It consists of a number of horizontal interconnected coiled tubes that are slowly rotated around their axis in a gravitational field. The stationary phase is retained by gravity in large-diameter coils rotated at speeds up to 100 rpm.

The instruments discussed above are capable of separating samples of several grams in periods of 4 to 12 h with an efficiency of 400-1500 theoretical plates. They are suitable for separating simple mixtures containing

approximately ten or fewer components. Countercurrent chromatography is
particularly attractive for preliminary fractionation of complex mixtures prior
to further analysis or for the isolation of compounds which are unsuitable for
separation by liquid chromatography due to undesirable interaction with the
stationary phase (e.g., irreversible adsorption). For these reasons most of the
proven applications in the literature involve the isolation of natural products
or biochemical compounds from crude samples.

6.6 <u>References</u>

1. P. E. Barker, in C. E. H. Knapman (Ed.),"Developments in Chromatography",
    Applied Science Publishers, London, Vol. 1, 1978, p. 41.
2. R. Bonmati and G. Guiochon, Perfum. Flavor., 3 (1978) 17.
3. R. G. Bonmati, G. Chapelet-Letourneux, and J. Reed Margulis,
    Chem. Engin., March 24 (1980) 70.
4. J. Guyimesi and L. Szepsey, Chromatographia, 9 (1976) 195.
5. K. I. Sakodynskii, S. A. Volkov, Yu. A. Kovan'Ko, V. Yu. Reznikov,
    and V. A. Averin, J. Chromatogr., 204 (1981) 167.
6. G. Schomburg, H. Kotter, and F. Hack, Anal. Chem., 45 (1973) 1236.
7. D. A. Leathard and B. C. Shurlock, "Identification Techniques in Gas
    Chromatography", Wiley, New York, 1970, p. 191.
8. G. Magnusson, J. Chromatogr., 109 (1975) 393.
9. A. A. Casselman and R. A. B. Bannard, J. Chromatogr., 90 (1974) 185.
10. H. T. Badings, J. J. G. van der Pol, and J. G. Wassink, Chromatographia,
    8 (1975) 440.
11. G. W. A. Rijnders, Adv. Chromatogr., 3 (1966) 215.
12. K. P. Hupe, Chromatographia, 1 (1968) 462.
13. M. Verzele, in E. S. Perry (Ed.), "Progress in Separation and
    Purification", Wiley, New York, 1 (1968) 83.
14. A. Zlatkis and V. Pretorius (Eds.), "Preparative Gas Chromatography",
    Wiley, New York, 1971.
15. B. Roz, R. Bonmati, G. Hagenbach, P. Valentin, and G. Guiochon,
    J. Chromatogr. Sci., 14 (1976) 367.
16. J. R. Conder, J. Chromatogr., 256 (1983) 381.
17. J. Albrecht and M. Verzele, J. Chromatogr. Sci., 8 (1970) 586.
18. D. A. Craven, J. Chromatogr. Sci., 8 (1970) 540.
19. J. Albrecht and M. Verzele, J. Chromatogr. Sci., 9 (1971) 745.
20. C. E. Reese and E. Grushka, Chromatographia, 8 (1975) 85.
21. E. Geeraert and M. Verzele, Chromatographia, 11 (1978) 640.
22. L. S. Ettre, in C. Horvath (Ed.), "High-Performance Liquid Chromatography",
    Academic Press, New York, Vol. 1, 1980, p. 1.
23. C. J. O. R. Morris and P. Morris, "Separation Methods in Biochemistry",
    Wiley, New York, 1976.
24. O. Mikes (Ed.), "Laboratory Handbook of Chromatographic and Allied
    Methods", Wiley, New York, 1979.
25. E. Heffman (Ed.), "Chromatography. Fundamentals and Applications of
    Chromatographic and Electrophoretic Methods", Elsevier, Amsterdam, Parts A
    and B, 1983.
26. M. Verzele and E. Geereart, J. Chromatogr. Sci., 18 (1980) 559.
27. B. Loev and M. M. Goodman, in E. S. Perry and C. K. Van Oss (Eds.),
    "Progress in Separation and Purification", Wiley, New York, Vol. 3, 1970,
    p. 73.
28. F. M. Rabel, in M. Zief and R. Speights (Eds.), "Ultrapurification",
    Dekker, New York, 1972, p. 157.
29. B. Low and K. Snader, Chem. Ind. (London), (1965) 15.
30. B. Low and M.Goodman, Chem. Ind. (London), (1967) 2026.
31. B. Bowden, J. Coll, S. Mitchell, and G. Stokie, Aust. J. Chem., 31 (1978)
    1303.

32. J. Coll, S. Mitchell, and G. Stokie, Aust. J. Chem., 30 (1977) 1859.
33. N. M. Targett, J. P. Kilcoyne, and B. Green, J. Org. Chem., 44 (1979) 4962.
34. W. C. Still, M. Kahn, and A. Mitra, J. Org. Chem., 43 (1978) 2923.
35. A. I. Meyers, J. Slade, R. K. Smith, E. D. Mihelich, F. M. Merchenson, and C. D. Liang, J. Org. Chem., 44 (1979) 2247.
36. H. Loibner and G. Seidl, Chromatographia, 12 (1979) 600.
37. M. Radke, H. Willsch, and D. H. Welte, Anal. Chem., 52 (1980) 406.
38. F. Eisenbeib and H. Henke, J. High Resolut. Chromatogr. Chromatogr. Commun., 2 (1979) 733.
39. W. N. Musser and R. E. Sparks, J. Chromatogr. Sci., 9 (1971) 116.
40. J. J. De Stefano and J. J. Kirkland, Anal. Chem., 47 (1975) 1103A.
41. A. Wehrli, Z. Anal. Chem., 277 (1975) 289.
42. L. R. Snyder and J. J. Kirkland, "Introduction to Modern Liquid Chromatography", Wiley, New York, 2nd Edn., 1979, p. 615.
43. C. E. Reese, in C. F. Simpson (Ed.), "Techniques in Liquid Chromatography", Wiley, New York, 1982, p. 97.
44. J. P. Wolf, Anal. Chem., 45 (1973) 1248.
45. J. J. De Stefano and H. C. Beachell, J. Chromatogr. Sci., 10 (1972) 654.
46. J. L. DiCesare and F. L. Vandemark, Paper No. 701 presented at the Pittsburgh Conference for Analytical Chemistry and Applied Spectroscopy, Atlantic City, New Jersey, 1981.
47. A. W. J. De Jong, H. Poppe, and J. C. Kraak, J. Chromatogr., 148 (1978) 127.
48. K. P. Hupe and H. H. Lauer, J. Chromatogr., 203 (1981) 41.
49. J. Klawiter, M. Kaminski, and J. S. Kowalczyk, J. Chromatogr., 243 (1982) 207.
50. M. Kaminski, J. Klawiter, and J. S. Kowalczyk, J. Chromatogr., 243 (1982) 225.
51. E. Geereart and M. Verzele, Chromatographia, 12 (1979) 50.
52. D. J. Pietrzyk, W. J. Cahill, and J. D. Stodola, J. Liq. Chromatogr., 5 (1982) 443.
53. D. J. Pietrzyk and J. D. Stodola, Anal. Chem., 53 (1981) 1822.
54. P. A. Bristow, J. Chromatogr., 122 (1976) 277.
55. K.-P. Hupe, H. H. Lauer, and K. Zech, Chromatographia, 13 (1980) 413.
56. D. Berger and B. Gilliard, J. Chromatogr., 210 (1981) 33.
57. R. P. W. Scott and P. Kucera, J. Chromatogr., 119 (1976) 467.
58. B. Coq, G. Cretier, C. Gonnett, and J. L. Rocca, Chromatographia, 12 (1979) 139.
59. B. Coq, G. Cretier, and J. L. Rocca, J. Chromatogr., 186 (1979) 457.
60. J. N. Little, R. L. Cotter, J. A. Prendergast, and P. D. McDonald, J. Chromatogr., 126 (1976) 439.
61. D. J. Pietrzyk and W. J. Cahill, J. Liq. Chromatogr., 5 (1982) 781.
62. E. Godbille and P. Devaux, J. Chromatogr. Sci., 12 (1974) 564.
63. E. Godbille and P. Devaux, J. Chromatogr., 122 (1976) 317.
64. G. Cretier and J. L. Rocca, Chromatographia, 16 (1983) 32.
65. A. J. P. Martin, I. Halasz, H. Engelhardt, and P. Sewell, J. Chromatogr., 186 (1979) 15.
66. H. Kalasz and C. Horvath, J. Chromatogr., 215 (1981) 295.
67. C. Horvath, A. Nahum, and J. F. Frenz, J. Chromatogr., 218 (1981) 365.
68. H. Kalasz and C. Horvath, J. Chromatogr., 239 (1982) 423.
69. L. C. Craig, W. Hausmann, P. Ahrens, and E. J. Harfenist, Anal. Chem., 23 (1951) 1326.
70. L.C. Craig and D. Craig, in A. Weissberger (Ed.), "Techniques of Organic Chemistry", Wiley, New York, 2nd. Edn., 1956, p. 149.
71. A. E. O'Keeffe, M. A. Dolliver, and E. T. Stiller, J. Amer. Chem. Soc., 71 (1949) 2452.
72. N. B. Mandava, Y. Ito, and W. D. Conway, Amer.Lab., October (1982) p. 62.
73. N. B. Mandava, Y. Ito, and W. D. Conway, Amer.Lab., November (1982) p. 48.
74. T. Tanimura, J. J. Pisano, Y. Ito, and R. L. Bowman, Science, 54 (1970) 169.

75. K. Hostettmann, Adv. Chromatogr., 21 (1983) 165.
76. K. Hostettmann, M. Hostettmann-Kaldas, and O. Sticher, J. Chromatogr.,
    186 (1979) 529.
77. Y. Ogihara, O. Inoue, H. Otsuka, K.-I. Kawai, T. Tanimura, and S. Shibata,
    J. Chromatogr., 128 (1976) 218.
78. K. Hostettmann, M. Hostettmann-Kaldas, and K. Nakanishi, J. Chromatogr.,
    170 (1979) 355.
79. W. Murayama, T. Kobayashi, Y. Kosuge, H. Yano, Y. Wunagaki, and
    K. Nunogaki, J. Chromatogr., 239 (1982) 643.
80. L. A. Sutherland and Y. Ito, Anal. Biochem., 108 (1980) 367.
81. Y. Ito, J. Chromatogr., 188 (1980) 33.
82. Y. Ito, J. Chromatogr., 188 (1980) 43.
83. Y. Ito and G. J. Putterman, J. Chromatogr., 193 (1980) 37.
84. Y. Ito and R. L. Bowman, J. Chromatogr., 147 (1978) 22.
85. Y. Ito, Anal. Biochem., 100 (1979) 271.
86. Y. Ito and R. L. Bowman, J. Chromatogr., 136 (1977) 189.
87. Y. Ito and R. L. Bowman, Anal. Biochem., 85 (1978) 230.
88. Y. Ito, J. Chromatogr., 196 (1980) 295.
89. Y. Ito and R. Bhatnagar, J. Chromatogr., 207 (1981) 171.

Chapter 7

SAMPLE PREPARATION FOR CHROMATOGRAPHIC ANALYSIS

430

7.1 <u>Introduction</u>

In most instances, mixtures obtained from biological or environmental
sources are often too complex, too dilute, or are incompatible with the
chromatographic system to permit analysis by direct injection. Preliminary
fractionation, isolation, and concentration of the sample components of interest
are needed prior to analysis. Figure 7.1 illustrates a flow diagram for a
typical organic analysis. A representative sample is obtained and then
fractionated and concentrated by sample extraction and cleanup procedures to
ensure effective separation, detection, and system compatibility in the final

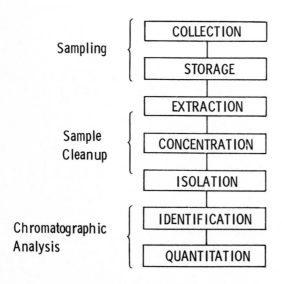

Figure 7.1 Flow diagram of the procedures used to prepare samples for
chromatographic analysis.

chromatographic determination.

Obtaining a representative sample is an important part of any analysis since errors or faults made at the sampling stage cannot be corrected at any later point in the analytical scheme [1-3]. Sampling may be a problem because of the size or lack of homogeneity of the sample. If one desired samples from a lake, then size would preclude total analysis, and subsamples would have to be taken. Any single sample may represent water at that location or depth and may not be representative of all water in the lake. Also the sample would contain two phases – the liquid phase and suspended matter. The sample may not be representative of the whole body of water as the concentration of suspended matter may vary with location and depth. Solutions to the latter problem include multiple or integrated sampling. It may be necessary to sample the two phases separately; adsorption techniques would be convenient for organics in water and filtration for a particulate sample. Large samples such as vegetable matter are invariably inhomogeneous and in a form not suited to chromatographic analysis. Sampling in this case may involve operations such as drying, cutting, grinding, mincing, pressing, or sieving to reduce the size and nonuniformity of the sample. As a dried, fine, free-flowing powder the sample could be remixed to provide an homogeneous subsample. The above examples are illustrative of problems often encountered in analysis as nearly all samples represent some problem which requires a solution before the analysis commences. Additional problems may arise from contamination due to careless collection and handling or alteration of the sample during storage. At trace levels almost any surface which the sample comes into contact becomes a possible source of contamination. Chemical processes that occur in the sample between the time of sample collection and analysis can invalidate the analytical results. Examples of such processes are photodecomposition, adsorption, vaporization loss, thermal decomposition, microbial action, and chemical reaction. Samples being held for extended periods prior to organic analysis should generally be stored in the dark in glass containers and maintained at subzero temperatures, since these conditions tend to inhibit the porcesses listed above [4-6]. Tissue and food samples when macerated may release enzymes capable of changing the sample composition upon storage. It may be preferable to store these samples whole, in the frozen state under nitrogen and to macerate them just prior to analysis [7]. Recommended practices for sampling biological materials for pesticide residues have been discussed by Leng [8]. Automation of the sample preparation process is discussed by Burns [9]. General reviews of sample preparation for gas chromatographic [10,11] and liquid chromatographic [12] analysis are available.

7.2 Isolation and Concentration Techniques using Physical Methods

The most frequently used methods of sample isolation and concentration for organic compounds involve distillation, extraction, and adsorption techniques. Some of these methods are summarized in Table 7.1 for the analysis of trace organic compounds in water [10,13]. These methods will be elaborated on below and in subsequent sections of this chapter.

Distillation is a suitable technique for the isolation of volatile organic compounds from liquid samples or the soluble portion of solid samples [7,11,14]. The physical basis of separation depends on the distribution of constituents between the liquid mixture and the vapor in equilibrium with that mixture. The more volatile constituents are concentrated in the vapor phase which is collected after condensation. The effectiveness of the separation is dependent on the physical properties of the components in the mixture, the equipment used, and the method of distillation. Several types of distillation columns can be used and are listed in Table 7.2. The single-stage concentric tube, rotary film, climbing-film, and falling-film evaporators operated at atmospheric or reduced pressure are the most useful for isolating organic volatiles. The fast-action climbing and falling-film evaporators are more efficient, easier to handle, and less likely to produce artifacts than the solvent distillation equipment found in most organic synthesis laboratories. Contact times of the solutes with heated surfaces are very short in such arrangements and concentration increases up to several hundred-fold can be obtained, depending on the product and the desired degree of isolation. Volatile polar organics such as alcohols, nitriles, aldehydes, and ketones are particularly suited to isolation by distillation, and frequently a clean extract is obtained that can be analyzed without further purification. Peters has described a small, all-glass distillation-concentration apparatus which is particularly well-suited for isolating concentrates of volatile polar organics for subsequent chromatographic analysis, Figure 7.2A [15]. Designed for aqueous samples, the apparatus consists of a round-bottomed flask, condenser, condensate collection (distillate) chamber, steam/water contact column, and an overflow return tube for returning a portion of the condensate to the distillation flask. In practice, the volatile polar organics will azeotropically distill into the distillate chamber and be preferentially retained there. The overflow of condensate back to the distillation flask is stripped by the rising steam and the volatile polar organics are recycled back into the distillate chamber. Two factors are important in judging the efficiency of the apparatus: the boiling point difference between water and the compound or its azeotrope and the time required to establish equilibrium between the vapor phase and condensate.

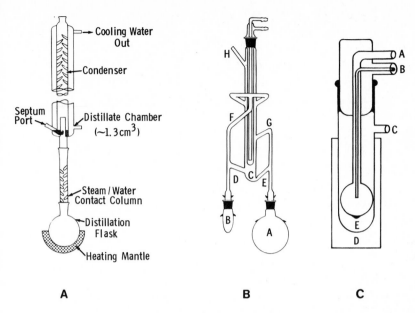

A   **B**   **C**

Figure 7.2   Apparatus used for sample preparation involving distillation or sublimation. A, distillation apparatus for organic volatiles in water; B, micro apparatus for continuous steam distillation with continuous liquid–liquid extraction of the distillate; C, micro sublimation apparatus.

At least a 1°C boiling point difference is required and an equilibration time less than one hour is preferred. The absolute recovery of volatile organics in the distillate chamber is typically 80%, with the remaining 20% of the compounds entrained in the reflux condenser. By continually withdrawing the organic concentrate from the distillate chamber after equilibrium has been established, virtually 100% recovery can be obtained. Concentration factors of 300- to 400-fold are possible with this apparatus.

      Steam distillation, a special distillation method, is used for the isolation of less volatile organic compounds. Vaporization of the sample is achieved by either continuously feeding steam into the mixture or generating steam in situ by heating the mixture with water to a sufficiently high temperature. Volatile and moderately volatile sample components are entrained in the steam vapor and carried over into the condenser, leaving involatile fractions and solid residues behind. For large sample sizes, phase separation occurs after condensation but solvent extraction is required for the recovery of small or dilute samples. Likens and Nickerson have described a widely-used apparatus for continuous steam distillation with continuous liquid–liquid extraction of the distillate [16]. Verzele has designed a micro version of this

TABLE 7.1

COMPARISON OF REPRESENTATIVE METHODS FOR THE CONCENTRATION OF TRACE ORGANIC COMPOUNDS IN WATER

Method	Principle	Scope	Comments
Freeze Concentration	Water sample is partially frozen, concentrating the dissolved substances in the unfrozen portion.	All sample types	Minimizes sample losses due to volatilization or chemical modification. Principal losses occur due to occlusion, adsorption, evaporation, and channelling the in ice layer. Limited sample size.
Lyophilization	Water sample is frozen and pure water is removed by sublimation under vacuum.	Nonvolatile organics	Can handle large sample volumes. Selective loss of volatile organics. Inorganic constituents concentrated simultaneously.
Vacuum Distillation	Water is evaporated at reduced pressure and at or near ambient temperature.	Nonvolatile organics	Slow process when sample volumes are large. Inorganics also concentrated. Sample contamination is low but sample may be modified due to thermal degradation or chemical and microbial reactions.
Reverse Osmosis	Water is forced through a membrane by application of pressure thereby enriching the water sample in constituents which ordinarily cannot pass through the membrane.	Molecular weight > 200	Compounds of small size are not concentrated. Inorganic materials may contaminate sample. Membranes may either adsorb constituents or release impurities into the sample.
Ultra-filtration	Water sample is filtered under pressure through a membrane that will pass molecular constituents below a certain size and retain those above that size.	Large molecules	Porosity of membrane determines the size of molecules concentrated. Usually used for compounds > 1000 molecular weight. Can concentrate large sample volumes at low temperatures.

Solvent Extraction	Aqueous sample is partitioned with an immiscible organic solvent. Extraction efficiency depends on the affinity of the solute for the organic solvent.	All sample types	Samples with a high affinity for water are not extracted. Extractions can be performed by a simple single equilibration or by multiple equilibrations with fresh solvent. Solvent impurities concentrated along with sample.
Surface Adsorption	Water sample is passed through a column of the adsorbent and the adsorbed organic constituents subsequently eluted with a smaller volume of organic solvent.	All sample types	Adsorbents include charcoal, macroreticular resins, polyurethane foams, bonded phases, and ion-exchangers. Generally have high capacity but sample discrimination may be a problem. Sample modification and incomplete recovery are further possible problems.
Gas-stripping	An inert purge gas sweeps over or sparges through the aqueous sample thereby transporting volatile constituents from the liquid phase to the gas phase, which permits them to be subsequently trapped cryogenically or on adsorbents.	Volatile samples	Examples include closed-loop stripping, purge-and-trap, and headspace techniques. Limited to small sample volumes. Extraction efficiency depends on the physical and chemical properties of the solute.

TABLE 7.2

GENERAL PROPERTIES OF DISTILLATION COLUMNS

Column	Types	General Features	Comments
Empty	Straight, smooth tubes	Small pressure drop and low hold-up	Concentric tube columns; very efficient, used for micro-, high-temperature, and vacuum distillation.
Packed	Straight or smooth tubes packed with spirals to increase surface area	High pressure drop and hold-up; in general, packing size about 1/10 diameter of column	Packings such as Raschig rings, spirals, helices, spheres, saddles, etc.; random packing not very efficient
Plate	Bubble plate, bubble cap, and sieve plate	High pressure drop and large liquid hold-up	Special-purpose columns for repetitive distillations where high purity distillate is required; Oldershaw column or sieve plate type is favored
Stationary element	Continuous glass or wire gauze, spiral, wire helix, etc.	Moderate pressure drop and hold-up	Heli-Pak a favored packing because of extremely low HETP even at high loads
Rotating	Spinning band, rotating concentric tube, or wire spiral	Moderately small pressure drop and hold-up	Spinning band columns useful for microvacuum distillation

apparatus which provides a concentrated extract suitable for chromatographic analysis [17-20]. The apparatus can handle 10-100 ml of aqueous solution or 1-20 g of solid material blended with water and provides an extraction of the distillate into 1.0 ml of pentane or methylene chloride, Figure 7.2B. The sample is placed in Flask A and the extraction solvent in flask B. To commence operation, the phase separator, C, is charged with a mixture of pentane and water, ice water is circulated through the cold-finger condenser, and pentane reflux is established by immersing flask B in a hot water bath. After 5 min, steam is generated by applying heat to flask A. The vapor channels F and G are insulated. The construction of the apparatus shown in Figure 7.2B is such that the low-density solvent (pentane) returns through arm D to flask B and the high-density solvent (water) returns through arm E to flask A. Reversing the positions of flasks A and B enables the extraction to be performed with a heavier-than-water solvent such as methylene chloride. After enrichment is complete, normally one hour is sufficient, steam generation is stopped while solvent extraction is continued for another 20 min to ensure that all the steam-distillable material is collected. An internal standard is added to the extraction solvent to allow for evaporative and entrainment losses during operation. For aroma constituents and pesticide compounds recoveries of 90-105% were typically obtained.

Entrainment in the gas or vapor phase forms the basis of three methods related to steam distillation. Zlatkis isolated an essential oil fraction from plant samples by co-distillation and solvent extraction using a continuous stream of nitrogen to carry over volatile organics from a mixture of solid sample and boiling water [21]. The aqueous condensate was extracted with a 2.0 ml volume of solvent prior to analysis. Jennings used reflux trapping for the isolation of volatile organic compounds from headspace vapors [22]. The dynamic headspace sample is passed into the base of a reflux condenser containing refluxing Freon 12. The components of lowest volatility in the headspace sample are condensed and migrate to the boiling flask where they accumulate. Freon 12 has a boiling point of -30°C and is easily removed from the extract by evaporation or fractional distillation. Sweep co-distillation has been used for the isolation of pesticide compounds in crude extracts [23-25]. The crude sample extract is injected into a glass column packed with glass wool and heated to 150-170°C. The volatile pesticides are swept through the packed column by a stream of nitrogen and several consecutive injections of a solvent such as ethyl acetate. The tube effluent is condensed in a cooled receiver and is generally sufficiently clean to permit direct chromatographic analysis. Non-volatile co-extractants remain on the glass wool.

Sublimation, the direct vaporization of a solid without first passing through a liquid phase, is a useful method of sample cleanup for those compounds which possess this ability, for example the polycyclic aromatic hydrocarbons [26,27]. The extraction yield with vacuum sublimation is dependent on the vapor pressure of the components at a given temperature and pressure and also on the nature of the sample matrix. Thus, different times are required to reach constant extraction yields for different compounds, and for the same compounds in different sample matrices. For polycyclic aromatic hydrocarbons adsorbed on carbonaceous materials extraction times vary from approximately 1 h to greater than 24 h. Under the most favorable circumstances sublimation is a rapid, solvent-free extraction procedure, giving a high yield of sample with a minimum of sample handling.

Lyophilization (freeze-drying), the removal of water by vacuum sublimation of ice, is a convenient method for concentrating involatile organic compounds in aqueous solution [28]. High concentration factors may be obtained as large volumes of aqueous solutions containing traces of organic substances can be concentrated. Sample losses, particularly of volatile compounds, occur when the bulk of the water has been removed and the temperature rises while the sample remains under a relatively high vacuum. Volatile acids and bases can be converted to their salt forms to minimize losses. Equipment for lyophilization is commercially available and external cooling with ice–salt eutectic mixtures can be used to control sample temperature and volatility losses.

Freeze concentration has been used for the concentration of aqueous solutions of organic volatiles and substances which are heat labile [7,29,30]. For successful results, the contact layers between the liquid and solid phases should be continuously disturbed by stirring or shaking and part of the solution should remain unfrozen at the end of the concentrating procedure. Artifact formation is uncommon by this procedure, but sample losses due to occlusion, adsorption, evaporation, and channelling in the ice layer may occur. Evaporation is the principal source of losses for organic volatiles if the apparatus is not enclosed. On the laboratory scale, a 2 liter sample volume can be concentrated to a 40–50 ml volume in about nine hours with a 90–100% recovery of the organic species.

7.3 Isolation and Concentration Techniques using Solvent Extraction

Extractions using liquids to solubilize all or part of the sample matrix are widely used. The techniques can be applied to vapors, liquids, and solids and have the following advantages: a large selection of pure solvents providing a wide range of solubility and selectivity are available, equipment requirements are simple, and solution phase samples are convenient and compatible with the sampling requirements of chromatographic instruments, Table 7.3 [10]. Liquid extractions are usually carried out discontinuously with the attainment of equilibrium between two immiscible phases or continuously under conditions where equilibrium is not necessarily obtained. Examples of discontinuous liquid extractions are liquid–liquid partition and countercurrent distribution which may be used for either sample isolation or cleanup. Examples of continuous extraction include the heavier–than or lighter–than liquid–liquid extractors, Soxhlet extractors, and countercurrent separators such as droplet, centrifugal, or rotating disk devices. Some guidelines for predicting the results from partitioning a sample between two immiscible liquid phases are summarized in Table 7.4 [31,32]. The efficiency of an extracting solvent depends primarily on the affinity of the solute for the extracting solvent (K_D), the phase ratio (V), and the number of extractions (n). For simple batchwise extractions K_D should be large, as there is a practical limit to the phase volume of the extracting solvent and the number of extractions that can be performed before the method becomes tedious and results in a very dilute sample extract. For extractions in which K_D is small, countercurrent distribution techniques should be used so that n can be made very large. This usually requires the use of automated equipment, for example the Craig distribution train. For some systems the value of K_D may

TABLE 7.3

THE ISOLATION OF ORGANIC COMPOUNDS BY LIQUID EXTRACTION

Sample Type	Equipment Requirement	Comments
Organic vapors in air or other gaseous mixtures	Impinger, gas wash bottle, or similar device	·Solvent acts as a selective extractant, retaining the sample because of its higher affinity for the solvent compared to the gas. ·Solutions of chemical or physical complexing agents may be used to improve the extraction efficiency. ·Solubility of the extractants may be adjusted by changes in the temperature of the extracting solvent.
Aqueous solutions	Separatory funnel Continuous extractor Countercurrent distribution apparatus	·In its simplest form an aliquot of the aqueous solution is shaken with an equal volume of an immiscible organic solvent. ·Limited to small sample volumes and solutes with large distribution constants. ·Several extractions are required when the distribution coefficient is small. The addition of salts, pH adjustment, ion-pairing reagents, etc., can be used to improve the distribution of organic solutes into the extracting solvent. ·When the distribution coefficient is very small continuous liquid-liquid extraction or countercurrent distribution apparatus is required. ·Large sample volumes may also be extracted using continuous and countercurrent distribution methods.
Organic liquids	Mixing device	·Selective extraction by mixing with an organic solvent. Trituration of semi-liquid samples. ·Precipitation or freezing used to remove coextractants.
Solid samples	Shaker Homogenizer Soxhlet extractor	·Solid samples such as soil are usually mechanically shaken with an appropriate solvent for a set period of time. ·Extraction efficiency may be improved by warming the solvent or by heating to reflux. ·Bulky samples such as plant materials are dried, cut, ground, pressed, or milled and sieved prior to extraction to promote even and efficient extraction. ·Tissue samples are homogenized in the presence of a water miscible organic solvent to promote efficient extraction. ·Samples which are difficult to extract efficiently with a few solvent exchanges can be extracted continuously at room temperature or at the boiling point of the solvent in a Soxhlet apparatus.

be made more favorable by adjusting pH to prevent ionization of acids or bases, by forming ion-pairs with ionizable solutes, by forming lipophilic complexes with metal ions, or by adding neutral salts to the aqueous phase to diminish the solubility of the organic solute. The selectivity of the extraction process is both a strength and a weakness when applied as a sampling technique. With multicomponent samples a single solvent is unlikely to extract all components, causing discrimination and hence a biased analysis [33]. This discrimination may be useful if the solvent discriminates against the extraction of solutes which are not of interest in the analysis. Solvent selection for the extraction of organics from water is discussed in references [33-35].

TABLE 7.4

BASIC RELATIONSHIPS FOR PREDICTING SOLUTE DISTRIBUTION IN LIQUID-LIQUID PARTITION

Nernst Distribution Law: Any neutral species will distribute between two immiscible solvents such that the ratio of the concentrations remains a constant

$$K_D = \frac{[A]_o}{[A]_{aq}}$$

K_D = distribution constant

$[A]_o$ = concentration of A in organic phase

$[A]_{aq}$ = concentration of A in aqueous phase

The fraction of A extracted is given by θ where:

$$\theta = \frac{[A]_o V_o}{[A]_o V_o + [A]_{aq} V_{aq}} = \frac{K_D V}{1 + K_D V}$$

V = phase ratio V_o/V_{aq}

V_o = Volume of organic phase

V_{aq} = Volume of aqueous phase

The fraction extracted in n successive extractions

$$\theta = 1 - \left[\frac{1}{1 + K_D V} \right]^n$$

$K_D V = 10$, 99% of solute extracted with n = 2

$K_D V = 1$, 99% of solute extracted with n = 7

$K_D V = 0.1$, 50% of solute extracted with n = 7

Countercurrent Distribution: The relationship of the distribution constant K_D of the solute in a CCD process to the concentration in the various separatory funnels or stages is given by

$$[\theta + (1 - \theta)]^n = 1$$

The fraction $T_{n,r}$ of the solute present in the rth stage for n transfers is given by:

$$T_{n,r} = \frac{n!}{r! \, (n-r)!} \cdot \frac{(K_D V)}{(1 + K_D V)^n}$$

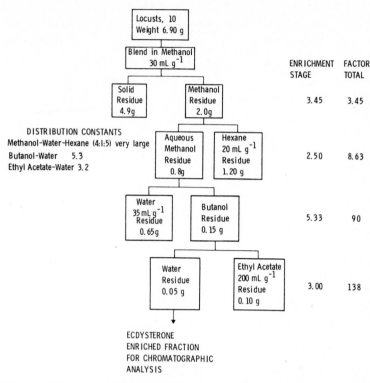

Figure 7.3   Extraction scheme used to isolate insect moulting hormone
(ecdysterone) from the desert locust.

Figure 7.3 is an example of the use of solvent extraction to isolate and
concentrate the insect moulting hormone ecdysterone from the desert locust
[36,37].   Ecdysterone, $2\beta,3\beta,14\alpha,20,22,25$-hexahydroxy-$5\beta$-cholest-7-en-6-one, is
a polar steroid with a high solubility in methanol.   The anaesthetized insects
are macerated in a blender with methanol and the solid residue is discarded.
Liquid-liquid partition is used to further isolate the ecdysterone fraction as
indicated in Figure 7.3.   A nonpolar solvent is used to remove fats and lipids
from the polar fraction and a selective partition with butanol isolates
ecdysterone from water-soluble coextractants.

Figure 7.4 is an example of a general solvent extraction scheme for
isolating neutral, base, weak acid, and strong acid fractions from a sample that
is soluble in methylene chloride [38].   Note that the acid and base fractions
are extracted by changing the pH of the aqueous phase to manipulate the
selectivity (K_D values) of the solvent system.   Neutral solutes are not affected
by changes in pH but further fractionation is possible using selective chemical
reagents.   Sodium bisulfite or Girard T and P reagents can be used to extract

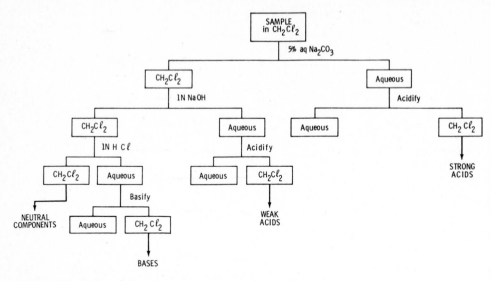

Figure 7.4 Solvent fractionation scheme for isolating neutral, base, and
acid fractions from a sample soluble in methylene chloride.

aldehydes and ketones as water-soluble derivatives [38]. The chemistry and
properties of Girard T (trimethylethylammonium acetyl hydrazide chloride) and
Girard P (pyridinium acetyl hydrazide chloride) reagents are reviewed in
reference [39]. These reagents produce water-soluble derivatives of aldehydes
and ketones which can be easily hydrolyzed back to the parent compound at high
pH.

Dimethyl sulfoxide provides a high extraction efficiency for compounds that
contain hydrogen bonding or π-electron rich sites [40]. The solvent system
pentane-dimethyl sulfoxide has been shown to be useful for the fractionation of
polar and unsaturated solutes, notably the separation of aliphatic and
polycyclic aromatic hydrocarbons, Figure 7.5. Partitioning between pentane and
dimethyl sulfoxide provides a separation of polar and nonpolar solutes. The
dilution of the dimethyl sulfoxide phase with water diminishes the extent of
π-electron attraction with the solute without dramatically diminishing hydrogen
bonding interactions. Back extraction of the aqueous dimethyl sulfoxide phase
with pentane provides a separation between hydrogen bonding and neutral polar
solutes.

Figure 7.6 is an example of the use of liquid extraction for the analysis
of trace levels of di-(2-ethylhexyl)phthalate (DEHP), a commercial plasticizer,
in human placenta [41]. The frozen placenta samples were dissected into small
pieces and then extracted with acetonitrile (1.0 ml/g) in a Wearing blender.
The suspension was filtered under vacuum, the residue re-extracted, and the
combined acetonitrile extracts diluted with an equal volume of water prior to

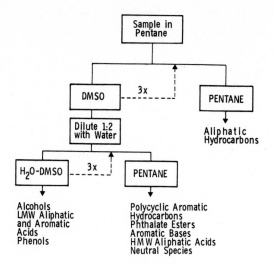

Figure 7.5 Selective solvent partition scheme for isolating polar and unsaturated solutes.

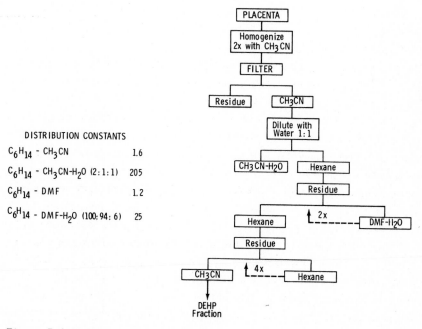

DISTRIBUTION CONSTANTS

C_6H_{14} - CH_3CN 1.6

C_6H_{14} - CH_3CN-H_2O (2:1:1)   205

C_6H_{14} - DMF   1.2

C_6H_{14} - DMF-H_2O (100:94:6)   25

Figure 7.6 Solvent partition scheme for the isolation of di-(2-ethylhexyl)-phthalate from human placenta.

extraction with hexane. The distribution constant for the water-acetonitrile-hexane extraction is very large and permits the isolation of DEHP in a small

volume of hexane. The residue from the hexane phase is further concentrated by
partition between hexane-dimethylformamide-water (100:94:6). Water was added to
the dimethylformamide to improve the distribution ratio of DEHP in favor of
hexane while still permitting the removal of most of the polar contaminants
which interfere in the final chromatographic analysis. Final sample cleanup was
performed by partition between hexane and acetonitrile saturated with hexane.
Since the distribution constant is small, multiple extractions were required to
provide a high extraction efficiency.

Although solvent extraction is simple and does not require complex
equipment, it is not entirely free of practical problems. The formation of
emulsions can be a problem if they cannot be readily broken up by conventional
techniques such as filtration through a glass wool plug, centrifugation,
refrigeration, addition of a trace amount of methanol, or by salting out. The
rate and extent of extraction may be different for a solute in a test system
than in a practical sample. Partial association of drug substances with
proteins in plasma samples has been recognized as one instance when the
extraction efficiency may vary substantially from that obtained using water as a
model system [42]. Dilution of the plasma sample with water or the formation of
a homogeneous extract prior to forming the two phase system will often solve
this problem. Sample preparation for the determination of organic compounds in
biological fluids has been reviewed by Reid [5].

Continuous liquid extraction techniques are used when the sample volume is
large, the distribution constant is small, or the rate of extraction is slow.
Numerous continuous liquid-liquid extractors using lighter-than-water and
heavier-than-water solvents have been described [43-49]. Generally, either the
lighter or heavier density liquid is boiled, condensed, and allowed to percolate
repetitively through an immiscible companion solvent in which the sample is
dissolved. A laboratory scale heavier-than-water liquid-liquid extractor is
shown in Figure 7.7A [47]. Other designs are capable of handling large sample
volumes (approximately 10 1/h) by coupling several units in series [43,48], can
be used for on-site sampling [44], or can be easily modified to work with
variable sample volumes [45,46]. Extractions may be performed over several
hours and concentration factors up to 100,000 obtained. The continuous liquid
extraction of solids is carried out in a Soxhlet extractor. The solvent is
vaporized, condensed, and allowed to percolate through the solid sample
contained in an extraction thimble. The return of solvent to the boiling flask
is discontinuous; working on the syphon principle, the solvent is returned only
when a certain volume of extraction solvent is reached in the extraction
chamber. Soxhlet extractors are available for milligram to kilogram sample
quantities and for extractions at either room temperature or near the boiling
point of the solvent [50]. Unattended operation is possible. The main

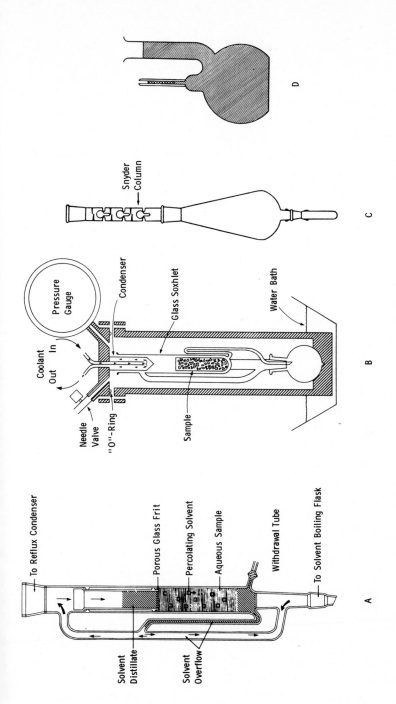

Figure 7.7   Apparatus used for sample preparation involving solvent extraction.
A, laboratory scale heavier-than-water liquid-liquid extractor;
B, pressurized Soxhlet extractor for use with supercritical fluids;
C, Kuderna-Danish evaporative concentrator;
D, micro extraction apparatus.

disadvantage is that the compounds extracted must be stable at the boiling point of the extracting solvent, as eventually they are accumulated in the boiling flask. Jennings has described a pressurized system that permits the use of liquid carbon dioxide with a standard glass Soxhlet extractor, Figure 7.7B, for the extraction of solid samples and for the recovery of organic volatiles from adsorbents [22]. Supercritical fluids, and in particular carbon dioxide, are excellent solvents for most organic compounds [51]. This extractor, which can utilize dry ice as a source of carbon dioxide, has the advantages of achieving high extraction efficiencies under an inert atmosphere and at low temperatures. The carbon dioxide is allowed to dissipate at subzero temperatures, yielding a solvent-free extract for analysis.

Liquid extraction uses large solvent volumes compared to the sample volumes that can be analyzed chromatographically. This dilution of the sample is often too great for direct determination without a preconcentration step. Large sample volumes may be evaporated using a rotary evaporator [52] or a Kuderna-Danish evaporative concentrator (Figure 7.7C) [53]. Rotary evaporators are available in most laboratories and provide a convenient means of solvent evaporation under reduced pressure. Volatile compounds are largely lost during concentration with this apparatus and even the recovery of less volatile materials may be lower than expected due to entrainment of the sample in the solvent vapors, adsorption on the glass walls of the flask and apparatus, as well as from uncontrolled expulsion from the flask due to uneven evaporation. For the concentration of compounds of reasonable volatility the Kuderna-Danish evaporative concentrator is preferred. The evaporative concentrator is operated at atmospheric pressure under partial reflux conditions using a three-ball Snyder column with an efficiency of about 2.7 theoretical plates. Condensed vapors in the Snyder column are returned to the boiling flask, washing down organics from the sides of the glassware; the returning condensate also contacts the rising vapors and helps to recondense volatile organics. Although the Kuderna-Danish concentrator provides a slower rate of evaporation than rotary evaporators it generally provides higher recoveries of trace organic components [52]. It is generally not possible to reduce sample volumes of several hundred milliliters to less than one milliliter in a single apparatus. Samples are usually concentrated to 5-10 ml and then transferred to a micro Kuderna-Danish evaporator [54] or to a controlled-rate evaporative concentrator [55,56]. The latter devices combine the advantages of the rotary evaporator and vacuum evaporation with the gas blow-down method. The gas blow-down method is suitable for the evaporation of volatile solvents of less than 25 ml volume. A gentle stream of pure gas is passed over the surface of the extract contained in a conical-tipped vessel or culture tube partially immersed in a warm water bath.

The solvent evaporation rate is a function of the gas flow rate, the position of the gas inlet tube relative to the refluxing solvent, the water bath temperature, and the solvent surface area. High gas flow rates must be avoided to prevent loss of sample by nebulization. The choice of gas is important to avoid contaminating the sample. Cylinder gases such as prepurified nitrogen or helium should be used and further purification effected if contamination is a problem. Two convenient high capacity methods of gas decontamination are passage through a U-tube filled with a mixture of molecular sieve and Carbosieve immersed in a dry ice-acetone coolant bath [57] or by using a femtogas purification train [58]. The latter consists of five 10 x 0.5 cm I.D. stainless steel tubes filled with different packings and connected in series. Tubes 1 and 2 are packed with a bifunctional gas-solid adsorption and gas-liquid partitioning material (prepared from 3 g Carbopack B, 3 g actived charcoal, 3 g Apiezon L, and 21 g of Chromosorb W-AW), tube 3 with 13-X molecular sieve, tube 4 with 20% sulfuric acid on silica, and tube 5 with Carbosieve S. Both methods are capable of producing very pure gases suitable for trace analysis.

Contamination of extracts from solvents and glassware or sample losses due to adsorption are frequently encountered problems in trace analysis. Glassware should be thoroughly cleaned and rinsed with pure solvent and, if stored, maintained in a dust-free area. Heating in a muffle furnace or high temperature vacuum oven is often a convenient method of decontaminating glassware [41]. Solution or vapor phase silylation can be used to deactivate clean glass surfaces [59,60]. Whenever possible the sample should contact only those materials which may be easily cleaned such as stainless steel, Teflon, and glass. The solvents used in liquid extraction are a potentially rich source of contaminants, particularly as relatively large solvent volumes are concentrated along with the sample. Needless to say, solvents of the highest available purity should be used. A 200 ml solvent extract concentrated to 100 microliters represents a 2000-fold concentration, and thus even very minor impurities in the extracting solvent will be present in fairly high concentration after evaporation. In trace analysis, sample blanks are run frequently to estimate the extent of contamination in a particular method.

As the solvent evaporation process is slow and the possibility of contamination from the solvent and glassware high, micromethods using small solvent and sample volumes are attractive and compatible with the sample size requirements of chromatographic techniques. Figure 7.7D is an example of a micro-extraction apparatus which enables up to 980 ml of an aqueous sample to be extracted with 200 microliters of hexane [61]. The sample is extracted by shaking, and after allowing sufficient time for phase separation, additional water is added through the side arm, forcing the lighter organic phase into a capillary tube where it can be sampled by syringe. About 50 microliters of

solvent are recovered, the remainder is accounted for by dissolution in the
aqueous phase and by evaporation. Typical recoveries of hydrocarbons,
chlorinated pesticides, and phthalate esters at trace levels averaged 40-70% for
one extraction and 89-99% for three consecutive extractions. Similarly, glass
syringes with a luer-slip fitting can be used to extract sample volumes of 1 to
50 ml with solvent volumes of 20 microliters and greater [62]. Extraction takes
place in the syringe barrel and, after phase separation, the organic layer is
easily dispensed through a narrow-bore glass transfer tube attached by a PTFE
sleeve to the luer fitting. Rapid extractions of aqueous solutions of various
volumes have been described using a polytron to homogenize the sample and
solvent followed by refrigerated centrifugation to separate the emulsion formed
[63,64]. In favorable circumstances, extraction efficiencies of 90-100% can be
obtained. This makes it possible to minimize the volume of extracting solvent
used, greatly reducing the time required to perform the extraction. Dunges and
Zlatkis have described the use of simple apparatus for extracting microliter
volumes of organics in biological fluids [65-67]. As the volume of extracting
solvent was small only a minimum of concentration was needed prior to analysis.

7.4 Sample Cleanup Using Liquid-Solid Chromatography

After solvent extraction, liquid-solid chromatography is frequently used to
further fractionate the extract on the basis of polarity. Either column or
thin-layer techniques may be used, but column methods are generally preferred as
sample recovery is more straightforward. Many different adsorbents have been
used in liquid-solid chromatography; their general properties are given in Table
7.5 [32,68-70]. Further discussion can be found in Chapter 4.10, as well as in
specific reviews for florisil [71-73], hydroxylapatite [74,75], and lipophilic
Sephadex [76]. The controlled porosity gels are not adsorbents but materials
for size-exclusion chromatography, and are included here for completeness. For
sample cleanup silica gel, alumina, florisil, and controlled porosity gels are
most frequently used. The philosophy of the separation requirements for sample
cleanup is different to classical liquid chromatography. The emphasis is placed
on fractionation rather than individual component resolution. Component
resolution is better left to subsequent stages in which high resolution
chromatographic techniques are used. Column sample cleanup is used to isolate
the components of interest by removing components of different polarity or size,
that would either interfere directly in the chromatogram by masking the
components of interest or prevent the continuous operation of the instrument by
changing the performance of the analytical column. Adsorption chromatography
can also be used as a concentration technique by applying the sample to the
column in a large volume of a noneluting solvent and then eluting the adsorbed

sample components with a small volume of a strong eluting solvent. In a typical procedure, the sample is applied to the column in a small volume of a weak solvent and the separation is affected by eluting with a series of solvents of increasing polarity. The components of interest are eluted in a small number of fractions. The dimensions of the column used are dictated by the size of the sample and the resolution required. Wide columns are used for large samples and long columns for difficult separations. The activity of the adsorbent is an important consideration as this will affect the fraction in which the components of interest will be found. Adsorbent activity, and therefore retention, is controlled by the purposeful addition of a known amount of water to the dry adsorbent. Adsorbent activity can be estimated by the Brockmann scale, Table 7.6, which, although based on the elution characteristics of a series of test dyes, provides a reasonable general indication of sample retention [32,73]. As many sample extracts contain some water they should be dried with anhydrous sodium sulfate or molecular sieves prior to passage through the column to avoid changes in column activity during the analysis. A 1-2 cm plug of anhydrous sodium sulfate may be added to the top of the column packing as a further precaution. General sources concerning the use of different adsorbents for pesticide residue analysis [77-79], biological fluids [5], and environmental analysis [80,81] should be consulted for further details.

TABLE 7.6

STANDARDIZATION OF ADSORBENT ACTIVITY

Brockmann Activity Grade	Percentage of Water (w/w)		
	Alumina	Silica Gel	Florisil
I	0	0	0
II	3	5	7
III	6	15	15
IV	10	25	25
V	15	38	35

Adsorbents are also used as support materials for stationary liquid phases or chemical reagents. Liquid-liquid partition chromatography is rarely used for sample cleanup although it is still used for the isolation of natural products. Hydrophobic celite or silanized silica is used to support a stationary liquid phase while an immiscible mobile phase, saturated with the stationary phase, is used for sample elution. Its principal advantage compared to liquid extraction is its higher separation efficiency.

Silica or alimuna can be coated with such chemical reagents as sulfuric acid, silver nitrate, alkaline potassium permanganate, sodium hydroxide, etc. to

TABLE 7.5

ADSORBENT MATERIALS USED FOR SAMPLE CLEANUP AND CONCENTRATION

Type	Composition	Comments and Applications
Silica Gel	$SiO_2 \cdot xH_2O$	Most widely used general adsorbent. May irreversibly bind some strongly basic substances. For details of adsorption mechanism, see Chapter 4.10.
Alumina	$Al_2O_3 \cdot xH_2O$	Prepared by low-temperature ($< 700°C$) dehydration of alumina trihydrate and is a mixture of γ-alumina with perhaps a small amount of the less active form (α-alumina) and sodium carbonate. Commerically available in 3 forms. Neutral alumina, pH 6.9 to 7.1, is the most widely used and is suitable for the separation of hydrocarbons, esters, aldehydes, ketones, quinones, alcohols, and weak organic acids and bases. Basic alumina, pH 10 to 10.5, is used to separate acid-labile substances. In aqueous or partially aqueous solutions it exhibits strong cation-exchange properties. Acid alumina, pH 3.5 to 4.5 (an acid-washed preparation of neutral alumina), acts as an anion exchanger and is used to separate inorganic compounds and acidic organic compounds.
Florisil	Magnesium silicate	Separation properties intermediate between silica and alumina. Widely used in the cleanup of organochlorine pesticides. Variation in chromatographic properties associated with the presence of variable amounts of sodium sulfate. Basic nitrogen compounds may not be eluted due to chemisorption.
Carbon macroreticular resins	Carbon polystyrene polyacryl- amide	Mainly used for batchwise adsorption of organics in water. Large affinity, high capacity, and low selectivity are their principal properties. Details given elsewhere in this section.
Hydroxyl- apatite	$Ca_5(PO_4)_3OH$	Form of calcium phosphate used primarily for the fractionation and purification of medium and high molecular weight substances encountered in biochemistry.

Polyamides	Polycapro-lactam Nylon	Used for the separation of polar molecules by a combination of reversed-phase and hydrogen bonding mechanisms. Further details are given in Chapter 9.13.
Celite	Diatomaceous earth	Weak adsorbent used as a support in liquid-liquid partition chromatography and as a filter aid.
Cellulose		Retains compounds by a combination of adsorption, partition and ion-exchange. Mainly used to separate molecules of high molecular weight. Further details given in Chapter 9.18.
Controlled-porosity gels	Cross-linked dextran, Polystyrene, Polyacryl-amide, agrose	Used for size-exclusion chromatography. Water-compatible gels are used to separate large biochemical molecules. For sample cleanup the organic solvent compatible gels (Sephadex LH-20, Bio-Beads SX) are used to separate small molecules from a high molecular weight sample matrix.

improve the selectivity of the cleanup stage [82–84]. These modified adsorbents are useful for retaining chemically active coextractants which would otherwise be eluted from the untreated adsorbent column.

Controlled porosity gels can be used to fractionate extracts based on molecular size with aqueous or organic mobile phases. They are particularly useful when a small molecule such as an organochlorine pesticide is to be determined in a matrix comprised of high molecular weight oils, fats, and polymers. By size-exclusion the large sample molecules are eluted ahead of the pesticide, providing an efficient cleanup. In this respect the method is complementary to the use of adsorbents which provide a separation based on polarity. A particular advantage of the size-exclusion columns is that since the sample rarely contaminates the column (total elution is achieved) it may be used repetitively. The method can thus be easily automated and equipment for this purpose has been described [85,86].

7.5 The Use of Internal Standards

An internal standard is a substance that is added to the sample at the earliest possible point in the analytical scheme to compensate for sample losses occurring during extraction, cleanup, and final chromatographic analysis. The properties desired of an ideal internal standard are summarized in Table 7.7. Rarely will an internal standard meet all of these requirements. They are most closely approached when a restricted range of similar substances are to be determined. Numerous examples can be found for the analysis of drugs in biological fluids where structurally similar analogs of the drug are available for use as internal standards [87,88]. The opposite extreme is represented by multicomponent environmental samples where the identity of all components is not known and for which an impractically large number of internal standards

TABLE 7.7

PROPERTIES REQUIRED OF AN IDEAL INTERNAL STANDARD

The internal standard should resemble the analyte as closely as possible in terms of chemical and physical properties.

The internal standard should not be a normal constituent of the sample.

The internal standard should be incorporated into the matrix in exactly the same way as the analyte. A situation which is rarely achieved.

The internal standard and analyte should be resolved chromatographically to baseline (except for isotopically labelled samples when mass discrimination or radioactive counting are used for detection), elute close together, respond to the detection system in a similar way, and be present in nearly equal concentrations.

---

would be required to account for the different properties of the substances present in the sample [89].

Substances used as internal standards include analogs, homologs, isomers, enantiomers, and isotopically labelled analogs of the analyte to be determined. Analogs and homologs are perhaps the most widely used substances simply because they are more likely to be available. Amino acids can be determined with high precision by gas chromatography using a chiral column and the unnatural enantiomers as internal standards [90,91]. Isotopically labelled internal standards are frequently used in gas chromatography-mass spectrometry where the mass discriminatory power of the mass spectrometer can be taken advantage of to differentiate between the analyte and internal standard.

As well as an internal standard, a carrier substance may be added to the sample. This technique is applied when the analyte is in low concentration and is likely to be lost by adsorption or binding to the sample matrix or the apparatus. The carrier substance should have a similar structure to the analyte and is added to the sample in large excess compared to the analyte concentration. It must be easily separated from the analyte in the final chromatographic separation.

7.6 Sorption Techniques for Trace Enrichment of Dilute Solutions

Substances present in trace amounts generally do not permit direct determination by the most appropriate instrumental method. Therefore, the sample concentration must be amplified by trace enrichment techniques, many of which involve a phase transfer. In this case, trace enrichment can be described as the process by which the substances of interest are preconcentrated by their selective removal from the bulk sample matrix to reach a concentration sufficient for detection. The enrichment technique can be general, for example the adsorption of organics from water, or it may be selective, for instance the ion-exchange adsorption of acidic compounds from water without collection of neutral or basic substances. The latter example represents both trace enrichment and sample cleanup. Both processes are frequently employed in the same analytical scheme; of course, if the concentration of the substance of interest is sufficient for detection then only the removal of interfering substances by sample cleanup is necessary for analysis.

In this section we will consider solid-phase sorbents as concentrating agents for the analysis of dilute solutions [12,92-96]. Commonly used sorbents are charcoal, macroreticular porous polymers, polyurethane foams, bonded-phase packings, and ion-exchange resins. The characteristics which govern a sorbent's usefulness for a particular problem are its sample capacity and breakthrough volume. The breakthrough volume is defined as the volume of sample that can be

passed through a sorbent bed before elution of the investigated compound
commences. The larger the breakthrough volume, the greater the sample volume
that can be processed, and the greater the enrichment factor obtained. Values
for the breakthough volume may be calculated or determined experimentally from
model systems. In either case, procedures are tedious and frequently
inaccurate. The values obtained for model systems may not be appropriate for
real samples where desorption or adsoprtion by other components of the sample
matrix may radically affect these values.

7.6.1 Carbon Adsorbents

Graphitized charcoal, activated carbon, and carbon molecular sieves are
three forms of carbon used chromatographically. Carbon molecular sieves have
small pores and very large surface areas and are thus suited more for the
separation of gases by gas chromatography than for use as a sorbent in water
treatment. Graphitized charcoal is a relatively pure form of carbon prepared at
temperatures in excess of 1000°C. Because of its low surface area (6-30 m^2/g)
it is mainly used as an adsorbent or support in gas chromatography. Partially
graphitized carbon black (e.g., Carbopack B) has a surface area of about 100
m^2/g and a pH = 10.25 in aqueous suspension. It has been used to adsorb a wide
range of organic species from water [97]. Activated carbon is prepared by the
low temperature oxidation of vegetable charcoals. These materials have large
surface areas (300-1000 m^2/g), a wide pore diameter distribution, and a
heterogeneous surface containing active functional groups. Standard methods for
the extraction of organic compounds from water using activated carbon filters
have been described [98-101]. Very large volumes, such as 1,200 liters, may be
sampled at a rate of 120 ml/min, requiring about one week for completion.
Precleaned carbon, 30 mesh, in a 18 x 3 inch diameter glass tube is used for
sample collection. The carbon adsorbent is then removed from the cartridge,
spread on a glass tray, and dried at 35-40°C until free flowing. The adsorbed
organics are removed by Soxhlet extraction with 1.5 liters of chloroform or
ethanol. For clean surface waters the extracted organic residue is usually less
than 50 micrograms per liter.

Belfort has explained the adsorption properties of carbon using solvophobic
theory as follows, "In aqueous solute adsorption on activated carbon,
non-covalent (physical) interactions are usually considered dominant. Of these
interactions, ion and hydrogen bonding are caused by positive attractive forces
between the adsorbate and the adsorbent. Hydrophobic interactions, on the other
hand, originate from a net repulsion between the water and the nonpolar regions
of the activated carbon as well as the nonpolar (and nonpolar moieties of the)
solute or adsorbate. Thus, as a consequence of the very high cohesive energy

density of the water, nonpolar molecules and nonpolar regions or surfaces are 'squeezed out' of the water phase and can associate if in close proximity to each other." [102]. The breakthrough volume of organic compounds on carbon is thus a complex function of the properties of the adsorbate, the sorbent, and the solution conditions. As a general rule, the breakthrough volume declines as the polarity and water solubility of the sample increases.

Although carbon adsorption is convenient and to some extent standardized, there are several problems with the technique which can affect the identification and quantitation of specific compounds in water. There is considerable variability in the collection efficiency for different compounds and for different experimental conditions, the desorption of compounds from the adsorbent is not always total, and the desorbed compounds are not always identical to those originally extracted. It is well know that activated carbon can act as a catalyst for oxidation and other chemical reactions. With the possible exception of low molecular weight polar organic compounds, most studies have concluded that macroreticular porous polymer adsorbents provide a higher recovery of organics from water than carbon, Table 7.8 [93,97,103-106]. However, the performance of carbon as an extractant is acceptable for many analytical problems [93,97,104,106].

TABLE 7.8

COMPARISON OF RECOVERIES WITH AMBERLITE XAD-2 AND ACTIVATED CARBON (FILTRASORB 300)

	Average Recovery (%)	
	Amberlite XAD-2	Filtrasorb 300
Alkanes	5	15
Esters	61	49
Alcohols	73	47
Phthalate esters	82	24
Phenols	45	7
Chlorinated hydrocarbons	43	55
Chlorinated aromatic compounds	70	11
Aromatic compounds	68	6
Aldehydes and Ketones	54	24
Pesticides	34	16
Carboxylic acids	1	2

7.6.2 Macroreticular Porous Polymer Adsorbents

Macroreticular resins are prepared by suspension polymerization of, for example, styrene-divinylbenzene copolymers in the presence of a substance which

is a good solvent for the monomer but a poor swelling agent for the polymer
[107]. The macroreticular resins consist of agglomerates of randomly packed
microspheres permeated by a network of holes and channels. This results in
higher mechanical stability and greater surface areas than
styrene–divinylbenzene gels. Characteristic properties of the Amberlite XAD
series of mcaroreticular resins manufactured by Rohm and Haas are given in Table
7.9 [93,108]. The XAD–1, –2, and –4 resins are aromatic in character, very
hydrophobic, and possess no ion-exchange capacity. Amberlite XAD–7 and –8 are
acrylic ester resins, non-aromatic in character, and possess very low
ion-exchange capacity, on the order of 0.01 milliequivalents per gram for· XAD–8
[108]. The acrylic ester resins are more hydrophilic than the
styrene–divinylbenzene copolymers; accordingly, they are more easily wetted and
adsorb more water. The adsorption mechanism of organic compounds from water on
the XAD-resins is similar to that on carbon, although recoveries are generally
higher and catalytic activity is greatly diminished. The recovery of neutral
organic molecules from water is universally high for a wide range of sample
types, Table 7.10. Partially dissociated organic compounds (e.g., phenols,
carboxylic acids, etc.) are generally weakly sorbed by the resin and the
percentage retained diminishes with dilution due to a higher degree of
dissociation. Adjusting the pH of the sample leads to increased capacity for
these compounds but average recoveries at low concentration are often not
comparable to those of undissociated solutes [93,108-110]. Acidic or basic

TABLE 7.9

PHYSICAL PROPERTIES OF AMBERLITE XAD RESINS

Type	Composition[a]	Character	Average Pore Diameter, Å	Specific Surface Area, m^2/g	Specific Pore ml/g	Solvent Uptake g/g of dry resin
XAD–1	STY–DVB	hydrophobic	200	100	0.69	– – – – –
XAD–2	STY–DVB	hydrophobic	85–90	290–300	0.69	0.65–0.70
XAD–4	STY–DVB	hydrophobic	50	750	0.99	0.99–1.10
XAD–7	methacry-late polymer	moderately hydrophilic	80	450	1.08	1.89–2.13
XAD–8	methyl methacry-late polymer	moderately hydrophilic	250	140	0.82	1.31–1.36

[a] STY–DVB = styrene-divinylbenzene copolymer

TABLE 7.10

TYPICAL RECOVERIES FOR ORGANIC COMPOUNDS AT 10-100 ppb (PESTICIDES ppt) IN
WATER ON AMBERLITE XAD-2

Compound Type	Typical Recoveries (%)
Alcohols	90 - 100
Aldehydes	90 - 100
Ketones	90 - 100
Esters	80 - 100
Acids[a]	90 - 100
Phenols[a]	80 - 95
Ethers	80 - 95
Halogen compounds	75 - 95
Polycyclic aromatic hydrocarbons	85 - 100
Alkylbenzenes	80 - 90
Nitrogen and sulfur compounds	85 - 95
Pesticides	80 - 95

[a] After pH adjustment

compounds may be eluted from the resin with an aqueous solution of appropriate
pH without displacing neutral molecules, effecting enrichment and cleanup stages
simultaneously.

Macroreticular porous polymers are widely used for the removal of trace
organic compounds from fresh water, waste water, salt water, and from aqueous
biological samples. The high affinity of these polymers for neutral organic
compounds, high sample capacity, low water retention, chemical inertness, and
ease of sample recovery are features which have contributed to their wide
acceptance. Relatively large sample volumes, 1-100 liters, may be pumped
through a short adsorbent column, 6 x 1.5 cm, at high flow rates, 4-7 column bed
volumes per minute [93,103,109-112]. The adsorbed sample is recovered by
elution with a small volume of organic solvent such as methanol, ether, acetone,
ethyl acetate, etc. After evaporating the eluting solvent to a small volume
enrichment factors of 10^3-10^7 can be obtained. For sample sizes less than 1
liter, a 1.2-1.8 x 25 mm mini-column, can be used without sample loss. Only
50-100 microliters of solvent is required for sample elution, eliminating the
need for evaporation [104]. Thermal desorption and Soxhlet extraction have been
used as alternative methods for sample recovery [93,113]. Thermal desorption
avoids the use of elution solvents and therefore provides the highest degree of

sample enrichment. Unfortunately, the thermal stability of the polymers is not very high and therefore limits the technique to easily desorbed molecules [114,115]. Polymers of the Amberlite type produce a high chromatographic background when heated to temperatures of 250-275°C. Naphthalene, ethylbenzene, and benzoic acid have been identified as the principal contaminants. Tenax is more thermally stable than the Amberlite polymers and produces relatively little chromatographic background at desorption temperatures. However, Tenax has a low surface area, 19-30 m^2/g, limiting its adsorption capacity and resulting in lower recoveries and a restriction of sample size. In spite of these disadvantages its use in studies of water pollution involving trace concentrations of organic compounds has been demonstrated [116-118]. The analytical methodology involves column sampling in the usual way, centrifugation and vacuum desiccation to remove entrapped water, and thermal desorption at 280°C for gas chromatography.

Breakthrough volumes for the macroreticular polymers can be calculated theoretically from a knowledge of the molar solubility of the compound [108] or determined experimentally [119]. In either case, the values obtained for model systems may be inaccurate for real samples where solution conditions and the presence of coextractants will have some effect. Model systems are usually prepared by adding the test sample, in a small volume of organic solvent, to water. Erroneous results may occur for test samples added to water in concentrations that exceed their water solubility and for volatile compounds with a higher density than water. The latter compounds collect at the surface and evaporate from the open sample reservoir unless the air space above the reservoir is minimized [109].

Macroreticular porous polymers generally require careful purification prior to use in trace analysis [120]. Fines are removed by slurrying in methanol and monomers, contaminants, and anti-bacterial agents are removed by sequential solvent extraction for 8 h each with methanol, acetonitrile, and ether [109]. Acrylic Amberlite resins are usually rinsed with dilute sodium hydroxide solution prior to solvent extraction [108]. The solvent-purified resins are stored in glass-stoppered bottles under methanol until used. Purified resins do not store well in the dry state. Monomers and other minor contaminants are apparently trapped interstitially within the resin during the polymer bead formation process. In the dry state, mechanical or spontaneous cracking of the beads leads to recontamination.

7.6.3 Polyurethane Adsorbents

Polyurethane as an open-pore or foam material, has been used as an adsorbent for neutral organics in water [121-126]. Open-pore polyurethanes are

prepared in situ by reacting a solution of an isocyanate with a polyol inside
the adsorbent column. The open-pore polyurethanes are composed of agglomerated
spherical particles, 1-10 micrometers in diameter, bonded to one another in a
rigid and highly permeable structure [121]. Cellular (foamed) polyurethanes are
commerically available and used in large quantities for many manufactured
products. Both types function as adsorbents with a high degree of efficiency
for the collection of polynuclear aromatic hydrocarbons and pesticides
[122-124]. They also exhibit some weak base ion-exchange capacity. Their area
of application and method of use are very similar to the macroreticular porous
polymers, over which no significant analytical advantages have been
demonstrated. For a general review of the chromatographic properties of
polyurethane foams and rubbers see references [125,126].

7.6.4 Bonded-Phase Adsorbents

Bonded-phase adsorbents for off-line sample preparation were introduced in
about 1978. Chemically they are similar to the column packings used in HPLC
except that a larger average pacticle diameter, 40 micrometers, is used to
facilitate the sampling process. Silica gel, nonpolar and polar bonded-phases,
and ion-exchange sorbents are available, Table 7.11, packed into disposable
polyethylene columns in amounts from 100 to 500 mg. The packing is retained by
two 20 micrometer polyethylene frits. A vacuum manifold is used to facilitate
rapid sampling and solvent elution of the retained organics. With the 500 mg
column, sample sizes as large as 100 ml may be used although most applications
reported have used much smaller volumes, particularly for biological and
physiological samples [127-132]. Average recoveries, in the range 80-100%, at
flow rates up to 33 ml/min have been obtained. The low cost, high sorptive
capacity, wide range of sorbent selectivity, and simplicity of use will continue
to contribute to recent interest in these materials.

7.6.5 Ion-Exchange Resins

Ion-exchange resins can be used for the efficient collection of acidic and
basic substances. Typical examples include the selective recovery of phenols,
acids, and amides from water [133-135] and steroids [129], organic acids
[136,137], and amino acids [138,139] from biological fluids. The recovery of
these materials using nonpolar adsorbents is often variable and incomplete.
Ion-exchange is particularly attractive for the above applications since neutral
molecules, which may interfere in the final chromatographic analysis, are easily
washed from the ion-exchanger without affecting the recovery of the ionized
components. For the complete recovery of phenols from water it has been shown
that the sample pH should be at least two units higher than the pK_a value of the

TABLE 7.11

CHARACTERISTICS OF SILICA BASED BONDED-PHASE SORBENTS

Sorbent Type	Sample Type	Typical Application
Octadecyl	Reversed-phase extraction of nonpolar compounds	Drugs, Essential oils, Food Preservatives, Vitamins, Plasticizers, Pesticides, Steroids, Hydrocarbons
Octyl	Reversed-phase extraction of moderately polar compounds. Compounds which bind too tightly to octadecyl silica	Priority Pollutants, Pesticides
Phenyl	Reversed-phase extraction of nonpolar compounds. Provides less retention of hydrophobic compounds	Not widely used
Cyanopropyl	Normal phase extraction of polar compounds	Amines, Alcohols, Dyes, Vitamins, Phenols
Silica Gel	Adsorption of polar compounds	Drugs, Alkaloids, Mycotoxins, Amino Acids, Flavinoids, Heterocyclic Compounds, Lipids, Steroids, Organic Acids, Terpenes, Vitamins
Diol Functionality	Normal phase extraction of polar compounds (similar to silica gel)	Proteins, Peptides, Surfactants
Aminopropyl	Weak anion-exchange extraction	Carbohydrates, Peptides, Nucleotides, Steroids, Vitamins
Dimethylaminopropyl	Weak anion-exchange extraction	Amino Acids
Aromatic Sulfonic Acid Functionality	Strong cationic-exchange extraction and reversed-phase extraction (eliminates ion-pairing when used in place of octydecyl silica)	Amino Acids, Catecholamines, Nucleosides, Nucleic Acid Bases
Quaternary Amines	Strong anion-exchange extraction	Antibiotics, Nucleotides, Nucleic Acids

phenols, and a reducing agent (e.g., sodium bisulfite) should be added to avoid oxidation during sampling [134]. It is also suggested that barium hydroxide be used to precipitate inorganic acids (e.g., sulfate, phosphate) in biological fluids prior to sampling. This avoids interference in the recovery and final analysis of the organic acids [137]. Organic acids can be eluted from DEAE-Sephadex with pyridinium acetate, which is easily removed from the extract prior to analysis by lyophilization [136,137].

7.7 On-Line Trace Enrichment by HPLC

A suitable arrangement for on-line trace enrichment using standard HPLC equipment, a multifunctional valve, and a trace enrichment column is shown in Figure 7.8. Compared to off-line sampling techniques, greater accuracy can be expected by minimizing sample losses from adsorption to container walls, degradation upon heating, and sample manipulation, as well as minimizing interferences from solvent impurities [95,140-142]. The apparatus in Figure 7.8 can also be automated for unattended operation, increasing sample throughput. The analytical column itself could be used for trace enrichment, the proviso being that the solvent for the sample is a non-eluting solvent for the column [92,143-145]. The sample would then be concentrated at the head of the column and a different mobile phase selected for the separation. The disadvantage of this approach is that any part of the sample which remains on the column will lead to a degradation in its performance. Therefore, it is preferrable to separate the enrichment process from the separation process as indicated in Figure 7.8.

For trace enrichment very short columns, 2-5 mm long, packed with good quality HPLC sorbents are used. Longer columns, perhaps equal in length to the analytical column, are employed for sample cleanup. The analytical column and the precolumn may contain the same sorbent [127,141,146-150] or sorbents of different selectivity for which mobile phase incompatibility is not a problem [127,151,152]. Details of column switching techniques for sample cleanup are discussed in Chapter 4.19. Here we will describe the use of short precolumns for trace enrichment.

The requirements of the precolumn for trace enrichment are that its diameter should be comparable or smaller than that of the analytical column, its length should be as short as possible, and the sorbent used should be of similar size and quality to that selected for the analytical column. The minimum acceptable length for the precolumn is a complex parameter since it will depend on the absolute detection limit of the solute, the concentration of solute in the solution to be analyzed, and the magnitude of the retention of the solute in the system under investigation [142,153]. The ultimate enrichment factor

A. FILL SAMPLE LOOP

B. INJECT SAMPLE ONTO TRACE ENRICHMENT COLUMN TO WASTE

A. BACKFLUSH TRACE ENRICHMENT COLUMN TO HPLC COLUMN TO DETECTOR

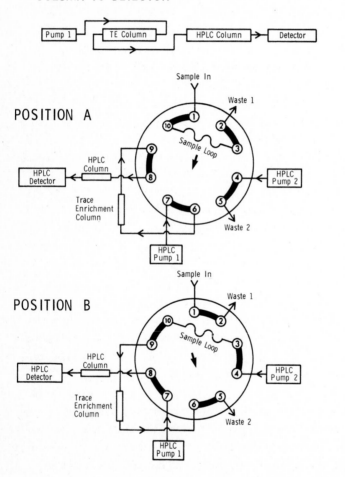

Figure 7.8 Apparatus for on-line trace enrichment by HPLC using a multi-functional valve. (Reproduced with permission from Valco Instruments, Inc.)

obtainable is determined by the breakthrough volume of the solute on the precolumn and the solubility of the solute in the mobile phase used with the analytical column. The breakthrough volume can be measured experimentally or predicted from chromatographic data [153]. The first approach becomes tedious if multiple solutes are to be determined and the latter is subject to large errors. The parameters required to calculate breakthrough volumes are the precolumn dead volume (V_o), theoretical plate count (n), and the capacity factor for the solute in the sample solution, usually water. For short precolumns the influence of extracolumn band broadening can be very large, making the measurements of V_o and n subject to large errors. Capacity factor measurements are subject to error because many organic compounds have long retention times and poor peak shapes on bonded-phase columns when water alone is used as the mobile phase.

As previously mentioned, precolumns for trace enrichment are generally very short; consequently, pressure drops are very low and the sample solution may be pumped through the precolumn at high flow rates, typically 5-25 ml/min. Sampling times are thus short, even for sample volumes of 0.1-10.0 liters, which are normal sample sizes for water analysis. The retention of acidic or basic organic solutes can be enhanced by adjusting the pH of the sample solution to suppress ionization. Under favorable circumstances enrichment factors of 10^4 may be obtained, considerably extending the scope of HPLC for trace analysis. A disadvantage of trace enrichment with sorbents of low selectivity is that many sample constituents are concentrated on the precolumn and the final sample complexity may be too great to permit detection after separation. Under these circumstances selective detectors or multidimensional chromatographic techniques may have to be employed.

7.8 Static Headspace Analysis

Gas chromatographic headspace analysis indirectly determines volatile constituents in liquid or solid samples by analyzing the vapor phase that is in thermodynamic equilibrium with the sample in a closed system [154]. It is used predominantly for the determination of trace concentrations of volatile substances in samples which are difficult to analyze by conventional chromatographic means [154-162]. Examples include dilute solutions where the matrix would obscure the components of interest, damage the column, or require excessively long analysis times due to the presence of late eluting peaks; inorganic or high molecular weight polymers which can not be volatilized or solubilized under normal conditions; and inhomogeneous mixtures such as blood, sewage, colloids, etc., which require extensive sample cleanup prior to analysis. In the above situations the advantages of the headspace method are economy of effort and the attainment of a sample which is relatively free from its matrix and the problems associated with the chromatographic properties of the matrix. As will become obvious later, the main disadvantage of quantitative headspace analysis is the need for careful and extensive calibration.

A quantitative expression relating the concentration of analyte in thermodynamic equilibrium between fixed gas and liquid phases can be developed in two ways. In terms of partial pressure, the chromatographic peak area (A_B) of compound B in the vapor phase above a liquid is proportional to its partial vapor pressure, P_B', equation (7.1) [163].

$$A_B = C_B P_B' \tag{7.1}$$

A_B = chromatographic peak area response of B in the gas phase

C_B = compound-dependent proportionality constant

P_B' = partial vapor pressure of B

According to Henry's law the partial vapor pressure of the analyte above the solution depends on the mole fraction, X_B, of the analyte B and the saturated vapor pressure, $P_B°$, of the pure compound B at a given temperature, corrected for any deviation from ideality by the activity coefficient γ_B, equation (7.2).

$$P_B' = X_B \gamma_B P_B° \tag{7.2}$$

Combining equations (7.1) and (7.2) and eliminating P_B' yields an expression relating X_B to the vapor phase composition of the sample, equation (7.3).

$$X_B = \frac{A_B}{C_B \gamma_B P_B°} \tag{7.3}$$

Alternatively, the equilibrium concentration of the analyte B in the gas phase above a liquid is proportional to the partition coefficient, K_B, equation (7.4) [155,164].

$$K_B = C_L/C_G \tag{7.4}$$

K_B = partition coefficient of B between the liquid and gas phase at a fixed temperature

C_L = concentration of B in the liquid phase

C_G = concentration of B in the gas phase

The equilibrium condition is also satisfied by equation (7.5).

$$C_L{}^\circ V_L = C_L V_L + C_G V_G \qquad\qquad (7.5)$$

$C_L{}^\circ$ = initial concentration of B in the liquid phase

V_L = volume of the liquid phase

V_G = volume of the gas phase

Eliminating C_L from equations (7.4) and (7.5) leads to equation (7.6).

$$C_L{}^\circ = \frac{C_G(K_B V_L + V_G)}{V_L} \qquad\qquad (7.6)$$

From equations (7.3) and (7.5) it can be seen that the concentration of B in the headspace above a liquid in equilibrium with a vapor phase will depend on the volume ratio of the gas and liquid phases and the compound-specific partition coefficient, K_B, which in turn is matrix dependent $(\gamma_B=1)$.

From the experimental point of view the headspace sampling technique is very simple. The sample, either solid or liquid, is placed in a glass vial of appropriate size and closed with a rubber septum, Figure 7.9A [154,165]. The vial is carefully thermostatted until equilibrium is established. The gas phase is sampled by syringe for manual procedures or with an electropneumatic dosing system in automated headspace analyzers. With the automated analyzer, a hypodermic needle penetrates the rubber septum on the sample vial and the gas phase is pressurized until it is equal to the column head pressure [166]. At this point a valve switches off the carrier gas flow, automatically releasing the excess pressure in the sample vial via the column. The volatile constituents in the sample are simultaneously carried into the column. At the end of this sampling sequence the valve on the automatic injector closes and the carrier gas supply to the column is restored. When automatic headspace sampling is used with capillary columns, it is recommended that a viscous carrier gas such as helium, narrow bore capillary columns, and long column lengths (independent of the resolution needed for the separation) be used to generate a sufficiently high pressure differential for successful operation of the automatic sampler [161].

For occasional manual headspace analysis a large glass hypodermic syringe, with a two-way valve attached to its luer tip, is a convenient sample chamber and can be adjusted to control the phase volume ratios [167]. After equilibration the headspace is sampled through the valve using a second syringe.

The sensitivity of the headpsace sampling method can be improved in some instances by adjusting the pH, salting out, or raising the temperature of the sample. As the concentration of the analyte in the gas phase is proportional to the concentration of the undissociated part of the analyte in solution,

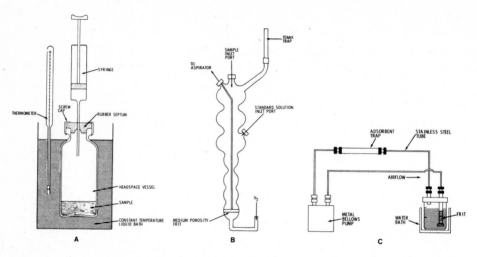

Figure 7.9   Apparatus used for gas chromatographic headspace analysis.
A, static headspace sampling apparatus; B, gas phase stripping
apparatus; C, closed-loop gas phase stripping apparatus.

adjusting solution pH so that the analyte exists mainly in the nondissociated
form will improve sensitivity [155,164,168,169]. Salting out by adding an
inorganic salt to an aqueous solution [155,164,168] or a non-electrolyte such as
water to a miscible organic solvent [169] can produce an increase in sensitivity
of over 100-fold in favorable cases. Raising the temperature of the sample
increases the saturated vapor pressure of the analyte and hence its
concentration in the gas phase. However, enhancement values are usually small,
due to the limited temperature range accessible in most cases [154]. The
principal sources of error in headspace analysis are usually associated with
adsorption of the analyte onto the rubber septum or premature analysis of
systems which equilibrate slowly [154,159,160,164].

Functional group analysis can also be performed by headspace analysis. The
normal sample headspace is run, a derivatizing reagent capable of forming an
involatile derivative with compounds containing a particular functional group is
then added to the liquid phase, and a second headspace sample is run. The
difference between the two chromatograms indicates which volatile components

contain that particular functional group.

A unique feature of headspace analysis is that the information obtained from the experiment, the chromatographic peak area of a substance in the gas phase, is an indirect measure of the concentration of that same substance in the original liquid phase. The liquid and gas phase concentrations are related to each other by the partition coefficient and an experimentally-derived proportionality constant. The partition coefficient is matrix dependent and unknown for most analyses; it must therefore be accounted for by calibrating the sampling system. Suitable calibration methods are summarized in Table 7.12. If the sample matrix can be obtained in a pure form, then calibration is simply performed by adding known amounts of the analyte to the matrix and analyzing the calibration standards in a manner identical to the original samples [154,170]. If the sample matrix can not be duplicated then the method of standard additions is used [154,159,160,171]. Here it is assumed that the addition of a small, known amount of analyte to the analyte already present in the sample will not change either the partition coefficient or the ratio of the gas and liquid phases. If, in an independent experiment, the relationship between detector response and the substance to be determined has been established, then the original amount of the analyte in the liquid phase can be calculated from equation (7.7) [154,159,160,171].

$$W_B = \frac{W_S - W_B}{(C'_{BG}/C_{BG}) - 1} \tag{7.7}$$

W_B = original amount of sample in the liquid phase

W_S = amount of standard added after having withdrawn and analyzed a sample of the gaseous phase

W_B = amount of equilibrated analyte in the sample measured before the addition of standard ($W_S \gg W_B$)

C_{BG} and C_{BG}' = concentration of analyte B in the gaseous phase of the original sample before and after the addition of the internal standard

Solid samples which form homogeneous solutions may be analyzed by either of the above methods depending on whether or not the sample matrix can be exactly duplicated. The solvent must be free of volatile impurities and preferably of low volatility so that it elutes after the sample components in the chromatogram and does not interfere in their determination [154,156]. For insoluble solid samples, and for samples for which matrix duplication or the standard additions method are inappropriate, the method of multiple headspace extraction is utilized [162,172-176]. This method employs a controlled, stepwise sampling of the headspace from a single sample. After each extraction the sample is

TABLE 7.12

CALIBRATION METHODS FOR QUANTITATIVE HEADSPACE ANALYSIS

Sample Type	Calibration Method	Principle	Examples
Homogeneous Solutions	Model System	Matrix available in pure form. Known amounts of substance added to pure matrix and determined by the same experimental procedure as samples	Volatile organic compounds in drinking water, beverages, vegetable oils, mineral oils, etc.
Inhomogeneous samples (liquid + solid)	Model System Standard Additions	Model systems may be used if a pure sample matrix is available (e.g., blood, milk). In the standard addition method the sample is analyzed twice, the second time after the addition of a known amount of the substance to be determined. The sample must be re-equilibrated after addition of the standard.	Residual monomers in polymer dispersions, alcohol and toxic substances in blood olfactory substances in milk.
Soluble Solids	Model System Standard Additions	For solid samples forming homogeneous solutions the model system may be used if pure sample matrix materials are available; otherwise, the standard additions method is used.	Inorganic, organic, and polymeric solids and salts. Monomers in polymers.
Insoluble Solids	Multiple Headspace Extraction	MHE can be used for substances of high volatility with a small partition coefficient. Method is based on a stepwise gas extraction at equal time intervals. Normal headspace chromatogram is run, a fraction of the gas phase exhausted, and a second headspace chromatogram is run. The difference in peak areas provides a measure of the total peak area of the analyte.	Volatiles and monomers in insoluble polymers. Aroma volatiles from foodstuffs, fruits, spices, tobacco, etc. Residual solvents in pharmaceuticals and printed films.

re-equilibrated and a further fraction of headspace is removed. The
concentration of analyte in the headspace becomes smaller after each extraction,
while the partition coefficient remains constant. Repetition of the process
would eventually result in the complete stripping of the analyte from the matrix
and the original amount of analyte could be obtained from the sum of all the
partial chromatographic peak areas for each extraction. Fortunately, exhaustive
extraction is not required to calculate the total amount of substance present.
In most cases two measurements suffice to estimate the total area for the
analyte using equation (7.8) [173].

$$\Sigma A_n = A_1^2 \ (A_1 - A_2) \tag{7.8}$$

ΣA_n = total of all partial peak areas

A_1 = peak area obtained in the first headspace analysis

A_2 = peak area obtained in the second headspace analysis

7.9 Dynamic Headspace and Gas Phase Stripping Analysis

Substances which are low in concentration and have unfavorable partition
coefficients can not be readily determined by the static headspace method.
Dynamic headspace or gas and vapor phase stripping techniques, combined with
sorbent or cryogenic trapping, are required to increase the amount of sample
available for analysis [159,160,177]. When a gas is used to strip volatile
organic substances from a liquid, the rate at which a substance is removed from
solution is given by equations (7.9) and (7.10) [159,160].

$$\frac{dW_B}{dt} = -W_B \left[\frac{F}{V_G + K_B V_L} \right] \tag{7.9}$$

$$\frac{W_B}{W_B^{\circ}} = \exp - \left[\frac{Ft}{V_G + K_B V_L} \right] \tag{7.10}$$

W_B = instantaneous amount of W_B in the sample

W_B° = amount of W_B present at the start of the experiment

F = stripping gas flow rate

V_G = volume of gas passed through the liquid in time t

V_L = sample volume

t = stripping time

Assuming that all the material stripped from solution is retained by the trap,
the amount of analyte stripped will depend primarily on the substance-specific

partition coefficient and experimental variables such as flow rate, time, and total volume of stripping gas passed through the solution. Only in those instances when the organic compound has low water solubility ($<$ 2% w/w) and is relatively volatile (b.p. $<$ 200°C) can quantitative extraction be expected [178]. In most circumstances the methods described in this section provide only semiquantitative information unless calibration with model systems is also employed.

In the main, the dynamic headspace and gas phase stripping techniques have been used to determine volatile organic compounds in water suspected of containing environmental contaminants and/or potential carcinogens, or for the profiling of biological fluids to detect volatile marker substances suitable for the early detection of disease in man. The organic volatile fraction in the latter case is comprised of a chemically diverse group of substances of widely different polarity; most are alcohols, ketones, aldehydes, O- and N-heterocyclic compounds, isocyanates, sulfides, and hydrocarbons [177]. They contain from 1 to 12 carbon atoms and have boiling points generally less than 300°C.

The dynamic headspace method is a simple extension of the static headspace method. A continuous flow of gas is swept over and above the surface of a liquid, carrying the headspace volatiles to a trap where they are accumulated prior to analysis. The method is suitable for determining the concentration of organic volatiles in urine and waste water, where the sample volume available for analysis is not a restriction [177,179-184]. A suitable apparatus is shown in Figure 7.10 [177,179]. In a typical experiment, an aliquot of 24-hour urine, ammonium sulfate (200 g/l), and a magnetic stirrer bar are placed into a 500 ml sampling bottle, which has a ground-glass joint at its neck. The sample bottle is fitted with a condenser connected to a thermostatically controlled water circulator. Single or multiple sorbent traps, usually filled with Tenax, are attached to the other end of the condenser. A flow of helium is established through the apparatus at 20 ml/min. The organic volatile fraction is collected by heating the rapidly stirred sample in a water bath set at 90°C; the sampling time usually being one hour.

Because the organic volatiles in urine samples equilibrate slowly with the gas phase above the fluid, the minimum sample volume of urine needed for a detailed profile is 20.0 ml. The passage of organic volatiles from urine into the gas phase is also influenced by their solubility in water, the possibility that selective adsorption to high molecular mass biological molecules will diminish their vapor pressure, and the area of contact between the gas and liquid phase. Transfer of organic volatiles to the gas phase is favored by adding a salt to the sample, maintaining an elevated temperature, and vigorously stirring the sample so as to create a vortex. Because the equilibration of volatiles between the gas and liquid phases is slow, sampling times are

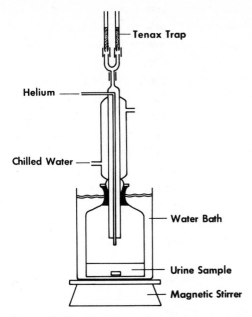

Helium

Chilled Water

Tenax Trap

Water Bath

Urine Sample

Magnetic Stirrer

Figure 7.10   Apparatus for dynamic headspace analysis of urine with sorbent
              trapping. (Reproduced with permission from ref. 177. Copyright
              American Association of Clinical Chemists).

comparatively long; thus it is possible to trap several samples sequentially
during the same experiment. These samples may differ quantitatively, a
reflection of the fact that the volatiles are removed gradually and
incompletely.

    Gas phase stripping techniques can improve the yield of organic volatiles
from water or biological fluids by facilitating the transfer of volatiles from
the liquid to the gas phase; it is also more suitable than dynamic headspace
analysis when the sample volume is restricted. Sometimes called the
purge-and-trap technique, it is used routinely in many laboratories, and
automated equipment is available [178,185-190]. A suitable apparatus for gas
phase stripping is shown in Figure 7.9B [187]. Gas phase stripping differs from
the dynamic headspace method in that the sampling gas is introduced below the
liquid level through a fritted orifice; the finely dispersed bubbles provide
maximum surface contact between the gas and liquid phases. As the organic
volatiles move into the gas phase they are rapidly and continuously carried
away, thus favoring the stripping process. The apparatus shown in Figure 7.9B
has baffled walls to diminish sample carryover due to foaming, has an inlet port
so that standard solutions may be added below the level of the sample by
syringe, and contains a center tube which facilitates emptying and cleaning the
apparatus. For volatile halocarbons in waste water the stripping process is

essentially complete and detection limits below the microgram/1 level can be obtained. Biological fluids often foam excessively, a problem which can sometimes be controlled by using a column packed with glass beads for dispersion in place of the glass frit [177,185].

Gas phase stripping with adsorption of the volatiles on a small charcoal trap in a closed-loop system was popularized by Grob for determining very low levels of organic volatiles (ng/l) in drinking water [34,191-195]. A version of this apparatus is shown in Figure 7.9C [196]. It provides high reproducibility, sensitivity, and low blank values [197]. Low and medium molecular weight compounds, up to eicosane in the n-alkane series, may be analyzed. The trapped organic volatiles are removed from the miniature charcoal trap by elution with a small volume of carbon disulfide and analyzed directly by capillary column gas chromatography. An influent preheater (not shown in Figure 7.9C) is generally used to warm the stripping gas to about 40°C to prevent condensation of water in the charcoal trap.

Generally speaking, the purge-and-trap technique is the method of choice for determining organic volatiles in water because of its ease of operation. If greater sensitivity is required, the closed-loop stripping apparatus should be used.

Zlatkis and co-workers have described a transevaporator sampling apparatus for the solvent stripping of organic volatiles from small (5-500 microliter) samples of biological fluids with collection and separation of the organic volatiles from the extracting solvent on an adsorbent column [57,177,198-201]. The transevaporator apparatus is suitable for the semiquantitative analysis of organic volatiles in serum, urine, saliva, cerebrospinal fluid, breast milk, amniotic fluid, sweat, and tissue homogenates. With reference to Figure 7.11, the sample is injected by syringe into the base of a microcolumn packed with Porasil E. The microcolumn retains most of the water, the high molecular weight polar organics, and the inorganic salts present in biological fluids. It also distributes the sample as a film over a large surface area and thus improves the efficiency with which the organic volatiles are extracted. The stripping gas passes through the microcolumn, removing the organic volatiles which are then collected on a Tenax trap, providing a modified headspace sample. A solvent-extraction profile is obtained using gas pressure to force a volatile solvent such as 2-chloropropane through the microcolumn; this solvent carries the solvent-elutable organic volatiles to the glass bead trap, where the volatiles are collected and the solvent is evaporated. Thermal desorption is used to transfer the organic volatiles from the trap to a capillary column for separation by gas chromatography. Figure 7.12 is a representative chromatogram of the organic volatile fraction from 50 microliters of serum using the transevaportor in the solvent-elution mode.

TRANSEVAPORATOR

MODIFIED HEADSPACE   SOLVENT EXTRACTION MODE

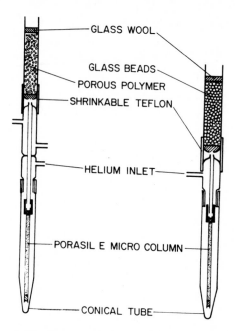

Figure 7.11   Transevaporator sampling apparatus for isolating organic volatiles
from small sample volumes. (Reproduced with permission from
ref. 177. Copyright American Association of Clinical Chemists).

7.10   Sampling Methods for Volatile Organic Compounds in Air

In general, three classes of organic compounds can be distinguished as
normal constitutents or environmental contaminants in air [202-206]. There are
substances which are normally gases at room temperature or have high vapor
pressures (e.g., C_1 - C_5 carbon compounds), substances having sufficient
volatility to yield measurable vapor concentrations at room temperature (e.g.,
C_5 - C_{20} carbon compounds), and compounds of intermediate or restricted
volatility attached to solid particles. No single sampling method or
chromatographic column is capable or resolving the entire spectrum of organic
compounds found in air. Analytical methodology is thus conventionally divided
into procedures which are geared toward the determination of gases, of organic
compounds in the intermediate volatility range, and of materials which are
commonly associated with particulate matter [207].

Gaseous hydrocarbons and some very volatile halocarbon compounds are
usually collected by cryogenic condensation using liquid oxygen as a coolant to

474

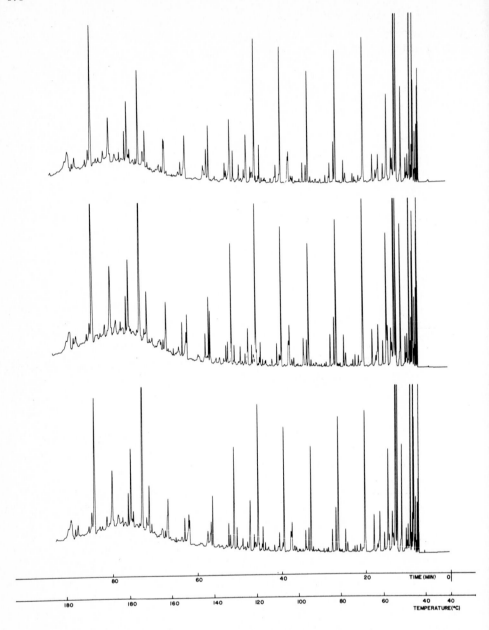

Figure 7.12   Profiles of organic volatiles from 50 microliters of human serum
using the transevaporator sampling apparatus and capillary column
gas chromatography.   Replicate chromatograms indicate sampling
reproducibility.

prevent excessive condensation of air in the sampler [208,209]. As air contains a large concentration of water, microfog formation or blocking of the trap with ice may limit the collection efficiency. Grab sampling into evacuated containers with processing in the laboratory is often preferred to cryogenic trapping in the field. Plastic bags or metal or glass containers, pre-evacuated or used with pumps, are most frequently employed for grab sampling. Grab sampling may also be used for sampling organic volatiles in the intermediate volatility range. Its principal disadvantages are the possibility of adsorption or catalytic modification on the container wall, contamination of the sample by the container, and modification of the sample by container leakage. Extensive cleaning and testing of the containers between use is also required and tends to detract from the routine application of the method.

Particulate organic matter is generally collected using impactor systems, electrostatic precipitators, or high-volume filtration samplers [210,211]. The high-volume sampler is a large surface area filter made from glass fiber, porous Teflon, or some similarly inert material, through which air is drawn by a pump. It is the method most frequently used, and the method for which approved operating procedures have been established by the EPA for monitoring atmospheric pollution. Approximately 2000 m^3 of air per 24 h can be sampled, yielding approximately 0.1 g of particulates. Cascade impactors use more complex equipment than high-volume samplers and provide a size fractionation of the particles. The particles themselves are comprised largely of inorganic material from which the organic fraction is isolated by extraction.

Organic compounds of intermediate volatility are usually collected by drawing filtered air through a bed of an appropriate sorbent. Porous polymers, charcoal, or inorganic adsorbents such as silica gel or alumina are most frequently used. Of these, the porous polymers, and in particular Tenax, are the most important [203,206,212-217]. Tenax will efficiently adsorb a wide range of organic compounds, has a low efficiency for the retention of water, and is thermally stable, permitting the use of high desorption temperatures. However, compounds of molecular weight lower than about C_7 are not quantitatively retained at room temperature. A more active sorbent such as charcoal, graphitized carbon black, or Ambersorb is required to trap these compounds [203,218,219]. Although the collection efficiency of carbon sorbents is generally higher than Tenax at room temperature, the use of these materials is not without problems. Their higher affinity for water, greater catalytic activity, and incomplete sample recovery are the principal problems. Dual traps containing Tenax in the front portion and a carbon sorbent in the rear can be used for the quantitative collection of samples which cover a wide volatility range. Inorganic adsorbents can trap polar compounds or act as supports for specific derivatizing reagents. They are generally used for applications where

476

their high affinity for water and reactivity with reactive gaseous pollutants
(e.g., CO_2, SO_2, etc.) are not of overriding concern. The retained compounds
are stripped from the adsorbent by solvent elution, leading to a large dilution
of the sample compared to direct thermal desorption employed with the porous
polymers. With reference to problems in sampling air pollutants, Simmonds has
described a selective membrane separator for drying air in a dynamic
flow-through system [167,220,221].

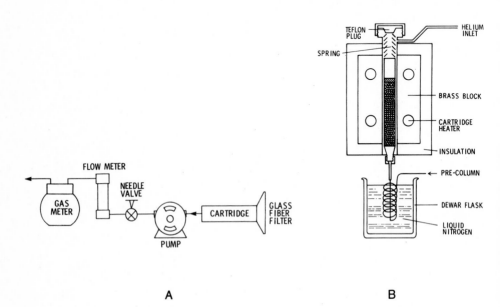

A   B

Figure 7.13   Apparatus for sampling airborne organic samples. A, trapping of
organic volatiles in air using sorbent traps; B, thermal
desorption chamber.

A schematic diagram of an apparatus used for trapping organic volatiles in
air with Tenax is shown in Figure 7.13A [216]. Although sorbent traps of
different dimensions are frequently used, a sorbent bed of 1.5 x 6.0 cm is
generally sufficient for sampling air volumes of 5 to 200 liters at 10–200
ml/min. The Tenax sorbent may require decontamination before use by Soxhlet
extraction overnight with pentane and then methanol, followed by thermal
conditioning of the dried sorbent in a stream of purified helium at 250–350°C
for 24 h. After conditioning, the Tenax traps can be stored in culture tubes
with Teflon-lined caps until needed. The collected samples are analyzed by
thermal desorption from the trap into a cryogenically–cooled capillary
precolumn. Alternatively, the first few loops of the analytical column may be
cooled in a Dewar flask of liquid nitrogen. A thermal desorption chamber for

sorbent traps is shown in Figure 7.13B. The desorption temperature and time required to strip the sample from the sorbent depend on the properties of the sample, but temperatures of 250-350°C and times from 5-30 min are common. As a general guide, a purge flow of helium at 15 ml/min and a temperature of 270°C for 8 min is sufficient to recover most of the organic compounds trapped from air. As the desorption process is kinetically slow, the desorbed sample must be reconcentrated to provide a narrow sample plug for separation by capillary column gas chromatography. This is achieved by cryogenic focussing as illustrated in Figure 7.13B. Passing carrier gas through the heated precolumn trap rapidly flushes the sample onto the analytical column. The above desorption process can be performed manually or automated for unattended operation.

The collection/desorption/separation process is relatively trouble free. Excess water, however, may cause sample loss in the desorption step due to microfog formation and/or blockage of the precolumn. Passing a dry stream of gas through the cartridge for a short time or storing it in a sealed container with desiccant will usually eliminate these problems.

The parameter which characterizes the collection efficiency of a particular sorbent for a particular analyte is the breakthrough volume. As there is no clear concensus on the definiton of this parameter or on the preferred method of measurement, numerous conflicting values can be found in the literature. The breakthrough volume is usually assumed to be the volume of gas, containing analyte, that can be passed through the sorbent bed until its concentration at the outlet reaches some fraction of its inlet concentration. This fraction has been variously defined as 100%, 50%, 5%, or 1% of the analyte concentration at the inlet. Several methods have been used to measure the breakthrough volume of sorbent traps [222,223]. These include purging a trap directly into a flame ionization detector [222,223], field sampling with dual traps in series [203,222], loading traps with a known amount of analyte which is then purged with a volume of gas and the amount of sample remaining determined [222], or by estimation from the measurement of chromatographic retention volume data at different temperatures [203,216,224-227]. An approximately linear relationship exists between the logarithm of the specific retention volume of a substance and column temperature. Therefore, the retention volume of a substance can be measured at several higher column temperatures and the value at 20°C or any other temperature can be obtained by extrapolation. Some representative breakthrough volumes on Tenax measured in this way are given in Table 7.13.

The breakthrough volume depends mainly on the affinity of the analyte for the sorbent (V_g), the efficiency (measured in theoretical plates) of the sorbent traps, and the trapping temperature (see Table 7.13). Within reasonable

TABLE 7.13

BREAKTHROUGH VOLUMES (1/g) OF SOME COMMON ATMOSPHERIC POLLUTANTS ON TENAX-GC

Under field sampling conditions a cartridge, containing about 2.2g (7.96 ml) of Tenax is commonly used.

Chemical Class	Compound	Boiling Point(°C)	Adsorbent Trap Temperature (°C)		
			21.1	26.7	32.2
Hydrocarbons	hexane	68.7	7.7	5.5	4.1
	heptane	98.4	34.1	25.0	17.7
	1-heptene	93.6	61.4	42.3	29.1
	cyclohexane	80.7	11.8	8.6	6.4
	benzene	80.1	24.5	17.3	12.3
	toluene	110.6	111.4	78.6	55.4
	ethylbenzene	136.2	315.0	221.4	156.4
	biphenyl	256	14381.0	10228.0	7293.6
Halogenated	vinyl chloride	13	0.57	0.45	0.36
hydrocarbons	1,2-dichloropropane	95	52.3	36.8	26.4
	1,3-dichloropropane	121	83.6	60.9	44.1
	2,3-dichloropropane	94	69.1	47.3	32.3
	bromobenzene	155	490.0	347.0	246.0
Alcohols	methanol	64.7	0.36	0.27	0.18
	propanol	97.5	6.4	4.5	3.2
	ethylene glycol	197	30.4	21.4	15.0
Amines	dimethylamine	7.4	1.8	1.4	0.91
	pyridine	115	85.9	60.9	43.2
	aniline	184	1724.0	1176.0	802.7
Aldehydes	acetaldehyde	20	0.91	0.45	0.41
	benzaldehyde	179	1594.0	1083.0	737.3
Ketones	acetone	56	5.4	3.6	2.7
	methyl ethyl ketone	81	17.7	12.3	8.6
	acetophenone	202	1258.0	909.0	654.0
Esters	ethyl acetate	77	32.7	21.8	14.5
	methyl acrylate	80	34.1	22.7	15.4
	methyl methacrylate	100	144.5	95.0	62.3

experimental limits, the breakthrough volume is independent of normal variations in humidity and the concentration of organic pollutants (<100 ppm) in air [222,225,228]. As a wide variety of sampling conditions is employed for collecting organic volatiles on sorbent traps, the idea of a safe sampling volume has been proposed [225]. This has been defined as the volume of air containing a particular vapor contaminant that may be sampled under a variety of circumstances without significant breakthrough occurring. A value of about 50% of the specific retention volume of the analyte was proposed for the safe sampling volume. This generalization has been criticized because it ignores the contribution of the sorbent trap efficiency to the collection efficiency [223,226,227]. The safe sampling volume was therefore redefined to be less than the sampling volume calculated from equation (7.11).

$$V = V_g (1 - 2/\sqrt{n}) \tag{7.11}$$

V = maximum sampling volume per gram of sorbent

V_g = specific retention volume

n = number of theoretical plates for the sorbent trap

A more explicit expression for the breakthrough volume, which describes the shape of the breakthrough curve, is given by equation (7.12) [223].

$$R(V) = C \text{ erfc.} \frac{[(n/2)^{1/2}(1 - V/V_g)]}{2} + C \text{ erfc.} \frac{[(n/2)^{1/2}(1 + V/V_g)]}{2} \tag{7.12}$$

$R(V)$ = analyte concentration at the outlet of the sorbent trap

C = analyte concentration

Analysis of equation (7.12) indicates that the collection efficiency of the sorbent will increase as n increases and collection will be nonquantitative for small values of n. For example, if n = 2, it would be impossible to collect a sample with an efficiency greater than 85% since, even at the start of sampling, 15% of the analyte is passing through the sorbent without being collected. This applies independent of the value for V_g, which may be very large. For a collection efficiency of 99.9%, a sorbent trap with a minimum of 10.7 theoretical plates is required regardless of the sampling volume or analyte being collected. Thus, the sorbent trap should conform to a minimum length and be packed homogeneously with sorbent of a well-defined mesh range compatible with obtaining a minimum pressure drop across the trap at normal sampling flow rates.

During the sampling operation, part of the sample is immobilized on the sorbent while air is continuously drawn through the sorbent trap. There is, therefore, a possibility that reactive gases such as ozone, sulfur dioxide, halogens, etc., although normally present in low concentrations in air, might modify the sample composition during collection. Model experiments, in which air was doped with known concentrations of potentially reactive agents, did not, however, reveal any significant problems in this respect [222,229,230]. Artifact formation is more likely to occur at the high temperatures required for desorption.

Numerous reports of the identification of volatile organic compounds in air using capillary column chromatography-mass spectrometry and sorbent trapping have appeared [203-206,212-218,231]. Methane, with a mixing ratio of about 1.6 ppm, is by far the most abundant hydrocarbon in unpolluted air [208]. Alkanes and substituted aromatic compounds account for the majority of organic volatiles other than methane identified, but are present in much lower concentrations (ppt-ppb), Table 7.14 [206]. Changes in sample site, time of day over which the sample was collected, and meteorological conditions can lead to changes in concentration of about 20-fold for the above compounds. The use of coal and oil to generate energy is the principal source of volatile organic compounds in air.

TABLE 7.14

AVERAGE CONCENTRATIONS OF SOME ORGANIC VOLATILES IN URBAN AIR SAMPLES

Compound	Concentration (ppb)
Benzene	1.3 - 15
n-Nonane	1.6 - 4.4
Toluene	0.3 - 9.7
n-Decane	1 - 2.7
Ethylbenzene	3.1 - 4.5
p-Xylene	2.1 - 3.4
m-Xylene	5.4 - 7.8
o-Xylene	3.0 - 4.8
Methylethylbenzene	1.5 - 4.0
Limonene	0 - 5.7

7.11 Personal Sampling and Occupational Hygiene

The levels of toxic chemicals are approximately 10 to 100-fold higher indoors than in the surrounding outdoor atmosphere [232]. In many instances this can be related to the use of better insulation practices for energy conservation, which diminish the exchange of indoor and outdoor air, and to the increasing use of synthetic materials in construction and furniture. Wherever volatile substances are manufactured, processed, transported, or handled in any way, the possibility of exposure to toxic substances can be very great [233]. Because of the concern of responsible agencies, guidelines are now in effect which limit occupational exposure to certain chemicals in the workplace, and these have been extended to include the use of certain materials in the home. Standards for acceptable levels of exposure can be established by combining analytical data with results from animal toxicity studies and human epidemiological investigations. Standards for individual substances are usually set to reflect ceiling values, a threshold concentration value never to be exceeded even momentarily, and also time-weighted average values, which reflect exposure over a normal working period, for example 8 h. Threshold limit values are set conservatively to allow for individual susceptibilities resulting from differences in age, weight, health, respiration rate, rates of absorption, and personal habits such as smoking or drinking. They are believed to be safe concentrations, below which the worker is assumed to be protected from short term effects such as chronic toxicity and environmental nuisances and long term effects such as cancer.

Three methods of personal monitoring are widely used to measure the concentration of airborne organic volatiles in the region of the mouth. These are bubblers, vapor adsorption tubes, and passive samplers. If the monitoring device is to be worn throughout the work period, it is important that the device is lightweight, unobtrusive, silent, and permits normal movement. For static monitors, weight and size present less of a problem.

Passive samplers may be worn in the form of a badge and have the advantages of low cost and convenience. They contain an adsorbent, such as charcoal, in a protective polymeric support [234,235]. The adsorbent is in contact with a stagnant air layer, across which the contaminants diffuse and are trapped by the adsorbent at a rate proportional to their concentration in air. The concentration of adsorbed materials is determined by solvent elution and gas chromatography in the usual way. In general, field trials with passive samplers have not been as reliable as anticipated from laboratory studies and further work is required.

Bubblers or impingers have several disadvantages for personal monitoring. They are clumsy to wear, there is a large dilution factor in the adsorbent solution, high air sampling rates and hence large pumps are needed, and spray and evaporative losses must be controlled or accounted for in some way. Because of the above disadvantages, bubblers are used mainly for the collection of very high boiling, reactive, or polar substances that can not be quantitatively recovered from solid sorbents [236].

The most common method of personal monitoring is a sorbent cartridge in conjunction with a small pump to maintain a fixed flow of air through the cartridge [237,238]. Carbon adsorbents are used for trapping volatile organic compounds [239]. Typical sorbent traps are glass tubes (5-6 cm long and 4 mm I.D.) that have been divided with polyethylene plugs; they contain approximately 100 mg of sorbent in the front section and 50 mg in a back section. The two sorbent sections are analyzed separately by eluting the trapped organic volatiles with carbon disulfide, followed by gas chromatography [240-244]. Dimethylformamide has been recommended as a better solvent than carbon disulfide for the recovery of polar organic compounds [245]. Adding a few percent of an alcohol to carbon disulfide or a biphasic system, water-carbon disulfide, may also improve the recovery of polar compounds [238,241]. The recovery of compounds from charcoal may be nonquantitative due to irreversible binding, for example 2-nitropropane, indene, cyclohexane, or because of slow equilibration of the sample with the eluent leading to inconveniently long analysis times. Careful calibration is required when recovery from the sorbent is incomplete. Silica, alumina, and chemical reagents supported on adsorbents are also used in specific cases [238,246].

482

Semivolatile organic compounds, for example pesticides, polychlorinated biphenyls, etc., are sampled with a low-volume sampler containing a Tenax-packed cartridge between two plugs of polyurethane foam [232,247]. The sample is recovered by Soxhlet extraction overnight using hexane-diethyl ether (95:5). The sorbent trap can then be reused. Polyurethane foam shows a fairly high efficiency for collecting semivolatile compounds, is lightweight, and has a low flow resistance, permitting high sampling rates. Pesticide compounds can be determined in the concentration range 1 ng/m^3 to 1 microgram/m^3 in a 3 m^3 air sample collected over a 12 h period [247].

7.12 Methods for Preparing Standard Mixtures of Volatile Organic Compounds in Air

The preparation of synthetic mixtures of gases or volatile organic liquids in gases is more difficult and potentially less accurate than methods for preparing liquid mixtures [248,249]. This is because gases cannot be easily weighed, volumes may change during handling, and temperature and pressure effects must be considered. Many methods have been used for preparing mixtures

TABLE 7.15

METHOD FOR PREPARING STANDARD MIXTURES OF VOLATILE ORGANIC COMPOUNDS IN A GAS

Method	Principle
Single Rigid Chamber	A known amount of the compound is introduced into a single rigid chamber of known dimensions. Magnetic stirrer or similar device used for homogeneous mixing.
Exponential Dilution Flask	A known amount of pure component or standard mixture is introduced into the vessel, which is stirred for effecient mixing. The vessel is continuously flushed with a steady stream of pure gas causing the concentration of the vapor to decrease with time.
Gas Stream Mixing	Two or more gas streams flowing at a known rate are mixed to give the desired concentration. Multiple dilution stages may be used to give lower concentrations.
Permeation Tubes	A volatile liquid when enclosed in an inert plastic tube, may escape by dissolving in and permeating through the walls of the tube at a constant and reproducible rate. The permeation rate depends on the properties of the tube material, its dimensions, and on temperature.
Diffusion Systems	The liquid whose vapor is to be the contaminant of the gas phase is contained in a reservoir maintained at a constant temperature. The liquid is allowed to evaporate and the vapor diffuses slowly through the capillary tube into a flowing gas stream. If the rate of diffusion of the vapor and the flow-rate of the diluent gas are known, the vapor concentration in the resultant gas mixture can be calculated.

with reasonable accuracy, Table 7.15, and these methods can be broadly
classified as static or dynamic. Static methods involve preparing and storing
the mixture in a closed vessel, for example a cylinder, flask, or plastic bag.
The sample volume is thus limited to that of the container. Cylinders must be
used to store mixtures at high pressures. Static systems are preferred when
comparatively small volumes of mixtures are required at moderately high
concentration levels, but losses of components to the vessel walls may occur.
Dynamic systems generate a continuous flow of mixture and can produce large
volumes, with lower surface losses, owing to an equilibrium between the walls
and the flowing gas stream. Whether static or dynamic systems are employed, the
methods used to create homogeneous mixing of the gas and vapor are important
considerations and some provision for creating forced convection is incorporated
into most devices.

The single rigid chamber is the simplest of the mixing devices. The
concentration of the standard mixture produced is given by equation (7.13).

$$C = C_o \exp (-V_W/V) \tag{7.13}$$

C = instantaneous concentration
C_o = initial concentration
V = container volume
V_W = volume of sample withdrawn

The exponential dilution flask, Figure 7.14A, is a hybrid static/dynamic system,
first described by Lovelock [250] and later modified and further characterized
by others [251-253]. This simple, reliable device is commercially available and
has been widely used. It produces an outlet concentration described by equation
(7.14).

$$C = C_o \exp (-Qt/V) \tag{7.14}$$

Q = volumetric gas flow rate
t = time after sample introduction

Its principal problems arise from surface adsorption losses, mechanical wear of
the mixing device, and the difficulty of accurately measuring the initial sample
concentration. However, it is capable of providing adequate accuracy and
precision for most analytical applications involving readily volatilized
substances. Very low gas phase concentrations can be prepared by mixing the
output from the flask with diluent gas.

Permeation tube devices are now very popular for generating standard vapor
concentrations. The permeation tube contains a volatile liquid sealed in an
inert permeable membrane, usually Teflon or a fluorinated copolymer of ethylene

and propylene, through which it diffuses at a fixed and controlled rate. The driving force behind the process is the difference in partial pressures between the inner and outer walls of the tube. This depends on the dissolution of the vapor in the membrane, the rate of diffusion through the membrane wall, and the rate at which the vapor is removed from the outer surface of the membrane [253-257]. The mass permeation rate per unit tube length can be expressed by equation (7.15).

$$G = \frac{730 \, PM}{\log (d_2/d_1)} \, P_1 \tag{7.15}$$

G = mass permeation rate in micrograms/min per cm of tube length
P = permeation constant for the vapor through the membrane in cm^3 (STP)
M = molecular weight of the vapor
P_1 = gas pressure inside the tube, mmHg
d_2 = outside diameter of the tube
d_1 = inside diameter of the tube

For samples with low vapor pressures at room temperature, elevated temperature ovens are used to raise permeation rates to yield desirable values. Gases or vapors with high membrane permeability require devices other than the standard single-walled tubes, for example multiwalled tubes, microbottles, or permeation wafer devices, to yield reasonable lifetimes. Commercially available permeation tubes have lifetimes of several months and provide a simple and inexpensive method of calibration for laboratories interested in determining only a few substances, or for those who need to perform measurements infrequently.

The diffusion system, Figure 7.14B, is a useful and simple apparatus for preparing mixtures of volatile and moderately volatile vapors in a gas stream [224]. The method is based on the constant diffusion of a vapor from a tube of accurately known dimensions, producing a gas phase concentration described by equation (7.16).

$$S = \frac{DMPA}{RTL} \, \ln \left[\frac{P}{P - p} \right] \tag{7.16}$$

S = diffusion rate of vapor out of tube
D = diffusion coefficient
M = molecular weight
P = pressure of the diffusion cell
A = cross-sectional area of the diffusion tube
R = gas constant
T = temperature
L = length of the diffusion tube
p = partial pressure of the diffusing vapor

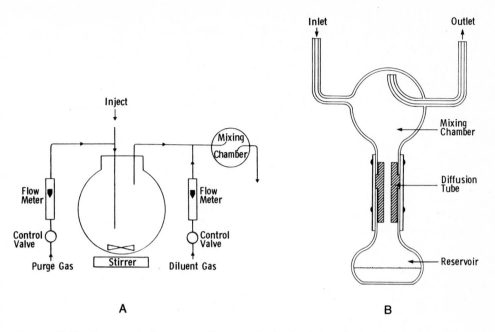

Figure 7.14   Apparatus for preparing standard mixtures of organic vapors.
A, exponential dilution flask; B, diffusion tube system.

Within limits, broad concentration ranges can be prepared by varying the tube
dimensions and/or the flow rate of the diluent gas. Diffusion tube systems are
preferable to permeation tubes when the latter are not commercially available.

Alternative methods for preparing standard gas mixtures, such as
evaporative, electrolytic, pyrolytic, and chemical methods, are discussed in
references [248,249,258].

7.13   Derivatization Techniques for Gas Chromatography

Gas chromatography is the technique of choice for the separation of
thermally stable volatile organic and organometallic compounds. Unfortunately
many compounds of biomedical and environmental interest, particularly those of
high molecular weight and/or containing polar functional groups, are thermally
labile at the temperatures required for their separation. Derivatization, in
effect a microchemical organic synthesis, is used to improve the thermal
stability of such compounds which would otherwise not be suitable for gas
chromatographic analysis. In most instances, derivatization reactions are
performed to convert protonic functional groups into thermally stable non-polar
groups. Thus, as well as improving the thermal stability of a compound, the
derivatized compound often exhibits improved peak shape with a minimization of
undesirable column interactions which could lead to irreversible adsorption and

skew peak formation. As a wide variety of derivatizing reagents are available the opportunity also exists to purposefully adjust the volatility of a compound to eliminate peak overlaps in the chromatogram. Derivatization can also be used to enhance the detectability of a compound through the introduction of detector-oriented tags or organic groups. In particular, reagents predisposing a substance to detection by the sensitive and selective electron-capture detector have been well-used.

The anatomy of a typical derivatizing reagent is shown in Figure 7.15. It has two halves: the organic portion which controls volatility and, in the case of reagents used with the electron-capture detector, provides the detector-oriented response; and the reactive group which provides the means by which the organic chain is attached to the substrate. The choice of reactive group controls the range of application of the reagent to different functional groups, the selectivity of the reagent towards certain functional groups in the presence of others, and the rate and extent of the reaction. The choice of the organic chain will influence the detection characteristics of the derivative, the rate and completeness of the derivatization reaction (resulting from steric and electronic effects), and the volatility of the derivatized molecule.

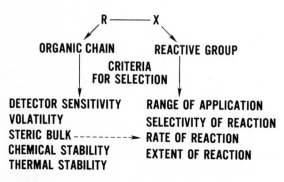

Figure 7.15 Anatomy of a derivatizing reagent.

Equipment requirements for derivatization reactions are generally simple; typically, glass tapered reaction vials or culture tubes with Teflon-lined plastic screw-caps are used. As the sample requirement for gas chromatography is only a few microliters, reaction vessels tend to be small. Tapered reaction vials are available in sizes from 0.1 to 10.0 ml, with the smaller sizes being the most useful. Reagents and soluble samples are usually measured and handled by syringe. Reaction solutions are mixed by hand agitation, vortex mixing, or with miniature Teflon-coated stirring paddles driven by a magnetic stirrer.

Slow reactions are usually accelerated by heating the reaction mixture with a convection oven, hot plate, oil bath, or drilled block heater. Screw-capped culture tubes are particularly suitable for reactions occurring under reflux with the air-cooled top section of the tube acting as a condenser. Mininert valves and vacuum hydrolysis tubes are other useful devices. A wide selection of equipment for derivatization reactions is available from chromatographic supply companies.

The reagents and conditions used for derivatization reactions are so varied that no attempt at a comprehensive coverage can be presented here. Only general reactions, common reagents, and a few typical applications will be discussed. Extensive compilations of methods and conditions can be found in references [259-263]. Subject reviews covering the general selection of reagents [264], trialkylsilyl reagents [265], cyclizing reagents for bifunctional compounds [266,267], alkylation reactions [268,269], and reagents for the preparation of electron-capturing derivatives [262,270,271] are available. Reviews of derivatization reactions for particular compound classes, for example pharmaceutical products [262,272-274], amino acids [275], insecticide and herbicide residues [276-279], and food additives [280] are also available.

7.13.1 Aklylsilyl Derivatives

The most versatile and universally applicable derivatizing reagents for polar molecules containing protonic functional groups are the alkylsilyl reagents. Nearly all functional groups which present a problem in gas chromatography can be converted to alkylsilyl ethers or esters, Figure 7.16. The most common derivatizing reagents are the trimethylsilyl (TMS) reagents. On occasion, higher alkyl homologs or halogen-containing alkyl or aryl substituted analogs of the TMS derivatives are used to impart greater derivative hydrolytic stability, improved separation characteristics, increased sensitivity when used with selective detectors, or to provide mass spectra containing greater diagnostic information [265]. These newer reagents and their particular areas of application will be discussed later.

The structures of the most widely used trimethylsilylating reagents are given in Figure 7.17. Reactions are usually carried out under anhydrous conditions in glass vials with Teflon-lined screw-caps. Many reactions occur instantaneously at room temperature; slower reactions are accelerated by raising the temperature and/or by adding a catalyst, usually trimethylchlorosilane (TMCS). In the absence of a detailed model for the silylation mechanism [263], it is possible to rank the silylating reagents of Figure 7.17 according to their "silyl donor ability" and the functional groups of Figure 7.16 according to their "silyl acceptor ability". For the trimethylsilyl reagents the approximate

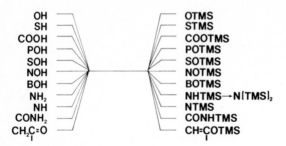

Figure 7.16 Functional groups forming trimethylsilyl derivatives.

order of "silyl donor ability" is: trimethylsilylimidazole (TMSIM) >
N,O-bis-(trimethylsilyl)trifluoroacetamide (BSTFA) > N,O-bis-(trimethylsilyl)acetamide (BSA) > N-methyl-N-(trimethylsilyl)trifluoroacetamide (MSTFA) >
N-trimethylsilyldiethylamine (TMSDEA) > N-methyl-N-(trimethylsilyl)acetamide
(MSTA) > trimethylchlorosilane (TMCS) with base > hexamethyldisilazane (HMDS).
For organic functional groups the approximate order of "silyl acceptor ability"
is: alcohols > phenols > carboxylic acids > amines > amides. Primary functional
groups react faster or more completely than secondary functional groups, which
in turn are more reactive than tertiary functional groups. Within the above
framework, the reaction between a good "silyl donor" and a good "silyl acceptor"
is likely to be facile and quantitative under mild conditions. There seem to be
few applications for which the use of weak "silyl donors" is either necessary or
desirable. Other important considerations for the selection of the correct
reagents for a particular application are summarized in Table 7.16. The
strongest silylating reagent of all is a mixture of TMSIM-BSTFA-TMCS (1:1:1).

The rate of the silylation reaction is also affected by steric factors, the
use of catalysts, the choice of solvent, and the reaction temperature. The
trimethylsilyl group has similar geometry to the t-butyl group and is slightly
larger in size. Thus, impeded access of reagent to the functional groups to be
derivatized can be the rate-determining step. For example, consider the
reaction conditions required for the quantitative derivatization of steroid
hydroxyl groups in the different steric environments given in Table 7.17.
Different reaction rates are observed, due firstly to the nature of the
functional group (primary, secondary, or tertiary) and secondly to changes in
the steric environment of the different groups. Reaction rates are also
influenced to a lesser extent by the choice of solvent. Polar solvents, such as
pyridine, dimethylformamide, acetonitrile, dioxane, or tetrahydrofuran, promote

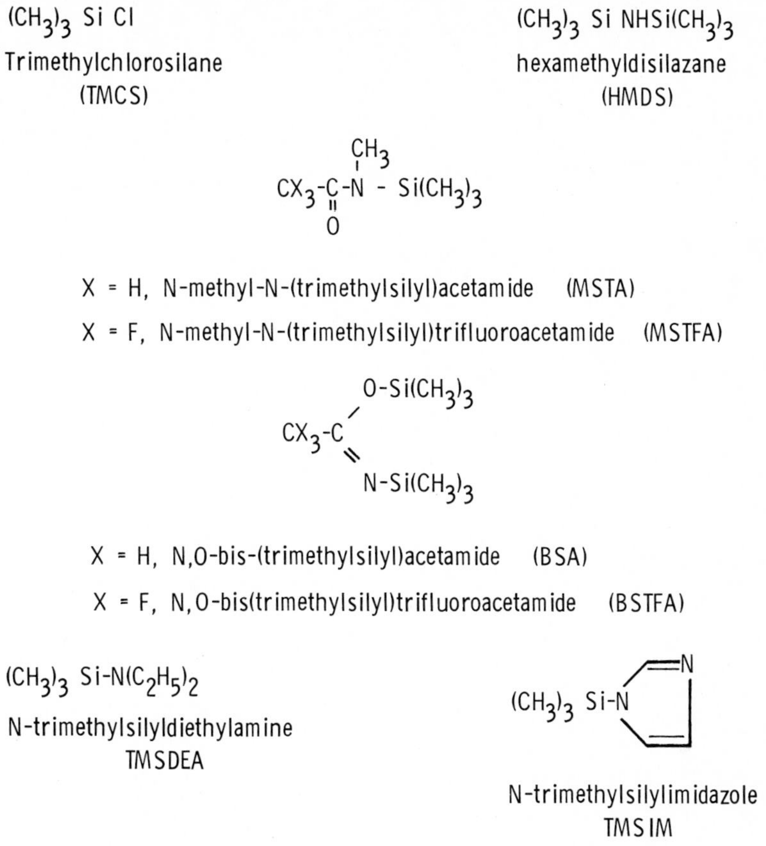

(CH₃)₃ Si Cl

Trimethylchlorosilane

(TMCS)

(CH₃)₃ Si NHSi(CH₃)₃

hexamethyldisilazane

(HMDS)

$$CX_3\text{-}\overset{\overset{\displaystyle CH_3}{|}}{\underset{\underset{\displaystyle O}{\|}}{C}}\text{-N - Si(CH}_3)_3$$

X = H, N-methyl-N-(trimethylsilyl)acetamide   (MSTA)

X = F, N-methyl-N-(trimethylsilyl)trifluoroacetamide   (MSTFA)

$$CX_3\text{-}C\begin{smallmatrix}\diagup O\text{-Si(CH}_3)_3\\ \diagdown N\text{-Si(CH}_3)_3\end{smallmatrix}$$

X = H, N,O-bis-(trimethylsilyl)acetamide   (BSA)

X = F, N,O-bis(trimethylsilyl)trifluoroacetamide   (BSTFA)

(CH₃)₃ Si-N(C₂H₅)₂

N-trimethylsilyldiethylamine

TMSDEA

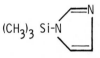

(CH₃)₃ Si-N

N-trimethylsilylimidazole

TMSIM

Figure 7.17   Structures of the most commonly used trimethylsilylating reagents.

the reaction and are generally preferred. The silylating reagents themselves
have good solubilizing properties for many compounds and can be used without
additional solvent. The primary criteron for selecting a solvent is that it
must solubilize both substrate and reagents. Increasing the temperature of the
reaction will often improve substrate solubility and enhance the rate of
reaction. Some difficult-to-derivatize functional groups will only react to
completion at elevated temperatures. Normal reaction temperatures are room
temperature, 60°C, 100°C, and 150°C; in a few instances, temperatures as high as
200°C have been used.

The optimum conditions for a particular reaction are achieved by
appropriate changes in the parameters discussed above. It is obvious that

several sets of conditions will give the same results and the governing feature
is the need for a quantitative reaction in a convenient time.

TABLE 7.16

OBSERVATIONS ON THE USE OF TRIMETHYLSILYL REAGENTS

Reagent	Property	Comments
TMSIM	Does not react with amino groups. Can be used to form TMS derivatives of carbohydrates in aqueous solution. Does not promote enol-ether formation with unprotected ketone groups.	Most generally useful reagent, preferred for most applications. Exceptions are the formation of N-TMS derivatives and the separation of low molecular weight TMS derivatives
BSTFA or BSA	Reagents of choice for the formation of N-TMS derivatives.	May promote the formation of enol-TMS ethers unless ketone groups are protected.
BSTFA	Produces volatile by-products which do not interfere with the GC analysis of low molecular weight compounds.	Reactivity is similar to BSA and is generally preferred to BSA for most applications.
MSTFA	Most volatile by-products of all.	Particularly useful for the separation of low molecular weight derivatives in the presence of excess reagent.
TMCS	A poor silylating reagent unless used in the presence of base (e.g., pyridine, diethylamine). Causes extensive enol-ether formation with unprotected ketone groups.	Mainly used to catalyze the reaction of other silylating reagents.

Besides the wide applicability and ease of use of the TMS reagents, the
fact that most reactions occur cleanly without artifact or by-product formation
adds to the attraction of these reagents. Secondary products can be formed
under normal conditions for silyl ether formation with compounds containing
unprotected ketone groups. These products arise from the formation of enol-TMS
ethers (below), the yield of the latter being high, and in some cases
quantitative in the presence of an acid catalyst such as TMCS.

$$\underset{\underset{O}{\|}}{RHC_2-C-R} \longrightarrow \underset{\underset{OH}{|}}{R-CH=C-R} - - - - -> \underset{\underset{OTMS}{|}}{R-CH=C-R}$$

Keto-enol equilibrium   enol-TMS ether

In general, enol-TMS ethers are hydrolytically and thermally unstable, making
them less desirable reaction products even when formed in quantitative yield.

TABLE 7.17

CONDITIONS FOR THE FORMATION OF TMS ETHERS OF STEROID HYDROXYL GROUPS

Hydroxyl group environment	Quantitative reaction with TMSIM	Selective reaction	
		BSA	BSA–TMCS(4:1)[a]
Mammalian Steroids			
All primary and normal secondary OH groups (less hindered than 11β–OH), phenolic groups, and unhindered tertiary OH	Less than 1 h at room temp. Often instantaneous	1–4 h at room temp. 1 h at 60°C	As for TMSIM
11β–OH	Less than 1 h at 100°C	Does not react	5 h room temp.
CH_2OTMS $C=N{\sim}OCH_3$ $C\text{'''}OH$	2 h at 100°C	– – –	30 h at 60°C
CH_3 CHOTMS C–OH	8 h at 150°C	– – –	Nonquantitative reaction
Insect Steroids			
All primary and normal secondary OH groups and some unhindered tertiary OH groups	Less than 1 h at room temp.	Overnight at room temp.	– – –
OTMS OH	4 h at 100°C		
hindered tertiary (e.g., 14α–OH)	12 h at 140°C	– – –	– – –

[a] Ketone groups must be protected as their methoximes

To avoid the formation of enol-TMS ethers a reagent which does not promote enol
formation, such as TMSIM, or prior protection of the ketone group by conversion
to a methoxime derivative is used. Methoxime-trimethylsilyl derivatives are
widely used for the analysis of biological extracts and in metabolic profiling
studies. Two peaks for each compound may appear in the chromatogram due to the
separation of the of the syn- and anti-methoxime isomers formed upon

derivatization.

Multiple products are frequently observed for the separation of TMS-sugar derivatives. At equilibrium reducing sugars can exist in more than one isomeric form known as anomers. Formation of their TMS derivatives followed by gas chromatography will result in multiple peaks corresponding in composition to the equilibrium anomeric mixture [259].

The TMS reagents are generally compatible with other reagents for the formation of mixed derivatives of polyfunctional molecules, provided that the TMS ethers are formed as the last step in the reaction sequence. The TMS ethers exhibit moderate hydrolytic stability and are cleaved under conditions normally used for the formation of alkyl esters, acetates and amides, oximes, cyclic boronic esters, acetonides, and most reactions employing acid catalysis. The TMS reagents, with a few exceptions, do not normally cleave any of the common protecting groups used in gas chromatography.

The mass spectra of TMS ethers are characterized by weak or absent molecular ions; the $[M-15]^+$ ion formed by cleavage of a methyl to silicon bond is generally more abundant. This ion can be used to determine the molecular weight provided that it is not mistaken for the molecular ion itself. Dissociation of the molecular ion often results in prominent secondary fragment ions containing the ionized dimethylsiloxy group attached to a hydrocarbon portion of the molecule. In common with alkyl ethers, -cleavage of the bond adjacent to oxygen is favored. Characteristic ions of hydroxy TMS ethers are:

$$[(CH_3)_3Si]^+ \qquad [HO=Si(CH_3)_2]^+ \qquad [(CH_3)_2Si=O-Si(CH_3)_3]^+$$

$$m/z = 73 \qquad\qquad m/z = 75 \qquad\qquad\qquad m/z = 147$$

The m/z 73 ion is prominent in virtually all TMS spectra and is often the base peak. The m/z 147 ion is common in polyhydroxy TMS compounds containing two or more TMS groups in close proximity. The TMS group undergoes a prolific number of intramolecular migrations and rearrangements (including McLafferty rearrangements) to give prominent silicon-containing ions. Perdeuterated silyl reagents are available to aid in elucidating specific fragmentation and rearrangement processes based on a comparison of the TMS and d_9-TMS spectra. The use of the perdeuterosilyl reagents either alone or in conjunction with isotopic labeling (e.g., ^{18}O, ^{15}N, ^{13}C) is one of the principal tools used to probe the fragmentation processes of TMS derivatives.

New reagents which are either homologs or analogs of the trimethylsilylating reagents have been developed to improve the hydrolytic stability or detectability of the silyl ether derivatives [265]. Homologous trialkylsilyl reagents containing ethyl, propyl, isopropyl, butyl, or hexyl groups have been used to improve the separation of complex mixtures on columns of low resolving power. The formation of TMS ethers of low molecular weight substances alters their volatility very little and complex samples containing either none, one, two, etc. functional groups are often incompletely resolved on packed columns. The use of homologous trialkylsilyl ethers of higher molecular weight than the trimethylsilyl ethers enables a better fractionation of the sample by removing the silyl ethers from the region occupied by substances which do not form derivatives and also by magnifying the separation between components containing different numbers of functional groups. The mass spectra of the trialkylsilyl ethers are characterized by weak or absent molecular ions with a prominent ion due to cleavage of an alkyl group from silicon [281,282]. The higher mass end of the spectrum contains a relatively abundant series of ions produced by elimination of C_nH_{2n} fragments from the alkyl groups bonded to silicon. Elimination of dialkylsilanol (R_2HSiOH) from ions such as $[M-R]^+$ was observed to be the major mode of fragmentation [by contrast TMS ethers eliminate $(CH_3)_3SiOH$]. Samples containing two derivatized groups have abundant doubly-charged ions due to cleavage of an alkyl group from both silicon centers.

Alkyldimethylsilyl ethers containing an ethyl, propyl, isopropyl, t-butyl, or allyl substituent are more volatile than the trialkylsilyl derivatives and have many of the same advantages. The isopropyldimethylsilyl and t-butyldimethylsilyl ethers are approximately 10^2 to 10^3 and 10^4 times more stable, respectively, to solvolysis than the TMS ethers. Consequently they find widespread use in organic synthesis as well as chromatography [283,284]. The bulkiness of the derivative group, which explains the hydrolytic stability of these derivatives, also influences the rate and extent of the derivative reaction. The t-butyldimethylsilyl ethers have good gas chromatographic properties with retention times 2 or 3 methylene units higher than the TMS derivatives, are relatively insensitive to moisture, and are stable to adsorption chromatography using columns or thin-layer plates. The N-t-butyldimethylsilyl derivatives are more labile than the O-t-butyldimethylsilyl ethers but more stable than the O-trimethylsilyl ethers [285]. The allyldimethylsilyl ethers were introduced in the hope of providing a derivative more hydrolytically stable than the TMS ethers and more reactive and volatile than the t-butyldimethylsilyl ethers [286,287]. The susceptibility of allyl groups to nucleophilic displacement limits the general use of this derivative. The mass spectra of the alkyldimethylsilyl ethers have weak molecular ions with an abundant $[M-R]^+$ ion due to cleavage of the bond between silicon and the larger alkyl group. This ion is often the base peak of the spectrum, particularly when R = t-butyl [286,288]. When the positive charge is localized on the siloxy group, the $[M-R]^+$ ion acts as the precursor ion for the remaining silicon-containing fragments in the spectrum [289].

To improve the detectability of silyl ethers, silylating reagents containing electron-capturing groups have been prepared [262,265,270,290]. These reagents contain either a halomethyl or pentafluorophenyl substituent in place of one of the TMS methyl groups. The structures of these reagents are shown below:

$$CH_2X-\underset{\underset{CH_3}{|}}{\overset{\overset{CH_3}{|}}{Si}}-Y$$

halomethyldimethylsilyl reagents

X = Cl, Br, I

Y = Cl, $N(C_2H_5)_2$, $NHSi(CH_3)_2CH_2X$

$$C_6F_5-\underset{\underset{R}{|}}{\overset{\overset{CH_3}{|}}{Si}}-Y$$

flophemesyl reagents

R = CH_3 Y = Cl, NH_2,$N(C_2H_5)_2$

alkylflophemesyl reagents

R = $CH(CH_3)_2$ Y = Cl

R = $C(CH_3)_3$ Y = Cl

halomethylflophemseyl reagents

R = CH_2Cl Y = Cl

The flophemesyl ethers have hydrolytic stability similar to the TMS ethers. The alkylflophemesyl reagents have improved stability [291,292].

The response of the electron-capture detector towards halogen-containing compounds follows the order I > Br > Cl >> F; this is the reverse order of the gas chromatographic volatility of their compounds, Table 7.18. Although fluorocarbon compounds capture electrons poorly, multiple substitution is favored by the exceptional volatility of perfluorocarbon compounds upon gas chromatography. However, with the exception of the flophemesyl reagents, none of the fluorocarbon compounds in Table 7.18 have adequate detection sensitivity to be used with the electron-capture detector. If the high volatility of perfluorocarbon substituents is to be exploited in the preparation of electron-capturing derivatives then a secondary center in the molecule is required to synergistically enhance detector response. An example of this principle, to be elaborated upon later, is the heptafluorobutyl and ketone groups in the heptafluorobutyryl and heptafluorobutyramide derivatives. Further details of the response of the electron-capture detector to organic molecules is contained in Chapter 3.4.1.

The halomethyldimethylsilyl derivatives are usually prepared by the reaction of 1,3-bis-(halomethyl)-1,1,3,3-tetramethyldisilazane and the halomethyldimethylchlorosilane or by the addition of an aliquot of the reagent resulting from the reaction of excess halomethyldimethylchlorosilane with diethylamine in hexane [293,294]. The rate and extent of reaction are limited

TABLE 7.18

RELATIVE VOLATILITY AND ELECTRON-CAPTURE DETECTOR SENSITIVITY OF
HALOCARBON-CONTAINING $RR_1(CH_3)Si$-CHOLESTEROL ETHERS

R	R_1	Relative Retention Time[a]	Least Detectable Amount (ng) (cholesterol)	Least Detectable Amount (pg) (octanol)
CH_3	CH_3	1.00	- - -	
$CF_3(CH_2)_2$	CH_3	1.26	1500	
$CF_3(CF_2)_2(CH_2)_2$	CH_3	1.37	115	
$ClCH_2$	CH_3	2.10	75	
C_6F_5	CH_3	3.14	4	4.0
C_6F_5	$CH(CH_3)_2$	4.57		5.0
CH_2Br	CH_3	5.13	0.5	
C_6F_5	CH_2Cl	6.30		0.9
C_6F_5	$C(CH_3)_3$	6.30		6.0
CH_2I	CH_3	12.82	0.005	

[a] Determined on a 1.0 m x 2.0 mm I.D. nickel column packed with 1% OV-101
on Gas Chrom Q (100-120 mesh), temperature 250°C, nitrogen flow-rate
75 ml/min.

by the "silyl donor" power of the reagents and by steric factors. The
iodomethyldimethylsilyl reagents are unstable and cannot be stored
conveniently. Iodomethyldimethylsilyl derivatives are prepared in situ by
halide ion-exchange between the chloro- or bromomethyldimethylsilyl ethers and
sodium iodide. Vigorous reaction conditions can result in the displacement of
either the halogen atom or halomethyl group to give a non-electron-capturing
derivative [265,296]. As an approximate guide to the sensitivity of the
halomethyldimethylsilyl ethers to the electron-capture detector, the
chloromethyldimethylsilyl ethers have a similar response to the electron-
capture detector as observed with the flame ionization detector; the
bromodimethylsilyl ethers are several-fold more responsive with detection limits
below the nanogram level; while, the iodomethyldimethylsilyl ethers are the most
responsive of all and can be determined at the low picogram level. The mass
spectra of the halomethyldimethylsilyl ethers have weak molecular ions and an
abundant characteristic daughter ion at $[M-CH_2X]^+$ [297,298]. As the halogen, X,
decreases in electronegativity, so the bond from the silicon atom to the carbon
bearing X becomes stronger and the loss of CH_2X less likely.

The "silyl donor" power of the flophemesyl reagents in pyridine was
established as flophemesylamine > flophemesyl chloride > flophemesyl-
-diethylamine > flophemesyldisilazane >> flophemesylimidazole [290,299,300].
Flophemesylamine is a particularly useful reagent that selectively reacts with
unhindered primary and secondary hydroxyl groups in the presence of unprotected
ketone groups. The addition of a catalyst (flophemesyl chloride) changes the
above order and mixtures of flophemesyldiethylamine and flophemesyl chloride
provide the most potent "silyl donor" media. The reaction conditions used for
the quantitative derivatization of steroid hydroxyl groups are summarized in
Table 7.19 [290]. The isopropylflophemesyl and t-butylflophemesyl reagents have
similar properties to the flophemesyl reagents unless reaction conditions are
subject to steric control [291,292]. These derivatives have greater hydrolytic
stability than the flophemesyl ethers and are better suited to trace analysis
when extensive sample manipulation and cleanup is required prior to
determination by gas chromatography. The flophemesyl reagents can be detected
with an electron-capture detector in the low nanogram to picogram range. The
mass spectra of the flophemesyl ethers often have prominent molecular ions with
a higher percentage of the total ion current carried by hydrocarbon fragments
than is observed with the TMS ethers [301,302]. The principal ions of lower m/z
are dominated by the presence of fluorosilane ions (m/z 47 $[SiF]^+$, m/z 77
$[Si(CH_3)_2F]^+$, and m/z 81 $[Si(CH_3)F_2]^+$) and fluorocarbon ions originating from
fluorocarbon tropylium ions (m/z 181 $[C_7H_2F_5]^+$, m/z 163 $[C_7H_3F_4]^+$, m/z 159
$[C_8H_6F_3]^+$, and m/z 145 $[C_7H_4F_3]^+$) [303]. The fluorocarbon tropylium ions arise
by fluorine-alkyl exchange between the pentafluorophenyl ring and silicon.
Their further decomposition produces a characteristic series of fluorocarbon
fragments. The mass spectra of the isopropylflophemesyl and
chloromethylflophemesyl derivatives are characterized by weak or absent
molecular ions, very few dominant silicon-containing ions, and a relatively
abundant series of fluorocarbon ions [291]. Loss of the alkyl or halomethyl
group from the molecular ion is often the base peak of the spectrum [291,292].

7.13.2 Haloalkylacyl Derivatives

The haloalkyl acid chlorides and anhydrides are probably the most studied
reagents for the introduction of an electron-capturing group into compounds with
protonic functional groups (except carboxylic acids), Figure 7.18. Although
they contain halogen atoms to provide a response to the electron-capture
detector, they are equally suitable for use with the flame ionization detector,
and have almost entirely displaced the hydrocarbonacyl derivatives from general
use. In particular, the perfluorocarbonacyl reagents (trifluoroacteyl,
pentafluoropropionyl, heptafluorobutyryl) produce stable, volatile derivatives

TABLE 7.19

CONDITIONS FOR THE FORMATION OF FLOPHEMESYL STEROID ETHERS
A = 3 h at 60°C; B = 0.25 h at room temp.; C = 6 h at 85°C; NR = no reaction;
NQ = none quantitative; CP = cyclic product; IR = not all hydroxyl groups
react; except for flophemesylamine ketone groups are converted to methoximes.

Steroid	Flophemesyl Chloride	Flophemesyl-amine	Flophemesyl-diethylamine: Flophemesyl Chloride(10:1)	Flophemesyl-diethylamine: Flophemesyl Chloride (1:1)
Cholesterol	A	B	B	B
Ergosterol	A	B	B	B
Cholestanol	A	B	B	B
2β,3β–dihydroxy-5α–cholestanol	A	B	B	B
2β,5α,6β–trihydroxy-cholestane	A[a]	B[a]	B[a]	B[a]
2β,5α–dihydroxy-cholestan-6-one	A[b]	B[b]	B[b]	NQ
2β,3β,14α–trihydroxy-cholest-7-en-6-one	A[c]	B[c]	B[c]	B[c]
3α,20α–dihydroxy-5β–pregnane	A	B	B	B
17α–methyl–17β–hydroxy-androst-4-en-3-one	NQ	NR	B	B
11β–hydroxyandrost-4-en-3,17-dione	NR	NR	NR	A
17α–hydroxypregn-4-en-3,20-dione	NR	NR	NR	C
3α,17α,20α–trihydroxy-5β–pregnane	CP	CP	CP	CP
17α,21–dihydroxypregn-4-en-3,11,20-trione	IR	IR	IR	– –
17β,11β,21–trihydroxy-pregn-4-en-3,20-dione	IR	IR	IR	NQ

[a] 2β and 6β groups react; [b] 2β group reacts; and [c] 2β and 3β groups react.

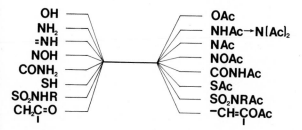

Figure 7.18 Functional groups forming acylated derivatives.

which often elute earlier and with better peak shape than the hydrocarbonacyl derivatives.

For derivative formation the appropriate anhydride, sometimes in the presence of an acid acceptor such as pyridine, leads to a rapid, quantitative reaction of all but the most sterically hindered groups. The use of the anhydride can lead to undesirable side reactions (dehydration, enolization, etc.) with sensitive molecules due to the strong acid conditions prevailing in the reaction medium. For these compounds, acylation can be performed using the acylimidazole reagents in which the by-product of the reaciton is the weakly amphoteric imidazole. The perfluoroalkylacyl derivatives have good gas chromatographic properties and high volatility. The chloroacteyl and bromoacetyl derivatives have long retention times often accompanied by poor peak shape and poor thermal stability compared to the perfluoroacyl derivatives.

The relative response of the electron-capture detector to some haloalkylacyl derivatives is summarized in Table 7.20 [304-309]. In general terms, the monochloroacteyl and chlorodifluoroacteyl derivatives are more sensitive than the trifluoroacteyl derivatives. Increasing the fluorocarbon chain length of the fluorocarbonacyl derivatives increases their electron-capture detector response without inconveniently increasing their retention times. The heptafluorobutyryl derivatives are considered to be the best compromise between detector sensitivity and volatility for most applications.

The mass spectra of the halocarbonacyl derivatives frequently have abundant

TABLE 7.20

RELATIVE RESPONSE OF THE ELECTRON-CAPTURE DETECTOR TO SOME HALOALKYLACYL DERIVATIVES

Derivative	Amphetamine	Testosterone	Thymol	Diethyl-stilbesterol	Benzylamine
Acetyl		1.0			1.0
Monofluoroacetyl			0.007		
Monochloroacetyl	1.0	40	0.3	2.7	750
Chlorodifluoro-acetyl		340			
Dichloroacetyl				2.6	
Trichloroacetyl	540			2.1	
Trifluoroacetyl	< 0.1	4		1.7	200
Pentafluoro-propionyl	40	50	1.3	15	5725
Heptafluorobutyryl	90	190	1.0	23	17875
Perfluorooctonyl	230	600		21	
Pentafluoro-benzoyl	770		6.9		

ions at high m/z values. Consequently they have often found use in studies using single-ion monitoring for detection and quantitatiion of the derivatives. The abundance of the molecular ion varies and depends primarily on the nature of the compound derivatized. In most cases, ions corresponding to $[C_nF_{2n+1}]^+$ occur abundantly in the mass spectra of the perfluorocarbonacyl derivatives. The $[C_nF_{2n+1}CO]^+$ ion is usually also fairly prominent and a loss of $(C_nF_{2n+1}CO_2)$ occurs readily in the mass spectra of alcohols and phenols. Aliphatic amines tend to give abundant ions corresponding to $[C_nF_{2n+1}CONHCH_2]^+$.

7.13.3 Esterification

Esterification is used to derivatize carboxylic acids and other acidic functional groups [259-261]. In a typical reaction the carboxylic acid is dissolved in an excess of an alcohol which contains a few percent of an acid catalyst such as hydrogen chloride, sulfuric acid, thionyl chloride, boron trifluoride etherate, boron trichloride, or an ion-exchange resin [310]. Many esterification reactions are slow and elevated temperatures are frequently used. As the esterification reaction is an equilibrium reaction it may be necessary to remove the water formed as a product during the reaction to obtain a quantitative yield of derivative. A Dean-Stark trap, reflux in a Soxhlet apparatus containing a desiccant in the thimble, or a solvent which reacts with water (e.g., 2,2-dimethoxypropane) may be used for this purpose [311]. To derivatize carboxylic acids with low solubility in higher alcohols the technique of transesterification is used. For example, cystine and the basic amino acids are virtually insoluble in n-butanol. To prepare their butyl esters, the methyl esters are formed first, the methylating reagents removed by evaporation, and the methyl esters then converted to the butyl esters by refluxing the residue with an excess of n-butanol and acid catalyst [138,312]. The yield of butyl esters is usually quantitative compared to a few percent in the case of direct reaction. Electron-capturing derivatives can be prepared using alcohols such as 2,2,2-trifluoroethanol, 2,2,3,3,3-pentafluoro-1-propanol, 1-chloro-1,1,3,3,3-pentafluoro-2-propanol, and hexafluoro-2-propanol [313-318].

7.13.4 Alkylation

The term "alkylation" covers a variety of techniques in which an active hydrogen atom is replaced in a chemical reaction by an alkyl or, sometimes, an aryl group [259,267,268,272]. Functional group types which can be alkylated are shown in Figure 7.19. A variety of reagents and methods are used in their preparation, of which the most important are the reaction with an alkyl halide and catalyst, diazoalkanes, N,N'-dimethylformamide dialkyl acetals, extractive alkylation, pyrolytic alkylation, and arylation. Older reactions, such as the

$$R - COOH \longrightarrow R - COOR'$$

$$R - SO_2OH \longrightarrow R - SO_2OR'$$

$$R - OH \longrightarrow R - OR'$$

$$R - SH \longrightarrow R - SR'$$

$$R - NH_2 \longrightarrow R - N(R')_2$$

$$R - NHR_2 \longrightarrow R_1R_2NR'$$

$$R - CONH_2 \longrightarrow R - CON(R')_2$$

$$R - SO_2NH_2 \longrightarrow R - SO_2N(R')_2$$

$$R_1 - COCH_2CO - R_2 \longrightarrow R_1 - COCH = \underset{\overset{|}{OR'}}{C} - R_2$$

$$R' = alkyl\ group$$

Figure 7.19 Functional groups which can be alkylated.

use of dimethyl sulfate with base catalyst (sodium hydroxide, potassium carbonate), are infrequently used as the reactions tend to be slower and less quantitative than the reagents mentioned above. Reagents such as 3-alkyl-1-p-tolyltriazenes, alkyl fluorosulfonates, and trialkyloxonium fluoroborates have been used in organic synthesis but remain little explored for analytical purposes.

Alkyl halides in the presence of silver oxide will convert any non-hindered carboxylic acid (or its salt) to the corresponding alkyl ester in minutes, and phenolic or thiol groups will also be alkylated rapidly [259]. Hydroxyl groups are alkylated slowly and not always to completion. The alkyl halides most frequently used are the lower molecular weight aliphatic bromides and iodides (e.g., methyl, ethyl, propyl, isopropyl, butyl, etc.) or benzyl and substituted benzyl bromides. The reaction has been extensively investigated for the alkylation of sugars. Oxidative degradation of free sugars and migration of O-acetyl groups can lead to unexpected by-products. N-acetyl groups are usually stable to the reaction conditions. In a typical reaction, silver oxide is added to a solution of the substrate in an excess of alkyl halide and the mixture is shaken in the dark until reaction is complete. Slow reactions are usually monitored at intervals by any suitable chromatographic technique and fresh portions of silver oxide are added as necessary.

Barium oxide and sodium hydride are more potent catalysts than silver
oxide. With barium oxide catalysis, reactions occur more rapidly but O-acetyl
migration is promoted. With sodium hydride, even sterically hindered groups may
be quantitatively alkylated but unwanted C-alkylation instead of, or in addition
to, O-alkylation is a possibility.

When sodium hydride is used as a catalyst in dimethyl sulfoxide, the
methylsulfinyl anion is formed. Usually, the methylsulfinyl anion is prepared
initially by heating sodium hydride and dimethyl sulfoxide together under
anhydrous conditions. When the evolution of hydrogen has ceased, a solution of
the substrate in dimethyl sulfoxide is added and the mixutre is stirred and
cooled prior to the slow dropwise addition of the alkyl halide. The reaction
conditions are suitable for derivatizing extremely small amounts of material
such as those required for sequencing polysaccharides and peptides by mass
spectrometry [319,320]. Under the above reaction conditions, O-acyl groups are
completely replaced by O-alkyl groups and N-acyl groups are converted to
N-alkyl-N-acylamido derivatives:

$$ROCOR_1 \longrightarrow ROR'$$

$$RNHCOOR_1 \longrightarrow RN(R')COR_1$$

R' = alkyl group

Base-sensitive substances will undergo side-reactions during treatment with this
reagent [321].

Diazoalkane alkylating reagents include diazomethane, diazoethane,
diazopropane, diazobutane, diazoisobutane, and phenyldiazomethane [272];
diazomethane is most frequently used. The other reagents are used principally
to improve chromatographic separation when the methyl derivatives are unsuitable
or to avoid loss of volatile methyl derivatives of low molecular weight
substances during the removal of excess reagents. Reaction conditions are very
simple: a solution of the diazoalkane is added to a solution of the substrate
at or below room temperature until a faint yellow color persists and the
evolution of nitrogen gas ceases. Excess reagent may be removed by evaporation
or destroyed by addition of acid (e.g., acetic acid).

Diazoalkanes alkylate acidic and enolic groups rapidly and other groups
with replaceable hydrogens slowly. Carboxylic and sulfonic acids, phenols, and
enols are alkylated virtually instantaneously when treated with this reagent.
Lewis acid catalysts (e.g., $BF_3 \cdot Et_2O$) are used to promote the reaction of
substances containing hydrogen atoms of low reactivity (e.g., alcohols).
Because of the large difference in reaction rates, many substances containing
carboxylic acid and phenolic groups can be selectively alkylated in the presence
of less reactive functional groups. Alternatively, methanol can be used to

advantage as a solvent for the alkylation of carboxylic acids. When Lewis-acid catalysis is used, acid-labile substances may undergo undesirable transformations induced by the catalyst while O-acylated sugars with free hydroxyl groups are alkylated with O-acyl migration. The diazoalkanes are versatile synthetic reagents in preparative organic chemistry. As well as alkylation, they also undergo a wide range of addition and cyclization reactions which may result in unexpected products with multifunctional compounds [272].

The diazoalkanes are toxic substances and may explode on contact with rough surfaces. Consequently, many workers prefer not to make large quantities of these materials when only small quantities are needed for derivatization reactions. Simple micro-diazoalkane generators capable of rapidly preparing small quantities of the reagents, as required, have been described [322,323].

Various N,N'-dimethylformamide dialkyl acetals, $(CH_3)_2NCH(OR')_2$, are commercially available in which $R' = CH_3$, C_2H_5, C_3H_7, and C_4H_9. They react rapidly with carboxylic acids, phenols, and thiols to give the corresponding alkyl derivatives [324-327]. Free amines and amides give the N-dimethylaminomethylene and N-alkyl derivatives, respectively [324,326-330]. Simple amino acids are fully derivatized to the N-dimethylaminomethylene alkyl esters [324]:

$$RNH_2 \xrightarrow{(CH_3)_2NCH(OR')_2} RN=CHN(CH_3)_2$$

$$RCONHR_1 \longrightarrow RCONR_1R'$$

$$\underset{\underset{NH_2}{|}}{RCH-COOH} \longrightarrow \underset{\underset{N=CHN(CH_3)_2}{|}}{RCH-COOR'}$$

Although the reaction conditions may be similar, the products generated with these reagents may not be the same and this must be taken into consideration. The dimethylformamide dialkyl acetals can condense with compounds containing active methylene groups and with N-heterocyclic compounds; alkylation or an exchange reaction to generate a new acetal may also occur.

Derivatization reactions should be performed under scrupulously dry conditions as the reagents readily hydrolyze to dimethylformamide and the appropriate alcohol. Under favorable conditions, the reagents act as their own

solvent and the reaction is complete as soon as the sample dissolves. Less
reactive compounds may require the addition of solvent and/or heating at 50 to
60°C for about 15 min.

Extractive alkylation is used to derivatize acids, phenols, alcohols, or
amides in aqueous solution [272,331-334]. The pH of the aqueous phase is
adjusted to ensure complete ionization of the acidic substance which is then
extracted as an ion-pair with tetraalkylammonium hydroxide into a suitable
immmiscible organic solvent. In the poorly solvating extractant, the substrate
anion possesses high reactivity and the nucleophilic displacement reaction with
an alkyl halide occurs under favorable conditions.

$$ROH + Bu_4NOH \longrightarrow (Bu_4N^+RO^-) + H_2O$$
$$(Bu_4N^+RO^-) + R'Br \longrightarrow ROR' + Bu_4NBr$$

Various tetraalkylammonium ions are used to form extractable ion-pairs
including tetrabutyl-, tetrapentyl-, and tetrahexylammonium. The choice of
cation depends on the efficiency with which the ion-pair will extract from
aqueous solution and then undergo alkylation. Almost any alkyl halide can be
used as the alkylating agent; selection is based on the separation required in
the chromatogram and the degree of sensitivity needed for detection purposes.
The lower members of the homologous series of alkyl bromides and iodides are
used with the flame-ionization detector and pentafluorobenzyl bromide with the
electron-capture detector.

Derivative formation is usually rapid at room temperature in the organic
phase, although elevated temperatures may occasionally be required. On
completion of the reaction, the derivative can be isolated by selective
extraction if the substrate contains other functional groups which make this
possible or, more generally, by evaporating the organic extract and taking up
the residue in a hydrocarbon solvent. A non-polar solvent is recommended since
the reaction by-product, the alkylammonium halide, is almost insoluble in such
solvents. The removal of the alkylammonium halide from the final extract prior
to gas chromatography is important to avoid the problems of column
contamination, wide solvent fronts, interference with the detector response, and
by-product formation.

Pyrolytic alkylation is the process whereby a volatile alkyl derivative of
an acidic compound is formed by thermal decomposition of a quarternary
N-alkylammonium salt of the acid in the heated injection port of a gas
chromatograph [267,268]. The alkyl derivative and the other volatile products
of the reaction are then swept onto the column by the carrier gas and eluted in
the usual manner. Pyrolytic alkylation can be used for the analysis of a
variety of organic compounds containing acidic NH and OH functional groups.

These include carboxylic acids, phenols, barbiturates, sulfonamides, and heterocyclic nitrogen compounds such as purines, pyrimidines, and xanthines. For the preparation of methyl derivatives, aqueous or methanolic solutions of tetramethylammonium hydroxide, phenyltrimethylammonium hydroxide, trimethylanilinium hydroxide, or m-trifluoromethylphenyltrimethylammonium hydroxide are usually used. Higher alkyl homologs are prepared with tetraethyl-, tetrapropyl-, or tetrabutylammonium hydroxides, etc.

The derivative forming process involves two sequential reactions: deprotonation of the acidic substrate in aqueous solution by the strongly basic tetraalkylammonium ion and the thermal decomposition of the quarternary N-alkylammonium salt formed to give a tertiary amine and alkyl derivative.

$$RCOOH + C_6H_5(CH_3)_3N^+OH^- \longrightarrow RCOO^- + C_6H_5N^+(CH_3)_3 + H_2O$$

$$RCOO^- + C_6H_5N^+(CH_3)_3 \longrightarrow RCOOCH_3 + C_6H_5N(CH_3)_2$$

For some weak acids both processes may occur virtually simultaneously in the injector oven of the gas chromatograph.

The tetraalkylammonium hydroxides deprotonate the more acidic organic acids ($pK_a < 12$) rapidly and virtually qunatitatively (e.g., carboxylic acids pK_a 4-5; phenols, pK_a 9-12). The weaker organic acids ($pK_a > 12$) are not extensively deprotonated under such conditions and consequently do not form derivatives by pyrolotic alkylation (e.g., aliphatic alcohols, pK_a 16-19; amides, pK_a about 25).

The thermal decomposition of the quarternary N-alkylammonium salt occurs by a nucleophilic attack of the anion upon the quarternary N-alkylammonium cation in the injector oven of the gas chromatograph. The optimum pyrolysis temperature depends primarily upon the structure of the cation and is established by trial and error. The usual operating range is 220 - 375°C. Any excess tetraalkylammonium hydroxide that is co-injected with the sample is thermally decomposed to give volatile products (the tertiary amine with small quantities of the corresponding alkyl alcohols and dialkyl ether). Methanolic trimethylanilinium hydroxide also produces small quantities of anisole as a by-product.

Despite its great utility, pyrolytic alkylation has not been without problems. One major defect of the method is that certain base- and heat-sensitive compounds can be isomerized or degraded under the conditions of high alkalinity and high temperature employed. For example, phenobarbitone undergoes base-catalyzed cleavage and partial degradation to form N-methyl-α-phenylbutyramide as a by-product [272,335].

$$\text{(barbiturate structure)} \xrightarrow{\;C_6H_5(CH_3)_3N^+OH^-\;} \text{(methylated product)} \;+\; \text{(amide by-product)}$$

However, by-product formation can be controlled by adjusting the pH of the extract prior to injection. This is done by co-injection with an acid or by using an alkylating agent of lower basicity (e.g., trimethylanilinium hydroxide). Similar observations have been made for other substances belonging to different compound classes. Another problem is that the efficiency of the thermal decomposition reaction is often strongly affected by the rate at which the sample solution is injected. In many instances, slow injection (2–5 seconds per microliter) yields better results in terms of solvent response, peak height, and peak resolution than rapid injection.

By far the most important reagent for arylation is 2,4-dinitrofluorobenzene [267,272]. This reagent will derivatize amine, thiol, and phenol groups in buffered aqueous solution or in non-aqueous solvents. Heating is usually required and reactions may be non-quantitative, particularly for sterically-hindered groups. 2,4-Dinitrobenzenesulfonic acid is preferred for some applications due to its higher selectivity for amines and the fact that the acidic derivatizing reagent can be easily separated from the derivative in the reaction medium. The principal limitation of the dinitrophenyl derivatives is their poor gas chromatographic volatility. However, they provide a high response towards the electron-capture detector and can be used for trace analysis [336,337].

7.13.5 Pentafluorophenyl-Containing Derivatives

Reagents for the preparation of pentafluorophenyl-containing derivatives are summarized in Table 7.21. They can be used to derivatize a broad spectrum of organic compounds, including tertiary amines. In some cases these reagents offer a high degree of specificity, for example the determination of primary amines with pentafluorobenzaldehyde and carboxylic acids with pentafluorobenzyl alcohol. The pentafluorophenyl-containing derivatives are generally easy to prepare, are volatile with good gas chromatographic properties, and provide a high response to the electron-capture detector. Their reactions proceed smoothly and few unexpected by-products have been identified. With secondary amines, pentafluorobenzaldehyde can react to form mixtures of ring-substituted products involving hydrogen fluoride elimination and cyclic derivatives can be formed with α-hydroxylamines [346,347].

TABLE 7.21

REAGENTS FOR THE INTRODUCTION OF THE PENTAFLUOROPHENYL GROUP INTO ORGANIC COMPOUNDS

Reagent	Functional group type	Reference
Pentafluorobenzoyl chloride	Amines, phenols, alcohols	338, 339
Pentafluorobenzyl bromide	Carboxylic acids, phenols, mercaptans, sulfonamides	339
Pentafluorobenzyl alcohol	Carboxylic acids	340
Pentafluorobenzaldehyde	Primary amines	341
Pentafluorobenzyl chloroformate	Tertiary amines	342, 343
Pentafluorophenacetyl chloride	Alcohols, phenols, amines	338
Pentafluorophenoxyacetyl chloride	Alcohols, phenols, amines	338
Pentafluorophenylhydrazine	Ketones	344
Pentafluorobenzylhydroxylamine	Ketones	345

Pentafluorobenzyl bromide is also a potent lachrymatory agent and is unstable under some conditions used for extractive alkylation [348]. Pentafluorophenylhydrazine and pentafluorobenzyl hydroxylamine derivatives of ketone-containing compounds can give two peaks due to the formation of syn- and anti-geometrical isomers [345,349,350].

7.13.6 Reagents for the Selective Derivatization of Bifunctional Compounds

Bifunctional compounds are characterized by the presence of at least two reactive functional groups on a molecular framework that places these groups in close proximity to one another. As such they do not constitute a defined chemical class of substances but are widely distributed among all classes of functionalized molecules (e.g., steroids, lipids, carbohydrates, nucleosides, catecholamines, prostaglandins, amino acids, etc.). In general terms, bifunctional compounds are compounds containing alkyl chains with functional groups on carbon atoms 1,2-, 1,3-, or 1,4- with respect to one another, or aromatic rings with ortho-substituted functional groups. Specific reagents can react with these groups to form stable cyclic derivatives [259,266,267].

Reagents used to form cyclic derivatives can be divided into two groups: those which can be used to derivatize a wide range of functional groups and reagents which are highly selective for specific functional groups or compounds (e.g., acetone, phenylenediamine, dimethyldiacetoxysilane). Of the general

reagents the most important are the cyclic boronic ester derivatives introduced
by Brooks and co-workers [351,352].

$$
\begin{array}{ccc}
\begin{array}{c}
\text{RCH--XH} \\[2pt]
\mid \\[2pt]
(\text{CH}_2)_n \\[2pt]
\mid \\[2pt]
\text{RCH--XH}
\end{array}
\;+\;
\begin{array}{c}
\text{HO} \\
\quad\diagdown \\
\quad\quad \text{BR}_1 \\
\quad\diagup \\
\text{HO}
\end{array}
\;\longrightarrow\;
\begin{array}{c}
\text{RCH--X} \\
\quad\quad\quad\diagdown \\
(\text{CH}_2)_n \quad \text{BR}_1 \\
\quad\quad\quad\diagup \\
\text{RCH--X}
\end{array}
\;+\; \text{H}_2\text{O}
\end{array}
$$

Bifunctional compound	boronic acid	cyclic boronic ester

X = O, N, S, CO_2, CO
n = 0, 1, 2

Reference [266] contains a complete compilation of all published applications on
the use of boronic acids through 1979. Their dominant position as derivatizing
reagents for bifunctional compounds is a consequence of their broad range of
application, ease of reaction, good thermal and GC properties, and their useful
mass spectral features. Disadvantages include the poor hydrolytic stability
exhibited by many derivatives and the ease of solvolysis observed in multiple
derivatization procedures. The boronate group may be partially or completely
displaced when the selective reaction of a remote functional group is required
to improve chromatographic properties. Boronate derivatives can be prepared
from compounds having two functional groups in close proximity such as alkyl
1,2-diols, 1,3-diols, 1,4-diones, 1,2-enediols, 1,2-hydroxyacids,
1,3-hydroxyacids, 1,2-hydroxylamines, 1,3-hydroxylamines, and aromatic compounds
with ortho-substituted phenol, amine, and carboxylic acid groups. Mild
conditions are usually sufficient for derivative formation and a typical
reaction involves mixing the boronic acid and substrate in an anhydrous solvent
at room temperature for a short time (1.0 to 30 min). In some cases excess
boronic acid may be required to force the equilibrium reaction to completion
and, for those derivatives which are exceptionally moisture sensitive, a means
of removing the water produced in the reaction is required (e.g., molecular
sieves can be added to the reaction medium, 2,2-dimethoxypropane can be added as
a water scavenger, or periodic azeotropic evaporation with benzene or
dichloromethane can be used). Direct injection of boronate derivatives with
remote unprotected polar funtional groups in the presence of excess boronic acid
invariably results in poor chromatographic performance, exemplified by tailing
peaks of reduced peak height. Sequential derivatization of the various
functional groups is required in this case and special attention must be paid to
the possibility that strong reaction conditions could result in loss of the
boronate group.

The boronic acids, methaneboronic acid, butaneboronic acid, t-butaneboronic acid, cyclohexaneboronic acid, and benzeneboronic acid, have all been used to prepare derivatives for gas chromatography. The cyclohexaneboronates and benzeneboronates have long retention times in comparison to the other boronate derivatives and can be inconvenient for the analysis of poly-bifunctional or high molecular weight compounds. The t-butaneboronates are surprisingly volatile on silicone stationary phases of low polarity with retention times closer to those of the methaneboronates than the n-butaneboronates. Unfortunately the reagent and derivatives have poor hydrolytic and air stability which limits their practical use. The methaneboronates are very volatile and the small molecular weight increment formed by derivatization is useful in the mass spectrometry of high molecular weight compounds. The butaneboronate derivatives provide a convenient compromise between volatility and stability making them the most studied derivatives. The stability of the boronate derivatives to TLC and other hydrolysis conditions is variable, depending both on local stereochemistry of the bifunctional group and the individual boronic acid used to prepare the derivative.

The boronate derivatives have useful mass spectral properties with prominent molecular ions, or quasi $[M + 1]^+$ molecular ions in the case of chemical ionization mass spectrometry. The boronate group is not strongly directing in influencing the mode of fragmentation as charge localization invariably occurs at a center remote from the boronate group due to the electrophilic character of the boron atom. This has the advantage that the abundant ions in the mass spectrum are characteristic of the parent molecule and not the derivatizing reagent. The natural isotopic abundance of boron ($^{10}B{:}^{11}B$ = 1:4.2) aids in the identification of boron-containing fragments in the low resolution mass spectra of the boronate derivatives. The boron-isotope distribution is a disadvantage when the mass spectrometer is operated as a single ion gas chromatographic detector, since the ion current carried by the boron-containing fragment is divided in the same ratio as the isotope distribution with a consequent reduction in sensitivity.

Boronic acids containing electron-capturing substitutents were developed by Poole and co-workers, Table 7.22 [353,354]. The 3,5-bis(trifluoromethyl)-benzeneboronates are remarkably volatile, with retention times significantly shorter than the benzeneboronate derivatives. The 4-iodobutaneboronates have retention times approximately 1.8 times those of the benzeneboronates. The 3-nitrobenzeneboronates and naphthaleneboronates have inconveniently long retention times for many applications. With the exception of the naphthaleneboronates, all of the derivatives show a moderate detector response which is enhanced by the introduction of halogen atoms. The position of the chlorine atom in the benzene ring influences the response of the detector and the highest response is obtained with 2,4-dichlorobenzeneboronate. The response of the 3,5-bis(trifluoromethyl)benzeneboronate derivatives is markedly temperature-dependent; it reaches a maximum at low detector temperatures and declines rapidly as the temperature is increased. It was recommended that the four boronic acids, 2,4-dichlorobenzeneboronic acid, 4-bromobenzeneboronic acid, 4-iodobutaneboronic acid, and 3,5-bis(trifluoromethyl)benzeneboronic acid, would provide a suitable selection of reagents for most applications [354]. The mass spectra of the derivatives formed with the above reagents almost always have prominent molecular ions, often the base peak or one of the most abuandant ions in the mass spectrum [355-357]. The simple loss of a halogen atom (Cl, Br, F) from the molecular ion or principal daughter ions is observed in all spectra but is rarely prominent. Daughter ions containing boron are generally among the more abundant ions in the mass spectra.

TABLE 7.22

A COMPARISON OF THE VOLATILITY AND ECD SENSITIVITY OF THE ELECTRON-CAPTURING BORONIC ACIDS

Boronic Ester	Relative Retention	Minimum Detectable Quantity (pg of pinacol)	Optimal Detector Temperature (°C)
3,5-Bis(trifluoromethyl) benzeneboronate	0.3 ± 0.05	3.0	180
Benzeneboronate	1.0	200.0	150
4-Iodobutaneboronates	1.8 ± 0.5		
4-Bromobenzeneboronates	3.9 ± 0.8	3.0	350
2,6-Dichlorobenzeneboronates	4.3 ± 2.0	18.0	380
2,4-Dichlorobenzeneboronates	4.7 ± 1.7	4.0	380
3,5-Dichlorobenzeneboronates	5.0 ± 1.1	11.0	380
2,4,6-Trichlorobenzeneboronates	6.9 ± 1.8	4.0	380
3-Nitrobenzeneboronates	11.7 ± 3.4	4.0	300
Naphthaleneboronates	18.5 ± 4.6	2550.0	350

Ethylphosphonothioic dichloride (EPTD) reacts with bifunctional compounds containing OH, NH_2, or CO_2H groups in the presence of triethylamine to form cyclic derivatives as shown below [358].

$$
\begin{array}{ccc}
\text{RCH-XH} & \text{Cl} \quad \text{S} & \\
\vert & \quad \backslash\vert\vert & \\
\text{(CH}_2)_n & + \quad \text{P-C}_2\text{H}_5 & \xrightarrow{\text{(C}_2\text{H}_5)_3\text{N}} \\
\vert & \quad / & \\
\text{RCH-XH} & \text{Cl} &
\end{array}
\quad
\begin{array}{c}
\text{RCH-X} \quad \text{S} \\
\vert \quad \backslash\vert\vert \\
\text{(CH}_2)_n \quad \text{P-C}_2\text{H}_5 \\
\vert \quad / \\
\text{RCH-X}
\end{array}
$$

XH = OH, NH_2, CO_2H
n = 0, 1

EPTD also reacts with ortho-substituted aromatic compounds as well as with some compounds containing enolizable ketone groups. Some derivatives give two peaks on gas chromatography due to the formation of geometric isomers. Low-level detection of the derivatives is possible using phosphorus- or sulfur-selective detectors. The mass spectra provide abundant molecular ions with a characteristic elimination of ethyl sulfide to form a daughter ion which is often the base peak of the spectrum. Other ions characteristic of the derivatives are $[C_2H_5PS]^+$ and $[PS]^+$.

Dimethyldichlorosilane and dimethyldiacetoxysilane form cyclic siliconides [359,360] while acetone forms acetonides with cis-diol groups [266]. Incomplete reaction, possible formation of by-products, and poor hydrolytic stability limit the applications of the siliconides. The acetonides, a specific example of the formation of acetal and ketal derivatives formed with aldehydes and ketones, are free from most of the problems observed with siliconides. The introduction of a halogen substituent into the alkyl portion of acetone enhances the acidic character of the carbonyl group and promotes the formation of a series of stable monofunctional adducts and cyclic derivatives with α-substituted carboxylic acids not observed with acetone itself [361].

X = NH   1,3-oxazolidin-5-one
X = O   1,3-dioxolan-5-one
X = S   1,3-oxathiolan-5-one

The oxazolidinone derivatives formed with hexafluoroacetone or 1,3-dichlorotetrafluoroacetone have been used for the separation of amino acids after silylation or acylation of side chain functional groups [362,363].

In acidic solution 1,2-diaminobenzene selectively reacts with α-keto acids to form thermally-stable cyclic quinozalinol derivatives. The phenolic hydroxy group is usually further derivatized to the trimethylsilyl ether to improve the chromatographic performance of the quinozalinol [364,365]. Reagents containing nitro or halogen substituents have been used with the electron-capture detector [364-367].

Biguanides undergo two cyclizing reactions which are useful for their gas chromatographic analysis. Condensation with acetylacetone [368,369] or hexafluoroacetylacetone [369-371] produces substituted pyrimidine derivatives.

$$\underset{\text{NH}}{\overset{\|}{R-C}}-NH_2 \;+\; (CX_3CO)_2CH_2 \;\longrightarrow\; R-C$$

X = H or F

The derivatives can be formed in aqueous or physiological solution (biguanides are an important group of pharmacologically-active compounds). With anhydrides, biguanides form cyclic 2,4-disubstituted-2-6-amino-1,3,5-s-triazine derivatives which have good gas chromatographic properties [372-374].

$$\underset{\text{NH}}{\overset{\|}{R-C}}\diagdown_{\underset{H}{N}}\diagup\underset{\text{NH}}{\overset{\|}{C}}-NH_2 \quad (R_1CO)_2O \;\longrightarrow\;$$

The properties of further miscellaneous cyclizing reagents have been compiled in reference [266].

7.14 Derivatization Techniques for High Performance Liquid Chromatography

Derivatives in liquid chromatography are invariably prepared to improve the response of a substance towards a particular detector [259,260,375-377]. Other applications include aiding in the isolation of a substance or improving the chromatographic separation obtained with mixtures having overlapping peaks. However, the versatility and wide range of separation methods and conditions available in HPLC makes this latter point more one of potential than of actual practice. Also, from an historical view point, HPLC was promoted as the technique to free the analyst from the need to derivatize polar compounds prior to their analysis, the established practice in gas chromatography at the time. The last few years have witnessed an explosive growth in the use of reversed-phase chromatographic systems for most routine analyses. This, coupled with the availability of detectors with adequate sensitivity and the commerical availability of water miscible solvents with little absorbance at 210 nm, enables many substances to be detected at this generally non-selective wavelength.

Fluorescence and electrochemical derivatives and postcolumn derivatization reactions are potentially of greater interest in HPLC than precolumn UV sensitive derivatives. They provide high sensitivity with detection limits down to the picogram level, coupled with derivative selectivity. This combination of selectivity and sensitivity is necessary for trace analysis of substances in complex matrices. This is an area which is likely to expand in the near future as the acceptance of fluorescence and electrochemical detectors becomes more widely established.

7.14.1 Derivatives for UV–Visible Detection

The "workhorse" detector in HPLC is the fixed wavelength, 254 nm, UV absorbance detector. It has adequate sensitivity for most routine applications if the sample has a reasonable absorbance in the region of the measuring wavelength. For compounds lacking a UV-chromophore in this region, but having a reactive functional group, derivatization provides a convenient means of introducing into the molecule a chromophore suitable for its detection. To some extent, this approach was made less essential by the wider use of variable wavelength detectors with a usuable range of 180 to 800 nm for detection at the absorbance maximum of the substance of interest. However, these detectors tend to be more common in research laboratories and on newer instruments than in quality control laboratories.

The majority of reagents used to introduce UV–chromophores into functionalized molecules contain a reactive group which controls the chemical aspects of reactivity and selectivity and a substituted aromatic moiety which provides the chromophore for detector response. Table 7.23 summarizes the most widely used chromophoric groups, their wavelength of maximum absorbance, and their absorbance measured at 254 nm [378]. Those reagents with an $\varepsilon_{max} > 10,000$ at 254 nm can provide detection limits for the derivatized molecule at the low nanogram level. Ideally, the chromophoric group should be small and nonpolar so as not to dominate the chromatographic separation. These demands are fairly well met by the substituted aromatic chromophores shown in Table 7.23.

Functional groups which can be derivatized and suitable reagents for this purpose are summarized in Table 7.24. Many of the reactions and reagents are the familiar ones used in qualitative analysis for the characterization of organic compounds by physical means. Alcohols are converted to esters by reaction with an acid chloride in the presence of a base catalyst (e.g., pyridine, tertiary amine). If the alcohol is to be recovered after the separation, then a derivative which is fairly easy to hydrolyze, such as p-nitrophenylcarbonate, is convenient. If the sample contains labile groups, phenylurethane derivatives can be prepared under very mild reaction conditions.

$$ROH + ONCC_6H_5 \longrightarrow ROCONHC_6H_5$$

Alcohols in aqueous solution can be derivatized with 3,5-dinitrobenzoyl chloride.

TABLE 7.23

CHROMOPHORES OF INTEREST FOR ENHANCED UV DETECTABILITY

Structure	Chromophore	λ_{max}	ε_{254}
	2,4-dintrophenyl	–	$>10^4$
	benzyl	254	200
	p-nitrobenzyl	265	6200
	3,5-dinitrobenzyl	–	$>10^4$
	benzoate	230	low
	p-nitrobenzoate	254	$>10^4$
	toluoyl	236	5400
	p-chlorobenzoate	236	6300
	anisyl	262	16000
	phenacyl	250	$\simeq 10^4$
	p-bromophenacyl	260	18000
	2-naphthacyl	248	12000

TABLE 7.24

REAGENTS FOR THE INTRODUCTION OF CHROMOPHORES INTO FUNCTIONALIZED MOLECULES

Functional Group Reacted	Derivatizing Reagent	Reference
Alcohols	3,5-Dinitrobenzyl Chloride	379
	Pyruvoyl Chloride	380
	p-Iodobenzenesulfonyl Chloride	380
	Benzoyl Chloride	381
	p-Nitrobenzoyl Chloride	382
	p-Nitrophenyl Chloroformate	383
	Phenyl Isocyanate	384
	Trityl Chloride	385
	p-Dimenthylaminophenyl Isocyanate	386
Amines	3,5-Dinitrobenzoyl Chloride	375
	Pyruvoyl Chloride	380
	p-Methoxybenzoyl Chloride	387
	n-Succinimidyl-p-nitrophenylacetate	378
	Benzoyl Chloride	378
	p-Nitrobenzoyl Chloride	388
	p-Toluenesulfonyl Chloride	389
	2-Naphthacyl Bromide	390
	2,4-Dinitrofluorobenzene	391
Ketones and Aldehydes	2,4-Dinitrophenylhydrazine	392
	p-Nitrobenzylhydroxylamine Hydrochloride	378
	2-Diphenylacetyl-1,3-indandione-1-hydrazone	393
Carboxylic Acids	p-Bromophenacyl Bromide	394
	p-Nitrobenzyl-N,N'-diisopropylisourea	395
	1-(p-Nitro)benzyl-3-p-tolyltriazine	396
	Phenacyl Bromide	397
	Benzyl Bromide	394
	N-Chloromethyl-4-nitrophthalimide	398
	Naphthyldiazomethane	399
Epoxides	Diethyldithiocarbamate	400
Isocyanates	N-p-Nitrobenzyl-N-n-propylamine	401
Mercaptans	Pyruvoyl Chloride	380
Phenols	Pyruvoyl Chloride	380
	3,5-Dinitrobenzoyl Chloride	375
	p-Iodobenzoylsulfonyl Chloride	380
	p-Nitrobenzenediazonium Tetrafluoroborate	402
	Diazo-4-aminobenzonitrile	377

    Amines are usually derivatized via the acid chlorides to form phenyl-substituted amides. Sanger introduced the use of 2,4-dinitro-1-fluorobenzene for the identification of N-terminal amino acid residues in proteins [259]. This reagent has also been widely used for the derivatization of amine-containing compounds in general.

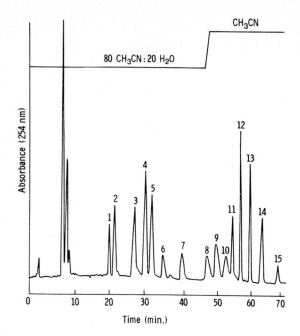

Figure 7.20 Separation of phenacyl ester derivatives of long-chain fatty acids by reversed-phase chromatography. Components: 1, lauric; 2, myristoleic; 3, linolenic; 4, myristic; 5, archidonic; 6, linoleic; 7, eicosatrienoic; 8, palmitic; 9, oleic; 10, petroselinic; 11, eicosadienoic; 12, stearic; 13, arachidic; 14, behenic; and 15, lignoceric. (Reproduced with permission from ref. 397. Copyright American Chemical Society).

N-Succinimidyl-p-nitrophenylacetate can be used to derivatize amines under mild conditions without the use of a catalyst.

Methods for the derivatization of carboxylic acids are based on well-known alkylation reactions. Figure 7.20 shows a separation of some fatty acids as their phenacyl ester derivatives by reversed-phase chromatography [397]. Alkylation reactions can be carried out smoothly in the presence of phase transfer catalysts such as the crown ethers [394]. The crown ether solubilizes the potassium carbonate catalyst in the organic solvent, helping to promote the alkylation reaction. Aryl-tolyltriazines have been used as derivatizing reagents for carboxylic acids, but a question of safety as well as the generation of interfering by-products has curtailed their general use. They have largely been superseded by the use of p-nitrobenzyl-N,N'-diisopropyl-isourea. This reagent has the advantage of promoting facile reactions without the need for a base catalyst.

$$O_2N-\langle\ \rangle-CH_2OC\overset{NCH(CH_3)_2}{\underset{NHCH(CH_3)_2}{<}} + HOOCR \longrightarrow O_2N-\langle\ \rangle-CH_2OOCR + O=C\overset{NHCH(CH_3)_2}{\underset{NHCH(CH_3)_2}{<}}$$

Aldehydes and ketones may be selectively derivatized with either 2,4-dinitrophenylhydrazine or p-nitrobenzylhydroxylamine hydrochloride without affecting other functional groups present in the molecule. Epoxides may be converted to -hydroxyldithiocarbamoyl esters by reaction with sodium diethyldithiocarbamate. Isocyanates can be converted to urethane derivatives with p-nitrobenzyl-N-n-propylamine [383].

$$\underset{R}{\overset{H_2C}{\underset{HC}{|}}}>O + \underset{C_2H_5}{\overset{C_2H_5}{>}}N C\overset{S}{\overset{||}{S^-}}\ Na^+ \longrightarrow \underset{\overset{|}{HCOH}}{\overset{CH_2SCN}{\overset{S}{\overset{||}{}}}}\overset{C_2H_5}{\underset{C_2H_5}{<}}$$

7.14.2 Derivatives for Fluoroescence Detection

The measurement of fluorescence has certain advantages as a detection technique for liquid chromatography. The most important of these are high sensitivity, wide linear response, and a relatively high substance specificity. The substance specificity arises from the fact that only a small percentage of organic molecules are naturally fluorescent and, among those that do fluoresce, some distinction is possible by wavelength discrimination.

There is only a limited selection of reagents available for the introduction of fluorescent groups into organic molecules. Many of these have been devised for use in either protein sequencing or for the detection of amino acids in automated amino acid analyzers. Thus, there are several reagents available for reaction with amino groups but the selection is more restricted for other functional groups. The structure of the most frequently used derivatizing reagents are shown in Figure 7.21. The functional groups which can be derivatized with the various reagents are summarized in Table 7.25. Several reviews of the chromatographic applications of fluorescence derivatization techniques are available [259,260,419-421].

Fluorescence is a molecular property, subject to certain variations with changes in the molecular environment. In chromatographic terms, the most

TABLE 7.25

REAGENTS FOR THE PRECOLUMN PREPARATION OF FLUORESCENT DERIVATIVES IN HPLC

Reagent	Abbreviation	Application	Reference
1-Ethoxy-4-(dichloro-s-triazinyl)naphthalene	EDTN	Primary and secondary alcohols and phenols	403
4-Dimethylamino-1-naphthoyl nitrile		Primary and secondary but not tertiary alcohols	404
4-Bromomethyl-7-methoxycoumarin	Br-Mmc	Carboxylic acids	405
4-Bromomethyl-6,7-dimethoxycoumarin		Carboxylic acids	406
9-(Chloromethyl)anthracene	9-ClMA	Carboxylic acids	407
9-Anthradiazomethane		Carboxylic acids	408
9,10-Diaminophenanthrene		Carboxylic acids	409
Dansyl hydrazine	DnS-H	Aldehydes and ketones	410
2-Diphenylacetyl-1,3-indandione-1-hydrazone		Aldehydes and ketones	411
5-Dimethylaminonaphthalene 1-sulfonylaziridine		Thiols	412
N-(9-Acridinyl)maleimide		Thiols	413
Dansyl chloride	DnS-Cl	Primary and secondary amines, phenols, amino acids, and imidazoles	12
2,5-Di-n-butylamino-naphthalene-1-sulfonyl chloride	Bns-Cl	as for Dns-Cl	12
Fluorescamine	Fluram	Primary amines and amino acids	12
o-phthaldialdehyde	OPT	Amines and amino acids	413
7-Chloro-4-nitrobenzo-2-oxa-1,3-diazole	Nbd-Cl	Primary and secondary amines, phenols, and thiols	414
9-Fluorenylmethyl chloroformate		Primary and secondary amines	415
9-Isothiocyanatoacridine		Primary and secondary amines	416
Pyridoxal		Amino acids	375
Phenanthreneboronic acid		Bifunctional compounds	417
5,5-Dimethyl-1,3-cyclohexanedione		Aldehydes	418

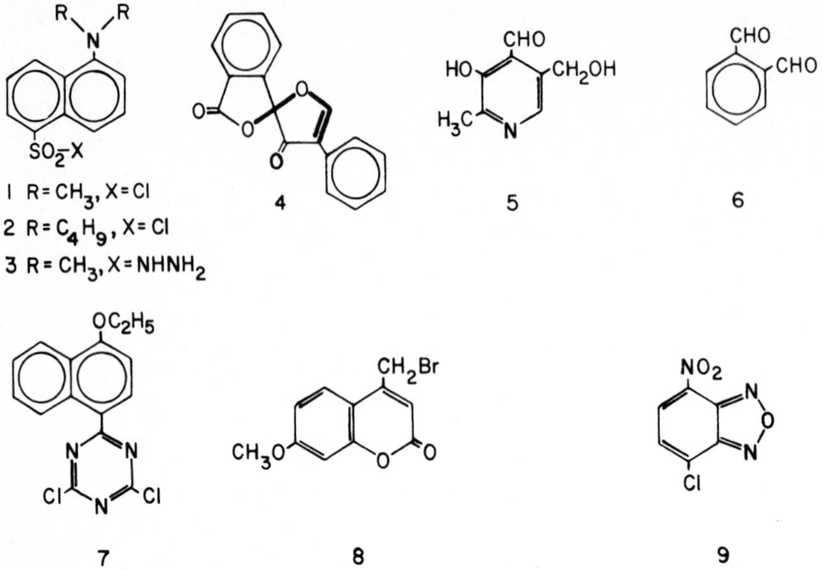

Figure 7.21 Reagents for the formation of fluorescent derivatives. 1, Dansyl chloride; 2, dabsyl chloride; 3, dansyl hydrazine; 4, fluorescamine; 5, pyridoxal; 6, o-phthaldialdehyde; 7, 1-ethoxy-4-(dichloro-s-triazinyl)naphthalene; 8, 4-bromomethyl-7-methoxycoumarin; and 9, 7-chloro-4-nitrobenzo-2-oxa-1,3-diazole.

important of these are changes in fluorescence intensity and shifts in the optimum wavelength of fluorescence excitation and emission with the mobile phase composition. Water and other strongly hydrogen-bonding solvents sometimes have a quenching effect on fluorescence. Compounds which fluoresce in aqueous solution may be dramatically influenced by solution pH. Compounds fluorescing in organic solvents may show a shift in fluorescence maxima and a change in intensity with solvent polarity. All of these effects must be examined along with the needs of the chromatographic system in developing a suitable analytical method.

Dansyl chloride is the most widely used of the derivatizing reagents. It forms derivatives with primary and secondary amines readily, less rapidly with phenols and imidazoles, and very slowly with alcohols. The reaction medium is usually an aqueous-organic mixture (e.g., 1:1 acetone-water) adjusted to a pH of 9.5 - 10. Dansyl chloride has two major application areas. It is used to determine small amounts of amines, amino acids, and phenolic compounds as well as basic drugs and their metabolites in tissues and biological fluids. It finds further use in biochemistry for peptide sequencing and for the fluorogenic labelling of proteins and enzymes. Several analogs of dansyl chloride are also in use. The butyl analog (Figure 7.21) has better storage properties and forms derivatives which can be more readily extracted into organic solvents. Its reaction and fluorescence properties are very similar to dansyl chloride. Dansyl hydrazine (Figure 7.21) is a selective reagent for the analysis of primary amines. It also reacts with secondary amines, alcohols, and water but these reactions do not lead to fluorescent products. Fluorescamine is widely used for the detection of amines, amino acids, and peptides after ion-exchange chromatography, particularly in conjunction with automated amino acid analyzers. Another widely used reagent for this purpose is o-phthaldialdehyde. o-Phthaldialdehyde produces strongly fluorescent derivatives with amino acids and biogenic amines in alkaline medium that contains a reducing agent (e.g., 2-mercaptoethanol) [422]. Figure 7.22 shows the separation of the o-phthaldialdehyde amino acid derivatives from a protein digest using reversed-phase chromatography [423]. 7-Chloro-4-nitrobenzo-2-oxa-1,3-diazole reacts with primary and secondary amines to form intensely fluorescent derivatives. It also reacts with anilines, phenols, and thiols to form derivatives which do not fluoresce or are only weakly fluorescent. Although the reagent is non-fluorescent, it nevertheless interferes in the fluorescence of its products and must be separated from them prior to measurement. Recent studies have indicated that $Nbd-OCH_3$ is formed as a reactive intermediate under normal reaction conditions in methanol. It reacts faster than Nbd-Cl and is

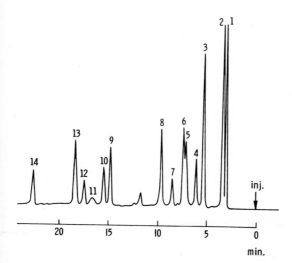

Figure 7.22 Separation of the o-phthaldialdehyde amino acid derivatives from a protein hydrolysate by reversed-phase chromatography. Components: 1, Asp; 2, Glu; 3, Ser; 4, His; 5, Thr; 6, Gly; 7, Arg; 8, Ala; 9, Val; 10, Phe; 11, NH_4^+; 12, Ile; 13, Leu; and 14, Lys. (Reproduced with permission from ref. 423. Copyright American Chemical Society).

probably the principal reactive species under these conditions [414].
7-Fluoro-4-nitrobenzo-2-oxa-1,3-diazole has also been shown to be a more
reactive analog of Nbd-Cl towards primary and secondary amines [424]. Pyridoxal
(Figure 7.21) undergoes a condensation reaction with amino acids in alkaline
medium to form a Schiff base complex which can be converted into a stable
fluorescent product by reduction with sodium borohydride.

Amino Acid   Schiff Base   Fluorescent Product
Complex

1-Ethoxy-4-(dichloro-s-triazinyl)naphthalene (Figure 7.21) reacts with primary
hydroxyl and phenol groups to form fluorescent derivatives [403]. It is not
clear whether it also reacts with secondary hydroxyl and amino groups.
Recently, 4-dimethylamino-1-naphthoyl nitrile has been introduced for the base
catalyzed reaction of primary and secondary hydroxyl groups at room temperature
[404]. Triethylamine was used as a catalyst. Sterically hindered secondary and
tertiary hydroxyl groups do not react. 5-Dimethylaminonaphthalene-1-sulfonyl-
aziridine is a selective reagent for the derivatization of sulfhydryl groups in
aqueous solution at pH 8.2 [412]. Under other reaction conditions, weaker
nucleophiles such as amines and alcohols may also react.
4-Bromomethyl-7-methoxycoumarin in anhydrous acetone smoothly alkylates
carboxylic acids in the presence of potassium carbonate solvolysed by a crown
ether catalyst [405,425]. Derivatives of saturated fatty acids prepared in this
way show a lower fluorescent yield as the concentration of organic solvent in
the mobile phase is increased [426,427]. Thus, with gradient elution
reversed-phase separations the response of later eluting derivatives is much
lower than those at the start of the chromatogram.

     All fluorescent derivatives absorb UV-light. Thus, the reagents in Table
7.25 are also suitable for use with a UV-absorbance detector, although the
sensitivity and selectivity may be less than that obtained with a fluorescence
detector.

7.14.3 Derivatives for Electrochemical Detection

     Any reagent in Table 7.24 which results in the formation of derivatives
containing a nitroaromatic chromophore for UV detection could also be used as a

derivative for electrochemical detection [428]. The detection principle is based on the electrochemical reduction of the aromatic nitro group. p-Aminophenol derivatives of carboxylic acids [429] and p-dimethylaminophenyl isocyanate derivatives of arylhydroxylamines [386] are also suitable for electrochemical detection. However, precolumn derivatization techniques for electrochemical detection are not without problems. The detector response is adversely affected by traces of the derivatizing reagents and/or by-products which may have to be removed prior to injection. The choice of mobile phase for the separation is limited by the requirements of the detector. For example, in reversed-phase chromatography the amount of organic modifier which can be used to control retention is limited by the detector preference for an aqueous mobile phase. This may result in poor mixture separation or complete retention of the hydrophobic derivative on the column.

7.14.4 Reaction Detectors

Reaction detectors are a convenient means of performing on-line postcolumn derivatization in HPLC. The derivative reaction is performed after the separation of the sample by the column and prior to detection in a continuous reactor. The mobile phase flow is not interrupted during the analysis and reaction, although it may be augmented by the addition of a secondary solvent to aid the reaction or to conform to the requirements of the detector. The advantage of this system, apart from its ease of automation, is that both the separation and detection process can be optimized separately. Reaction detectors are finding increasing application for the analysis of trace components in complex matrices where both high detection sensitivity and specificity are needed. Many suitable reaction techniques have been published for this purpose [377,430-432].

Advantageous features of postcolumn derivatization are that artefact formation is rarely a problem and it is also not essential that the reaction employed for derivatization go to completion or be well defined, as long as it is reproducible. Some general problems with postcolumn reaction detectors are associated with mobile phase compatibility and reagent selection. Very seldom will the optimal chromatographic eluent also provide an ideal reaction medium. This is particularly true for electrochemical detectors which function correctly only within a restricted range of aqueous solvents, pH, ionic strength and organic modifier concentrations [428]. The reagent must not be detectable under the same conditions as the derivative. Finally, the reaction must be fast enough that column resolution is not destroyed by diffusion in the reaction device. This sets an upper limit of about 20 minutes for the slowest reaction that may be employed. The time required for reaction most profoundly influences

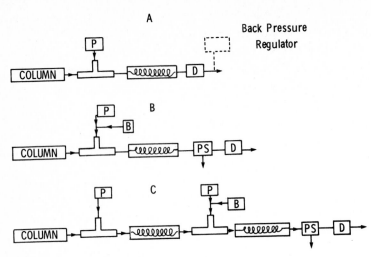

Figure 7.23   Schematic diagrams of some typical reaction detectors used in
HPLC. A, non-segmented tubular reactor; B, segmented tubular
reactor; C, extraction segmented reaction detector. P, pump;
PS, phase separator; B, device for introducing bubbles; and
D, detector.

the design of the reaction detector.

The reaction detector system must provide for the continuous addition of
controlled volumes of one or more reagents to the column effluents, followed by
mixing of the column effluent-reagent mixture and incubation at some time and
temperature governed by the needs of the reaction. Detection of the reaction
products is usually by colorimetry, fluorescence, or electrochemical
techniques. Some typical reaction detector configurations are shown in Figure
7.23. The addition of reagent to the column effluent takes place at fairly low
pressures and peristaltic pumps can be used for this purpose. It is
particularly important that the reagent pump provide a pulse-free flow or the
performance of the detector will be compromised. Nearly all devices contain a
mixing-tee or some similar component for contacting and homogeneously mixing the
column eluent and reagent stream [433,434]. The mixing-tee contributes to band
broadening by dispersion and should be miniaturized. Volumes of 150-600
microliters are typical for these devices. If the segmentation principle is
employed, a device for introducing a bubble of liquid or gas into the column
eluent is required [433-437]. This results in the separation or segmentation of
the column eluent into a series of reaction compartments whose volume is
governed by the dimensions of the transfer tube and the frequency of bubble
introduction. Prior to detection a phase separator is needed to remove the
segmentation agent. Dispersion in the phase separator can be a serious
limitation when column resolution is low. A phase separator is also used when

ion-pair formation or extraction is part of the detection system. Reagents or by-products which could interfere with detector performance are removed in this way [438-440]. Phase separators are not one hundred percent efficient and a fraction of the detection stream (approximately 20-40%) goes to waste. For reactions using solid-phase reagents, immobilized enzymes, etc., a packed bed tubular reactor is used. The column eluent flows through the reactor with or without the addition of further reagent, pH buffer, etc. Reaction occurs on the surface of the packing, generating a product that can be detected downstream of the reactor [441,442]. Reactions that are slow at room temperature may be accelerated by thermostatting the reactor to a higher temperature. A cooling coil may be required prior to detection to prevent interference in the operation of the detector or to avoid bubble formation. Models have been devised to predict the influence of postcolumn dispersion on the column separation for the various common reaction detector designs [431-435,443,444]. The influence of the reaction rate on detector design will be qualitatively discussed below.

For fast reactions (i.e., < 30 s), tubular reactors are commonly used. They simply consist of a mixing-tee and a coiled stainless steel or Teflon capillary tube of narrow bore. The length of the capillary tube and the flow rate through it control the reaction time. The design of the mixing-tee is a critical part of the apparatus, particularly for mixing column eluents and reagent solutions of dissimilar densities. Reagents such as fluorescamine and o-phthaldialdehyde are frequently used in this type of detector to determine primary amines, amino acids, indoles, hydrazines, and aminophosphoric acids in biological or environmental samples [377]. In its simplest form, a tubular reactor may alter the pH or compostion of the mobile phase to enhance detection sensitivity [445,446], while at the other extreme, several tubular reactors may be employed in series to perform multiple chemical reactions and extraction prior to detection [447]. Non-segmented flow is generally used with short tubular reactors; however, recent work suggests that even for very fast reactions, band broadening can be reduced by using the segmentation principle [431].

For reactions of intermediate kinetics (i.e., those with reaction times from 0.3 to several minutes), and in particular, for reactions employing solid-phase reagents, a packed bed reactor is preferred. The packed bed reactor is constructed from a length of column tubing packed with an inert material of small diameter such as glass beads. As band broadening is controlled by axial diffusion and convective mixing, packing the reactor bed should be given the same attention as packing the analytical column. Packed bed reactors have been used with fluorescamine and ninhydrin for the detection of amino acids [377]. Deelder et al. have described a postcolumn reaction that utilizes two reagent solutions pumped separately into one mixing-tee. This reagent mixture is then

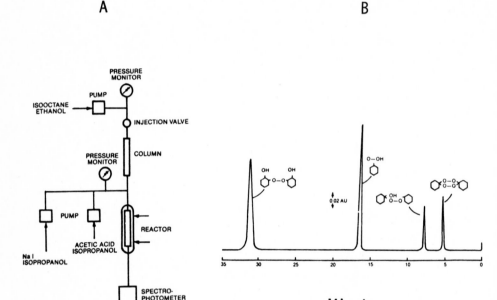

Figure 7.24 A, packed bed reaction detector for the determination of peroxides. B, separation of a test mixture or peroxides using detector A. (Reproduced with permission from ref. 448. Copyright Elsevier Scientific Publishing Co.)

mixed separately with the HPLC eluent containing the peroxides to be detected (Figure 7.24) [448]. The final homogeneous solution enters the packed bed reactor where reaction between the acidic iodide reagent and peroxides occurs to liberate molecular iodine, which is spectrophotometrically detected. The packed bed reactor is particularly useful for reactions involving solid-phase reagents such as catalysts, immobilized enzymes, and metallic reducing agents [377,441,442,449]. Regnier has described the use of several enzyme-based detectors for the determination of isoenzymes in biological fluids [442]. These detectors provide very high detection specificity, minimizing sample preparation.

For slow reactions (reaction times from 5 to 20 minutes), air or liquid segmented reactors are used. These work on the same principle as the auto-analyzers used in many clinical laboratories. The effluent stream is split into segments by the introduction of air or a immiscible liquid bubble at fixed time intervals. The optimal conditions for operation involve small liquid segments with a high frequency of segmentation, short reaction tubes of low internal diameter, high flow rates, and a well-designed phase separation unit just upstream from the detector. Predictions of theory must be tempered with practical needs of the reaction, which controls the length of the reaction coil. Band broadening occurs mainly by axial diffusion with a small contribution from sample transfer between segments by wetting of the tube wall. The air segmented reactor is probably the most widely applied reaction detector in HPLC. This is due to a wide selection of suitable reactions, many of which were described originally for use with the auto-analyzer. Major application areas include the analysis of amino acids, sugars, phenols, amines, pharmaceutical products, pesticides, and environmental samples [377,430-435].

One limition of postcolumn reaction detectors is that the reagent, always present in large excess, should not interfere with detection of the newly-formed derivative. There are several useful reactions, particularly involving ion-pair formation or complexation, for which reagent interference are commonly observed. The reaction products are usually much lower in polarity than the reagent, permitting an extractive separation of the two. Extraction in the continuous flow mode by segmenting with immiscible organic solvent plugs (the extraction solvent) or by air segmentation with an additional flow of extraction solvent and a mixing device can be used. A suitable experimental arrangement for ion-pair extraction using air segmentation is shown in Figure 7.25 [450]. Solvent segmentation is more convenient for extraction detection and is becoming more widely used. Applications include the ion-pair detection of antibiotics [439], pharmaceutical compounds possessing a tertiary amine structure not easily derivatized by other techniques [431,438-440], and alkylsulfonate detergents [451].

The photochemically-induced change of a compound in a flowing stream can be the basis of a reaction detector. Photolysis may be used to convert a substance into a more readily detected product, for example the introduction of fluorescence or conductance. It can also be used to decompose the sample, generating a fragment which can be coupled with an appropriate reagent for detection. As an example of the latter approach, nitrosamines can be photolytically degraded to nitriles which are then determined colorimetrically by the Griess reaction [431]. Photochemical reaction detectors are fairly simple in design, comprised of a high-powered lamp in a reflective housing. The Teflon or quartz reaction coil is wound around the lamp [437,453]. Teflon has excellent transparency to UV light and is the preferred material for the

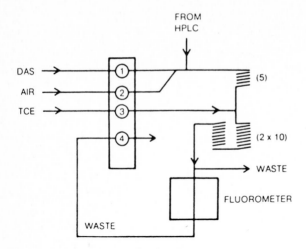

Figure 7.25 Postcolumn ion-pair extraction detector. DAS is the sodium salt
of 9,10-dimethoxyanthracene-2-sulfonate and forms ion pairs with
basic compounds. TCE is tetrachloroethane and is used to extract
the ion pair. (Reproduced with permission from ref. 450.
Copyright Dr. Alfred Huethig Publishers).

reaction coil. The reaction time, controlled by the length of the coil, is
generally maximized by trial and error. Overexposure may result in the
decomposition of the product to be detected. The high specificity of the
photochemical reaction detector is useful for particular applications
[377,437,454]. As few compounds are photolytically sensitive, however, its
range of applications is rather limited.

7.14.5 Ion-Pair Chromatography

Ion-pair chromatography is a derivatization technique of relatively recent
origin that has quickly gained wide acceptance as a versatile and efficient
method for the separation of ionized and ionizable molecules by high pressure
liquid chromatography. Substances of this type have poor separation
characteristics on most column types with the exception of ion-exchange
columns. Ion-exchange methods are often plagued by poor column efficiencies,
short column operation life and batch to batch irreproducibilities in
manufacture compared to other HPLC column types. This, coupled with the limited
selection of ion-exchange packings available, has done much to promote the
general interest in the ion-pair chromatographic technique. Ion-pair
chromatography makes use of existing high efficiency, small particle normal or
reversed-phase packings. When chemically bonded reversed-phase packings are
used, the selectivity of the separation system can be easily altered by making
all changes to the composition of the mobile phase. An important advantage of

ion-pair chromatography is its ability to simultaneously separate samples containing neutral and ionized molecules.

The origins of ion-pair chromatography can be traced to the use of ion-pairing reagents for the liquid extraction of ionic compounds in aqueous solution with an immiscible organic solvent. This topic will not be discussed here, except as it relates to the separation mechanism in HPLC. A detailed review of ion-pair liquid-liquid extraction has been given by Schill [455].

Superficially, it is thought that the basis for ion-pair chromatography is the formation of a lipophilic complex by the association of two ions of opposite charge. Ion-pair formation occurs in the aqueous phase and the net result is the transfer of the ionic solute into an organic phase. Chromatographic separations result principally from the dynamic exchange of the ionized solute between the mobile and stationary phases either directly as the ion-pair or after dissociation into its constituent parts. As will be shown in the next section, the above picture is a gross oversimplification of the actual situation. Indeed, the separation can be explained in terms of ionic interactions which do not require the formation of a formal ion-pair complex.

Ion-pair chromatography can be performed with normal or reversed-phase chromatographic techniques. In the normal phase mode, the ion-pair reagent, in an aqueous buffered solution, is coated onto silica gel to form the stationary phase and organic solvents are used as the mobile phase. Column retention and selectivity are regulated by changing the composition of the organic mobile phase. Perhaps the most important advantage of this system is the possibility of using detector oriented counter ions of high sensitivity to improve the detection of compounds with little UV absorption or fluorescence. The main limitation is that the sample must be transferred to the organic phase prior to analysis; many ionic hydrophilic substances have low solubility in lipophilic solvents. An example of the use of normal phase ion-pair chromatography for the separation of a mixture of dipeptides is shown in Figure 7.26 [456].

Reversed-phase systems employing permanently bonded alkyl layers are the most convenient to use experimentally. There is no column preparation needed, and retention and separation selectivity are controlled by adjusting the composition of the mobile phase. Changes in the nature and concentration of the ion-pair reagent, the buffer composition and pH value, and the type and amount of the organic modifier in the mobile phase are easily made. Since the stationary phase is chemically bonded to the support, gradient elution techniques can be used to further optimize the separation system. As ion-pair chromatography is used for the separation of ionic or easily ionizable substances, there are obvious advantages to having an aqueous mobile phase as far as sample compatibility is concerned. Samples are prepared by dissolution in the mobile phase prior to injection. Figure 7.27 shows the separation of a

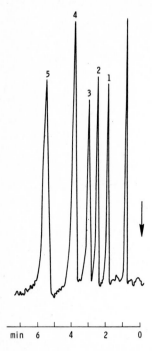

Figure 7.26 Separation of dipeptides by ion-pair chromatography. Components:
1, leu-leu; 2, phe-val; 3, val-phe; 4, leu-val; and 5, met-val.
Mobile phase CHCl$_3$-pentanol (9:1); stationary phase LiChrospher
Si 100 coated with naphthalene-2-sulfonate 0.1 M, pH 2.3.
(Reproduced with permission from ref. 456. Copyright Elsevier
Scientific Publishing Co.)

mixture of benzoic and sulphonic acids by reversed-phase partition
chromatography with n-pentanol coated onto LiChrosorb RP-2 as the stationary
phase [457]. Figure 7.28 is a separation of some antihistamine and decongestant
drugs by reversed-phase ion-pair chromatography using a chemically bonded alkyl
stationary phase [458].

    The mechanism by which separation takes place in ion-pair chromatography is
not without controversy. Three models have been proposed to explain different
aspects of the separation process. Each model is in effect an extreme situation
or boundary condition, and actual separations probably involve a mixed model
mechanism or occur by a mechanism yet to be defined. In this section a
phenomological approach to the description of the various models will be
adopted. As a rigorous mathematical treatment is complex, only a few important
and practically useful equations will be presented. For further information,
detailed reviews of the ion-pair chromatographic process should be consulted
[458-464].

    The ion-pair model, the dynamic ion-exchange model, and the ion-interaction

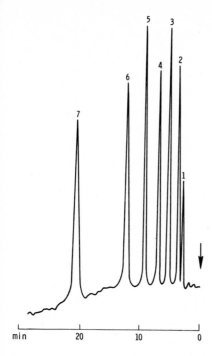

Figure 7.27 Separation of benzoic and sulfonic acids by reversed-phase
ion-pair chromatography. Components: 1, 4-hydroxybenzoic acid;
2, 3-aminobenzoic acid; 3, 4-hydroxybenzoic acid; 4, 3-hydroxy-
benzoic acid; 5, benzenesulfonic acid; 6, benzoic acid;
7, toluene-4-sulfonic acid. Mobile phase tetrabutylammonium,
0.03 M, pH 7.4; stationary phase 1-pentanol on LiChrosorb RP-2.
(Reproduced with permission from ref. 457. Copyright Elsevier
Publishing Co.)

model have been proposed to explain the ion-pair separation mechanism. The
ion-pair model (also known as the partition model) presumes that ion-pair
formation occurs in the mobile phase prior to the adsorption or partition of the
ion-pair complex into the stationary phase. This model most closely explains
the experimental results obtained with non-bonded reversed-phase columns (e.g.,
n-pentanol coated onto silica gel), in which the stationary phase behaves as a
bulk liquid. The ion-pair model is, however, unable to explain ion-pair
interactions with chemically bonded reversed-phase columns. The bonded alkyl
groups, with a film thickness on the order of a monolayer, have restricted
motion compared to a bulk liquid; a dynamic ion-exchange model is more
appropriate in this situation. The hypothesis presumes that the ion-pairing
reagent is adsorbed onto the stationary phase surface where it behaves as a
liquid ion-exchanger. Separation and retention are due to ionic interactions
between the ionized solute molecules and the counterions adsorptively bound to
the stationary phase. The ion-interaction model, does not require that ion-pair

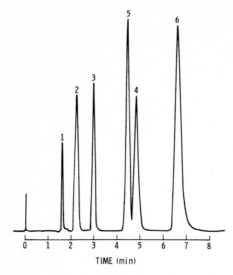

TIME (min)

Figure 7.28   Separation of antihistamine and decongestant drugs by reversed-phase ion-pair chromatography. Components: 1, maleic acid; 2, phenylephrine; 3, phenylpropanolamine; 4, naphazoline; 5, phenacetin; and 6, pyrilamine. Mobile phase methanol-water (1:1) containing 5 mM hexanesulfonate and 1% acetic acid, flow rate 3 ml/min. (Reproduced with permission from ref. 458. Copyright Preston Publications, Inc.)

formation occur in either phase nor is it based on classical ion-exchange chromatography. The ion-interaction model presumes that an electrical double layer is formed at the stationary phase surface as a result of the dynamic adsorption of the lipophilic ion-pairing reagent. Sample retention is assumed to be due to the coulombic attraction of the solute at the electrical double layer resulting from the surface charge density provided by the counterion, accompanied by an additional "sorption" effect from the attraction of the lipophilic portion of the solute for the non-polar stationary phase surface. The net result is a pair of ions (not necessarily an ion-pair) has been adsorbed onto the stationary phase. The partition or adsorption of the counterion to the stationary phase is a dynamic process and so can result in a chromatographic separation.

For convenience, the boundary conditions of the ion-pair model can be assumed to be the formation of an ion-pair AB from the interaction between a hypothetical monovalent cationic species, A^+, and a counterion, B^-. The ion-pair is formed initially in the aqueous phase and is extracted into the organic phase. An identical treatment can be applied to ion-pair formation for anionic solutes with cationic counterions. Under the above conditions, the following equilibria apply:

$$A_{aq}^{+} + B_{aq}^{-} \xrightleftharpoons{K_1} AB_{aq} \tag{7.17}$$

$$AB_{aq} \xrightleftharpoons{K_2} AB_{org} \tag{7.18}$$

$$AB_{org} \xrightleftharpoons{K_3} A_{org}^{+} + B_{org}^{-} \tag{7.19}$$

The experimental conditions in ion-pair chromatography are adjusted so that K_2 is large and K_3 is comparatively small. The extraction constant, $E_{AB(org)}$, for the equilibrium condition can be related to the counterion concentration $[B^-]$ by equation (7.20).

$$E_{AB(org)} = \frac{[AB]_{org}}{[A^+]_{aq} [B^-]_{aq}} \tag{7.20}$$

The bulk liquid distribution constant D_{AB} ($[AB]_{org}/[A^+]_{aq}$) can be substituted into equation (7.20) to give:

$$D_{AB} = E_{AB(org)} [B^-]_{aq} \tag{7.21}$$

To account for the influence of side reactions (e.g., protonation, aggregation, dissociation, etc.) on the ion-pair extraction process, the distribution ratio, D_{AB}, can be redefined in terms of the conditional equilibrium constant, E_{AB}^{*}, such that

$$D_{AB} = E_{AB}^{*} C_{B}^{-} \tag{7.22}$$

where $E_{AB}^{*} = \dfrac{C_{AB}}{C_{A}^{+} C_{B}^{-}}$

C_{AB} = total molar concentration of A^+ extracted into an organic phase as an ion-pair with B^-

C_{A}^{+} and C_{B}^{-} = total molar concentrations of A^+ and B^- in the aqueous phase

In normal phase partition chromatography, the capacity factor of solute A is given by

$$k = \frac{\theta}{E_{AB}^{*} [B^-]_{aq}} \tag{7.23}$$

while for reversed-phase systems

$$k = E_{AB}^{*} [B^-]_{aq} \theta \tag{7.24}$$

θ = phase volume ratio = V_s/V_m

The chromatographic retention time of a compound can be expressed conveniently in terms of the capacity factor, k, by

$$t_R = \frac{L}{u}(1 + k) \tag{7.25}$$

t_R = elution time of the retained substance

u = linear velocity of the eluent

L = colume length

The retention time for the normal phase and reversed-phase partition systems, respectively, can be related to the counterion concentration, $[B^-]$, by the expressions

$$t_R = \frac{L}{u}\left[1 + \frac{\theta}{E_{AB}^* \, [B^-]_{aq}}\right] \tag{7.26}$$

and

$$t_R = \frac{L}{u}(1 + \theta \, E_{AB}^* \, [B^-]_{aq}) \tag{7.27}$$

As mentioned previously, the ion-pair model is applicable to normal and reversed-phase columns coated with bulk liquids, but is not suitable for a description of chemically bonded reversed-phase separations. The dynamic ion-exchange model is more appropriate for explaining ion-pair retention when the thickness of the stationary phase is small. The ionic solute, A^+, is assumed to interact with a counterion, B^-, to form a complex AB which can bind to the nonpolar ligand L (the stationary phase film) to form a moiety LAB. It is assumed that this binding occurs reversibly and that the interaction is due to hydrophobic interactions and not electrostatic or hydrogen-bonding effects. Other equilibria that can occur include the binding of B^- initially to the staionary phase (thus generating a liquid ion-exchanger) and subsequently to A^+ to form the adduct LAB, or the free species A^+ and B^- may individually undergo hydrophobic binding to form complexes of the type LA^+ or LB^-. Thus, sample retention is a product of the equilibrium reactions (7.28) through (7.33).

$$[A^+]_{aq} + [L]_{org} \underset{\longleftarrow}{\overset{K_1}{\longrightarrow}} [LA^+]_{org} \tag{7.28}$$

$$[A^+]_{aq} + [B^-]_{aq} \underset{\longleftarrow}{\overset{K_2}{\longrightarrow}} [AB]_{aq} \tag{7.29}$$

$$[L]_{org} + [B^-]_{aq} \underset{\longleftarrow}{\overset{K_3}{\longrightarrow}} [LB^-]_{org} \tag{7.30}$$

$$[AB]_{aq} + [L]_{org} \underset{\longleftarrow}{\overset{K_4}{\longrightarrow}} [LAB]_{org} \tag{7.31}$$

$$[LB^-]_{org} + [A^+]_{aq} \underset{}{\overset{K_5}{\rightleftharpoons}} [LAB]_{org} \tag{7.32}$$

$$[LA^+]_{org} + [B^-]_{aq} \underset{}{\overset{K_6}{\rightleftharpoons}} [LAB]_{org} \tag{7.33}$$

The capacity factor for compound A^+ can be related to the counterion concentration, $[B^-]$, by the expression

$$k = \frac{(k_o + \beta[B^-])}{(1 + Y[B^-])} \tag{7.34}$$

k_o = capacity factor for A in the absence of B

β = product of the equilibrium constants $K_2 \cdot K_4$, $K_3 \cdot K_5$ or $K_1 \cdot K_6$ in equations (7.28) through (7.33)

Y = K_2 or K_3

The capacity modification factor, δ_{CMF}, can be used to express the extent to which the presence of B^- modulates the retention of solute A^+. It is defined as

$$\delta_{CMF} = \frac{k}{k_o} = \frac{1 + \beta/k_o[B^-]}{1 + Y[B^-]} \tag{7.35}$$

Based on the above equation, it is possible to define several boundary conditions for the ion-pair process and to relate them to the concentration of the counterion B^-. These conditions are summarized in Table 7.26. Small polar counterions such as $H_2PO_4^-$ and ClO_4^- do not tend to bind significantly to the hydrophobic stationary phase and ion-pair formation in the aqueous phase is favored (Table 7.26, $K_2 \gg K_4$; K_3 small). Reagents of intermediate polarity, such as the alkylsulfonic acids, favor the formation of ion pairs in the aqueous phase with strong binding of the ion pair to the stationary phase (Table 7.26, $K_4 \gg K_2$; K_3 small). The long chain alkylsulfonates are readily adsorbed onto the ligand surface by hydrophobic attraction and tend to behave as liquid ion-exchangers (Table 7.26, K_3 large).

There are many parameters which can influence retention in ion pair chromatography. Some of the more important of these are summarized in Table 7.27. At first sight, the long list might suggest that optimizing the separation system for a particular sample in ion pair chromatography is a difficult task. However, this is not the case in practice and a few trial experiments will yield the desired results for simple separations.

Important decisions that will influence the chance of success must be made before attempting a separation. The first decision is the choice of column type; a convenient starting point is to select a chemically bonded

TABLE 7.26

LIMITING CONDITIONS FOR THE RETENTION OF A$^-$, AS AN ION PAIR, AB, BY A CHEMICALLY
BONDED ALKYL STATIONARY PHASE

Boundary Condition	Remarks
$K_2 \gg K_4$; K_3 small	Ion-pair formation in the aqueous phase is favored. Ion-pair complex can bind less readily to the stationary phase than the free solute A$^-$ or its conjugate base, ($K_1 \gg K_4$). k will be less than k_o
$K_4 \gg K_2$; K_3 small	Ion-pair formation occurs in the mobile phase prior to the complex binding rapidly and more strongly to the stationary phase than the free solute. k will be greater than k_o
K_3 large	Modification of the stationary phase by the counterion is important with the stationary phase assuming the characteristics of a dynamic liquid ion-exchanger

TABLE 7.27

VARIABLE PARAMETERS FOR REVERSED-PHASE ION-PAIR CHROMATOGRAPHY

Variable	Effect
Type of counterion	Retention increases with the ability of the counterion to form an ion pair.
Size of counterion	An increase in the size of the counterion will increase retention.
Concentration of counterion	Increasing concentration increases retention up to a limit.
pH	Effect is dependent upon nature of the solute. Retention increases as pH maximizes concentration of ionic form of solute.
Type of organic modifier	Retention decreases with increasing lipophilic nature.
Concentration of organic modifier	Retention decreases with increasing concentration.
Temperature	Retention decreases as temperature increases.
Stationary phase	Retention increases with the lipophilic character of the stationary phase.

reversed-phase column, then to optimize the separation via changes in the mobile
phase composition. Chemically bonded reversed-phase columns are commercially
available, can be used without any further treatment, and are compatible with

gradient elution techniques. Normal phase and non-bonded reversed-phase columns must be prepared in the laboratory prior to use. As the stationary phase is not bonded to the support, some precautions will be required to prevent stripping of the stationary phase during the chromatographic run. In normal phase chromatography, it is difficult to optimize the counterion concentration or pH of the stationary phase, as a new column must be prepared for each experiment. For some applications, a major advantage of the normal phase technique is that it enables the use of counterions with a high detector response to enhance the detection capabilities of solutes which otherwise have poor responses.

To select a counterion for a particular separation, the most important consideration is charge compatibility. The counterion should ideally be univalent, aprotic, soluble in the mobile phase (reversed-phase methods), nondistructive to the chromatographic system, and should not undergo aggregation or other secondary equilibria. Some of the compounds generally used as counterions are summarized in Table 7.28. For the separation of anionic solutes such as carboxylic acids and sulfonates, the tetraalkylammonium salts in the chloride or phosphate forms are generally used. Protonated amines are another possibility, but these are generally less useful than the tetraalkylammonium salts. Inorganic anions, alkyl or aryl sulfonates, and alkyl sulfates are generally used for the separation of cationic solutes. Hydrophobic amino compounds are best separated with small hydrophilic counterions such as dihydrogenphosphate, bromide, or perchlorate. For cations of intermediate polarity, hydrophobic counterions such as naphthalene sulfonate, picrate, or bis-(2-ethylhexyl)phosphate are used. The reversed-phase separation of small or polar cations requires the use of lipophilic counterions such as alkyl

TABLE 7.28

TYPICAL COUNTERIONS

Type	Major Applications
Quarternary amines, e.g., tetramethyl, tetrabutyl, palmityltrimethylammonium ions	For strong and weak acids, sulfonated dyes, and carboxylic acids.
Tertiary amines, e.g., trioctylamine	Sulfonates.
Alkyl and aryl sulfonates, e.g., methane- or heptanesulfonate, camphorsulfonic acid	For strong and weak bases, benzalkonium salts, and cathecholamines.
Perchloric acid	Forms very strong ion pairs with a wide range of basic solutes.
Alkyl sulfates, lauryl sulfate	Similar to sulfonic acids, but yields different selectivity.

sulfonates, sulfamates, or sulfates. With these counterions, increasing the length of the alkyl chain leads to an increase in retention of the solute but not necessarily to an increase in the selectivity of the separation. When selectivity is the overriding goal of the separation, the counterion should be selected so as not to dominate the chromatographic behavior of the ion pair. Comprehensive bibliographies of ion-pair applications are available to assist in the selection of suitable counterions [461,462].

There is no general rule for predicting the optimum concentration of a counterion for a particular separation. For low counterion concentrations, solute retention usually increases fairly rapidly with increasing counterion concentration until a plateau region is reached, at which point retention changes slightly but unpredictably with further addition of counterion. For large counterions a mobile phase concentration of 0.005 M is fairly common. With small counterions much higher concentrations are generally used (e.g., 0.01 - 0.50 M). For any particular separation, the optimum concentration of the counterion is established by experiment.

In reversed-phase chromatography, retention and selectivity are also governed by the pH, ionic strength, and percentage of organic modifier in the mobile phase, as well as by the choice of counterion. The pH of the mobile phase is chosen so that both the sample and the counterion are completely ionized. The pH of the mobile phase is usually controlled by making the aqueous component a buffer solution. The composition and concentration of the buffer can thus influence the chromatographic properties of the solute. In general terms, pH can have a large influence on retention and separation by ion-pair chromatography. Some guidelines for selecting the optimum pH for a given sample are summarized in Table 7.29. For chemically bonded reversed-phase columns, the working range is between pH 2.0 to 8.0. Outside this range, physical damage to the column (e.g., corrosion, cleavage of the bonded phase, dissolution of silica, etc.) can occur.

The most convenient and useful method for controlling retention and selectivity in reversed-phase chromatography is by adjusting the concentration of the organic modifier in the mobile phase. This is also true for ion-pair chromatography. Generally, either methanol or acetonitrile is used as the organic modifier in such concentration that under gradient elution conditions, the least polar solvent used is able to completely dissolve the ion-pair reagent. The more common ion-pair reagents show much higher solubilities in methanol than acetonitrile.

With a little experience and practice, excellent results can be obtained by ion-pair chromatography for the separation of ionized molecules. Although it

TABLE 7.29

SELECTION OF pH

Type of Solute	Example	pH (for RP-IPC)	Comment
Strong acid $(pK_a < 2)$	Sulfonated dyes	2 - 7.4	These solutes are ionized through-out the pH range; actual pH selected is dependent upon other types of solutes present.
Weak acids $(pK_a > 2)$	Amino acids, Carboxylic acids	6 - 7.4	Solutes ionized; retention dependent upon the nature of the ion pair.
		2 - 5	Ionization of solutes is suppressed; retention dependent upon the nature of solute (not ion pair).
Strong bases $(pK_a > 8)$	Quarternary amines	2 - 8	Solutes are ionized throughout pH range; similar to strong acids.
Weak bases $(pK_a < 8)$	Catechol-amines	6 - 7.4	Ionization is suppressed; retention dependent upon the nature of the solute.
		2 - 5	Solutes are ionized; retention dependent upon the nature of the ion pair.

would seem that there are many parameters to optimize, for simple separations this is not a problem, as changes in just one or two of the parameters discussed above while maintaining the others within sensible ranges will yield the desired results. For the analysis of complex mixtures, the added degree of flexibility, is important for success compared to the rather restricted possibilities in ion-exchange chromatography. For routine work, much time can be saved by using TLC to scout for approximate conditions for the separation prior to transferring the method to HPLC [462].

7.15 Derivatives for the Chromatographic Separation of Inorganic Cations and Anions

Gas and, later, liquid chromatographic methods have had an enormous impact on the analysis of organic mixtures. Quite naturally this has led to investigations of chromatographic methods for the separation of inorganic compounds [465-467]. Some of these techniques, for example ion-exchange and ion chromatography, are now well established techniques for inorganic analysis and were described in detail in Chapter 4. Gas chromatography has been used to separate many inert, thermally stable organometallic compounds. Organometallic compounds of Li, Be, Mg, Zn, Hg, B, Ga, Al, Si, Ge, Sn, Pb, As, Sb, Cr, Mo, Se, Te, Mn, Fe, Co, Cu, Ni, Ru, containing alkyl, aryl, carbonyl, hydride, halide, nitrile, and alkoxide groups have been separated by employing the standard conditions used to analyze organic compounds. Certain metal hydrides and halides have sufficient volatility to be separated by gas chromatography, but only with difficulty, after certain instrument modifications have been made [468]. Reactive volatile inorganic compounds, for example SO_2Cl, UF_6, ClO_2F, SF_5Cl, etc., can be separated by perfluorocarbon stationary phases coated on Teflon supports [469,470]. These columns are used because of their resistance to chemical attack but they are generally very inefficient in the chromatographic sense (100–300 theoretical plates per meter). Porous polymer adsorbents have been used successfully to analyze some reactive compounds with reasonable column efficiencies.

Many metals, cations, and anions can be made sufficiently thermally and hydrolytically stable for separation by gas, liquid, or thin-layer chromatography by derivatization. It is this aspect of inorganic analysis that will be discussed here. Most of these methods provide the possibility of multi-element or multi-ion determinations. As far as elemental analysis is concerned, some overlap exists with spectroscopic techniques; for elemental analysis spectroscopic techniques are generally preferred. However, another important area of inorganic analysis is speciation. In this case chromatographic methods are required to separate the individual components prior to detection. Element-tunable or element-specific spectroscopic detectors are often used in these studies to provide a selective profile of the substances of interest.

The preparation of neutral chelates for the separation of metals by gas chromatography has been studied for many years [259,468,471–474]. The principal limit to the success of this approach has been the paucity of suitable reagents which can confer the necessary volatility, thermal stability, and chemical inertness on the metal ion. The class of metal chelating reagents which have been most studied are the β-diketonates, including their halo- and alkyl-substituted analogs, as well as the corresponding thio and amine forms. Enolization and subsequent ionization of β-diketones occurs as indicated below:

$$R-\underset{\underset{O}{\|}}{C}-\underset{2}{CH_2}-\underset{\underset{O}{\|}}{C}-R' \underset{H^+}{\overset{-H^+}{\rightleftharpoons}} R-\underset{\underset{O}{\|}}{C}-CH=\underset{\underset{O^-}{|}}{C}-R'$$

In many instances the complexing enolate anion forms neutral chelates with metals whose preferred co-ordination number is twice their oxidation state; the resultant complexes are effectively co-ordinatively saturated, thus precluding further adduction by solvent or other ligand species. Stable complexes are formed with ions such as Be(II), Al(III), and Cr(III), which conform to the above rule. Other ions such as Ni(II), Co(II), Fe(II), and La(III), which readily adduct additional neutral ligands and assume a co-ordination state greater than twice their oxidation state, are difficult to gas chromatograph. Hydrates which lower the volatility and increase the polarity of the derivative may be formed, resulting in undesirable column behavior, or non-solvated chelates may polymerize or may react on-column with active sites to give excessive peak broadening or even irreversible adsorption, particularly at low chelate concentrations. Numerous examples exist of metal chelate derivatives that cannot be quantitatively eluted at low levels by gas chromatography. Non-fluorinated chelates usually have only marginally suitable thermal and chromatographic stability, often requiring elution temperatures that are too high for thermal degradation to be completely absent. The most frequently employed complexing ligands are 1,1,1-trifluoropentane-2,4-dione (HTFA) and 1,1,1,5,5,5-hexafluoropentane-2,4-dione (HHFA). Only in the case of the HTFA derivatives of beryllium, aluminum, and chromium have trace level concentrations of the elements in environmental samples been determined [262,475-477]. Other ligands studied include the β-thioketones, salicylaldimines, dialkyldithiophosphates, and dialkylthiocarbamates [467,473].

Liquid chromatographic methods are well suited to the separation and determination of metal chelates that can be extracted into organic solvents. Many chelates also absorb strongly in the UV or visible regions, facilitating detection. The most widely used derivatives are the acetylacetonates, Schiff base chelates, hydrazones, dithizonates, and metal dithiocarbamates [478-485]. It might be expected that many chelates would be more stable in HPLC than GC but generally this has not been the case. Many complexes are kinetically labile and decompose on the column. For example, the acetylacetonate chelates of Mn(II), Be(II), Co(III), Cr(III), Rh(III), Ir(III), Pd(I), and Pt(II) are stable to reversed-phase chromatography while the derivatives of Fe(III), Co(II), Zn(II), and Pb(II) are degraded on the column [479]. Ligand-substitution reactions may occur between the metal parts of the column and the metal complexes being chromatographed [480,486,487]. This is most pronounced for the diethyldithiocarbamate derivatives of Bi(III), Zn(II), Cd(II), and Pb(II), since these complexes are relatively unstable and may readily undergo exchange with nickel from the stainless steel column components. To minimize these interactions PTFE or glass-lined tubing and plastic radial compression columns have been used with partial success. Conditioning the column with a concentrated solution of chelates or incorporating disodium ethylenediaminetetraacetic acid into the mobile phase may also be required [480].

The facile alkylation of mercury in the environment coupled with the high sensitivity of alkylmercurial compounds to electron-capture detection have prompted studies on the use of alkylating reagents as a means of derivatizing inorganic mercury for gas chromatographic analysis. Among the more commonly used alkylating reagents are the arylsulfinates [488,489], $[Co(III)(CN)_5CH_3]^{3-}$ [490], sodium tetraphenylborate [490,491], and tetramethyltin [492].

$$C_6H_5SO_2H + HgCl_2 \xrightarrow{HCl} C_6H_5HgCl + SO_2 + HCl$$

$$[Co(III)(CN)_5R]^{3-} + Hg^{2+} \longrightarrow RHg^+$$

$$(CH_3)_4Sn + Hg^{2+} \xrightarrow{2Cl^-} CH_3HgCl + (CH_3)_3SnCl$$

$$NaB(C_6H_5)_4 + 2HgCl_2 + 3H_2O \longrightarrow 2Hg(C_6H_5)_2 + 3HCl + NaCl + B(OH)_3$$

The alkyl and aryl mercurials are thermally stable with good gas chromatographic properties. The electron-capture detector responds to ppb concentrations of the derivatives. Alkylation has also been used for the simultaneous determination of trace quantities of bromide and chloride by gas chromatography with electron-capture detection [493]. The anions were derivatized with dimethyl sulfate.

Selenium(IV) reacts with substituted 1,2-diaminobenzene or 2,3-diaminonaphthalene in acidic solution to form stable cyclic derivatives which can be extracted into an organic solvent and analyzed by gas or liquid chromatography [494-497].

With chloro, bromo, or nitro substituents the piazselenols can be determined by an electron-capture detector at the low picogram level. Se(VI) does not form piazselenol derivatives so the reaction with diaminobenzene can be used to determine the concentration of Se(IV) and Se(VI). Selenium(VI) may be reduced to Se(IV) with boiling hydrochloric acid, providing a value for total Se and a value for Se(VI) by difference.

The most general method for the simultaneous analysis of oxyanions by gas chromatography is the formation of trimethylsilyl derivatives [259,498,499].

$$AsO_3^{3-} \longrightarrow (TMS)_3AsO_3$$
$$BO_3^{3-} \longrightarrow (TMS)_3BO_3$$

$$CO_3^{2-} \longrightarrow (TMS)_2CO_3$$

$$SO_4^{2-} \longrightarrow (TMS)_2SO_4$$

Trimethylsilyl derivatives of silicate, carbonate, oxalate, borate, phosphite, phosphate, orthophosphate, arsenite, arsenate, sulfate, and vanadate, usually as their ammonium salts, are readily prepared by reaction with BSTFA-TMCS (99:1). Fluoride can be derivatized in aqueous solution and extracted by trimethylchlorosilane into an immiscible organic solvent [500].

$$(C_2H_5)_3SiOH + F^- \longrightarrow (C_2H_5)_3SiF$$

A number of methods are available for the derivatization of cyanide and thiocyanate [501-503].

$$CN^- + Chloramine\text{-}T \longrightarrow ClCN + products$$

$$CN^- + Br_2 \longrightarrow BrCN + Br^-$$

$$SCN^- + 4Br_2 + 4H_2O \longrightarrow BrCN + SO_4^{2-} + 7Br^- + 8H^+$$

The cyanogen halides can be detected at trace levels using headspace gas chromatography with electron-capture detection. Thiocyanate can be quantitatively determined in the presence of cyanide by the addition of an excess of formaldehyde which converts the cyanide ion to the unreactive cyanohydrin.

Iodine, in acid solution, reacts with acetone to form monoiodoacetone, which can be detected at low concentrations by electron-capture detection [504,505]. Strong acid conditions are required to inhibit the formation of the diiodoacetone derivative. The reaction is specific for iodine. Iodide can be determined after oxidation with potassium iodate. The other halide anions are not oxidized by iodate to the reactive molecular form and thus do not interfere in the derivatization reaction.

The nitrate anion can be determined in aqueous solution by reaction with a mixture of benzene and sulfuric acid [506-509]. Nitrobenzene is formed in 90 ± 8% yield for nitrate concentrations in the range 0.12 - 62 ppm. The nitrite anion can be determined after oxidation to nitrate with potassium permanganate.

7.16 Visualization Techniques for Thin-Layer Chromatography

Qualitative and quantitative measurements in high performance thin-layer chromatography (HPTLC) are made by in situ scanning densitometry (Chapter 9.10). Instrumental measurements provide an accurate record of the separation in the form of a chromatogram. As many organic compounds absorb UV-visible radiation or are fluorescent, they are readily determined without resorting to derivatization. Visualization techniques are used in TLC for two principal reasons: to enable colorless compounds to be detected by eye and to increase the selectivity of the detection process by reaction of the separated compounds with a reagent having chemical specificity for a particular functional group or compound class. Visualization techniques are widely used in conventional TLC where the separation efficiency is much lower than HPTLC.

Several hundred visualization reagents have been described for use in TLC [510-512]. They can be divided into several different types. Reagents are classified as reversible or destructive, depending on the type of interaction they undergo with the separated compounds, and as selective or universal, based on the specificity of the reaction. Most methods involve spraying the plate or dipping the plate into a chemical reagent followed by a conditioning period, for example heating in an oven, during which time a color is developed at positions where sample components are present. The most common nondestructive methods employ iodine vapor, water, fluorescein, bromine, or pH indicators as the visualizing reagents [513]. In the iodine vapor method the dried plate is enclosed in a chamber containing a few crystals of iodine; components on the chromatogram are stained more rapidly than the background and appear as yellow-brown spots on a light yellow background. As little as 0.1 to 0.01 micrograms of sample can be visualized in this way and the reaction can be reversed by simply removing the plate from the visualization chamber and allowing the iodine to evaporate. Compounds containing unsaturated hydrocarbon or phenolic groups may irreversibly react with iodine. Mercaptans are oxidized to disulfides. Spraying a TLC plate with water reveals hydrophobic compounds as white spots on a translucent background when the water-moistened plate is held against the light. A TLC plate sprayed with 0.05% aqueous fluorescein, followed by exposure to bromine vapor while still damp, becomes uniformly colored (due to formation of eosin) except at positions where separated components are located. Solutions of pH indicators, for example bromocresol green and bromophenol blue, are widely used for the detection of acidic and basic compounds. The above methods are all fairly universal and reversible so that they may be used for detection when sample recovery for further studies is required.

Destructive spray reagents may be used when sample recovery is not important. Again, numerous reagents are known [510,511] but among the more generally used are solutions of sulfuric acid, sulfuric acid-acetic anhydride (1:4), and sulfuric acid-sodium dichromate. After spraying, the plate is usually heated in an oven for some time at 110-115°C. Organic compounds are converted to carbon and visualized as black spots on a white-grey background. Alternatively, ammonium bisulfite can be incorporated into the adsorbent layer and, after the development process, sulfuric acid is generated in situ by heating the plate in an oven. When the reaction conditions are carefully controlled, charring reactions can provide reliable quantitative data [514]. Heating a sample on a TLC plate to a temperature of 100–150°C for 1–12 h in the presence of ammonium hydrogen carbonate vapors induces fluorescence without charring in most organic compounds [515]. Regardless of the structure of the compound treated, the in situ fluorescence characteristics of the corresponding derivatives are very similar, all showing excitation maxima at approximately 380 nm and emission maxima at 455 and/or 475 nm. The fluorescent derivatives obtained are very stable for several weeks and are not significantly influenced by the presence of atmospheric oxygen or moisture. Fluorescence may also be induced by heating in the presence of acid vapors (e.g., HCl, HBr, HNO_3, $HClO_4$), silicon tetrachloride, ammonium hydrogen sulfate, and zirconyl salts [516,517]. Alternatively, Shanfield has recommended the use of a gaseous electrical discharge for fluorescence induction [518,519]. The discharge treatment produces a higher yield of fluorescent material, is more reproducible, and requires an exposure time of only several minutes compared to several hours needed by the thermal method.

It has been observed that many fluorescent compounds adsorbed onto silica gel produce a diminished fluorescence response which may be accompanied, in some cases, by a shift in the emission maximum to longer wavelengths. Impregnating the developed plate by dipping or spraying with a viscous liquid such as liquid paraffin [520,521], triethanolamine [520,522], Triton X-100 [520,523-525], or Fomblin Y-Vac [525] prior to measurement will enhance the fluorescence signal sufficiently for trace level detection. Fluorescence enhancement values from 10- to 200-fold have been observed in favorable cases. It is assumed that adsorption onto silica gel provides additional nonradiative pathways for loss of the fluorescent excitation energy; these pathways are relieved by transfer of the adsorbed solute to the liquid state when the plate is impregnated with a nonvolatile liquid. However, once in solution spot broadening due to diffusion becomes important. The poly(perfluoroalkyl ether) Fomblin Y-Vac was found to be particularly effective in minimizing spot broadening in the detection of polycyclic aromatic hydrocarbons [525].

Numerous reagents are available which form colored products with either specific functional groups or compounds distinguished by class [510-512]. They are useful for identifying specific compounds in complex mixtures. Those reactions which are sufficiently specific may be used to qualitatively identify compounds which are incompletely resolved on the plate. Quantitative analysis may also be possible if the product can be formed reproducibly and adsorbs or fluoresces at a wavelength where interferences are absent.

7.17 Separation of Stereoisomers and Diastereoisomers

Stereoisomers have identical molecular formulas but differ from one another in that their atoms are arranged differently in space. Such molecular arrangements can give rise to a center or plane of asymmetry with the ability to rotate a beam of polarized light. Two stereoisomers related as non-superimposable mirror images are called enantiomers. They possess identical physical properties except that they rotate polarized light in opposite directions.

Chromatographic interest in enantiomeric resolution is, however, not due to their unique interaction with plane polarized light. The spatial dissimilarity which effects the way a beam of polarized light is rotated is also predominant in the way biological molecules interact in living systems. In most cases optical enantiomers exhibit different biological activities due to differences in their strength of interaction with the corresponding receptor, have different transport mechanisms, and may be metabolized by different routes. Thus, chiral recognition is an inherent feature of the biological mechanism by which enzymes, chemical agents, and drugs perform their prescribed functions. For this reason a knowledge of the enantiomeric purity, compositions and type of natural products and xenobiotics is of considerable importance in biology and medicine.

There are two approaches to the separation of enantiomers by chromatographic means. The direct route employs a chiral phase to distinguish the spatial difference between enantiomeric forms. The indirect route involves the synthesis of diastereomeric derivatives using an optically pure derivatizing reagent and subsequent separation in a nonchiral chromatographic system. The advantages and disadvantages of each approach will be discussed below.

7.17.1 Separation of Enantiomers as Their Diastereomeric Derivatives

Enantiomeric mixtures containing a reactive functional group can be derivatized with an optically pure reagent to produce a mixture of diastereomers. Unlike enantiomers, diastereomers have different physical properties and can be separated by a wide range of techniques including fractional crystallization, selective precipitation, distillation, solvent partition, and chromatographic methods. Simply forming diastereomeric derivatives does not in itself guarantee resolution, it merely makes resolution possible in a nonchiral separation system. The magnitude of the physical difference between diastereomers and the selectivity and efficiency of the chromatographic system all influence the extent of resolution. Often the resolution of diastereomeric derivatives is enhanced when the chiral centers of the enantiomer and the reagent are in close proximity in the derivative and when the reagent is comprised of conformationally immobile groups or contains bulky groups attached directly to the chiral center. Two potential sources of error in establishing the enantiomeric composition of a mixture are that the reagent may cause a degree of racemization during reaction or may react with the enantiomers at different rates because they exhibit energetically different diastereomeric transition states. However, the prinicpal limitation of the method is the requirement that the derivatizing reagent be readily available in an optically pure form. If the reaction is to be used for preparative purposes, ready reconversion of the diastereoisomers into their respective enantiomeric moieties is also a necessary condition. Some representative derivatizing reagents for enantiomeric resolution by gas chromatography are given in Table 7.30. These reagents must form derivatives that are thermally stable and free

TABLE 7.30

TYPICAL CHIRAL DERIVATIZING REAGENTS FOR GAS CHROMATOGRAPHIC ENANTIOMERIC RESOLUTION

Reagent	Functional Group Reacted	Reference
N-Trifluoroacetyl-prolyl chloride	Amines	526,527
α-Chloroisovaleryl chloride	Amines	528
α-Methoxy-α-(trifluoromethyl)-phenylacetyl chloride	Amines	530
Menthyl chloroformate	Amines	530
	Alcohols	531
Teresantalinyl chloride	Amines	532
Drimanoyl chloride	Amines	533
	Alcohols	533
trans-Chrysanthemoyl chloride	Amines	533
	Alcohols	533
α-Phenylbutyric anhydride	Amines	534
3β-Acetoxy-Δ^5-etienic acid chloride	Alcohols	535
1-Phenylisocyanate	Alcohols	536
2-Phenylpropionyl chloride	Alcohols	537
2-Butanol	Carboxylic acids	538
3-Methyl-2-butanol	Carboxylic acids	539
Menthyl alcohol	Carboxylic acids	540
2-Octanol	Carboxylic acids	541
2,2,2-Trifluoro-1-pentylethylhydrazine	Ketones	542

from racemization during the chromatographic process. Although these derivatives may be separated on nonchiral stationary phases, preferably using capillary columns, the resolution of diastereoisomers is often greater on a chiral stationary phase [528].

A selection of reagents for enantiomeric resolution by HPLC is shown in Figure 7.29. Liquid-solid chromatography is often used for separating diastereoisomers on account of its unique lock-key separation mechanism, which is particularly useful for separating molecules of different shape. Many of the reagents employed for enantiomeric resolution by HPLC also contain a chromophore to improve the detection characteristics of the derivatives. Reagents containing a 4-dimethylamino-1-naphthyl chromophore can be used with fluorescence detection for trace analysis [543]. Without the

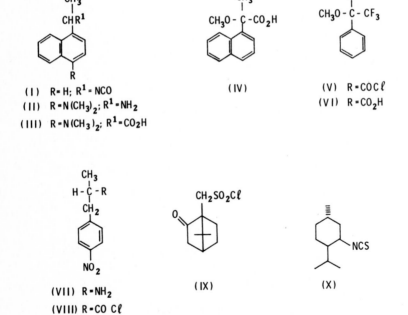

(I)   R=H; R^1=NCO
(II) R=N(CH$_3$)$_2$; R^1=NH$_2$
(III) R=N(CH$_3$)$_2$; R^1=CO$_2$H

(IV)

(V) R=COCℓ
(VI) R=CO$_2$H

(VII) R=NH$_2$
(VIII) R=CO Cℓ

(IX)

(X)

Figure 7.29   Representative chiral derivatizing reagents for enantiomer separation by HPLC. (I), a-naphthalylethylisocyanate; (II), L-1-(4-dimethylamino-1-naphthyl)ethylamine; (III), L-1-(4-dimethylamino-1-naphthyl)acetic acid; (IV), a-methoxy-a-methyl-1-naphthaleneacetic acid; (V), (+)-a-methoxy-a-trifluoromethylphenylacetyl chloride; (VI), carboxylic acid form of (V); (VII), (R)-(+)-a-methoxy-p-nitrobenzylamine; (VIII), acid chloride form of (VII); (IX), (+)-camphor-10-sulfonyl chloride; (X), (+)-neomenthyl isothiocyanate. Where not explicitly stated the reagent must be in a single enatiomeric form to be of use as a resolving agent.

4-dimethylamino-substituent the fluorescence properties are poor and the
naphthyl-derivatives are usually determined by UV detection [544,545]. The
reactive functional group of the reagent determines its area of application.
For example, α-methoxy-α-methyl-1-naphthaleneacetic acid was used to derivatize
the amino functionality [545] and 1-aminoethyl-4-dimethylaminonaphthalene the
carboxylic acid functionality [546] of racemic amino acids.
N-trifluoroacetyl-1-prolyl chloride [546], 2,3,4,6-tetra-O-acetyl-β
-glucopyranosyl isothiocyanate [547], 10-camphorsulfonyl chloride [548], and
1,7-dimethyl-7-norbornyl isothiocyanate and neomethyl isothiocyanate [549] have
been used to derivatize primary and secondary amines in amino acids and
pharmaceutical products. α-Methyl-p-nitrobenzylamine was used to resolve
racemic carboxylic acids [550], while α-methoxy-α-trifluoromethylacetyl chloride
[551] and α-naphthylethyl isocyanate [552] are used to separate racemic
alcohols.

Practical details of the preparation of optically pure reagents and the
conditions for the formation and separation of diastereomeric derivatives are
reviewed in references [259,260,553,554].

7.17.2 Chiral Stationary Phases for Enantiomer Resolution

The direct separation of enantiomers without prior conversion to
diastereoisomers would realize several advantages. It could be applied to those
enantiomers which do not contain reactive functional groups and could not be
limited by the need for reagents of high optical purity. Sample manipulation,
preparation, and separation would also be simplified. However, just as there is
no universal resolving reagent for diastereoisomer separations there is no
single chiral stationary phase for the resolution of all enantiomer mixtures.

The separation of enantiomers using a chiral stationary phase in gas
chromatography occurs because interactions between the enantiomers and the
stationary phase results in the formation of diastereomeric association
complexes having different solvation enthalpies and hence different retention
characteristics [555]. A number of chiral phases, including amides
($RCONHCHR_1R_2$), diamides ($RCONHCHR_1CONHR_2$), dipeptides (N-trifluoroacetylamino
acid-amino acid-R), and polysiloxane polymers with chiral substituents have been
described in the literature [556,557]. Some examples are shown in Figure
7.30. The chiral phase in Figure 7.30 (I) was used to separate the enantiomeric
N-trifluoroacetylamines, N-trifluoroacetylamino acid esters, and
α-alkylcarboxylic acid amides [558]. N-(1R,3R)-trans-chrysanthemoyl-
(R)-1-(α-naphthyl)ethylamine, which contains two chiral centers, was used to
successfully resolve racemic pyrethroid insecticides [559]. The diamide phases,
of which Figure 7.30 (II) is an example, show high stereoselectivity for the

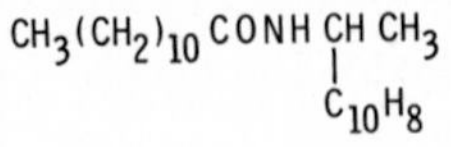

$$CH_3(CH_2)_{10}\ CONH\ CH\ CH_3$$
$$\underset{C_{10}H_8}{|}$$

(I)

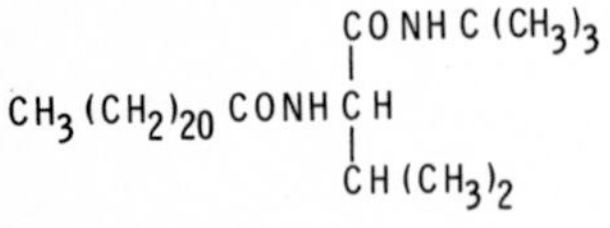

$$CH_3(CH_2)_{20}\ CONH\ \overset{CO\ NH\ C(CH_3)_3}{\underset{CH(CH_3)_2}{\overset{|}{C}\ H}}$$

(II)

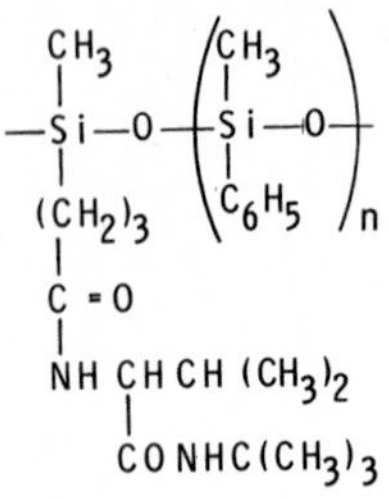

$$-\overset{CH_3}{\underset{(CH_2)_3}{\overset{|}{Si}}}-O-\left(\overset{CH_3}{\underset{C_6H_5}{\overset{|}{Si}}}-O-\right)_n$$
$$\underset{C = O}{|}$$
$$\underset{NH\ CH\ CH\ (CH_3)_2}{|}$$
$$\underset{CO\ NHC(CH_3)_3}{|}$$

(IV)

$$\overset{(CH_3)_2\ CH}{\underset{F_3C\ CO\ NH\ CH\ CO\ NH\ CH\ COO\ C_6H_{11}}{\quad |\qquad\qquad |}}\ \overset{CH(CH_3)_2}{\quad}$$

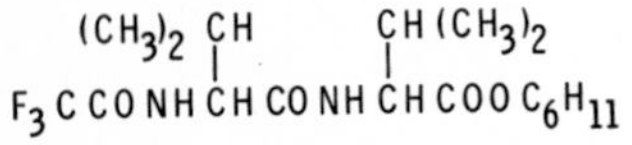

(III)

Figure 7.30   Some representative chiral stationary phases for gas chromatography. (I), N-lauroyl-S-α-(1-naphthyl)ethylamine; (II), N-docosanoyl-L-valine-t-butylamide; (III), N-trifluoroacetyl-L-valyl-valine cyclohexyl ester; (IV), polymeric chiral phase prepared from polycyanopropylmethyl-phenylmethyl silicone and L-valine-t-butylamide.

separation of enantiomeric amides derived from amino acids, amines, amino alcohols, α-hydroxy acids, and esters of aromatic diols [559-562]. The most thermally stable of these phases, N-docosanoyl-L-valine-2-(2-methyl)-n-heptadecylamide, has an upper temperature limit of 200°C. Other analogs have lower operating temperature limits which can be a problem for the elution of the least volatile of the N-trifluoroacetyl-α-amino acid methyl and isopropyl esters. Many dipeptide stationary phases, of which Figure 7.30 (III) is an example, have been evaluated for the separation of enantiomeric amino acid derivatives [563-567]. Phases providing the highest resolution of enantiomeric pairs contain a trifluoroacetyl group, bulky side groups and similarly bulky ester groups, and are operated at the lowest feasible temperature. Problems in the practice of enantiomeric resolution on these phases include poor resolution of some derivatives, particularly those of proline and aspartic acid, the overlap of D- and L-enantiomers of different amino acids, and long retention times for the least volatile amino acid derivatives due to the upper temperature operating limit (different phases are stable to 110 - 170°C). Polymeric chiral phases, for example Figure 7.30 (IV), have been synthesized to provide chiral phases with good chromatographic performance and higher operating temperature limits. All phases are prepared from a polycyanopropylmethyldimethylsiloxane [568-570] or polycyanopropylmethylphenylmethylsiloxane [571,572] by hydrolysis

of the cyano group to the carboxylic acid, which is then coupled with
L-valine-t-butylamide or L-valine-S-α-phenylethylamide [573,574] to introduce
the chiral center. Separation of each chiral center by several dimethylsiloxane
or methylphenylsiloxane units (> 7) seems to be essential for good enantiomeric
resolving power and thermal stability. The upper temperature limit of these
phases is about 230°C; at higher temperatures racemization of the chiral centers
becomes significant. Nearly all the enantiomeric protein amino acid
N-pentafluoropropionylamide isopropyl ester derivatives can be separated in a
single analysis [568,571]. Racemic mixtures of α-amino alcohols, glycols,
aromatic and aliphatic α-hydroxycarboxylic acids, and amines can also be
separated on these phases [568-572]. Racemic hydroxyacids and carbohydrate
derivatives were separated on the L-valine-S-α-phenylethylamide-containing phase
[573]. The commerically available chiral polymeric phases discussed above are
the most widely used chiral phases for gas chromatography. The general approach
used for their synthesis should allow the interested researcher to synthesize
"tailor-made" phases to suit particular requirements.

For HPLC separations, chiral stationary phases can be prepared simply by
impregnating an adsorbent such as silica gel with a chiral reagent. The solvent
for development is chosen such that the chiral reagent is not eluted from the
column. For example, the separation of the carbohelicenes shown in Figure 7.31
was obtained using a silica gel column impregnated with 25% w/w
R(-)-2-(2,4,5,7-tetranitro-9-fluorenylideneaminooxy)propionic acid (TAPA)
[575]. TAPA is a well known resolving agent for optically active aromatic and
unsaturated hydrocarbon compounds capable of interacting with the former through
charge transfer complexation. Since these charge transfer complexes are really
diastereomeric complexes, they have different equilibrium constants for their
formation and dissociation on the silica gel support. The need for the
resolving agent to be completely insoluble in the mobile phase is an unnecessary
limitation on the selection of mobile phase for the separation. Permanently
bonded phases offer greater versatility and can be conveniently prepared by
condensation reactions employing the commercially available 3-aminopropylsilica
phase as substrate. Gil-Av has shown how TAPA can be chemically bonded to this
phase for the resolution of carbohelicenes and aromatic diols [576]. In a
related application, Lochmuller [577] prepared a bonded phase by condensation of
L-alanine with 3-aminopropylsilica, followed by reaction with
1-fluoro-2,4-dinitrobenzene to introduce a charge transfer group. This phase
was used to separate aza-helicene enantiomers. The structures of some
representative bonded chiral stationary phases, all prepared from
3-aminopropylsilica, are shown in Figure 7.32 [575,578-581]. The chiral phase,
Figure 7.32 (II), can be conveniently prepared by passing a solution of
(R)-N-(3,5-dinitrobenzoyl)phenylglycine through a prepacked column of

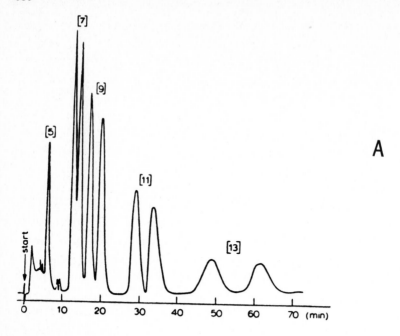

A

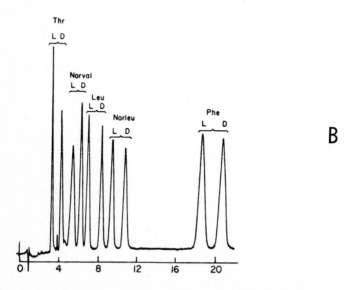

B

Figure 7.31 HPLC separation of enantiomeric mixtures using a chiral phase.
A, separation of a mixture of racemic carbohelicenes using silica
gel impregnated with 25% R-(-)-TAPA as stationary phase and
dichloromethane-cyclohexane (1:3) as mobile phase. B, separation
of D,L-dansyl amino acids by reversed-phase chromatography using
0.65 mM L-2-isopropyl-dien-Zn(II) and 0.17 M NH$_4$Ac, pH 9.0, in
acetonitrile-water (35:65) as the mobile phase.[4]

3-aminopropyl bonded silica [578]. This ionically-bonded phase is suitable for use with relatively nonpolar solvents. The chiral phase shown in Figure 7.32 (I) is probably the most widely investigated phase and has wide applicability, particularly for samples containing π-acid substituents [579,581]. Solutes lacking π-acid substituents can often be derivatized to incorporate such a functionality. For example amines, alcohols, and thiols can be derivatized with 2,4-dinitrofluorobenzene or 3,5-dinitrobenzoyl chloride to provide derivatives with better separation characteristics than the parent solute. The N-n-valeryl-L-valyl-aminopropylsilica phase, Figure 7.32 (III), was used to separate racemic N-acetyl-α-amino acid methyl esters [580]. The mechanism of chiral recognition was assumed to involve selective complexation between the phase and the amino acid derivative amide group, and to some extent the steric effect of the chiral alkyl group in the localized solute molecule.

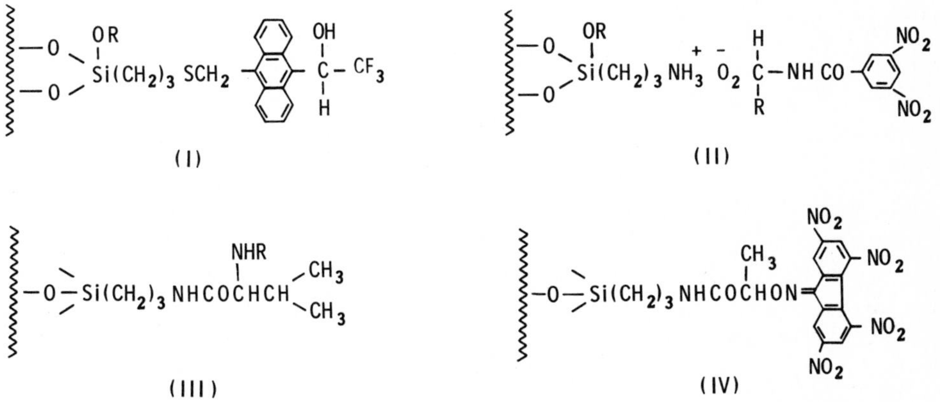

(I)   (II)

(III)   (IV)

Figure 7.32   Representative bonded chiral phases for HPLC. (I), Pirkle phase; (II), N-(3,5-dinitrobenzoyl)phenylglycine ionically bonded to 3-aminopropylsilanized silica; (III), N-n-valeryl-L-valyl-3-aminopropylsilanized silica; (IV), R-(-)-2-(2,4,5,7-tetranitro-9-fluorenylideneaminooxy)pripionamidepropyl silanized silica.

A general model, known as the three-point model, has been used to enumerate the requirements of the stationary phase for the separation of enantiomers [578-582]. The principal tenet of this model is that a minimum of three simultaneous interactions are needed to distinguish the handedness of a chiral object. In molecular terms, at least three significant interactions between one of the enantiomers and the chiral stationary phase are required, and at least one of these interactions must be stereochemically dependent. Lochmuller has suggested that "environmental chirality" is the principal requirement for the separation of enantiomers and direct "attachment" is not required [577].

7.17.3 Chiral Mobile Phases for Enantiomer Resolution

Chiral stationary phases are difficult to prepare reproducibly, are
sometimes of lower chromatographic efficiency than expected, and, in practice,
optimizing separation conditions is restricted by the fixed nature of the chiral
centers. Chiral mobile phases are free from many of these problems,
optimization of the separation is more convenient, and conventional
reversed-phase columns may be used, Table 7.31. Gil-Av devised an ion-exchange
method for the separation of D- and L-amino acids using optically active
copper(II)-proline complexes as the chiral mobile phase additive [583]. When an
L-proline ligand was used, the L-enantiomer eluted ahead of the D-enantiomer and
vice versa when D-proline was used. With a racemic mixture of the proline
enantiomers separation was not obtained. The stereoselectivity was ascribed to
differences in the stability constants of diastereomeric species formed by the
L-proline-copper(II) complex with an L- or D-amino acid. The racemic
dansyl-amino acid derivatives can be separated by reversed-phase chromatography
using L-2-R-4-octyldiethylenetriamine-M(II) (R = ethyl, isopropyl, isobutyl; M =
Zn or Cd) [584], proline and arginine-Cu(II) complexes [585], or

TABLE 7.31

CHIRAL MOBILE PHASE ADDITIVES

Mobile Phase Additive	Samples Separated	Chromatographic System	Reference
1) Metal Chelates			
Proline-Cu(II)	Amino acids	ion-exchange	583
L-2-R-4-octyldiethylene-triamine-M(II) R = ethyl, isopropyl, isobutyl M = Zn or Cd	Dansyl amino acids	reversed-phase	584
Proline-Cu(II) Arginine-Cu(II)	Dansyl amino acids	reversed-phase	585
L-Prolyl-n-octylamide-Ni(II)	Dansyl amino acids	reversed-phase	586
L-Aspartylcyclohexyl-amide-Cu(II)	Amino acids	reversed-phase	587
2) Non-metal chelates			
N-(2,4-Dinitrophenyl)-L-alanine-n-dodecyl ester	1-Azahexahelicenes	reversed-phase	588
(+)-10-Camphorsulfonate	Amines	ion-pair	589

L-prolyl-n-octylamide-Ni(II) [586] as the chiral component of the mobile phase.
A typical separation is shown in Figure 7.31B [584]. Underivatized racemic
amino acids were separated using reversed-phase chromatography with
L-aspartylcyclohexylamide-copper(II) as the chiral component of the mobile phase
[587]. The resolved amino acid–copper complexes could be detected at 230 nm,
thus eliminating the need for a postcolumn reaction system or for derivative
formation. The uncomplexed non-aromatic amino acids have very poor UV-detection
characteristics.

Above an upper concentration maximum (about 20 mM) the separation of the
racemic dansyl amino acids was independent of concentration of the chiral
complex but could be further optimized by varying the pH buffer concentration
and the amount of organic modifier in the mobile phase. The maximum separation
was generally found to lie at fairly high pH values, resulting in column
degradation with continuous use. Evidence has also been presented which
indicates that the chiral metal chelate is adsorbed onto the column packing in a
manner somewhat similar to ion-pairing reagents [586]. Thus the actual
separation mechanism might involve the adsorption of the chiral metal complex to
form a chiral bonded phase. The separation might then involve the interaction
between the chiral metal complex, immobilized on the stationary phase, and the
exchangeable amino acid ligands in the mobile phase.

Two alternatives to metal-complex mobile phase additives have also been
investigated. Lochmuller used N-(2,4-dinitrophenyl)-L-alanine-n-dodecyl ester
as a non-ionic chiral mobile phase additive for the resolution of
1-azahexahelicenes by reversed-phase chromatography [588]. The resolution
obtained was found to be a function of the mobile phase polarity and the
concentration of chiral additive used. Schill was able to separate enantiomeric
amines by ion-pair formation with a chiral counter ion, (+)-10-camphorsulfonate
[589]. Diastereomeric ion-pairs have structural differences resulting in
different distribution coefficients between the mobile and stationary phases.
Differences cannot be expected, however, if the ion-pair components are bound
only by electrostatic attraction between charged groups. For chiral recognition
a minimum of three interactions is needed to yield diastereomeric ion-pairs with
different mobile phase solvation or stationary phase adsorption. Ion-pair
formation of diastereomeric molecular complexes was also used to separate
racemic alkaloids [590].

7.18 Complexation Chromatography

The rapid and reversible formation of complexes between some metal ions and
organic compounds that can function as electron donors can be used to adjust
retention and selectivity in gas and liquid chromatography. Such coordinative

interactions are very sensitive to subtle differences in the composition or stereochemistry of the donor ligand, owing to the sensitivity of the chemical bond towards electronic, steric, and strain effects. Difficult to separate mixtures of constitutional, configurational, and isotopic isomers may be separated by complexation chromatography. Perhaps the most widely known example is the use of silver ions to complex organic compounds containing π-electrons in various kinds of double and triple bonds, and heteroatoms such as N, O, and S with lone pairs of electrons [591]. The selectivity of the silver nitrate-containing phases results from the marked effect that relatively small structural or electronic changes in the donor ligand have on the stability constants of the complexes. Some of these trends are summarized in Table 7.32. A limitation of silver ion complexation chromatography is the low upper temperature limit of the columns, 65°C or 40°C, as claimed by different authors [591]. Schurig has investigated the use of metal camphorate complexes for both isomeric and enantiomeric separations. Using either dicarbonyl rhodium-trifluoroacetyl-d-camphorate [591] or the dimeric 3-trifluoroacetyl- or 3-heptafluorobutyryl-(1R)-camphorates of Mn(II), Co(II), Ni(II), or Cu(II) dissolved in the noncoordinating stationary phase squalane, both analytical separations and thermodynamic constants can be obtained for σ-donor ligands [592-594]. As an example of the resolving power of these chiral complexing agents, consider the separation of racemic 2,2-dimethylchloroaziridine (Figure 7.33A) on a capillary column containing 0.113 M nickel(II)-bis-3-heptafluoro-butyryl-(1R)-camphorate (Figure 7.33B) in squalane. The remarkable resolution of the racemate (Figure 7.33C) is accounted for by the direct participation of the lone-electron pair of the chiral nitrogen at the enantiospecific coordination site of the nickel camphorate complex [595]. Other complexing metal ion salts, for example Hg(II), Cu(II), Pd(II), Pt(II), and Th(I), have been investigated but, in general, few advantages over the examples discussed above were found [596-598].

In reversed-phase HPLC, the formation of silver complexes with unsaturated ligands results in a decrease in retention due to an increase in the hydrophilic character of the complex compared to the parent ligand [599-602]. Varying the concentration of silver nitrate in the mobile phase enables the retention of the complexed species to be changed over a wide range, facilitating method development. For retinyl esters, which contain an unsaturated fatty acid side chain, a linear relationship was found between the logarithm of the capacity factor value and the silver ion concentration in the mobile phase [601].

Grushka bonded the dithiocarbamate ligand to silica and used the column to separate nucleotides and nucleosides in the presence of Mg(II) ions [603]. The diphosphate and triphosphate groups of the nucleotides bind Mg(II) very strongly and compete with the dithiocarbamate ligand of the stationary phase for the

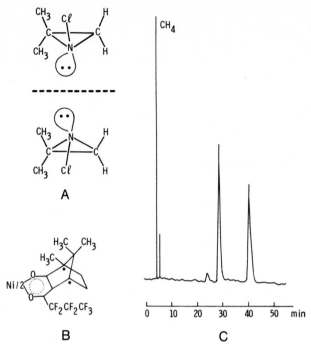

Figure 7.33 A, enantiomeric forms of 2,2-dimethylchloroaziridine;
B, nickel (II) bis-3-heptafluorobutyryl-(IR)-camphorate;
C, resolution of mixture A on a 100 m capillary column coated with
B dissolved in squalane.

TABLE 7.32

THE INFLUENCE OF OLEFIN STRUCTURE ON CHROMATOGRAPHIC RETENTION USING SILVER
NITRATE-CONTAINING STATIONARY PHASES

Substitution at the double bond decreases the retention volume.

A 1-alkyl compound has a lower retention volume than those of the 3- and
4-alkyl isomers

Olefins having a substituent in the 3- position have higher retention volumes
than those of the 4- isomers

Cyclobutenes have less tendency to form complexes than the corresponding 5-
and 6-membered cyclo-olefins

Cyclopentene derivatives have higher retention volumes than those of the
corresponding isomeric cyclohexenes

A conjugated double bond system has a lower complex-forming capacity than a
simple double bond

metal cation. The retention of the nucleotides was found to vary inversely with the magnesium ion concentration in the mobile phase and also exhibited a strong pH dependence. Likewise, the competition between the amino groups of an aminopropylsilica bonded phase and the amino groups of aminosugars or peptides for cadmium or zinc ions was used to enhance their resolution in a reversed-phase chromatographic system [604].

Karger investigated the use of a relatively hydrophobic chelating agent, namely 4-dodecyl-diethylenetriamine, in the presence of Zn(II) ions for the reversed-phase separations of dansyl amino acids, dipeptides, and aromatic carboxylic acids [605]. The metal-derived selectivity results from the formation of outersphere complexes. The formation and dissociation of these complexes is rapid, contributing to the high separation efficiencies observed. Innersphere complexes, involving the formation of direct metal-ligand bonds, are generally slower processes and result in a loss of chromatographic efficiency. The general usefulness of outersphere complexation to control selectivity in the separation of donor ligands remains to be fully investigated, but both theory and practice predict that this technique will find many future uses.

7.19 Ligand-Exchange Chromatography

In an extensive review of this topic, Davankov described ligand-exchange chromatography as a process in which interaction between the stationary phase and the species to be separated occurs during the formation of coordination bonds inside the coordination sphere of the complex-forming metal ion [606]. When a stationary phase that contains a chiral ligand metal complex interacts with enantiomeric mobile ligands, a discrimination of the latter can occur due to enantioselectivity in the formation of mixed-ligand sorption complexes. To take advantage of this difference in thermodynamic stabilities, the kinetics of the complex formation/dissociation reaction must be rapid. These criteria have been met for copper proline or copper hydroxyproline complexes, immobilized via chemical attachment to various polymeric resins, for the separation of enantiomeric amino acids and peptides [606-610]. The efficiency of these columns is often low for analytical purposes, but the high selectivity factors obtained can be exploited for preparative-scale separations. Copper(II)-modified silica gel was found to be a suitable stationary phase for the separation of small peptides by ligand-exchange chromatography [609]. Here the retention mechanism was ascribed to a combination of the hydrophobic character of the peptides in the polar eluent used for the separation and their ability to form complexes with the coordination sites on the stationary phase.

7.20 Qualitative Identification and Microreaction Techniques for Gas
Chromatography

Before the advent of modern hyphenated techniques (GC-MS, GC-FTIR),
numerous qualitative physical and chemical tests were devised for the
identification of peaks in a gas chromatogram [611]. With the exceptions of
retention index scales and the coincidence of retention of standard and sample
on two phases of different selectivity, these identification techniques have
largely passed into disuse. The greater access to capillary column gas
chromatography-mass spectrometry (GC-MS) and its higher information content are
primarily responsible for this trend.

The methods described in this section form a fragmented group of physical
and chemical techniques. For the most part, they are simple to perform,
inexpensive, require minimum instrument modification, and, in a few instances,
provide an elegant solution to an otherwise complex problem.

In many instances, a linear correlation exists between the retention volume
of a compound and its physical properties (e.g., vapor pressure, boiling point,
and carbon number) [611]. Vapor pressure and boiling point correlations are
really only useful for ideal solutions; as nonideal behavior is common, the
method is of limited utility. It can be used to predict the approximate
position of a peak in a chromatogram but should not be used to confirm the
identity of an unknown compound. For compounds forming an homologous series,
plots of the logarithm of the adjusted retention volume versus carbon number
often exhibit a high degree of linearity for all but the first new members of
the series. This correlation is useful for determining whether an unknown peak
could belong to a given homologous series, particularly when the sample already
contains several known members of that series to establish the basic linear
relationship.

The partition coefficient of a substance between several immiscible solvent
pairs can be combined with retention time data to confirm the identity of a
substance for which a pure standard exists [612-614]. Devised by Bowman and
Beroza, the substance specific partition coefficient ("p-value") was defined as
the fractional amount of substance partitioning into the nonpolar phase of an
equal-volume two-phase system. Only nanogram quantities of sample are required
for the measurement and p-values are often sufficiently characteristic to
distinguish between such closely related substances.

Peak shift techniques can also be used to aid in the identification of
compounds with reactive functional groups. When a compound is converted into a
derivative its chromatographic properties are usually altered, resulting in a
characteristic shift in its position in the chromatogram. If the derivatizing

reagent is specific for a particular functional group then only the retention times of compounds containing that group will be changed in the chromatogram. If two substances containing different functional groups coelute in the chromatogram, selective derivatization may be used to distinguish between them or to separate them. In general, observing changes in the chromatogram before and after derivatization provides a simple method of functional group identification.

Carbon–skeleton gas chromatography provides a means of identifying the hydrocarbon skeleton of a compound which contains functional groups [615]. The basis of the method is the catalytic hydrogenolysis or hydrogenation of the sample prior to separation. Hydrogenolysis involves the removal of heteroatoms such as halides, N, O, or S (usually present as functional groups, e.g., NH_2, OH, SH, etc.) such that the hydrocarbon species so generated retains either the carbon skeleton of the parent or the next lower homolog of the series, Table 7.33. Under similar conditions unsaturated hydrocarbons may be hydrogenated [616,617]. For certain compounds dehydrogenation and carbon–carbon bond fission can occur.

TABLE 7.33

PRODUCTS OF CARBON–SKELETON CHROMATOGRAPHY

Compound Class	Reaction
1) Compounds giving parent exclusively	
Paraffinic hydrocarbons	None
Unsaturated compounds	Multiple bonds saturated
Halogenated compounds	C–X bond cleaved
Alcohols (sec. or tert.)	C–O bond cleaved
Esters (alcohol part sec. or tert.)	C–O bond cleaved
Ethers (sec. or tert.)	C–O bond cleaved
Ketones	C=O bond cleaved
Amines (sec. or tert.)	C–N bond cleaved
Amides (NH attached to sec. or tert. C)	C–N bond cleaved
Sulfides	C–S bond cleaved
2) Compounds giving mainly next lower homolog	
Aldehydes	$RCHO \longrightarrow RH$
Acids	$RCOOH \longrightarrow RH$
Anhydrides	$(RCO)_2O \longrightarrow RH$
Alcohols (primary)	$RCH_2OH \longrightarrow RH$
Esters (C–O attached to primary C)	$R'COOCH_2R \longrightarrow RH, R'CH_3, R'H$
Amides (NH attached to primary C)	$R'CONHCH_2R \longrightarrow RH, RCH_3$

In a typical experimental arrangement, the injection heater of the gas chromatograph is used to heat a short catalyst bed containing platinum, palladium, copper, or nickel coated on a diatomaceous support. The catalyst bed can be the top portion of a packed column or a precolumn connected to a packed or capillary column. Hydrogen carrier gas flows through the heated catalyst bed (220-350°C) and then into the column. The sample is injected by syringe onto the catalyst bed where a rapid reaction occurs and the products are displaced onto the analytical column. Catalyst preparation, activation, and conditioning prior to use are important considerations [615,617,618]. A forked-column arrangement has been described for the quantitative analysis of complex mixtures in which the parent hydrocarbons are also present [619]. The two arms of the fork are identical except that one arm contains support and the other support plus catalyst. Injection of the sample alternatively into the two arms enables the natural concentration of the hydrocarbon to be subtracted from the amount found in the reduction mode. Dechlorination prior to capillary column gas chromatography has been performed in solution using nickel boride [$(Ni_2B)_2H_3$] as a reducing agent [620]. With further investigation this reagent may prove to be generally useful for a wide range of reactions.

An important application of carbon-skeleton gas chromatography is the simplification of the analysis of complex samples such as polychlorinated biphenyls, polybrominated biphenyls, and polychloroalkanes [618-621]. These complex mixtures of halogenated isomers produce multiple peaks when gas chromatographed, making quantitation difficult. The isomers have identical carbon skeletons, resulting in a very simple chromatogram after hydrodechlorination.

The location of the position of double bonds in alkenes or similar compounds is a difficult process when only very small amounts of sample are available. Mass spectrometry is often unsuited for this purpose; migration of the double bond occurs during ionization, resulting in the generation of similar spectra for different isomers. The position of double bonds can, however, be identified after oxidation to either the ozonide or cis-diol. Ozonolysis is simple to carry out and occurs sufficiently rapidly that reaction temperatures of -70°C are common [259,611,615,622]. Micro-ozonolysis apparatus are commercially available or can be readily assembled in the laboratory using standard equipment and a Tesla coil (vacuum tester) as the ozone generator. Reaction yields of ozonolysis products are typically 70 to 95%, although structures such as triple bonds and α, β-unsaturated nitriles are quite resistant to the process. After ozonolysis, either pyrolysis, or preferably, reduction with triphenyl phosphine, converts the ozonides to aldehydes (or ketones in the case of alkyl substituents attached directly to the carbon atom

of the double bond). Identification of the product by gas chromatography
enables the position of the original double bond to be established.

The stereospecific oxidation of a double bond to a cis-diol with osmium
tetroxide and subsequent conversion to a stable cyclic derivative (acetonide,
boronic ester) is a convenient method of double bond location by either gas
chromatography or gas chromatography-mass spectrometry [623,624]. Mass spectra
of the cyclic derivatives yield fragments characteristic of the location of the
double bond and abundance differences characteristic of the geometric
arrangement of substituents attached to it. The position of the double bond in
the molecule is indicated by a simple a-cleavage to form two ions containing
either of the end groups of the original double bond.

Methods have been developed by which compounds containing common functional
groups can be subtracted from complex mixtures prior to gas chromatography.
Known generally as reaction or subtraction chromatography, the reactions may
take the form of sample preparation techniques (derivatization), precolumn
reactions in loops or vessels, on-column or in-syringe reactions, or postcolumn
reactions [611,615,625-635]. The difference between the reacted and unreacted
chromatograms is used to identify which compounds contain a particular
functional group. Some representative reagents and their areas of application
are summarized in Table 7.34. In its simplest form the reagent is coated onto a
support and packed into a short column that is usually positioned between the
injection device and analytical column or, occasionally, between the column and
detector. Such columns expose a large surface area of reagent to the sample, so
that, provided the reaction is rapid, a short column ensures complete
subtraction at normal carrier gas flow rates without distorting the peak shape
of compounds which do not react. Some reactions are reversible, which affects
the operating temperature range for the reaction, others generate secondary
by-products which can confuse the identification, and, of course, all reagents
eventually become exhausted or deactivated with use. Special techniques for use
with capillary columns have been described [630,634,635].

A sophisticated form of reaction chromatography used for the fractionation
of complex mixtures by capillary column gas chromatography has been described by
Sievers [636,637]. A europium coordination polymer was used to selectively
retain nucleophilic compounds such as ketones, aldehydes, alcohols, and
carboxylic acids by complexation. As the complexation process is reversible, a
manifold was constructed that enabled capillary column chromatograms of the
total sample, the retained nucleophiles, and the sample without the retained
nucleophiles to be obtained.

As well as reaction chromatography, class reactions can be performed on
isolated fractions or on-line to identify compounds containing particular
functional groups [611,638-640]. The on-line method involves passing a fraction

TABLE 7.34

PRECOLUMN SUBTRACTION REAGENTS USED FOR QUALITATIVE IDENTIFICATION IN GAS CHROMATOGRAPHY

Subtraction Reagent	Compound Functionalities Subtracted
Molecular Sieve 5A	n-Alkanes and other straight-chain molecules in the presence of branched chain molecules
Salts of Ag, Cu, Hg often mixed with concentrated sulfuric acid	Alkenes, alkynes
Concentrated sulfuric acid	Aromatics, alkenes, alkynes, basic compounds
Maleic anhydride	Conjugated dienes
Boric acid 2-Nitrophthalic anhydride	Primary and secondary alcohols
Lithium aluminum hydride	Alcohols, aldehydes, ketones, esters, epoxides
Zinc oxide	Carboxylic acids, partial subtraction of alcohols and phenols
NaOH-quartz	Phenols
Phosphoric acid	Epoxides, bases
Versamid 900	Organic compounds containing active halogens
Bromine/Carbon tetrachloride	Unsaturated compounds
NaBr-alumina	Organic compounds with functional groups
Benzidine	Aldehydes, ketones, epoxides
Sodium borohydride Hydroxylamine Sodium bisulfite Semicarbazide 3,4,5-Trimethoxybenzylhydrazine	Ketones, aldehydes
FFAP o-Dianisidine	Aldehydes

of the column effluent into a solution of either a single or a series of chemically-specific reagents. The reagents used are similar to those employed in organic qualitative analysis and spot tests. Most have a precipitation or colorimetric end-point. The principal limitation is that 10-100 micrograms of sample is generally required to produce a positive test.

7.21   Pyrolysis Gas Chromatography

Pyrolysis gas chromatography is an indirect method of analysis in which heat is used to transform a compound into a series of volatile products that are characteristic of the original compound and the experimental conditions. The initiation reactions involved in the pyrolysis of most organic materials are accounted for by the generation of free radicals from the cleavage of a single bond or by the unimolecular elimination of simple molecules such as water or carbon dioxide. The subsequent reactions of these species via abstraction or combination reactions, or by diffusion, produce products characteristic of the original sample. Gas chromatography is the analytical technique used for product identification and is often directly coupled with mass spectrometry when more information than a comparative fingerprint (pyrogram) of the sample is required. Pyrolysis may be on-line in the form of a modified sample inlet for the gas chromatograph, or off-line, in which case the pyrolysis products are usually cryogenically collected and injected either directly as a gas or as a liquid dissolved in a solvent via the standard inlet. Pyrolysis gas chromatography finds many applications in polymer chemistry, geochemistry, biomolecular analysis, and taxonomy for the identification and, under some circumstances, quantitative analysis of intractable samples such as rubbers, polymers, sediments, bacteria, etc. [641-646]. Many of these samples are mixtures with differing degrees of complexity, often of high molecular weight substances, or substances having limited volatility or solvent solubility, which would be difficult to analyze by conventional gas chromatographic methods. However, the application of pyrolysis gas chromatography is not limited to such samples alone; small molecules such as sulfonamides can be readily detected in pharmaceutical products or biological fluids by pyrolysis gas chromatography. Upon pyrolysis, sulfonamides extrude sulfur dioxide which normally leads to two main components via fragmentation and proton rearrangement as shown below:

sulfapyridine   aniline   2-aminopyridine

Aniline is produced by pyrolysis of all medicinal sulfonamides and acts as an internal reference standard, while the other fragment is an aromatic amine characteristic of the specific drug under investigation [647].

Three methods of pyrolysis - continuous mode, pulsed mode, and laser disintegration - are commonly used. Continuous mode pyrolyzers include a wide range of tube furnaces and microreactors which may be operated on- or off-line [641,645,648,649]. Pyrolysis is normally performed by inserting the smaple into a continuously-heated zone. The sample may be supported on a boat and dropped or pushed into the heated zone, or is encapsulated in a quartz tube which, after pyrolysis, is broken to release the products. When pyrolysis is performed on-line a valving arrangement is usually required to divert the carrier gas or auxiliary gas flow through the furnace to flush the pyrolysis products onto the column. The principal disadvantage of continuous pyrolysis is that the temperature of the heated wall of the microfurnace is much higher than the temperature of the sample, and heat transfer is relatively slow. Additionally, primary pyrolysis products released from the surface of the sample expand into the hot zone of the furnace, thus increasing the probability of secondary reactions. It will be recognized later that this process differs from that occurring in pulsed-mode pyrolyzers and may result either in the formation of different products or in the formation of the same products, but with a different distribution between components. Pulse-mode pyrolyzers include resistively-heated electrical filaments or ribbons [643,648-650] and radio frequency induction-heated wires [641,645,648,651]. The filament or ribbon-type pyrolyzers are simple to construct, Figure 7.34A, and typically consist of an inert wire or ribbon (Pt or Pt-Rh alloy) connected to a high-current power supply. Samples soluble in a volatile solvent are applied to the filament in the form of a thin film. Insoluble materials are placed in a crucible or quartz tube, heated by a basket-like shaped or helical wound filament [648]. The coated filament is contained within a low dead volume chamber through which the carrier gas flows, sweeping the pyrolysis products onto the column. The pyrolysis temperature is controlled by the current passed through the wire. An inductively-heated Curie point pyrolyzer is shown in Figure 7.34B. The sample, as a thin-film coated on a ferromagnetic wire or a solid contained within a recess forged into the wire, is heated inductively in a radio frequency field to its Curie point, the temperature at which the alloy becomes paramagnetic and ceases to absorb radio frequency energy. This pyrolysis technique provides a highly reproducible pyrolysis temperature since the Curie-point temperature of a ferromagnetic material depends only on the composition, of the alloy. By selecting wires of different alloy compositions temperatures in the range 300 to 1000°C may be obtained. The principal advantage of the pulse-mode pyrolyzers is that, as the sample and heat source are in direct contact, the primary pyrolysis products are quenched as they rapidly expand away into the cooler regions, diminishing the possibility of secondary products forming. Heating rates are rapid (milliseconds), reproducible, and controllable by adjusting the filament-heating current or alloy composition and radio frequency energy of the

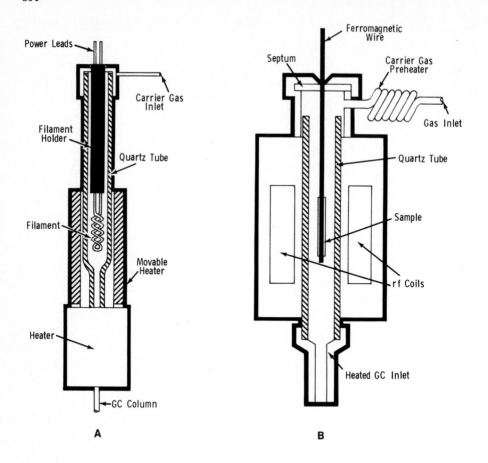

Figure 7.34   Apparatus for pyrolysis gas chromatography. A, filament or
               ribbon-type pyrolyzer and B, Curie point pyrolyzer. (Reproduced
               with permission from ref. 648. Copyright American Chemical
               Society).

Curie point pyrolyzer. Lasers are particularly suitable for controlled
pyrolysis as very high energies can be directed at a small portion of the sample
in the form of a series of pulses of short duration (100 microseconds)
[642,645,652,653]. Laser pyrolysis conditions differ considerably from those of
thermal pyrolysis, typically result in ionization and the formation of a plasma
plume. Quenching of the high temperature plasma, combined with the products
generated from thermal shock waves transmitted through the sample, constitute
the origin of the products observed in the pyrogram. In general, laser
pyrolysis results in fewer, if different, products than thermal pyrolysis. If
the laser radiation is not absorbed by the sample (e.g., colorless samples with
visible lasers) a substance which is capable of functioning as an absorption
center, for example powdered carbon or nickel, is intimately mixed with the

sample to be pyrolyzed. Alternatively, an infrared carbon dioxide laser can be used. Laser pyrolysis devices may be more widely used in the future, but the filament and Curie point pyrolyzers are currently more important.

Many investigations of the optimum conditions for thermal pyrolysis have been reported [645]. In general, good inter-laboratory reproducibility of pyrograms can be obtained but intra-laboratory comparisons have often been poor [654]. The major factors which influence reproducibility are the type of pyrolyzer, the pyrolysis temperature and temperature rise-time, sample size and homogeneity, gas chromatographic column and separation conditions, and the pyrolyzer-gas chromatographic interface. Optimization of the pyrolyzer using reference standards is one solution to this problem. The temperature rise-time, the time required to reach the final pyrolysis temperature, is influenced by the geometry of the filament, the sample amount, and the design of the power supply [654]. Rise times are on the order of tens of milliseconds for filament and Curie point pyrolyzers. To avoid the production of thermal gradients across the pyrolysis sample, thin samples are preferred, and only a few micrograms are needed for analysis.

The statistical analysis of the pyrograms requires a high reproducibility of duplicate runs. Pattern-recognition procedures are required for the comparison of complex pyrograms, a situation which is common for environmental or biological samples [655-658]. Also, in taxonomic identification of bacteria and fungi it is of paramount importance that the organism always exhibits the same chemical characteristics [644]. Factors such as sub-culturing, the growth time, and the medium upon which the organism is grown will affect the reproducibility of the pyrogram.

7.22 <u>References</u>

1. R. S. Smith and G. V. James, "The Sampling of Bulk Materials", The Royal Society of Chemistry, London, 1981.
2. C. J. Kirchner, Environ. Sci. Technol., 17 (1983) 174A.
3. B. Kratochvil and J. K. Taylor, Anal. Chem., 53 (1981) 924A.
4. H. S. Hertz, W. E. May, S. A. Wise, and S. N. Chesler, Anal. Chem., 50 (1978) 428A,
5. E. Reid, Analyst, 101 (1976) 1.
6. Q. V. Thomas, J. R. Stork, and S. L. Sammert, J. Chromatogr. Sci., 18 (1980) 583.
7. C. Weurman, J. Argic. Food Chem., 17 (1969) 370.
8. M. L. Leng, in H. A. Moye (Ed.), "Analysis of Pesticide Residues", Wiley, New York, 1981, p. 395.
9. D. A. Burns, Anal. Chem., 53 (1981) 1403A.
10. C. F. Poole and S. A. Schuette, J. High Resolut. Chromatogr. Chromatogr. Commun., 6 (1983) 526.
11. W. G. Jennings and A. Rapp, "Sample Preparation for Gas Chromatographic Analysis", Huthig, Heidelberg, 1983.
12. J. F. Lawrence, "Organic Trace Analysis by Liquid Chromatography", Academic Press, New York, 1981.
13. R. L. Jolley, Environ. Sci. Technol., 15 (1981) 874.

14. R. Teranishi, I. Hornstein, P. Issenberg, and E. L. Wick, "Flavor Research: Principles and Techniques", Dekker, New York, 1971.
15. P. L. Peters, Anal. Chem., 52 (1980) 211.
16. G. B. Nickerson and S. T. Likens, J. Chromatogr., 21 (1966) 1.
17. M. Godefroot, P. Sandra, and M. Verzele, J. Chromatogr., 203 (1981) 325.
18. M. Godefroot, M. Stechele, P. Sandra, and M. Verzele, J. High Resolut. Chromatogr. Chromatogr. Commun., 5 (1982) 75.
19. J. Rijks, J. Curvers, Th. Noy, and C. Cramers, J. Chromatogr., 279 (1983) 395.
20. C. Bicchi, A. D'Amato, G. M. Nano, and C. Frattini, J. Chromatogr., 279 (1983) 409.
21. K. R. Kim, A. Zlatkis, J.-W. Park, and U. C. Lee, Chromatographia, 15 (1982) 559.
22. W. G. Jennings, J. High Resolut. Chromatogr. Chromatogr. Commun., 2 (1979) 221.
23. R. W. Storherr and R. R. Watts, J. Assoc. Off. Anal. Chem., 48 (1965) 1154.
24. B. A. Karlhuber and D. O. Eberle, Anal. Chem., 47 (1975) 1094.
25. J. Kovac, V. Batora, A. Hankova, and A. Szokolay, Bull. Environ. Contam. Toxicol., 13 (1975) 692.
26. U. R. Stenberg and T. E. Alsberg, Anal. Chem., 53 (1981) 2067.
27. D. T. Kaschani, in W. Bertsch, S. Hara, R. E. Kaiser and A. Zlatkis (Eds.), "Instrumental HPTLC", Huthig, Heidelberg, 1980, p. 185.
28. R. A. Chalmers and R. W. E. Watts, Analyst, 97 (1972) 224.
29. J. Shapiro, Anal. Chem., 39 (1967) 280.
30. R. E. Kepner, S. van Straten, and C. Weurman, J. Agric. Food Chem., 17 (1969) 1123.
31. W. R. Robins, Anal. Chem., 51 (1979) 1860.
32. C. J. O. R. Morris and P. Morris, "Separation Methods in Biochemistry", 2nd Edn., Halstead Press, New York, 1976.
33. R. G. Webb, Int. J. Anal. Chem., 5 (1978) 239.
34. W. L. Budde and J. W. Eichelberger, "Organic Analysis using GC-MS", Ann Arbor Science, Ann Arbor, 1979.
35. R. A. Hites, Adv. Chromatogr., 15 (1977) 69.
36. E. D. Morgan and C. F. Poole, J. Insect Physiol., 22 (1976) 885.
37. E. D. Morgan and C. F. Poole, Adv. Insect Physiol., 12 (1976) 17.
38. S. G. Colgrove and H. J. Svec, Anal. Chem., 53 (1981) 1737.
39. D. H. Wheeler, Chem. Revs., 62 (1962) 205.
40. D. F. S. Natusch and B. A. Tomkins, Anal. Chem., 50 (1978) 1429.
41. C. F. Poole and D. G. Wibberley, J. Chromatogr., 132 (1977) 511.
42. J. Vessman, Methodological Surveys Series A, 10 (1981) 284.
43. M. C. Goldberg, L. DeLong, and M. Sinclair, Anal. Chem., 45 (1973) 89.
44. M. Ahnoff and B. Josefsson, Anal. Chem., 46 (1974) 658.
45. W. A. Hoffman, Anal. Chem., 50 (1978) 2158.
46. V. M. Buchar and A. K. Agrawal, Analyst, 106 (1981) 620.
47. E. Stephanou and W. Giger, Environ. Sci. Technol., 16 (1982) 800.
48. M. C. Goldberg and E. R. Weiner, Anal. Chim. Acta, 115 (1980) 373.
49. T. L. Peters, Anal. Chem., 54 (1982) 1913.
50. G. F. Griffin and E. A. DeLwiche, Anal. Chem., 54 (1982) 2616.
51. G. M. Schneider, E. Stahl, and G. Wilke, "Extraction with Supercritical Gases", Verlag Chemie, Weinheim, 1980.
52. F. W. Karasek, R. E. Clement, and J. A. Sweetman, Anal. Chem., 53 (1981) 1050A.
53. F. A. Gunther, R. C. Blinn, M. J. Kolbezen, and J. H. Barkley, Anal. Chem., 23 (1951) 1835.
54. N. P. Sen and C. Dalpe, Analyst, 97 (1972) 216.
55. G. D. Price and D. A. Carlson, Anal. Chem., 53 (1981) 554.
56. C. E. Higgins and M. R. Guerin, Anal. Chem., 52 (1980) 1984.
57. A. Zlatkis, C. F. Poole, R. S. Brazell, K. Y. Lee, F. Hsu, and S. Singhawangcha, Analyst, 106 (1981) 352.

58. T. J. Nestrick and L. L. Lamparski, Anal. Chem., 53 (1981) 122.
59. D. C. Fenimore, C. M. Davis, J. H. Whitford, and C. A. Harrington, Anal. Chem., 48 (1976) 2289.
60. S. O. Farwell and S. J. Gluck, Anal. Chem., 52 (1980) 1968.
61. D. A. J. Murray, J. Chromatogr., 177 (1979) 135.
62. J. F. J. van Rensberg and A. J. Hasset, J. High Resolut. Chromatogr. Chromatogr. Commun., 5 (1982) 574.
63. V. Lopez-Avila, R. Northcutt, J. Onstot, M. Wickham, and S. Billets, Anal. Chem., 55 (1983) 881.
64. A. J. Burgasser and J. F. Colaruotolo, Anal. Chem., 51 (1979) 1588.
65. W. Dunges, "Pre-Chromatographic Methods", Huthig, Heidelberg, 1984.
66. W. Dunges, Anal. Chem., 45 (1973) 963.
67. M. Stafford, M. G. Horning, and A. Zlatkis, J. Chromatogr., 126 (1976) 495.
68. O. Mikes (Ed.), "Laboratory Handbook of Chromatographic and Allied Methods", Wiley, New York, 1979.
69. C. H. Giles and I. A. Easton, Adv. Chromatogr., 3 (1966) 67.
70. L. R. Snyder, "Principles of Adsorption Chromatography", Dekker, New York, 1968.
71. L. R. Snyder, J. Chromatogr., 12 (1963) 488.
72. N. S. Radin, Method. Enzymol., 14 (1969) 317.
73. P. A. Mills, J. Assoc. Off. Anal. Chem., 51 (1968) 29.
74. M. Spencer and M. Grynpas, J. Chromatogr., 166 (1978) 423.
75. M. Spencer, J. Chromatogr., 166 (1978) 435.
76. J. Ellingborne, E. Nystrom, and J. Sjovall, Method. Enzymol., 14 (1969) 317.
77. D. E. Wells and S. J. Johnstone , J. Chromatogr., 140 (1977) 17.
78. R. C. Chapman and C. R. Harris, J. Chromatogr., 166 (1978) 513.
79. W. Horwitz (Ed.), "Official Methods of Analysis", Assoc. Off. Anal. Chem., Washington, D. C., 1975.
80. K. H. Altgett and T. H. Gouw, Adv. Chromatogr., 13 (1975) 71.
81. M. L. Lee, M. Novotny, and K. D. Bartle, "Analytical Chemistry of Polycyclic Aromatic Compounds", Academic Press, New York, 1981.
82. L. L. Lamparski, T. J. Nestrick, and R. J. Stehl, Anal. Chem., 51 (1979) 1453.
83. M. L Langhorst and L. A. Shadoff, Anal. Chem., 52 (1980) 2037.
84. L. L. Lamparski and T. J. Nestrick, Anal. Chem., 52 (1980) 2045.
85. D. L. Stalling, R. C. Tindle, and J. L. Johnson, J. Assoc. Off. Anal. Chem., 55 (1972) 32.
86. L. D. Johnson, R. H. Waltz, J. P. Ussary, and F. E. Kaiser, J. Assoc. Off. Anal. Chem., 59 (1976) 174.
87. J. Vessman, Methodological Surveys Series A, 10 (1981) 341.
88. A. P. de Leenheer and H. J. C. F. Nelis, Analyst, 106 (1981) 1025.
89. S. N. Chesler, H. S. Hertz, W. E. May, S. A. Wise, and F. R. Guenther, Int. J. Environ. Anal. Chem., 5 (1978) 259.
90. H. Frank, G. J. Nicholson, and E. Bayer, J. Chromatogr., 167 (1978) 187.
91. H. Frank, N. Vujtovic-Ockenga, and A. Rettenmeier, J. Chromatogr., 279 (1983) 507.
92. J. J. Kirkland, Analyst, 99 (1974) 858.
93. M. Dressler, J. Chromatogr., 165 (1979) 167.
94. J. F. Lawrence, Anal. Chem., 52 (1980) 1122A.
95. R. W. Frei and U. A. Th. Brinkman, Trends Anal. Chem., 1 (1981) 45.
96. E. Reid (Ed.), "Trace-Organic Sample Handling", Ellis Horwood, London, 1981.
97. A. Bacaloni, G. Goretti, A. Lagana, B. M. Petronio, and M. Rotatori, Anal. Chem., 52 (1980) 2033.
98. R. W. Beulow, J. K. Carswell, and J. M. Symons, J. Amer. Water Works Assoc., 65 (1973) 57.
99. R. W. Beulow, J. K. Carswell, and J. M. Symons, J. Amer. Water Works Assoc., 65 (1973) 195.

568

100. J. J. McCreary and V. L. Snoeyink, J. Amer. Water Works Assoc., 69 (1977) 437.
101. D. K. Chow and M. M. David, J. Amer. Water Works Assoc., 69 (1977) 555.
102. G. Belfort, Environ. Sci. Technol., 13 (1979) 939.
103. P. van Rossum and R. G. Webb, J. Chromatogr., 150 (1978) 381.
104. A. Tateda and J. S. Fritz, J. Chromatogr., 152 (1978) 329.
105. D. C. Kennedy, Environ. Sci. Technol., 7 (1973) 138.
106. C. D. Chriswell, R. L. Ericson, G. A. Junk, K. W. Lee, J. S. Fritz, and H. J. Svec, J. Amer. Water Works Assoc., 69 (1977) 669.
107. K. A. Kunk and R. Kunin, J. Polymer Sci., Part A1, 6 (1968) 2689.
108. G. R. Aiden, E. M. Thurman, R. L. Malcolm, and H. E. Walton, Anal. Chem., 51 (1979) 1799.
109. G. A. Junk, J. J. Richard, M. D. Grieser, Đ. Witiak, J. L. Witiak, M. D. Arguello, R. Vick, H. J. Svec, J. S. Fritz, and G. V. Calder, J. Chromatogr., 99 (1974) 745.
110. H. A. Stuber and J. H. Leenheer, Anal. Chem., 55 (1983) 111.
111. P. R. Musty and G. Nickless, J. Chromatogr., 89 (1974) 185.
112. V. N. Mallet, J. M. Francoeur, and G. Volpe, J. Chromatogr., 150 (1978) 381.
113. B. Olufsen, Anal. Chim. Acta, 113 (1980) 393.
114. W. V. Ligon and R. L. Johnson, Anal. Chem., 48 (1978) 481.
115. W. E. May, S. N. Chesler, S. P. Cram, B. H. Gump, H. S. Hertz, D. P. Enangonio, and S. M. Dyszel, J. Chromatogr. Sci., 13 (1975) 535.
116. V. Leoni, G. Puccetti, R. J. Colombo, and A. M. D'Ovidio, J. Chromatogr., 125 (1976) 399.
117. J. F. Pankow and L. M. Isabelle, J. Chromatogr., 237 (1982) 25.
118. J. F. Pankow, L. M. Isabelle, and T. Kristensen, J. Chromatogr., 245 (1982) 31.
119. A. Przyjazny, W. Janicki, W. Chrzanowski, and R. Staszewski, J. Chromatogr., 245 (1982) 256.
120. G. Hunt and N. Pangaro, Anal. Chem., 54 (1982) 369.
121. J. D. Navratil, R. E. Sievers, and H. F. Walton, Anal. Chem., 49 (1977) 2260.
122. P. R. Musty and G. Nickless, J. Chromatogr., 100 (1974) 83.
123. D. K. Basu and J. Saxena, Environ. Sci. Technol., 12 (1978) 791.
124. H. D. Gesser, A. B. Sparling, A. Chow, and C. W. Turner, J. Amer. Water Works Assoc., 65 (1973) 220.
125. G. J. Moody and J. D. R. Thomas, Analyst, 104 (1979) 1.
126. G. J. Moody and J. D. R. Thomas, "Chromatographic Separation and Extraction with Foamed Plastic and Rubbers", Dekker, New York, 1982.
127. G. J. de Jong and J. Zeeman, Chromatographia, 15 (1982) 453.
128. W. A. Saner, J. R. Jadamec, R. W. Sager, and T. J. Kileen, Anal. Chem., 51 (1979) 2180.
129. M. Tetsuo, H. Ericksson, and J. Sjovall, J. Chromatogr., 239 (1982) 287.
130. R. Lafont, J.-L. Pennetier, M. Andrianjafintrimo, J. Claret, J.-M. Modde, and C. Blais, J. Chromatogr., 236 (1982) 137.
131. S. L. Pallante, M. Stognlew, M. Colvin, and D. J. Liberto, Anal. Chem., 54 (1982) 369.
132. L. C. Ramirez, C. Millot, and B. F. Maume, J. Chromatogr., 229 (1982) 267.
133. C. D. Chriswell, R. C. Chang, and J. S. Fritz, Anal. Chem., 47 (1975) 1325.
134. J. J. Richard and J. S. Fritz, J. Chromatogr. Sci., 18 (1980) 35.
135. J. J. Richard, C. D. Criswell, and J. S. Fritz, J. Chromatogr., 199 (1980) 143.
136. F. A. Chalmers and R. W. E. Watts, Analyst, 97 (1972) 958.
137. J. A. Thompson and S. P. Markey, Anal. Chem., 47 (1975) 1313.
138. C. F. Poole and M. Verzele, J. Chromatogr., 150 (1978) 439.
139. R. W. Zumwalt, D. Roach, and C. W. Gehrke, J. Chromatogr., 53 (1970) 171.
140. J. F. K. Huber and R. R. Becker, J. Chromatogr., 142 (1977) 765.

141. J. Lankelma and H. Poppe, J. Chromatogr., 149 (1978) 587.
142. H. P. M. van Vliet, Th. C. Bootsman, R. W. Frei, and U. A. Th. Brinkman, J. Chromatogr., 185 (1979) 483.
143. J. B. Reust and V. R. Meyer, Analyst, 107 (1982) 673.
144. P. Guinebault and M. Broquaire, J. Chromatogr., 217 (1981) 509.
145. B. L. Karger, M. Martin, and G. Guiochon, Anal. Chem., 46 (1974) 1640.
146. R. J. Dolphin, F. W. Willmott, A. D. Mills, and L. P. J. Hoogeveen, J. Chromatogr., 122 (1976) 259.
147. J. F. K. Huber, I. Fogy, and C. Fioresi, Chromatographia, 13 (1980) 408.
148. F. Erni, H. P. Keller, C. Morin, and M. Schmitt, J. Chromatogr., 204 (1981) 65.
149. J. C. Gfeller and M. Stockmeyer, J. Chromatogr., 198 (1980) 162.
150. G. J. De Jong, J. Chromatogr., 183 (1980) 203.
151. J. A. Apffel, T. V. Alfredson, and R. E. Majors, J. Chromatogr., 206 (1981) 43.
152. T. V. Alfredson, J. Chromatogr., 218 (1981) 715.
153. C. E. Werkhoven-Goewie, U. A. Th. Brinkman, and R. W. Frei, Anal. Chem., 53 (1981) 2072.
154. H. Hachenberg and A. P. Schmidt, "Gas Chromatographic Headspace Analysis", Heyden, London, 1977.
155. A. G. Vitenberg, B. V. Ioffe, and V. N. Borisov, Chromatographia, 7 (1974) 610.
156. B. Kolb (Ed.), "Applied Headspace Gas Chromatography", Heyden, London, 1977.
157. G. Charalambous (Ed.) "Analysis of Food and Beverages, Headspace Techniques", Academic Press, New York, 1978.
158. P. Gagliardi and G. R. Verga, J. Chromatogr., 279 (1983) 323.
159. J. Drozd and J. Novak, J. Chromatogr., 165 (1979) 141.
160. J. Novak, Adv. Chromatogr., 21 (1983) 303.
161. L. S. Ettre, J. E. Purcell, J. Widonski, B. Kolb, and P. Pospisil, J. Chromatogr. Sci., 18 (1980) 116.
162. B. Kolb, P. Pospisil, and M. Auer, J. Chromatogr., 204 (1981) 371
163. B. Kolb, J. Chromatogr., 122 (1976) 553.
164. A. G. Vitenberg, L. M. Kuzenetsoua, I. L. Butseva, and M. D. Inshakov, Anal. Chem., 49 (1977) 129.
165. J. Drozd, J. Novak, and A. J. Rijks, J. Chromatogr., 158 (1978) 471.
166. B. Kolb, P. Pospisil, T. Borath, and M. Auer, J. High Resolut. Chromatogr. Chromatogr. Commun., 2 (1979) 283.
167. P. G. Simmonds and E. Kerns, J. Chromatogr., 178 (1979) 863.
168. S. L. Friant and I. H. Suffet, Anal. Chem., 51 (1979) 2167.
169. R. J. Steichen, Anal. Chem., 48 (1976) 1398.
170. E. A. Dietz and K. F. Singley, Anal. Chem., 51 (1979) 1809.
171. W. J. Khazal, J. Vejrosta, and J. Novak, J. Chromatogr., 157 (1978) 125.
172. B. Kolb, M. Auer, and P. Pospisil, J. Chromatogr., 279 (1983) 341.
173. B. Kolb, Chromatographia, 15 (1982) 587.
174. B. Kolb and P. Pospisil, Chromatographia, 10 (1977) 705.
175. B. V. Ioffe and A. G. Vitenberg, Chromatographia, 11 (1978) 282.
176. W. M. Dong, Chromatographia, 14 (1981) 441.
177. A. Zlatkis, R. S. Brazell, and C. F. Poole, Clin. Chem., 27 (1981) 789.
178. T. A. Bellar and J. J. Lichtenberg, J. Amer. Water Works Assoc., 66 (1974) 739.
179. A. Zlatkis, H. A. Lichtenstein, and A. Tishbee, Chromatographia, 6 (1973) 67.
180. H. M. Liebich, A. Al-Babbili, A. Zlatkis, and K. Kim, Clin. Chem., 21 (1975) 1294.
181. L. C. Michael, M. D. Erickson, S. P. Parks, and E. D. Pellizzari, Anal. Chem., 52 (1980) 1836.
182. A. B. Robinson, D. Partridge, M. Turner, R. Teraniski, and L. Pauling, J. Chromatogr., 85 (1973) 19.
183. K. E. Murrey, J. Chromatogr., 135 (1977) 49.
184. M. L. McConnell and M. Novotny, J. Chromatogr., 112 (1975) 559.

185. R. E. Sievers, R. M. Barkley, G. A. Eiceman, R. H. Shapiro, H. F. Walton, K. J. Kolonko, and L. R. Field, J. Chromatogr., 142 (1977) 745.
186. T. A. Bellar, J. J. Lichtenberg, and J. W. Eichelberger, Environ. Sci. Technol., 10 (1976) 962.
187. H. C. Hu and P. H. Weiner, J. Chromatogr. Sci., 18 (1980) 333.
188. B. S. Middleditch, J. Chromatogr., 239 (1982) 159.
189. M. Thomason, M. Shoults, W. Bertsch, and G. Holzer, J. Chromatogr., 158 (1978) 437.
190. W. Bertsch, E. Anderson, and G. Holzer, J. Chromatogr., 112 (1975) 701.
191. K. Grob and F. Zurcher, J. Chromatogr., 117 (1976) 285.
192. K. Grob, K. Grob, and G. Grob, J. Chromatogr., 106 (1975) 299.
193. W. E. Coleman, J. W. Munch, R. W. Slater, R. G. Melton, and F. C. Kopfler, Environ. Sci. Technol., 17 (1983) 571.
194. R. Hites, Adv. Chromatogr., 15 (1977) 69.
195. M. M. Thomason and W. Bertsch, J. Chromatogr., 279 (1983) 383.
196. R. S. Narang and B. Bush, Anal. Chem., 52 (1980) 2076.
197. H. Boren, A. Grimvall, and R. Savenhed, J. Chromatogr., 252 (1982) 139.
198. A. Zlatkis, C. F. Poole, R. S. Brazell, D. A. Bafus, and P. S. Spencer, J. Chromatogr., 182 (1980) 137.
199. A. Zlatkis, C. F. Poole, R. S. Brazell, K. Y. Lee, and S. Singhawangcha, J. High Resolut. Chromatogr. Chromatogr. Commun., 2 (1979) 423.
200. K. Y. Lee, D. Nurok, and A. Zlatkis, J. Chromatogr., 158 (1978) 377.
201. A. Zlatkis and K. Kim, J. Chromatogr., 126 (1976) 475.
202. J. P. Mieure and M. W. Dietrich, J. Chromatogr. Sci., 11 (1973) 559.
203. G. Holzer, H. Shanfield, A. Zlatkis, W. Bertsch, P. Juarez, H. Mayfield, and H. M. Liebich, J. Chromatogr., 142 (1977) 755.
204. T. J. Hughes, E. D. Pellizzari, L. Little, C. Sparacino, and A. Kolber, Mut. Res., 76 (1980) 51.
205. W. Bertsch in W. G. Jennings (Ed.), "Applications of Glass Capillary Gas Chromatography", Dekker, New York, 1981, p. 71.
206. W. Bertsch, R. C. Chang, and A. Zlatkis, J. Chromatogr. Sci., 12 (1974) 175.
207. W. Bertsch, "Sampling of Organic Volatiles", Huthig, Heidelberg, 1984.
208. J. Rudolph, D. H. Ehhalt, A. Khedim and C. Jebsen, J. Chromatogr., 217 (1981) 301.
209. H. B. Singh, L. J. Salas, and R. E. Stiles, Environ. Sci. Technol., 16 (1982) 872.
210. T. L. Chan, P. S. Lee, and J. S. Siak, Environ. Sci. Technol., 15 (1981) 89.
211. D. Schuetzle, Environ. Hlth. Persp., 47 (1983) 65.
212. W. Bertsch, A. Zlatkis, H. M. Liebich, and H. J. Schneider, J. Chromatogr., 99 (1974) 673.
213. E. D. Pellizzari, J. E. Bunch, B. H. Carpenter, and E. Sawicki, Environ. Sci. Technol., 9 (1975) 552.
214. E. D. Pellizzari, R. H. Carpenter, J. E. Bunch, and E. Sawicki, Environ. Sci. Technol., 9 (1975) 556.
215. E. D. Pellizzari, J. E. Bunch, R. E. Berkley, and J. McRae, Anal. Chem., 48 (1976) 803.
216. K. J. Krost, E. D. Pellizzari, S. G. Walburn, and S. A. Hubbard, Anal. Chem., 54 (1982) 810.
217. A. I. Clark, A. E. McIntyre, J. N. Lester, and R. Perry, J. Chromatogr., 252 (1982) 147.
218. P. Ciccioli, G. Bertoni, E. Brancaleoni, R. Fratarcangeli, and F. Bruner, J. Chromatogr., 126 (1976) 757.
219. A. Raymond and G. Guiochon, J. Chromatogr. Sci., 13 (1975) 173.
220. B. E. Foulger and P. G. Simmonds, Anal. Chem., 51 (1979) 1089.
221. W. F. Burns, D. T. Tingey, R. C. Evans, and E. H. Bates, J. Chromatogr., 269 (1983) 1.
222. E. D. Pellizzari, J E. Bunch, R. E. Berkeley, and J. McRae, Anal. Letts., 9 (1976) 45.
223. G. I. Senum, Environ. Sci. Technol., 15 (1981) 1073.

224. J. Namiesnik, L. Torres, E. Kozlowski, and J. Mathieu, J. Chromatogr., 208 (1981) 239.
225. R. H. Brown and C. J. Purnell, J. Chromatogr., 178 (1979) 79.
226. T. Kanaka, J. Chromatogr., 153 (1978) 7.
227. L. D. Butler and M. F. Burke, J. Chromatogr. Sci., 14 (1976) 117.
228. J. Janak, J. Ruzickova, and J. Novak, J. Chromatogr., 99 (1974) 689.
229. J. E. Bunch and E. D. Pellizzari, J. Chromatogr., 186 (1979) 811.
230. A. Venema, N. Kampstra, and J. T. Sukkel, J. Chromatogr., 269 (1983) 179.
231. D. Schuetzle, in G. R. Waller and O. C. Dermer (Eds.) "Biochemical Applications of Mass Spectrometry", Interscience, New York, 1980, p. 969.
232. R. G. Lewis and K. E. Macleod, Anal. Chem., 54 (1982) 310.
233. D. T. Coker, Analyst, 106 (1981) 1036.
234. C. J. Purnell, M. D. Wright, and R. H. Brown, Analyst, 106 (1981) 590.
235. R. W. Coutant and D. R. Scott, Environ. Sci. Technol., 16 (1982) 410.
236. D. Grosjean and K. Fung, Anal. Chem., 54 (1982) 1221.
237. E. C. Gunderson and E. L. Fernandez, in G. Choudhary (Ed.), "Chemical Hazards in the Workplace," American Chemical Society Symposium Series No. 149, American Chemical Society, Washington, D. C., 1981, p. 179.
238. R. G. Melcher, P. W. Langvardt, M. L. Langhorst, and S. A. Bouyoucos, in G. Choudhary (Ed.), "Chemical Hazards in the Workplace", American Chemical Society Symposium Series, No. 149, American Chemical Society, Washington, D. C., 1981, p. 155.
239. F. Mangani, A. Mastrogiacomo, and O. Marras, Chromatographia, 15 (1982) 712.
240. L. D. White, D. G. Tayler, P. A. Mauer, and R. E. Kupel, Amer. Ind. Hyg. Ass. J., 31 (1970) 225.
241. NIOSH Manual of Analytical Methods, U. S. Department of HEW, Public Health Service, Center for Disease Control, NIOSH, Cincinnati, 2nd Edn., 1977.
242. A. T. Saalwaechter, C. S. McCammon, C. P. Roper, and K. S. Carlberg, Amer. Ind. Hyg. Ass. J., 38 (1977) 476.
243. S. Crisp, Ann. Occup. Hyg., 23 (1980) 47.
244. H. G. Baxter, R. Blakemore, and J. P. Moore, Ann. Occup. Hyg., 23 (1980) 117.
245. I. Johansen and J. F. Wendelboe, J. Chromatogr., 217 (1981) 317.
246. R. G. Melcher, W. L. Garner, L. W. Severs, and J. R. Vaccaro, Anal. Chem., 50 (1978) 251.
247. R. G. Lewis and M. D. Jackson, Anal. Chem., 54 (1982) 594.
248. G. O. Nelson, "Controlled Test Atmospheres, Principles and Techniques" Ann Arbor Science Publishers, Ann Arbor, MI, 1971.
249. R. S. Barratt, Analyst, 106 (1981) 817.
250. J. E. Lovelock, Anal. Chem., 33 (1961) 162.
251. H. P. Williams and J. D. Winefordner, J. Gas Chromatogr., 4 (1966) 271.
252. J. J. Ritter and N. K. Adams, Anal. Chem., 48 (1976) 612.
253. H. Nozoye, Anal. Chem., 50 (1978) 1727.
254. R. P. Scaringelli, A. E. O'Keefe, E. Rosenberg, and J. P. Bell, Anal. Chem., 42 (1970) 871.
255. B. E. Saltzman, W. R. Burg, G. Ramaswamy, Environ. Sci. Technol., 5 (1971) 1121.
256. A. E. O'Keefe, Anal. Chem., 49 (1977) 1278.
257. J. Namiesnik, Chromatographia, 17 (1983) 47.
258. J. S. Marhevka, G. D. Johnson, D. F. Hagen, and R. D. Danielson, Anal. Chem., 54 (1982) 2607.
259. K. Blau and G. S. King (Eds.), "Handbook of Derivatives for Chromatography", Heyden, London, 1978.
260. D. R. Knapp, "Handbook of Analytical Derivatization Reactions", Wiley, New York, 1979.
261. J. Drozd, "Chemical Derivatization in Gas Chromatography", Elsevier, Amsterdam, 1981.
262. A. Zlatkis and C. F. Poole (Eds.), "Electron Capture: Theory and Practice in Chromatography", Elsevier, Amsterdam, 1982.

263. A. E. Pierce, "Silylation of Organic Compounds", Pierce Chemical Company, Rockford, IL, 1968.
264. J. Drozd, J. Chromatogr., 113 (1975) 303.
265. C. F. Poole and A. Zlatkis, J. Chromatogr. Sci., 17 (1979) 115.
266. C. F. Poole and A. Zlatkis, J. Chromatogr., 184 (1980) 99.
267. W. C. Kossa in R. W. Frei and J. F. Lawrence (Eds.), "Chemical Derivatization in Analytical Chemistry", Plenum, New York, Vol. 1, 1981, p. 99.
268. W. C. Kossa, J. MacGee, S. Ramachandran, and A. J. Webber, J. Chromatogr. Sci., 17 (1979) 177.
269. A. Hutshoff and A. D. Forch, J. Chromatogr., 220 (1981) 275.
270. C. F. Poole and A. Zlatkis, Anal. Chem., 52 (1980) 1002A.
271. C. F. Poole, Chem. Ind. (London), (1976) 479.
272. J. D. Nicholson, Analyst, 103 (1978) 1.
273. J. D. Nicholson, Analyst, 103 (1978) 193.
274. S. Ahuja, J. Pharm. Sci., 65 (1976) 163.
275. P. Husek and K. Macek, J. Chromatogr., 113 (1975) 139.
276. S. U. Kahn, Res. Revs., 59 (1975) 21.
277. W. P. Cochrane and R. B. Maybury, in A. Zlatkis and C. F. Poole (Eds.), "Electron Capture: Theory and Practice in Chromatography", Elsevier, Amsterdam, 1981, p. 205.
278. W. P. Cochrane, J. Chromatogr. Sci., 13 (1975) 246.
279. W. P. Cochrane, J. Chromatogr. Sci., 17 (1979) 124.
280. H. B. S. Conacher and B. D. Page, J. Chromatogr. Sci., 17 (1979) 188.
281. D. J. Harvey and W. D. M. Patton, J. Chromatogr., 109 (1975) 73.
282. D. J. Harvey, Org. Mass Spectrom., 12 (1977) 473.
283. E. J. Corey and A. Venkateswarlu, J. Amer. Chem. Soc., 94 (1972) 6190.
284. K. K. Ogilvie, S. L. Beaucage, and D. W. Entwistle, Tet. Lett., 16 (1976) 1255.
285. M. A. Quilliam and J. B. Westmore, Anal. Chem., 50 (1978) 59.
286. G. Phillipou, J. Chromatogr., 129 (1976) 384.
287. I. A. Blair and G. Phillipou, J. Chromatogr. Sci., 15 (1977) 478.
288. D. J. Harvey, Biomed. Mass Spectrom., 4 (1977) 265.
289. M. A. Quilliam and J. B. Westmore, Steroids, 29 (1977) 579.
290. C. F. Poole, A. Zlatkis, W.-F. Sye, S. Singhawangcha, and E. D. Morgan, Lipids, 15 (1980) 734.
291. C. F. Poole, W.-F. Sye, S. Singhawangcha, F. Hsu, A. Zlatkis, A. Arfwidsson, and J. Vessman, J. Chromatogr., 199 (1980) 123.
292. C. F. Poole, S. Singhawangcha, L.-E. Chen Hu, W.-F. Sye, R. S. Brazell, and A. Zlatkis, J. Chromatogr., 187 (1980) 331.
293. C. Eaborn, C. A. Holder, D. R. M. Walton, and B. S. Thomas, J. Chem. Soc. (C), (1969) 2502.
294. B. S. Thomas, J. Chromatogr., 56 (1971) 37.
295. E. K. Symes and B. S. Thomas, J. Chromatogr., 116 (1976) 163.
296. J. B. Brooks, J. A. Liddle, and C. C. Alley, Anal. Chem., 47 (1975) 1960.
297. C. J. W. Brooks and B. S. Middleditch, Anal. Lett., 5 (1972) 611.
298. J. R. Chapman and E. Bailey, Anal. Chem., 45 (1973) 1636.
299. E. D. Morgan and C. F. Poole, J. Chromatogr., 104 (1975) 351.
300. A. J. Francis, E. D. Morgan, and C. F. Poole, J. Chromatogr., 161 (1978) 111.
301. C. F. Poole and E. D. Morgan, Org. Mass Spectrom., 10 (1975) 537.
302. A. J. Francis, E. D. Morgan, and C. F. Poole, Org. Mass Spectrom., 13 (1978) 671.
303. C. F. Poole, W.-F. Sye, S. Singhawangcha, and A. Zlatkis, Org. Mass Spectrom., 15 (1980) 486.
304. E. Anggard and A. Hankey, Acta Chem. Scand., 23 (1969) 3110.
305. L. Dehennin, A. Reiffstock, and R. Scholler, J. Chromatogr. Sci., 10 (1972) 224.
306. N. K. McCallum and R. J. Armstrong, J. Chromatogr., 78 (1973) 303.
307. J. F. Lawrence, J. J. Ryan, and R. Leduc, J. Chromatogr., 147 (1978) 398.
308. J. J. Ryan and J. F. Lawrence, J. Chromatogr., 135 (1975) 117.

309. D. D. Clarke, S. Wilk, and S. E. Gitlow, J. Gas Chromatogr., 4 (1966) 310.
310. P. K. Kadaba, J. Pharm. Sci., 63 (1974) 1333.
311. N. B. Lorette and J. H. Brown, J. Org. Chem., 24 (1959) 261.
312. R. W. Zumwalt, D. Roach, and C. W. Gehrke, J. Chromatogr., 53 (1970) 171.
313. S. Mierzwa and S. Witek, J. Chromatogr., 136 (1977) 105.
314. C. C. Alley, J. B. Brooks, and G. Choudary, Anal. Chem., 48 (1976) 387.
315. E. Watson, B. Travis, and S. Wilk, Life Science, 15 (1974) 2167.
316. D. D. Godse, J. J. Warsh, and H. C. Stancer, Anal. Chem., 49 (1977) 915.
317. S. Wilk and M. Orlowski, Anal. Biochem., 69 (1975) 100.
318. D. A. Davis, D. A. Durden, P. Pun-Li, and A. A. Boulton, J. Chromatogr., 142 (1977) 517.
319. H. R. Morris, D. H. Williams, and R. P. Ambler, Biochem. J., 125 (1971) 189.
320. H. Nikaido, Eur. J. Biochem., 15 (1970) 57.
321. D. M. W. Anderson, I. C. M. Dea, P. A. Maggs, and A. C. Munroe, Carbohyd. Res., 5 (1967) 489.
322. H. M. Fales, T. M. Jaouni, and J. F. Babashak, Anal. Chem., 45 (1973) 2302.
323. I. Hazai and G. Alexander, J. High Resolut. Chromatogr. Chromatogr. Commun., 5 (1982) 583.
324. J. P. Thenot and E. C. Horning, Anal. Letts., 5 (1972) 519.
325. J. P. Thenot, E. C. Horning, M. Stafford, and M. G. Horning, Anal. Lett., 5 (1972) 217.
326. N. Nose, S. Kobayashi, A. Hirose, and A. Watanabe, J. Chromatogr., 123 (1976) 167.
327. D. Karashima, S. Takahaski, A. Shigematsu, T. Furukawa, T. Kuhara, and I. Matsumoto, Biomed. Mass Spectrom., 3 (1976) 41.
328. M. W. Scoggins, J. Chromatogr. Sci., 12 (1975) 146.
329. A. A. Arbin and P.-O. Edlund, Acta Pharm. Suec., 11 (1974) 249.
330. D. G. Bailey, H. L. Davis, and G. E. Johnson, J. Chromatogr., 121 (1976) 263.
331. A. Arbin, H. Brink, and J. Vessman, J. Chromatogr., 170 (1979) 25.
332. O. Gyllenhaal, H. Brotell, and P. Hartvig, J. Chromatogr., 129 (1976) 295.
333. J. Dockx, Synthesis, (1973) 441.
334. A. Brandstrom, "Preparative Ion Pair Extraction", Swedish Academy of Pharmaceutical Sciences, Stockholm, 1974.
335. R. Osiewicz, J. Chromatogr., 88 (1974) 157.
336. I. C. Cohen, J. Norcup, J. J. A. Ruzicka, and B. B. Wheals, J. Chromatogr., 44 (1969) 251.
337. A. C. Moffat and E. C. Horning, Anal. Lett., 3 (1970) 205.
338. A. Zlatkis and B. C. Pettit, Chromatographia 2 (1969) 484.
339. S. B. Margin and M. Worland, J. Pharm. Sci., 61 (1972) 1235.
340. A. J. F. Wickramasinghe and R. S. Shaw, Biochem. J., 141 (1974) 179.
341. J.-C. Lhuguenot and B. F. Maume, J. Chromatogr. Sci., 12 (1974) 411.
342. P. Hartvig and J. Vessman, J. Chromatogr. Sci., 12 (1974) 722.
343. P. Hartvig and J. Vessman, Anal. Lett., 12 (1974) 223.
344. R. A. Mead, G. C. Hattmeyer, and K. B. Eik-Nes, J. Chromatogr. Sci., 7 (1969) 554.
345. K. T. Koshy, D. G. Kaiser, and A. L. Van der Silk, J. Chromatogr. Sci., 13 (1975) 97.
346. A. C. Moffat, E. C. Horning, S. B. Martin, and M. Rowland, J. Chromatogr., 66 (1972) 255.
347. P. O. Edlund, J. Chromatogr., 187 (1980) 161.
348. O. Gyllenhall, J. Chromatogr., 153 (1978) 517.
349. K. Kobayashi, M. Tanaka, and S. Kawi, J. Chromatogr., 187 (1980) 413.
350. Y. Hoshika and G. Huto, J. Chromatogr., 152 (1978) 224.
351. G. M. Anthony, C. J. W. Brooks, I. MacLean, and I. Sangster, J. Chromatogr. Sci., 7 (1969) 623.
352. C. J. W. Brooks and I. MacLean, J. Chromatogr. Sci., 9 (1971) 18.

353. C. F. Poole, S. Singhawangcha, and A. Zlatkis, J. Chromatogr., 158 (1978) 33.
354. C. F. Poole, S. Singhawangcha, and A. Zlatkis, J. Chromatogr., 186 (1979) 307.
355. S. Singhawangcha, C. F. Poole, and A. Zlatkis, Org. Mas Spectrom., 15 (1980) 505.
356. S. Singhawangcha, C. F. Poole, and A. Zlatkis, J. Chromatogr., 183 (1980) 433.
357. C. F. Poole, L. Johansson, and J. Vessman, J. Chromatogr., 194 (1980) 365.
358. C. F. Poole, S. Singhawangcha, L.-E. Chen Hu, and A. Zlatkis, J. Chromatogr., 178 (1979) 495.
359. R. W. Kelley, J. Chromatogr., 43 (1969) 229.
360. T. A. Baillie, C. J. W. Brooks, and B. S. Middleditch, Anal. Chem., 44 (1972) 30.
361. H. D. Simmonds and D. W. Wiley, J. Amer. Chem. Soc., 82 (1960) 2288.
362. P. Husek and V. Felt, J. Chromatogr., 152 (1978) 546.
363. P. Husek, V. Felt, and M. Matucha, J. Chromatogr., 234 (1982) 381.
364. K. H. Nielsen, J. Chromatogr., 10 (1963) 463.
365. U. Langenbeck, A. Hoinowski, K. Mantel, and U.-H. Mohring, J. Chromatogr., 143 (1977) 39.
366. T. Hayashi, T. Sigura, H. Terada, S. Kawai, and T. Ohno, J. Chromatogr., 118 (1976) 403.
367. A. Frigerio, P. Martelli, K. M. Baker, and P. A. Bondi, J. Chromatogr., 81 (1973) 139.
368. M. S. Lennard, J. H. Silas, A. J. Smith, and G. T. Tucker, J. Chromatogr., 133 (1977) 161.
369. J. G. Allen, P. B. East, R. J. Francis, and J. L. Haigh, Drug Metabol. Disp., 3 (1975) 332.
370. P. Erdtmansky and T. J. Goehl, Anal. Chem., 47 (1975) 750.
371. S. L. Malcolm and T. R. Marten, Anal. Chem., 48 (1976) 807.
372. T. Kawabata, H. Ohshima, T. Ishibashi, M. Matsui, and T. Kitsuwa, J. Chromatogr., 140 (1977) 47.
373. S. B. Martin, J. H. Karam, and P. H. Forsham, Anal. Chem., 47 (1975) 545.
374. D. Alkalay, J. Volk, and M. F. Bartlett, J. Pharm. Sci., 65 (1976) 525.
375. J. F. Lawrence and R. W. Frei, "Chemical Derivatization in Liquid Chromatography", Elsevier, Amsterdam, 1976.
376. S. Ahuja, J. Chromatogr. Sci., 17 (1979) 168.
377. R. W. Frei and J. F. Lawrence (Eds.), "Chemical Derivatization in Analytical Chemistry", Vol. 1, Plenum Press, New York, 1981.
378. T. Jupille, J. Chromatogr. Sci., 17 (1979) 160.
379. M. A. Carey and H. W. Persinger, J. Chromatogr. Sci., 10 (1972) 537.
380. R. W. Ross, J. Chromatogr. Sci., 14 (1976) 505.
381. J. Lehrfield, J. Chromatogr., 120 (1976) 141.
382. F. Nachtmann, H. Spitzy, and R. W. Frei, J. Chromatogr., 122 (1976) 293.
383. C. R. Clark and J. L. Chan, J. Chromatogr., 83 (1973) 471.
384. B. Bjorkuist and H. Tolvonen, J. Chromatogr., 153 (1978) 265.
385. Y. Suzuki and K. Tani, Buneseki Kagaku, 28 (1979) 610.
386. D. G. Musson and L. A. Sternson, J. Chromatogr., 188 (1980) 159.
387. C. R. Clark and M. M. Wells, J. Chromatogr. Sci., 16 (1975) 332.
388. L. L. Needham and M. M. Kochhar, J. Chromatogr., 114 (1975) 220.
389. T. Sugiura, T. Hayashi, S. Kawai, and T. Ohno, J. Chromatogr., 110 (1975) 385.
390. A. Hulshoff, H. Rosenboom, and J. Renema, J. Chromatogr., 186 (1980) 535.
391. G. B. Cox, J. Chromatogr., 83 (1973) 471.
392. F. A. Fitzpatrick, S. Siggia, and J. Dingman, Anal. Chem., 44 (1972) 2211.
393. R. A. Braun and W. A. Mosher, J. Amer. Chem. Soc., 80 (1958) 3048.
394. H. D. Durst, M. Milano. E. J. Kikta, S. A. Connelly, and E. Grushka, Anal. Chem., 47 (1975) 1797.
395. D. R. Knapp and S. Krueger, Anal. Lett., 8 (1975) 603.

396. I. R. Politzer, G. W. Griffin, B. J. Dowty, and L. J. Laseter, Anal. Lett., 6 (1973) 539.
397. R. F. Borch, Anal. Chem., 47 (1975) 2437.
398. W. Linder, J. Chromatogr., 198 (1980) 367.
399. D. P. Matthees and W. C. Purdy, Anal. Chim. Acta, 109 (1979) 161.
400. D. Munger, L. A. Sternson, A. J. Repta, and T. Higuchi, J. Chromatogr., 143 (1977) 375.
401. K. L. Dunlap, R. L. Sandridge, and J. Keller, Anal. Chem., 48 (1976) 497.
402. K. Kuwata, M. Uebori, and Y. Yamazaki, Anal. Chem., 52 (1980) 857.
403. M. Schafer and E. Mutscher, J. Chromatogr., 164 (1979) 247.
404. J. Gogo, S. Komatsu, M. Goto, and T. Nambara, Chem. Pharm. Bull. Japan, 29 (1981) 899.
405. S. Lam and E. Grushka, J. Chromatogr., 158 (1978) 207.
406. R. Farinotti, Ph. Siard, J. Bourson, S. Kirkiacharian, B. Valeur, and G. Mohuzier, J. Chromatogr., 256 (1983) 243.
407. W. D. Korte, J. Chromatogr., 243 (1982) 153.
408. N. Nimura and T. Kinoshita, Anal. Lett., 13 (1980) 191.
409. J. B. F. Lloyd, J. Chromatogr., 189 (1980) 359.
410. G. Avigad, J. Chromatogr., 139 (1977) 343.
411. D. J. Pietrzyk and E. P. Chan, Anal. Chem., 42 (1970) 37.
412. E. P. Lankmayr, R. W. Budna, K. Muller, F. Nachtmann, and F. Rainer, J. Chromatogr., 222 (1981) 249.
413. H. Umagat, P. Kucera, and L. F. Wen, J. Chromatogr., 239 (1982) 463.
414. M. Ahnoff, I. Grundevik, A. Arfwidsson, J.Fonselius, and B.-A. Persson, Anal. Chem., 53 (1981) 484.
415. H. A. Moye and A. J. Boning, Anal. Lett., 12 (1979) 25.
416. A. P. de Leenheer, J. E. Sinsheimer, and J. H. Bruckhalter, J. Pharm. Sci., 62 (1972) 1370.
417. C. F. Poole, S. Singhawangcha, A. Zlatkis, and E. D. Morgan, J. High Resolut. Chromatogr. Chromatogr. Commun., 1 (1978) 96.
418. K. Mopper, W. L. Stahovec, and L. Johnson, J. Chromatogr., 256 (1983) 243.
419. R. W. Frei and J. F. Lawrence, J. Chromatogr., 83 (1973) 321.
420. J. F. Lawrence and R. W. Frei, J. Chromatogr., 98 (1974) 253.
421. J. F. Lawrence, J. Chromatogr. Sci., 17 (1979) 147.
422. T. Skaaden and T. Greibrokk, J. Chromatogr., 247 (1982) 111.
423. P. Lindroth and K. Mopper, Anal. Chem., 51 (1979) 1667.
424. Y. Watanabe and K. Imai, J. Chromatogr., 239 (1982) 723.
425. W. Dunges, Anal. Chem., 49 (1977) 442.
426. W. Voelter, R. Huber, and K. Zech, J. Chromatogr., 217 (1981) 491.
427. J. B. F. Lloyd, J. Chromatogr., 178 (1979) 249.
428. P. T. Kissinger, K. Bratin, G. C. Davis, and L. A. Pachla, J. Chromatogr. Sci., 17 (1979) 137.
429. S. Ikenoya, O. Hiroshima, M. Ohmae, and K. Kawabe, Chem. Pharm. Bull. Japan, 28 (1980) 2941.
430. G. Schwedt and E. Reh, Chromatographia, 13 (1980) 779.
431. R. W. Frei, Chromatographia, 15 (1982) 161.
432. R. S. Deelder, A. T. J. M. Kuijpers, and J. H. M. Van den Berg, J. Chromatogr., 255 (1983) 545.
433. A. H. M. Scholten, U. A. Th. Brinkman and R. W. Frei, Anal. Chem., 54 (1982) 1932.
434. A. H. M. Scholten, U. A. Th. Brinkman, and R. W. Frei, J. Chromatogr., 218 (1981) 3.
435. G. Schwedt and E. Reh, Chromatographia, 14 (1981) 317.
436. J. C. Gfeller, G. Frey and R. W. Frei, J. Chromatogr., 142 (1977) 271.
437. A. H. M. Scholten, U. A. Th. Brinkman, and R. W. Frei, J. Chromtogr., 205 (1981) 229.
438. J. T. Stewart, Trends Anal. Chem., 1 (1982) 170.
439. K. Tsuji, J. Chromatogr., 158 (1978) 337.
440. J. F. Lawrence, U. A. Th. Brinkman, and R. W. Frei, J. Chromatogr., 185 (1979) 475.

441. J. F. Studebaker, J. Chromatogr., 185 (1979) 497.
442. T. D. Schlabach and F. E. Regnier, J. Chromatogr., 158 (1978) 349.
443. J. F. K. Huber, K. M. Jonker, and H. Poppe, Anal. Chem., 52 (1980) 2.
444. R. S. Deelder, M. G. F. Kroll, A. J. B. Beeren, and J. H. M. Van den Berg, J. Chromatogr., 149 (1978) 669.
445. H. Yoshida, T. Sumida, T. Masujima, and H. Imai, J. High Resolut. Chromatogr. Chromatogr. Commun., 5 (1982) 509.
446. S. H. Lee, L. R. Field, W. N. Howald, and W. F. Trager, Anal. Chem., 53 (1981) 467.
447. H. Nakamura and Z. Tamura, Anal. Chem., 54 (1982) 1951.
448. R. S. Deelder, M. G. F. Kroll, and J. H. M. Van den Berg, J. Chromatogr., 125 (1976) 307.
449. E. P. Lankmayr, B. Maichin, G. Knapp, and J. Nachtmann, J. Chromatogr., 224 (1981) 239.
450. R. W. Frei, J. F. Lawrence, U. A. Th. Brinkman, and I. Honigberg, J. High Resolut. Chromatogr. Chromatogr. Commun., 2 (1979) 11.
451. F. Smedes, J. C. Kraat, C. F. Werkhoven-Goewie, U. A. Th. Brinkman, and R. W. Frei, J. Chromatogr., 247 (1982) 123.
452. A. H. M. Scholten, P. L. M. Welling, U. A. Th. Brinkman, and R. W. Frei, J. Chromatogr., 199 (1980) 239.
453. U. A. Th. Brinkman, P. L. M. Welling, G. De Vries, A. H. M. Scholten, and R. W. Frei, J. Chromatogr., 217 (1981) 463.
454. C. Baiocchi, E. Campi, M. Gennaro, E. Mentasi, and P. Mirti, Chromatographia, 15 (1982) 660.
455. G. Schill, in J. A. Mirnsky and Y. Marcus (Eds.), "Ion-Exchange and Solvent Extraction", Dekker, New York, Vol. 6, 1974, p. 1.
456. J. Crommen, B. Fransson, and G. Schill, J. Chromatogr., 142 (1977) 283.
457. B. Fransson, K. G. Wahlund, I. M. Johansson, and G. Schill, J. Chromatogr., 125 (1976) 327.
458. B. A. Bidlingmeyer, J. Chromatogr. Sci., 18 (1980) 525.
459. R. Gloor and E. L. Johnson, J. Chromatogr. Sci., 15 (1977) 413.
460. E. Tomlinson, T. M. Jefferies, and C. M. Riley, J. Chromatogr., 159 (1978) 315.
461. M. T. W. Hern, Adv. Chromatogr., 18 (1980) 59.
462. G. Schill, R. Modin, K. O. Berg, and B.-A. Persson, in K. Blau and G. S. King (Eds.), "Handbook of Derivatives for Chromatography", Heyden, London, 1978, p. 550.
463. R. H. A. Sorel and A. Hulshoff, Adv. Chromatogr., 21 (1983) 87.
464. B. A. Bidlingmeyer, Liq. Chromatogr. HPLC Magz., 1 (1983) 345.
465. G. Schwedt, "Chromatographic Methods in Inorganic Analysis", Huthig, Heidelberg, 1981.
466. T. R. Crompton, "Gas Chromatography of Organometallic Compounds", Plenum Press, New York, 1982.
467. G. Guiochon and C. Pommier, "Gas Chromatography of Inorganics and Organometallics", Ann Arbor Science Publishers, Ann Arbor, 1973.
468. J. A. Rodriguez-Vazquez, Anal. Chim. Acta, 73 (1974) 1.
469. R. S. Juvet and R. L. Fisher, Anal. Chem., 38 (1966) 1860.
470. W. S. Pappas and J. G. Million, Anal. Chem., 40 (1968) 2176.
471. C. F. Poole and A. Zlatkis, in A. Zlatkis and C. F. Poole (Eds.), "Electron Capture: Theory and Practice in Chromatography", Elsevier, Amsterdam, 1981, p. 191.
472. R. W. Moshier and R. E. Sievers, "Gas Chromatography of Metal Chelates", Pergamon, Oxford, 1965.
473. P. C. Uden and D. E. Henderson, Analyst, 102 (1977) 889.
474. C. A. Burgett, Sep. Purif. Methods, 5 (1976) 1.
475. C.-J. Kuo, I.-H. Lin, J.-S. Shih, and Y.-C. Yeh, J. Chromatogr. Sci., 20 (1982) 455.
476. W. D. Ross and R. E. Sievers, Environ. Sci. Technol., 6 (1972) 155.
477. L. C. Hansen, W. G. Schribner, T. W. Gilbert, and R. E. Sievers, Anal. Chem., 43 (1971) 349.
478. G. Schwedt, Chromatographia, 12 (1979) 613.

479. R. C. Gurira and P. W. Carr, J. Chromatogr. Sci., 20 (1982) 461.
480. S. R. Hutchins, P. R. Haddad, and S. Dilli, J. Chromatogr., 252 (1982) 185.
481. D. E. Henderson, R. Chaffee, and F. P. Novak, J. Chromatogr. Sci., 19 (1981) 79.
482. O. Liska, J. Lehotay, E. Brandsteterova, G. Guiochon, and H. Colin, J. Chromatogr., 172 (1979) 384.
483. T. H. Risby and C. A. Tollinche, J. Chromatogr. Sci., 16 (1978) 448.
484. A. M. Bond and G. G. Wallace, Anal. Chem., 53 (1981) 1209.
485. B. R. Willeford and H. Veening, J. Chromatogr., 251 (1982) 61.
486. N. Haering and K. Ballschmiter, Talanta, 27 (1980) 873.
487. Y.-T. Shih and P. W. Carr, Talanta, 28 (1981) 411.
488. P. Jones and G. Nickless, J. Chromatogr., 76 (1973) 295.
489. P. Jones and G. Nickless, J. Chromatogr., 89 (1974) 201.
490. P. Zarnegar and P. Mushak, Anal. Chim. Acta, 69 (1974) 389.
491. V. Luckow and H. R. Russel, J. Chromatogr., 150 (1978) 187.
492. C. J. Cappon and J. C. Smith, Anal. Chem., 49 (1977) 365.
493. K. Funazo, T. Hirashima, H.-L. Wu, M. Tanaka, and T. Shono, J. Chromatogr., 243 (1982) 85.
494. Y. Shimoishi, J. Chromatogr., 136 (1977) 85.
495. C. F. Poole, N. J. Evans, and D. G. Wibberley, J. Chromatogr., 136 (1977) 73.
496. T. Stijve and E. Cardinale, J. Chromatogr., 109 (1975) 239.
497. K. Kurahashi, S. Inoue, S. Yonekura, Y. Shimoishi, and K. Toei, Analyst, 105 (1980) 690.
498. W. C. Butts and W. T. Rainey, Anal. Chem., 43 (1971) 538.
499. D. J. Harvey and M. G. Horning, J. Chromatogr., 79 (1973) 65.
500. K. Ranfit, Z. Anal. Chem., 269 (1974) 18.
501. G. Nota, C. Improta, and V. R. Miraglia, J. Chromatogr., 242 (1982) 359.
502. J. C. Valentour, V. Aggarwal, and I. Sunshine, Anal. Chem., 46 (1974) 924.
503. G. Nota and R. Polombari, J. Chromatogr., 84 (1973) 37.
504. R. A. Hasty, Mikrochim. Acta, (1973) 621.
505. St. Grays, J. Chromatogr., 100 (1974) 43.
506. D. J. Glover and J. C. Hoffsommer, J. Chromatogr., 94 (1974) 334.
507. J. C. Hoffsommer, D. J. Glover, and D. Krubose, J. Chromatogr., 103 (1975) 182.
508. R. L. Tanner, R. J. Fajer, and J. Gaffney, Anal. Chem., 51 (1979) 865.
509. J. W. Tesch, W. R. Rehg, and R. E. Sievers, J. Chromatogr., 126 (1976) 743.
510. J. G. Kirchner, "Thin-Layer Chromatography", Wiley, New York, 2nd. Edn., 1978.
511. J. C. Touchstone and M. F. Dobbins, "Practice of Thin-Layer Chromatography ", Wiley, New York, 1978.
512. E Stahl, "Thin-Layer Chromatography", Springer-Verlag, New York, 1969.
513. G. D. Barrett, Adv. Chromatogr., 11 (1974) 145.
514. K. Y. Lee, D. Nurok, and A. Zlatkis, J. Chromatogr., 174 (1979) 187.
515. R. Segura and A. M. Gotto, J. Chromatogr., 99 (1974) 643.
516. L. Zhou, H. Shanfield, F.-S. Wang, and A. Zlatkis, 217 (1981) 329.
517. R. Segura and X. Navarro, J. Chromatogr., 217 (1981) 329.
518. H. Shanfield, F. Hsu, and A. J. P. Martin, J. Chromatogr., 126 (1976) 457.
519. H. Shanfield, K. Y. Lee, and A. J. P. Martin, J. Chromatogr., 142 (1977) 387.
520. W. Funk, R. Kerler, J.-Th. Schiller, V. Damman, and F. Arndt, J. High Resolut. Chromatogr. Chromatogr. Commun., 5 (1982) 534.
521. S. Uchiyama and M. Uchiyama, J. Chromatogr., 153 (1978) 135.
522. A. Oztunc, Analyst, 107 (1982) 585.
523. S. Uchiyama and M. Uchiyama, J. Liq. Chromatogr., 3 (1980) 681.
524. S. Uchiyama and M. Uchiyama, J. Chromatogr., 262 (1983) 340.

525. S. S. J. Ho, H. T. Butler, and C. F. Poole, J. Chromatogr., 281 (1983) 330.
526. E. A. Hoopes, E. T. Peltzer, and J. L. Bada, J. Chromatogr. Sci., 16 (1978) 556.
527. R. W. Souter, J. Chromatogr., 108 (1975) 265.
528. W. A. Konig, K. Stoelting, and K. Kruse, Chromatographia, 10 (1977) 444.
529. S. Gal, J. Pharm. Sci., 66 (1977) 169.
530. J. W. Westley and B. Halpern, J. Org. Chem., 33 (1968) 3978.
531. R. G. Arnett and P. K. Stumpf, Anal. Biochem., 47 (1972) 638.
532. T. Nambara, J. Goto, K. Taguchi, and T. Iwata, J. Chromatogr., 100 (1974) 180.
533. C. J. W. Brooks, M. T. Gilbert, and J. D. Gilbert, Anal. Chem., 45 (1973) 896.
534. M. T. Gilbert, J. D. Gilbert, and C. J. W. Brooks, Biomed. Mass Spectrom., 1 (1974) 274.
535. M. W. Anders and M. J. Cooper, Anal. Chem., 43 (1971) 1093.
536. W. Pereira, V. A. Bacon, W. Patton, and B. Halpern, Anal. Lett., 3 (1970) 23.
537. S. Hammarstrom and M. Hambert, Anal. Biochem., 52 (1973) 169.
538. B. Halpern and J. W. Westley, Aust. J. Chem., 19 (1966) 1533.
539. W. A. Konig, W. Rahn, and J. Eyme, J. Chromatogr., 133 (1977) 141.
540. J. P. Kamerling, G. J. Gerwig, J. F. G. Vliengenthart, M. Duran, D. Ketting, and S. K. Wadman, J. Chromatogr., 143 (1977) 117.
541. A. Murano, Agric. Biol. Chem., 36 (1972) 2203.
542. W. E. Pereira, M. Salomon, and B. Halpern, Aust. J. Chem., 24 (1971) 1103.
543. J. Goto, N. Goto, and T. Nambara, J. Chromatogr., 239 (1982) 559.
544. J. Goto, N. Gogo, A. Hikichi, T. Hishimaki, and T. Nambara, Anal. Chim. Acta, 120 (1980) 187.
545. J. Goto, M. Hasegawa, S. Nakamura, K. Shimada, and T. Nambara, J. Chromatogr., 152 (1978) 413.
546. J. Hermansson and C. Von Bahr, J. Chromatogr., 221 (1980) 109.
547. N. Nimura, H. Ogura, and T. Kinoshita, J. Chromatogr., 202 (1980) 375.
548. H. Furukawa, Y. Mori, Y. Takeuchi, and K. Ito, J. Chromatogr., 136 (1977) 428.
549. T. Nambara, S. Ikegawa, M. Hasegawa, and J. Goto, Anal. Chim. Acta, 101 (1978) 111.
550. D. Valentine, K. K. Chan, C. G. Scott, K. K. Johnson, K. Troth, and G. Saucy, J. Org. Chem., 41 (1976) 62.
551. M. Koreeda, G. Weiss, and K. Nakanishi, J. Amer. Chem. Soc., 95 (1973) 239.
552. W. H. Pirkle and R. W. Anderson, J. Org. Chem., 39 (1974) 3901.
553. I. S. Krull, Adv. Chromatogr., 16 (1978) 175.
554. E. Gil-Av and D. Nurok, Adv. Chromatogr., 10 (1974) 99.
555. C. H. Lochmuller and R. W. Souter, J. Chromatogr., 113 (1975) 283.
556. W. A. Konig, J. High Resolut. Chromatogr. Chromatogr. Commun., 5 (1982) 588.
557. R. H. Liu and W. W. Ku, J. Chromatogr., 271 (1983) 309.
558. S. Weinstein, B. Feibush, and E. Gil-Av, J. Chromatogr., 126 (1976) 97.
559. N. Oi, T. Doi, H. Kitahara, and Y. Inda, J. Chromatogr., 237 (1982) 297.
560. R. Charles, U. Beither, B. Feibush, and E. Gil-Av., J. Chromatogr., 112 (1975) 121.
561. W. A. Bonner and N. E. Blair, J. Chromatogr., 169 (1979) 153.
562. R. Charles and E. Gil-Av, J. Chromatogr., 195 (1980) 317.
563. W. A. Konig and G. J. Nicholson, Anal. Chem., 47 (1975) 951.
564. F. Andrawes, R. S. Brazell, W. Parr, and A. Zlatkis, J. Chromatogr., 112 (1975) 197.
565. R. S. Brazell, W. Parr, F. Andrawes, and A. Zlatkis, Chromatographia, 9 (1976) 57.
566. I. Abe, T. Kohno, and S. Musha, Chromatographia, 11 (1978) 393.

567. K. Grohman and W. Parr, Chromatographia, 5 (1972) 18.
568. H. Frank, G. J. Nicholson, and E. Bayer, J. Chromatogr. Sci., 15 (1977) 174.
569. H. Frank, G. J. Nicholson, and E. Bayer, J. Chromatogr., 146 (1978) 197.
570. H. Frank, G. J. Nicholson, and E. Bayer, J. Chromatogr., 167 (1978) 187.
571. T. Saeed, P. Sandra, and M. Verzele, J. Chromatogr., 186 (1979) 611.
572. T. Saeed, P. Sandra, and M. Verzele, J. High Resolut. Chromatogr. Chromatogr. Commun., 3 (1980) 35.
573. W. A. Konig, I. Benecke, and S. Sievers, J. Chromatogr., 217 (1981) 71.
574. W. A. Konig and I. Benecke, J. Chromatogr., 269 (1983) 19.
575. F. Mikes, G. Boshart, and E. Gil-Av, J. Chromatogr., 122 (1976) 205.
576. Y. H. Kim, A. Tishbee, and E. Gil-Av, J. Chem. Soc. Chem. Comm., (1981) 75.
577. C. H. Lochmuller and R. R. Ryall, J. Chromatogr., 150 (1978) 511.
578. W. H. Pirkle, J. M. Finn, J. L. Schreiner, and B. C. Hamper, J. Amer. Chem. Soc., 103 (1981) 3964.
579. W. H. Pirkle, D. W. House, and J. M. Finn, J. Chromatogr., 192 (1980) 143.
580. S. Hara and A. Dobashi, J. Chromatogr., 186 (1979) 543.
581. W. H. Pirkle and J. M. Finn, J. Org. Chem., 46 (1981) 2935.
582. R. Audebert, J. Liq. Chromatogr., 2 (1979) 1063.
583. P. E. Hare and E. Gil-Av, Science, 204 (1979) 1226.
584. J. N. Le Page, W. Linder, G. Davies, D. E. Seitz, and B. L. Karger, Anal. Chem., 51 (1979) 433.
585. S. Lam, F. Chow, and A. Karmen, J. Chromatogr., 199 (1980) 295.
586. Y. Tapuhi, N. Miller, and B. L. Karger, J. Chromatogr., 205 (1981) 325.
587. C. Gilon, R. Leshem, and E. Grushka, Anal. Chem., 52 (1980) 1206.
588. C. H. Lochmuller and E. C. Jensen, J. Chromatogr., 216 (1981) 333.
589. C. Pettersson and G. Schill, J. Chromatogr., 204 (1981) 179.
590. G. Szepesi, M. Gazdag, and R. Ivancsics, J. Chromatogr., 244 (1982) 33.
591. V. Schurig, R. C. Chang, A. Zlatkis, E. Gil-Av, and F. Mikes, Chromatographia, 6 (1973) 223.
592. V. Schurig, R. C. Chang, A. Zlatkis, and B. Feibush, J. Chromatogr., 99 (1974) 147.
593. V. Schurig, Chromatographia, 13 (1980) 263.
594. V. Schurig and R. Weber, J. Chromatogr., 217 (1981) 51.
595. V. Schurig, W. Burkle, A. Zlatkis, and C. F. Poole, Naturwissenschaften, 66 (1979) S423.
596. H. Traitler and M. Rossier, J. High Resolut. Chromatogr. Chromatogr. Commun., 5 (1982) 189.
597. W. Szczepaniak, J. Naworcki, and W. Wasiak, Chromatographia, 12 (1979) 559.
598. G. E. Baiulescu and V. A. Ilie, "Stationary Phases in Gas Chromatography", Pergamon Press, Oxford, 1975.
599. T. J. Tscherne and G. Capitano, J. Chromatogr., 136 (1977) 337.
600. B. Vonach and G. Schomburg, J. Chromatogr., 149 (1978) 417.
601. M. G. M. de Ruyter and A. P. de Leenheer, Anal. Chem., 51 (1979) 43.
602. P. L. Phelan and J. R. Miller, J. Chromatogr. Sci., 19 (1981) 13.
603. E. Grushka and F. K. Chow, J. Chromatogr., 199 (1980) 283.
604. V. K. Dua and C. A. Bush, J. Chromatogr., 244 (1982) 128.
605. N. H. C. Cooke, R. L. Viavattene, R. Eksteen, W. S. Wong, G. Davies, and B. L. Karger, J. Chromatogr., 149 (1978) 391.
606. V. A. Davankov, Adv. Chromatogr., 18 (1980) 139.
607. I. A. Yamskov, B. B. Berezkin, V. A. Davankov, Yu. A. Zolotarev, I. N. Dastavalov, and N. F. Myasoedov, J. Chromatogr., 217 (1981) 539.
608. G. Gubitz, W. Jellenz, and W. Santi, J. Chromatogr., 203 (1981) 377.
609. F. Guyon, A. Foucault, and M. Caude, J. Chromatogr., 186 (1979) 677.
610. K. Sugden, C. Hunter, and J. G. Lloyd-Jones, J. Chromatogr., 192 (1980) 228.
611. D. A. Leathard and B. C. Shurlock, "Identification Techniques in Gas Chromatography", Wiley, New York, 1970.

612. M. C. Bowman and M. Beroza, J. Ass. Off. Agr. Chem., 48 (1965) 943.
613. M. Beroza and M. C. Bowman, Anal. Chem., 37 (1965) 291.
614. M. C. Bowman and M. Beroza, Anal. Chem., 38 (1966) 1544.
615. M. Beroza and M. N. Inscoe, in L. S. Ettre and W. H. McFadden (Eds.), "Ancillary Techniques of Gas Chromatography", Wiley, New York, 1969, p. 89.
616. L. Neelakantan and H. B. Kostenbauder, Anal. Chem., 46 (1974) 452.
617. M. R. Tirgan and N. Sharifi-Sandjani, Analyst, 105 (1980) 441.
618. M. Cooke, G. Nickless, A. M. Prescott, and D. J. Roberts, J. Chromatogr., 156 (1978) 293.
619. M. Cooke, G. Nickless, and D. J. Roberts, J. Chromatogr., 187 (1980) 47.
620. P. A. Kennedy, D. J. Roberts, and M. Cooke, J. Chromatogr., 249 (1982) 257.
621. M. Cooke and D. J. Roberts, J. Chromatogr., 193 (1980) 437.
622. M. Beroza and B. A. Bierl, Anal. Chem., 39 (1967) 1131.
623. J. A. McCloskey and M. J. McClelland, J. Amer. Chem. Soc., 87 (1965) 5090.
624. V. Dommes, F. Wirtz-Pietz, and W.-H. Kunau, J. Chromatogr. Sci., 14 (1976) 360.
625. V. G. Berezkin, "Analytical Reaction Gas Chromatography", Plenum Press, New York, 1968.
626. M. Beroza, J. Chromatogr. Sci., 13 (1975) 314.
627. A. Simon, J. Palagyi, G. Speier, and Z. Furedi, J. Chromatogr., 150 (1978) 135.
628. J. H. Tumlinson and R. R. Heath, J. Chem. Ecol., 2 (1976) 87.
629. A. B. Attygalle and E. D. Morgan, Anal. Chem., 55 (1983) 1379.
630. E. D. Morgan, R. P. Evershed, and R. C. Tyler, J. Chromatogr., 186 (1979) 605.
631. B. P. Moore and W. V. Brown, J. Chromatogr., 121 (1976) 279.
632. M. S. Black, W. R. Rehg, R. E. Sievers, and J. J. Brooks, J. Chromatogr., 142 (1977) 809.
633. P. Kabo, J. Chromatogr., 205 (1981) 39.
634. B. J. Gudzinowicz, M. J. Gudzinowicz, and H. F. Martin, "Fundamentals of Integrated GC-MS", Dekker, New York, 1976, Part I, p. 182.
635. N. Gelsomini, J. Chromatogr., 279 (1983) 473.
636. J. E. Picker and R. E. Sievers, J. Chromatogr., 203 (1981) 29.
637. J. E. Picker and R. E. Sievers, J. Chromatogr., 217 (1981) 275.
638. C. Merritt, in L. S. Ettre and W. H. McFadden (Eds.), "Ancillary Techniques in Gas Chromatography", Wiley, New York, 1969, p. 325.
639. T. S. Ma and A. S. Ladas, "Organic Functional Group Analysis by Gas Chromatography", Academic Press, London, 1976.
640. R. C. Crippen, "The Identification of Compounds with the Aid of Gas Chromatography", McGraw-Hill, New York, 1973.
641. R. L. Levy, J. Chromatogr., 8 (1966) 49.
642. C. E. Roland Jones and C. A. Cramers (Eds.), "Analytical Pyrolysis", Elsevier, Amsterdam, 1977.
643. R. W. May, E. F. Pearson, and D. Scothern, "Pyrolysis Gas Chromatography", Royal Chemical Society, London, 1977.
645. V. G. Berezkin, CRC Crit. Revs. Anal. Chem., 11 (1981) 1.
646. W. J. Irwin, "Analytical Pyrolysis: A Comprehensive Guide", Dekker, New York, 1982.
647. W. J. Irwin and J. A. Slack, J. Chromatogr., 139 (1977) 364.
648. C. J. Wolf, M. A. Grayson, and D. L. Fanter, Anal. Chem., 52 (1980) 348A.
649. Y. Sugimura and S. Tsuge, Anal. Chem., 50 (1978) 1968.
650. J. Chih-An Hu, Anal. Chem., 49 (1977) 537.
651. J. P. Schmid, P. P. Schmid, and W. Simon, Chromatographia, 9 (1976) 597.
652. O. F. Falmer and L. V. Azarrage, J. Chromatogr. Sci., 7 (1969) 665.
653. R. L. Hanson, D. Brookins, and N. E. Vanderborgh, Anal. Chem., 48 (1976) 2210.
654. J. Q. Walker, J. Chromatogr. Sci., 15 (1977) 267.

655. J. J. R. Mertens, E. Jacobs, A. J. A. Callaerts, and A. Buekens, Anal. Chem., 54 (1982) 2620.
656. G. Blomquist, E. Johansson, B. Soderstrom, and S. Wold, J. Chromatogr., 173 (1979) 19.
657. E. Renier and F. L. Bayer, J. Chromatogr. Sci., 16 (1978) 623.
658. S. L. Morgan and C. A. Jacques, Anal. Chem., 54 (1982) 741.

Chapter 8

HYPHENATED METHODS FOR IDENTIFICATION AFTER CHROMATOGRAPHIC SEPARATION

8.1   Introduction

A chromatogram provides information regarding the complexity (number of components), quantity (peak height or area), and identity (retention parameter) of the components in a mixture. Of these parameters the certainty of identification based solely on retention is considered very suspect, even for simple systems; a peak eluting at a certain position in a chromatogram may be a substance other than the one anticipated. When the identity can be firmly established the quantitative information from the chromatogram is very good. The reverse situation applies to spectroscopic techniques which provide a rich source of qualitative information from which substance identity may be inferred with a reasonably high degree of certainty. Spectroscopic instruments have, however, two practical limitations: it is often difficult to extract quantitative information from their signals and, for reliable identification, they require pure or single-component samples. Chromatographic and spectroscopic techniques thus provide complementary information about the identity of the components and their concentration in a sample. Their tandem operation provides more information about a sample than the sum of the information gathered by either instrument independently [1]. This provides the driving force for the inception of combined instruments, often referred to as "hyphenated" systems. The principal hyphenated techniques are gas chromatography interfaced with mass spectrometry (GC-MS), Fourier transform infrared spectrometry (GC–FTIR), and optical emission spectroscopy (GC–OES); and liquid chromatography combined with mass spectrometry (LC–MS), Fourier transform infrared spectrometry (LC–FTIR), and nuclear magnetic resonance spectroscopy (LC–NMR). Our discussion will be limited to the problems of interfacing the various spectroscopic instruments to chromatographic systems and the use and management of the information obtained. Under the heading of interfacing there are two important aspects: the mechanical coupling of the two or more instruments and the control of the total system by computer.

8.2 Instrumentation for Mass Spectrometry

· A mass spectrum is a histogram of the relative abundance of individual ions having different mass-to-charge ratios (m/z) generated from a sample of neutral molecules. The important processes in obtaining a mass spectrum are vaporization, ionization, separation of the ions in a vacuum according to their m/z ratio, detection of these ions, and data management to reduce the information to an easily readable form. A mass spectrometer is a sophisticated measurement instrument and the processes involved are quite complex. Many texts describe the operating principles of mass spectrometry in detail [2–12]. Here we will discuss only those aspects of instrumentation of primary interest to chromatographers.

Ionization of organic molecules may be achieved in many ways, for example, by electron impact, chemical ionization, atmospheric pressure ionization, field emission/desorption, laser ablation, and californium-252 plasma desorption [3, 10, 13]. Of these methods, electron impact, chemical ionization, and to a lesser extent, atmospheric pressure ionization are the techniques most frequently used in GC–MS and LC–MS. Several factors contribute to the popularity of electron impact ionization: stability, ease of operation, simple construction, precise beam intensity control, relatively high efficiency of ionization, and narrow kinetic energy spread of the ions formed. The ionization source consists of a heated, evacuated cavity in which a beam of electrons with a narrow energy distribution is generated from a heated tungsten or rhenium filament. The energy of the ionizing electrons is controlled by the accelerating voltage established between the cathode and the source housing (normally). Electrons with energies in the range of 5 to 100 eV may be used but, unless otherwise stated, a value of 70 eV is standard practice. Since most organic compounds have ionization potentials of 7 to 20 eV, the energy transferred on collision between the electron and a neutral molecule is sufficient to cause both ionization and extensive fragmentation. The majority of ions formed by this process are singly-charged parent ions or molecular fragments, as well as a few multiply-charged ions and some negatively-charged ions. In most cases the number of negatively-charged ions is only a small fraction of the total number of ions formed. Ionization takes place at a temperature sufficient to maintain the sample in the vapor phase at a pressure below 10^{-5} Torr, which is sufficient to ensure that the average mean free path of the ion is large enough for it to escape the source without undergoing a significant number of ion-molecule collisions. The positive ions are extracted from the ion source by a repeller electrode having a small positive potential which directs them towards the analyzer slits. Alternatively, they may be extracted by the field penetration of the high voltage on the focussing electrodes. In both instances the ion beam is usually focused, collimated, and accelerated to provide a beam of narrow energy dispersion that is capable of traversing the analyzer section of the mass spectrometer. In modern mass spectrometers the ionization source and analyzer sections are usually differentially pumped, allowing the source to operate at a distinctly higher pressure than the analyzer unit, for which pressures below 10^{-7} Torr are common.

The bombardment of a neutral molecule with a beam of high energy electrons results in the formation of a molecular (or parent) ion radical with excess energy. The molecular ion provides useful information concerning the structure of the molecule as it defines the molecular weight. Unfortunately under electron impact conditions, this ion may be too unstable to be present in the spectrum even when beams of comparatively low energy (e.g., 15-20 eV) are used. In these circumstances a softer ionization method, such as chemical ionization, may be used to obtain molecular weight information. As well as providing information complementary to electron impact generated mass spectra, chemical ionization methods are blessed with other characteristics which makes them suitable for use in the trace level analysis of organic compounds.

Chemical ionization mass spectra are produced by ion-molecule reactions between neutral sample molecules and a high pressure (0.2-2 Torr) reagent gas ion plasma [13-15]. Fundamentally and practically several differences exist between chemical and electron impact ionization. In the chemical ionization source the concentration of reagent gas molecules exceeds that of the sample by several orders of magnitude, typically $1:10^3$ or $1:10^4$. Thus the electron beam ionizes the reagent gas with little direct ionization of the sample molecules. As the source is operated at high pressures compared to electron impact sources, ion-molecule reactions are now favored, and sample ionization results. These collisional processes are gentler than electron impact ionization and produce stable molecular ions or molecular ion adducts with little additional fragmentation. The exact mechanism of this process is discussed in the next paragraph. Chemical ionization sources are operated at much higher electron energies (200-500 eV) to ensure penetration of the ionizing electrons into the active source volume and require a "tight source" which is differentially pumped because of the large pressure differential established between the source and analyzer sections of the mass spectrometer. The high pressure conditions in the chemical ionization source also favor the production of thermal electrons which may be captured by molecules with a high electron affinity to form negatively-charged molecular ions or molecular ion fragments after dissociation [2,15,16]. The negative ion beam may be two or three orders of magnitude more intense than that generated under electron impact conditions, making negative chemical ionization mass spectrometry a viable analytical technique. As many compounds which readily form negative ions are biologically active or toxic substances, negative chemical ionization mass spectrometry is rapidly becoming an important technique in biomedical and environmental studies. Its areas of general application parallel those of the electron-capture detector. The detection of negative ions generally requires certain changes in a positive-ion mass spectrometer. For magnetic sector instruments this involves reversing the polarity of the accelerating voltage and magnet current. Quadrupole mass spectrometers can be operated in the pulsed positive and negative ion chemical ionization mode (PPNICI) where quasi-simultaneous recording of positive and negative ions is accomplished by switching the polarity of the ion source and focussing lens potentials at a rate of approximately 10 Hz. The alternative "packets" of positive and negative ions ejected from the ion source traverse the ion filter and are detected by either of two continuous dynode electron multipliers held at opposite potentials. Tandem chemical ionization/electron impact and combined sources which can be rapidly switched from one ionization mode to the other are available on some instruments [15,17].

Several reagent gases are used in chemical ionization mass spectrometry [3,10,14]. By way of example consider the processes that occur with methane, one of the most widely used regeant gases. A methane plasma is generated by a combination of electron impact and ion-molecule reactions which can be summarized by the following reaction scheme.

$$CH_4 + e^- \longrightarrow [CH_4]^+ + 2e^-$$

$$[CH_4]^+ + CH_4 \longrightarrow [C_2H_5]^+ + CH_3$$

$$[CH_3]^+ + CH_4 \longrightarrow [C_2H_5]^+ + H_2$$

$$[CH_2]^+ + CH_4 \longrightarrow [C_2H_4]^+ + H_2$$

$$[CH_2]^+ + CH_4 \longrightarrow [C_2H_3]^+ + H_2 + H$$

$$[CH]^+ + CH_4 \longrightarrow [C_2H_2]^+ + H_2 + H$$

$$[C_2H_5]^+ + CH_4 \longrightarrow [C_3H_7]^+ + H_2$$

$$[C_2H_3]^+ + CH_4 \longrightarrow [C_3H_5]^+ + H_2$$

At a source pressure of approximately one Torr, the ions formed from methane consist mainly of $[CH_5]^+$ (48%), $[C_2H_5]^+$ (40%), and $[C_3H_5]^+$ (6%). In the gas phase, $[CH_5]^+$ and $[C_2H_5]^+$ ions function as Bronsted acids and can transfer a proton to the sample molecule as shown below.

$$[CH_5]^+ + M \longrightarrow [MH]^+ + CH_4$$
$$[C_2H_5]^+ + M \longrightarrow [MH]^+ + C_2H_4$$

The $[C_2H_5]^+$ ion can also function as a Lewis acid, forming collision-stabilized complexes or molecular ion-adducts:

$$[C_2H_5]^+ + M \longrightarrow [M(C_2H_5)]^+$$

The product ions generated by chemical ionization are stable even-electron species with relatively little excess energy compared to direct electron impact. The chemical ionization mass spectrum is characterized, therefore, by the presence of a few intense molecular ion adducts with very little further fragmentation. With some experience or knowledge of the sample type, the sample molecular weight is readily identified from the m/z value of the molecular ion adducts. The nature of the molecular ion adduct formed depends primarily on the sample composition and the selection of reagent gas. Some typical reagent gases and the predominant reactive ions formed in the chemical ionization plasma are summarized in Table 8.1 [10]. Water and ammonia are relatively mild protonating reagents. With tetramethylsilane, trimethyl- silylation rather than protonation occurs while dimethylamine appears to selectively ionize carbonyl compounds, with protonated dimethylamine apparently attached to the carbonyl group. Nitric oxide reacts with ketones, esters, and carboxylic acids to give primarily $[M + NO]^+$ ions. With aldehydes and esters hydride abstraction occurs. The rare gases such as He, Ar, and Xe cause ionization by a charge transfer mechanism. The reagent ion abstracts an electron from the sample which results in the formation of an odd electron molecular ion, the same product as would be expected from electron impact, except that it is formed under less energetic conditions.

$$[He]^+ + M \longrightarrow [M]^+ + He$$

The energetics and thermodynamics of the chemical ionization process have been reviewed by Richter and Schwarz [14]. In general, only exothermic ion-molecule reactions proceed at high rates. This condition is nearly always fulfilled for proton transfer reactions under chemical ionization conditions if the proton affinity of the sample exceeds that of the conjugated base of the relevant reagent ion. A second, no less important consequence of the energetics of the ionization process is that the exothermicity controls to a considerable degree the stability of the $[MH]^+$ ions and their subsequent fragmentation. The temperature of the reagent gas must also be considered; as it contributes to the vibrational energy of the ionized particle, it may control the extent of fragmentation. The relative abundance of ions in chemical ionization mass spectra are often markedly temperature dependent.

TABLE 8.1

PRINCIPAL REAGENT IONS FORMED UNDER CHEMICAL IONIZATION CONDITIONS

Reagent Gas	Predominant Reagent Ions at ca. one Torr
Methane	CH_5^+, $C_2H_5^+$, $C_3H_5^+$
Propane	$C_3H_7^+$, $C_3H_8^+$
Isobutane	$C_4H_9^+$
Hydrogen	H_3^+
Ammonia	NH_4^+, $(NH_3)_2H^+$, $(NH_3)_3H^+$
Water	H_3O^+
Tetramethylsilane	$(CH_3)_3Si^+$
Dimethylamine	$(CH_3)_2NH_2^+$, $[(CH_3)_2NH]_2H^+$, $C_3H_8N^+$
Helium	He^+

Although atmospheric pressure ionization sources are somewhat novel at present they possess certain advantages in terms of simplicity and sensitivity for coupling to chromatographic equipment [18-20]. Operating at atmospheric pressure, an inert gas is bombarded with electrons from a Nickel-63 foil or a high voltage corona discharge to form a plasma of electrons and ions resulting from ionization and ion-molecule reactions of trace impurities such as oxygen, water, or solvent molecules in the carrier gas. Ionization of sample molecules results from ion-molecule reactions of these impurity ions and cluster ions as well as from direct electron attachment reactions. Both positive and negative ions are generated; the most common ions result from hydride abstraction or addition, charge transfer, electron-capture, and complex formation with cluster ions. The ions are sampled through an orifice (25 micrometer diameter pinhole) separating the source from the analyzer section of the mass spectrometer.

Ions leaving the source of the mass spectrometer must be separated according to their mass-to-charge ratio prior to detection. This is achieved by imposing an external electric or magnetic field on the ion beam to effect dispersion (resolution). The resolving power of a mass spectrometer is a measure of its ability to distinguish between two neighboring masses. Formally, the mass resolution of an instrument (R) is defined as the highest m/z value for which an ion can be separated from another ion of adjacent m/z value and identical abundance with a valley (between the ions) no greater than 10% of the height of the ions on the recorded spectrum. In general, mass spectrometers operate in two resolution ranges, low to medium resolution with values of R = 500–2000 and high resolution with values of R = 10,000–75,000. Low resolution mass spectrometry is routinely used in GC–MS and LC–MS applications. High resolution measurements provide accurate elemental composition data for all ions in the spectra or for a single ion such as the molecular ion. High resolution mass spectrometers are often referred to as double focussing, as the ion beam from the source is first focussed by an electrostatic analyzer which diminishes the energy dispersion of the ion beam and then by a magnetic analyzer which provides dispersion of the ion beam according to the mass-to-charge ratio of the ions. The combined process permits very accurate mass measurement (to within 10 ppm). Double-focussing mass spectrometers are more expensive and, when operated at high resolution, less sensitive and slower scanning (typically 6 s/decade) than single-focussing instruments. Their general use in GC–MS is less frequent than single-focussing instruments and we will not discuss them further. Burlingame has discussed the general problems in the design and implementation of a high resolution GC–MS system [21].

Rapid scanning mass spectrometers providing unit mass resolution are routinely used in GC–MS and LC–MS. Ion separation based on the mass-to-charge ratio of the ions is accomplished using either a magnetic sector, quadrupole filter, or time-of-flight mass analyzer. In a magnetic field, ions of different mass-to-charge ratio will be focussed at the detector according to the conditions represented by equation (1).

$$m/z = \frac{H^2 r^2}{2V} \tag{8.1}$$

m = mass of the ion
z = charge on the ion
H = magnetic field strength
r = radius of the circular path into which the ions are deflected
V = ion accelerating voltage

The position of the detector and the radius of curvature of the flight tube are usually fixed so that the complete mass spectrum is recorded by scanning (i.e., sequentially bringing all ions of different m/z into focus at the detector). Scanning is accomplished by changing the magnetic field (H) at a constant accelerating voltage (V) or by holding the magnetic field constant and varying the accelerating voltage. For most applications it is preferable to vary the magnetic field at a constant accelerating potential to record the mass spectrum. This is partly because the efficiency with which ions are transmitted through the analyzer depends on their momentum; at lower accelerating potentials relatively fewer ions are transmitted than at higher accelerating potentials, and thus scanning by varying the accelerating potential results in mass discrimination. Practically, this results in fewer ions of high m/z being recorded compared to those of low m/z. Scan times of 1-2 s per decade of mass (e.g., 50-500 amu) are required to accurately record the mass spectrum of early eluting peaks from a capillary column. This is possible with modern laminated magnets, although it should be kept in mind that, because of magnetic hysteresis, a certain amount of time is required prior to each scan for the magnetic field to reset and stabilize. Typical conditions might involve a scan time of 1.5 s and a reset time of 1.5 s. Effectively, therefore, scans can only be initiated every 3 s and chromatographic peaks with peak widths less than about 15 s may be erroneously recorded under these conditions due to the rapidly changing sample concentration over the scan period. Some magnetic sector instruments incorporate a Hall effect probe which, in conjunction with a data system, improves the accuracy of mass marking and reduces scan cycle times.

Quadrupole mass filter have enjoyed rapidly increasing popularity over magnetic sector instruments in the last few years except for applications involving the detection of masses at high m/z or mass resolution greater than about 1000. Earlier quadrupole designs employing circular rods have been superseded by instruments with hyperbolic poles, thus diminishing the problem of mass discrimination which limited the mass range and sensitivity of older instruments [22,23]. The quadrupole mass filter consists of four parallel hyperbolic rods in a square array such that the inside radius of the array (field radius) is equal to the smallest radius of curvature of the hyperbolae. Diagonally opposite rods are electrically connected to radio frequency and direct current voltages. For a given radio frequency/direct current voltage ratio only ions of a specific m/z value are transmitted by the filter and reach the detector. Ions with a m/z different than the transmitted ion are deflected away from the principal axis of the system and strike the rods; these ions are not transmitted by the quadrupole filter. To scan the mass spectrum the frequency of the radio frequency voltage and the ratio of the alternating current/direct current voltages are held constant while the magnitude of the alternating current and direct current voltages are varied. The transmitted ions of mass m/z are then linearly dependent on the voltage applied to the quadrupoles, producing a m/z scale which is linear with time. The voltages applied to the rods are usually chosen to give equal peak widths over the entire mass range. Modern quadrupole mass spectrometers with a mass range of 2 to 1500 amu and minimum peak widths of 0.3 amu are commercially available.

Quadrupole mass spectrometers have several attractive features when compared to magnetic sector instruments. They are smaller, less expensive, and more flexible. The low source voltages present fewer problems for sample introduction or for high pressure operation than chemical ionization and atmospheric pressure ionization methods. Electric fields can be precisely controlled much more easily than magnetic fields so that tuning desired m/z values for selected ion monitoring is more easily accomplished. Quadrupoles may also be scanned at faster rates, 0.1 s/decade, with a delay time of 3 ms between scans, and are thus well-suited for use with fast eluting peaks from capillary columns. In fact the boundary condition for the scan time is due to ion statistics, and related to the dwell time necessary at each mass in the spectrum to provide a reproducible measure of the ion current at each m/z value. If this condition is not met the mass spectrum will be distorted. Scan times of approximately 1 s/decade are therefore more common for practical reasons.

Time-of-flight mass spectrometers are rarely used for general organic analysis. The basic instrument comprises a source and detector separated by an evacuated drift tube usually 1 to 2 m in length. A voltage pulse of a few nanoseconds duration is used to eject ions from the source through a series of grids which provide an accelerating voltage. The "bunch" of ions so accelerated theoretically experiences the same force and therefore each ion acquires a velocity which is related to the square root of its mass. The ions are thus separated in time as they traverse the length of the drift tube. By operating the detector synchronously with the switching of the source voltage (i.e., injection of ions into the drift tube) a mass spectrum is produced. Mass scans can be made in a few milliseconds but spectrum reproducibility is limited by ion statistics. Resolution values of 500-600 may be obtained.

For ion detection a Faraday cage collector or, more commonly, an electron multiplier is used. The Faraday cage collector consists of a collector plate housed inside a cup at an angle such that impinging ions can not escape. The total ion current measured is therefore proportional to the total number of ions collected and their charge. The response is independent of the mass and energy of the ions and mass discrimination is low. Obtaining measurable currents requires large amplification circuits which are slow and prevent fast scanning. Alternatively, the electron multiplier detector produces current amplification of 10^3 to 10^8 with little noise and/or dark/background curent. The response time of this detector is very fast, leading to negligible signal broadening. Working on the principle of secondary electron emission, an ion impinging on the conversion dynode ejects several electrons which are accelerated to another electrode (dynode) where additional electrons are ejected. The discrete stage electron multiplier may have as many as 10 to 40 stages (dynodes), with each stage connected to a successively higher potential by a voltage divider, and the final collector (anode) connected to a solid-state amplifier. Some instruments employ a Channeltron electron multiplier as opposed to the discrete stage version. The Channeltron electron multiplier is a hollow glass tube coated with semiconducting material which functions both as an electron emitting dynode surface and as a voltage divider to establish the electrostatic field required to accelerate the secondary electrons. As for the discrete stage electron multiplier, positive ions are attracted towards and impact on the negative input edge of the Channeltron. The collisions release electrons which are accelerated down the tube by the electrostatic field. After a short distance these electrons collide with the sides of the Channeltron and release additional electrons. This process if repeated down the whole length of the Channeltron. The overall gain is similar to the discrete stage electron multiplier. With both devices modifications may be required to facilitate the detection of negative ions as discussed previously. The electron multiplier detectors show poor gain stability, a gain that is dependent upon the charge, mass, angle of incidence, and nature of the impacting ion, as well as the work function and surface condition of the impacted surface. Saturation effects limit the dynamic range of the detector and are usually significant when the output current exceeds 10^{-8} A. These characteristics are obviously not desirable in an ion detector but the high gain of the electron multiplier makes it indispensible and its short-comings must be tolerated.

Instrument design in mass spectrometry is a constant evolutionary process. At the time of writing ion cyclotron resonance and Fourier transform mass spectrometry may provide breakthroughs in low cost, rapid scanning, high resolution GC-MS [24]. Combination instruments employing GC-MS-FTIR may allow more positive sample identification [25]. Inexpensive mass detectors employing simple quadrupole systems have been introduced as selective gas chromatographic detectors for selected ion monitoring. The latter technique is discussed in a subsequent section.

8.3   Interfacing a Gas Chromatograph to a Mass Spectrometer

The general problem of interfacing a gas chromatograph to a mass spectrometer arises from the material flow requirement differences between the two instruments. The gas chromatograph typically operates at atmospheric pressure while source pressures in the mass spectrometer are in the range of 2 to 10^{-5} Torr for chemical and electron impact ionization. The interface should thus be capable of providing an adequate pressure drop between the two instruments, should maximize the throughput of sample molecules while maintaining a gas flow rate compatible with the source operating pressure, should not introduce excessive dead volume at the column exit, and should not degrade or otherwise modify the chemical constitution of the sample. Several interface designs are available and may be used in different circumstances. Here we will review those interface methods which have been adopted as standard methods on commercial GC-MS systems [4,5,8-10,26,27].

Modern differentially pumped mass spectrometers can accept column flow rates of 1-2 ml/min (adjusted to STP) into the ion source. This is also the optimum flow rate range for open tubular capillary columns. Coupling such columns to a modern mass spectrometer with high-capacity vacuum pumps therefore presents the fewest problems. Because of their flexibility, inertness, and low bleed characteristics fused silica columns with immobilized phases are often connected directly to the ion source with a simple vacuum-tight flange coupling and column support. This allows the position of the column end within the ion source to be adjusted for maximum ion yield [27-32]. The sample is quantitatively transferred to the ion source but the performance of the capillary column may be compromised by the large pressure drop placed across the column. Operation in this mode usually results in a substantial decrease in column retention times but a much less dramatic reduction in resolution. Another direct coupling device employs a fixed inlet restrictor to the mass spectrometer and a low-dead-volume needle valve which acts as a flow diverter. This interface is more versatile as it allows a wider selection of column flow rates, permits columns to be changed without instrument shutdown, and provides a by-pass by which large solvent volumes or corrosive derivatizing reagents can be vented away from the mass spectrometer. Disadvantages are the introduction of dead volume and the possibility of transforming labile compounds on the hot metal surfaces of the valve.

In many respects the use of an open split coupling is preferred over the valve interface [33-36]. For open split coupling the interface essentially consists of the capillary column or outlet tubing of the chromatograph and the inlet tubing of the mass spectrometer separated by a narrow slit, which is scavenged by helium at atmospheric pressure. The amount of the sample entering the mass spectrometer depends on the ratio of the inlet flow rate of the ion source (which is fixed for given inlet capillary dimensions, source conductance, temperature, and carrier gas) to the column flow rate. The total sample is transferred to the mass spectrometer only if the outlet flow rate of the column equals the inlet flow rate of the ion source; hence, no split occurs. If the flow rate of the column is larger than that of the source inlet, part of the sample is split off at the interface and does not reach the mass spectrometer. If the column flow rate is less than the inlet flow rate of the source, the sample is diluted by make-up gas and the mass flow of sample to the ion source is reduced. The splitting or dilution of the sample at the interface can be

594

determined by equation (8.2) [35].

$$C_{mo} = C_{cl} \left[\frac{W_{cl} - W_s}{W_{mo}} \right] \tag{8.2}$$

C_{mo} = sample concentration at the mass spectrometer inlet
C_{cl} = sample concentration at the column outlet
W_{cl} = flow rate at column outlet
W_s = flow rate of make-up gas
W_{mo} = flow rate at mass spectrometer inlet

Activating a valve in the make-up gas line provides a means of diverting solvent or reagent species away from the mass spectrometer and then rapidly restoring the analytical conditions after a preset time. The open split interface can be made of inert materials. It has been reported that Pt/Ir transfer tubing may cause chemical transformation of certain compounds when hydrogen is used as a carrier gas [37]. The advantages of the open split interface are that atmospheric pressure is maintained at the end of the column, column resolution is not affected, it is versatile with respect to the use of different column types and flow rates, and it permits easy change or replacement of the analytical column without affecting the operation of the mass spectrometer.

The problems associated with coupling wide bore columns to a mass spectrometer are more severe than those encountered with capillary columns. Packed columns are operated at much higher flow rates, 20 to 60 ml/min, and although this diminishes the influence of dead volumes in the interface on sample resolution, it poses a problem due to the pressure and volume flow rate restriction of the mass spectrometer. The interface must provide a pressure drop between column and mass spectrometer source on the order of 10^4 to 10^6, it must reduce the volumetric flow of gas into the mass spectrometer without diminishing the mass flow of sample by the same amount, and it must also retain the integrity of the sample eluting from the column in terms of the separation obtained and its chemical constitution [8,9,26,27]. To meet the above requirements the interface must thus function as a molecular separator.

The performance of any type of separator is characterized in terms of its separation factor (enrichment) N and separator yield (efficiency) Y [38]. The separator yield is defined as the ratio of the amount of sample entering the mass spectrometer (Q_{ms}) to that entering the separator (Q_{GC}), usually expressed as a percentage.

$$Y = \frac{Q_{ms}}{Q_{GC}} \times 100\% \tag{8.3}$$

It represents the ability of the device to allow organic material to pass into the ion source of the mass spectrometer. The separation factor is defined as the ratio of sample concentration in the carrier gas entering the mass spectrometer to the sample concentration in the carrier gas entering the separator.

$$N = \frac{Q_{ms}/V_{ms}}{Q_{GC}/V_{GC}} = \frac{Q_{ms}V_{GC}}{Q_{GC}V_{ms}} = \frac{(P_s/P_{He})_{ms}}{(P_s/P_{He})_{GC}}$$

(8.4)

Q_{ms} = amount of sample entering the mass spectrometer
V_{ms} = volume of carrier gas entering the mass spectrometer
Q_{GC} = amount of sample entering the separator
V_{GC} = volume of carrier gas entering the separator
P_s = sample partial pressure
P_{He} = partial pressure of carrier gas (helium)

The separation factor varies greatly depending on the type of separator, the gas chromatographic flow, the vacuum system efficiency, and the molecular weight of the sample. The separation factor N and the separator yield Y are algebraically related to one another by the expression:

$$N = \frac{Y}{100} \times \frac{V_{GC}}{V_{ms}}$$

(8.5)

Several types of molecular separators have been described; we will consider only three of these in detail. The effusion separator, the jet separator, and the membrane separator, Figure 8.1, are the principal molecular separators supplied with commercial GC-MS systems for packed column work. Other useful separators are the Teflon separator [39,40] and the silver-palladium separator [41-43]. The Teflon separator preferentially removes helium through a very thin Teflon capillary mounted in a heated, vacuum chamber. Separator yields and enrichment factors are reasonably high but problems such as peak broadening, a narrow operating temperature range (270-330 °C), restricted carrier gas flow rates, and functional group discrimination have resulted in the discontinuation of its use. The silver-palladium separator is based on a unique property of this alloy: it has a high permeability to hydrogen while remaining impermeable to other gases and organic compounds at 250°C. It is capable of providing the highest efficiency and enrichment of all separators but is little used for organic trace analysis due to the catalytic activity of the alloy. However, its simple design and lack of support vacuum system makes it ideally suited for special applications such as extraterrestrial GC-MS.

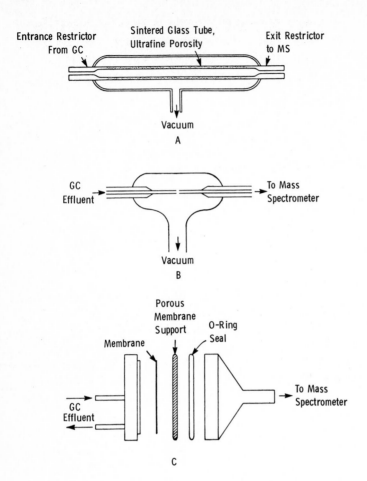

Figure 8.1 Common interfaces for GC–MS. A, effusion interface; B, jet
        interface; C, membrane interface.

The Watson–Biemann separator is an example of an effusion separator (Figure
8.1). It consists of an ultrafine-porosity sintered glass tube enclosed in a
vacuum envelope with glass capillaries at the entrance and exit to provide flow
restriction [9,44]. The average pore size of the frit is approximately one
micrometer. The function of the separator is to achieve an enrichment of sample
in the carrier gas while simultaneously producing a pressure drop from
atmospheric at the column exit to the working pressure of the vacuum system of
the mass spectrometer. The diameter of the entrance capillary is important in
affecting a pressure drop sufficient to satisfy the conditions of molecular flow
in the pressure reduction system. For the condition of molecular flow to apply,
the mean free path of the gas must be large compared to the pore diameters of
the frit. Thus, the corresponding pressure in the glass frit is on the order of

a few Torr (typically 1–10 Torr). Under these conditions the rate at which a gas effuses to the exhaust vacuum will be inversely proportional to the square root of the molecular weight and directly proportional to the partial pressure of each component. The quantity of any gas, Q, going through the porous glass frit is given by equation (8.6).

$$Q = Kp \, (1/M)^{1/2} \qquad\qquad (8.6)$$

K = constant depending on the conductance of the porous tube
p = partial pressure of the component
M = molecular weight

The ratio of the quantity of sample Q_s to the quantity of helium carrier gas V_{He} that goes through the frit is:

$$(Q_s/V_{He})_{frit} = (P_s/P_{He}) \, (M_{He}/M_s)^{1/2} \qquad\qquad (8.7)$$

Consequently, a fractionation of sample to carrier gas is obtained which depends upon the inverse ratio of the square root of the molecular weights. The exit constriction diameter controls the sensitivity of the apparatus; the diameter is limited only by the maximum pressure that can be tolerated in the ion source of the mass spectrometer. If sample fractionation is to be avoided then the mean free path of the gas must be small compared to the diameter of the exit capillary. Thus, under conditions of viscous flow, the flow is independent of molecular weight and all components are transported with equal velocity. The quantity of a given component entering the mass spectrometer is given by equation (8.8).

$$(Q_s/V_{He})_{ms} = P_s/P_{He} \qquad\qquad (8.8)$$

and the ratio of the quantity of sample to the quantity of helium that enters the mass spectrometer depends on the partial pressure of each component in the fritted tube. The diameter of the entrance and exit capillary tubing is set by trial and error. Once optimized it is not normally altered. The variable parameters of the Watson–Biemann separator, carrier gas flow and temperature, are optimized by experiment. The optimum operating temperature for each sample provides the maximum amount of sample entering the mass spectrometer and the production of symmetrical peaks. At either extreme of this temperature operating plateau both the enrichment and peak symmetry will decrease dramatically. The carrier gas flow rate affects the efficiency of the separator and the performance of the mass spectrometer. An upper limit to the flow rate is established by the conductance of the ion source and the optimum separator flow rate usually lies close to this upper limit. At lower flow rates peak shape and spectrometer response are poor.

Probably the most popular separator for use with packed columns is the jet separator, Figure 8.1. The effluent from the gas chromatograph is throttled through a fine orifice where it rapidly expands into a heated vacuum chamber. During this expansion the lighter helium gas molecules rapidly diffuse away from the core of the supersonic jet which becomes enriched in the heavier sample molecules. The core is received by a second jet and enters the mass spectrometer. The alignment, orifice diameter, and relative spacing of the expansion and collection jets are crucial. Normally these are fixed by the manufacturer so that maximum enrichment will only be obtained over a narrow flow rate range. Clogging of the expansion jet by particulate matter is a frequent problem whose remedy often tests the ingenuity of the operator [45,46]. Mechanical proding with a fine wire, solvent flushing under vacuum, brief treatment with dilute hydrofluoric acid, a high voltage discharge, and conditioning in an annealing oven are among the more popular methods of wizardry.

The silicone membrane separator, Figure 8.1, works on the principle that the transmission of organic compounds across the membrane may be two orders of magnitude greater than that of an inert gas [47]. The effluent from the gas chromatograph is exposed to the high pressure side of the membrane. It is either channeled along a short (several cm) spiral path over the membrane or simply allowed to flow into a small dead-volume chamber which is exposed to the membrane. The membrane is usually a dimethylsilicone rubber about 0.02-0.04 mm thick with a surface area usually much less than one cm^2. To maintain column resolution the dead volume of the cavity on the high pressure side of the membrane must be minimized. The low pressure side of the membrane leads directly to the mass spectrometer source. The surface area, thickness, and contact time are critical design parameters and temperature is an important operating parameter. The fundamental factor limiting transmission of organic compounds through a membrane is the permeability of the membrane. Membrane permeability is a function of solubility S, diffusivity D, pressure difference across the membrane $P_1 - P_2$, exposed area A, and membrane thickness L. The amount of sample transmitted by the membrane Q_{ms} is therefore given by equation (8.9).

$$Q_{ms} = DS \ (P_1 - P_2)/L \tag{8.9}$$

A vital consideration in the use of the membrane separator is the temperature dependence of the transmission of species across the membrane. Total transmission depends on both solubility and diffusivity. For organic solutes membrane solubility is the most important parameter. Solubility increases with decreasing temperature so that the highest enrichment possible is obtained at the lowest temperature commensurate with maintaining the integrity of the column separation.

No single separator is ideal for all situations. As can be seen from Table 8.2, the membrane separator would at first sight seem to be the most useful. It provides the highest sample enrichment and essentially allows the column to be operated separately from the mass spectrometer so that both systems can be individually optimized. Unfortunately, the performance of the membrane separator is critically influenced by temperature. There is an optimum temperature for the transmission of each organic compound across the membrane; at less than optimum temperature conditions peak distortion may occur and the relatively low upper operating temperature precludes its use with organic compounds of low volatility. The jet and effusion separators require auxiliary vacuum equipment and only operate efficiently over a narrow range of flow and temperature conditions. Careful deactivation of the effusion separator may be required to avoid sample decomposition at high temperature.

TABLE 8.2

COMPARISON OF MOLECULAR SEPARATORS

Typical Properties	Effusion (Watson–Bieman)	Jet	Membrane
Yield (%)	20–30	30–70	30–80
Enrichment	4–7	6–14	10–30
Carrier gas flow range (ml/min)	10–30	15–25	1–60
Temperature (approximate upper limit, °C)	350	300	255

With modern differentially pumped mass spectrometers single-stage separators are adequate. Hybrid two-stage combinations may be required with older instruments having inefficient pumping systems. The enrichment factor of a two-stage separator is the product of the enrichment factors for each stage.

8.4 Interfacing a Liquid Chromatograph to a Mass Spectrometer

Interfacing a gas chromatograph to a mass spectrometer is relatively simple compared to the problems encountered in LC–MS. Here the mass flow created by vaporizing the eluent from the liquid chromatograph is several orders of magnitude greater than the conductance of the mass spectrometer. For example, a

liquid chromatograph operating at a flow rate of 1.0 ml/min would produce nearly
200-1000 ml/min of vapor at STP, of which the mass spectrometer could accept 1-2
ml/min The sample portion of this vapor plume is so small that an enrichment
factor on the order of 10^5 is required, a value many times greater than that
needed for GC-MS. Also, liquid chromatography is frequently selected for the
separation of samples which are thermally sensitive and cannot be volatilized
without pyrolysis. Those ionization techniques which require volatilization in
a heated source prior to ionization would thus not be suitable for thses
compounds. Some mobile phases, such as those containing inorganic salts or
buffers, may present additional problems, such as clogging of the interface, as
they are generally not volatile or volatalizable substances. LC-MS is thus not
truly domesticated and, although commercial LC-MS interfaces exist, they
represent a compromise situation which retains some of the inherrent problems
discussed above [27,48-50].

The principal methods of interfacing a liquid chromatograph and a mass
spectrometer are direct introduction into an atmospheric pressure ionization
source [51], direct introduction into a chemical ionization source [52,53], the
expanding jet interface [54,55], and the moving belt interface [56,57]. By
virtue of atmospheric pressure operation the atmospheric pressure ionization
source can accept much higher gas or solvent vapor flow rates than conventional
high-vacuum sources. The column eluent is mixed with a stream of preheated
carrier gas in a heated evaporator tube and carried directly into the ion
source. An ion plasma is generated by bombardment with electrons from a corona
discharge and by ion-molecule reactions. Both positive and negative ions are
generated in large numbers. A fraction of the ion plasma is sampled directly by
the mass spectrometer through a 25 micrometer orifice. Detection limits are
lower than might be anticipated due to consumption of the ionizing electrons by
solvent molecules and, in particular, by electron-capturing solvent impurities.
The chemical ionization spectrum is dependent upon the mobile phase composition
and will therefore be different for different solvents and may even change for
binary solvent mixtures of different composition. The choice of solvent is
essentially limited to those having a lower proton affinity than the solute.

A modified chemical ionization source, Figure 8.2A, can accept a fraction
of the eluent from a liquid chromatograph. This fraction represents a flow of
about 10-70 microliters/min, and is thus only a small fraction of the normal
flow of a standard liquid chromatograph (less than 1%). The column eluent
enters the ion source directly through a cooled insertion probe. The interface
itself is a variable diaphragm orifice, 5-15 micrometers in diameter, which is
changed to match the viscosity of the solvent to the conductance of the ion
source. Because of the large quantities of solvent that go into the mass
spectrometer, a cryogenic trap (cold finger) is provided in the source region to

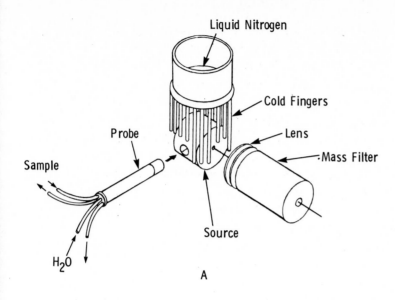

Liquid Nitrogen

Cold Fingers

Probe

Lens

Mass Filter

Sample

H₂O

Source

A

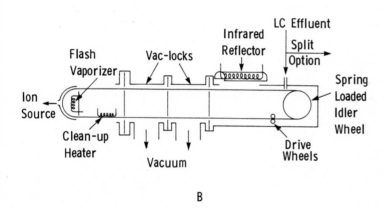

LC Effluent

Infrared
Reflector

Split
Option

Flash
Vaporizer

Vac-locks

Spring
Loaded
Idler
Wheel

Ion
Source

Clean-up
Heater

Vacuum

Drive
Wheels

B

Figure 8.2   Interfaces for LC-MS. A, direct liquid introduction interface;
          B, moving belt interface.

supplememt the normal chemical ionization pumping system. Within the source,
the vaporized solvent acts as the chemical ionization reagent and thus directly
influences the mass spectra produced. Problems arise from clogging of the micro
orifice by particulate matter, the need to "deice" the cold trap at regular
intervals, and the need to maintain the liquid state of the effluent in the
insertion probe (the reason for the supply of cooling water) while inside the
heated source block. Perhaps the most significant drawback of the direct
introduction interface (of which there are designs other than the one described

above, see reference [48]) is the low sample utilization rate, which rarely exceeds 1% under normal operating conditions. The use of microbore HPLC columns may solve this problem to some extent, although extracolumn dead volumes seriously limit the interface design.

The expanding jet and moving belt interfaces represent current attempts to increase sample utilization by enrichment. There are two versions of the jet interface proposed by Vestal. In an earlier design, the effluent from the liquid chromatograph forms a liquid jet which is irradiated by a focussed laser beam (50 watt CO_2 laser) to increase the rate of solvent vaporization. At the interface, which has its own vacuum system, the rapidly expanding solvent vapor is pumped away, leaving an enriched primary molecular beam which enters the mass spectrometer source through a micro-orifice. The source area, which is an integral part of the interface design, incorporates various modifications for focusing the primary molecular beam and for producing either electron impact or chemical ionization mass spectra. In a more recent design, Vestal has employed an electrically-heated vaporizer which generates a supersonic jet at the column/source interface [55]. The jet normally contains a mist of fine particles or droplets which traverse the ion source of the mass spectrometer and directly enter the entry port of the source vacuum pumping system. The system will accept flow rates of up to 2.0 ml/min and can be adapted to any commercial quadrupole chemical ionization mass spectrometer by relatively simple modifications to the ion source.

As there often exists a substantial difference in volatility between the solutes typically analyzed by liquid chromatography and the solvents used for their separation, it should be possible to use selective solvent vaporization as an enrichment technique in LC-MS. This principle is employed on-line in the moving belt interface, Figure 8.2B. The column eluent is brought either into direct contact with the belt or a zero-volume splitter, depending on the ability of the belt to hold the solvent. The solvent flow that can be accommodated depends on the dimensions of the belt and also on the solvent type. The belts are typically about 0.3 cm wide, are made of a metal mesh or polyimide material, and have a capacity of about 1.5 ml/min for volatile nonpolar solvents and about 0.1-0.2 ml/min for water-containing mobile phases. The thin film of eluent on the belt is carried along under a focussed infrared lamp which effects a rapid preliminary vaporization of solvent. The solute and risidual solvent then pass through two serial vacuum locks and into the mass spectrometer vacuum envelope. Heating the belt in a chamber adjacent to the ion source flash vaporizes the sample into the source. On the return journey a cleanup heater is provided to remove solute residues that might cause ghost peaks on recycle.

The operating characteristics of the LC-MS interfaces discussed above are compared in Table 8.3. No single system is ideal. The moving belt interface is probably the most widely characterized device and does provide reasonable enrichment for samples with boiling points greater than about 170–180°C. Either electron impact or chemical ionization mass spectra may be obtained for those compounds which can be thermally vaporized from the belt surface. In the other methods, chemical ionization occurs in the presence of the solvent. The choice of solvent will thus influence the nature of the ions generated. This mechanism is not completely understood but involves the formation of solvent cluster ions as the ionizing reagent gas. Details can be found in references [55,58,59]. The abundance and mass of the ions formed depends on the solvent composition and source pressure and temperature. Greater attention to such details may be necessary to generate reproducible reference spectra for interlaboratory identification purposes.

TABLE 8.3

CHARACTERISTICS OF LC-MS INTERFACES

Property	Moving Belt	Atmospheric Pressure Ionization	Expanding Jet	Direct Liquid Inlet
Accepted flow of solvent (ml/min)				
Nonpolar	1–2	1	1	0.01–0.07
Polar	0.1–0.3	1	0.5	0.01–0.07
Sample yield (%)	30–40		1	0.3–1.0
Mass spectrometer mode	EI/CI	CI	EI/CI	CI
Low mass limit (amu)	90–100		40/90	150
Technical complexity	high	medium	very high	low
Adaptable to existing instruments	yes	yes	no	yes
Detection limit (g)				
Single ion recording	10^{-12}	10^{-12}	10^{-9}	10^{-12}
Full scan	10^{-9}	10^{-9}	10^{-6}	10^{-9}

8.5 Computer Acquisition of Mass Spectral Data

The amount of data generated during GC–MS runs is immense and no one would consider attempting to handle this data manually today, although this was common practice only a few years ago. Computers are now less expensive and readily available. Virtually all modern GC–MS instruments employ computers and/or sophisticated microprocessors for several purposes. The most important of these are data acquisition and instrument control; data processing, including background subtraction, mass marking, normalization, formatting, printing, and display; and identification of mass spectra by library search methods or with interpretive programs employing self-training algorithms [2,60-63]. We will consider the role of the computer in data acquisition, data handling, and data interpretation separately.

Consider Figure 8.3. The computer may receive analog signals from the detector, the ion source, or, for magnetic sector instruments, a Hall effect probe. The latter is used to sense the magnetic field during scanning and provides a time base for mass marking the spectrum. The total ion current monitor is positioned between the ion source and mass analyzer in such a way that it intercepts a few percent of the total ion beam. It thus provides a continuous monitor of the rate of ion production in the source which can be correlated with the rate of sample mass entering the source. The total ion current monitor then provides a representation of the chromatogram which can be used to identify peaks of interest for mass spectral scanning. Other methods for reconstructing the chromatogram under continuous scan control are available and these will be discussed later. The total ion monitor is often used with magnetic scanning instruments to optimize the source operating conditions prior to sample analysis.

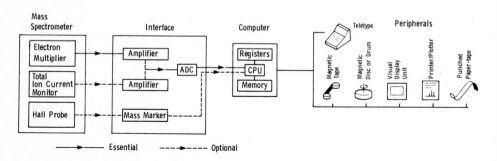

Figure 8.3 System for the control and acquisition of data from a mass spectrometer.

The signal output from the electron multiplier detector is an analog voltage proportional to the number of ions arriving at the detector and varies during mass scanning. As this signal is not suitable for computer acquisition an interface, an analog-to-digital converter (ADC), is needed. The critical parameters of the ADC are the sampling frequency and the dynamic range (ratio of the largest to smallest voltages handled). Recording mass spectra usually requires a dynamic range of at least 10^6, which is not possible with a fast ADC. The effective dynamic range of the system must be increased by dividing the mass spectrometer output among a number of amplifiers having different gains, and then sampling these channels in turn through a multiplexer. Alternatively a programmable amplifier under the control of the data system might be used. The scaled, digitized data is usually preprocessed at the interface or by the data system to remove noise spikes, single ion peaks, and metastable peaks using one of several peak recognition and threshold routines. Afterwards the digitized data is converted into position and intensity measurements for each peak recognized. Peak position is usually estimated as the centroid time or peak maxima time for the digital data array and intensity is determined by the maximum peak height or peak area. The final process is mass conversion; the conversion of the time axis into mass increments. This is usually effected through a calibration of the instrument based on the position of peaks of known mass from a standard compound such as perfluorokerosene or heptacosafluorotributylamine. For low resolution mass spectrometry external calibration is used to establish the calibration at the start of the day and then repeated as required to check for drift. In addition to time, the Hall voltage in magnetic scanning instruments or the applied voltage in quadrupole instruments can be used to define the mass scale. Whichever method is used, reproducible scan behavior is required, as succeeding spectra are mass converted by interpolation of the calibration data. The mass spectra are usually normalized so that either the most intense peak or the sum of all mass peaks is assigned a value of 100%. If the total ion current signal is recorded simultaneously with mass data it can be used to correct for fluctuations in ion current during the course of a scan. Other algorithms are available to subtract out background contributions, for example from column bleed, or to eliminate interferences from overlapping chromatographic peaks. For on-line recording of mass spectral data during a chromatographic run the raw data is usually stored directly on tape and then manipulated at the end of the run. The operator may then view part or all of this data, deciding which portions to keep and which data to erase. A video display unit is usually used for this purpose. Spectra of interest may be printed from the screen onto a digital plotter to provide a hard copy of the ion chromatogram or mass spectrum.

The computer does not merely passively accept data from the mass spectrometer. Conversation is bimodal and the computer continuously monitors and modifies the operating status of the instrument by actuating electronic switches and adjusting voltages in accordance with a menu of preset operating instructions, selected at the beginning of the analysis. Control of the scanning function of the mass spectrometer is one variable that has a direct bearing on the methods used for data processing. Two scanning techniques, continuous scanning and selected ion monitoring, are used routinely. In the continuous scanning mode the computer initiates a scan of the mass spectrum at fixed time intervals and repeats this process throughout the analysis. An advantage of this technique is that a complete record of the sample analysis is available that may be searched retrospectively if the object of the analysis should change at a future date. The immediate problem is to decide which of the many recorded spectra are most relevant to the problem at hand. In most cases a summary of the total analysis in the form of a total ion current profile (TICP) is generated [64]. The TICP is defined as a normalized plot of the sum of the ion abundance measurements in each member of a series of mass spectra as a function of the serially indexed spectrum number. Some authors refer to this as a reconstructed chromatogram. A scan containing significant information about the sample is then easily located from the TICP and a copy of the mass spectra of interest requested by identifying the appropriate scan number. When specific information is required about the analysis of a single component or a few components whose mass spectral properties are known or can be estimated, then mass chromatography may be used. In this case a mass chromatogram, a display against time of the intensity recorded at a specific mass value, is used to identify the substance of interest. The specificity is increased by simultaneously displaying the mass chromatograms of several characteristic ions of the substance of interest. The mass chromatograms should maximize in the same scan sequence if the desired substance is present in the sample. The scan number at which this occurs can be identified and the full mass spectra obtained to ensure the substance is correctly identified.

When the object of the analysis may be clearly defined, it is not necessary to record the full mass spectra of all substances present in the sample. As discussed above, a few ions characteristic of the substance or substances of interest may be monitored throughout the experiment. Known as selected ion monitoring (SIM), it is defined as the dedication of the mass spectrometer to the acquisition of ion abundance data at only selected masses in real time as components emerge from the chromatographic system [3,7,63,64]. It differs from the continuous scan mode in that only preselected ions are recorded during the analysis by rapid switching of the accelerating voltage for magnetic sector or the quadrupole voltage for quadrupole mass spectrometers. Its principal advantage is high sensitivity, as only the ions of interest are measured at the detector and the dwell time over which the ion current is recorded for each ion is increased compared to the continuous scan mode. It lends itself to problems in quantitative organic trace analysis of complex mixtures, a topic discussed in the next section. Compared to the continuous scan mode a disadvantage is that the full range of ions is not recorded so that, if at a later date information different to that originally sought is required, it is not available without repeating the analysis.

As well as processing data into a form convenient for use by a mass spectroscopist the computer can also help to identify the mass spectrum of an unknown compound by retrieval or interpretive procedures [66-69]. Retrieval processes using library search procedures are available on most GC-MS instruments. Commercial spectral libraries containing some 30,000-60,000 reference mass spectra of a general nature or for specific compound types such as EPA priority pollutants, common drugs of abuse, etc. are readily available. In this technique spectra from a magnetic tape or disc file are read in turn and compared with the unknown spectrum. The computer should correctly identify the authentic spectrum if present, and otherwise should pick similar spectra, guiding the analyst towards the correct identification. Selection and encoding of certain mass peaks or features from a spectrum reduces both the storage and time requirements of a search system. For instance, two common methods of library searching use the eight most abundant peaks in the mass spectrum or the two most abundant peaks present every fourteen mass units to abbreviate the mass spectrum. The degree of correspondence between the library and unknown spectra are then calculated and ranked according to the degree of fit using one of several pattern recognition methods. As there are many more organic compounds than there are reference mass spectra an unknown compound may remain unknown after a library search. Interpretive procedures using artificial intelligence methods may be useful. Here, in simple terms, the computer is armed with the basic knowledge of the mass spectroscopist concerning the most common modes of fragmentation, the most frequent neutral losses, etc., and with this information it deduces possible molecular structures in the same way that a trained mass spectroscopist would. The programs for this process are rather large and elaborate, requiring a main frame computer for their execution. Generally, the mass spectroscopist does not have this facility available in his laboratory and the programs are accessed from one of the regional mass spectrometry centers by telephone hookup.

8.6 Quantitative Selected Ion Monitoring

The mass spectrometer can be operated as a very sensitive and mass selective chromatographic detector. The mass spectrometer can be tuned to monitor a single ion or to switch at high frequency between several preset m/z values throughout the chromatographic separation. Selectivity is obtained because mass is a substance-characteristic parameter. When several masses are recorded sequentially the specificity is increased as the probability of two different substances having in common the same retention time and the same characteristic mass ion ratios is very small. High sensitivity is obtained because the dwell time at which the ion current for each mass is measured is increased compared to the normal scan mode. The sensitivity of selected ion monitoring is from 100 to 1000 times greater than is obtained in the scan mode, providing detection limits in the range 10^{-9} to 10^{-15}g. This limit will depend on the ionization efficiency of the compound, the fraction of the total ion current carried by the ion monitored, the contribution of background ions to the signal, mass spectrometer resolution, and the sensitivity of the detector. Measurements may be made using electron impact or chemical ionization methods with positive or negative ion detection. Selected ion monitoring is now a widely used technique in the areas of environmental [64,70-73] and biomedical analysis [3,7,11,65,73]. The advent of relatively inexpensive quadrupole mass ion detectors for gas chromatography has led to the greater accessibility of selected ion monitoring techniques to many laboratories lacking sophisticated multipurpose scanning instruments.

Another feature of selected ion monitoring is the use of stable isotope analogs as internal standards [74]. Differing from the substance to be determined only in mass, they are ideally suited to the multiple ion monitoring techniques. They can be added to the sample at an early stage to correct for losses at all stages of sample workup and analysis [75]. Suitable ^{2}H, ^{13}C, ^{15}N, ^{18}O, etc., analogs may not be available and their synthesis in high isomeric purity may be difficult. When a stable isotope analog is not available then an homologous compound containing the same m/z ions as the substance to be determined is the next best choice and may be more readily available or easier to synthesize. Further comments on the selection of internal standards for quantitative selected ion monitoring are summarized in Table 8.4.

8.7 On-Line Coupling of a Gas or Liquid Chromatograph with a Fourier Transform Infrared Spectrometer

Molecular vibrational transitions give rise to the absorption of energy in the infrared region of the electromagnetic spectrum. These characteristic absorption frequencies can be correlated with the structure of the molecule and are particularly useful for indicating the presence of functional group sub-units. However, except by comparison with an authentic spectrum, it is rarely possible to identify an unknown compound entirely on the basis of its infrared spectrum. Infrared spectrometry is particularly poor for distinguishing between homologs without additional molecular weight information. This information may be obtained by mass spectrometry. The two techniques are complementary as a knowledge of the infrared spectra of a substance helps in the interpretation of the mass fragmentation pattern. Attention has thus been given to the construction of hybrid GC-FTIR-MS instruments for on-line structural elucidation with computer interpretation [1,76].

TABLE 8.4

INTERNAL STANDARDS FOR QUANTITATIVE SELECTED ION MONITORING

Type of Standard	Comments
Stable isotope analog of the substance quantified.	Has similar physical and chemical properties as the unlabelled compound. Amount of unlabelled material present must be minimal if linear calibration curves are to be obtained. If a deuterated analog is used, a trideuterated molecule is the minimum analog which will give a linear calibration curve. May be expensive. Labelled derivatizing reagents available.
Homologous compound whose mass spectrum contains a reasonably intense ion at the same mass as the ion being determined.	Synthesis often easier and less expensive than labelled analog. Physical and chemical properties close to those of the substance being determined. Must have the same m/z ion as the substance being determined.
Compound with similar physical properties to the substance being measured but whose mass spectrum has no significant ions in common with it.	Least likely to provide accurate data but the most likely to be available.

Until a few years ago dispersive (grating or prism) instruments were used to acquire infrared spectra. These instruments were relatively slow and insensitive, which made recording spectral information on-the-fly from chromatographic instruments all but impossible except under conditions which severely compromised the performance of the chromatograph. They could of course be used as selective single-wavelength detectors for gas chromatography [77] in which case no spectral information beyond the presence of a functional group absorbing at the measuring wavelength was obtained. This situation improved dramatically with the advent of interferometric recording instruments and algorithms for the fast Fourier transform of the interferogram into an IR absorption or transmission spectrum. Data acquisition and transformation require a minicomputer as part of the instrument ensemble. The infrared spectrum is recorded by a Michelson interferometer which consists of two mirrors at right angles to one another and a beam splitter which bisects the angle between the two mirrors. One mirror is stationary and the other may be moved at a constant velocity. After reflection the two beams recombine at the beam splitter and, for any particular wavelength, constructively or destructively interfere depending on the difference in optical paths between the two branches of the interferometer. For a broad band emission source each frequency comes in and out of phase at a characteristic optical path difference, and the superposition of all frequencies produces the observed interferogram. The combined beam then traverses the sample compartment and is focused on the detector. The detector signal is sampled at very precise intervals during the mirror scan. Both the sampling rate and mirror velocity are precisely controlled by modulation of an auxiliary reference beam from a helium-neon laser. After data collection the dedicated computer executes a Fourier transform of the data to provide a single beam spectrum which may then be ratioed against a background spectrum to provide the customary transmittance (or absorbance) versus wavenumber infrared spectrum [78-79]. The advantages of Fourier transform infrared spectrometry (FTIR) viz. a vis. dispersive measurements are that all frequencies are recorded at all times throughout the experiment (Fellgett advantage), as programmable slits are not required the energy throughput of the instrument is higher (Jacquinot advantage), higher frequency accuracy (0.01 cm^{-1}) is possible because of the use of the frequency lock mechanism provided by the reference source, and there is no stray light contribution to background noise as in dispersive instruments. For on-line coupling to chromatographic instruments the principal advantages are those of greater sensitivity and rapid scanning ability. The infrared spectrum may be scanned in less than a second and individual scans summed during the elution of a chromatographic peak to provide sensitivity two orders of magnitude greater than those obtained with dispersive instruments. Computer aquisition provides attendant advantages such as background subtraction, peak deconvolution, and library search facilities,which add to the attraction of FTIR methods when combined with chromatographic instruments [80-81].

Before considering on-line techniques for acquiring infrared spectral data we will briefly review some of the off-line procedures [82-84]. Although generally tedious, these methods are likely to be used in those laboratories requiring infrequent access to infrared data or those lacking access to an FTIR spectrometer. It should be recalled that spectra can be obtained off-line with either a dispersive or Fourier transform instrument, but by using the spectrum accumulation feature of the latter the sensitivity may be two orders of magnitude greater [85].

Gas chromatographic effluents are usually trapped in a mini-column containing a few milligrams of potassium bromide or potassium chloride [86,87] or in a microcell cooled to -70°C [88,89]. Samples trapped on potassium bromide mini-columns are usually formed into micro-disc pellets and analyzed in the conventional manner using a beam condenser to concentrate the source energy into the small area occupied by the sample. With a dispersive instrument about 1-10 micrograms of sample is usually required to obtain a recognizable spectrum. Suitable microcells of 1-2 microliter volumes and made of a material transparent to infrared radiation (e.g., silver chloride) are commercially available. The sample is usually collected in the cell by cryogenic condensation, often in the presence of a solvent to improve trapping efficiency. Sensitivity is similar to the micro-disc method. Both methods usually allow only a fraction of the sample to be trapped and are generally used with packed columns and high sample loadings. For liquid chromatography the effluent is generally collected at the detector exit and added to a few milligrams of potassium bromide. Excess solvent is evaporated and the sample formed into a micro-disc. This procedure is rarely successful with mobile phases containing water, for which solvent extraction of the sample into a volatile organic solvent is made prior to addition to the potassium bromide. With preparative HPLC sufficient sample may be collected for analysis without resorting to micro techniques.

On-line GC-FTIR techniques have been ably reviewed by Erickson [90]. Several interface designs of varying degrees of complexity are available but the most widely used approaches are the flow-through light pipe [91-93], stop-flow light pipe [92-94], and cryogenic matrix isolation [95]. A stand alone FTIR spectrometer can be interfaced to a gas chromatograph by a light pipe. The gas-chromatographic effluent passes through a heated transfer line to the light pipe which is in the optical path. The effluent may, depending on the instrument design, be split to or passed through a standard gas-chromatographic detector which signals the arrival of the peak to the data system. In other instruments the chromatogram is reconstructed from the interferograms stored in the data system and an external detection system is not required. The light pipe usually consists of a gold-coated, heated glass tube with potassium bromide or zinc selanide windows affixed to either end. For optimum performance the volume of the light pipe should be less than the volume of gas containing the peak from the chromatograph, the flow of gas should be laminar to avoid intermingling of the peaks in the cell and transfer lines, it must transmit infrared radiation with a minimum of signal loss, and it must be inert to the sample at elevated temperatures. Typical dimensions are 0.1 to 0.3 cm internal diameter and 20 to 100 cm in length, giving volumes of approximately 0.8 to 5.0 ml. The transmitted infrared energy is usually detected by a high sensitivity mercury-cadmium telluride detector cooled with liquid nitrogen. In some instruments the transfer lines are equipped with a set of air-actuated valves that can be used to trap a particular peak in the light pipe. Signal averaging provides improved sensitivity but at the expense of not recording other peaks which must be by-passed. Depending on the properties of the substance and the method of detection somewhere between 10 nanograms and 1 microgram of sample is required for useful spectral data.

Matrix isolation provides an alternative method of interfacing a gas chromatograph to a FTIR spectrometer [95]. The effluent from the gas chromatograph is co-condensed with an inert carrier gas at cryogenic temperatures to form a solid matrix from which an infrared spectrum can be directly obtained. The interface device, termed a cryocollector, thus serves as a sample collector, concentrator, cell, and storage device. Continuous sample collection and recording are performed using a stepping carousel controlled by an upstream flow-through detector, which actuates the drive mechanism of the carousel as each peak above a certain threshold concentration elutes. The cryogenic spectra, measured at 15 K, are of high quality and relatively free of band broadening or pertubations due to intermolecular associations. From 1-10 ng of sample is required to record a spectrum and approximately 160 pg is needed for the detection of a strong absorbing band.

After the data are collected the dedicated computer executes a Fourier transform of the data into a single beam spectrum. The raw or transformed data can also be used to deconvolute co-eluting chromatographic peaks, to construct the gas chromatogram, to perform selected functional group monitoring, to quantify each peak in the chromatogram, and to compare unknown spectra with library spectra [80,81,90,91]. Data processing in FTIR is not as refined or as versatile as in mass spectrometry but this situation will conceivably change with time.

In GC-FTIR the mobile phase is essentially nonabsorbing and the transmission of energy at all frequencies occurs unabated. This is not the case with HPLC-FTIR where the solvents commonly used often show appreciable energy absorption. In a conventional HPLC-FTIR system, the effluent from the column is passed through a flow cell, and interferograms are continually monitored and stored during the entire chromatographic run [97-100]. At the end of the separation, the solution spectra are computed and the absorption bands due to the mobile phase are subtracted out. In order to prevent solvent bands from "blocking out" excessively wide regions of the spectrum, the pathlength of the flow cell must be quite small. Typical cells have path lengths of 0.1 to 0.2 mm and cell volumes of 3 to 12 microliters. Even under these conditions there are several disadvantages. No information about the solute can be obtained in spectral regions where the mobile phase has appreciable absorption. Subtraction of the mobile phase contribution to the signal is not possible for all solvents; since this applies to water reversed-phase chromatography is not feasible. Solvent subtraction cannot be satisfactorily performed under gradient elution conditions, limiting HPLC-FTIR to isocratic separations. Compared to GC-FTIR the sensitivity of HPLC-FTIR is at least 20-fold less and detection limits similar to refractive index detection have often been quoted. The data system allows up to five spectral windows to be monitored for selective spectral monitoring and this seems to be the most popular method, using the detector in a manner somewhat similar to selected ion monitoring in GC-MS [99]. Microbore

HPLC–FTIR provides higher detectability and makes feasible the use of costly solvents with large spectral windows. However, its general usefulness in this respect requires further validation [100,101].

Two interfaces have been described which enable the infrared spectrum of a solute to be obtained after solvent removal [102–103]. Both are complex mechanical devices which require the solute to be less volatile than the solvent as enrichment is achieved by selective evaporation. For example, the interface described by Griffiths [102] consists of a carousel containing 32 cups holding a few milligrams of potassium chloride supported on a metal screen. The eluent from the liquid chromatograph is sprayed at a controlled rate into a heated concentrator tube using a stream of nitrogen; the gas flow rate and temperature are selected so that about 90% of the solvent is evaporated. When an upstream UV detector signals the elution of a solute peak the effluent from the concentrator tube is switched by a solenoid valve and allowed to drip into one of the cups on the carousel. A microprocessor controls the entire switching process and after solute collection moves the carousel cup to the second position where a slow stream of air is drawn through the cup to eliminate the remaining solvent. A further movement positions the sample in the spectrometer and an infrared spectrum is measured by diffuse reflectance. Submicrogram detection limits were observed for all samples of low volatility, and chromatographic resolution was maintained even for closely separated peaks. The technique is not recommended for reversed–phase chromatography due to the high surface tension and latent heat of vaporization of water, which precludes its complete removal under normal operating conditions.

8.8 Interfacing a Liquid Chromatograph to a Nuclear Magnetic Resonance Spectrometer

Interest in coupling a liquid chromatograph to a nuclear magnetic resonance (NMR) spectrometer is relatively recent [104–107]. For although NMR spectra contain useful information from which structural and stereochemical information are readily obtained the technique requires relatively large sample amounts or long spectrum acquisition times, which are not in keeping with the experimental conditions common to HPLC. Because of the low sensitivity of the measuring technique only proton spectra can be obtained. For occasional measurements, the peak may be collected at the detector exit and its spectrum taken in special microcells after solvent exchange if the eluting solvent is not compatible with the normal operation of the spectrometer. Whether on- or off-line, solvent incompatibility is a serious problem as deuterated solvents are generally too expensive to use as mobile phases. Microcells have volumes of 0.03-0.05 ml and are commercially available. Using FTNMR and signal averaging a usable proton spectrum can be obtained from sub-microgram amounts of sample.

For on-line coupling the design of the flow cell is an important consideration. The geometry of the measuring cell must allow for laminar flow, the cell volume should not exceed the elution volume of a peak but otherwise should be as large as possible to ensure the presence of the maximum amount of sample during measurement, and the residence time of a peak within the measuring volume should exceed three times the longnitudinal relaxation time of the solute protons. Both the intensity and peak width of the NMR signal, therefore, depend on the residence time of the solute in the flow cell and, consequently, the column flow rate. With a superconducting magnet resolution of approximately 0.5 Hz is possible compared to 1-3 Hz with an iron magnet. Spectra may be obtained by continuous flow or in the stop-flow mode. Stop-flow has the advantage that signal averaging may be used to enhance sensitivity but at the expense of increasing the analysis time. With a superconducting high field magnet detection limits in the lower micromolar range have been obtained in the continuous flow mode and high nanomolar range in the stop-flow mode. Sensitivity is approximately an order of magnitude less with instruments having a permanent iron magnet. Perhaps a viable technique with large diameter preparative columns, it has many limitations for analytical use, not least of which is the high cost of superconducting high field FTNMR equipment. In this case the advantages of on-line coupling are not so well-founded and off-line techniques appear to be just as useful and present fewer compromises in the operation of both the chromatograph and the spectrometer.

8.9 Gas Chromatography-Optical Emission Spectroscopy

Plasma sources and, to a lesser extent, flames and carbon furnaces are readily coupled to a gas chromatograph [108-112]. The only interface required in most cases is little more than a direct connection with an auxiliary flow of make-up gas to minimize dead volume effects. Plasma sources are capable of exciting intense emission from all elements in the periodic table and are thus uniquely suited to the detection of C,H,D,O,N,S,P, and the halogens, elements of particular importance for the analysis of organic compounds. The plasma consists of a mass of predominantly ionized gas at a temperature of 4000-10,000 K. This state can be maintained directly by an electrical discharge through the gas (d.c. plasma) or indirectly via inductive heating of the gas. The latter is established by an electromagnetic field using power generated at radio frequencies (inductively coupled plasma) or microwave frequencies (microwave induced plasma). When organic compounds enter the plasma molecular breakdown occurs due to absorption of energy from the plasma and emission spectra are produced which are characteristic of the atoms involved. The intensity of these emissions is measured by a direct-reading optical emission spectrometer coupled to suitably placed photomultiplier tubes. All elements are excited simultaneously and, at plasma temperatures, molecular breakdown is complete; the measured response for each element is thus proportional to the number of atoms in the plasma and independent of the structure of the parent compound. With a multichannel instrument (commercial models presently have 12 channels) as many elements as there are channels can be measured simultaneously. Ratioing the response of each channel during the passage of a chromatographic peak will, after response standardization with compounds of known elemental composition, enable the empirical formula of unknown compounds to be directly determined [108-117]. Combining retention data with a knowledge of the empirical formula of a substance is often sufficient for identification. Also, when identifying an unknown compound, knowledge of its empirical formula is a useful aid in the interpretation of other spectroscopic data.

Instrument developments in the area of plasma detectors for both gas and liquid chromatography are proceeding at a rapid pace. Most early work was done with direct current or inductively coupled argon or low pressure helium plasmas. Helium is the preferred plasma gas as it provides a simpler background spectrum, higher excitation energy, and improved linear response range. Many systems now use the atmospheric helium microwave induced plasma [114]. This seems to be an excellent source for gas chromatography but, because of its limited sample consumption rate, is difficult to interface to a liquid chromatograph without diverting all but a few percent of the column eluent away from the detector [118]. Detection limits thus suffer. Direct current and inductively coupled plasmas can be interfaced with few problems to a liquid chromatograph in which the column eluent is nebulized directly into the plasma.

Despite the attractive features of plasma chromatographic detectors they remain little used. One reason is the high cost of multi-element spectroscopic equipment. As much effort is being directed at simplifying all aspects of plasma emission spectroscopy for elemental analysis, this cost disadvantage could diminish in the future. Element sensitivity is already adequate for many applications and detection limits suitable for trace analysis have been demonstrated [108-111].

8.10 References

1. T. Hirschfeld, Anal. Chem., 52 (1980) 297A.
2. J. T. Watson, in G. R. Waller and O. C. Dermer (Editors), "Biomedical Applications of Mass Spectrometry", First Supplementary Volume, Wiley, New York, 1980, p. 3.
3. B. J. Millard, "Quantitative Mass Spectrometry", Heyden, London, 1978
4. A. M. Greenway and C. F. Simpson, J. Phys. E, 13 (1980) 1131.
5. R. D. Craig, R. H. Bateman, B. N. Green, and D. S. Millington, Phil. Trans. R. Soc. Lond. A, 293 (1979) 135.
6. N. Gochman, L. J. Bowie, and D. N. Bailey, Anal. Chem., 51 (1979) 525A.

616

7. W. A. Garland and M. L. Powell, J. Chromatogr. Sci., 19 (1981) 392.
8. M. C. Ten Noever De Brauw, J. Chromatogr., 165 (1979) 207.
9. W. H. McFadden, "Techniques of Combined Gas Chromatography-Mass Spectrometry: Applications in Organic Analysis", Wiley, New York, 1973.
10. B. J. Gudzinowicz, M. J. Gudzinowicz and H. F. Martin, "Fundamentals of Integrated Gas Chromatography-Mass Spectrometry", Dekker, New York, Part II (1976) and Part III (1977).
11. B. Halpern, CRC Crit. Revs. Anal. Chem. 11 (1981) 49.
12. B. S. Middleditch (Editor), "Practical Mass Spectrometry. A Contemporary Introduction", Plenum, New York, 1979.
13. R. M. Milberg and J. C. Cook, J. Chromatogr. Sci., 17 (1979) 43.
14. W. J. Richter and H. Schwarz, Angew. Chem. Int. Ed. Engl., 17 1978) 424.
15. R. E. Mather and J. F. J. Todd, Int. J. Mass Spectrom. Ion Phys., 30 (1979) 1.
16. R. C. Dougherty, Anal. Chem., 53 (1981) 625A.
17. G. P. Arsenault, J. J. Dolhun, and K. Biemann, Anal. Chem., 43 (1971) 1720.
18. D. I. Carroll, I. Dzidic, R. N. Stillwell, M. G. Horning, and E. C. Horning, Anal. Chem., 46 (1974) 706.
19. E. C. Horning, D. I. Carroll, I. Dzidic, K. D. Haegele, M. G. Horning, and R. N. Stillwell, J. Chromatogr., 99 (1974) 13.
20. M. W. Siegel and W. L. Fite, J. Phys. Chem., 80 (1976) 2871.
21. J. Meili, F. C. Walls, R. McPherron, and A. L. Burlingame, J. Chromatogr. Sci., 17 (1979) 29.
22. P. H. Dawson (editor), "Quadrupole Mass Spectrometry and its Application", Elsevier, Amsterdam, 1976.
23. K. Feser and W. Kogler, J. Chromatogr. Sci., 17 (1979) 57.
24. R. L. White and C. L. Wilkins, Anal. Chem., 54 (1982) 2443.
25. R. W. Crawford, T. Hirschfeld, R. H. Sanborn, and C. M. Wong, Anal. Chem., 54 (1982) 817.
26. J. T. Watson, in L.S. Ettre and W. H. McFadden (editors), "Ancillary Techniques of Gas Chromatography", Wiley, New York, 1969, p. 145.
27. W. H. McFadden, J. Chromatogr. Sci., 17 (1979) 2.
28. F. Vangaever, P. Sandra, and M. Verzele, Chromatographia, 12 (1979) 153.
29. T. E. Jensen, R. Kaminsky, B. D. Veety, T. J. Wozniak and R. A. Hites, Anal. Chem., 54 (1982) 2388.
30. A. Friedli, J. High Resolut. Chromatogr. Chromatogr. Commun., 4 (1981) 495.
31. C. Sunol and E. Gelpi, J. Chromatogr., 142 (1977) 559.
32. K. Rose, J Chromatogr., 259 (1983) 445.
33. D. Henneberg, U. Henrichs, and G. Schomburg, Chromatographia, 8 (1975) 449 .
34. D. Henneberg, U. Henrichs, and G. Schomburg, J. Chromatogr., 112 (1975) 343.
35. J. F. K. Huber, E. Matisova, and E. Kenndler, Anal. Chem., 54 (1982) 1297.
36. E. Wetzel, Th. Kuster, and H.-Ch. Curtius, J. Chromatogr., 239 (1982) 107.
37. G. C. Jamieson, J. High Resolut. Chromatogr. Chromatogr. Commun., 5 (1982) 633.
38. M. A. Grayson and C. J. Wolf, Anal. Chem., 39 (1967) 1438.
39. S. R. Lipsky, C. G. Horvath, and W. J. Murray, Anal. Chem., 38 (1966) 1585.
40. J. E. Arnold and H. M. Fales, J. Gas Chromatogr., 3 (1965) 131.
41. P. G. Simmonds, G. R. Schoemake, and J. E. Lovelock, Anal. Chem., 42 (1970) 881.
42. J. E. Lovelock, P. G. Simmonds, and G. R. Schoemake, Anal. Chem., 42 (1970) 969.
43. D. P. Lucero, J. Chromatogr. Sci., 9 (1971) 105.
44. J. T. Watson and K. Biemann, Anal. Chem., 37 (1965) 844.

45. W. E. Wentworth, Y-Chi. Chen, A. Zlatkis, and B. S. Middleditch, Anal. Chem., 54 (1982) 1895.
46. F. L. Cardinall and L. E. Lowe, Anal. Chem., 54 (1982) 1454.
47. C. C. Greenwalt, K. J. Voorhees, and J. H. Futrell, Anal. Chem., 55 (1983) 468.
48. P. J. Arpino and G. Guiochon, Anal. Chem., 51 (1979) 682A.
49. D. E. Games, Adv. Chromatogr., 21 (1983) 1.
50. G. Guiochon and P. J. Arpino, J. Chromatogr., 271 (1983) 13.
51. D. I. Carroll, I. Dzidic, R. N. Stillwell, K. D. Haegele, and E. C. Horning, Anal. Chem., 47 (1975) 2369.
52. M. Dedieu, C. Juin, P. J. Arpino, and G.Guiochon, Anal. Chem., 54 (1982) 2375.
53. P. J. Arpino, J. P. Bounine, M. Dedieu, and G. Cuiochon, J. Chromatogr., 271 (1983) 43.
54. C. R. Blakley, M. J. McAdams, and M. L. Vestal, J. Chromatogr., 158 (1978) 261.
55. C. R. Blakley and M. L. Vestal, Anal. Chem., 55 (1983) 750.
56. R. P. W. Scott, C. G. Scott, M. Munroe, and J. Hess, J. Chromatogr., 99 (1974) 395.
57. W. H. McFadden, H. L. Schwartz, and S. Evans, J. Chromatogr., 122 (1976) 389.
58. R. D. Voyksner, J. R. Hass, and M. M. Bursey, Anal. Chem., 54 (1982) 2465.
59. R. D. Voyksner, C. E. Parker, J. R. Hass, and M. M. Bursey, Anal. Chem., 54 (1982) 1583.
60. J. R. Chapman, "Computers in Mass Spectrometry", Academic Press, New York, 1978.
61. A. Carrick, "Computers and Instrumentation", Heyden, London, 1979.
62. M. J. E. Hewlins, Chem. Brit., 12 (1976) 341.
63. J. R. Chapman, J. Phys. E, 13 (1980) 365.
64. W. L. Budde and J. W. Eichelberger, J. Chromatogr., 134 (1977) 147.
65. C. Fenselau, Anal. Chem., 49 (1977) 563A.
66. R. Buchi, J. T. Clerc, J. H. Koenitzer, and D. Wegmann, Anal. Chim. Acta, 103 (1978) 21.
67. F. W. McLafferty, Anal. Chem., 49 (1977) 1441.
68. F. W. McLafferty and R. Venkataraghavan, J. Chromatogr. Sci., 17 (1979) 24.
69. S. C. Gates, M. J. Smisko, C. L. Ashendel, N. D. Young, J. F. Holland, and C. C. Sweeley, Anal. Chem., 50 (1978) 433.
70. L. H. Kieth, J. Chromatogr. Sci., 17 (1979) 48.
71. B. N. Colby, P. W. Ryan, and J. E. Wilkinson, J. High Resolut. Chromatogr. Chromatogr. Commun., 6 (1983) 72.
72. A. D. Sauter, P. E. Mills, W. L. Fitch, and R. Dyer, J. High Resolut. Chromatogr. Chromatogr. Commun., 5 (1982) 27.
73. A. P. De Leenheer and A. A. Cryyl, in G. R. Waller and O. C. Dermer (Editors), "Biomedical Applications of Mass Spectrometry", First Supplementary Volume, Wiley, New York, 1980, p. 1170.
74. T. A. Baillie, Pharmacol Rev., 33 (1981) 81.
75. M. Claeys, S P. Markey, and W. Maenhaut, Biomed. Mass Spectrom., 4 (1977) 122.
76. C. L. Wilkins, G. N. Giss, R. L. White, G. M. Brissey, and E. C. Onyiriuka, Anal. Chem., 54 (1982) 2260.
77. H. H. Hausdorff, J. Chromatogr., 134 (1977) 131.
78. D. A. Hanna, G. Hangac, B. A. Hohne, G. W. Small, R. C. Wieboldt, and T. L. Isenhour, J. Chromatogr. Sci., 17 (1979) 423.
79. P. R. Griffiths, "Chemical Infrared Fourier Transform Spectroscopy", Wiley, New York, 1975.
80. D. A. Hanna, J. C. Marshall, and T. L. Isenhour, J. Chromatogr. Sci., 17 (1979) 434.
81. M. F. Delaney and P. C. Uden, J. Chromatogr. Sci., 17 (1979) 428.

618

82. S. K. Freeman, in L. S. Ettre and W. H. McFadden (Eds.), "Ancillary Techniques of Gas Chromatography", Wiley, New York, 1969, p. 227.
83. D. A. Leathard and B. C. Shurlock, "Identification Techniques in Gas Chromatography", Wiley, New York, 1970.
84. H. J. Sloan and R. J. Obremski, Appl. Spec., 31 (1977) 506.
85. R. Cournayer, J. C. Shearer, and D. H. Anderson, Anal. Chem., 49 (1977) 2275.
86. J. T. Chen, J. Assoc. Off. Anal. Chem., 48 (1965) 380.
87. W. J. de Klein and K. Ulbert, Anal. Chem., 41 (1969) 682.
88. G. D. Price, E. C. Sunas, and J. F. Williams, Anal. Chem., 39 (1967) 138.
89. A. S. Currey, J. F. Read, C. Brown, and R. W. Jenkins, J. Chromatogr., 38 (1969) 200.
90. M. D. Erickson, Appl. Spec. Revs., 15 (1979) 261.
91. D. R. Mattson and R. L. Julian, J. Chromatogr. Sci., 17 (1979) 416.
92. R. H. Shaps, M. J. Flanagan, and A. Varano, J. Chromatogr. Sci., 17 (1979) 454.
93. K. Krishnan, R. Curbelo, P. Chiha, and R. C. Noonan, J. Chromatogr. Sci., 17 (1979) 413.
94. L. V. Azarraga and C. A. Potter, J. High Resolut. Chromatogr. Chromatogr. Commun., 4 (1981) 60.
95. S. Bourne, G. T. Reedy, and P. T. Cunningham, J. Chromatogr. Sci., 17 (1979) 460.
96. D. T. Sparks, R. B. Lam, and T. L. Isenhour, Anal. Chem., 54 (1982) 1922.
97. D. W. Vidrine and D. R. Mattson, Appl. Spectrosc., 32 (1978) 502.
98. D. W. Vidrine, J. Chromatogr. Sci., 17 (1979) 477.
99. C. Combellas, H. Bayart, B. Jasse, M. Caude, and R. Rosset, J. Chromatogr. 259 (1983) 211.
100. C. C. Johnson and L. T. Taylor, Anal. Chem., 55 (1983) 436.
101. K. Jinno and Ch. Fujimoto, J. High Resolut. Chromatogr. Chromatogr. Commun., 4 (1981) 532.
102. D. T. Kuehl and P. R. Griffiths, J. Chromatogr. Sci., 17 (1979) 471.
103. D. T. Kuehl and P. R. Griffiths, Anal. Chem., 52 (1980) 1394.
104. J. F. Haw, T. E. Glass, and H. C. Dorn, Anal. Chem., 53 (1981) 2332.
105. J. F. Haw, T. E. Glass, and H. C. Dorn, Anal. Chen., 55 (1983) 22.
106. E. Bayer, K. Albert, M. Nieder, E. Grom, and T. Keller, J. Chromatogr., 186 (1979) 497.
107. E. Bayer, K. Albert, M. Nieder, E. Grom, G. Wolf, and M. Rindlisbacher, Anal. Chem., 54 (1982) 1747.
108. J. W. Carnahan, K. J. Mulligan, and J. A. Caruso, Anal. Chim. Acta, 130 (1981) 227.
109. A. I. Zander and G. M. Hieftje, Appl. Spectrosc. 35 (1981) 357.
110. V. Rezl, J. Chromatogr., 251 (1982) 35.
111. T. H. Risby and Y. Talmi, CRC Crit. Rev. Anal. Chem., 14 (1983) 231.
112. R. M. Dagnall, T. S. West, and P. Whitehead, Anal. Chem., 44 (1972) 2074.
113. C. Feldman and D. A. Batistoni, Anal. Chem., 49 (1977) 2215.
114. J. Bonnekessel and M. Klier, Anal. Chim. Acta, 103 (1978) 29.
115. D. L. Windsor and M. B. Denton, J. Chromatogr. Sci., 17 (1979) 492.
116. D. B. Quimby, M. F. Delaney, P. C. Uden, and R. M. Barnes, Anal. Chem., 52 (1980) 259.
117. S. A. Estes, P. C. Uden, and R. M. Barnes, Anal. Chem., 58 (1981) 1829.
118. H. A. H. Billiet, J. P. J. van Dalen, P. J. Schoenmakers, and L. De Galan, Anal. Chem., 55 (1983) 847.

Chapter 9

HIGH PERFORMANCE THIN-LAYER CHROMATOGRAPHY

9.1   Introduction

    Thin-layer chromatography (TLC) is methodologically perhaps the simplest of
the chromatographic separation techniques. The mechanisms employed for
separation are identical to those of high pressure liquid chromatography
(HPLC). The important difference between the two techniques is one of practice
rather than of the physical phenomena from which separation is derived. In TLC,
the stationary phase consists of a thin layer of sorbent coated on an inert,

rigid backing material. The sample to be separated is applied to the surface of the sorbent layer as a spot, a short distance from the bottom edge of the plate. The separation is carried out in an enclosed chamber by contacting the bottom edge of the plate with the mobile phase, which advances through the stationary phase by capillary forces. A separation of the sample is produced by the differential migration of the sample components in the direction traveled by the mobile phase.

Many analysts believe that thin-layer chromatography is a simple, rapid qualitative technique for the separation of simple mixtures. The methodological ease of this technique, however, belies its separating power. Accumulated experiences employing the unfavorable practices of conventional TLC have assigned a limited role to TLC in complex mixture analyses. Thus, there remains an impression among many analysts that TLC is a poor alternative to high pressure liquid chromatography for quantitative analyses of all but the simplest of mixtures.

Recent changes in the practice of TLC have created a renaissance of interest in this technique and led to its wider acceptance as a powerful separating tool for multicomponent trace level analysis. This changed role and the accompanying expectations have spawned a new expression - high performance thin-layer chromatography (HPTLC), analogous to the name change applied to liquid chromatography when it underwent a similar expansion in performance capabilities. The performance breakthrough in TLC was not a result of any specific advance in instrumentation or materials, but was rather a culmination of improvements in practically all of the operations of which TLC is comprised. In this chapter we will describe the experimental parameters which constitute the HPTLC process and contrast these methods with their conventional TLC counterparts. Two monographs [1,2] and a review [3] have been devoted to describing the HPTLC process. There are many sources to the more extensive conventional TLC literature, of which references [4-8] are a good starting point for both the practice and general application areas of TLC.

9.2 Comparison of Conventional and High Performance Thin-Layer Chromatography

Improvements in the quality of the adsorbent layer and methods of sample application, and the availability of scanning densitometers for in situ quantitative analysis were all important developments in the evolution of HPTLC. The new HPTLC plates are prepared from specially purified silica gel with average particle diameters of between 5 to 15 micrometers with a narrow particle size distribution. These plates give HETP values of about 12 micrometers and a maximum of about 5000 usable theoretical plates in any separation. By contrast, conventional TLC plates prepared from silica gel with average particle diameters of 20 micrometers and a rather broad particle size distribution gave HETP values of about 30 micrometers and a maximum of approximately 600 usable theoretical plates. The new HPTLC plates can provide an increase in performance approaching an order of magnitude over conventional TLC plates. Thus, it is possible to carry out separations on HPTLC plates that were not possible by conventional TLC and also to carry out those separations which are presently carried out by conventional TLC in much shorter times.

Due to the lower sample capacity of the HPTLC layer the amount of sample applied to the layer is reduced. Sample volumes of 100–200 nl, which give starting spots of 1.0–1.5 mm diameter, are typical. After developing a distance of 3–6 cm separated spots with 3–6 mm diameters are obtained. As spots are more compact than in conventional TLC, the lower sample capacity of the plate does not present any problems and results in detection limits ten times better than conventional TLC. The compact starting spots in HPTLC also allow an increase in the number of samples which may be applied along the edge of the plate. Depending on the method of development and the size of the plate, as many as 18 (10 x 10 cm plate) or 36 (10 x 20 cm plate) samples can be simultaneously separated. Linear development from two directions (i.e., simultaneously from opposite edges towards the middle) doubles the number of samples which can be separated.

Although the mobile phase velocity is greater for conventional TLC than for HPTLC plates, longer migration distances are required to obtain an equivalent separation. Therefore, analysis times are considerably shorter for HPTLC. Typical migration distances in HPTLC are 3–6 cm while development times are an order of magnitude lower than in conventional TLC.

In summary, HPTLC is a more rapid, efficient and sensitive technique than conventional TLC. Because of these advantages, HPTLC represents a considerable advance in the practice of TLC. Table 9.1 provides a summary of the experimental parameters defining conventional versus high performance thin–layer chromatography.

9.3 Comparison of High Performance Liquid and Thin–Layer Chromatography

Since the areas of application for high performance liquid chromatography (HPLC) and HPTLC, and hence the growth curves for interest and purchase of equipment, overlap to some extent, interest has heightened in comparing the performance of the two techniques. It is relatively easy to demonstrate that HPLC is a more efficient chromatographic technique than HPTLC if the only criteria for comparison are based upon HETP values and separation numbers, etc. The conventionally packed HPLC columns used in most analytical laboratories are

TABLE 9.1

A COMPARISION OF THE TECHNIQUES OF CONVENTIONAL AND HIGH PERFORMANCE
THIN-LAYER CHROMATOGRAPHY

Parameter	Conventional TLC	HPTLC
Plate Size	20 x 20 cm	10 x 10 cm 10 x 20 cm

Particle Size
    Average
    Distribution
	
	20 micrometers
	10-60 micrometers
		
		5 micrometers
		tight

Adsorbent Layer Thickness	100-250 micrometers	200 micrometers
Plate Height	30 micrometers	12 micrometers
Total Number of Usable Theoretical Plates	< 600	< 5000
Separation Number	7-10	10-20
Sample Volume	1-5 microliters	0.1-0.2 microliters
Starting Spot Diameter Diameter of Separated Spots	3-6 mm 6-15 mm	1-1.5 mm 2-5 mm
Solvent Migration Distance	10-15 cm	3-6 cm
Separation Time	30-200 min	3-20 min

Detection Limits
    Absorption
    Fluorescence
	
	1-5 ng
	0.05-0.1 ng
		
		0.1-0.5 ng
		0.005-0.01 ng

Tracks per Plate	10	18 or 36

capable of providing approximately 10,000 theoretical plates with separation numbers of 20-40. Special HPLC columns, such as capillary and microbore columns, are capable of even greater efficiency. However, HPTLC gains certain advantages because it is an open bed while HPLC is a closed bed system. These advantages are outlined below.

In HPLC the mobile phase velocity is electronically controllable up to a limit. This limit is established by the maximum pressure gradient that can be maintained across the column. For HPTLC, the mobile phase velocity is governed by the capillary forces which transport the solvent through the sorbent bed. This disadvantage is partially offset by the fact that the development process is different for the two techniques. In HPLC the development occurs by elution chromatography; each sample component must travel the complete length of the column bed and the total separation time is determined by the time required for the slowest moving component to reach the detector. The HPTLC process is governed by development chromatography. The total time for the separation is the time required for the solvent front to migrate a fixed or predetermined distance and is independent of the migration distance of the sample components. For most analyses, only a few components of the sample are of interest and the mobile phase is selected so as to provide the necessary separation. The remainder of the sample material can be left at the origin or moved away from the region of maximum resolution. This results in considerable savings of time. There is no possibility of "poisoning" a TLC plate, which is far too easy to do with an HPLC column. Such poisoning results in a considerable loss of time while components accumulated at the head of the column are completely eluted. This may result in permanent damage (i.e., loss of resolution) to the column if these samples cannot be removed. TLC plates are disposed of at the conclusion of each separation and are thus immune to the above problems.

Detection in HPTLC, unlike HPLC, is a static process, being completely separated from chromatographic development. Consequently, the selection of the mobile phase does not limit the choice of detector. For example, UV-absorbing solvents cannot be used with UV detectors in HPLC. In HPTLC the solvent is completely evaporated between development and measurement so it does not influence the detection process.

Perhaps the most important feature of development chromatography is that the sample is separated by distance rather than time. This freedom from time constraints permits the utilization of any of a variety of techniques to enhance the sensitivity of detection, such as reactions which increase light absorbance or fluorescence emission and wavelength selection for optimum response of each component measured. The separation can be scanned as many times as desired at a variety of wavelengths, and a complete UV-visible or fluorescence spectrum can be easily plotted out for each component. Thus, the detection process in HPTLC is more flexible and variable than for HPLC. Detection limits under optimum conditions are approximately the same for both techniques.

Because of the nature of the method of development, analysis in HPLC is by necessity performed in a sequential manner. Each sample must individually undergo the same sequence of injection, separation, detection, and column re-equilibration. For a series of n samples, the total time for the analysis will be n multipled by the time for each individual analysis. HPTLC techniques permit simultaneous sample analysis with the possibility of substantially reducing the time required for the analyses of a large group of samples. Conceivably as many as 72 samples (36 along two opposite edges of a 20 x 10 cm plate developed horizontally in a linear development chamber) could be separated simultaneously. In contrast to HPLC, the time required to analyze n samples is reduced by a factor of n/72 per sample. In practice, this savings in time is somewhat reduced, as allowance must be made for the fact that some of the available tracks are occupied by calibration standards. In addition, sample application and the scanning of each separation requires a finite length of time. Even after making suitable allowances for the above points, HPTLC can provide significant reductions in time when multiple samples of a similar type are analyzed.

In conclusion, for individual samples HPLC can provide greater separating power than HPTLC. The greater methodological simplicity (selection of operating variables) and detection flexibility of HPTLC offset this disadvantage to some extent. When a high thoughput of similar samples is required, then HPTLC can provide considerable savings in time. In essence, the two techniques complement one another and therefore selection should be based on the type of problem to be solved. The relative merits of the two techniques are summarized in Table 9.2.

TABLE 9.2

A COMPARISON OF THE SCOPE OF HIGH PERFORMANCE LIQUID (HPLC) AND THIN-LAYER (HPTLC) CHROMATOGRAPHY

Parameter	HPTLC	HPLC
Plate Height (micrometers)	12	2-5
Number of Plates	< 5000	6-10,000
Separation Number	10-20	20-40
Type of Chromatography	Open Bed	Closed System
Development Technique	Development Chromatography	Elution Chromatography
Control of Mobile Phase Velocity	Fixed by Capillary Forces[a]	Easily adjusted
Sample Simultaneously Available for Detection	Yes	No
Multiple Scan of Same Separation	Yes	No
Plot out UV-Visible/ Fluorescence Spectrum	Yes	Yes (diode array detector)
System Equilibration Time	Small	Small to Large [b]
Simultaneous Sample Analysis	Yes	No
Sample Capacity (throughput)	High	Low
Methodological Simplicity	Good	Not so Good
Detection Limits	Approximately the Same	

[a] May be controlled in the U-chamber or overpressured-chamber
[b] Re-equilibration time in gradient elution HPLC can be quite long

9.4   Simplified Theory of High Performance Thin-Layer Chromatography

The prevailing experimental conditions in a TLC separation are difficult to define accurately or to control experimentally. As a consequence, a precise model of the separation process would be overly complex and inappropriate to the needs of the analytical chemist. Under the usual experimental conditions employed in TLC, both the stationary and mobile phases remain ill defined and exist in a state of flux with the vapor phase, all of which may change continuously during the separation process. Contact with vapors, both prior to and during the chromatographic process, modifies the properties of the stationary phase. For adsorbents, the amount of activation and deliberate or accidental deactivation changes their chromatographic behavior. Once in the developing chamber, the adsorbent layer progressively adsorbs the vapor phase prior to being wetted by the mobile phase. The porosity of the stationary phase changes, and with it the apparent speed of the mobile phase. In addition, the mobile phase does not evenly penetrate the adsorbent layer. Capillary forces are stronger in the narrow inter-particle channels, leading to more rapid advancement of the mobile phase. Larger pores below the solvent front are filled at a slower rate and result in an increased thickness of the mobile phase layer. During the chromatographic process a solvent gradient in the mobile phase is produced as the solvent front migrates through the adsorbent layer. This is particularly true for mixed mobile phases where the more polar component is selectively adsorbed. Even single solvent systems exhibit impurity gradients in the direction of the solvent front. If the vapor and mobile phase are not in equilibrium, evaporation will cause a loss of mobile phase from the plate surface.

Gaining control of the above phenomena is experimentally difficult as TLC separations are normally practiced. The U-chamber described by Kaiser most closely approaches the experimental requirements for true reproducibility and theoretical validation of TLC data [9]. For more detailed discussions of the process, references [10-20] should be consulted.

The fundamental parameter used to characterize the position of a spot in a TLC chromatogram is the retardation factor, or R_F value. It represents the ratio of the distance migrated by the sample compared to that traveled by the solvent front.

$$R_F = \frac{Z_{substance}}{Z_{mobile\ phase}} = \frac{Z_s}{Z_f} \qquad (9.1)$$

Z_s = distance migrated by the sample (spot) from its origin

Z_f = distance moved by the mobile phase from the sample origin to the solvent front

Boundary conditions: $1 > R_F > 0$

$R_F = 0$, spot does not migrate from the origin

$R_F = 1$, spot is not retained by the stationary phase and migrates with the solvent front

Systematic errors in the measurement of R_F values arise from the difficulty in locating the precise position of the solvent front. If the adsorbent layer, mobile phase, and vapor phase are not in equilibrium then condensation of the vapor phase or evaporation of the mobile phase in the region of the solvent front will give an erroneous R_F value. This value may be either too high or too low, depending on the prevailing conditions. The R_F values for linear, circular, and anticircular development are related by equations (9.2) and (9.3):

$$R_{F(L)} = [R_{F(c)}]^2 \qquad (9.2)$$

$$R_{F(L)} = 1 - [1 - R_{F(ac)}]^2 \qquad (9.3)$$

$R_{F(L)}$ = R_F value for a component determined by linear development

$R_{F(c)}$ = R_F value for the same component determined by circular development

$R_{F(ac)}$ = R_F value for the same component determined by anticircular development

Equation (9.2) assumes that the point of sample application and the mobile phase entry position are located exactly in the center of the circular chromatogram [9,19]. If the points of sample application and solvent entry are not identical but separated by a distance r, the correct equation becomes

$$R_{F(c)} = [R_{F(L)} (1 + 2r) + r^2]^{1/2} - r \qquad (9.4)$$

The capacity factor, k, of a spot is defined as the ratio of the time spent by the substance in the stationary phase compared to that spent in the mobile phase.

$$k = \frac{t_s}{t_m} = \frac{\text{retention time in the stationary phase}}{\text{retention time in the mobile phase}} \qquad (9.5)$$

The capacity factor and R_F value are related by the expression

$$R_F = \frac{t_m}{t_m + t_s} = \frac{1}{k + 1} \tag{9.6}$$

which can be rearranged to give

$$k = \frac{1 - R_F}{R_F} \tag{9.7}$$

For optimum resolution in the various development modes, the preferred values

for k are:

$k \simeq 100$ for circular development
$k \simeq 10$ for linear development
$k \simeq 1$ for anticircular development

Thus, circular development provides good separation for compounds with low R_F values while anticircular development is preferred for the separation of compounds with high R_F values.

When an equilibrium exists between the adsorbent layer, the mobile phase and the vapor phase, the speed with which the solvent front ascends a TLC plate is given by equation (9.8).

$$(Z_f)^2 = K t \tag{9.8}$$

Z_f = migration distance of the solvent front in cm after t seconds

K = mobile phase velocity constant in cm^2/s
t = time in seconds

If true equilibrium does not exist during the development process, then rather complex correction factors must be applied to equation (9.8). These have been discussed in some detail by Guiochon and Siouffi [14]. The velocity constant, K, is related to the experimental conditions by equation (9.9).

$$K = 2 K_o d_p \frac{\tau}{\eta} \cos \Theta \tag{9.9}$$

K_o = permeability constant
d_p = mean particle diameter
τ = surface tension of the mobile phase
η = viscosity of the mobile phase
Θ = contact angle

The permeability constant, K_o, is a dimensionless constant that takes into account the profile of the external pore size distribution, the effect of porosity on the permeability of the adsorbent layer, and the ratio of the bulk liquid velocity to the solvent front velocity. For silica gel HPTLC plates it has a value of about 7.9×10^3 [18]. The contact angle for most organic solvents on silica gel is generally close to zero (cos $\Theta = 1$). For water organic solvent mixtures the contact angle on reversed-phase plates may be much greater than zero and therefore the rate of solvent front migration is very slow.

The mobile phase velocity can be related to both the characteristics of the TLC bed and the properties of the mobile phase by equation (9.10), derived by Belenki et al. [20].

$$u = \frac{dZ_f}{dt} = \left[\frac{K_o d_p}{\eta Z_f}\right] \cos \Theta - Z_f \left[\frac{d_p \rho g}{12}\right] \tag{9.10}$$

u = velocity of the solvent front
ρ = mobile phase density
g = gravitational field constant

HPTLC plates are coated with particles of small diameter; hence the rate of solvent advancement is slower than for conventional TLC plates prepared from particles of larger diameter. The more uniform, tightly packed layers used in HPTLC result, however, in higher K_o values, which may partially compensate for this particle size effect. In practice, HPTLC is claimed to be a faster separation technique than TLC (Table 9.1), since the migration distance required to obtain a separation equivalent to conventional TLC is much shorter. As shown by equation (9.8), the time required for the solvent front to reach a certain position on the plate is approximately proportional to $(Z_f)^2$. Solvents of low viscosity and density are preferred in TLC when short analysis times are required, equation (9.10).

The ultimate chromatographic performance of a TLC plate, and therefore resolution, is dependent upon the following parameters: the velocity constant of the mobile phase, the diffusion coefficient of the substance in the mobile phase, the mean particle diameter, and the particle size distribution of the stationary phase. As in all modes of chromatography, the mobile phase velocity is determined by the particles of small size while chromatographic efficiency (HETP) is dependent upon the coarser particles. In all systems performance is improved by using particles of a narrow size distribution. Modern HPTLC plates are prepared from particles of small diameter and, more importantly, of a very narrow size distribution. Separations by HPTLC are characterized by a series of compact symmetrical spots, with the exception of components eluting close to the solvent front. Consequently, as far as HPTLC separations are concerned, spot broadening is controlled only by molecular diffusion. Resistance to mass transfer phenomena can be ignored at normal mobile phase velocities. In conventional TLC, elongated and irregularly shaped spots are not uncommon and here the contribution of mass transfer kinetics to spot broadening cannot be ignored.

Performance in TLC can be evaluated in terms of such familiar chromatographic parameters of efficiency as the number of theoretical plates (n), height equivalent of a theoretical plate (HETP), and separation number (SN). Before establishing the equations to calculate these parameters, it is necessary to highlight some of the special features of the TLC process which differ from the column chromatographic systems, for which the above parameters have been more widely adopted. In column chromatographic techniques, all substances travel the same migration distance (the length of the column) but have different diffusion times (retention times on the column). This is opposite to TLC where all substances have the same diffusion time (the plate is developed for a fixed time) but migration distances vary. The chromatographic measures of performance in TLC (e.g., n, HETP, SN) are all correlated to the migration distance of the substance. Their numerical values are evaluated for a specific value of R_F and are thus dependent on their position in the chromatogram.

In HPTLC, the effect of diffusion is the same for all substances and spot broadening is a function of the distance migrated. The distribution of sample within a spot is thus essentially Gaussian and the number of theoretical plates, n, is given by equation (9.11).

$$n = 16 \left[\frac{Z_s}{w_b} \right]^2 \tag{9.11}$$

w_b = peak width at the base

As $Z_s = R_F Z_f$ from equation (9.1), then

$$n = 16 \left[\frac{R_F Z_f}{w_b} \right]^2 \tag{9.12}$$

The height equivalent to a theoretical plate, HETP, is given by equation (9.13).

$$\text{HETP} = \frac{Z_s}{n} = \frac{R_F Z_f}{n} = \frac{w_b^2}{16 R_F Z_f} \tag{9.13}$$

Note that the values of n and HETP depend on the migration distance, R_F.

Assuming that the local plate height can be represented by Knox's equation for the reduced plate height in HPLC (Chapter 4.6.2).

$$h = \frac{B}{v} + A v^{1/3} + C v \tag{9.14}$$

where $h = \dfrac{\text{HETP}}{d_p}$ and $v = \dfrac{u d_p}{D_m}$

the A term is a measure of the quality of the packing, the B term axial diffusion in the packed bed, and the C term the resistance to mass transfer in that bed. For well-prepared plates the A term is approximately unity and the C term is very small (0.01-0.03). If the contribution to mass transfer is neglected, then the average plate height can be described by equation (9.15) [15,16,21].

$$H = b (Z_f + Z_o) + a \left[\frac{(Z_f^{2/3} - Z_o^{2/3})}{(Z_f - Z_o)} \right] \tag{9.15}$$

H = average plate height
Z_o = distance of the sample spot above the solvent level in the tank

$$b = \frac{2 \cdot \gamma D}{K R_F} \qquad\qquad a = \frac{3}{2} A d_p \left[\frac{d_p}{2D} \right]^{1/3}$$

Assumptions: $\gamma_m = \gamma_s = \gamma$ = tortuosity factor

$D_m = D_s = D$ = diffusion coefficient

subscripts m and s refer to the mobile phase and stationary phase, respectively

Equation (9.15) predicts that the efficiency of conventional TLC plates (coarse particles) increases with development length and, for long migration distances, should exceed that of HPTLC plates (fine particles). As Z_f increases the mobile phase velocity declines dramatically for HPTLC plates and diffusional broadening of the spots limits the separation efficiency. Thus separations in HPTLC should be performed using short migration distances, in agreement with accepted practice.

The separation number (SN) is useful for comparing the separating power of different chromatographic systems. For TLC it is defined as the number of substances completely separated between R_F = 0 and R_F = 1 by a homogeneous mobile phase (i.e., no solvent gradients in the direction of development). Two substances are considered to be completely separated when the distance between the two adjacent peak maxima and the sum of both peak widths at half height are the same. The separation number can then be calculated according to the procedure devised by Kaiser [9]. Under diffusion-controlled conditions peak broadening is a linear function of the migration distance. By extrapolation of the measured peak widths at half height, the half height peak widths of an imaginary compound eluted at the solvent front (b_1) and at the sample origin (b_0) can be calculated. The calculated value of b_0 is slightly larger than the measured half height peak width prior to development by an amount equal to the "rest diffusion value". The separation number is calculated from equation (9.16).[a]

$$SN = \frac{Z_f}{b_o + b_1} - 1 \qquad (9.16)$$

Typical separation numbers for HPTLC cover the range 10-20 and are 7-10 for conventional TLC. These values are in keeping with the experimentally verified separating power of the two techniques. Using the same approach, the number of "real" plates can also be calculated from equation (9.17).

$$N_{real} = 5.54 \left[\frac{Z_s}{b_s - b_o} \right]^2 \qquad (9.17)$$

b_s = peak width at half height for substance s

N_{real} and n differ by a correction term which takes into account the extrapolation value of the peak width at half height of the starting spot.

The real plate number is useful for comparing different TLC coatings and for optimizing separation systems. For this purpose the value of N_{real} at R_F = 1 or R_F = 0.5 is usually calculated. The expression for N_{real} at R_F = 1, $(N_1)_{real}$ becomes

$$(N_1)_{real} = 5.54 \left[\frac{Z_f}{b_1 - b_o} \right]^2 \qquad (9.18)$$

The number of real plates for conventional TLC is generally less than 600 and less than 5000 for HPTLC. This provides plate height values, H_{real}, of about 30 micrometers for conventional TLC and 12 micrometers for HPTLC.

The resolution, R_s, between two closely migrating spots is defined as

$$R_s = 2 \left[\frac{Z_{s1} - Z_{s2}}{w_{b1} + w_{b2}} \right] \qquad (9.19)$$

Z_{s1} = migration distance of sample 1 with base width w_{b1}

Z_{s2} = migration distance of sample 2 with base width w_{b2}

Boundary conditions: $Z_{s1} > Z_{s2}$ (i.e., Z_{s1} is the spot of higher

R_F value for any pair)

---

[a] This is an approximation and gives a value slightly less than the correct value. The true value is given by

$$SN = \frac{\log (b_o/b_1)}{\log \left[\dfrac{1 - b_1 + b_o}{1 + b_1 - b_o} \right]}$$

For two spots with similar R_F values, $w_{b1} = w_{b2} = w_b$, substituting $Z_s = R_F Z_f$
into equation (9.19) gives

$$R_s = Z_f \frac{R_{F1} - R_{F2}}{w_b} = \frac{Z_f}{w_b} \Delta R_F \qquad (9.20)$$

$$\Delta R_F = R_{F1} - R_{F2}$$

The peak width at the base, w_b, can be expressed in terms of the number of
theoretical plates, n, by equation (9.11) and substituted into equation (9.20)
to give

$$R_s = \frac{\sqrt{n}}{4} \frac{\Delta R_F}{R_{F1}} \qquad (9.21)$$

The plate number, n, the equilibrium constants embodied in R_F, and the
denominator of equation (9.21) are all dependent upon the R_F value. Thus, the
relationship between resolution, R_s, and the R_F value is complex. The TLC
analog of the classical resolution equation used in column chromatography
(Chapter 1.5) is given by

$$R_s = \underbrace{\frac{1}{4} \left[\frac{k_1 - 1}{k_2} \right]}_{1} \underbrace{\left[\frac{n}{\bar{k} + 1} \right]^{1/2}}_{2} \underbrace{\left[\frac{\bar{k}}{\bar{k} + 1} \right]}_{3} \qquad (9.22)$$

where k_1 and k_2 are the capacity factor constants for spots 1 and 2 and $\bar{k}$ is
their mean value. The first term in equation (9.22) defines the selectivity of
the system. If k_1 and k_2 do not differ, separation is impossible (i.e., R_s =
0). The second term measures the quality of the stationary phase while the
third term takes into account the relative position of the two spots in the
chromatogram. Figure 9.1 illustrates the relationship between resolution for
two closely migrating spots and the R_F value for the spot of higher R_F. Maximum
resolution occurs for a R_F value of approximately 0.3. The resolution does not
change significantly with R_F values varying between 0.2 and 0.5; within this
range, the resolution is greater than 92% of the maximum value (75% between R_F =
0.1 to 0.6). Outside this range, the resolution declines rapidly again,
illustrating the strong correlation between separating power in HPTLC and the
position of the spots in the chromatogram.

Thus far the theory of TLC has been discussed in terms of mobile phase
development through the application of capillary forces. A new method of
development, overpressured TLC (OPTLC), uses a mechanical pump to force the
mobile phase through the sorbent bed, which is sandwiched between the sorbent
backing plate and a pressurized elastic membrane [22,23]. The mobile phase

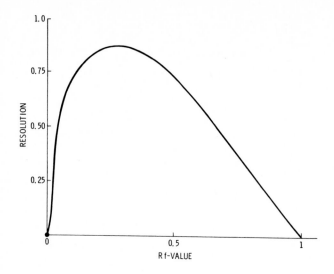

Figure 9.1 Change in resolution of two closely migrating spots as a function of the R_F value of the faster moving spot.

velocity is independent of the migration distance (assuming that a constant flow pump is used) and a linear relationship exists between the migration distance of the mobile phase and time, equation (9.23).

$$Z_f = u_o t \qquad\qquad (9.23)$$

t = mobile phase velocity in OPTLC
u_o = development time

In contrast to ascending TLC, there is a fairly regular linear increase in the number of theoretical plates, the separation number, and resolution with solvent migration distance. This major difference is attributed to the fact that spot broadening by diffusional processes plays a minor role at the relatively high linear mobile phase velocities commonly employed in OPTLC. This is shown in Figure 9.2, in which the average plate height ($\overline{H}$) is plotted against migration distance for conventional and high performance TLC plates using normal and overpressured development [22].

The discussion of development in this section has been limited to those processes that occur in an enclosed chamber with a fixed migration distance. Other development methods, principally continuous and multiple development, can produce higher resolution than the conventional single development mode already discussed. A detailed discussion of these methods is presented in section 9.8.

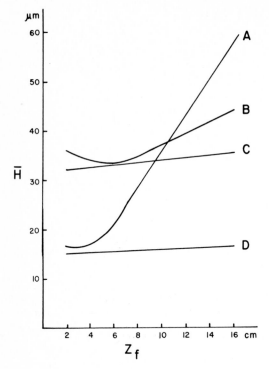

Figure 9.2   Variation of the average plate height with migration distance in
normal and overpressured thin-layer chromatography. A, HPTLC
plate, normal development; B, conventional TLC plate, normal
development; C, conventional TLC plate, overpressured development;
and D, HPTLC plate, overpressured development. (Reproduced with
permission from ref. 22. Copyright Elsevier Scientific
Publishing Co.)

9.5   High Performance Thin-Layer Chromatographic Plates

The exacting experimental conditions required for the reproducible
preparation of HPTLC plates are more readily obtained in the manufacturing
setting than in the research laboratory. Conventional TLC plates are also
commercially available or may be prepared in the laboratory by well-tried
standard procedures [4-6]. The commercial TLC plates available have layers
ranging in thickness from 100 to 2000 micrometers of silica gel, alumina,
cellulose, polyamide, keiselguhr, Sephadex, etc., which are supported by either
glass, aluminum, or plastic backing sheets. Most plates also contain a binder
such as gypsum, starch, or polyesters to impart the desired mechanical strength,
durability, and abrasion resistance to the adsorbent layer. If fluorescence
quenching is used for detection purposes, a UV-indicator such as
manganese-activated zinc silicate may be added to the adsorbent layer.

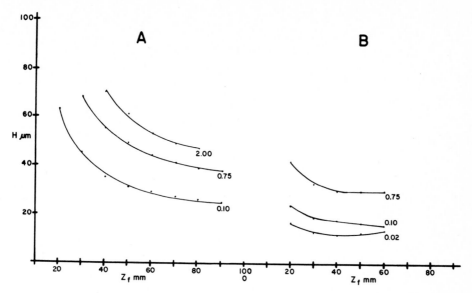

Figure 9.3 Plate height as a function of migration distance for conventional (A) and high performance (B) silica gel TLC plates. Sample, lipophilic dyes; sample loading in micrograms. (Reproduced with permission from ref. 24. Copyright Elsevier Scientific Publishing Co.)

    The same adsorbents used in the manufacture of conventional TLC plates are used to prepare HPTLC plates; the principal differences are the particle size and particle size distribution. Conventional TLC plates are prepared from silica gel with a mean particle diameter of approximately 20 micrometers and a particle size distribution of 5-30 micrometers, while particles with a mean diameter of 5 micrometers and a very tight size distribution are used in the preparation of HPTLC plates. These smaller particles are easier to pack as dense, homogeneous beds, which add to the chromatographic efficiency of the plate. A plot of plate height against solvent migration distance for both conventional TLC and HPTLC silica gel plates is shown in Figure 9.3. Both plates were prepared from the same type of silica gel but with different mean particle diameters and particle size distributions. This diagram clearly indicates the greater efficiency of the HPTLC plates. Note also that the HETP minimum occurs at a shorter migration distance for the HPTLC plate [24,25]. Because the efficiency of the layer is influenced by the sample size, the sample capcity of the HPTLC plate is considerably lower than the conventional TLC plate. In conventional TLC, microgram amounts of sample are used routinely, but these amounts should be scaled down to nanogram quantities to maintain optimum separation conditions in HPTLC. Thus, sample application equipment for HPTLC is designed to operate in the nanoliter range, compared to the microliter sample volumes employed in conventional TLC.

The physical properties of two commercial silica gels used to prepare HPTLC plates are summarized in Table 9.3. The particular separation characteristics of silica gel are controlled by its pore system (and via this pore network the available surface area), the texture of the skeletal substance (the ratio of silanol groups to siloxane groups), and the deactivation of the silica surface by adsorption of polar modifying substances (e.g., the uptake of moisture). These features govern the selectivity of the adsorbent as they determine the nature and extent of the adsorption mechanism with the sample components [26].

TABLE 9.3

PHYSICAL PROPERTIES OF SILICA GELS USED TO PREPARE HPTLC PLATES

Property	Merck Si 60	Whatman HP-K
Mean Pore Diameter (A)	60	80
Pore Volume (ml/g)	0.82	0.70
Specific Surface Area (m^2/g)	550	300
Mean Particle Size (micrometers)	5	5
Layer Thickness (micrometers)	200	200
Stable pH Range	1 $\longrightarrow$ 8	1 $\longrightarrow$ 8
pH of a 10% Aqueous Suspension	7.0	7.0-7.2

HPTLC plates with a concentrating zone of an inert synthetic silica gel are also commercially available. They are most useful when large sample amounts are applied, when very dilute sample solutions are used, or when crude samples (e.g., biological fluids) are applied directly to the plate. HPTLC plates with concentrating zones can also be used in laboratories lacking the sophisticated equipment necessary for spotting nanoliter volumes onto normal HPTLC plates. However, HPTLC plates without concentrating zones are preferred for the separation of small sample sizes, for obtaining maximum resolution, and for the greatest quantitative accuracy [27].

HPTLC plates with concentrating zones are prepared from two layers of silica gel having different properties. The two layers abut each other, parallel to one edge, forming a very narrow interface. The separating zone is prepared from the same silica gel used to coat normal HPTLC plates and occupies most of the plate area. The concentrating zone is prepared from porous silica gel of larger pore diameter and extremely low surface area [28]. The concentrating zone simplifies the process of sample application since microliter

volumes can be applied either as spots or streaks to any position on the
concentrating zone. Alternatively, the entire zone can be immersed in a dilute
solution of the sample. During development, the sample migrates out of the
concentrating zone and is focussed at the interface as a narrow band. This
sample band may be less than 1 mm wide, resulting in excellent chromatographic
efficiency. However, since the distribution of the sample may not be even
within the band, the quantitative accuracy of densitometric measurements may be
lowered.

HPTLC would be needlessly limited as a separation tool if the only
stationary phase material available were silica gel, since not all sample types
are conveniently separated on this medium. Silica gel may lack the necessary
selectivity for a particular separation or the sample may contain components
which bind too strongly to the surface and thus remain close to the origin in
all solvent systems. Poor migration characteristics, especially streaking, may
also be a problem. Assuming that a mobile phase cannot be found to correct
these problems, a change in stationary phase is indicated. The simplest
solution is to change the properties of the stationary phase by impregnating the
silica gel layer with a solid or liquid of different properties [29]. For
example, hydrophilic layers can be prepared by impregnation with
dimethylformamide, dimethyl sulfoxide, ethylene glycol, and polyethylene glycol
while lipophilic layers are prepared with liquid paraffin, undecane, silicone
oils, ethyl oleate, etc. These layers are easily prepared by the horizontal or
vertical immersion of the plate into a solution of the impregnating reagent
dissolved in a volatile solvent. The plate is removed, excess solution is
allowed to drain away, and the volatile solvent is evaporated, leaving an even
coating of the modifying agent. A disadvantage of this preparation method is
that the mobile phase must be selected so as not to destroy the homogeneity of
the impregnated layer. The ultimate solution is to chemically bond the
modifying agent to the silica gel, forming a new layer which is non-extractable
by the mobile phase. This approach will be described later.

The selectivity of the silica gel layer can also be modified by the
incorporation of chemically selective reagents. Those most widely used are
silver nitrate for the separation of saturated and unsaturated compounds by
selective bonding of silver ions with the π-electrons of double bonds, boric
acid for the separation of vicinal bifunctional isomers by chelation, and picric
acid for the separtion of polycyclic aromatic hydrocarbons by complex
formation. The above represent just a few examples of the many reagents and
reactions described in the literature [4,6].

The chemically-bonded layers mentioned above are available commercially as
reversed-phase plates. These plates have lipophilic layers permanently bonded
to the silica surface via $\equiv$Si-O-Si-R bonds. The Merck products suffixed RP-2,

RP-8, and RP-18 contain ethyl, octyl, and octadecyl ($R = C_2H_5-$, $C_8H_{17}-$ and $C_{18}H_{37}-$) alkyl groups, respectively. According to the manufacturer, the reversed-phase plates are prepared from a medium pore silica gel of particle size 5-7 micrometers with a high conversion of silanol groups to the chemically bonded alkyl substituents [30]. Another series of plates, given an 'S' designation, differs from the RP-8 and -18 plates in that they are prepared from silica gel of a larger particle diameter (ca. 11-12 micrometers) and are only partially silyltated [31,32]. Machery-Negel supplies octadecyl-coated reversed-phase plates prepared from silica gel with 5-10 micrometers particle diameter and 50%, 75%, or 100% silylation. It is not specified as to what the percentage silylation refers, but it is unlikely that it represents the percentage of silanol groups reacted. The KC_{18} plates manufactured by Whatman are prepared from silica gel of mean particle diameter of about 20 micrometers and with an "optimum" coverage of octadecyl groups, corresponding to a carbon load of about 12% [33]. Unreacted silanol groups are endcapped by silylation with an unspecified reagent. High performance reversed-phase plates of acceptable quality can be prepared in the laboratory by reacting an alkyltrichlorosilane or alkyldimethylchlorosilane in situ with precoated silica gel HPTLC plates [34-36]. HPTLC plates prepared by coating silanized silica gel onto glass plates without a binder give soft, fragile layers of poor quality [37]. This fragility is due to the poor adhesion of the silanized silica gel particles to one another and to the glass backing plate.

Silica gel shows an amphoteric hydrophilic/lipophilic character due to the simultaneous existence of silanol and siloxane groups. Virtually all solvents used in TLC can wet this surface with a contact angle of, or close to, zero. This is not true of the alkyl bonded reversed-phase plates which are not generally wetted by organic solvents containing more than about 20-40% water. Consequently, the selection of the solvent for sample application should be considered carefully, as water-rich solvents may not adequately penetrate the plate surface for sample deposition [37]. Generally, the best results are achieved by using an organic solvent that wets the layer, dissolves the solute, and has a low enough solvent strength to prevent predevelopment of the applied spot.

The water content of the mobile phase also has a dramatic influence on the mobile phase velocity of reversed-phase plates [18,31,37-40]. Table 9.4 illustrates the dramatic changes in mobile phase velocity observed with changes in the water content of water-ethanol mobile phase mixtures [18]. Even when using a relatively non-viscous organic solvent such as acetonitrile, only 20-30% water can be tolerated while still preserving reasonable analysis times on the Merck RP bonded phases. The Whatman KC_{18} plates have mobile phase velocities three to four times greater than the Merck plates. It should be noted that with mobile phases containing more than about 40% water, the precoated KC_{18} layer tends to swell and/or flake off, thus preventing normal development of the plate. This drawback can be overcome by adding inorganic salts to the mobile phase (e.g., 3% w/v sodium chloride or ammonium acetate solution rather than water). The polarity of the mobile phase can then be varied from 5-80% sodium chloride solution in methanol and 25-80% in acetonitrile without greatly changing in the mobile phase velocity. Nearly all separations recorded have used water-methanol or water-acetonitrile mixtures as the mobile phase. Studies of the influence of mobile phase composition on selectivity and resolution, the retention mechanism, and the development rate are lacking [18,30,39].

TABLE 9.4

VELOCITY CONSTANT AND WETTING ANGLE OF RP-18 REVERSED-PHASE PLATES WITH WATER-ETHANOL MIXTURES

Water Concentration (% v/v)	$K \times 10^3$ (cm^2/s)	$\cos \Theta$
0	16.6	0.87
4	13.4	0.78
10	12.3	0.76
20	7.8	0.61
30	5.5	0.48
40	3.5	0.34
50	1.5	0.14

HPTLC silica gel plates silylated with 3-aminopropylsilyl groups are available from Merck [41]. The chromatographic properties of the aminoalkyl-bonded plates are largely controlled by the nature of the amine group, with some additional contributions from hydrophobic interactions with the organic surface and polar interactions with unreacted silanol groups. For ionic or ionizable substances the dominant retention mechanism is due to the weakly basic ion-exchange behavior of the amino group. When the mobile phase is an organic solvent mixture these plates act as weak adsorbents and affect separations predominantly by the number of polar functional groups present in each component. The aminoalkyl-bonded plates are wet by all solvents, including pure water. Various authors have prepared aminoalkyl-bonded plates by the in situ reaction of 3-aminopropylchlorosilanes and other substituted aminoalkyltrialkoxysilanes with silica gel HPTLC plates [42]. 3-Cyanopropyl-bonded phases have been prepared by a similar reaction technique [35,43]. The cyanoalkyl-bonded plates exhibit weak adsorption behavior with some lipophilic character. They can be used in both normal and reversed-phase separations.

Precoated TLC layers of both native and microcrystalline cellulose have been used for several years, primarily for the separation of very polar compounds (e.g., carbohydrates, carboxylic acids, amino acids, nucleic acid derivatives, phosphates, etc.). HPTLC plates coated with a special grade of microcrystalline cellulose as well as with a type of porous silica (silica gel 50,000) with a large pore diameter (5000 nm) and very low surface area (ca. 0.5 m^2/g), are available for the separation of polar compounds [28]. The silica gel 50,000 is identical to the material used to prepare the concentrating zone of those HPTLC plates with such zones.

The HPTLC cellulose plates possess advantages similar to those exhibited by HPTLC silica gel plates with regard to the nature of the plate surface, the particle size range, the particle size distribution, the homogeneity of the layer and the associated increase in chromatographic separation efficiency and detection sensitivity. The HPTLC silica gel 50,000 and cellulose plates also show very similar separation properties for polar samples. Compared to cellulose, however, silica gel plates are non-swelling in organic solvents and they can be used with aggressive visualization reagents without darkening or reaction.

9.6   Sample Application in High Performance Thin-Layer Chromatography

The introduction of the sample into the adsorbent layer is a critical process in HPTLC. In order to achieve optimum resolution and sample detectability, the initial starting spot should be of minimal size. However, the overall efficiency of the chromatographic system shows a stronger correlation with the size of the developed spots and depends less on the starting spot diameter [44]. Although this does not advocate the use of starting spots with large diameters, as spots of this type exhibit an elliptical profile upon development with the HPTLC plates currently available, the optimum initial diameter of the sample spot is about 1.0 mm and there seems to be no added advantage in applying spots of a smaller diameter.

The size of a spot deposited as a solution will depend on both the sample volume and the R_F value of the least retained component in the sample. The R_F value here refers to the expected value if the sample solvent were used as the mobile phase. Generally, one chooses the solvent of lowest eluting strength which completely dissolves the sample and provides a quantitative transfer of the sample from the applicator. It is also important that the application solvent wets the sorbent; otherwise the sample will not penetrate the layer. This was discussed in the previous section and is most important when reversed-phase layers are used.

For most quantitative work, a platinum-iridium capillary of fixed volume, either 100 or 200 nl, sealed into a glass support capillary of larger bore, provides a convenient and durable spotting device. The capillary tip is polished to provide a smooth, planar surface of small area (ca. 0.05 mm^2) which,

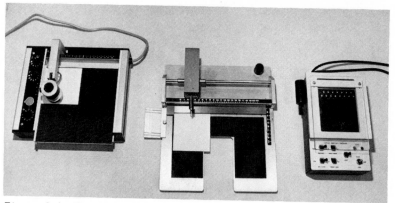

Figure 9.4   From left to right, Nanomat, "rocker" applicator, and contact
spotter.

when used with a mechanical applicator, does not seriously damage the plate
surface. Spotting by hand invariably damages the surface. Mechanical
application is made possible by attaching a metal collar to the glass support
capillary so that it can be held by a magnet and lowered to contact the plate's
surface. The simplest form is the "rocker" applicator: an arm houses a magnet
at one end to hold the dosimeter and a counter-weight at the other to control
the force with which the dosimeter strikes the plate's surface, Figure 9.4. The
dosimeter is both lowered to and removed from the plate's surface by a tipping
action of the applicator arm about its fulcrum. A somewhat more sophisticated
capillary applicator is the Camag Nanomat, which holds and lowers the dosimeter
electromagnetically, Figure 9.4. The height from which the dosimeter is
lowered, the time it spends in contact with the plate's surface and the number
of repetitions needed for complete sample transfer can be controlled
electronically. Both applicators use a click-stop grid mechanism to aid in the
even spacing of the samples and to provide a frame of reference for sample
location during scanning densitometry. The Nanomat is also equipped with an
accessory for spotting circular chromatograms.

As an alternative to the dosimeter, samples can be spotted with a
microsyringe similar to those used in gas chromatography. For accurate
dispension of nanoliter volumes, the syringe plunger is controlled by a
micrometer screw gauge. This micrometer microsyringe can be operated to provide
various precisely-selected volumes or, via a fixed lever mechanism, a repetitive
constant sample volume. An advantage of the microsyringe over the dosimeter is
that it delivers the sample volume by displacement rather than capillary action
and thus does not deform the plate surface. The microsyringe needle is brought
just close enough to the plate's surface for the convex sample drop of the

642

ejected liquid to touch it. Damage to the plate surface can be completely
avoided in this way.

As well as in the form of spots, samples can be applied to the HPTLC plate
as either streaks or narrow bands. The Camag Linomat can provide streaks of any
desired length by mechanically controlling the distance the plate moves beneath
a fixed sample syringe. A controlled nitrogen atomizer sprays the sample from
the syringe, forming narrow, homogeneous bands on the plate surface.

Any of the above methods are suitable for spotting solutions of low
viscosity, or viscous solutions by repetitive application after dilution.
However, samples from biological sources invariably yield viscous residues;
dilution and repetitive sample application can be tedious and time consuming for
these samples. The contact spotter, Figure 9.4, provides a simple means of

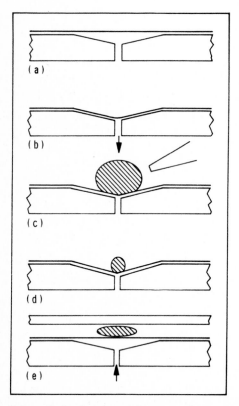

Figure 9.5   Sample application by contact spotting. A specially treated
fluoropolymer film is pulled into a series of depressions in a
metal plate by application of vacuum (a and b). Sample solution
is delivered by pipette (c) and, after evaporation, a residue
remains (d), which is transferred to the HPTLC plate by replacing
the vacuum with slight pressure (e). (Reproduced with permission
from ref. 3. Copyright American Chemical Society).

spotting viscous samples or large sample volumes (up to 100 microliters) onto
HPTLC plates. This apparatus is designed for the solventless sample application
of evaporated residues of several samples simultaneously at precise locations on
the thin-layer plate [45]. The transfer medium is a fluorinated
ethylene-propylene film, coated with perfluorokerosene, positioned over a series
of depressions in a metal platform. With the application of vacuum through
orifices in the center of each depression the film conforms to the contour of
the platform's surface. The sample solutions are pipetted into these
depressions and the solvent is evaporated by gentle heat and a flow of
nitrogen. The HPTLC plate is then positioned over the film, adsorbent side
down, and with slight pressure replacing the vacuum, the spots are transferred
simultaneously to the plate, Figure 9.5. Problems can arise due to the
non-quantitative transfer of highly crystalline samples. This is solved by the
addition of a small amount of a non-volatile solvent (e.g., methyl myristate) to
the solution. Transfer then occurs quantitatively.

9.7 Linear and Radial Development

Development in thin-layer chromatography is the process by which the mobile
phase moves through the sorbent layer, thereby inducing differential migration
of the sample components. In a closed bed technique such as HPLC only linear
development of the chromatogram is possible. An open bed technique like HPTLC
does not suffer from this limitation and the chromatogram can be developed in
the radial as well as the linear direction. The principal development modes for
use in HPTLC are linear, circular, and anticircular development.

The linear development mode represents the simplest situation and is the
most widely used of the three techniques. The samples are spotted along one
edge of the plate and the solvent migrates to the opposite edge, effecting a
separation. Viewed in the direction of development, the chromatogram consists
of compact symmetrical spots of increasing diameter. Spots eluting close to the
solvent front may be distorted and ellipitcal or rod-like in shape. The usable
separation area of the plate is defined by an R_F range of approximately 0.1 to
0.8.

If the point of sample application and the point of entry of the mobile
phase are at the center of the plate then this mode of development is called
circular chromatography. This method is a powerful technique for separating
samples having low R_F values. Spots near the origin remain symmetrical and
compact while those nearer the solvent front are compressed in the direction of
devlopment and elongated at right angles to this direction, Figure 9.6A. Spots
near the solvent front may be diffuse and elliptical.

In anticircular development the sample is applied along the circumference

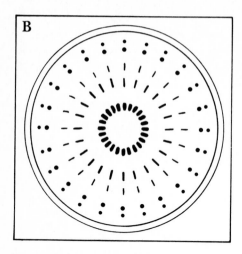 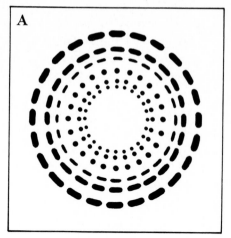

Figure 9.6   Circular development with the point of solvent entry at the plate
center (A).   Anticircular development from the outer circle towards
the center (B).   (Reproduced with permission from ref. 3.
Copyright American Chemical Society).

of an outer circle and developed toward the center of the plate. The unique
features of this method are high speed development and large sample capacity.
The mobile phase velocity remains constant across the plate as a consequence of
the equivalence between the quadratic decrease in mobile phase velocity and a
similar reduction in the plate area to be wetted. Anticircular development is
thus the fastest method of development, both practically and theoretically. As
many as 48 samples can be placed along the outer edge of the development circle
of a 10 x 10 cm HPTLC plate, providing the highest sample capacity obtainable in
TLC.

Anticircular development is useful for separating compounds with high R_F
values. Spots near the origin remain compact while those towards the solvent
front are considerably elongated in the direction of migration, but changed very
little in width when viewed at right angles to the direction of development,
Figure 9.6B. Although sample resolution is optimum with high R_F values, much of
the potential separating power is lost due to spot elongation. Unfortunately,
this elongation of the sample spots is unavoidable. It arises as a consequence
of the lateral compression induced by the mobile phase flow through a
continuously decreasing plate area.

9.8 Continuous and Multiple Development

The literature describes many examples where the conventional solvent
development of a TLC plate does not provide the necessary resolving power to

separate a sample. Typical examples are samples containing components with very similar R_F values and those in which a single solvent development is optimum for the separation of some components but leaves others unseparated in the vicinity of the origin or the solvent front. The solution to these and similar problems is to improve the resolving power of the TLC system by the techniques of continuous and multiple development. These methods increase the separating power of TLC by allowing the use of more selective mobile phases or by increasing the number of theoretical plates employed in the separation, or both.

In the continuous development technique, the mobile phase traverses the TLC plate to a predetermined fixed distance, at which point the mobile phase is allowed or induced to continuously evaporate. Thus the development of the TLC plate does not stop at the conventional solvent front, since a mechanism is provided for the complete evaporation of the solvent at this point. The mobile phase velocity and thus the sample migration can be controlled for any time period desired. Under these circumstances, components with low R_F values can be made to traverse the entire length of the TLC plate and their resolution is improved by maximizing the number of theoretical plates available for the separation. The principal disadvantages of this approach are that the resolution between spots increases only proportionally to $\sqrt{n}$, the spots become more diffuse (spot broadening is linearly related to migration distance), and, if long bed lengths are employed for the separation, the analysis time is greatly increased. An alternative approach to improving the resolution obtainable by continuous development TLC is to optimize mobile phase selectivity and to reduce the plate length required for the separation to the minimum value that provides the necessary number of theoretical plates [46]. This is easily achieved with the short-bed continuous development chamber, which will be described later. In general terms, selectivity in TLC can usually be increased by a reduction in the solvent strength of the mobile phase. If the selectivity of the mobile phase can be increased sufficiently by the use of a mobile phase of low R_F value, then the number of theoretical plates required for a particular separation can be reduced. Thus short bed lengths at high mobile phase velocities can be employed to minimize the analysis time. With modern HPTLC plates this latter approach is generally preferred.

The rate of solvent migration in continuous development TLC is constant and can be described by equation (9.24) [47–52].

$$u_c = \frac{K}{2L} \qquad (9.24)$$

K = mobile phase velocity constant
L = length of plate traversed by the solvent
u_c = mobile phase velocity in the continuous development mode

The total analysis time is the sum of two components: the time during which the solvent front traverses the TLC plate and the time during which continuous development occurs, equation (9.25).

$$t_L = t_1 + t_2 \qquad (9.25)$$

t_L = total analysis time
t_1 = time during which the solvent front traverses the TLC plate
t_2 = time during which continuous development occurs

Similarly, the total distance migrated by each solute is the sum of the distances that the solute migrates during times t_1 and t_2, such that the following approximations are valid, equations (9.26) to (9.28).

$$M_D = d_1 + d_2 \qquad (9.26)$$

$$d_1 = R_F \, (L - x) \qquad (9.27)$$

$$d_2 = R_F \, \frac{K}{2L} \, t_2 = R_F \left[\frac{t_L}{2L} - \frac{L}{2} \right] \qquad (9.28)$$

M_D = total distance migrated by each solute

d_1 = distance migrated during time t_1

d_2 = distance migrated during time t_2

x = distance between the sample origin and the height of the mobile phase in the tank

Combining the above equations and eliminating x leads to the generally useful expression for M_D, equation (9.29) [52].

$$M_D = R_F \left[\frac{L^2 - 2LK + Kt_L}{2L} \right] \qquad (9.29)$$

Equation (9.29) can be used to predict the center-to-center spot separation of any two solutes by the difference in their M_D values. With binary mobile phase mixtures, K can be replaced by a quadratic expression, equation (9.30), which contains the mole fraction of the stronger solvent as a variable.

$$K = a_1 + a_2 x_s + a_3 (x_s)^2 \qquad\qquad (9.30)$$

a_1, a_2, and a_3 = experimentally-derived constants
x_s = mole fraction of the stronger solvent

Using equations (9.29) and (9.30) and the method of overlapping resolution maps for critical solute pairs, Nurok has shown that it is possible to predict separations by continuous development TLC [52]. The resolution domain of the optimization triangle is bounded by the plate length, development time, and mole fraction of the stronger solvent, as shown in Figure 9.7. The optimum region for separation is the unshaded portion of the triangle, while the shaded area represents combinations of experimental variables for which acceptable resolution does not exist. Nurok has shown that for a given separation, continuous development will always yield a shorter analysis time than conventional development [49].

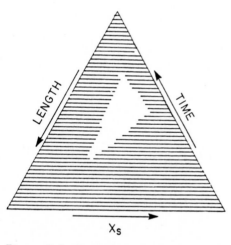

Figure 9.7   Optimization triangle for continuous development TLC. The experimental variables bounding the triangle are plate length, time, and mole fraction of the stronger solvent of a binary pair. The shaded area is generated by the method of overlapping resolution maps. The unshaded area is the optimized region for the separation. (Reproduced with permission from ref. 52. Copyright Elsevier Scientific Publishing Co.)

A technique similar to that of continuous development is "evaporative TLC". Here the TLC plate is heated by external means to induce controlled evaporation of the mobile phase from the developing plate [53]. The net result is a solvent velocity profile which decreases from the solvent reservoir to the solvent front. With this technique modest improvements in separation rates have been claimed due to increased integral flow. There exists an important difference between the methods of continuous development and evaporative TLC. In the former, the separation takes place in an ideally saturated atmosphere, whereas in the latter, the vapor phase is not saturated and the development may be performed in an open chamber.

The term "multiple development" encompasses a variety of TLC techniques involving the repeated development of a TLC plate in either one or two dimensions with the same or different solvents for constant or varying distances. In its simplest form, the chromatogram is developed in a given solvent, the plate is removed from the chamber, and the solvent is allowed to evaporate. The plate is then returned to the same solvent and developed for a second time to the same solvent front. The sequence of development and solvent removal can be repeated any number of times until the desired separation is achieved. Provided that the experimental regime remains as described above, then the apparent R_F value for the second of two components migrating close together, R_{Fa}, after n developments is given by equation (9.31)

$$(1 - R_F)^n = 1 - R_{Fa} \qquad (9.31)$$

and the resolution of the two bands by equation (9.32).

$$R_s = C \, [1 - (1 - R_F)^n]^{1/2} \, (1 - R_F) \qquad (9.32)$$

R_F = single development value

R_{Fa} = apparent R_F value after n developments

C = constant dependent on the value of R_{Fa} for the second of two closely migrating components. It embodies the values for the capacity factor and the number of theoretical plates.

The condition for maximum resolution with n identical developments is given by equation (9.33).

$$R_F = 1 - [2/(n + 2)]^{1/2} \qquad (9.33)$$

If the solvent system has been optimized for the separation then the maximum resolution obtainable in n developments is given in Table 9.5.

A unique feature of the multiple development technique is the existence of an interface, the solvent front, which periodically traverses the stationary bed and any sample components within it. The result of this process is a natural spot reconcentration phenomenon [54]. As the solvent front advances across the spot, the solute molecules behind the front advance in the direction of solvent flow toward those motionless solute molecules still beyond the front. The net result is a compression of the spot in a bottom-to-top direction as illustrated in Figure 9.8. For a spot of initial width, w_i, the spot width w after n developments is given by equation (9.34).

TABLE 9.5

MAXIMUM RESOLUTION IN MULTIPLE DEVELOPMENT THIN-LAYER CHROMATOGRAPHY

Number of Developments (n)	Optimum R_F		Relative Resolution[a]
	Single Development (R_F)	Position of band after multiple development (R_{Fa})	
1	0.33	0.33	1.00
2	0.29	0.50	1.31
3	0.26	0.61	1.52
5	0.22	0.71	1.73
10	0.16	0.82	2.00

[a] R_s measured divided by R_s for a single development (R_F = 0.33)

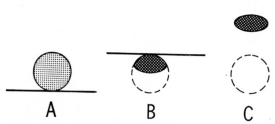

Figure 9.8 Spot reconcentration mechanism in multiple development.
A, advancing solvent front contacts lower edge of spot; B, solvent
front traverses spot, producing a compression in the bottom-to-top
direction; C, spot is developed normally after being reconcentrated.

$$w = w_i (1 - R_F)^n \tag{9.34}$$

The conventional development of a TLC plate results in a linear increase in spot
width with the distance migrated by the spot. The spot reconcentration
phenomenon counteracts this broadening mechanism, as shown in Figure 9.9. The
dotted line, which takes on a zig-zag shape, indicates the change in spot
diameter during each cycle of the multiple development process. Two features
are clearly visible: the sharp decrease in spot width as the solvent front
traverses the spot (reconcentration mechanism) and the general increase in spot
width by the normal processes taking place during development of the
chromatogram. This latter broadening of the spot is of course equivalent to
that which occurs in normal TLC.

The problem of separating mixtures containing components with very
different R_F values can be circumvented in the multiple development technique.

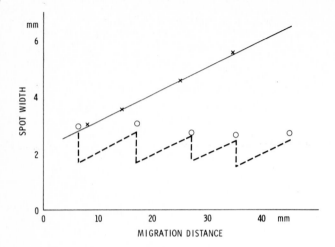

Figure 9.9   Spot broadening as a function of development distance. The solid
line represents the constant rate of spot broadening by normal
development.   The broken line illustrates the influence of multiple
development techniques on the spot width via the reconcentration
mechanism. (Reproduced with permission from ref. 54. Copyright
American Oil Chemists Society).

By successively changing the solvent strength of the mobile phase at any or all
development steps, the k value for each component, or groups of adjacent
components, can be optimized. The advent of scanning densitometers for the in
situ detection of the separated spots has had a considerable impact on the use
of the multiple development technique for the separation of mixtures containing
weakly and strongly retained components. It is no longer necessary to have the
entire sample simultaneously separated on the plate and available for
measurement. The preferred method is to fractionate the sample, a few
components at a time, based on their relative mobilities and to make
quantitative measurements at the conclusion of any development cycle in which
adequate separation has been achieved. With HPTLC plates a combination of
continuous and multiple development techniques provides the means to optimize
the resolution of complex mixtures while minimizing the time required for each
analysis. The experimental parameters that must be optimized for each cycle in
the multiple development sequence are plate length, development time, and
solvent selectivity [55].

The features of both multiple development and "evaporative TLC" can be
combined in a completely instrumentalized high resolution TLC technique called
programmed multiple development [56-58]. Programmed multiple development (PMD)
has been defined as the repeated development of a TLC plate with the same
solvent in the same direction for gradually increasing distances. Between

developments the solvent is removed from the thin-layer bed by controlled evaporation while the plate remains in contact with the solvent reservoir. A unique feature of this development method is that the spot reconcentration mechanism occurs twice per development cycle, first during solvent advance, as with all multiple development techniques, and then during solvent removal. The solvent front recedes slowly by forced evaporation, by the application of either heat to the back of the plate or a stream of inert gas across the front. Thus, as the solvent front recedes through the spot, the molecules at the top of the spot are deposited in the bed while those molecules at the bottom of the spot continue to move upwards until the solvent front recedes completely below the spot. The result is a top-to-bottom compression of the spot diameter. The solvent front then recedes until a steady-state situation is reached between the evaporation of solvent from all wetted areas and the solvent flux entering the bed. The next cycle commences by terminating the heat or gas flow used to induce forced evaporation. As the number of cycles is increased, the spots become more and more rod-like with very narrow top-to-bottom diameters as a consequence of this spot reconcentration mechanism. The PMD instrument allows up to 99 solvent advance and evaporation cycles to be performed automatically and apparent plate numbers in excess of 100,000 have been reported using a program of 68 developments lasting 72 hours.

So far the discussion of multiple development TLC has dealt exclusively with unidimensional development. This method has the advantage of preserving the ability of the TLC plate to handle multiple samples. When sample throughput is not the main concern and the required separation needs only a modest increase in resolution, it is possible to use two-dimensional TLC [59]. In two-dimensional TLC, the sample is developed along one edge of the plate, the plate is removed from the chamber, and the solvent is evaporated prior to returning the plate to the chamber and developing a second time in a direction orthogonal to the first. If the same mobile phase is used for both developments, the bed length and thus the total number of theoretical plates are doubled and the resolution is improved by up to a factor of $\sqrt{2}$. Two-dimensional development is often used with samples containing some components which have similar k values for any set of separation conditions, such that some overlap of sample spots always occurs with normal development. The use of different separation conditions in each development permits the separation of one group of sample components in the first development and some other group of components in the second. Two-dimensional TLC can also be used in conjunction with specific chemical reagents which alter either the sample (e.g., reduction, oxidation, derivatization) or the plate (e.g., impregnation with silver nitrate) after the first development. The principal limitation of two-dimensional TLC is that scanning densitometry cannot be used for quantitation.

A technique related to two-dimensional TLC, two-dimensional column chromatography, has recently been described by Guiochon [60,61]. Here, the column is essentially a thin layer of sorbent, similar to the sorbent bed in TLC, except that it is retained in a rectangular trough. The trough is pressurized by an elastic membrane covering the top surface. The column is developed alternately in orthogonal directions, employing an intermediate gas or solvent flushing step to completely eliminate the first solvent prior to starting the second development [62]. The mobile phase is forced through the sorbent bed by a mechanical pump. Detection is performed on-line with a photodiode array detector strip located along one edge of the column opposite the entry position of the second mobile phase. The principal drawbacks to two-dimensional column chromatography are the necessity of selecting two different retention mechanisms for the separation of complex mixtures, the possible interference of the solvent used for the first development in the second, and, perhaps most importantly, the detection of the separated components. In theory, the sample capacity for two-dimensional column chromatography should exceed several thousand components, compared to a few hundred in HPLC. This has maintained interest in the technique despite the formidable instrumental problems.

9.9 Development Chambers for HPTLC

To obtain reproducible R_F values in thin-layer chromatography it is essential that the development process takes place in a chamber designed to enable the plate, mobile phase, and vapor phase to reach equilibrium [63]. For conventional TLC, large volume chambers, lined with paper that has been saturated by the mobile phase, are routinely used. Such chambers are adequate for qualitative work but fail to provide the controlled environment necessary for accurate and reproducible measurement of TLC data. Equilibration of the vapor phase with the sorbent layer of the plate is easily achieved in the compact sandwich chamber. This sandwich chamber is formed by clamping the TLC plate and a glass plate together, separated by a U-shaped spacer a few millimeters thick. Development is carried out by placing the open end of the chamber in a solvent trough containing the mobile phase.

Vapor phase equilibration is equally important in HPTLC. Development chambers based on the experience gained in conventional TLC are commercially available. The twin-trough chamber, Figure 9.10, is the simplest of the HPTLC developing chambers. It consists of a standard developing tank with a raised, wedge-shaped bottom to minimize solvent consumption. The wedged bottom divides the tank into two compartments so that it is possible to either develop two plates simultaneously or to use one compartment to precondition the sorbent

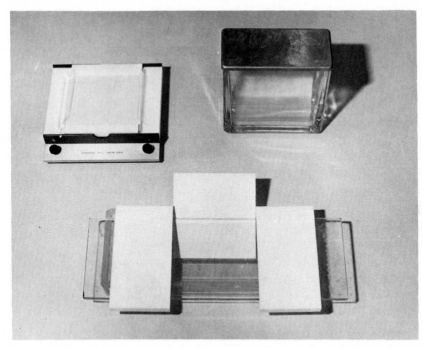

Figure 9.10   TLC chambers used for HPTLC. Front, short-bed continuous
development chamber. Back left, linear development chamber and
right, twin-trough chamber.

layer prior to development [64]. The plate is placed in one compartment and the
mobile phase or the preconditioning solvent in the other. Examples of sorbent
preconditioning procedures that are easily carried out in the twin-trough
chamber are control of humidity with water-sulfuric acid mixtures and plate
deactivation with aqueous ammonia solutions for the separation of basic
components. The mobile phase can be added directly to the plate compartment or,
when the mobile phase is first used for preconditioning, the chamber can be
tilted to allow the transfer of solvent into the compartment for development.
The twin-trough chamber is widely used for routine quality control
applications.

The linear development chamber, Figure 9.10, is a sandwich-type chamber for
the horizontal development of HPTLC plates. The samples can be developed either
from edge-to-edge, thereby maximizing the number of theoretical plates available
for the separation, or from both ends simultaneously toward the plate center,
hence maximizing the sample capacity of the plate. The mobile phase is
transported from the reservoir to the sorbent layer by two glass microscope

slides; the liquid rises by surface tension and capillary forces. The mobile phase then travels through the sorbent layer by capillary action. When a plate is developed from both edges simultaneously, the chamber must be leveled to allow the two solvent fronts to migrate at the same speed and meet precisely in the middle. At this point the capillary forces balance out and the chromatographic development ceases.

Related to the linear development chamber is the Vario-KS chamber. The plate is also developed in the horizontal position in a sandwich configuration. However, this chamber is supplied with a variety of conditioning trays, divided into lanes or square segments, and up to five mobile phase reservoirs. Thus the plate may be segregated into five environmentally separated lanes that can be developed with five different mobile phases simultaneously. The condition or activity of the adsorbent layer can be adjusted as desired for each of the five sample lanes independently or any desired activity gradient can be formed in the direction of mobile phase migration for each lane. The Vario-KS chamber is used primarily for scouting the conditions necessary for optimum separation of a mixture and also to assess the influence of preloading the sorbent layer with different solvent vapors [65]. It is used less often for carrying out multiple separations of the same sample. A heating block accessory is available for carrying out separations by continuous development.

The short-bed continuous development chamber (SB/CD chamber) is designed specifically for use with continuous and multiple development techniques, Figure 9.10. The chamber has a low profile and a wide base to permit development close to the horizontal positon and to minimize the chamber volume. The base contains four glass ridges running nearly the entire length of the chamber. The four ridges, along with the back wall of the chamber, provide five positions which permit a choice of the plate length used for a particular separation. One end of the HPTLC plate protrudes from the chamber, enabling continuous evaporation of the mobile phase at the junction formed by the plate surface and the cover lid of the chamber.

The U-chamber, shown in Figure 9.11, is designed for optimal circular HPTLC development. The mobile phase velocity is electronically controlled by a stepping motor that drives a captive syringe. The syringe feeds the mobile phase directly onto the horizontal plate via a platinum-iridium capillary. The chamber volume is miniaturized to provide optimum conditions for the saturation of the vapor phase in contact with the plate. The sorbent layer can be equilibrated prior to and also during development with any desired volatile vapor. An injection valve is provided to introduce the sample onto the plate surface via the mobile phase. When the sample is applied onto the pre-wetted layer in this mode, the chromatogram consists of a series of concentric rings of varying circumference. A circular heating block is available as an accessory to

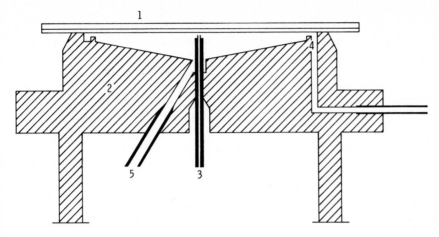

Figure 9.11   Cross-sectional diagram of the Camag U-chamber for circular
chromatography.

evaporate the mobile phase at the solvent front for use in the continuous
development mode. The U-chamber provides the most controlled environment for
obtaining reproducible HPTLC data [1].

The U-chamber was the forerunner of the pressurized ultramicro chamber
(PUM) used for overpressured TLC [22,66,67]. A unique feature of this
development chamber is the use of an elastic membrane under external pressure
which is in contact with the surface of the TLC plate. In this way the vapor
phase above the sorbent layer is entirely eliminated. The mobile phase is
pumped through the sorbent layer, permitting either linear or circular
development. A wide selection of flow rates may be employed with the proviso
that the input pressure of the mobile phase should not exceed the external
pressure on the membrane.

The anticircular development chamber contains many of the features found in
the U-chamber. The plate is developed horizontally, the chamber volume is
small, and the vapor phase can be controlled by external means [68]. An outer
circle is cut from the adsorbent layer to allow the solvent to migrate in only
the desired direction. This adsorbent-free annulus provides a barrier that
prevents solvent migration towards the outer edge of the plate. A very narrow
circular channel in direct contact with the solvent reservoir is contacted with
the adsorbent layer sealing the chamber to start the development process. The
mobile phase moves by capillary forces through the bed and cannot be controlled
by external means. Mobile phase consumption in both circular and anticircular
development modes is very low.

9.10 Quantitative Evaluation of Thin-Layer Chromatograms

Commercial instruments for performing quantitative evaluation of TLC chromatograms by direct photometric measurements first appeared in about 1967. Such instruments have played an important role in the evolution of modern TLC; without such equipment the exquisite resolution obtained by HPTLC would be to no avail and TLC would have remained a semi-quantitative technique. At best, the visual scanning of a TLC plate allows detection of about 1-10 micrograms of colored components with a reproducibility of no better than 10-30%. Excising the separated spots, eluting the substance from the silica gel, and measuring their concentration by solution photometry is time consuming and fairly insensitive. For compounds having average UV absorbance, 10-100 micrograms per spot is needed for detection purposes, an amount far in excess of the sample capacity of HPTLC plates. Since the human eye is incapable of accurately detecting the edge of a spot, scraping a spot off a plate requires a generous boundary of silica gel between the separated spots for accurate quantitative data. Incomplete elution of the sample from the silica gel and nonspecific background absorbance due to colloidal silica gel in the analytical solution compound the problem. In situ detection is essential for the accurate measurement of both spot size and location, for a true measure of inter-spot separation, and for rapid, accurate quantitation.

In situ measurements of substances on TLC plates can be made by a variety of methods: reflectance, transmission, simultaneous reflectance and transmission, fluorescence quenching, and fluorescence [1,69,70]. Light striking the plate surface is both transmitted and diffusely scattered by the layer. Light striking a spot on the plate will undergo absorption so that the light transmitted or reflected is diminished in intensity at those wavelengths, forming the absorption profile of the spot. The measurement of the signal diminution in the transmission or reflectance mode due to this absorption by the spot provides the mechanism for in situ quantitation. For a given compound, measurements in the transmission mode provide greater peak heights, up to a factor of two, accompanied by significantly higher baseline noise when compared to reflectance [71]. In reflectance most of the scattered light arises from layers close to the surface and is less significantly influenced by changes in the thickness of the TLC medium, which generates much of the background noise in transmission measurements. Transmission measurements are limited to those wavelengths greater than 320 nm due to strong absorption by the glass backing plate and by the silica gel itself at shorter wavelengths. Reflectance measurements can be made at any wavelength ranging from the UV to the near infrared (185-2500 nm). For spots with absorption maxima greater than 320 nm, the optimum signal-to-noise ratio is obtained by simultaneously recording both

the reflectance and the transmission signals at the same point on the plate
[72,73]. If the reflectance and transmission signals are combined by a
corresponding weighting factor, baseline fluctuations can be dramatically
reduced and the signal-to-noise ratio improved by up to two orders of
magnitude.

An exact theoretical description of the optical behavior of a thin-layer
plate is nearly impossible. Since TLC plates are opaque and scatter light
strongly, absorption measurements cannot be expected to obey the Beer-Lambert
law. However, a simplified theory is available and is adequate for the
situation pertaining to chromatographic measurements. Known as the Kubelka-Munk
theory, its derivation is beyond the scope of this section and the interested
reader is referred to references [74-76]. The Kubelka-Munk equation takes the
form

$$\frac{(1 - R)^2}{2R} = \varepsilon \frac{c}{s} \tag{9.35}$$

R = reflectance
ε = absorption coefficient
s = scatter coefficient
c = spot concentration

Reflectance measurements are made by scanning the plate with a slit-shaped
measuring area. Under these conditions the calibration curves show an initial
linear region passing through the origin but curving toward the concentration
axis at higher concentrations. Some typical calibration curves recorded under
the above conditions are shown in Figure 9.12 [77]. The curved upper branch of
the calibration graph can often be linearized by a parabolic approximation

$$A^2 = f(c) \tag{9.36}$$

or a two-fold logarithmic approximation

$$\log A = f(\log c) \tag{9.37}$$

where A is the area enclosed by the reflectance curve. If an internal standard
with absorption properties similar to the compound to be determined is used in
the analysis, then the ratio of the peak areas of the compound to the internal
standard plotted against concentration is usually linear over one to two orders
of magnitude.

UV-absorbing compounds can be measured by fluorescence quenching as well as
by reflectance. The fluorescence quenching technique provides a means of
visualizing spots absorbing UV-light on special TLC plates incorporating a
fluorescent indicator. When such a plate is exposed to UV light of short
wavelength, the UV-absorbing spots appear dark against the brightly fluorescing

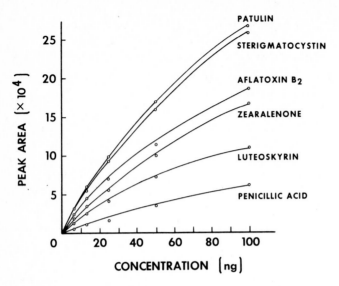

Figure 9.12   Calibration curves for some mycotoxins measured by reflectance in the absorption mode. (Reproduced with permission from ref. 77. Copyright American Chemical Society).

background of lighter color. The UV-absorbing spots behave similarly to an optical filter, absorbing a portion of the fluorescence excitation radiation and thus diminishing the fluorescence emission intensity. Compared to the measurement of reflectance, fluorescence quenching exhibits severe background fluctuations as a result of the inhomogeneous distribution of the fluorescent indicator in the adsorbent layer. The fluorescence quenching method can only be applied to those substances whose absorption spectra overlap the excitation spectrum of the fluorescent indicator. The fluorescent indicators in common use have a maximum absorption around 280 nm and virtually no absorbance below 240 nm. Fluorescence quenching measurements are thus less specific and less sensitive than reflectance measurements.

When applicable, the measurement of natural fluorescence is the detection mode of choice. The signal strength does not depend on the reflectance of the TLC medium since the fluorescence observed is dependent solely on the fluorescence of the individual sample molecules. When compared with absorption measurements, a much greater linear response range with a concentration profile independent of spot shape is obtained. For compounds which fluoresce strongly, detection limits of 10 to 100 times lower than the detection limits in the absorption mode are possible. As in solution photometry, there is an exponential correlation between the fluorescence intensity, I_{FL}, and the quantity of substance, c:

$$I_{FL} \approx I_{o\lambda} (1 - 10^{-\varepsilon_\lambda c})$$ (9.38)

ε_λ = absorption coeffcient at the excitation wavelength

$I_{o\lambda}$ = intensity of the excitation wavelength

For low concentrations, the product $\varepsilon_\lambda c$ is small, and the linear approximation may be applied [78].

$$I_{FL} \simeq I_{o\lambda} \varepsilon_\lambda c$$ (9.39)

The reproducibility of absorbance and fluorescence measurements by scanning densitometry is 1-3%. A sagging solvent front or variations in its homogeneity can seriously deteriorate this reproducibility. For quantitative work in which a high degree of reproducibility is desired, the influence of spot location on reproducibility can be largely eliminated by using the data pair technique [79].

9.11   Operating Requirements for Scanning Densitometers

The scanning densitometers currently available share many features. They are capable of measuring both absorption and fluorescence in either the transmission or reflectance modes and absorption in the fluorescence quenching mode. Different lamps must be used as light sources in order to cover the entire UV-visible range. Halogen or tungsten lamps are used for the visible range and deuterium lamps for the UV range. For fluorescence measurements, high intensity mercury or xenon lamps are used. Some filter densitometers use mercury lamps for absorption measurements in the UV range. Filter densitometers are inexpensive but limited in selectivity to the characteristic line spectrum of the source. Either grating or prism monochromators are used to select the desired measuring wavelength with deuterium and tungsten sources. For fluorescence measurements, a monochromator or filter is used to select the excitation wavelength. A cut-off filter which transmits the wavelength of emission but attenuates the wavelength of excitation is placed between the detector and the plate. Interference filters can be substituted if better selectivity is required; however, sensitivity decreases, since emission spectra usually consist of broad bands. The monochromatic light reflected or transmitted by the TLC plate and the fluorescence emission signal is measured by either a photomultiplier or a phototransistor. Photomultipliers are preferred as they have a wide wavelength range of operation and also provide a linear output signal as a function of excitation energy.

At this time there are three optical arrangements predominantly employed in scanning densitometers, Figure 9.13. The single-beam arrangement is capable of producing excellent quantitative results but spurious background noise resulting from fluctuations in the source output, inhomogeneity in the distribution of

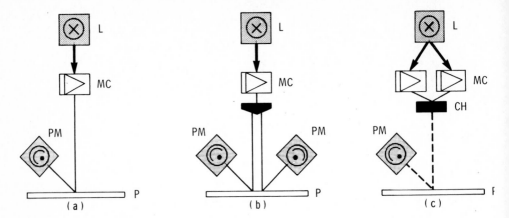

Figure 9.13 Schematic diagrams showing the optical arrangements of different types of scanning densitometers (a) single beam, (b) double beam in space, (c) double beam in time: L = source; MC = monochromator; P = TLC plate; and PM = photomultiplier. (Reproduced with permission from ref. 3. Copyright American Chemical Society).

extraneous adsorbed impurities, and irregularities in the plate surface can be troublesome. Background disturbances are compensated for in the double-beam operating mode by exposing the plate to two beams and recording the difference of the two signals. The two beams can be either separated in time at the same point on the plate or separated in space and recorded simultaneously by two detectors. In the single-beam, dual-wavelength mode, fluctuations caused by scattering at a light absorbing wavelength (λ_1) are compensated for by substracting the fluctuations at a different wavelength (λ_2), at which the spot exhibits no absorption but experiences the same scatter [2,80]. The two beams are altered by a chopper and recombined into a single beam to provide the difference signal at the detector. Because the scatter coefficient is wavelength dependent to some extent, the correction is better when λ_1 and λ_2 are close together. Background correction of spurious absorption by plate impurities can be very good in this mode, as is indicated in Figure 9.14 [81]. The single-beam, dual-wavelength mode provides a well-defined baseline, whereas the baseline is less clearly defined when single wavelength operation is used -- a condition unfavorable for quantitative analysis.

The double beam in space optical arrangement divides a single beam of monochromatic light into two beams which scan different portions of the plate. One beam scans the sample lane while the other traverses the blank region between sample lanes. The two beams are subsequently detected by matched photomultipliers and a difference signal is fed to the recorder; fluctuations in the output of the source are corrected in this way. As the two beams impinge on different areas of the plate, however, small irregularities in the plate surface

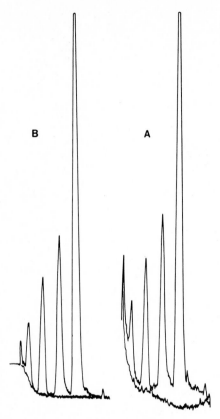

B   A

Figure 9.14   Use of background correction to improve baseline stability in the
analysis of a mixture containing metoprolol and some potential
contaminants.   A was measured using the single wavelength mode:
λ = 280 nm and B by the single-beam, dual-wavelength mode:
λ_1 = 280 nm, λ_2 = 300 nm.   (Reproduced with permission
from ref. 81.   Copyright Elsevier Scientific Publishing Co.)

and undesired background contributions from impurities in the adsorbent layer
may still pose problems.   In practice, the quality and surface homogeneity of
HPTLC plates are generally very good, and single-beam, single-wavelength
operation is predominantly used.   The lower quality of conventional and homemade
TLC plates generally requires the use of background correction to obtain a
stable baseline.

In all densitometers the position of the beam is fixed.   The plate is
scanned by mounting it on a movable stage controlled by stepping motors.   The
motor-driven scanning stage transports the plate through the beam in the
direction perpendicualr to the slit length; the stage is manually operated or
motor driven in the orthogonal direction.   Each scan represents a lane whose
length is defined by the sample migration distance and whose width is specified

by the slit setting. Alternately, each lane can be scanned in a zig-zag motion, a method which effectively provides reproducible readings for spots of irregular shape. Irregularly shaped spots are encountered infrequently in HPTLC and usually indicate chromatographic problems.

Circular and anticircular chromatograms are scanned by radial or peripheral scanning. Scanning in the direction of sample migration is radial scanning and the distance to the neighboring track assumes the dimensions of an angle in degrees. Scanning at right angles to the direction of development is termed peripheral scanning, analogous to the stylus tracking the groove in a record. Radial scanning is usually preferred over peripheral scanning.

To establish the optimum wavelength for maximum sensitivity and to select characteristic wavelength response ratios to confirm sample identity or establish the presence of unresolved interferents, most densitometers make some

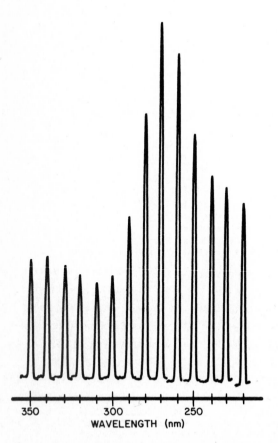

Figure 9.15 Absorption spectra for PTH-L-asparagine obtained by repetitively scanning through the spot on a HPTLC plate at different wavelengths.

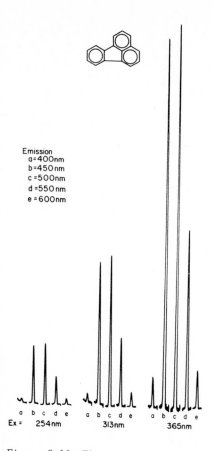

Figure 9.16   Fluorescence spectra of fluoranthene measured as in Figure 9.15, using the fluorescence mode and filters between the plate and detector.

provision for recording the UV–visible or fluorescence spectrum of any desired spot. This may be done either manually or automatically by scanning repetitively across the spot while changing the absorption wavelength or a series of fluorescence emission filters, as shown in Figures 9.15 and 9.16. A line drawn through the peak maxima gives the characteristic absorbance envelope of the compound as a function of wavelength. For identification purposes, characteristic absorbance or fluorescence emission response ratios can be determined by sequential wavelength scanning [55,77,82,83]. The wavelength response ratios are independent of concentration within the linear operating range of the detector. Sequential wavelength scanning may also be used to enhance the resolution of a mixture. This requires that two separate wavelengths be used for measurement. These wavelengths must be chosen such that neither unresolved component has significant absorption or fluorescence

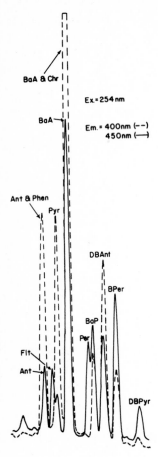

Figure 9.17   Separation of a mixture of polycyclic aromatic hydrocarbons by
reversed-phase continuous multiple development HPTLC. Sequential
wavelength scanning at two emission wavelengths provides
complementary data on the complexity of the mixture: excitation
= 254 nm, emission = 400 nm (- - - -) and 450 nm (————).

where the other component is to be determined. An example is shown in Figure
9.17 for an incomplete separation of a mixture of polycyclic aromatic
hydrocarbons. Sequential wavelength scanning at two different emission
wavelengths provides information about the complexity of the mixture that is
more complete than found in either of the individual scans. Recording second
and higher order derivatives of the densitometer signal allows partially
separated components to be resolved and quantified [84-86]. This is a
potentially powerful technique for enhancing the separating power of HPTLC.

    An important consideration when assessing the performance of a scanning
densitometer is how faithfully it transforms the separation on the plate into a

strip chart chromatogram. Here the parameters of interest are component resolution, dynamic signal range, and detectability; the last is measured by the signal-to-noise ratio. The slit dimensions controlling the size of the measuring beam, the scan rate, and the total electronic time constants of the instrument and recording device are the most important experimental variables directly affecting these parameters. Optimization of these variables for recording HPTLC chromatograms in the reflectance [71,87], transmission [71], and fluorescence [83,88-90] modes is discussed in the literature. No absolute measure of resolution is available, and this makes a comparison of the performance of individual densitometers difficult except for side-by-side comparison of the same separation with both instruments set to optimum measuring conditions. Sample detectability is more easily determined and a protocol has been suggested to standardize this parameter for HPTLC measurements [91,92].

9.12 Application Highlights from the HPTLC Literature

Examples from the HPTLC literature have been selected to illustrate different aspects of the modern practice of TLC. The first example, the separation of the PTH-amino acid derivatives, illustrates the flexibility of the continuous multiple development technique for the separation of a complex mixture. Chromatographically, the PTH-amino acids represent a worse case situation; the mixture covers a broad polarity range and simultaneously contains components of similar structure or near identical separation properties. For this separation a series of solvents of increasing polarity is needed to optimize the resolution of the mixture. The second set of examples deals with the analysis of drugs in blood. It is in the area of clinical analysis that HPTLC has had its greatest impact, due mainly to its ability to analyze multiple samples simultaneously. Complete clinical assays are described to give the reader an idea of the sample treatment required prior to separation by HPTLC. Comparisons with other chromatographic methods for therapeutic drug monitoring are also included. The remaining two examples are taken from the environmental area. The separation of mycotoxins in admixture and polycyclic aromatic hydrocarbons in particulate samples were selected to illustrate the advantage of sequential wavelength scanning to detect unresolved compounds on an HPTLC plate and for the measurement of spectral parameters for use in qualitative compound identification.

9.12.1 Separation of PTH-Amino Acids

Mammalian proteins and peptides are comprised of amino acid residues connected by amide bonds. There are 20 common amino acids principally found in mammalian proteins; individual proteins and peptides differ only in the number

666

and sequence of these amino acids. The amino acid sequence can be determined in a number of ways, one of the most common is the Edman degradation reaction. With this automated method the protein structure can be determined by identifying the N-terminal amino acid, as its phenylthiohydantoin derivative (PTH-amino acid), at the conclusion of each cycle of the sequenator. GC, HPTLC, TLC, and MS have been used to identify the PTH-amino acids, but no single technique has proven completely adequate for this purpose [93].

Continuous multiple development HPTLC has recently been applied to this problem. 18 of the 20 common protein PTH-amino acids have been separated by five sequential developments with three changes of mobile phase [93]. Alanine and tryptophan were not separated in this scheme but could be resolved to baseline in a different solvent system. Thus, all 20 PTH-amino acids can be identified with an analysis time of well under one hour. By using unidimensional development, the high sample capacity of the HPTLC plate is preserved; samples and standards can be run simultaneously to improve the acccuracy of identification.

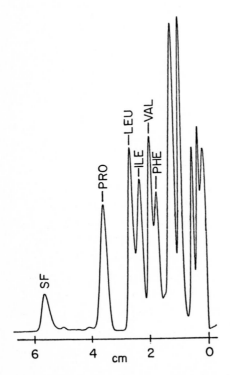

Figure 9.18   Separation of PTH-amino acids.   Second development in the solvent system methylene chloride-isopropanol (99:1), 10 minutes. (Reproduced with permission from ref. 93.   Copyright Elsevier Scientific Publishing Co.)

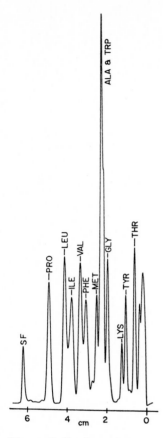

Figure 9.19   Separation of PTH–amino acids. Third development in the solvent
               system methylene chloride–isopropanol (99:1), 10 minutes.
               (Reproduced with permission from ref. 93. Copyright Elsevier
               Scientific Publishing Co.)

The separation is carried out on HPTLC silica gel plates in a short-bed
continuous development chamber. The plate was developed in either position 2
(plate length 3.5 cm) or position 4 (plate length 7.5 cm). The first
development was made with methylene chloride for 5 minutes in position 2 of the
development chamber. The function of this development step is to provide an
initial ordering of the derivatives in the region of the origin, thereby
improving the resolution of the mixture in subsequent development steps. For
the second development, the mobile phase was changed to methylene chloride–
isopropanol (99:1) and the plate developed for 10 minutes in position 4 of the
development chamber. Scanning the plate at the end of this stage enables any of
the derivatives of proline, leucine, isoleucine, valine, or phenylalanine to be
identified, Figure 9.18. The third continuous development is a repeat of step
2. It provides better resolution of the derivatives separated previously as

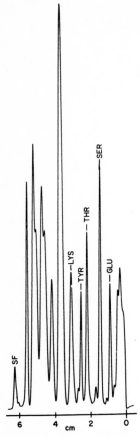

Figure 9.20 Separation of PTH-amino acids. Fourth development in the solvent
           system methylene chloride–isopropanol (97:3), 10 minutes.
           (Reproduced with permission from ref. 93. Copyright Elsevier
           Scientific Publishing Co.)

well as enabling the derivatives of methionine, alanine/tryptophan, glycine,

lysine, tyrosine, and threonine to be identified, Figure 9.19. In this

development sequence, tryptophan often appears as a shoulder on the alanine

peak, but it is not adequately resolved for identification purposes. The

alanine and tryptophan derivatives can be separated almost to baseline in a 10

minute development with hexane–tetrahydrofuran (9:1) in position 2 of the

short-bed continuous development chamber. For the fourth development step, the

mobile phase is changed to methylene chloride–isopropanol (97:3) and the plate

redeveloped for 10 minutes in position 4. This improves the separation of the

derivatives of lysine, tyrosine, and threonine, and, in addition, serine and

glutamine may be identified, Figure 9.20. Figure 9.20 also illustrates the

advantage of being able to make measurements with the densitometer at any step

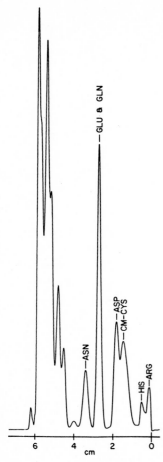

Figure 9.21 Separation of PTH-amino acids. Fifth development in the solvent
system ethyl acetate-acetonitrile-glacial acetic acid
(74.3:25:0.7), 10 minutes. (Reproduced with permission from ref.
93. Copyright Elsevier Scientific Publishing Co.)

in the development sequence. The separation of the PTH-derivatives of lysine,
tyrosine, threonine, serine, and glutamine is to baseline at this stage and
their identification is easily achieved. However, the derivatives of
leucine/isoleucine, valine/phenylalanine, and tryptophan/alanine/glycine have
started to merge together and could not be determined individually, although
they were well-separated in earlier steps. At the fourth development stage the
very polar PTH-amino acid derivatives remain essentially unresolved in the
region of the origin. A much more polar mobile phase is selected for their
resolution in the final development step. A 10 minute development in position 4
of the short-bed chamber with ethyl acetate-acetonitrile-glacial acetic acid
(74.3:3:25.7) separates the remaining PTH-amino acid derivatives of arginine,

histidine, S-(carboxymethyl)-cysteine and aspargine, except for glutamic acid/glutamine, Figure 9.21. However, glutamic acid is separated from glutamine in the fourth development step so the two are easily differentiated from one another.

9.12.2 Analysis of Drugs in Blood

The quantitative analysis of drugs in blood samples is complicated by the low concentrations normally encountered and by interferences from metabolic products which may closely resemble the parent drug. Thus, analytical procedures furnishing both sensitivity and selectivity are usually required for these determinations; most successful methods involve chromatographic separations. The major disadvantage with both gas and liquid chromatographic procedures for routine clinical determinations is the considerable instrument time involved in sequential separations - an unavoidable consequence of any column based separation system. Thin-layer chromatography, on the other hand, permits simultaneous separation of many samples but is hindered by relatively low sensitivity and resolution. High performance thin-layer chromatography does not suffer from these latter limitations and is well-suited to therapeutic drug monitoring in a busy clinical laboratory.

The suitability of HPTLC for the separation of phenothiazine drugs is illustrated in Figure 9.22. This separation is almost to baseline and would be more than adequate for drug identification purposes. One of these drugs, chlorpromazine, is a widely administered psychopharmacologic agent in mental health facilities. Estimates of the amount of chlorpromazine present in blood serum during therapy range from about 10 ng/ml to 300 ng/ml. A simple procedure for the direct extraction of tricyclic antidepressant drugs has been devised. The drugs are extracted from a 1.0 ml serum sample at high pH into isoamyl alcohol-heptane (1.5:98.0), back extracted in to acid, basified, and then re-extracted with organic phase. Solvent evaporation to a small volume completes the sample cleanup sequence. The extracts are applied directly to the HPTLC plate with a dosimeter [94-96]. The recovery and reproducibility of the assay was improved by adding internal standards, butaperazine for chlorpromazine and loxapine for the tricyclic antidepressants, as well as a carrier substance, perphenazine, to the serum sample. Perphenazine was selected as it is structurally similar to the drugs being determined but has a low R_F value in the separation system and could be used in large excess without interfering with the compounds being measured. Its influence is most dramatically seen on the recovery and reproducibility of the assay for the drugs at low concentrations

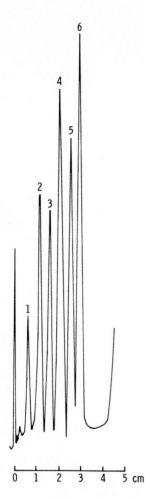

Figure 9.22   HPTLC separation of 1, acetophenazine; 2, perphenazine;
              3, trifluoroperazine; 4, promazine; 5, thioridazine; and
              6, chlorpromazine. Silica gel plate using multiple development
              in the mobile phase benzene–acetone–ammonium hydroxide
              (80:20:0.2). (Reproduced with permission from ref. 94. Copyright
              American Association for Clinical Chemistry).

(10 ng/ml) but it still provides a useful improvement at higher levels, Table
9.6. Figure 9.23 shows a chromatogram of a patient's blood serum containing 45
ng/ml of chlorpromazine. The metabolic products of drugs are usually more
difficult to determine than the parent compound because enzymatic modification
results in their increased polarity. Figure 9.24 shows an HPTLC chromatogram of
chlorpromazine and its major metabolic products, resolved almost to baseline in
a single chromatographic step.

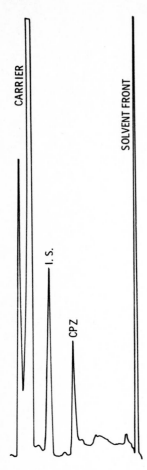

Figure 9.23   HPTLC chromatogram of a serum sample from a patient receiving
chlorpromazine (CPZ). The concentration of chlorpromazine was
45 ng/ml. (Reproduced with permission from ref. 94. Copyright
American Association for Clinical Chemistry).

The extraction and work-up procedure for the tricyclic antidepressants
amitriptyline, nortiptyline, imipramine, and desipramine is the same as used for
chlorpromazine. To maximize the sensitivity, measurements are made at two
different wavelengths, 240 and 275 nm, corresponding to the maximum absorption
regions of nortriptyline and amitriptyline and of desipramine and imipramine,
respectively. Figure 9.25 shows the scans at two different wavelengths for the
separated tricyclic antidepressants at therapeutic levels in serum. The lower
trace is the serum blank and indicates that the analysis is free from any
co-extracted interferences.

A microanalytical procedure utilizing HPTLC has been described for
determining concentrations of the anticonvulsant drugs phenobarbital, phenytoin,
primidone, and carbamazepine in plasma [97]. The attractive features of this

673

TABLE 9.6

RECOVERY AND REPRODUCIBILITY OF THE HPTLC ASSAY FOR PSYCHOPHARMACOLOGICALLY
ACTIVE DRUGS IN SERUM USING PERPHENAZINE AS A CARRIER SUBSTANCE

Drug	Concentration in Serum µg/l	Recovery, %		Reproducibility (CV)	
		With Carrier	Without Carrier	With Carrier	Without Carrier
Chlorpromazine	100	96	87	3.3	2.6
	10	80	54	8.0	34.2
Amitriptyline	100	99	87	2.6	7.9
	10	95	71	7.8	11.3
Nortriptyline	100	77	73	3.1	12.4
	10	83	62	8.5	18.5
Imipramine	100	91	86	4.0	7.9
	10	86	77	4.2	11.3
Desipramine	100	70	62	6.0	13.5
	10	82	72	10.7	17.7

assay are its small sample requirement (50 microliters of plasma), the use of a
simple and rapid single solvent extraction as the isolation step, and the use of
the contact spotter to simultaneously concentrate and apply 15 samples to the
HPTLC plate. The method lends itself to the analysis of numerous samples as
well as to emergency situations, where a rapid and accurate assay is required.

The extracts are separated by developing with chloroform in a
pre-equilibrated chamber containing a separate beaker of concentrated ammonium
hydroxide. This system moves the carbamazepine to an R_F of 0.4 and leaves the
remaining anticonvulsants at the origin. The carbamazepine is determined by
reflectance densitometry at 285 nm without interference from plasma
co-extractants, Figure 9.26. The plate is then re-developed in a more polar
solvent system of chloroform-isopropanol-ammonium hydroxide (80:20:1) with
ammonium hydroxide present in a separate beaker. Two 7-minute developments are
required for the complete separation of the remaining anticonvulsants, Figure
9.26. The plate is scanned at 215 nm for quantitative measurements. The
extraction efficiency for all four anticonvulsants is greater than 95% and the
minimum detectable amount of drug on the plate varies from 1 to 5 ng.

Using the above assay, a single plasma sample can be extracted,
chromatographed, and quantified within two hours. A much larger number of
samples can be processed in approximately the same period of time since the
extraction and chromatographic steps are carried out concurrently. The
densitometric scanning and quantitation is a sequential process which requires
approximately one additional minute per sample.

674

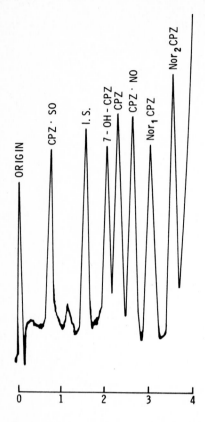

ORIGIN

CPZ · SO

I.S.

7 - OH - CPZ

CPZ

CPZ · NO

Nor₁ CPZ

Nor₂CPZ

0	1	2	3	4

Figure 9.24 HPTLC separation of chlorpromazine and its major metabolites with
clozapine as internal standard. Mobile phase ethyl acetate–acetic
acid–water–acetone–isopropanol (40:5:5:2.5:2.5), migration
distance 4 cm, three multiple developments and scanned at 250 nm.
(Reproduced with permission from ref. 94. Copyright American
Association for Clinical Chemistry).

The antiarryhthmia drugs lidocain, procainamide, propranolol, and quinidine
are widely used in clinical practice for the treatment of cardiac disorders.
Dipehnylhydantoin (dilantin) is an antiepileptic drug occasionally also used to
treat cardiac disorders. A simple HPTLC method has been devised for the
determination of any one or all five antiarrhthymia drugs in 0.5 ml of serum
[98]. The neutral drugs, lidocain and diphenylhydantoin, can be extracted with
benzene while the basic drugs, procainamide, propranolol, and quinidine, are
extracted with the same solvent after first adjusting the pH of the sample to a
high value. Clozapine can be used as an internal standard for all the drugs.
The normal practice with this assay is to combine the benzene extracts,
evaporate them to a residue, and redissolve this residue in a small volume of
benzene for sample application. The sample recovery varies with the drug but is
fairly constant and reproducible at the therapeutic level, Table 9.7.

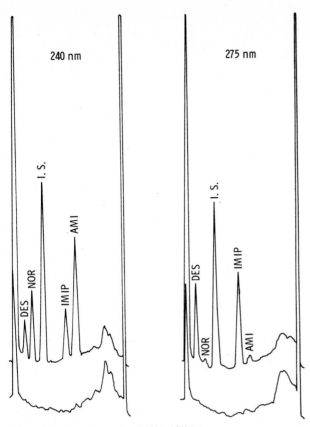

Figure 9.25   HPTLC separation of tricyclic antidepressants scanned at 240 nm
and 275 nm. Mobile phase ethyl acetate-ammonium hydroxide (97:3)
and migration distance 4 cm. Amitriptyline (AMI), nortriptyline
(NOR), imipramine (IMIP), and desipramine (DES). Internal standard
(IS) is loxapine. (Reproduced with permission from ref. 95.
Copyright Elsevier Scientific Publishing Co.)

TABLE 9.7

REPRODUCIBILITY AND RECOVERY TEST FOR SPIKED SERUM (n = 5)

Drug	Concentration (µg/ml)	Relative Standard Deviation (%)	% Recovery
Dilantin	3	7.3	54.8
	12	5.2	72.5
Lidocaine	1.5	13.1	78.3
	6	6.8	81.1
Procainamide	1.2	7.9	46.5
	4.8	6.3	51.8
Propranolol	0.3	12.4	89.3
	1.2	7.4	93.5
Quinidine	2.4	6.6	97.5
	9.6	3.2	103.5

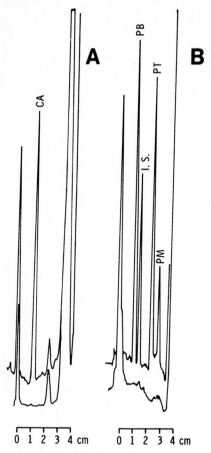

Figure 9.26 HPTLC separation of anticonvulsant drugs. A, solvent system
chloroform, carbamazepine (CA) 5 mg/1, determined at 285 nm, lower
trace is a plasma blank; B, second development in chloroform-
isopropanol-ammonium hydroxide (80:20:1), two times, scanned at
215 nm. Phenobarbital (PB) 40 mg/1, p-tolylbarbital (IS)
phenytoin (PT) 20 mg/1, and primidone (PM) 10 mg/1. (Reproduced
with permission from ref. 97. Copyright Elsevier Scientific
Publishing Co.)

It is possible to obtain a baseline separation of all five drugs and the
internal standard, clozapine, using a two-step development procedure requiring
about 30 minutes. Solvent system 1, benzene-ethyl acetate-methanol (4:4:1),
gives a baseline separation of lidocaine and dilantin, Figure 9.27. The solvent
front migrates 5 cm in about 7 minutes and the drugs are detected by scanning
densitometry at 220 nm after first drying the plate. The other drugs remain at
the origin in solvent system 1 while the internal standard, clozapine, migrates
only a short distance. A small peak, identified as caffeine, was found in most
serum samples studied, but did not interfere in the analysis. A second

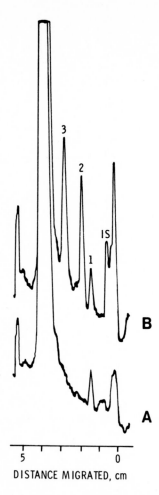

DISTANCE MIGRATED, cm

Figure 9.27   HPTLC separation of caffeine (1), lidocaine (2), diphenylhydantoin
               (3) in the solvent system benzene-ethyl acetate-methanol (4:4:1).
               A, serum blank; B, serum and drugs. (Reproduced with permission
               from ref. 98. Copyright Elsevier Scientific Publishing Co.)

development (5 cm in 13 minutes) is performed with the more polar solvent system
benzene-ethyl acetate-methanol-pyridine (4:2:3:3) to give a baseline separation
of procainamide, propranolol, quinidine, and the internal standard, clozapine,
Figure 9.28. In this solvent system lidocaine and dilantin migrate to the
solvent front. The drugs are determined by scanning densitometry at 290 nm.
The detection limit for clozapine, quinidine, lidocaine and dilantin is about 10
ng and about 2 ng for propranolol and procainamide. Twenty frequently
prescribed basic drugs were screened for interference in this assay. It was
found that only barbital, carbamazepine, and ethosuximide would present any
interference problems, indicating the excellent selectivity of this method for
the antiarrhythmia drugs studied.

678

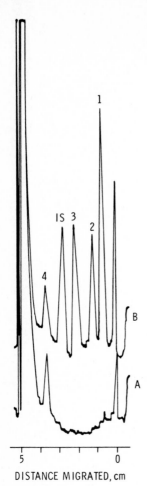

DISTANCE MIGRATED, cm

Figure 9.28   HPTLC separation of procainamide (1); propranolol (2); quinidine
(3); caffeine (4) after further development of the chromatogram
in Figure 9.27 with benzene-ethyl acetate-methanol-pyridine
(4:2:2:3). A, serum blank; B, serum and drugs. (Reproduced
with permission from ref. 98. Copyright Elsevier Scientific
Publishing Co.)

Apart from HPTLC, gas chromatography is the only method that has been used
to determine all five of these antiarrhythmia drugs. The time requirements of
HPTLC compare favorably to those of gas chromatography for a single analysis.
However, since samples can be analyzed simultaneously by HPTLC, the time
required on a per sample basis is much lower for HPTLC. The precision of HPTLC
and gas chromatography were shown to be similar. A major advantage of HPTLC is
that all 5 drugs or any combination of the 5 may be determined simultaneously in
a routine single assay. By gas chromatography, different separation conditions
and derivatization techniques, etc. are required for the individual drugs,

adding considerably to the labor, time, and instrument requirements for the analysis.

9.12.3 Identification of Mycotoxins in Admixture

Mycotoxins are toxic fungal metabolites found as contaminants in some harvested agricultural products. The control of these hazardous contaminants in food stocks requires an adequate analytical method for each mycotoxin. Usually, several mycrotoxin-producing fungi are isolated from the same foodstuff. Therefore, there is no real basis for deciding which mycotoxin to look for in any particular sample showing heavy contamination with molds. Since it is time consuming to analyze all the mycotoxins separately, a rapid screening method for their simultaneous analysis has obvious advantages. As many mycotoxins are thermally labile, liquid and thin-layer chromatography are the available options. The latter is preferable due to its greater methodological simplicity. This method also provides greater detection flexibility since both the mode and wavelength can be easily optimized to improve selectivity and sensitivity.

With HPTLC it is possible to determine 13 mycotoxins simultaneously in one sample and also to separate multiple samples at the same time [55,77,99]. The method uses seven continuous multiple developments and two solvent systems of different polarity to achieve baseline resolution of the mixture. It is possible to detect and/or quantify sterigmatocystin, zearalenone, citrinin, ochratoxin A, patulin, penicillic acid, luteoskyrin, and the aflatoxins B_1, B_2, G_1, G_2, M_1, and M_2 in an individual sample. The separation is carried out on silica gel HPTLC plates impregnated with ethylenediaminetetracetic acid disodium salt (EDTA) to eliminate the streaking of citrinin and luteoskyrin. Sequential wavelength scanning eliminates the problem of sample overlap and maximizes sensitivity. With in situ scanning densitometry, detection limits in the low nanogram range (0.2-2 ng) were obtained by UV-visible absorption measurements and in the low picogram (2-50 pg) by fluorescence.

The separation is performed by continuous multiple development using the separation sequence outlined in Table 9.8. The variables requiring optimization are the plate length, the mobile phase composition, the number of development stages, the time for each development, and the mode and wavelength used for determining the separated components at each sequence. The mixture is never completely separated on the plate at any one time, but each component is separated to baseline and recorded at some stage in the development sequence. The first development stage requires 5 minutes and yields a baseline separation of sterigmatocystin, zearalenone, and citrinin, Figure 9.29. The other mycotoxins remain near the origin. After second and third continuous

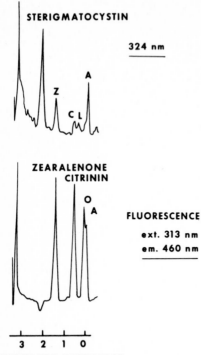

STERIGMATOCYSTIN

324 nm

Z

A

C L

ZEARALENONE
CITRININ

O

A

FLUORESCENCE

ext. 313 nm
em. 460 nm

3 2 1 0

MIGRATED DISTANCE, cm

Figure 9.29   HPTLC separation of mycotoxins. First continuous development,
sterigmatocystin determined by absorption (top) and zearalenone
and citrinin by fluorescence (bottom). Solvent system toluene-
ethyl acetate–formic acid (30:6:0.5). (Reproduced with permission
from ref. 77. Copyright American Chemical Society).

developments, ochratoxin A was separated sufficiently from the aflatoxins to be
determined, Figure 9.30. After a fourth continuous development, penicillic
acid, patulin, and luteoskyrin are determined, Figure 9.31. Penicillic acid and
patulin are separated to baseline but luteoskyrin and patulin still overlap.
Patulin, however, shows no absorption at the wavelength maximum for luteoskyrin
(440 nm) and, consequently, wavelength discrimination can be used to
differentiate between these two mycotoxins. The aflatoxins still remain close
to the origin and a more polar solvent system is required for their separation.
After three further continuous developments in the second solvent system, the
six aflatoxins are baseline resolved, Figure 9.32.

In general, the tentative identification of a compound by TLC is based on
the measurement of several R_F values in solvents of different polarity. With a
scanning densitometer, the UV-visible or fluorescence spectrum of the component
can be obtained and this information, along with a knowledge of the migration
distance of the component, can confirm the identity of the spot and establish
the presence of any contaminants. Spectral parameters for the tentative

TABLE 9.8

DEVELOPMENT STAGE AND SPECTROSCOPIC METHOD USED FOR THE DETECTION OF MYCOTOXINS

Development Stage	Time (min)	Mycotoxin Detected	Spectral Characteristic Used for Detection
toluene–ethyl acetate–formic acid (30:6.0:0.5)			
1st development	5.0	Sterigmatocystin Zearalenone Citrinin	Reflectance, 324 nm Fluorescence, Ex = 313 nm, Em = 460 nm
2nd development	5.0	no measurement	
3rd development	6.0	Ochratoxin A	Fluorescence, Ex = 313 nm, Em = 460 nm
4th development	6.0	Penicillic acid Patulin Luteoskyrin	Reflectance, 240 nm Reflectance, 280 nm Reflectance, 440 nm
toluene–ethyl acetate–formic acid (30:14:4.5)			
5th development	8.0	no measurement	
6th development	8.0	no measurement	
7th development	8.0	Aflatoxins $B_1, B_2, G_1, G_2, M_1,$ and M_2	Fluorescence, Ex = 365 nm, Em = 430 nm

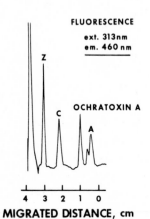

FLUORESCENCE
ext. 313nm
em. 460 nm

Z

C

OCHRATOXIN A

A

MIGRATED DISTANCE, cm
4 3 2 1 0

Figure 9.30 HPTLC separation of mycotoxins. Third continuous development, ochratoxin A determined by fluorescence. (Reproduced with permission from ref. 77. Copyright American Chemical Society).

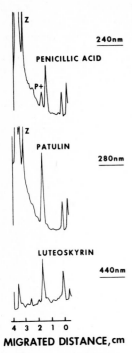

MIGRATED DISTANCE, cm

Figure 9.31   HPTLC separation of mycotoxins. Fourth continuous development,
penicillic acid and patulin are baseline resolved and measured
by absorption. Patulin and luteoskyrin overlap. Luteoskyrin
determined by selective wavelength scanning at 440 nm.
(Reproduced with permission from ref. 77. Copyright American
Chemical Society).

identification of mycotoxins can be developed by obtaining peak ratios at
several fixed wavelengths, Table 9.9. This procedure is more convenient than
recording the complete spectrum when many samples are to be determined. These
parameters are remarkably constant and independent of concentration provided
that measurements are made on the linear portion of the detector response
curve. By the use of standards, confidence limits can be established for each
spectral parameter. Values outside the confidence limits indicate either an
incorrect assignment of the mycotoxin or spectral overlap with a contaminant
which may affect the accuracy of the quantitative data.

9.12.4 Determination of Polycyclic Aromatic Hydrocarbons in Particulate Samples

Polycyclic aromatic hydrocarbons (PAHs) are well-documented environmental
contaminants and some are known human or animal carcinogens. They are formed
during the incomplete combustion of organic matter and thus enter the

TABLE 9.9

SPECTRAL PARAMETERS FOR TENTATIVE IDENTIFICATION OF MYCOTOXINS

Mycotoxin	Peak Height Ratio			
	UV-visible		Fluorescence	
	324/280	240/280	313/365	365/405
	Mean (% Relative Standard Deviation)			
Sterigmatocystin	4.80 (16.3)	4.70 (14.0)		
Zearalenone	0.60 (4.1)	1.20 (2.0)		
Patulin		0.22 (4.6)		
Citrinin			1.41 (5.1)	0.52 (3.6)
Aflatoxin B_1			0.31 (4.2)	0
Aflatoxin G_1			0.25 (5.6)	0.22 (4.0)
Aflatoxin M_1			0.29 (6.6)	0

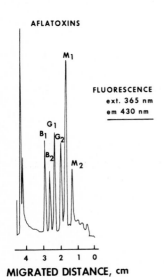

Figure 9.32   HPTLC separation of mycotoxins.   Seventh continuous development.
After the fourth development, the solvent system was changed to
toluene–ethyl acetate–formic acid (30:14:4:5).   Aflatoxins
determined by fluorescence.   (Reproduced with permission from
ref. 77.   Copyright American Chemical Society).

environment from a wide range of energy related sources.

Air particulate samples are usually collected by high volume filtration
samplers and then separated from inorganic material by sublimation or solvent
extraction, etc.   The soluble organic fraction of particulate samples is a
complex mixture of hydrocarbons, PAHs, and various functionalized polar aromatic
compounds.   It represents a formidable separation problem.   Typically, extensive
sample cleanup and fractionation are needed prior to separation by high

resolution gas or liquid chromatography. These procedures are laborious but essential for a detailed analysis of the sample. TLC techniques are useful for screening multiple samples to indicate the compounds which may be present and to provide approximate concentration levels [83,90,100-106].

The separating ability of TLC depends on its efficiency and selectivity. The chromatographic efficiency is optimized in the current practice of HPTLC. Selectivity in TLC is based on a combination of the properties of the separation system and the method of detection. The selection of the stationary phase is very important. Silica gel plates show little selectivity for the separation of PAHs and exhibit extensive quenching of the fluorescent signal [104,106]. In some cases catalytic decomposition of the sample may also occur. Satisfactory results are obtained using octadecylsilanized silica [83,90,105] or acetyl-cellulose [100,103] plates. Aqueous mixtures of methanol or acetonitrile are used as the mobile phase.

For PAH analysis, detection selectivity is of particular importance as many PAHs of similar structure and separation characteristics have easily distinguishable fluorescence spectra. Since TLC is an open-bed technique and the detection process is essentially static, the separated sample may be conveniently viewed several times with different detection conditions. This process is known as sequential wavelength scanning and is a useful aid in the identification or quantitation of components in a complex mixture. As an example of this technique, consider the separation of chrysene, pyrene, and fluoranthene shown in Figure 9.33. The three PAHs are not adequately separated for identification purposes by the chromatographic system. However, all three PAHs may be readily identified by fluorescence detection excited at either 254 nm (A) or 313 nm (B) and by measuring the fluorescence emission signal at 400 nm (solid line) and 500 nm (broken line). At 400 nm emission, only chrysene and pyrene are detected and fluoranthene is not observed. Changing the excitation wavelength does not affect the resolution but changes the emission intensity ratio of the chrysene and pyrene peaks at 400 nm. Fluoranthene is not significantly affected by the choice of excitation wavelength (254 or 313 nm) when its emission signal is measured at 500 nm. However, chrysene and pyrene do not fluoresce significantly at this wavelength and therefore do not interfere in the detection of fluoranthene. Thus, by scanning the plate at two different emission wavelengths all three PAHs may be identified without obtaining complete chromatographic resolution.

Figure 9.34 shows the sequential separations of a series of PAH standards and a crude Soxhlet extract of a diesel engine particulate sample. The resolution of the sample improves with an increase in the number of development sequences, as shown in Figure 9.34A to 9.34C. However, the separation system lacks the resolution necessary for identification purposes. Multiple scanning

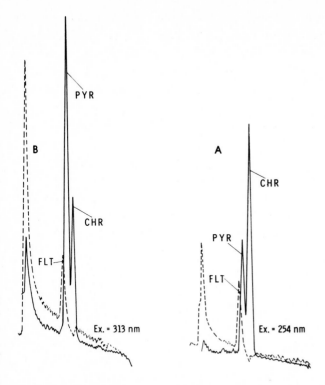

Figure 9.33   Sequential wavelength scanning used to determine chrysene (CHR),
               pyrene (PYR), and fluoranthene (FLT) unseparated on the HPTLC
               plate.   A, Ex = 254 nm; B, Ex = 313 nm.   Fluorescence emission,
               solid line, 400 nm and broken line, 500 nm.

of the separation at two different emission wavelengths, 400 nm (solid line) and

450 nm (broken line), aids in the identification of the PAHs, as indicated in

Figure 9.34C and 9.34D.   Figure 9.34D is a crude sample without cleanup and

illustrates the power of HPTLC as a rapid screening method for environmental

samples.   The method is adequate for providing preliminary data to decide

whether a more detailed analysis is required.   HPTLC is particularly well-suited

for this analysis as it has a high sample thoughput, is cost-effective, and can

be used with crude samples which would result in heavy contamination of an HPLC

or GC column.

HPTLC techniques can also be used for the quantitative determination of

PAHs in diesel engine particulate samples after sample cleanup [100,107].   The

method involves vacuum sublimation of the PAHs from the raw sample, separation

of the sublimate on columns of silica gel and alumina, and application of the

column fractions to the TLC plate.   Fourteen characteristic PAHs were determined

in the diesel samples with a detection limit of about 0.5 ng.   At the normal

concentrations of PAHs found in diesel samples a relative standard deviation of

1.4-6.0% was obtained. As simultaneous separations are possible with TLC, the
time required for a complete analysis is approximately 4 hours compared to one
day for the same number of samples analyzed using GC. Excellent agreement was
found between the TLC and GC method for determining these PAHs.

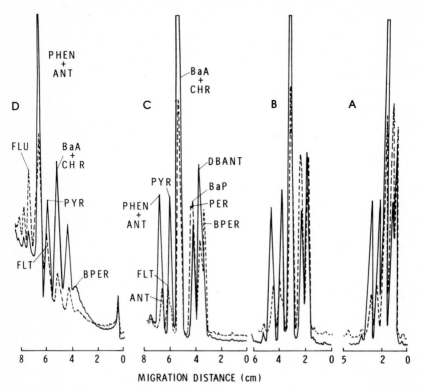

MIGRATION DISTANCE (cm)

Figure 9.34 Continuous multiple development and multiple wavelength scanning
in the fluorescence mode for the detection of PAHs in a diesel
engine particulate extract. A, B, and C are the separations of
reference standards and D is the particulate extract. The
separation conditions are octadecylsilanized silica plates
developed with methanol-water (20:4); A, 2x10 min; B, 3x10 +
1x13 min; C and D 3x10 + 2x13 + 1x15 min. [Last development
solvent changed to methanol-methylene chloride-water (20:3:3)].
Measurements made at Ex = 254 nm and Em = 400 nm solid line and
450 nm broken line. BPER, benzo[g,h,i]perylene; DBANT,
dibenzo[a,h]anthracene; BaP, benzo[a]pyrene; PER, perylene; BaA,
benzo[a]anthracene; CHR, chrysene; PYR, pyrene; FLT, fluoranthene;
ANT, anthracene; PHEN, phenanthrene; and FLU, fluorene.

9.13 Thin-Layer Chromatography on Polyamide Layers

Glass or solvent resistant polyester sheets precoated with thin layers of
polyamide are commercially available. The preferred method of preparing
polyamide TLC plates in the laboratory is the controlled evaporation of a formic

acid-polyamide solution from a horizontal glass plate [108]. This produces porous, durable layers consisting of small particles with acceptable chromatographic efficiency. Polyamide films can also be prepared by the conventional slurry-coating technique. As polyamide powders do not adhere readily to glass, a binder such as cellulose, gypsum, or synthetic glue is required.

Adsorption on polyamide layers is due mainly to the reversible formation of hydrogen bonds between the functional group of the sample and the carbonyl oxygen of the amide group. Separation occurs by the competition between the components of the sample and a polar mobile phase for the hydrogen bonding centers. A contribution to retention from a reversed-phase-type mechanism is also expected. Polyamide TLC plates have been used mainly for the separation of phenols, amino acid derivatives, heterocyclic nitrogen compounds, and carboxylic and sulfonic acids [108,109].

9.14 Thin-Layer Gel Chromatography

The published applications of thin-layer gel chromatography (TLG) are confined almost entirely to the separation of macromolecules of biochemical interest, and to proteins in particular. TLG is used to make comparisons of different protein fractions, to estimate the relative concentrations of components in a mixture, or to determine molecular weights [110-112]. The mechanism of separation is primarily size-exclusion as the sample migrates through a gel of controlled pore size swollen in a carefully selected buffer.

Soon after the widespread acceptance of column gel chromatography for the separation of macromolecules, it was shown that similar separations were possible using thin-layers of gel. However, interest in TLG remained at a low level for several years, due mainly to the difficulties involved in obtaining reproducible results. The principal reason for these difficulties was the variation in the migration of the solvent at different points on the plate. Nonuniform solvent migration results from non-equilibration of the stationary phase with the mobile phase, uneven spreading of the gel on the glass backing plate, evaporation of the solvent from the plate, and most important of all, a non-steady-state condition of the system at the working angle of tilt. All of these problems can be overcome or eliminated by adopting good laboratory practices [113].

Compared to column techniques, TLG has the advantage of being suitable for the separation of small quantities of samples (micrograms). In addition, this technique can compare numerous samples simultaneously and it is easily combined with other methods, such as isoelectric focusing and immunoprecipitation techniques. Column techniques are preferred for preparative applications and for the cleanup of crude samples to be determined by other analytical techniques.

Several types of gels suitable for use in TLG are commercially available. These include the Sephadex series of crosslinked dextrans, the Bio-Gel P series of crosslinked polyacrylamide gels, and the Enzacryl crosslinked poly(acryloylmorpholine) gels. All gels are available in a superfine grade (bead size approximately 25 microns) desirable for TLG. Prior to spreading on a clean glass plate, the gel is first swollen in an appropriate buffer/solvent system recommended by the manufacturer. Thin layers of 0.2 to 1.0 mm thickness are most commonly used. The plate is developed in a "sandwich box" chamber which has a fulcrum that enables the plate angle, and hence the flow rate, to be easily adjusted. Before spotting, the stationary phase must be equilibrated with the mobile phase used for the separation. This is best achieved by allowing the mobile phase to flow through the gel for several hours.

The samples are applied along the plate a few centimeters from one edge. Micropipettes are convenient for spotting 10 micrograms of sample in 1-5 microliters of solution. Although rectangular sample zones are more difficult to apply, they are preferred for densitometric evaluation. It is particularly important to minimize the disturbance of the gel surface during sample application. If the sample is not colored, it is usual to add one of several colored dyes to the sample prior to spotting [110]. The colored marker can be used to monitor the mobile phase flow rate through the gel and also acts as an internal standard for the calculation of relative migration distances. Most separations are performed with a migration speed of about 3 cm/h for a totally excluded compound, with a typical separation requiring about 4-6 hours.

The most widely used detection method in TLG is the paper-replica method. A piece of dry filter paper is placed over the wet gel as soon as the separation is complete and adsorbs over 90% of most separated solutes within a minute or two. The paper can then be removed from the gel and stained using the procedures common to zone electrophoresis. For colored solutes, reflectance densitometry can be used or a photographic evaluation can be made.

9.15 Sintered Thin-Layer Chromatography

Sintered TLC plates contain an adsorbent fixed in the three-dimensional space formed by the sintered glass powder and a glass supporting plate. The sintered glass powder binds the adsorbent particles without excessive damage to their surface structure and thus they retain all the properties expected of the adsorbent. Sintered plates are prepared by spreading a slurry that contains adsorbent and glass powder in an organic solvent on a soda-lime glass plate in the conventional manner [114]. The slurry is air dried and then heated in an

electric furnace to fuse the glass powder and adsorbent particles together. A
wide range of soda-lime, borosilicate, and ceramic glass powders have been used
as binders. Suitable adsorbents are silica gel, alumina, kieselguhr, florisil,
titania, and magnesia. Sintered sheets employing microparticulate polyolefins
as the binder rather than glass powder have also been prepared. The adsorbents
suitable for forming such organic sintered sheets are silica gel, alumina,
kieselguhr, florisil, synthetic porous polymers, ion-exchange resins, cellulose
ion-exchangers, dextran gels, and celluloses.

The sintered glass plates are highly porous and allow chromatographic
development to proceed normally. The separation mechanism is the same as for
conventionally prepared TLC plates, but compound retention is considerably
lower. When attempting separations previously obtained by conventional TLC on
sintered glass plates, it is usually necessary to reduce the polarity of the
mobile phase in order to obtain an equivalent separation. As they do not
contain any organic binder, the sintered glass plates are not charred when
sprayed and heated with such reagents as concentrated sulfuric acid, chromic
acid, and trichloroacetic acid. The sintered glass plates also have the
advantage of being reusable. Used plates can be regenerated by soaking in a
cleaning solution (e.g., chromic acid), washing with water, and then reactivated
by heating. Alternatively, the plates can be cleaned by heating at 400-450°C
for thirty minutes. Reusing a plate several times for the same separation
eliminates the influence that changes in the nature, activity, and layer
thickness of the adsorbent have on the reproducibility of R_F values. R_F values
obtained on recycled plates can be reproduced with a standard deviation of 0.05.

Thin-layer chromatography can be carried out on sintered glass rods as well
as on plates. These rods were developed specifically for use with a scanning
flame ionization detector. The motor-driven rod passes through the flame of a
hydrogen flame ionization detector where the separated zones are evaporated and
ionized. Detection limits for organic compounds range from 1 to 0.1 micrograms
per spot. The silica gel- and alumina-quartz sintered rods are prepared from a
special ceramic glass binder which produces a low background signal in the
detector. The sintered rods are thermally stable, acid resistant, and can be
used repeatedly without reactivation after passage through the flame. Many
applications of the commercially available equipment (Iatron) for the
quantitative determination of thermally labile or non-volatile organic compounds
of biomedical or environmental interest have been published [115-117].

9.16 Radiochromatography on Thin-Layer Plates

The stability, fate, metabolism, etc., of a compound in a biological or
environmental system are difficult to assess due to the multitude of mechanisms

available for its modification and the co-existence of many substances with similar chemical properties. Studies of this kind are simplified by incorporating a stable radioactive element into the test compound; this radioactive species serves to indentify the parent compound and the products derived from it in complex mixtures. Thin-layer chromatography is used in these studies to assess the radiochemical purity of the starting material and to identify the compound and the products derived from it.

The selection of a method to detect radiolabeled compounds in TLC depends largely on the frequency with which such measurements are made in the laboratory and the general availability of equipment for making radiochemical measurements. Expensive and dedicated radiochemical TLC scanners are commercially available and would be the choice for laboratories making frequent routine measurements of radiochromatograms. Methods employing less sophisticated equipment, or equipment generally available in radiochemical laboratories, would be the choice for those laboratories with only an occasional need for such determinations. The equipment available for thin-layer radiochromatography has been reviewed [118-120].

There are three principal methods for quantifying radioactive substances after separation by TLC. These are autoradiography, liquid scintillation counting, and direct scanning with radiation detectors. Autoradiography requires the direct contact of the TLC layer with a photographic emulsion (e.g., x-ray film). Autoradiography is inexpensive and does not require elaborate apparatus. Quantitative measurements can be made with a photodensitometer to measure the darkening of the developed film. The disadvantages are that the film may require extremely long exposure times, it is difficult to quantify, particularly for wide ranges of radioactivity, and it can show false images due to chemical interactions between the TLC layer and the film. For low energy beta emitters (e.g., ^{3}H), a large part of their energy is absorbed by the chromatographic layer and hence only a few electrons reach the photographic emulsion. The sensitivity of autoradiography for low energy beta emitters can be enhanced by incorporating into the TLC layer a scintillator which emits light upon absorbing beta particles. This light is absorbed to a lesser extent than the beta particles and can be registered by the film. The scintillator can be applied in a number of ways: by dipping, by spraying or incorporated in the slurry during the preparation of the TLC layer.

Liquid scintillation counting is the most sensitive technique for the quantitative analysis of separated spots on a TLC plate. It represents an indirect method since the adsorbent layer and the sample must usually be removed from the plate prior to measurement. Automatic equipment which can divide and remove the whole chromatogram as a series of narrow zones is commercially available. The compound to be counted is first eluted from the excised adsorbent and, after solvent removal, the residue is mixed with scintillator soltution and determined in the normal manner. Alternatively, the sample can be counted while adsorbed onto the silica gel without elution. To avoid self-quenching, a gelling agent is added to the scintillator cocktail to keep the adsorbent in suspension. Correction factors based on internal or external standards are required for quantitative measurements.

Scintillation counting is probably the most widely used method for quantifying TLC radiochromatograms, due in part to the general availability of scintillation counting equipment in bioanalytical·laboratories. The most convenient method, however, is direct scanning with radiation detectors. The necessary equipment is expensive and can be justified only by those laboratories with a high routine workload. Scanning is performed by moving the plate on a motor driven stage under, over, or between radiation detectors. The most suitable detectors are Geiger-Muller detectors or flow-through proportional counters. The technique is very rapid when the level of radioactivity is high and the data is provided in convenient form as a chromatogram on a strip-chart recorder. Its main disadvantage is a lack of sensitivity for detecting small quantities of low energy beta isotopes due to self-absorption problems within the adsorbent layer.

9.17 Preparative Thin-Layer Chromatography

The principal difference between analytical TLC and preparative TLC is one of scale and not of procedure or method. The importance of preparative TLC for the isolation of moderate quantities of pure materials has declined substantially in recent years, due firstly to the wider acceptance of dry column chromatography (Chapter 6.3) and more recently to the greater access many laboratories have to high performance preparative liquid chromatography. Dry column chromatography can perform any separation obtainable by preparative TLC, in addition to having a higher sample capacity and being less expensive and easier to perform. Preparative TLC plates range in size from 20 x 20 cm to 20 x 100 cm and are coated with an adsorbent layer 0.5 to 10.0 mm thick. The most commonly used layer thicknesses are 1.0 and 2.0 mm. The plates can be prepared in the laboratory by methods similar to those used to prepare conventional analytical TLC plates [121,122] or they may be obtained commercially. Manufactured plates are available in various sizes coated with silica gel, alumina, and silanized silica gel, the latter for reversed-phase chromatography. Fluorescent indicators are added to the adsorbent to aid in locating UV-absorbing compounds.

In preparative TLC the sample, a 5-10% solution in a volatile organic solvent, is applied as a streak along one edge of the plate. The maximum sample

load for a silica gel layer 1.0 mm thick is about 5 mg/cm. Cellulose and
reversed-phase layers have a lower sample capacity. Various mechanical devices
have been described for sample streaking since this is the preferred method of
sample application. Manual sample application must be performed with care to
avoid damaging the layer and producing irregular development. If the starting
line is not entirely even then removal of the separated zones will be
difficult. To some extent the sample zone can be refocussed by a short (e.g.,
1.0 cm) predevelopment step in a solvent of sufficient eluotropic strength to
cause the entire sample to move together. The plate is then dried and developed
in the solvent system used for the separation. When applying the sample to the
adsorbent layer, a margin of 2-3 cm is left at either side of the plate to avoid
uneven development.

After the separation the bands are visualized and the separated zones
carefully marked for removal. The zones are scraped off the plate with a
spatula or similar tool onto glassine weighing paper. A number of devices based
on the vacuum suction principle are commercially available for removing the
marked zones from the plate. These devices work like a vacuum cleaner and
collect the adsorbent material in a Soxhlet thimble or a glass chamber with a
fritted base. The sample is separated from the adsorbent by Soxhlet extraction,
elution, or solvent extraction. For extraction, a small quantity of water is
usually added to dampen the silica gel prior to shaking for several minutes with
several portions of an organic solvent. In an alternative method, enough water
is added to completely cover the adsorbent material and the aqueous suspension
is extracted several times with an immiscible organic solvent. Prior to solvent
evaporation, colloidal silica can be removed by filtration through a membrane
filter. A convenient and rapid method for this is a hypodermic syringe fitted
with a Swinney adapter and filter holder [123]. The possibility that recovery
will be non-quantitative by any of the above methods must always be kept in
mind.

High speed preparative separations can be achieved by centrifugal
preparative thin-layer chromatography using circular or anticircular development
[125]. The model CLC-5 centrifugal preparative liquid chromatograph (Hitachi)
can separate 2.0-10.0 g of sample per injection. The separation takes place by
radial thin-layer chromatography which uses centrifugal forces to move the
sample and solvent through the adsorbent layer. The sample and the solvent are
applied at the center of the spinning disk. As the sample components separate
they migrate as concentric rings of different circumference. When a component
reaches the end of the disk it is automatically detected and collected by a
fraction collector. Up to 500 theoretical plates may be generated in this
system and the throughput of purified material is about ten times greater than
preparative TLC.

9.18 Paper Chromatography

Paper chromatography was in use prior to the existence of TLC, but has largely been superseded by it in most analytical laboratories. Paper chromatography is now rarely used for the separation of lipophilic substances but still finds some applications, largely in biochemical laboratories, for the separation of hydrophilic compounds. A further decline in the use of paper chromatography was due to the availability of thin-layer plates coated with microcrystalline cellulose. The cellulose TLC plates have separation characteristics similar to paper but provide sharper spot shape and greater resolution. Separations by cellulose TLC are also 4 to 5 times faster than paper and provide significantly lower sample detection limits. For a more detailed treatment of paper chromatography than is presented here, the reader is directed to references [126-130].

Filter papers used for paper chromatography are prepared from specially purified cellulose (98-99% α-cellulose, 0.3-1.0% β-cellulose, 0.4-0.8% pentosans, and < 0.01 % mineral ash). Chromatographic papers are available in slow, standard, and fast grades corresponding to the degree of coarseness of the cellulose fibers used during manufacture and their packing density. They are characterized physically by their thickness, weight per unit area, and flow rate index, Table 9.10. Standard papers are most commonly used, representing the best compromise between resolution and the time required to develop the chromatogrm (10-20 h). Fast papers provide lower resolution but can be developed in a shorter time (4-6 h). They are used mainly for simple separations. Slow papers provide the highest resolution but are rarely used due to the long development times (20-50 h). The separation mechanism in paper chromatography is somewhat complex, involving adsorption on the cellulose fibers, ion-exchange with the free carboxylic acid groups formed during the manufacture of the paper (ca. 3-8 microequivalents/gram) and partition with the "water-cellulose complex" formed between chemically and physically bound water and the cellulose matrix of the paper.

Both impregnated and chemically modified papers are also commercially available. Impregnating agents include silica gel and alumina adsorbents, silicone oils for reversed-phase separations, and liquid ion-exchangers. Chemically modified papers are prepared mainly from ion-exchange celluloses similar to those used in column chromatography. Among the more widely used chemically modified papers are those prepared from carboxymethyl-(CM-), diethylaminoethyl-(DEAE), phosphate-(P-), and acetyl-(Ac-) celluloses.

Samples in a volatile solvent are applied to the chromatographic paper as spots or streaks of minimum size. Microcapillaries or microsyringes are used as sample applicators. Sample volumes are usually 5 microliters or less per spot;

TABLE 9.10

PHYSICAL PROPERTIES OF WHATMAN CHROMATOGRAPHY PAPERS

Type	Weight (g/m^2)	Thickness (mm)	Flow Index (water[a])	Comments
No. 1	87	0.16	130	Smooth, normally hard surface for general purpose chromatography.
No. 2	97	0.18	115	Smooth, normally hard surface, medium flow rate, recommended for optical or radiometric scanning.
No. 3	185	0.38	130	Thick paper, medium flow rate, rough surface, frequently used for the separation of inorganics and for electrophoresis.
No. 3mm	185	0.33	130	Similar to No. 3 but with smooth surface, generally a good paper for high sample loads.
No. 4	92	0.20	180	Smooth, normally hard surface, fast flow, recommended for routine work with small sample loadings.
No. 17	440	0.88	190	Extremely thick paper with a high flow rate, used for preparative chromatography.
No. 20	93	0.16	85	Smooth, normally hard surface, slow flow rate, very good resolving power.
No. 31 ET	190	0.53	225	Soft surface, fast flow rate, used mainly in electrophoresis of large molecules.

ADSORBENT LOADED PAPERS

Type	Weight (g/m^2)	Thickness (mm)	Flow Index (water[a])	Comments
SG81	100	0.25	110	Contains 22% SiO_2, pH 7, used for separations in which partition and adsorption are important. Particularly efficient for the separation of polar lipids, steroids, phenols, dyes, natural pigments, and keto acids.

SILICONE LOADED PAPERS

Type	Weight (g/m^2)	Thickness (mm)	Flow Index (water[a])	Comments
IPS				Paper is impregnated with a silicone oil, making it highly hydrophobic. Used for reversed-phase paper chromatography of lipophilic substances.

TABLE 9.10 (continued)

Type	Weight (g/m^2)	Thickness (mm)	Flow Index (water[a])	Comments
ION-EXCHANGE PAPERS				
DE81	85	0.17	20^b	Functional groups are diethylamino-ethyl residues, a weakly basic anion exchanger, 1.7 μequiv/cm^2. Used for the separation of anions, particularly nucleotides.
P81	85	0.17	10^b	Cellulose phosphate paper. Strong cation exchanger of high capacity, 18 μequiv/cm^2. Used for the separation of biogenic amines, antibiotics, histamimes, and some metals.
GLASS-FIBER PAPERS				
GF/A	52.5	0.25	2.5^c	Water absorption 25 μl/cm^2, general purpose medium
GF/B		0.73		Relatively slow flow rate, strong, used for electrophoresis.
GF/C	55	0.26	2.5^c	Water absorption 25 μl/cm^2, made from ultra-fine fibers.

[a] mm capillary rise in 30 minutes; [b] test conditions not specified;
[c] time for water to ascend 7.5 cm.

to avoid tailing effects, the sample amount should not exceed 100 micrograms. Once spotted, the paper is equilibrated with the vapors of the mobile phase, a process which may take several hours. The paper chromatograms can be developed in the linear or radial direction by the same methods described for TLC. Unlike TLC, descending development is frequently used as it is faster and more suitable for long paper sheets. It is important to maintain a saturated atmosphere in the chromatographic chamber, which should be gas tight and protected from temperature variations. After development, the chromatogram is air dried and the spots or bands are visualized. Colored compounds and those which fluoresce strongly when irradiated with UV-light are easily detected by eye and can be measured with a densitometer. Other compounds can be converted to colored or fluorescent zones by dipping in or spraying with a solution of a reactive, selective chemical reagent. For quantitative measurements it is difficult to obtain accuracy better than ± 1% in routine work. Fluorescence or radioactivity can be measured in situ but absorption measurements are usually made by cutting out the zones and eluting the separated components for measurment by solution photometry. The latter method is prone to all the errors common to detection by

elution in open bed chromatography. These are difficulty in precisely locating
sample zones, cross-contamination of zones, incomplete sample elution, and
contamination of the eluted solution by the chromatographic medium.

Preparative separations by paper chromatography are reserved for relatively
easy separations where alternative methods are more complicated. In most
instances column methods are found to be more suitable. Possible contamination
of the separated product with soluble material on the paper, as well as
cellulose fibers from the paper, must always be considered.

The most suitable method for preparative paper chromatography is the use of
wide sheets of thick paper. The sample is applied as a narrow streak along one
edge in loadings 10-20 mg/cm. Papers such as Whatman's No. 31ET offer higher
loadings of 20-40 mg/cm. After the development and drying of the chromatogram,
the position of the solute zones may be located on marker strips cut from the
middle and both sides of the paper. The zones can then be cut out for
individual elution by any appropriate method.

9.19 References

1. A. Zlatkis and R. E. Kaiser (Eds.), "HPTLC: High Performance Thin-Layer
    Chromatography", Elsevier, Amsterdam, 1977.
2. W. Bertsch, S. Hara, R. E. Kaiser, and A. Zlatkis (Eds.), "Instrumental
    HPTLC", Huthig, Heidelberg, 1980.
3. D. C. Fenimore and C. M. Davis, Anal. Chem., 53 (1981) 252A.
4. E. Stahl (Ed.), "Thin-Layer Chromatography", Springer-Verlag, New York,
    1969.
5. J. C. Touchstone and M. F. Dobbins, "Practice of Thin-Layer
    Chromatography", Wiley, New York, 1978.
6. J. G. Kirchner, "Thin-Layer Chromatography", 2nd Edn, Wiley, New York,
    1978.
7. J. G. Kirchner, J. Chromatogr. Sci., 13 (1975) 558.
8. H. J. Issaq and E. W. Barr, Anal. Chem., 49 (1977) 83A.
9. R. E. Kaiser, in A. Zlatkis and R. E. Kaiser (Eds.), "HPTLC: High
    Performance Thin-Layer Chromatogrpahy", Elsevier, Amsterdam, 1977, p. 15.
10. F. Geiss, "Die Parameter der Dunnschicht Chromatographia, eine moderne
    Einfuhrung in Grundlagen und Praxis", Friedr. Vieweg & Sohn, Wiesbaden,
    Germany, 1972.
11. T. H. Jupille, CRC Crit. Rev. Anal. Chem., 6 (1977) 325.
12. G. Guiochon, A. M. Siouffi, H. Engelhardt, and I. Halasz, J. Chromatogr.
    Sci., 16 (1978) 153.
13. G. Guiochon and A. M. Siouffi, J. Chromatogr. Sci., 16 (1978) 471.
14. G. Guiochon and A. M. Siouffi, J. Chromatogr. Sci., 16 (1978) 598.
15. G. Guiochon, F. Bressolle, and A. M. Siouffi, J. Chromatogr. Sci., 17
    (1979) 368.
16. A. M. Siouffi, F. Bressolle, and G. Guiochon, J. Chromatogr., 209 (1981)
    129.
17. G. Guiochon and A. M. Siouffi, J. Chromatogr., 245 (1982) 1.
18. G. Guiochon, G. Korosi, and A. M. Siouffi, J. Chromatogr. Sci., 18 (1980)
    324.
19. A. Li Wan Po and W. J. Irwin, J. High Resolut. Chromatogr. Chromatogr.
    Commun., 2 (1979) 623.
20. B. G. Belenki, V. I. Kolegov, and V. V. Nesterov, J. Chromatogr., 107
    (1975) 265.
21. J. P. Franke, W. Kruyt, and R. A. de Zeeuw, J. High Resolut. Chromatogr.
    Chromatogr. Commun., 6 (1983) 82.

697

22. E. Minscovics, E. Tyihak, and H. Kalasz, J. Chromatogr., 191 (1980) 293.
23. H. F. Hauck and W. Jost, J. Chromatogr., 262 (1983) 113.
24. J. Ripphahn and H. Halpaap, J. Chromatogr., 112 (1975) 81.
25. U. A. Th. Brinkman, G. De Vries, and R. Cuperus, J. Chromatogr., 198 (1980) 421.
26. H. Halpaap and J. Ripphahn, Chromatographia, 10 (1977) 613.
27. H. Halpaap and K.-F. Krebs, J. Chromatogr., 142 (1977) 823.
28. H. E. Hauck and H. Halpaap, Chromatographia, 13 (1980) 538.
29. H. Halpaap and J. Ripphahn, Chromatographia, 10 (1977) 613.
30. H. Halpaap, K.-F. Krebs, and H. E. Hauck, J. High Resolut. Chromatogr. Chromatogr. Commun., 3 (1980) 215.
31. U. A. Th. Brinkman and G. De Vries, J. Chromatogr., 258 (1983) 43.
32. U. A. Th. Brinkman and G. De Vries, J. High Resolut. Chromatogr. Chromatogr. Commun., 5 (1982) 476.
33. T. E. Beesley, in J. C. Touchstone (Ed.), "Thin-Layer Chromatography. Clinical and Environmental Applications", Wiley, New York, 1982, p. 1.
34. R. K. Gilpin and W. R. Sisco, J. Chromatogr., 124 (1976) 257.
35. M. Ericsson and L. G. Blomberg, J. High Resolut. Chromatogr. Chromatogr. Commun., 3 (1980) 345.
36. M. Ericsson and L. G. Blomberg, J. High Resolut. Chromatogr. Chromatogr. Commun., 6 (1983) 95.
37. A. M. Siouffi, T. Wawrzynowicz, F. Bressolle, and G. Guiochon, J. Chromatogr., 186 (1979) 563.
38. U. A. Th. Brinkman and G. De Vries, J. Chromatogr., 192 (1980) 331.
39. U. A. Th. Brinkman and G. De Vries, J. Chromatogr., 265 (1983) 105.
40. G. Grassini-Strazza and M. Cristalli, J. Chromatogr., 214 (1981) 209.
41. W. Jost and H. E. Hauck, J. Chromatogr., 261 (1983) 235.
42. M. Okamoto, F. Yamada, and T. Omori, Chromatographia, 16 (1983) 152.
43. T. Omori, M. Okamoto, and F. Yamada, J. High Resolut. Chromatogr. Chromatogr. Commun., 6 (1983) 47.
44. D. C. Fenimore, in W. Bertsch, S. Hara, R. E. Kaiser, and A. Zlatkis (Eds.), "Instrumental HPTLC", Huthig, Heidelberg, 1980, p. 81.
45. D. C. Fenimore and C. J. Meyer, J. Chromatogr., 186 (1979) 555.
46. J. A. Perry, J. Chromatogr., 165 (1979) 117.
47. D. Nurok and M. J. Richard, Anal. Chem., 53 (1981) 563.
48. D. Nurok, R. M. Becker, M. J. Richard, P. D. Cunningham, W. B. Gorman, and C. L. Bush, J. High Resolut. Chromatogr. Chromatogr. Commun., 5 (1982) 373.
49. D. Nurok, R. M. Becker, and K. A. Sassic, Anal. Chem., 54 (1982) 1955.
50. R. E. Tecklenburg, R. M. Becker, E. K. Johnson, and D. Nurok, Anal. Chem., 55 (1983) 2196.
51. R. E. Tecklenburg, B. L. Maidak, and D. Nurok, J. High Resolut. Chromatogr. Chromatogr. Commun., 6 (1983) 627.
52. R. E. Tecklenburg, G. H. Fricke, and D. Nurok, J. Chromatogr., 290 (1984) 75.
53. G. H. Stewart and C. T. Wendel, J. Chromatogr. Sci., 13 (1975) 105.
54. T. H. Jupille and J. A. Perry, J. Amer. Oil Chem. Soc., 54 (1976) 179.
55. K. Y. Lee, C. F. Poole, and A. Zlatkis, in W. Bertsch, S. Hara, R. E. Kaiser, and A. Zlatkis (Eds.), "Instrumental HPTLC", Huthig, Heidelberg, 1980, p. 245.
56. J. A. Perry, J. Chromatogr., 113 (1975) 267.
57. J. A. Perry, T. H. Jupille, and L. J. Glunz, Anal. Chem., 47 (1975) 65A.
58. J. A. Perry, T. H. Jupille, and A. B. Curtice, Sepn. Sci., 10 (1975) 571.
59. M. Zakaria, M.-F. Gonnord, and G. Guiochon, J. Chromatogr., 271 (1983) 127.
60. G. Guiochon, M.-F. Gonnord, M. Zakaria, L. A. Beaver, and A. M. Siouffi, Chromatographia, 17 (1983) 121.
61. G. Guiochon, L. A. Beaver, M.-F. Gonnord, A. M. Siouffi, and M. Zakaria, J. Chromatogr., 255 (1983) 415.

62. G. Guiochon, A. Krstulovic, and H. Colin, J. Chromatogr., 265 (1983) 159.
63. J. K. Rozylo and B. Oscik-Mendyk, J. High Resolut. Chromatogr. Chromatogr. Commun., 3 (1980) 291.
64. P. Petrin, J. Chromatogr., 123 (1976) 65.
65. F. Geiss and H. Schlitt, Chromatographia, 1 (1968) 392.
66. E. Tyihak, E. Minscovics, and H. Kalasz, J. Chromatogr., 174 (1979) 75.
67. E. Tyihak, E. Minscovics, H. Kalasz, and J. Nagy, J. Chromatogr., 211 (1981) 45.
68. R. E. Kaiser, J. High Resolut. Chromatogr. Chromatogr. Commun., 1 (1978) 164.
69. M. E. Coddens, H. T. Butler, S. A. Schuette, and C. F. Poole, Liq. Chromatogr. HPLC Magzn., 1 (1983) 282.
70. J. C. Touchstone and J. Sherma, "Densitometry in Thin-Layer Chromatography", Wiley, New York, 1979.
71. M. E. Coddens, S. Khatib, H. T. Butler, and C. F. Poole, J. Chromatogr., 280 (1983) 15.
72. L. R. Treiber, R. Nordberg, and S. Lindstedt, J. Chromatogr., 63 (1971) 211.
73. H. Jork, J. Chromatogr., 82 (1973) 85.
74. V. Pollock, Adv. Chromatogr., 17 (1979) 1.
75. S. Ebel and P. Post, J. High Resolut. Chromatogr. Chromatogr. Commun., 4 (1981) 337.
76. R. J. Hurtubise, "Solid Surface Luminescence Analysis. Theory, Instrumentation, Applications", Dekker, New York, 1981.
77. K. Y. Lee, C. F. Poole, and A. Zlatkis, Anal. Chem., 52 (1980) 837.
78. J. Goldman, J. Chromatogr., 78 (1973) 7.
79. H. Bethke, W. Santi, and R. W. Frei, J. Chromatogr. Sci., 12 (1974) 392.
80. H. Yamamoto, T. Kurita, J. Suzuki, R. Hira, K. Nakano, M. Makabe, and K. Shibata, J. Chromatogr., 116 (1976) 29.
81. M.-L. Cheng and C. F. Poole, J. Chromatogr., 257 (1983) 140.
82. L. Zhou, C. F. Poole, J. Triska, and A. Zlatkis, J. High Resolut. Chromatogr. Chromatogr. Commun., 3 (1980) 440.
83. H. T. Butler, M. E. Coddens, and C. F. Poole, J. Chromatogr., 290 (1984) 113.
84. J. Traveset, V. Such, R. Gonzalo, and E. Gelpi, J. High Resolut. Chromatog. Chromatogr. Commun., 5 (1982) 483.
85. V. Such, J. Traveset, R. Gonzalo, and E. Gelpi, J. Chromatogr., 234 (1982) 77.
86. J. Traveset, V. Such, R. Gonzalo, and E. Gelpi, J. Chromatogr., 204 (1981) 51.
87. H. T. Butler, S. A. Schuette, F. Pacholec, and C. F. Poole, J. Chromatogr., 261 (1983) 55.
88. H. T. Butler, F. Pacholec, and C. F. Poole, J. High Resolut. Chromatogr. Chromatogr. Commun., 5 (1982) 580.
89. H. T. Butler and C. F. Poole, J. High Resolut. Chromatogr. Chromatogr. Commun., 6 (1983) 77.
90. H. T. Butler and C. F. Poole, J. Chromatogr. Sci., 21 (1983) 385.
91. M. E. Coddens and C. F. Poole, Anal. Chem., 55 (1983) 2429.
92. M. E. Coddens and C. F. Poole, Liq. Chromatogr. HPLC Magzn., 2 (1984) 34.
93. S. A. Schuette and C. F. Poole, J. Chromatogr., 239 (1982) 251.
94. D. C. Fenimore, C. M. Davis, and C. J. Meyer, Clin. Chem., 24 (1978) 1386.
95. D. C. Fenimore, C. J. Meyers, C. M. Davis, F. Hsu, and A. Zlatkis, J. Chromatogr., 142 (1977) 399.
96. C. M. Davis and C. A. Harrington, J. Chromatogr. Sci., 22 (1984) 71.
97. C. M. Davis and D. C. Fenimore, J. Chromatogr., 222 (1981) 265.
98. K. Y. Lee, D. Nurok, A. Zlatkis, and A. Karmen, J. Chromatogr., 158 (1978) 403.

99. K. Y. Lee and A. Zlatkis, in J. C. Touchstone (Ed.)., "Advances in Thin-Layer Chromatogrphy. Clinical and Environmental Applications", Wiley, New York, 1982, p. 343.

100. J. Kraft, A. Hartung, K.-H. Lies, and J. Schulze, J. High Resolut. Chromatogr. Chromatogr. Commun., 5 (1982) 489.

101. D. Brocco, V. Cantuti, and G. P. Cartoni, J. Chromatogr., 49 (1970) 66.

102. R. C. Pierce and M. Katz, Anal. Chem., 47 (1975) 1743.

103. R. Tomingas, G. Voltmer, and R. Bednarik, Sci. Total Environ., 7 (1977) 261.

104. B. Seifert, J. Chromatogr., 131 (1977) 417.

105. W. A. Bruggeman, J. van der Steen, and O. Hutzinger, J. Chromatogr., 238 (1982) 335.

106. S. S. J. Ho, H. T. Butler, and C. F. Poole, J. Chromatogr., 281 (1983) 330.

107. D. T. Kaschani, in W. Bertsch, S. Hara, R. E. Kaiser, and A. Zlatkis (Eds.), "Instrumental HPTLC", Huthig, Heidelberg, 1980, p. 185.

108. K. T. Wang, Y.-T. Lin, and I. S. Y. Wang, Adv. Chromatogr., 11 (1974) 73.

109. L. Horhammer, H. Wagner, and K. Macek, J. Chromatogr., 9 (1967) 103.

110. J. N. Miller, in R. Epton (Ed.), "Chromatography of Synthetic Biological Polymers", Ellis Horwood, Chichester, 1978, p. 181.

111. L. Fischer, "An Introduction to Gel Chromatography", North Holland, Amsterdam, 1969, p. 316.

112. K. W. Williams, Lab. Pract. (London), 22 (1973) 306.

113. D. A. B. Young, Chromatographia, 12 (1979) 608.

114. T. Okumura, J. Chromatogr., 184 (1980) 37.

115. T. Okumura, T. Kadono, and A. Iso'O, J. Chromatogr., 108 (1975) 329.

116. R. G. Ackman, Methods in Enzymol., 72 (1981) 205.

117. C. C. Parrish and R. G. Ackman, J. Chromatogr., 262 (1983) 103.

118. T. R. Roberts, "Radiochromatography. The Chromatography and Electrophoresis of Radiolabelled Compounds", Elsevier, Amsterdam, 1978.

119. J. C. Touchstone and M. F. Dobbins, "Practice of Thin-Layer Chromatography", Wiley, New York, 1978, p. 275.

120. S. Prydz, Anal. Chem., 45 (1973) 2317.

121. J. G. Kirchner, "Thin-Layer Chromatography", 2nd. Edn., Wiley, New York, 1978, p. 279.

122. J. C. Touchstone and M. F. Dobbins, "Practice of Thin-Layer Chromatography", Wiley, New York, 1978, p. 311.

123. J. H. Nagel and J. C. Dittmer, J. Chromatogr., 42 (1969) 121.

124. J. R. DeDeyne and A. F. Vetters, J. Chromatogr., 103 (1975) 177.

125. E. Stahl and J. Muller, Chromatographia, 15 (1982) 493.

126. I. M. Hais and K. Macek, "Paper Chromatography", Academic Press, New York, 1963.

127. V. C. Weaver, Adv. Chromatogr., 7 (1968) 87.

128. C. J. O. R. Morris and P. Morris, "Separation Methods in Biochemistry", 2nd. Edn., Halsted Press, New York, 1976, p. 545.

129. Z. Prochazka, M. Hejtmanek, K. Sebesta, and V. Tomasek, in O. Mikes (Ed.), "Laboratory Handbook of Chromatography and Allied Methods", Ellis Harwood, Chichester, 1979, p. 64.

130. J. Sherma and G. Zweig, "Paper Chromatography", Academic Press, New York, 1971.

Subject Index

SAMPLE PREPARATION

J.T.Baker

cosmetic

petrochemical

clinical/biological

food + feed

pharmaceutical

environmental analysis

'BAKER'-10
EXTRACTION SYSTEM

J.T.Baker Chemical Co., Phillipsburg, NJ 08865

extraction, purification,
enrichment for analysis by
HPLC, GC, GC/ms, TLC, LSC,
RIA, electrophoresis and
spectrophotometry

J.T. Baker Chemicals B.V. - P.O. Box 1 - 7400 AA Deventer - Holland - Tel. 05700-11341 - Telex 49072

KONTRON's modular HPLC-components offer you full flexibility of perfomance AND price

1 Scanning UV-Detectors,
   UVIKON 720 LC/722 LC
2 Autosampler MSI-660 (also cooled)
3 Pump model 414
4 Variable UV-Detector UVIKON 730 LC
5 HPLC-Programmer, model 205
6 ANACOMP 220-Multitasking Computer
7 Variable Filter-Detector UVIKON 740 LC

Your partner in science and health

KONTRON
INSTRUMENTS

For further information please contact your local KONTRON Company

Austria (Vienna)	(0222) 670631	Japan (Tokyo)	(03) 2634801
France (Montigny le Br.)	(3) 0438152	Scandinavia (Stockholm)	(08) 7567330
Germany (Munich)	(08165) 771	Spain (Madrid)	(01) 7291155
Great Britain (St. Albans)	(0727) 66222	Switzerland (Zurich)	(01) 4354111
Italy (Milan)	(02) 50721	USA (Everett/Mass.)	(617) 3896400

International Marketing Services, KONTRON AG, Bernerstrasse Süd 169, 8048 Zurich/Switzerland
Telephone (01) 435 41 11, Telex 822 191